Advances in Marine Chitin and Chitosan II, 2017

Special Issue Editors

Hitoshi Sashiwa
David Harding

MDPI • Basel • Beijing • Wuhan • Barcelona • Belgrade

MDPI

Special Issue Editors

Hitoshi Sashiwa
Kaneka Co., Ltd
Japan

David Harding
Massey University
New Zealand

Editorial Office
MDPI AG
St. Alban-Anlage 66
Basel, Switzerland

This edition is a reprint of the Special Issue published online in the open access journal *Marinedrugs* (ISSN 1660-3397) from 2015–2017/ (available at: http://www.mdpi.com/journal/marinedrugs/special_issues/advance_chitosan).

For citation purposes, cite each article independently as indicated on the article page online and as indicated below:

Author 1, Author 2. Article title. *Journal Name*. **Year**. Article number/page range.

First Edition 2018

ISBN 978-3-03842-677-6 (Pbk)
ISBN 978-3-03842-678-3 (PDF)

Cover photo courtesy of Se-Kwon Kim

Table of Contents

About the Special Issue Editors vii

Preface to "Advances in Marine Chitin and Chitosan II, 2017" ix

Riccardo A. A. Muzzarelli, Mohamad El Mehtedi, Carlo Bottegoni, Alberto Aquili and
Antonio Gigante
Genipin-Crosslinked Chitosan Gels and Scaffolds for Tissue Engineering and Regeneration of
Cartilage and Bone
Reprinted from: *Mar. Drugs* **2015**, *13*(12), 7314–7338; doi: 10.3390/md13127068 1

Barbara Bellich, Ilenia D'Agostino, Sabrina Semeraro, Amelia Gamini and Attilio Cesàro
"The Good, the Bad and the Ugly" of Chitosans
Reprinted from: *Mar. Drugs* **2016**, *14*(5), 99; doi: 10.3390/md14050099 26

Pascal Viens, Marie-Ève Lacombe-Harvey and Ryszard Brzezinski
Chitosanases from Family 46 of Glycoside Hydrolases: From Proteins to Phenotypes
Reprinted from: *Mar. Drugs* **2015**, *13*(11), 6566–6587; doi: 10.3390/md13116566 58

Chunhua Wu, Liping Wang, Zhongxiang Fang, Yaqin Hu, Shiguo Chen, Tatsuya Sugawara
and Xingqian Ye
The Effect of the Molecular Architecture on the Antioxidant Properties of Chitosan Gallate
Reprinted from: *Mar. Drugs* **2016**, *14*(5), 95; doi: 10.3390/md14050095 77

María-Pilar Sánchez-Sánchez, Araceli Martín-Illana, Roberto Ruiz-Caro, Paulina Bermejo,
María-José Abad, Rubén Carro, Luis-Miguel Bedoya, Aitana Tamayo, Juan Rubio,
Anxo Fernández-Ferreiro, Francisco Otero-Espinar and María-Dolores Veiga
Chitosan and Kappa-Carrageenan Vaginal Acyclovir Formulations for Prevention of Genital
Herpes.
In Vitro and *Ex Vivo* Evaluation
Reprinted from: *Mar. Drugs* **2015**, *13*(9), 5976–5992; doi: 10.3390/md13095976 95

Niuris Acosta, Elisa Sánchez, Laura Calderón, Manuel Cordoba-Diaz, Damián Cordoba-Diaz,
Senne Dom and Ángeles Heras
Physical Stability Studies of Semi-Solid Formulations from Natural Compounds Loaded with
Chitosan Microspheres
Reprinted from: *Mar. Drugs* **2015**, *13*(9), 5901–5919; doi: 10.3390/md13095901 109

Seong-Chul Hong, Seung-Yup Yoo, Hyeongmin Kim and Jaehwi Lee
Chitosan-Based Multifunctional Platforms for Local Delivery of Therapeutics
Reprinted from: *Mar. Drugs* **2017**, *15*(3), 60; doi: 10.3390/md15030060 126

Sruthi Ravindranathan, Bhanu prasanth Koppolu, Sean G. Smith and David A. Zaharoff
Effect of Chitosan Properties on Immunoreactivity
Reprinted from: *Mar. Drugs* **2016**, *14*(5), 91; doi: 10.3390/md14050091 142

Weiai Zhang, Caijuan Ma, Zhengquan Su and Yan Bai
Resonance Rayleigh Scattering Spectra of an Ion-Association Complex of Naphthol Green
B–Chitosan System and Its Application in the Highly Sensitive Determination of Chitosan
Reprinted from: *Mar. Drugs* **2016**, *14*(4), 71; doi: 10.3390/md14040071 154

San-Lang Wang, Hsin-Ting Li, Li-Jie Zhang, Zhi-Hu Lin and Yao-Haur Kuo
Conversion of Squid Pen to Homogentisic Acid via *Paenibacillus* sp. TKU036 and the Antioxidant
and Anti-Inflammatory Activities of Homogentisic Acid
Reprinted from: *Mar. Drugs* **2016**, *14*(10), 183; doi: 10.3390/md14100183 165

Fernando Notario-Pérez, Araceli Martín-Illana, Raúl Cazorla-Luna, Roberto Ruiz-Caro,
Luis-Miguel Bedoya, Aitana Tamayo, Juan Rubio and María-Dolores Veiga
Influence of Chitosan Swelling Behaviour on Controlled Release of Tenofovir from Mucoadhesive
Vaginal Systems for Prevention of Sexual Transmission of HIV
Reprinted from: *Mar. Drugs* **2017**, *15*(2), 50; doi: 10.3390/md15020050 175

Anish Babu and Rajagopal Ramesh
Multifaceted Applications of Chitosan in Cancer Drug Delivery and Therapy
Reprinted from: *Mar. Drugs* **2017**, *15*(4), 96; doi: 10.3390/md15040096 191

Alexander J. Winkler, Jose Alfonso Dominguez-Nuñez, Inmaculada Aranaz,
César Poza-Carrión, Katrina Ramonell, Shauna Somerville and Marta Berrocal-Lobo
Short-Chain Chitin Oligomers: Promoters of Plant Growth
Reprinted from: *Mar. Drugs* **2017**, *15*(2), 40; doi: 10.3390/md15020040 210

Ya Gao, Yingbo Wang, Yimin Wang and Wenguo Cui
Fabrication of Gelatin-Based Electrospun Composite Fibers for Anti-Bacterial Properties and
Protein Adsorption
Reprinted from: *Mar. Drugs* **2016**, *14*(10), 192; doi: 10.3390/md14100192 231

Toril Andersen, Ekaterina Mishchenko, Gøril Eide Flaten, Johanna U. Ericson Sollid,
Sofia Mattsson, Ingunn Tho and Nataša Škalko-Basnet
Chitosan-Based Nanomedicine to Fight Genital Candida Infections: Chitosomes
Reprinted from: *Mar. Drugs* **2017**, *15*(3), 64; doi: 10.3390/md15030064 245

Lidong Cao, Xiuhuan Li, Li Fan, Li Zheng, Miaomiao Wu, Shanxue Zhang and Qiliang Huang
Determination of Inorganic Cations and Anions in Chitooligosaccharides by Ion Chromatography
with Conductivity Detection
Reprinted from: *Mar. Drugs* **2017**, *15*(2), 51; doi: 10.3390/md15020051 257

Khalid A. Ibrahim, Bassam I. El-Eswed, Khaleel A. Abu-Sbeih, Tawfeeq A. Arafat,
Mahmoud M. H. Al Omari, Fouad H. Darras and Adnan A. Badwan
Preparation of Chito-Oligomers by Hydrolysis of Chitosan in the Presence of Zeolite as Adsorbent
Reprinted from: *Mar. Drugs* **2016**, *14*(8), 43; doi: 10.3390/md14080043 266

Sara Skøtt Paulsen, Birgitte Andersen, Lone Gram and Henrique Machado
Biological Potential of Chitinolytic Marine Bacteria
Reprinted from: *Mar. Drugs* **2016**, *14*(12), 230; doi: 10.3390/md14120230 279

Cui Hao, Wei Wang, Shuyao Wang, Lijuan Zhang and Yunliang Guo
An Overview of the Protective Effects of Chitosan and Acetylated Chitosan Oligosaccharides
against Neuronal Disorders
Reprinted from: *Mar. Drugs* **2017**, *15*(4), 89; doi: 10.3390/md15040089 296

Laura De Matteis, Maria Alleva, Inés Serrano-Sevilla, Sonia García-Embid, Grazyna Stepien, María Moros and Jesús M. de la Fuente
Controlling Properties and Cytotoxicity of Chitosan Nanocapsules by Chemical Grafting
Reprinted from: *Mar. Drugs* **2016**, *14*(10), 175; doi: 10.3390/md14100175 311

Emilia Szymańska, Marta Szekalska, Robert Czarnomysy, Zoran Lavrič, Stane Srčič, Wojciech Miltyk and Katarzyna Winnicka
Novel Spray Dried Glycerol 2-Phosphate Cross-Linked Chitosan Microparticulate Vaginal Delivery System—Development, Characterization and Cytotoxicity Studies
Reprinted from: *Mar. Drugs* **2016**, *14*(10), 174; doi: 10.3390/md14100174 326

Daojiang Yu, Shan Li, Shuai Wang, Xiujie Li, Minsheng Zhu, Shai Huang, Li Sun, Yongsheng Zhang, Yanli Liu and Shouli Wang
Development and Characterization of VEGF165-Chitosan Nanoparticles for the Treatment of Radiation-Induced Skin Injury in Rats
Reprinted from: *Mar. Drugs* **2016**, *14*(10), 182; doi: 10.3390/md14100182 348

Zhiwen Li, Xige Yang, Xuesong Song, Haichun Ma and Ping Zhang
Chitosan Oligosaccharide Reduces Propofol Requirements and Propofol-Related Side Effects
Reprinted from: *Mar. Drugs* **2016**, *14*(12), 234; doi: 10.3390/md14120234 358

Yang Qu, Jinyu Xu, Haohan Zhou, Rongpeng Dong, Mingyang Kang and Jianwu Zhao
Chitin Oligosaccharide (COS) Reduces Antibiotics Dose and Prevents Antibiotics-Caused Side Effects in Adolescent Idiopathic Scoliosis (AIS) Patients with Spinal Fusion Surgery
Reprinted from: *Mar. Drugs* **2017**, *15*(3), 70; doi: 10.3390/md15030070 375

Hun Min Lee, Min Hee Kim, Young Il Yoon and Won Ho Park
Fluorescent Property of Chitosan Oligomer and Its Application as a Metal Ion Sensor
Reprinted from: *Mar. Drugs* **2017**, *15*(4), 105; doi: 10.3390/md15040105 390

About the Special Issue Editors

Hitoshi Sashiwa was born in Osaka, Japan, in 1963. He received his Ph.D. degree from Hokkaido University (Japan) under the supervision of Professor S. Tokura in 1991. He worked at Tottori University (Japan) as Assistant Associated Professor from 1988 to 2000. He worked with Profesor R. Roy at the University of Ottawa (Canada) for 2 years (1998–2000). He worked at AIST Kansai (Japan) as a postdoctoral scholar during 2000–2004. He has been affiliated with Kaneka Co., Ltd. (Japan) since April 2004. His research interests include chemical modification of chitin and chitosan and their biomedical applications. He is a member of The Society of Polymer Science, Japan, and the Japanese Society for Chitin and Chitosan. He is the sole author of 70 publications and co-author of 30 publications.

David Harding was born in London England in 1944. He received as BSc Honours from the University of Canterbury, Christchurch, New Zealand in 1966. He then spent over two years working for Eli Lilly (USA) at their research centre in the UK. In 1969, he went to Canada and completed a PhD in early 1973 at the University of Western Ontario. He returned to New Zealand in early 1973 as a postdoctoral fellow and was taken on staff to become in time a professor in 2012. Since returning to New Zealand his research interests have been several. Once significant series of projects in the 1990s involved inexpensive (large scale) protein purification for South San Francisco companies - Genentech and Genencor. The purification media involved was an excellent Czech cellulose. One of the products developed is still returning royalties in 2017 and is marketed by Pall for monoclonal antibody purification. In the late 1990s, this polysaccharide interest morphed into hydrogel research, very much including chitosan. This interest is still very much alive in 2017 and involves international collaborations as well support for Massey University's Institute of Veterinary and Biological Sciences' projects. In recent years, nanocellulose has been added to these programmes. He is a Fellow of the New Zealand Institute of Chemistry with patents, book chapters and over one hundred papers.

Preface to "Advances in Marine Chitin and Chitosan II, 2017"

In recent years, bio-base polymers from renewable resources have received increasing focus owing to the depletion of petroleum resources. Natural polysaccharides such as cellulose, hemicellulose, and starch are among the candidates from natural resources for biomass polysaccharide products including bioplastics. Although several kinds of anionic polysaccharides such as alginic acid, hyaluronic acid, heparin, and chondroitin sulfate exist in nature, natural cationic polysaccharides are quite limited. Chitin is second only to cellulose as the most natural abundant polysaccharide in the world. Chitosan, the product from the N-deacetylatation of chitin, appears to be the only natural cationic polysaccharide. Therefore, chitin and chitosan due to their unique properties are expected to continue to offer a vast number of possible applications for not only chemical or industrial use but also biomedical treatments. The research history on chitin, one of the most abundant natural polysaccharides on earth, started around 1970. Since the 1980s, chitin and chitosan research (including D-glucosamine, N-acetyl-D-glucosamine, and their oligomers) has progressed significantly over several stages in both fundamental research and industrial fields.

Previously, we have published "Advances in Marine Chitin and Chitosan" with quite an interesting and exciting issue. With the opening of this new book "Advances in Marine Chitin and Chitosan II, 2017", we have planned to produce a strong, very exciting issue that will encompass breakthroughs in highly valuable, scientific, and industrial research in this field. A large volume of chitin and chitosan research involves biomedical objectives, in particular controlled drug release. Nevertheless, this book covers recent trends in all aspects of basic and applied scientific research on chitin, chitosan and their derivatives.

Hitoshi Sashiwa and David Harding

Special Issue Editors

*marine
drugs*

MDPI

Review

Genipin-Crosslinked Chitosan Gels and Scaffolds for Tissue Engineering and Regeneration of Cartilage and Bone

Riccardo A. A. Muzzarelli [1,*], Mohamad El Mehtedi [2], Carlo Bottegoni [3], Alberto Aquili [3] and Antonio Gigante [3]

[1] Faculty of Medicine, Polytechnic University of Marche, Via Tronto 10/A, Ancona IT-60126, Italy
[2] Department of Industrial Engineering & Mathematical Sciences, Faculty of Engineering, Polytechnic University of Marche, Via Brecce Bianche, Ancona IT-60131, Italy; elmehtedi@univpm.it
[3] Clinical Orthopaedics, Department of Clinical and Molecular Sciences, Faculty of Medicine, Polytechnic University of Marche, Via Tronto 10/A, Ancona IT-60126, Italy; bottegonicarlo@gmail.com (C.B.); a.aquili@univpm.it (A.A.); a.gigante@univpm.it (A.G.)
* Correspondence: muzzarelli.raa@gmail.com; Tel.: +39-071-2206001

Academic Editor: Hitoshi Sashiwa
Received: 3 November 2015; Accepted: 2 December 2015; Published: 11 December 2015

Abstract: The present review article intends to direct attention to the technological advances made since 2009 in the area of genipin-crosslinked chitosan (GEN-chitosan) hydrogels. After a concise introduction on the well recognized characteristics of medical grade chitosan and food grade genipin, the properties of GEN-chitosan obtained with a safe, spontaneous and irreversible chemical reaction, and the quality assessment of the gels are reviewed. The antibacterial activity of GEN-chitosan has been well assessed in the treatment of gastric infections supported by *Helicobacter pylori*. Therapies based on chitosan alginate crosslinked with genipin include stem cell transplantation, and development of contraction free biomaterials suitable for cartilage engineering. Collagen, gelatin and other proteins have been associated to said hydrogels in view of the regeneration of the cartilage. Viability and proliferation of fibroblasts were impressively enhanced upon addition of poly-L-lysine. The modulation of the osteocytes has been achieved in various ways by applying advanced technologies such as 3D-plotting and electrospinning of biomimetic scaffolds, with optional addition of nano hydroxyapatite to the formulations. A wealth of biotechnological advances and know-how has permitted reaching outstanding results in crucial areas such as cranio-facial surgery, orthopedics and dentistry. It is mandatory to use scaffolds fully characterized in terms of porosity, pore size, swelling, wettability, compressive strength, and degree of acetylation, if the osteogenic differentiation of human mesenchymal stem cells is sought: in fact, the novel characteristics imparted by GEN-chitosan must be simultaneously of physico-chemical and cytological nature. Owing to their high standard, the scientific publications dated 2010–2015 have met the expectations of an interdisciplinary audience.

Keywords: chitosan; genipin; tissue engineering; biomedical uses; biochemical properties

1. Introduction and Scope

The most important applications of genipin in conjunction with chitosan are the preparation of elastic cartilage substitutes, the manufacture of carriers for the controlled release of drugs, the encapsulation of biological products and living cells, the biofabrication of tissues such as muscle and arterial walls, and the dressing of wounds in animals and humans. Genipin has definitely replaced glutaraldehyde and other crosslinkers mainly owing to the expanded biochemical significance of the genipin-crosslinked hydrogels (GEN-chitosan), but also owing to the advantages of stability, biocompatibility, well defined chemistry and general safety of the products whose manipulation, handling and quality assessment are currently done with advanced techniques and clearly defined protocols that guarantee absence of cytotoxicity.

1.1. Characteristic Properties of Genipin

The first review article on genipin was published in 2009 [1], but two early papers [2,3] on the isolation and structure of genipin deserve to be cited here because they are valid examples of exhaustive research and scientific soundness obtained with advanced equipment. Working in the early 1950s with Syntex S.A. in Mexico City, Carl Djerassi first synthesized 19-nor-17α-ethynyltestosterone (norethisterone). This steroid, derived from inedible yams of a wild plant *Dioscorea*, proved to be the most effective orally administered progestational agent discovered at that time. This was the start of a very fortunate research program that led to hundreds and hundreds of journal articles and patents. Syntex could boast of possessing the most advanced equipment such infrared and NMR spectrometers, at a time when neither the pharmaceutical industries, as Djerassi wrote, "nor my Alma Mater, the University of Wisconsin, had such equipment which proved to be enormously useful for steroid research" [4]. The work done paved the way to the first synthesis of a steroid contraceptive in 1953, "the Pill" that changed the habits of mankind [4]. In the frame of said research program, several other plants were investigated and several extracts were described scientifically with avalanches of data, thus starting the evolution of the empirical medicaments of the traditional medicine into scientifically assessed plant extracts, as it was the case of *Genipa americana* and *Gardenia jasminoides* Ellis that yielded commercial genipin. A more recent example is the food supplement from *Serenoa repens* (Permixon™ Pierre Fabre, Giem, France). Thus genipin is a part of the cultural legacy from Carl Djerassi.

Because it is recognized that genipin, rather than geniposide, is the main compound that exerts pharmacological activities [5], there is interest in its isolation and purification for use in therapy and in the manufacture of food commodities [6]. Genipin is choleretic; anti-depressant; antidiabetic; anticancer; antithrombotic; anti-inflammatory; antibacterial; gastro-, hepato-, and neuro-protective [7]; it prevents lipid peroxidation; and it protects the hippocampal neurons against the Alzheimer's amyloid beta protein [8].

The biochemical significance of genipin emerges in fact from a number of research projects in the areas of the therapies of vascular diseases, diabetes, hepatic dysfunctions, as well as biofabrication, dentistry, ophthalmology, wound healing and regeneration of nerve, tendon and other tissues, just to mention a few [9–20].

The main specifications of genipin (CAS 6902-77.8) are the following: white crystalline powder soluble in water, methanol, ethanol and acetone; chemical formula $C_{11}H_{14}O_5$; molar mass 226.226 g/mol; melting point 120–121 °C; UV (CH$_3$OH) λ_{max} 240 nm.

Although a minor molar ratio of genipin to chitosan is necessary for crosslinking the latter or other aminated polymers, genipin is expensive because during its preparation a large quantity is wasted owing to homopolymerization. Therefore *Fusarium solani* was screened as an efficient source of β-glucosidase for genipin preparation from geniposide by extraction with a 10-L ethyl acetate-water biphasic system. HPLC data indicated that immediately after hydrolysis genipin was extracted from the aqueous phase into ethyl acetate thus escaping homopolymerization that would have been unavoidable in the aqueous phase. With *Fusarium solani* ACCC 36223, genipin in the ethyl acetate phase was 15.7 g/L, corresponding to yields of 0.65 g·L^{-1}·h^{-1}. Efficient substrate conversion and side reactions elimination were the key aspects of the advances made; moreover genipin was easily purified via the sole recrystallization. These most recent conceptual and technical approaches will certainly permit a more convenient production at lower price [21]. The available methods for recovery of genipin and geniposide were described, as well as the methods for genipin and geniposide identification and quantification based on instrumental analyses. Analytical methods for genipin were implemented in view of effective recovery protocols [22–37].

1.2. Characteristic Properties of Chitosans

Chitins and chitosans of various origins along with some of their derivatives are today protagonists in the scenario of wound healing, tissue engineering, gene therapy, and other advanced biomedical areas, owing to their unique properties. Basic information on these polysaccharides, relevant to the title topic, can be found in books and review articles [38–49].

Being biocompatible, non-toxic, stable, sterilizable and biodegradable, chitosan exhibits most appreciated properties that enhance its versatility in the biomedical and biotechnological fields, such as immunostimulation, activation of macrophages, mucoadhesion, antimicrobial activity, and well assessed chemistry [50]. Moreover, chitosan can also be prepared in a variety of forms, namely hydrogels and xerogels, powders, beads, films, tablets, capsules, microspheres, microparticles, nanofibrils, textile fibers, and inorganic composites. Chitosan is today a protagonist in advanced fields, for example it is a high performing non-viral vector for DNA and gene delivery.

1.3. Genipin-Crosslinked Chitosan Hydrogels

Genipin reacts promptly with chitosan, as well as with proteins or amines in general [51], as a bi-functional crosslinking compound, thus producing blue-colored fluorescent hydrogels. The reaction between chitosan and genipin is well understood for a variety of experimental conditions and yields composites and complexes with no cytotoxicity for human and animal cells (Figure 1).

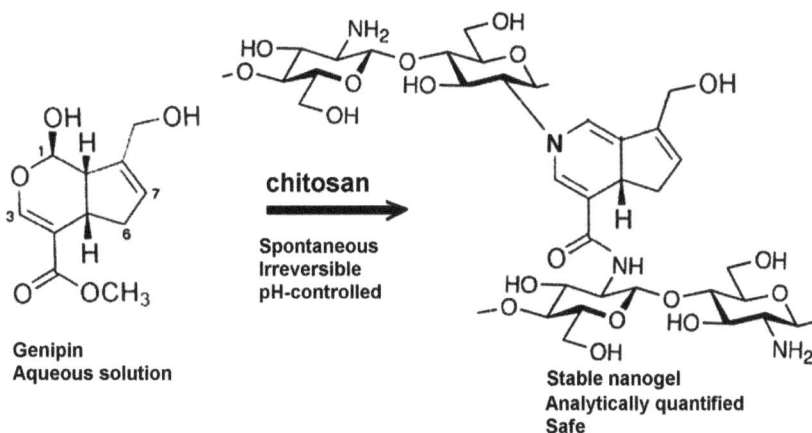

Figure 1. Genipin crosslinks chitosan spontaneously at a quite small molar ratio. On the right, two chitosan chains (represented by their structural units) react covalently with one mole of genipin to yield two newly formed chemical functions, namely the monosubstituted amide and the tertiary amine.

Chitosan nanoparticles crosslinked with genipin were prepared by reverse microemulsion that allowed obtaining highly monodisperse nanogels. Whilst ^{13}C· NMR provides evidence of the reaction as shown in Figure 2, the incorporation of genipin into chitosan was also confirmed and quantitatively evaluated by ^{1}H· NMR [52,53]. The hydrodynamic diameter of the genipin-chitosan nanogels ranged from 270 to 390 nm and no difference was found when the crosslinking degree was varied. The hydrodynamic diameters of the nanoparticles increased slightly at acidic pH. TEM data indicated that the nanoparticles had average diameters of from 3 to 20 nm and that they are spherical, have nearly uniform particle size distribution, and are not affected by particle agglomeration; these being interesting qualities for drug delivery. The progressive protonation of the amino groups as pH

3

decreases was confirmed by measuring the electrokinetic potential of the nanogels. The variation of water solubility of chitosan due to the crosslinking with genipin is a compromise between the decrease of crystallinity and the elastic force within the generated network. There was an insignificant variation of the average hydrodynamic diameter of the nanoparticles with pH, but a large progressive variation of zeta potential (from +30 to −7 mV) in the pH interval 4–9, indicative of the fact that these hydrogels are pH-sensitive [53].

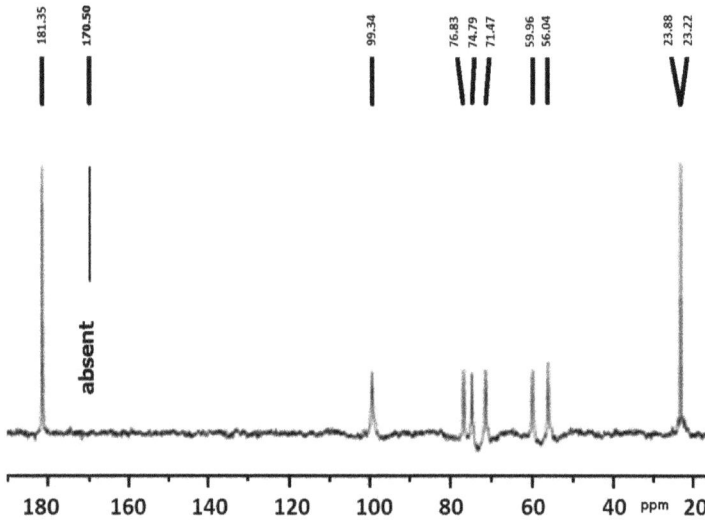

Figure 2. ^{13}C NMR spectrum of chitosan film crosslinked with genipin 0.10%. At 23.0 ppm the resonance signal of alkyl groups in the crosslinked chitosan was attributed to the chitosan + genipin linkage. The signal at 170.5 ppm, assigned to the ester group of plain genipin, disappeared as a consequence of the reaction, thus the resonance at 181.3 ppm is assigned to the amide generated by the reaction between the amino group of chitosan and the ester group of genipin.

Biodegradable polymers such as chitosan need to be crosslinked in order to modulate their general properties and to last long enough for delivering drugs over a desired period of time. Certain chemicals have been used for crosslinking chitosan such as glutaraldehyde, tripolyphosphate, ethylene glycol, diglycidyl ether and diisocyanate. However, the synthetic crosslinking reagents are all more or less cytotoxic and may impair the biocompatibility of a chitosan delivery system. Hence, efforts were made to provide crosslinking reagents that have low cytotoxicity and that form stable and biocompatible crosslinked products, for example tyrosinase was used to mediate quinone tanning of chitosans [54].

Chitosan can be used as a scaffold for tissue regeneration in porous or film form. However, as a porous scaffold it exhibits mechanical weakness: for example, when mouse fibroblasts are cultured on a porous chitosan scaffold, the narrow site of attachment and general weakness drastically depress the adhesion, and the cells tend to become round thus losing their prerogatives. On the other hand, when the cells are anchored to a surface endowed with stiffness, the cellular growth and differentiation rates are better, migration and aggregation become evident, and the cellular shapes favored by the support are those associated with proliferation, differentiation, and apoptosis.

A number of research teams are interested in using genipin to obtain stable and biocompatible chitosan hydrogels. Yao *et al.* indicated that the fibroblasts adhering to the GEN-chitosan scaffolds were 2.29 times more numerous compared to the fibroblasts on the pristine scaffold surface, the characteristic modulus of a genipin-crosslinked chitosan surface, ≈2.3 GPa, being nearly the double of the control [55]. A genipin crosslinked scaffold retains its own chemical composition while having significantly larger Young's modulus and hardness. Thus, the mechanical properties of a porous chitosan scaffold in film form are enhanced by genipin. In turn the enhanced general properties induce cell adhesion and proliferation in the modified porous scaffold. Interestingly, the pore size and mechanical properties of chitosan can be tuned for specific tissue regeneration.

Moreover, survival and proliferation of L929 fibroblasts were up-regulated after crosslinking with genipin, especially 0.5% genipin solutions. Analogous data were presented by Bao *et al.* for carboxymethylchitosan crosslinked with genipin in an article devoted to the mechanical properties of that class of hydrogels and their biocompatibility [56].

GEN-Chitosan hydrogels were prepared by incubation of solutions containing mixtures of genipin and chitosan in different ratios. They turned dark blue and became opaque, owing to exaggerated quantity of genipin. Upon lyophilization they yielded macroporous sponge-like scaffolds [57]. The *in vitro* cytocompatibility of hydrogels was demonstrated with L929 fibroblasts by the MTT method, in agreement with other authors [58]. The macroporous structure of the chitosan hydrogels could be tailored so that they enhanced their storage modulus, and also altered their hydrophilicity and swelling properties. The crosslinked hydrogels did not induce cytotoxic effects. Flow cytometry showed that fibroblasts possessed good viability on the surface of crosslinked gels (88.4%–90.9%) close to that on blank plates (93.7%) and chitosan films (92.8%). There was no quantitative difference in apoptotic or dead cells, thus crosslinking had little influence on viability, but the stiffness was the most important parameter influencing cell growth and made it possible to switch the cells either toward round or spreading shapes upon modulation of the hydrogel stiffness.

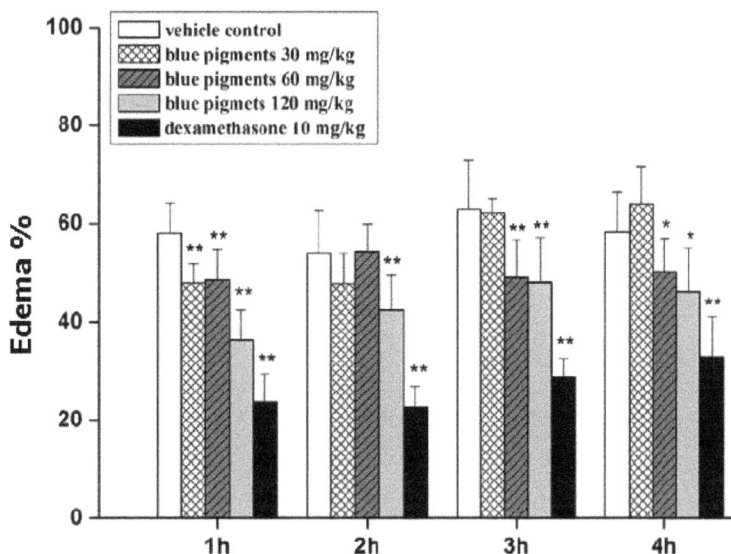

Figure 3. Anti-inflammatory effect of genipin + glycine blue pigment on edema in mice. Maximum edema depression was observed 1 h after edema induction. Notably, treatment with blue pigments at 120 mg/kg reduced edema by *ca.* 22% (from *ca.* 60% to 38%) at 1 h, whereas the positive control, dexamethasone (10 mg/kg) depressed the edema by *ca.* 35% at 1 h. The data are indicative of the safety of genipin which alleviates inflammation by exerting biochemical actions favorable to the organism. Reproduced from [33].

Safety of use was amply confirmed, thus chitosan composites have been taken into consideration in view of the production of biomaterials with desirable physicochemical and biological properties for tissue engineering. It is worth emphasizing that the safety of genipin has been demonstrated by a number of approaches, for example although the blue pigments derived from genipin and aminoacids have been used as value-added colorants for foods over the last 20 years in Eastern Asia, their biochemical significance has been explored as recently as in 2012 by Wang, Q.S. *et al.* who demonstrated that blue pigments did not only inhibit iNOS and COX-2 gene expression induced by LPS and subsequent production of NO and PGE_2, but reduced the production of cytokines (TNF-α, IL-6) induced by LPS in macrophages by the inhibition of signaling cascades leading to the activation of NF-κB [33]. Therefore, the results of recent studies provide strong scientific evidence for blue pigments to be developed as nutraceuticals for prevention and treatment of chronic inflammatory diseases. Nitric oxide is recognized as a mediator and regulator of inflammatory responses being produced in high amounts by iNOS in activated inflammatory cells. Blue pigments were found to inhibit LPS-induced NO production (Figure 3). Also the mRNA expression of iNOS was decreased by blue pigments, confirming the inhibitory effect of blue pigments on the NO production. That work also showed that blue pigments inhibited the expression of iNOS mRNA in LPS-stimulated macrophages. The effect of blue pigments on LPS-induced iNOS expression might result from the transcriptional inhibition of the iNOS gene. Further, the anti-inflammatory effect of blue pigments might be attributed to their inhibitory effect on PGE_2 production through blocking COX-2 gene and protein expression. Therefore, besides being safe, genipin is also beneficial owing to its positive action when present in functional foods.

1.4. Scope

The scope of this review article is therefore the evaluation of recent data on the title crosslinked hydrogels and scaffolds for (i) the description and appreciation of the experimental advances foreseen in the earlier review article by Muzzarelli [1], and actually performed in the most recent years; (ii) efficacy of said compounds in the upgrading of the chemical and biochemical characteristic properties and in their capacity to modulate the behavior of cells and stem cells *in vitro* and *in vivo*; (iii) treatments for the regeneration of the joint cartilage; and (iv) treatment and materials for enhanced osteogenesis and for the regeneration of bone, optionally including inorganic composites. Aspects related to sources of the raw materials, analytical chemistry, drug delivery, and economics are also considered. Understanding the synergy of the two ingredients of this class of composites in providing safety of use and efficacy in the pre-clinical trials is a further object of this work.

2. Therapies Based on the Genipin-Crosslinked Chitosan Alginate Complex

The encapsulation technology permits long-term delivery of desired therapeutic products to certain parts of the body without the use of immuno-suppressant drugs. In the study by Nayak *et al.* microcapsules composed of sericin and alginate micro bead as inner core with the outer chitosan shell were prepared for therapeutic applications [59]. The sericin-alginate micro beads were prepared via ionotropic gelation under high applied voltage and were coated with chitosan and crosslinked with genipin, their size (300–800 μm) depending on flow rate and applied voltage. Alamar Blue assay and confocal microscopy showed high cell viability and uniform cell distribution within the sericin-alginate-chitosan microcapsules indicative of the favorable internal microenvironment for the cells. In fact glucose consumption, urea secretion rate and intracellular albumin content increased in the microcapsules. The genipin crosslinked chitosan provided a fluorescent coating around the capsules, that appear light blue in the visible light. The coating is mandatory for the chemical stability of the capsules, particularly when dealing with *in vivo* delivery for therapeutic purposes: in fact, encapsulated hepatocytes generated enriched populations of metabolically and functionally active cells of therapeutic usefulness in acute liver failure.

Covalent crosslinking with genipin of chitosan alginate microcapsules provides significant enhancement of the microcapsule strength and resistance while maintaining the permeability. Aldana *et al.* reported the compatibility of genipin with other polymers such as polyvinylpyrrolidone and its suitability in making the soft, tough material for controlled drug release [60].

Moisture absorption can be modulated even in polyamide 6,6 fabrics when the surface is functionalized with the aid of GEN-chitosan hydrogels [61].

Scaffolds of chitosan-coated alginate were fabricated in a layer-by-layer fashion by Colosi *et al.* for drug delivery. A dispensing system based on two coaxial needles delivered alginate and calcium chloride solutions yielding alginate fibers according to designed patterns. Coating of the alginate fiber with chitosan and subsequent crosslinking with genipin assured the endurance of the scaffold. The crosslinking imparted to the scaffold a hierarchical chemical structure. Typical hepatic functions such as albumin and urea secretion and induction of CYP3A4 enzyme activity following drug administration were quite good [62].

Further chemical characteristics of the GEN-chitosan alginate combinations were reported by a number of authors: Silva *et al.* [63,64] further assessed the advantages of the LbL technique to generate functional biomimetic surfaces with tuned mechanical and chemical properties, and for the preparation of nanostructured multilayers tubes combining LbL and template leaching. Those works demonstrate the versatility and feasibility of LbL assembly to generate nanostructured devices including freestanding membranes with tunable permeability, besides mechanical and biological properties, by acting on the molar ratios of each polysaccharide and genipin.

Microcapsules with a calcium alginate core and a genipin-crosslinked chitosan alginate coating were prepared by Ranganath *et al.* [65] with good control over size, membrane thickness and density. Importantly, the authors interrelated membrane thickness, chitosan + alginate reaction rate constant, and diffusion coefficient. The large immunoglobulin and carbonic anhydrase were found to diffuse promptly. Compared to other microcapsules, the genipin treated microcapsules exhibited improved permselectivity of small nutrient compounds and proteins, while excluding antibodies.

3. Stem Cells in Regenerative Medicine

Many studies have indicated that human adipose-derived stem cells can easily be obtained from liposuction waste or arthroscopy, and maintained in a stable undifferentiated state during *in vitro* expansion [66]. Although ASC can be induced toward a chondrogenic phenotype with growth factors, that would make them suitable for cartilage regeneration, the use of exogenous growth factors may be impractical for clinical use owing to economic or regulatory issues. Instead, a bioactive scaffold exhibiting appropriate environmental signals may provide an alternative approach for inducing ASC chondrogenesis.

Stem cell transplantation has enormous potential in regenerative medicine [67,68]. Microencapsulation of stem cells is an efficient procedure for the preservation of viability and biochemical properties especially for the therapy of heart diseases. Paul *et al.* reported the use of microcapsules made of GEN-chitosan alginate for the delivery of human adipose stem cells (hASC) with the aim to increase the implant retention in the infarcted myocardium for maximum therapeutic benefit [69]. Under hypoxic conditions *in vitro*, the microencapsulated cells overexpressed higher amount of biologically active vascular endothelial growth factor (VEGF), thus the *in vivo* potential was investigated by using immunocompetent rats after induction of myocardial infarction. For this, rat groups received either empty control microcapsules, or 1.5×10^6 free hASC, or 1.5×10^6 microencapsulated hASC. Results showed significant retention (3.5-fold higher) of microencapsulated hASCs compared to free hASCs 10 weeks after transplantation. Microencapsulated hASC led to attenuated infarct size compared to the free hASC group and the empty microcapsule group, besides enhanced vasculogenesis and improved cardiac function. Therefore, the GEN-chitosan alginate microcapsules are deemed to be a valid aid for the significant improvement of the cardiac functions.

Porous cartilage-derived matrix (CDM) from porcine articular cartilage induced *in vitro* chondrogenic differentiation of adult human stem cells or chondrocytes without exogenous growth factors. Cheng *et al.* 2011 investigated CDM scaffolds crosslinked with genipin, seeded with ASC, and then cultured for four weeks [70]. By using a 0.05% genipin solution, a crosslinking degree of 50% was achieved (involving *ca.* one-half of the available lysine or hydroxylysine units in the cross linkage), and the ASC-seeded constructs exhibited no significant contraction during the culture. Moreover, the expression of cartilage-specific genes, the accumulation of cartilage-related macromolecules and the development of mechanical properties were comparable to the original CDM, thus making the cartilage-derived matrix crosslinked with genipin a contraction-free biomaterial suitable for cartilage tissue engineering [71]. Contraction of engineered cartilage *in vivo* invariably creates a gap between the construct and the nearby native cartilage. The fact that integration of engineered scaffolds with surrounding native tissue is crucial for both immediate functionality and long-term performance of the tissue enables one to appreciate the important contribution of genipin in solving this issue particularly crucial for articular cartilage repair because the surrounding native cartilage has scarce regeneration potential.

4. Genipin-Crosslinked Chitosan in Gastric Infections

Chitosan microspheres have been explored for pharmaceutical applications as drug delivery hydrogels in particular for the treatment of *Helicobacter pylori* gastric infection, owing to their mucoadhesive capacity. *H. pylori* is an important human pathogen that recognizes specific carbohydrate receptors, such as the fucose receptor, and produces the vacuolating cytotoxin, which induces inflammatory responses and modulates the cell junction integrity of the gastric epithelium. Nogueira *et al.* [72] proposed a different application of chitosan microspheres that capture and remove those bacteria from infected patients, taking advantage of their adhesive capacity for mucins and bacteria: they studied the effect of genipin on stability, size, charge and mucoadhesion of chitosan microspheres in acidic media. Chitosan microspheres (*ca.* 170 μm) were produced by ionotropic gelation and subsequently covalently crosslinked with genipin to various extents. Both the zeta potential and the swelling capacity of chitosan microspheres decreased with increasing crosslinking. When immersed in simulated gastric fluid with pepsin for seven days, the microspheres crosslinked with 10 mM genipin for 1 h presented an adequate balance between capacity to bind mucins, and free amino groups required for maintaining chitosan stability in acidic environment, and had gastric retention time *ca.* 2 h *in vivo*; they did not dissolve but simply doubled their size to *ca.* 345 μm. Although they maintained their *in vitro* mucoadhesion to soluble gastric mucins at pH 3.6 and 6.5 and presented an *in vivo* retention time of *ca.* 2 h in the stomach of mice, they were unable to lead to satisfactory results owing to the presence of pepsin [72]. Delmar *et al.* found that although the reaction between chitosan and genipin is apparently slow and might require up to four days for completion, the alteration of the pH within the small range of 4.00–5.50 dramatically affects the reaction, yielding hydrogels differing in appearance and properties [73]. The ability to manipulate the hydrogel properties, while adjusting the conditions slightly, provides a powerful and useful tool when designing chitosan hydrogels. Furthermore, the dependence of the properties on tiny pH modifications is crucial when reproducible and reliable results are sought.

On the other hand, Lin, Y.H. *et al.* [74] combined fucose-conjugated chitosan with genipin in genipin-crosslinked fucose-chitosan/heparin nanoparticles to encapsulate amoxicillin and straightforwardly make contact with the bacterium on the gastric epithelium. The nanoparticles effectively reduced drug release to gastric acids, and then released amoxicillin to inhibit *H. pylori* growth, and reduced disruption of the cell junction protein in the infected areas. Thus, with amoxicillin-loaded nanoparticles, a more complete *H. pylori* clearance effect was observed, and the *H. pylori* associated gastric inflammation in an infected animal model was definitely reduced. Thakur *et al.* also made use of highly stable GEN-chitosan beads in simulated gastric and intestinal fluids for the release of amoxicillin [75].

Further Aspects of Enhanced Antibacterial Efficacy

The antibacterial efficacy of GEN-chitosan has been well assessed by Wang R *et al.* who mixed the antifouling polymer poly(sulfobetaine methacrylate) and the bactericidal N-[(2-hydroxy-3-trimethylammonium) propyl] chitosan, in one coating onto a silicone surface, by using genipin [76]. Yu, S.H. *et al.* developed fucoidan-shelled chitosan beads with the purpose of oral delivery of berberine to inhibit the growth of bacteria [20]. Furthermore, a nanoparticles + beads complex was developed by incorporation of berberine-loaded chitosan + fucoidan nanoparticles in the fucoidan-shelled chitosan beads. It served as a drug carrier to delay the berberine release in simulated gastric fluid, with lag time of 2 h, and it effectively inhibited the growth of common clinical pathogens.

Drug administration via the oral mucosa is an attractive strategy owing to good patient compliance, prolonged localized drug effect, and avoidance of gastrointestinal drug metabolism and first-pass elimination. Oral drug delivery systems need to maintain an intimate contact with the mucosa lining in the wet conditions of the oral cavity for long enough to allow drug release and absorption. Chitosan and its derivatives have been examined for this purpose. In particular, the genipin treated carboxymethyl–hexanoyl chitosan, an amphiphilic chitosan derivative with quite good swelling ability, cytocompatibility and water solubility, was studied under physiological conditions [15]. Inspired by the wet adhesion of marine mussel adhesive protein, Xu, J.K. *et al.* [77] developed an oral drug delivery system using a catechol-chitosan hydrogel. The catechol functional groups were covalently linked to chitosan, and the resulting modified chitosan was crosslinked with genipin. Catechol groups significantly enhanced mucoadhesion *in vitro* when in contact with porcine mucosal membrane up to 6 h, whereas the chitosan hydrogels lost contact after 1.5 h. The new hydrogel systems sustained the release of lidocaine for about 3 h. *In vivo*, buccal patches adhered to rabbit buccal mucosa, thus lidocaine was monitored easily in the rabbit serum owing to the intimate contact provided by the highly mucoadhesive catechol-GEN-chitosan [77].

5. Genipin-Crosslinked Collagen/Gelatin for the Regeneration of the Cartilage

Collagen and gelatin have been treated with genipin in a number of instances with the intention to involve them in the treatment of cartilage: in their review Elzoghby *et al.* reported that the mechanism of crosslinking of proteins by genipin involves the free amino groups of lysine in the protein [78]. Recent advances on the regeneration of cartilage have been reviewed by Muzzarelli *et al.* [79] and Bottegoni *et al.* [80]. Because a crosslinker is necessary to improve and optimize mechanical strength, porosity and degradability of single biopolymers and their composites, Bi, L. *et al.* crosslinked chitosan + collagen scaffolds by using genipin [81]: the compressive strength was directly dependent on the genipin concentration in the interval 0.1% to 1.0% and on the crosslinking time. The pore size, degradation rate and swelling ratio changed significantly with different crosslinking conditions. For a similar genipin crosslinked chitosan + collagen material, Yan, L.P. *et al.* found that rabbit chondrocytes adhered well to the surface of the scaffolds and reached confluence, thus they suggested that the genipin crosslinked chitosan plus collagen may be a promising formulation for articular cartilage scaffolding [82].

Genipin-crosslinked recombinant human gelatin (preferred owing to its homogeneity in molecular weight and precisely defined properties) was efficiently internalized in the cells without inducing cytotoxicity. Genipin was also used to stabilize the structure of gelatin–dextran micelles encapsulating tea polyphenol to avoid disintegration after dilution: the crosslinked micelles were stable with no size change. Kuo *et al.* preferred bovine pituitary extract for study of the formation of neocartilage in chitosan/gelatin scaffolds, and cultured bovine knee chondrocytes in it over 28 days; collagen-II was synthesized in the constructs, thus demonstrating the chondrocytic phenotype of proliferated chondrocytes [83]. Yin *et al.* blended chitosan plus polylactide with collagen-II to fabricate layered composites potentially applicable in cartilage repair [84]. The manufacture of marine collagen porous structures crosslinked with genipin under high pressure CO_2 was investigated by Fernandes-Silva *et al.*: shark skin collagen was used to prepare prescaffolds by freeze-drying. Under dense CO_2 atmosphere, crosslinking of collagen with genipin was protracted for 16 h [85].

Modulation of the proliferation and matrix synthesis of chondrocytes by dynamic compression on genipin-crosslinked chitosan plus collagen scaffolds was also observed [86]. Dynamic compression is an important physical stimulus for the physiology of chondrocytes and engineering of the articular cartilage. Rabbit chondrocytes were seeded in genipin-crosslinked chitosan plus collagen and then cultured for three days prior to two weeks of cyclic compression of 40% strain, 0.1 Hz, and 30 min/day. The cell proliferation and the total GAG deposition was directly dependent on genipin quantity and dynamic compression.

While fully biocompatible gelatin microspheres for intra-articular drug delivery were prepared by Kawadkar *et al.* [87], emulsion-crosslinking was used by Kawadkar and Chauhan [88] to prepare chitosan microspheres with various concentrations of genipin and drug-to-polymer ratios for intra-articular delivery of flurbiprofen. The mean particle size was in the range 5.18–9.74 µm with drug entrapment up to 81%. The optimized microspheres were able to release the drug for more than 108 h. The biocompatibility of the microspheres in the rat knee joints was confirmed by histopathology. Pharmacokinetic data pointed out the extended release of flurbiprofen from microspheres in comparison with solution, so that GEN-chitosan qualified as an injectable drug vehicle. According to the *in vivo* data, the microspheres made of chitosan and genipin are safe for the synovia and maintain the drug concentration within the arthritic knee joint. In fact, Sarem *et al.* explained how genipin helps chitosan with gelatin scaffolds act as replacements of load-bearing soft tissues and concluded that the 1% genipin-crosslinked chitosan 40 with gelatin 60 scaffolds, prepared at room temperature for 24 h was a promising replacement of missing segments of load-bearing soft tissues. Owing to the hydrogel characteristics of said biopolymers, a significant amount of fluid can be retained in their structure. Hence, they can produce high compressive modulus comparable to native load-bearing soft tissues: these materials can be used for treatment or repair of articular cartilage and meniscus. This was attributed to the formation of polyelectrolyte complexes via ionic interactions between the amino groups of chitosan and the anionic groups in gelatin. Finally the presence of genipin depressed the depolymerization of chitosan by lysozyme, while still permitting an adequate degradation and ingrowth of newly formed tissues, *i.e.*, remodeling of tissues under loadbearing conditions [89].

The genipin-crosslinked chitosan + gelatin scaffolds containing bovine pituitary extract are quite effective in the regeneration of neocartilage. The histological and immunochemical staining showed chondrogenesis in the culture of bovine knee condrocytes using said scaffolds in a medium containing bovine pituitary extract. In addition, collagen-II was synthesized in the constructs, demonstrating the chondrocytic phenotype of proliferated bovine knee chondrocytes in said scaffolds over 28-day culture. In practice, the addition of the extract to the culture medium accelerated the regeneration of the articular cartilage [83].

Scaffolds made of chitosan, collagen and gelatin were prepared with the aid of carbon dioxide saturated solutions [90], the chitosan dissolution in carbonic acid being no longer a laboratory curiosity. Chitosan was dissolved upon saturation of an aqueous colloidal chitosan suspension with gaseous CO_2 under mild conditions: atmospheric pressure and room temperature. As CO_2 dissolves in water, the pH decreases owing to formation of carbonic acid. This is a fine demonstration that commonly used inorganic and organic acids are no longer indispensable for the dissolution of chitosan. Moreover, this approach simplifies and optimizes the preparation of wound dressing materials, where the presence of undesirable and cytotoxic counter ions such as acetate is avoided. The use of CO_2 for chitosan dissolution made the scaffold preparation more reproducible and economically sustainable. Porosity data are in Table 1; the values of other parameters were: dissolution degree (30%), lysozyme-induced degradation (5% after 168 h), good antioxidant properties, and especially absence of cytotoxicity against mouse NIH 3T3 fibroblasts, the viability being at the level of the control. The fibroblasts grew uniformly in the pores of the chitosan-protein structure owing to optimal swelling of the scaffold and even distribution of collagen, to which cells have high affinity. When the chitosan + protein scaffolds are treated with genipin, the color intensity reveals the extent of the crosslinking, as shown in Figure 4.

Table 1. Average cross-sectional areas of the pores and porosity of the chitosan-protein scaffolds crosslinked with different concentrations of genipin. The reaction time and temperature are not specified and depended on the protocol adopted.

Genipin Concentration (%, *w/w*)	Cross-Sectional Area of the Pores (μm^2)	Porosity (%)
0.0	187.9 ± 101.0	25.75 ± 1.47
0.5	274.6 ± 123.8	33.13 ± 1.30
1.0	533.9 ± 259.8	39.95 ± 1.25
2.0	1066.4 ± 396.7	44.75 ± 1.50

Data from [90], and set in novel tabular presentation.

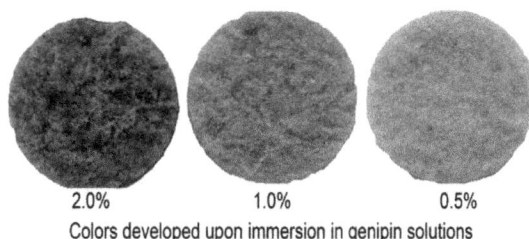

2.0%   1.0%   0.5%

Colors developed upon immersion in genipin solutions

Figure 4. Chitosan-protein scaffolds crosslinked with genipin at various concentrations, under identical conditions. The intensity of the blue color is an indication of the extent of crosslinking.

5.1. Fibrin, Poly-L-Lysine, Heparin, Hyaluronan

Fibrin is another biopolymer studied in conjunction with chitosan and genipin, in order to develop biocompatible microspheres. Human chondrocytes cultured on the composite substrate were viable during the culture period (28 days): at the end the composite substrate showed 41% more collagen-II and 13% higher production of sulfated glycosaminoglycans with respect to the amounts found at 14 days. The de-differentiated chondrocytes cultured in monolayer on the composite could re-acquire characteristics of differentiated cells without using three-dimensional substrates or chondrogenic media [91].

Nihn *et al.* established a quantitative framework for controlled release systems in order to deliver genipin into protein-based hydrogels. Covalent coupling between genipin and primary amines in fibrin gels obeys second-order kinetics in genipin concentration with an effective activation energy of -71.9 ± 3.2 kJ· mol^{-1}. Genipin-crosslinked fibrin clots are resistant to fibrinolytic degradation as measured by rheology. Interestingly, active genipin can be delivered from poly(D,L-lactide-co-glycolide) matrices to gels at rates that are comparable to the characteristic rate of incorporation in fibrin networks. Poly-L-lysine, the homopolymer of the essential aminoacid Lys, has been used in other works as a standard tissue culture coating to promote cellular adhesion. Films manufactured after blending it with chitosan enhance cellular attachment to chitosan. Genipin improves and maintains the stability of chitosan + poly-L-lysine blends [92]. In the article by Mekhail *et al.*, the viability of fibroblasts was enhanced more than six-fold after treating with genipin the gels containing nearly equal weight of the two polymers; the proliferation was enhanced up to five folds. Fibroblast viability was significantly enhanced after crosslinking the 60:40 and 50:50 gels; whilst it was not on 100 and 80:20 gels [93]. This is in agreement with data by other groups that reported that genipin improves the biocompatibility of various chitosan formulations [94–96]. Heparin was covalently crosslinked to the chitosan scaffolds by using genipin, which bound fibroblast growth factor-2 (FGF-2) while preserving its biological activity. At 1 µg/mL approximately 80% of the FGF-2 bound to the scaffold that showed good cytocompatibility and therefore could be used for the delivery of neural stem cells and growth factors for central nervous system repair [97]. Likewise, carrageenan and carboxymethylcellulose have been studied in conjunction with genipin [98,99].

Finally, a few words on hyaluronan, a polysaccharide possessing aspects of chemical similarity to chitosan. Jalani *et al.* [100] reported a method to produce tough, thermogelling, safe, injectable hydrogels made of chitosan and hyaluronan co-crosslinked with β-glycerophophate and genipin. The highly homogeneous gels form within half an hour, *i.e.*, faster than gels crosslinked with either genipin or β-glycerophosphate. The shear strength of co-crosslinked hydrogels was 3.5 kPa, higher than for any chitosan-based gel reported. Chondrocytes and nucleus pulposus cells thrive inside the gels and produce large amounts of collagen-II. The gelation took place *in vivo* within a short time after injection in rats and remained well localized for more than one week while the rats were healthy and active [100].

Swelling and degradation of the chitosan + hyaluronan complex could be controlled with the aid of genipin. Optimization of said parameters is important because cells need to adhere first to the scaffold and then proliferate. BMP-2 was immobilized in the complex by electrostatic attraction. Thus, high loading and sustained BMP-2 release were achieved. Reverse transcriptase PCR indicated that released BMP-2 facilitated osteogenesis at all stages, this being a key factor for bone regeneration [101].

5.2. Genipin Treatment of Tendon Cells and Matrix

Fessel *et al.* studied the effects of genipin treatment on tendon cells and their matrix, with a view to *in vivo* application to the repair of partial tendon tears [102]. They observed that post-treatment cell survival may be adequate to eventually repopulate and stabilize the tissue. Superficial blue pigmentation that qualitatively indicates genipin was documented. Inherent sample fluorescence (λ_{ex} 510–560 nm; λ_{em} 590 nm) was measured with the aid of fluorescence microscopy: uniform fluorescence throughout the cut sections indicated homogeneous crosslinking. According to a model the cell viability in tendon explants was concentration dependent, but cell survival was not time dependent. The model predicted that *ca.* 50% of cells would remain viable at genipin concentrations of 6.2 mM applied for 72 h, with decreasing cell viability after prolonged incubation. While many cells remained alive with genipin 5 mM or less for the time spans studied, effects on cell metabolism occurred at lower concentrations. The model predicted a 50% drop in metabolic activity at 0.4 mM after 72 h, this being consistent with reduced cell motility at similar concentrations. Although reduced collagen-I expression was also consistent with reduced metabolic activity, apoptosis markers and matrix degradation markers were not affected, this being a favorable finding regarding the potential of genipin for *in vivo* application. It appeared that 5 mM (and maybe lower concentrations) could induce relatively rapid crosslinking while leaving sub-populations of resident cells viable. In view of the clinical application of *in situ* crosslinking to arrest tendon tear propagation, it was deemed that the genipin concentration of 5 mM or slightly lower with an exposure of 72 h would be reasonable for *in vivo* studies.

It should be added that the said study focused only on acute and relatively rapid crosslinking effects, neither investigating whether long-term administration of genipin at lower doses (<1 mM) could possibly achieve a cumulative functional effect, nor observing the continuity of the documented effects. At a more basic level, so far the authors did not investigate how genipin crosslinks are actually formed or remain stable within the tissue, these aspects depending upon time and concentration [102].

6. Bone Regeneration

Chitosan respects the physiological bone formation and healing processes, and most importantly it enhances favorably the biochemical responses, owing to its inherent immunostimulating properties and susceptibility to lysozyme. Bone healing involves a sequence of events that should not be disturbed by the presence of a composite or scaffold. At the time of a fracture, the disruption of bone architecture and vascular network results in loss of mechanical stability and local decrease in oxygen and nutrients. The inflammatory response is accompanied by the activation of macrophages and infiltration of platelets that release various cytokines, which probably play a role in the initiation of the repair process by acting on various cells: post-fracture periosteal osteoprogenitor cells and osteoblasts differentiate to produce new bone. This process involves fibroblast growth factors and bone morphogenetic proteins. To provide crucial nutrient supplies to the cells, new blood vessels develop into the fracture callus. The matrix composed of various collagen isotypes develops, which may be important for presenting cytokines to receptive cells.

The chemical and technological versatility of chitosan enables researchers to prepare elaborated composites: for example, the research works on bone regeneration with the aid of bone cements have become more refined in terms of the effects of chitosan composites on the cells involved in the healing process. With the advent of nanotechnology the applications of fairly non-toxic nanocrystalline hydroxyapatite extends from bone repair and augmentation to the delivery of drugs, growth factors and genetic material to the bone: for this purpose, particles of uniform size with controlled morphology can be manufactured by using macromolecules as templates. A number of advantages have become evident, particularly when nano-hydroxyapatite is crystallized using biomimetic methods, or when the biopolymers are submitted to biomineralization. The hydroxyapatite nanoparticles influence favorably the morphology of attached cells, as a consequence of the adsorption of extracellular matrix proteins from serum, that in turn bind osteoblast precursors. Thus, an additional peculiarity of chitosan is emerging from most recent studies, namely the capacity to influence both the mineralization and the cell activity.

Chitosan, *N*-carboxymethyl chitosan, fibroin and poly(L-lactic acid) are at the basis of new strategies useful to stimulate stem cells to become osteoblasts, and to make co-cultures of osteoblasts and osteoclasts. With the aid of chitosan and sulfated chitosan, important advances have been made in the field of the delivery of human and recombinant bone morphogenetic proteins, in particular the morphogenetic protein-2 that exhibits a positive effect in every step of the bone regeneration. Again, the advances made in histology, cell culture, and cytology are accompanied by equally important contributions from material chemistry [103–105].

6.1. Enhanced Osteogenesis

The combined antibacterial efficacy and the modulation of the osteoblast behavior are very important for orthopedic applications. For example, Wu, F. *et al.* [106] loaded genipin crosslinked carboxymethylchitosan hydrogel with gentamycin and achieved enhanced adhesion, proliferation, and differentiation of osteoblasts besides full inhibition of *Staphylococcus aureus*. The degradation time of the CM-chitosan as well as the cellular responses depended on the genipin quantity. The loading of gentamycin increased of course the antibacterial efficiency, but it was also beneficial for the osteoblastic cell responses. Overall, the biocompatibility of the prepared hydrogel could be tuned by acting on the concentration of genipin and gentamycin, which interact with the available chemical groups of chitosan [106].

Bone defects surgically produced in sheep and rabbit models have been treated with freeze-dried modified chitosans because they promote direct endochondral ossification. Moreover, the pattern of bone regeneration has been studied in an osteoporotic experimental model with bone morphogenetic protein linked to chitosan [107]. The chitosan + collagen scaffolds had high proliferative effect if the degree of acetylation of chitosan was high, regardless of molecular weight. SEM demonstrated that MC3T3-E1 osteoblasts grew well on all tested scaffolds.

To provide a novel and effective drug delivery system that can enhance osteogenesis, Wang, G.C. *et al.* [108,109] evaluated the BMP-2 adsorption and release of bone morphogenetic protein-2 (BMP-2) on the superficial hydroxyapatite nanostructure of a coated GEN-chitosan that exhibited a loading efficiency of 65% (1.30 µg). The release of BMP-2 lasted for over 14 days in simulated body fluid, and induced an increase in alkaline phosphatase, indicative of osteogenic differentiation of seeded BMSCs. Hydroxyapatite + GEN-chitosan scaffolds also stimulated mRNA expression of osteogenic differentiation markers, namely osteopontin for three days, and osteocalcin for 14 days. Thus the superficial biomimetic HAp nanostructure within the composite scaffold promoted osteogenic differentiation *in vitro*. The hydroxyapatite nanostructure within the organic porous scaffold worked as a calcium source and absorption/release agent that suggested the design of bioactive scaffold for bone engineering. Nano-hydroxyapatite was included in GEN-chitosan films that were shown to be deprived of cytotoxicity against L929 cells [110].

Chitosan enhances bone and cartilage formation owing to its structural similarity to the extracellular matrix of bone cells. The pore sizes of the traditional scaffolds however, are seldom within the optimal range for cell ingrowth (100–400 μm), and the pores are not interconnected enough to allow cell infiltration. Moreover, high porosity and degradability of those chitosan scaffolds make them too weak for the purpose of bone repair. To qualify for these applications, membranes and scaffolds have to be stable in a wet environment, and should withstand simultaneously chemical and mechanical stresses, thus they need to be stabilized. Allowing cell infiltration and efficient nutrient and waste exchange has been an important goal in making good tissue scaffolds. Interconnected pores not only achieve this goal but also allow neovascularization that prevent the formation of a necrotic core in the scaffold. In practice, a microcomputer controls the movement of the robotic arm in *x* and *z* directions and the platform in y direction. The needle is raised 400 μm during each layer-step in the *z* direction. By raising the syringe needle one layer-step in the *z* direction, successive layers of hydrogel fibers are deposited onto previous layers in a 0°–90° pattern. According to the data obtained by Liu, I.H. *et al.*, osteoblasts secreted collagen-I in the first week and gradually differentiated into osteocytes; later on they expressed alkaline phosphatase that initiated the mineralization process by providing an alkaline environment and facilitating formation and nucleation of calcium phosphate in the GEN-chitosan 3D-plotted scaffolds [111]. In this respect it is interesting to note that the above mentioned article by Colosi *et al.* [62] contains valuable detailed information on a simplified technological approach to the manufacture of 3D-plotted scaffolds.

Loss of fibrous structure upon contact with aqueous solutions could limit practical utilization of 3D-plotted or electrospun chitosan nanofibers. To meet the demands for tissue engineering uses [112], post-electrospinning crosslinking may be performed to inhibit solubility and improve mechanical properties [113].

The invention by Lelkes and Frohbergh provided a scaffold comprising an electroprocessed, genipin-crosslinked mineralized chitosan nanofiber, definitely purified and capable of supporting the maturation of osteoblasts [114]. Osteoprogenitor cells, mesenchymal cells, stem cells, and osteocytes, are equally suitable. In this area of stimuli sensitive (smart) biomaterials that can facilitate regeneration of critical-size bone lesions, Frohbergh *et al.* [115] tested biomimetic scaffolds electrospun from chitosan expected to promote tissue repair in a critical size calvarial defect. Chitosan fibre mats are non-toxic and biocompatible, and therefore are convenient as filtration membranes and scaffolds for tissue engineering. They compared the *in vitro* ability of electrospun genipin-crosslinked chitosan to analogous scaffolds containing hydroxyapatite.

The cellular metabolic activity exhibited a biphasic behavior, indicative of initial proliferation followed by subsequent differentiation for all scaffolds. After three weeks in maintenance medium, ALP activity of mMSCs seeded onto GEN-chitosan + HAp scaffolds was approximately twice as much that of cells cultured on plain GEN-chitosan. Said mineralized scaffolds were also osseointegrative *in vivo*, as inferred from the enhanced bone regeneration in a murine model of critical size calvarial defects. Treatment of the lesions induced a 38% increase in the area of de novo generated mineralized tissue after three months, whereas plain scaffolds led to 10% increase. Mineralized scaffolds pre-seeded with mMSCs yielded 45% new mineralized tissue formation in the defects. The presence of HAp in the scaffolds significantly enhances their osseointegrative capacity: thus the mineralized GEN-chitosan may represents an unique biomaterial with possible clinical relevance for the repair of critical calvarial bone defects. In fact, autografts are seldom available, and alternative materials lead to poor integration with the host bone, owing to the absence of periosteum that contains osteoprogenitor cells and is crucial for growth and remodeling of bone tissue. The same authors [116] developed a one-step platform to electrospin nanofibrous scaffolds from chitosan, which also contain hydroxyapatite nanoparticles and are crosslinked with genipin, to stimulate osteoblast differentiation and maturation similar to the periosteum. The average fiber diameters of the electrospun scaffolds were 227 ± 154 nm as spun, and increased to 335 ± 119 nm after crosslinking with genipin. The Young's modulus of the composite fibrous scaffolds was 142 ± 13 MPa, which is similar to that of the natural periosteum [49].

Whilst genipin strongly improves the mechanical properties of composite rods, Pu *et al.* reported that the GEN-chitosan network made the hydroxyapatite composite rods much more stable than the controls against enzymatic depolymerization. The latter was tested with lysozyme for 72 h and the results indicated a weight loss rate of 4% for samples deprived of genipin, *versus* 0.5% for the genipin-containing samples. The bending strength and bending modulus of the crosslinked rods could reach 161 MPa and 7.2 GPa, respectively, with *ca.* 60% and 26% increase compared with un-crosslinked ones. Therefore, GEN-chitosan + hydroxyapatite composite rods with excellent mechanical properties are useful for internal fixation of bone fractures [117].

Rheological studies by Pandit *et al.* [118] demonstrated that the stiffness of hydrogels made of methylcellulose, chitosan and agarose increased upon crosslinking the chitosan with increasing amounts of genipin. Based on these results, gels crosslinked with 0.5% (w/v) genipin, having one third of the amino groups of chitosan crosslinked, exhibited a stiffness of 502 ± 64.5 Pa along with optimal characteristics to support bone regeneration. The gelling time decreased with increasing genipin concentrations. Again, these favorable chemical effects were accompanied by modulated cellular behaviors: in fact, proliferation of human umbilical vascular endothelial cells decreased by 10.7 times with increasing gel stiffness, in contrast to fibroblasts and osteoblasts, where it increased with gel stiffness by 6.37 and 7.8 times, respectively. Expression of differentiation markers by osteoblasts (osteocalcin, osteopontin and alkaline phosphatase) were significantly enhanced in the 0.5% (w/v) crosslinked gel, which also demonstrated enhanced mineralization by Day 25. Gels crosslinked with 0.5% (w/v) genipin still demonstrated significant bacterial inhibition [118].

Ge S.H. *et al.* explored the viability and differentiation of periodontal ligament stem cells on a nanohydroxyapatite-coated GEN-chitosan scaffold *in vitro* and *in vivo* [119]. Cell seeded scaffolds were used in a rat calvarial defect model, and new bone formation was assessed by hematoxylin and eosin staining at 12 weeks postoperatively. When seeded on said scaffolds the stem cells exhibited significantly greater viability, and up-regulated the bone-related markers to a greater extent than for controls, thus the calvarial bone repair was obtained.

6.2. Technical Advances, Novel Know-How and Improvements

Genipin spontaneously crosslinks chitosan gels, microspheres, and fibers (Figure 5), even in the presence of PVA, PVP, PEO, fibroin or gelatin. Data by Nwosu *et al.* indicate that being able to improve the stability of smart GEN-chitosan + PVP with solely physical manipulations without altering the initial chemical composition is an aspect of potential medical applications, for example, in wound dressing and drug delivery [120].

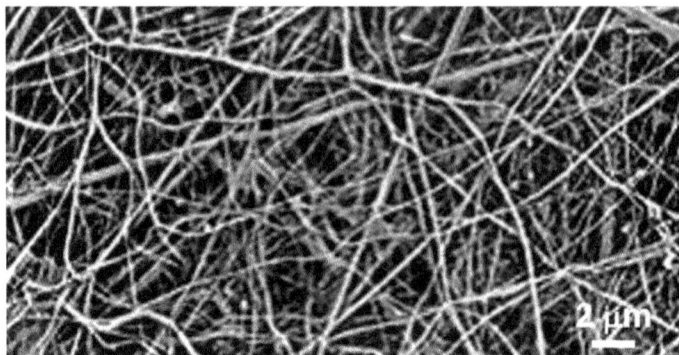

Figure 5. Typical genipin-crosslinked chitosan fiber mat obtained from electrospun chitosan mat subsequently treated with genipin: the nanofibrous form is preserved.

GEN-Chitosan + bioglass + PVP scaffolds were prepared by Yao Q.Q. *et al.* [121]. Improved resistance to enzymatic degradation of the scaffolds was obtained while enhancing biocompatibility and porosity of the bioglass scaffolds for good adhesion and proliferation of pre-osteoblasts MC3T3-E1 cells. The latter within the scaffolds showed well stretched F-actin bundles after four-day incubation. These structures, known as tunneling nanotubes, could mediate the intercellular transfer of organelles, plasma membrane components and cytoplasmic substances. Furthermore, those scaffolds qualified for the controlled delivery of the antibiotic vancomycin.

While several laboratories reported that that 0.025% genipin is sufficient to fully crosslink chitosan, Austero *et al.* [122] and Donius *et al.* [123] used 0.10% for two reasons: first, excess of the crosslinker was added to react with all the amino groups of chitosan; second, at concentrations less than 0.10%, the mats were soluble in acidic and neutral solutions. The 0.10% concentration enabled to make improvements in stability and versatility. Typical values for chitosan-genipin fiber mats were fiber diameter 176 ± 106 nm; fiber mat porosity $55.5\% \pm 6.8\%$; fiber-fiber contacts per unit volume $4.45 \ \mu m^{-3}$ [122,123]. The same research team [124] manufactured fibrous chitosan + hydroxyapatite composite scaffolds with 1 and 10 wt % mineral contents by electrospinning. The fibers, crosslinked with genipin, contained crystalline hydroxyapatite at 10% additive. Electrospun fibers had diameters 122–249 nm, in the range of those of fibrous collagen found in the extracellular matrix of bone. Young's modulus and ultimate tensile strength of the various crosslinked composite were in the range 2–15 MPa. Osteocytes seeded onto the mineralized fibers demonstrated good biocompatibility.

In a dynamic perfusion culture apparatus the flow rate of a culture medium through a chitosan scaffold influenced cell proliferation and expression of bone marker genes. The feasibility of culturing osteoblast-like MG-63 cells on chitosan + genipin scaffolds was demonstrated by Su, W.T. *et al.* [125] who confirm that flow perfusion cultures can improve cellular distribution and abundance in porous scaffolds. In fact, mineralized tissue is distributed throughout the entire area of the scaffolds cultured under flow perfusion; on the contrary on scaffolds cultured under static conditions fewer mineral depots are detected. These results suggest that osteoblast-like MG-63 cells seeded in chitosan scaffolds produce more calcium and phosphate in dynamic culture. In the latter, cells seeded into a scaffold receive mechanical stimulation provided by the mobile fluid, and undergo multilayered 3D growth and organization, thereby enhancing bone-related gene and phenotypic expression [126,127]; thus collagen-I and OCN gene expression are higher in dynamic culture than in static culture. A continuously pumping gas-permeable silicon tube allows for optimal exchange of gases, so that the O_2 partial pressure in the medium is higher than in a conventional culture plate. These results are in agreement with earlier articles, which reported that cultivation of osteoblast-like cells and rat bone marrow stem cells on 3D scaffolds in a perfusion culturing system enhances growth, differentiation, and mineralized matrix production *in vitro*.

Fully characterized scaffolds in terms of porosity, pore size, swelling, wettability, compressive strength, and mass loss were seeded with human mesenchymal stem cells and evaluated with respect to osteogenic differentiation with incubation time [128]. Experimental groups included GEN-chitosan + β-tricalcium phosphate that displayed interconnected honeycomb-like microstructures with porosity >65%. There was linear dependence of both water contact angle and pore size with crosslinker concentration. The metabolic activity of hMSCs seeded in those scaffolds was significantly higher than for controls, as well as their mineralization after 21 days of incubation in osteogenic medium.

Guided bone regeneration membranes prevent soft tissue infiltration into the graft space during dental interventions that involve bone grafting. Chitosan materials have shown promise in this area, owing to their biocompatibility and predictable biodegradability, but longer degradation time periods are needed for clinical applications. Chitosan membranes were electrospun using chitosan (70% deacetylated, 312 kDa, 5.5% *w/v*), with or without the addition of 5 or 10 mM genipin, in order to extend the degradation to meet the clinical time of four months [129]. Genipin addition resulted in median fibre diameters 144 nm, 154 nm respectively for 5 mm and 10 mm crosslinked, and 184 nm for uncrosslinked samples. The ultimate tensile strength of the mats was increased by 165% to 32 MPa with 10 mm crosslinking as compared to the uncrosslinked mats. Genipin-chitosan samples exhibited only 22% degradation based on mass loss, as compared to 34% for uncrosslinked mats at four months *in vitro*. Therefore electrospun chitosans may benefit from the reaction with genipin and can meet clinical degradation time frames for guided bone regeneration.

7. Conclusions

Recent works have produced a wealth of data on the advantages offered by new physical forms of chitosan stabilized with genipin [95,130–140]. The advances made become quite noticeable when the works dated 2010–2015 (amounting to >82% of the bibliography below) are compared to some most significant key articles [5,11,27,41,50,141–146] published in the 2000–2005 quinquennium.

The research topics currently considered span from the extraction of genipin to the preparation of advanced devices suitable for tissue engineering. The biochemical processes adopted in the preparation of the pigment aim at drastic improvement of the process yields, and cost abatement. Likewise, novel views on the manipulation of human cells permit to refine the preparation of scaffolds intended for the maximum performance: for example in the current year electrospun chitosan layers have been put on the market by Advanced BioMatrix, San Diego, CA, USA, to complete a catalog of analogous products based on collagen, gelatin, alginate, hyaluronan and more, and intended for cell culture and tissue engineering.

After the layer-by-layer scaffolds, today it is possible to generate functional biomimetic surfaces with tuned mechanical and biochemical properties. Simplified technological approaches to the manufacture of 3D-plotted scaffolds are providing exciting developments.

Authors unanimously recognize that electrospun nanofibrous chitosan scaffolds crosslinked with genipin are most attractive for tissue engineering. Nevertheless, the preparations described in the above cited articles most often omit indispensable details. It should be underlined that minor amounts of genipin are necessary when the latter is the ingredient of a scaffold, and that the final scaffold is expected to be light blue instead of black as shown in some illustrations. Besides that, it is becoming mandatory to express the calculations in terms of molar ratios between the components of a scaffolds whose quality has to be assessed by advanced instrumental analyses if they are expected to be recognized for purity, functionality and consistency.

Acknowledgments: The authors are grateful to Marilena Falcone and Simonetta Pirani, Central Library, Polytechnic University, Ancona, Italy, for assistance in handling the bibliographic information, and to Maria Weckx for help with the preparation of the manuscript.

Author Contributions: R.A.A. Muzzarelli, M. El Mehtedi., C. Bottegoni and A. Gigante contributed equally to the scientific elaboration of this review. A. Aquili provided assistance. The project was coordinated by R.A.A. Muzzarelli.

Conflicts of Interest: The authors declare no conflict of interest. This work stemmed from a personal initiative of the authors, who did not apply for financial support.

References

1. Muzzarelli, R.A.A. Genipin-crosslinked chitosan hydrogels as biomedical and pharmaceutical aids. *Carbohydr. Polym.* **2009**, *77*, 1–9. [CrossRef]

2. Djerassi, C.; Gray, J.D.; Kinci, F.A. Naturally occurring oxygen heterocycles. IX. Isolation and characterization of genipin. *J. Org. Chem.* **1960**, *25*, 2174–2177. [CrossRef]

3. Djerassi, C.; Nakano, T.; James, A.N.; Zalkow, L.H.; Eisenbraun, E.J.; Shoolery, J.N. Terpenoids. XLVII. The structure of genipin. *J. Org. Chem.* **1961**, *26*, 1192–1206. [CrossRef]

4. Djerassi, C. *This man's Pill: Reflections on the 50th Birthday of the Pill*; Oxford University Press: Oxford, UK, 2001.

5. Zheng, H.Z.; Dong, Z.H.; Yu, J. *Modern Research and Application of Chinese Traditional Medicine*; Academy Press: Beijing, China, 2000; Volume 4, pp. 3166–3172.

6. Yu, Y.; Feng, X.L.; Gao, H.; Xie, Z.L.; Dai, Y.; Huang, X.J.; Kurihara, H.; Ye, W.C.; Zhong, Y.; Yao, X.S. Chemical constituents from the fruits of *Gardenia jasminoides* Ellis. *Fitoterapia* **2012**, *83*, 563–567. [CrossRef] [PubMed]

7. Winotapun, W.; Opanasopit, P.; Ngawhirunpat, T. One enzyme catalyzed simultaneous plant cell disruption and conversion of released glycoside to aglycone combined with *in situ* product separation as green one-pot production of genipin from Gardenia fruit. *Enzym. Microbial Technol.* **2013**, *53*, 92–96. [CrossRef] [PubMed]

8. Manickam, B.; Sreedharan, R.; Elumalai, M. Genipin the natural water soluble crosslinking agent and its importance in the modified drug delivery systems: An overview. *Curr. Drug Deliv.* **2014**, *11*, 139–145. [CrossRef] [PubMed]

9. Hald, E.S.; Steucke, K.E.; Reeves, J.A.; Win, Z.; Alford, P.W. Long-term vascular contractility assay using genipin-modified muscular thin films. *Biofabrication* **2014**, *6*, 045005. [CrossRef] [PubMed]

10. Li, Y.H.; Cheng, C.Y.; Wang, N.K.; Tan, H.Y.; Tsai, Y.J.; Hsiao, C.H.; Ma, D.H.K.; Yeh, L.K. Characterization of the modified chitosan membrane crosslinked with genipin for the cultured corneal epithelial cells. *Colloids Surf.* **2015**, *126*, 237–244. [CrossRef] [PubMed]

11. Lin, C.K.; Lee, T.C.J.; Sung, H.W. Chemical Modification of Biomedical Materials with Genipin. U.S. Patent 6,608,040, 9 July 2003.

12. Wu, J.L.; Liao, C.Y.; Wang, Z.; Cheng, W.Z.; Zhou, N.; Wang, S.; Wan, Y. Chitosan-polycaprolactone microspheres as carriers for delivering glial cell line-derived neurotrophic factor. *React. Funct. Polym.* **2011**, *71*, 925–932. [CrossRef]

13. Wu, J.L.; Liao, C.Y.; Zhang, J.; Cheng, W.Z.; Zhou, N.; Wang, S.; Wan, Y. Incorporation of protein-loaded microspheres into chitosan-polycaprolactone scaffolds for controlled release. *Carbohydr. Polym.* **2011**, *86*, 1048–1054. [CrossRef]

14. Yang, Y.M.; Zhao, W.J.; He, J.H.; Zhao, Y.H.; Ding, F.; Gu, X.S. Nerve conduits based on immobilization of nerve growth factor onto modified chitosan by using genipin as a crosslinking agent. *Eur. J. Pharm. Biopharm.* **2011**, *79*, 519–525. [CrossRef] [PubMed]

15. Liu, T.Y.; Lin, Y.L. Novel pH-sensitive chitosan-based hydrogel for encapsulating poorly water-soluble drugs. *Acta Biomater.* **2010**, *6*, 1423–1429. [CrossRef] [PubMed]

16. Liu, Y.; Kim, H.I. Characterization and antibacterial properties of genipin-crosslinked chitosan/poly(ethylene glycol)/ZnO/Ag nanocomposites. *Carbohydr. Polym.* **2012**, *89*, 111–116. [CrossRef] [PubMed]

17. Liu, Y.G.; Xie, M.B.; Wang, S.B.; Zheng, Q.Y.; Chen, A.Z.; Deng, Q.J. Facile fabrication of high performances MTX nanocomposites with natural biomembrane bacterial nanoparticles using GP. *Mater. Lett.* **2013**, *100*, 248–251. [CrossRef]

18. Peng, C.; Zhao, S.Q.; Zhang, J.; Huang, G.Y.; Chen, L.Y.; Zhao, F.Y. Chemical composition, antimicrobial property and microencapsulation of mustard (*Sinapis alba*) seed essential oil by complex coacervation. *Food Chem.* **2014**, *165*, 560–568. [CrossRef] [PubMed]

19. Song, X.; Wu, H.; Li, S.; Wang, Y.; Ma, X.; Tan, M. Ultrasmall chitosan-genipin nanocarriers fabricated from reverse microemulsion process for tumor photothermal therapy in mice. *Biomacromolecules* **2015**, *16*, 2080–2090. [CrossRef] [PubMed]

20. Yu, S.H.; Wu, S.J.; Wu, J.Y.; Wen, D.Y.; Mi, F.L. Preparation of fucoidan-shelled and genipin-crosslinked chitosan beads for antibacterial application. *Carbohydr. Polym.* **2015**, *126*, 97–107. [CrossRef] [PubMed]

21. Zhu, Y.Y.; Zhao, B.T.; Huang, X.D.; Chen, B.; Qian, H. A substrate fed-batch biphasic catalysis process for the production of natural cross linking agent genipin with *Fusarium solani* ACCC 36223. *J. Microbiol. Biotechnol.* **2015**, *25*, 814–819. [CrossRef] [PubMed]

22. Bergonzi, M.C.; Righeschi, C.; Isacchi, B.; Bilia, A.R. Identification and quantification of constituents of *Gardenia jasminoides* Ellis (Zhizi) by HPLC-DAD-ESI-MS. *Food Chem.* **2012**, *134*, 1199–1204. [CrossRef] [PubMed]

23. Cano, E.V.; Echeverri-Lopez, L.F.; Gil-Romero, J.F.; Correa-Garces, E.A.; Zapata-Porras, S.P. Colorant Compounds Derived from Genipin or Genipin Containing Materials. U.S. Patent Application 2014/0350127 A1, 27 November 2014.

24. Chen, J.; Wu, H.; Xu, G.B.; Dai, M.M.; Hu, S.L.; Sun, L.L.; Wang, W.; Wang, R.; Li, S.P.; Li, G.Q. Determination of geniposide in adjuvant arthritis rat plasma by ultra-high performance liquid chromatography tandem mass spectrometry and its application to oral bioavailability and plasma protein binding. *J. Pharm. Biomed. Anal.* **2015**, *108*, 122–128. [CrossRef] [PubMed]

25. Chen, J.F.; Fu, G.M.; Wan, Y.; Liu, C.M.; Chai, J.X.; Li, H.G.; Wang, J.T.; Zhang, L.N. Enrichment and purification of gardenia yellow from *Gardenia jasminoides* var *radicans Makino* by column chromatography. *J. Chromatogr.* **2012**, *893–894*, 43–48. [CrossRef] [PubMed]

26. Gao, Y.; Sun, Y.; Wang, Y.; Zhang, J.; Xu, B.; Zhang, H.; Song, D. A practical and rapid method for the simultaneous isolation, purification and quantification of geniposide from the fruit of *Gardenia jasminoides* Ellis by MSPD extraction and UFLC analysis. *Anal. Methods* **2013**, *5*, 4112–4118. [CrossRef]

27. Lee, S.W.; Lim, J.M.; Bhoo, S.H.; Paik, Y.S.; Hahn, T.R. Colorimetric determination of amino acids using genipin from *Gardenia jasminoides*. *Anal. Chim. Acta* **2003**, *480*, 267–274. [CrossRef]

28. Ramos-de-la-Pena, A.M.; Renard, C.M.G.C.; Wicker, L.; Montanez, J.C.; García-Cerda, L.A.; Contreras-Esquivel, J.C. Environmental friendly cold-mechanical/sonic enzymatic assisted extraction of genipin from genipap (*Genipa americana*). *Ultrason. Sonochem.* **2014**, *21*, 43–49. [CrossRef] [PubMed]

29. Ramos-de-la-Pena, A.M.; Renard, C.M.G.C.; Montanez, J.C.; Reyes-Vega, M.; Contreras-Esquivel, J.C. A review through recovery, purification and identification of genipin. *Phytochem. Rev.* **2014**. [CrossRef]

30. Ramos-Ponce, L.M.; Vega, M.; Saldoval-Fabian, G.C.; Colunga-Urbina, E.; Rodriguez-Gonzalez, F.J.; Contreras-Esquivel, J.C. A simple colorimetric determination of the free amino groups in water soluble chitin derivatives using genipin. *Food Sci. Biotechnol.* **2010**, *19*, 683–689. [CrossRef]

31. Reich, M.S.; Akkus, O. Sporicidal efficacy of genipin: A potential theoretical alternative for biomaterial and tissue graft sterilization. *Cell Tissue Bank* **2013**, *14*, 381–393. [CrossRef] [PubMed]

32. Yang, Y.; Zhang, T.; Yu, S.; Ding, Y.; Zhang, L.; Qiu, C.; Jin, D. Transformation of geniposide into genipin by immobilized β-glucosidase in a two-phase aqueous-organic system. *Molecules* **2011**, *16*, 4204–4295. [CrossRef] [PubMed]

33. Wang, Q.S.; Xiang, Y.; Cui, Y.L.; Lin, K.M.; Zhang, X.F. Dietary blue pigments derived from genipin, attenuate inflammation by inhibiting LPS-induced iNOS and COX-2 espression via the NF-κB inactivation. *PLoS ONE* **2012**, *7*, e34122. [CrossRef] [PubMed]

34. Wu, S.W.; Horn, G. Genipin-Rich Material and Its Use. WO Patent 2013/070682 A1, 16 May 2013.

35. Zhang, M.; Ignatova, S.; Hu, P.; Liang, Q.; Wang, Y.; Sutherland, I.; Jun, F.W.; Luo, G. Cost-efficient and process-efficient separation of geniposide from *Gardenia jasminoides* Ellis by high performance counter-current chromatography. *Sep. Purif. Technol.* **2012**, *89*, 193–198. [CrossRef]

36. Zhou, M.; Zhuo, J.; Wei, W.; Zhu, J.; Ling, X. Simple and effective large-scale preparation of geniposide from fruit of *Gardenia jasminoides* Ellis using a liquid-liquid two-phase extraction. *Fitoterapia* **2012**, *83*, 1558–1561. [CrossRef] [PubMed]

37. Zhou, T.; Liu, H.; Wen, J.; Fan, G.; Chai, Y.; Wu, Y. Fragmentation study of iridoid glycosides including epimers by liquid chromatography-diode array detection/electrospray ionization mass spectrometry and its application in metabolic fingerprint analysis of *Gardenia jasminoides* Ellis. *Rapid Commun. Mass Spectrom.* **2010**, *24*, 2520–2528. [CrossRef] [PubMed]

38. Chiono, V.; Pulieri, E.; Vozzi, G.; Ciardelli, G.; Ahluwalia, A.; Giusti, P. Genipin-crosslinked chitosan/gelatin blends for biomedical applications. *J. Mater. Sci. Mater. Med.* **2008**, *19*, 889–898. [CrossRef] [PubMed]
39. Croisier, F.; Jerome, C. Chitosan based biomaterials for tissue engineering. *Eur. Polym. J.* **2013**, *49*, 780–792. [CrossRef]
40. Muzzarelli, R.A.A. *Chitin*; Pergamon Press: Oxford, UK, 1977.
41. Muzzarelli, R.A.A.; Muzzarelli, C. Chitosan chemistry: Relevance to the biomedical sciences. In *Advances in Polymer Science*; Heinze, T., Ed.; Springer Verlag: Berlin, Germany, 2005; pp. 151–209.
42. Muzzarelli, R.A.A. Chitins and chitosans for the repair of wounded skin, nerve, cartilage and bone. *Carbohydr. Polym.* **2009**, *76*, 167–182. [CrossRef]
43. Muzzarelli, R.A.A. Chitins and chitosans as immunoadjuvants and non-allergenic drug carriers. *Mar. Drugs* **2010**, *8*, 292–312. [CrossRef] [PubMed]
44. Muzzarelli, R.A.A. Chitin nanostructures in living organisms. In *Chitin Formation and Diagenesis*; Springer: Dordrecht, The Netherlands, 2011; Volume 34, pp. 1–34.
45. Muzzarelli, R.A.A. Chemical and technological advances in chitins and chitosans useful for the formulation of biopharmaceuticals. In *Chitosan-Based Systems for Biopharmaceuticals*; Sarmento, B., DasNeves, J., Eds.; Wiley: New York, NY, USA, 2012; pp. 3–22.
46. Muzzarelli, R.A.A. Nanochitins and nanochitosans, paving the way to eco-friendly and energy-saving exploitation of marine resources. In *Polymer Science: A Comprehensive Reference*; Elsevier: Amsterdam, The Netherlands, 2012; Volume 10, pp. 153–164.
47. Muzzarelli, R.A.A.; Baldassarre, V.; Conti, F.; Gazzanelli, G.; Vasi, V.; Ferrara, P.; Biagini, G. The biological activity of chitosan: An ultrastructural study. *Biomaterials* **1988**, *9*, 247–252. [CrossRef]
48. Muzzarelli, R.A.A.; Boudrant, J.; Meyer, D.; Manno, N.; DeMarchis, M.; Paoletti, M.G. A tribute to Henri Braconnot, precursor of the carbohydrate polymers science, on the chitin bicentennial. *Carbohydr. Polym.* **2012**, *87*, 995–1012. [CrossRef]
49. Muzzarelli, R.A.A.; El Mehtedi, M.; Mattioli-Belmonte, M. Emerging biomedical applications of nano-chitins and nano-chitosans obtained via advanced eco-friendly technologies from marine resources. *Mar. Drugs* **2014**, *12*, 5468–5502. [CrossRef] [PubMed]
50. Ravi Kumar, M.N.V.; Muzzarelli, R.A.A.; Muzzarelli, C.; Sashiwa, H.; Domb, A.J. Chitosan chemistry and pharmaceutical perspectives. *Chem. Rev.* **2004**, *104*, 6017–6084. [CrossRef] [PubMed]
51. Zhang, Y.N.; Yang, Y.; Guo, T.Y. Genipin-crosslinked hydrophobical chitosan microspheres and their interactions with bovine serum albumin. *Carbohydr. Polym.* **2011**, *83*, 2016–2021. [CrossRef]
52. Maggi, F.; Ciccarelli, S.; Diociaiuti, M.; Casciardi, S.; Masci, G. Chitosan nanogels by template chemical crosslinking in polyion complex micelle nanoreactors. *Biomacromolecules* **2011**, *12*, 3499–3507. [CrossRef] [PubMed]
53. Pujana, M.A.; Perez-Alvarez, L.; Iturbe, L.C.C.; Katime, I. Biodegradable chitosan nanogels crosslinked with genipin. *Carbohydr. Polym.* **2013**, *94*, 836–842. [CrossRef] [PubMed]
54. Muzzarelli, R.A.A.; Ilari, P.; Xia, W.; Pinotti, M.; Tomasetti, M. Tyrosinase mediated quinone tanning of chitinous materials. *Carbohydr. Polym.* **1994**, *24*, 294–300. [CrossRef]
55. Yao, C.K.; Liao, J.D.; Chung, C.W.; Sung, W.I.; Chang, N.J. Porous chitosan scaffold crosslinked by chemical and natural procedure applied to investigate cell regeneration. *Appl. Surf. Sci.* **2012**, *262*, 218–221. [CrossRef]
56. Bao, D.S.; Chen, M.J.; Wang, H.Y.; Wang, J.F.; Liu, C.F.; Sun, R.C. Preparation and characterization of double crosslinked hydrogel films from carboxymethylchitosan and carboxymethylcellulose. *Carbohydr. Polym.* **2014**, *110*, 113–120. [CrossRef] [PubMed]
57. Gao, L.; Gan, H.; Meng, Z.Y.; Gu, R.L.; Wu, Z.N.; Zhang, L.; Zhu, X.X.; Sun, W.Z.; Li, J.; Zheng, Y.; *et al.* Effects of genipin crosslinking of chitosan hydrogels on cellular adhesion and viability. *Colloids Surf.* **2014**, *117*, 398–405. [CrossRef] [PubMed]
58. Song, W.L.; Oliveira, M.B.; Sher, P.; Gil, S.; Noobrega, J.M.; Mano, J.F. Bioinspired methodology for preparing magnetic responsive chitosan beads to be integrated in a tubular bioreactor for biomedical applications. *Biomed. Mater.* **2013**, *8*, 045008. [CrossRef] [PubMed]
59. Nayak, S.; Dey, S.; Kundu, S.C. Silk sericin-alginate-chitosan microcapsules: Hepatocytes encapsulation for enhanced cellular functions. *Int. J. Biol. Macromol.* **2014**, *65*, 258–266. [CrossRef] [PubMed]

60. Aldana, A.A.; Gonzalez, A.; Strumia, M.C.; Martinelli, M. Preparation and characterization of chitosan/genipin/poly(*N*-vinyl-2-pyrrolidone) films for controlled release of drugs. *Mater. Chem. Phys.* **2012**, *134*, 317–324. [CrossRef]

61. Glampedaki, P.; Jocic, D.; Warmoeskerken, M.M.C.G. Moisture absorption capacity of polyamide 6,6 fabrics surface functionalised by chitosan-based hydrogel finishes. *Prog. Org. Coat.* **2011**, *72*, 562–571. [CrossRef]

62. Colosi, C.; Costantini, M.; Latini, R.; Ciccarelli, S.; Stampella, A.; Barbetta, A.; Massimi, M.; Devirgiliis, L.C.; Dentini, M. Rapid prototyping of chitosan-coated alginate scaffolds through the use of a 3D fiber deposition technique. *J. Mater. Chem.* **2014**, *2*, 6779–6791. [CrossRef]

63. Silva, J.M.; Duarte, A.R.C.; Caridade, S.G.; Picart, C.; Reis, R.L.; Mano, J.F. Tailored freestanding multi layered membranes based on chitosan and alginate. *Biomacromolecules* **2014**, *15*, 3817–3826. [CrossRef] [PubMed]

64. Silva, J.M.; Duarte, A.R.C.; Custodio, C.A.; Sher, P.; Neto, A.I.; Pinho, A.C.M.; Fonseca, J.; Reis, R.L.; Mano, J.F. Nanostructured hollow tubes based on chitosan and alginate multilayers. *Adv. Healthc. Mater.* **2014**, *3*, 433–440. [CrossRef] [PubMed]

65. Ranganath, S.H.; Tan, A.L.; He, F.; Wang, C.H.; Krantz, W.B. Control and enhancement of permselectivity of membrane-based microcapsules for favorable biomolecular transport and immunoisolation. *AIChE J.* **2011**, *57*, 3052–3062. [CrossRef]

66. Zhang, Z.J.; Zhang, H.; Kang, Y.; Sheng, P.Y.; Ma, Y.C.; Yang, Z.B.; Zhang, Z.Q.; Fu, M.; He, A.S.; Liao, W.M. miRNA expression profile during osteogenic differentiation of human adipose-derived stem cells. *J. Cell. Biochem.* **2012**, *113*, 888–898. [CrossRef] [PubMed]

67. Busilacchi, A.; Gigante, A.; Mattioli-Belmonte, M.; Muzzarelli, R.A.A. Chitosan stabilizes platelet growth factors and modulates stem cell differentiation toward tissue regeneration. *Carbohydr. Polym.* **2013**, *98*, 665–676. [CrossRef] [PubMed]

68. Debnath, T.; Ghosh, S.; Potlapuvu, U.S.; Kona, L.; Kamaraju, S.R.; Sarkar, S.; Gaddam, S.; Chelluri, L.K. Proliferation and differentiation potential of human adipose-derived stem cells grown on chitosan hydrogel. *PLoS ONE* **2015**, *10*, e0120803. [CrossRef] [PubMed]

69. Paul, A.; Chen, G.Y.; Khan, A.; Rao, V.T.S.; Shum-Tim, D.; Prakash, S. Genipin-crosslinked microencapsulated human adipose stem cells augment transplant retention resulting in attenuation of chronically infarcted rat heart fibrosis and cardiac dysfunction. *Cell Transplant.* **2012**, *21*, 2735–2751. [CrossRef] [PubMed]

70. Cheng, N.C.; Estes, B.T.; Young, T.H.; Guilak, F. Engineered cartilage using primary chondrocytes cultured in a porous cartilage-derived matrix. *Regen. Med.* **2011**, *6*, 81–93. [CrossRef] [PubMed]

71. Cheng, N.C.; Estes, B.T.; Young, T.H.; Guilak, F. Genipin-crosslinked cartilage-derived matrix as a scaffold for human adipose-derived stem cell chondrogenesis. *Tissue Eng.* **2013**, *19*, 484–496. [CrossRef] [PubMed]

72. Nogueira, F.; Goncalves, I.C.; Martins, M.C.L. Effect of gastric environment on Helicobacter pylori adhesion to a mucoadhesive polymer. *Acta Biomater.* **2013**, *9*, 5208–5215. [CrossRef] [PubMed]

73. Delmar, K.; Bianco-Peled, H. The dramatic effect of small pH changes on the properties of chitosan hydrogels crosslinked with genipin. *Carbohydr. Polym.* **2015**, *127*, 28–37. [CrossRef] [PubMed]

74. Lin, Y.H.; Tsai, S.C.; Lai, C.H.; Lee, C.H.; He, Z.S.; Tseng, G.C. Genipin-crosslinked fucose-chitosan/heparin nanoparticles for the eradication of *Helicobacter pylori*. *Biomaterials* **2013**, *34*, 4466–4479. [CrossRef] [PubMed]

75. Thakur, A.; Wanchoo, R.K.; Hardeep; Soni, S.K. Chitosan hydrogel beads: A comparative study with glutaraldehyde, epichlorohydrin and genipin as crosslinkers. *J. Polym. Mater.* **2014**, *31*, 211–223.

76. Wang, R.; Neoh, K.G.; Kang, E.T. Integration of antifouling and bactericidal moieties for optimizing the efficacy of antibacterial coatings. *J. Colloid Interface Sci.* **2015**, *438*, 138–148. [CrossRef] [PubMed]

77. Xu, J.K.; Strandman, S.; Zhu, J.X.X.; Barralet, J.; Cerruti, M. Genipin-crosslinked catechol-chitosan mucoadhesive hydrogels for buccal drug delivery. *Biomaterials* **2015**, *37*, 395–404. [CrossRef] [PubMed]

78. Elzoghby, A.O. Gelatin-based nanoparticles as drug and gene delivery systems: Reviewing three decades of research. *J. Control. Release* **2013**, *172*, 1075–1091. [CrossRef] [PubMed]

79. Muzzarelli, R.A.A.; Greco, F.; Busilacchi, A.; Sollazzo, V.; Gigante, A. Chitosan, hyaluronan and chondroitin sulfate in tissue engineering for cartilage regeneration: A review. *Carbohydr. Polym.* **2012**, *89*, 723–739. [CrossRef] [PubMed]

80. Bottegoni, C.; Muzzarelli, R.A.A.; Busilacchi, A.; Giovannini, F.; Gigante, A. Oral chondroprotection with nutraceuticals made of chondroitin sulphate plus glucosamine sulphate in osteoarthritis. *Carbohydr. Polym.* **2014**, *109*, 126–138. [CrossRef] [PubMed]

81. Bi, L.; Cao, Z.; Hu, Y.Y.; Song, Y.; Yu, L.; Yang, B.; Mu, J.H.; Huang, Z.S.; Han, Y.S. Effects of different crosslinking conditions on the properties of genipin-crosslinked chitosan/collagen scaffolds for cartilage tissue engineering. *J. Mater. Sci. Mater. Med.* **2011**, *22*, 51–62. [CrossRef] [PubMed]

82. Yan, L.P.; Wang, Y.J.; Ren, L.; Wu, G.; Caridade, S.G.; Fan, J.B.; Wang, L.Y.; Ji, P.H.; Oliveira, J.M.; Oliveira, J.T.; *et al.* Genipin-crosslinked collagen/chitosan biomimetic scaffolds for articular cartilage tissue engineering applications. *J. Biomed. Mater. Res.* **2010**, *95*, 465–475. [CrossRef] [PubMed]

83. Kuo, Y.C.; Wang, C.C. Effect of bovine pituitary extract on the formation of neocartilage in chitosan/gelatin scaffolds. *J. Taiwan Inst. Chem. Eng.* **2010**, *41*, 150–156. [CrossRef]

84. Yin, D.K.; Wu, H.; Liu, C.X.; Zhang, J.; Zhou, T.; Wu, J.J.; Wan, Y. Fabrication of composition-graded collagen/chitosan-polylactide scaffolds with gradient architecture and properties. *React. Funct. Polym.* **2014**, *83*, 98–106. [CrossRef]

85. Fernandes-Silva, S.; Moreira-Silva, J.; Silva, T.H.; Perez-Martin, R.; Sotelo, C.G.; Mano, J.F.; Duarte, A.R.C.; Reis, R.L. Porous hydrogels from shark skin collagen crosslinked under dense carbon dioxide atmosphere. *Macromol. Biosci.* **2013**, *13*, 1621–1631. [CrossRef] [PubMed]

86. Wang, P.Y.; Tsai, W.B. Modulation of the proliferation and matrix synthesis of chondrocytes by dynamic compression on genipin-crosslinked chitosan/collagen scaffolds. *J. Biomater. Sci. Polym. Ed.* **2013**, *24*, 507–519. [CrossRef] [PubMed]

87. Kawadkar, J.; Jain, R.; Kishore, R.; Pathak, A.; Chauhan, M.K. Formulation and evaluation of flurbiprofen-loaded genipin crosslinked gelatin microspheres for intra-articular delivery. *J. Drug Target.* **2013**, *21*, 200–210. [CrossRef] [PubMed]

88. Kawadkar, J.; Chauhan, M.K. Intra-articular delivery of genipin crosslinked chitosan microspheres of flurbiprofen: Preparation, characterization, *in vitro* and *in vivo* studies. *Eur. J. Pharm. Biopharm.* **2012**, *81*, 563–572. [CrossRef] [PubMed]

89. Sarem, M.; Moztarzadeh, F.; Mozafari, M. How can genipin assist gelatin/carbohydrate chitosan scaffolds to act as replacements of load-bearing soft tissues. *Carbohydr. Polym.* **2013**, *93*, 635–643. [CrossRef] [PubMed]

90. Gorczyca, G.; Tylingo, R.; Szweda, P.; Augustina, E.; Sadowska, M.; Milewski, S. Preparation and characterization of genipin crosslinked porous chitosan-collagen-gelatin scaffolds using chitosan-CO_2 solution. *Carbohydr. Polym.* **2014**, *102*, 901–911. [CrossRef] [PubMed]

91. Gamboa-Martinez, T.C.; Cruz, D.M.G.; Carda, C.; Ribelles, J.L.G.; Ferrer, G.G. Fibrin-chitosan composite substrate for *in vitro* culture of chondrocytes. *J. Biomed. Mater. Res.* **2013**, *101*, 404–412. [CrossRef] [PubMed]

92. Nihn, C.; Iftikhar, A.; Cramer, M.; Bettinger, C.J. Diffusion-reaction models of genipin incorporation into fibrin networks. *J. Mater. Chem. B* **2015**, *3*, 4607–4615.

93. Mekhail, M.; Jahan, K.; Tabrizian, M. Genipin-crosslinked chitosan/poly-L-lysine gels promote fibroblast adhesion and proliferation. *Carbohydr. Polym.* **2014**, *108*, 91–98. [CrossRef] [PubMed]

94. Mathew, A.P.; Oksman, K.; Pierron, D.; Harmand, M.F. Biocompatible fibrous networks of cellulose nanofibres and collagen crosslinked using genipin: Potential as artificial ligament/tendons. *Macromol. Biosci.* **2013**, *13*, 289–298. [CrossRef] [PubMed]

95. Norowski, P.A.; Mishra, S.; Adatrow, P.C.; Haggard, W.O.; Bumgardner, J.D. Suture pullout strength and *in vitro* fibroblast and RAW 264.7 monocyte biocompatibility of genipin crosslinked nanofibrous chitosan mats for guided tissue regeneration. *J. Biomed. Mater. Res.* **2012**, *100*, 2890–2896. [CrossRef] [PubMed]

96. Tseng, H.J.; Tsou, T.L.; Wang, H.J.; Hsu, S.H. Characterization of chitosan-gelatin scaffolds for dermal tissue engineering. *J. Tissue Eng. Regen. Med.* **2013**, *7*, 20–31. [CrossRef] [PubMed]

97. Skop, N.B.; Calderon, F.; Levison, S.W.; Gandhi, C.D.; Cho, C.H. Heparin crosslinked chitosan microspheres for the delivery of neural stem cells and growth factors for central nervous system repair. *Acta Biomater.* **2013**, *9*, 6834–6843. [CrossRef] [PubMed]

98. Hezaveh, H.; Muhamad, I.I. Controlled drug release via minimization of burst release in pH-response κ-carrageenan/polyvinyl alcohol hydrogels. *Chem. Eng. Res. Des.* **2013**, *91*, 508–519. [CrossRef]

99. Kaihara, S.; Suzuki, Y.; Fujimoto, K. *In situ* synthesis of polysaccharide nanoparticles via polyion complex of carboxymethyl cellulose and chitosan. *Colloids Surf.* **2011**, *85*, 343–348. [CrossRef] [PubMed]

100. Jalani, G.; Rosenzweig, D.H.; Makhoul, G.; Abdalla, S.; Cecere, R.; Vetrone, F.; Haglund, L.; Cerruti, M. Tough, *in situ* thermogelling, injectable hydrogels for biomedical applications. *Macromol. Biosci.* **2015**, *15*, 473–480. [CrossRef] [PubMed]

101. Nath, S.D.; Abueva, C.; Kim, B.; Lee, B.T. Chitosan-hyaluronic acid polyelectrolyte complex scaffold crosslinked with genipin for immobilization and controlled release of BMP-2. *Carbohydr. Polym.* **2015**, *115*, 160–169. [CrossRef] [PubMed]

102. Fessel, G.; Cadby, J.; Wunderli, S.; van Weeren, R.; Snedeker, J.G. Dose- and time-dependent effects of genipin crosslinking on cell viability and tissue mechanics: Toward clinical application for tendon repair. *Acta Biomater.* **2014**, *10*, 1897–1906. [CrossRef] [PubMed]

103. Muzzarelli, R.A.A.; Mattioli-Belmonte, M.; Tietz, C.; Biagini, R.; Ferioli, G.; Brunelli, M.A. Stimulatory effect on bone formation exerted by a modified chitosan. *Biomaterials* **1994**, *15*, 1075–1081. [CrossRef]

104. Muzzarelli, R.A.A. Chitosan scaffolds for bone regeneration. In *Chitin, Chitosan and Their Derivatives: Biological Activities and Applications*; Kim, S.K., Ed.; CRC Taylor & Francis: Boca Raton, FL, USA, 2010.

105. Muzzarelli, R.A.A. Chitosan composites with inorganics, morphogenetic proteins and stem cells, for bone regeneration. *Carbohydr. Polym.* **2011**, *83*, 1433–1445. [CrossRef]

106. Wu, F.; Meng, G.L.; He, J.; Wu, Y.; Wu, F.; Gu, Z.W. Antibiotic-loaded chitosan hydrogel with superior dual functions: Antibacterial efficacy and osteoblastic cell responses. *ACS Appl. Mater. Interfaces* **2014**, *6*, 10005–10013. [CrossRef] [PubMed]

107. Mattioli-Belmonte, M.; Biagini, G.; Muzzarelli, R.A.A.; Castaldini, C.; Gandolfi, M.G.; Krajewski, A.; Ravaglioli, A.; Fini, M.; Giardino, R. Osteoinduction in the presence of chitosan-coated porous hydroxyapatite. *J. Bioact. Compat. Polym.* **1995**, *10*, 249–257.

108. Wang, G.C.; Qiu, J.C.; Zheng, L.; Ren, N.; Li, J.H.; Liu, H.; Miao, J.Y. Sustained delivery of BMP-2 enhanced osteoblastic differentiation of BMSCs based on surface hydroxyapatite nanostructure in chitosan-HAp scaffold. *J. Biomater. Sci. Polym. Ed.* **2014**, *25*, 1813–1827. [CrossRef] [PubMed]

109. Wang, G.C.; Zheng, L.; Zhao, H.S.; Miao, J.Y.; Sun, C.H.; Ren, N.; Wang, J.Y.; Liu, H.; Tao, X.T. *In vitro* assessment of the differentiation potential of bone marrow-derived mesenchymal stem cells on genipin-chitosan conjugation scaffold with surface hydroxyapatite nanostructure for bone tissue engineering. *Tissue Eng.* **2011**, *17*, 1341–1349. [CrossRef] [PubMed]

110. Li, X.Y.; Nan, K.H.; Shi, S.; Chen, H. Preparation and characterization of nano-hydroxyapatite/chitosan crosslinking composite membrane intended for tissue engineering. *Int. J. Biol. Macromol.* **2012**, *50*, 43–49. [CrossRef] [PubMed]

111. Liu, I.H.; Chang, S.H.; Lin, H.Y. Chitosan-based hydrogel tissue scaffolds made by 3D plotting promotes osteoblast proliferation and mineralization. *Biomed. Mater.* **2015**, *10*, 035004. [CrossRef] [PubMed]

112. Schiffman, J.D.; Schauer, C.L. A review: Electrospinning of biopolymer nanofibers and their applications. *Polym. Rev.* **2008**, *48*, 317–352. [CrossRef]

113. Li, Q.; Wang, X.; Lou, X.; Yuan, H.; Tu, H.; Li, B.; Zhang, Y. Genipin-crosslinked electrospun chitosan nanofibers: Determination of crosslinking conditions and evaluation of cytocompatibility. *Carbohydr. Polym.* **2015**, *130*, 166–174. [CrossRef] [PubMed]

114. Lelkes, P.I.; Frohbergh, M. Electrospun Mineralized Chitosan Nanofibers Crosslinked with Genipin for Bone Tissue Engineering. U.S. Patent Application 2013/0274892, 17 October 2013.

115. Frohbergh, M.E.; Katsman, A.; Mondrinos, M.J.; Stabler, C.T.; Hankenson, K.D.; Oristaglio, J.T.; Lelkes, P.I. Osseointegrative properties of electrospun hydroxyapatite-containing nanofibrous chitosan scaffolds. *Tissue Eng.* **2015**, *21*, 970–981. [CrossRef] [PubMed]

116. Frohbergh, M.E.; Katsman, A.; Botta, G.R.; Lazarovici, P.; Schauer, C.L.; Wegst, U.G.K.; Lelkes, P.I. Electrospun hydroxyapatite-containing chitosan nanofibers crosslinked with genipin for bone tissue engineering. *Biomaterials* **2012**, *33*, 9167–9178. [CrossRef] [PubMed]

117. Pu, X.M.; Wei, K.; Zhang, Q.Q. *In situ* forming chitosan/hydroxyapatite rods reinforced via genipin crosslinking. *Mater. Lett.* **2013**, *94*, 169–171. [CrossRef]

118. Pandit, V.; Zuidema, J.M.; Venuto, K.N.; Macione, J.; Dai, G.H.; Gilbert, R.J.; Kotha, S.P. Evaluation of multifunctional polysaccharide hydrogels with varying stiffness for bone tissue engineering. *Tissue Eng.* **2013**, *19*, 2452–2463. [CrossRef] [PubMed]

119. Ge, S.H.; Zhao, N.; Wang, L.; Yu, M.J.; Liu, H.; Song, A.M.; Huang, J.; Wang, G.C.; Yang, P.S. Bone repair by periodontal ligament stem cell-seeded nanohydroxyapatite-chitosan scaffold. *Int. J. Nanomed.* **2012**, *7*, 5405–5414. [CrossRef] [PubMed]

120. Nwosu, C.J.; Hurst, G.A.; Novakovic, K. Genipin crosslinked chitosan-polyvinylpyrrolidone hydrogels: Influence of composition and postsynthesis treatment on pH responsive behavior. *Adv. Mater. Sci. Eng.* **2015**, *10*, 621289. [CrossRef]

121. Yao, Q.Q.; Li, W.; Yu, S.S.; Ma, L.W.; Jin, D.Y.; Boccaccini, A.R.; Liu, Y. Multifunctional chitosan/polyvinyl pyrrolidone/45S5 Bioglass® scaffolds for MC3T3-E1 cell stimulation and drug. *Mater. Sci. Eng.* **2015**, *56*, 473–480. [CrossRef] [PubMed]

122. Austero, M.S.; Donius, A.E.; Wegst, U.G.K.; Schauer, C.L. New crosslinkers for electrospun chitosan fibre mats. I. Chemical analysis. *J. R. Soc. Interface* **2012**, *9*, 2551–2562. [CrossRef] [PubMed]

123. Donius, A.E.; Kiechel, M.A.; Schauer, C.L.; Wegst, U.G.K. New crosslinkers for electrospun chitosan fibre mats. Part II. Mechanical properties. *J. R. Soc. Interface* **2012**, *10*, 20120946. [CrossRef] [PubMed]

124. Kiechel, M.A.; Beringer, L.T.; Donius, A.E.; Komiya, Y.; Habas, R.; Wegst, U.G.K.; Schauer, C.L. Osteoblast biocompatibility of premineralized, hexamethylene-1,6-diaminocarboxysulfonate crosslinked chitosan fibers. *J. Biomed. Mater. Res.* **2015**, *103*, 3201–3211. [CrossRef] [PubMed]

125. Su, W.T.; Wang, Y.T.; Chou, C.M. Optimal fluid flow enhanced mineralization of MG-63 cells in porous chitosan scaffold. *J. Taiwan Inst. Chem. Eng.* **2014**, *45*, 1111–1118. [CrossRef]

126. Radisic, M.; Marsano, A.; Maidhof, R.; Wang, Y.; Vunjak-Novakovic, G. Cardiac tissue engineering using perfusion bioreactor system. *Nat. Protoc.* **2008**, *3*, 719–738. [CrossRef] [PubMed]

127. Ichinohe, N.; Takamoto, T.; Tabata, Y. Proliferation, osteogenic differentiation, and distribution of rat bone marrow stromal cells in nonwoven fabrics by different culture methods. *Tissue Eng.* **2008**, *14*, 107–116. [CrossRef]

128. Siddiqui, N.; Pramanik, K.; Jabbari, E. Osteogenic differentiation of human mesenchymal stem cells in freeze-gelled chitosan/nano beta-tricalcium phosphate porous scaffolds crosslinked with genipin. *Mater. Sci. Eng.* **2015**, *54*, 76–83. [CrossRef] [PubMed]

129. Norowski, P.A.; Fujiwara, T.; Clem, W.C.; Adatrow, P.C.; Eckstein, E.C.; Haggard, W.O.; Bumgardner, J.D. Novel naturally crosslinked electrospun nanofibrous chitosan mats for guided bone regeneration membranes: Material characterization and cytocompatibility. *J. Tissue Eng. Regen. Med.* **2015**, *9*, 577–583. [CrossRef] [PubMed]

130. Bavariya, A.J.; Norowski, A.; Anderson, K.M.; Adatrow, P.C.; Garcia-Godoy, F.; Stein, S.H.; Bumgardner, J.D. Evaluation of biocompatibility and degradation of chitosan nanofiber membrane crosslinked with genipin. *J. Biomed. Mater. Res.* **2014**, *102*, 1084–1092. [CrossRef] [PubMed]

131. Du, M.C.; Zhu, Y.M.; Yuan, L.H.; Liang, H.; Mou, C.C.; Li, X.R.; Sun, J.; Zhuang, Y.; Zhang, W.; Shi, Q.; et al. Assembled 3D cell niches in chitosan hydrogel network to mimic extracellular matrix. *Colloids Surf.* **2013**, *434*, 78–87. [CrossRef]

132. Dimida, S.; Demitri, C.; De Benedictis, V.M.; Scalera, F.; Gervaso, F.; Sannino, A. Genipin-cross-linked chitosan-based hydrogels: Reaction kinetics and structure-related characteristics. *J. Appl. Polym. Sci.* **2015**, *132*. [CrossRef]

133. Reves, B.T.; Bumgardner, J.D.; Haggard, W.O. Fabrication of crosslinked carboxymethylchitosan microspheres and their incorporation into composite scaffolds for enhanced bone regeneration. *J. Biomed. Mater. Res.* **2013**, *101*, 630–639. [CrossRef] [PubMed]

134. Gaudiere, F.; Morin-Grognet, S.; Bidault, L.; Lembre, P.; Pauthe, E.; Vannier, J.P.; Atmani, H.; Ladam, G.; Labat, B. Genipin-crosslinked layer-by-layer assemblies: Biocompatible microenvironments to direct bone cell fate. *Biomacromolecules* **2014**, *15*, 1602–1611. [CrossRef] [PubMed]

135. Huang, J.J.; Yang, S.R.; Chu, I.M.; Brey, E.M.; Hsiao, H.Y.; Cheng, M.H. A comparative study of the chondrogenic potential between synthetic and natural scaffolds in an *in vivo* bioreactor. *Sci. Technol. Adv. Mater.* **2013**, *14*, 054403. [CrossRef]

136. Jayakumar, R.; Chennazhi, K.P.; Muzzarelli, R.A.A.; Tamura, H.; Nair, S.V.; Selvamurugan, N. Chitosan conjugated DNA nanoparticles in gene therapy. *Carbohydr. Polym.* **2010**, *79*, 1–8. [CrossRef]

137. Muzzarelli, R.A.A. Biomedical exploitation of chitin and chitosan via mechano-chemical disassembly, electrospinning, dissolution in imidazolium ionic liquids, and supercritical drying. *Mar. Drugs* **2011**, *9*, 1510–1533. [CrossRef] [PubMed]

138. Muzzarelli, R.A.A. New techniques for optimization of surface area and porosity in nanochitins and nanochitosans. In *Advances in Polymer Science: Chitosan for Biomaterials*; Jayakumar, R., Prabaharan, A., Muzzarelli, R.A.A., Eds.; Springer Verlag: Berlin, Germany, 2011; Volume 2, pp. 167–186.

139. Hurst, G.A.; Novakovic, K. A facile *in situ* morphological characterization of smart genipin-crosslinked chitosan-poly(vinyl pyrrolidone) hydrogels. *J. Mater. Res.* **2013**, *28*, 2401–2408. [CrossRef]

140. Moura, M.J.; Martins, S.P.; Duarte, B.P.M. Production of chitosan microparticles cross-linked with genipin: Identification of factors influencing size and shape properties. *Biochem. Eng. J.* **2015**, *104*, 82–90. [CrossRef]

141. Mi, F.L.; Tan, Y.C.; Liang, H.C.; Huang, R.N.; Sung, H.W. *In vitro* evaluation of a chitosan membrane cross-linked with genipin. *J. Biomater. Sci. Polym. Ed.* **2001**, *12*, 835–850. [CrossRef] [PubMed]

142. Chen, S.C.; Wu, Y.C.; Mi, F.L.; Lin, Y.H.; Yu, L.C.; Sung, H.W. A novel pH-sensitive hydrogel composed of *N,O*-carboxymethyl chitosan and alginate cross-linked by genipin for protein drug delivery. *J. Control. Release* **2004**, *96*, 285–300. [CrossRef] [PubMed]

143. Butler, M.F.; Ng, Y.F.; Pudney, P.D.A. Mechanism and kinetics of the crosslinking reaction between biopolymers containing primary amine groups and genipin. *J. Polym. Sci.* **2003**, *41*, 3941–3953. [CrossRef]

144. Mi, F.L.; Sung, H.W.; Shyu, S.S. Synthesis and characterization of a novel chitosan-based network prepared using naturally occurring crosslinker. *J. Polym. Sci. Polym. Chem.* **2000**, *38*, 2804–2814. [CrossRef]

145. Mi, F.L.; Sung, H.W.; Shyu, S.S. Drug release from chitosan-alginate complex beads reinforced by a naturally occurring cross-linking agent. *Carbohydr. Polym.* **2002**, *48*, 61–72. [CrossRef]

146. Mi, F.L.; Tan, Y.C.; Liang, H.F.; Sung, H.W. *In vivo* biocompatibility and degradability of novel injectable chitosan-based implant. *Biomaterials* **2002**, *23*, 181–191. [CrossRef]

marine drugs

MDPI

Review

"The Good, the Bad and the Ugly" of Chitosans

Barbara Bellich [1], Ilenia D'Agostino [2], Sabrina Semeraro [2], Amelia Gamini [1] and Attilio Cesàro [1,3,*]

[1] Laboratory of Physical and Macromolecular Chemistry, Department of Chemical and Pharmaceutical Sciences, University of Trieste, Via Giorgieri 1, 34127 Trieste, Italy; bbellich@units.it (B.B.); gamini@units.it (A.G.)

[2] Department of Life Sciences, University of Trieste, Via Giorgieri 1, 34127 Trieste, Italy; ilenia.d'agostino@phd.units.it (I.D.); ssemeraro@units.it (S.S.)

[3] Elettra-Sincrotrone Trieste, Strada Statale 14 km 163.5, Area Science Park, 34149 Trieste, Italy

* Correspondence: cesaro@units.it; Tel.: +39-040-558-3684

Academic Editors: Hitoshi Sashiwa and David Harding
Received: 14 March 2016; Accepted: 9 May 2016; Published: 17 May 2016

Abstract: The objective of this paper is to emphasize the fact that while consistent interest has been paid to the industrial use of chitosan, minor attention has been devoted to spread the knowledge of a good characterization of its physico-chemical properties. Therefore, the paper attempts to critically comment on the conflicting experimental results, highlighting the facts, the myths and the controversies. The goal is to indicate how to take advantage of chitosan versatility, to learn how to manage its variability and show how to properly tackle some unexpected undesirable features. In the sections of the paper various issues that relate chitosan properties to some basic features and to advanced solutions and applications are presented. The introduction outlines some historical pioneering works, where the chemistry of chitosan was originally explored. Thereafter, particular reference is made to analytical purity, characterization and chain modifications. The macromolecular characterization is mostly related to molecular weight and to degree of acetylation, but also refers to the conformational and rheological properties and solution stability. Then, the antimicrobial activity of chitosan in relation with its solubility is reviewed. A section is dedicated to the formulation of chitosan biomaterials, from gel to nanobeads, exploring their innovative application as active carrier nanoparticles. Finally, the toxicity issue of chitosan as a polymer and as a constructed nanomaterial is briefly commented in the conclusions.

Keywords: chitosan; physico-chemical properties; molecular weight; degree of acetylation; conformation; antimicrobial activity; from gel to nanobeads; drug delivery

1. Introduction

Chitosan is one of the most commonly cited polymers in the scientific research dealing with a wide range of biopharmaceutical and biomedical applications including food science and technology. It has been strongly indicated as a suitable functional material in view of its excellent biocompatibility, biodegradability, non-toxicity, and adsorption properties. Originally known as a marine polysaccharide from shrimps and crabs, chitosan was primarily thought to be an easily accessible substance from the food industry waste. The original samples of chitosan were often still raw materials, with few exceptions of purification applied to chitosan in medical and cosmetic uses. The main areas in its characterization were in terms of quality (either purity or harmfulness), intrinsic macromolecular properties and physical form. Independently of the final human use, the first and most important issue was, and is still, related to the presence of impurities, metals and other inorganics, proteins, pyrogenic, endotoxic and cytotoxic agents, bioburden, all of which have received over the years erratic attention. The macromolecular characterization is mostly related to molecular weight and to degree of acetylation, but also refers to the rheological properties and solution stability (solubility, aggregation, and filterability). Last, but not less important in many applications, is the physical form of the powder or other morphology of the sample. Since nowadays there is a real boom of interest in chitosan, it is worth mentioning that all these characteristics have been amply and often repeatedly reported in literature in original research and review articles. Indeed, more than a thousand papers (1150) were published in the decade 1981–1990, increasing to more than five thousand papers (5700) in the decade 1991–2000, to about 23,100 in the first decade of this century. A current search of "Chitin and Chitosan" in research article titles gives more than 1600 results, almost one-half of which were published in the last 15 years; so why write another review? The simple answer is that another view-point can still be worthwhile, provided that some properties are cross linked and related to some critical specificity of chitosan in a novel way, as reported schematically in Figure 1.

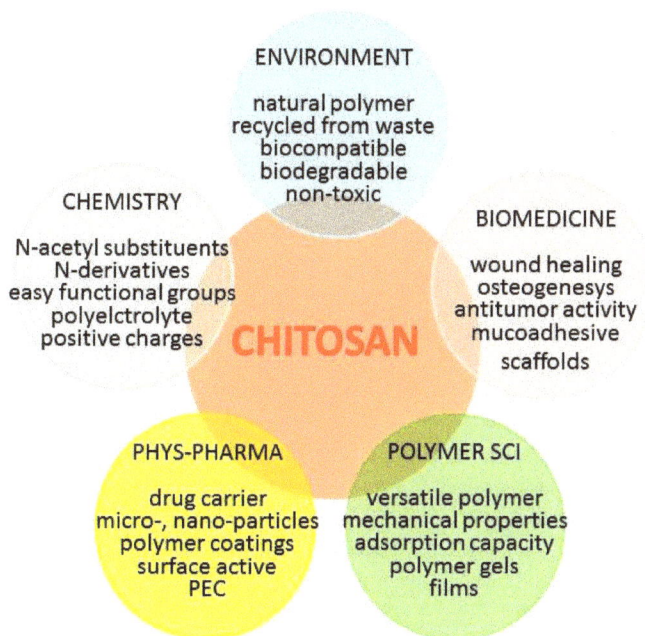

Figure 1. Schematic representation of relationship of chitosan properties with the pharmaceutical and biomedical applications.

Thus, having in mind to classify the "properties" of chitosan in categories as "the good, the bad and the ugly", a scrutiny of its characteristics is here overviewed in order to introduce the theme of this paper. Among all natural polysaccharides the presence of the monomer glucosamine, either fully acetylated or partially deacetylated, is unique in the chitin/chitosan polymers. Whether explicitly highlighted or only mentioned as a secondary aspect, this structural characteristic is probably the basis of its outstanding biophysical properties and the key for a rationale of its applications.

In summary, the most simple and salient characteristics that make chitosan a beneficial and valuable polysaccharide are:

1. Chitosan is a linear natural polymer of glucosamine/acetylglucosamine that behaves as a polyelectrolyte with positive charge density at low pH.
2. Among the industrial polysaccharides, chitosan is an exception, being the only high molecular weight cationic polyelectrolyte, while polysaccharides are generally either neutral or anionic.

3.   Chitosan is often claimed to be GRAS (Generally Recognized As Safe) and bioabsorbable.

The same versatile "good" characteristics are indeed also the reason for several undesired effects, starting from the most obvious one when dealing with copolymers, *i.e.*, the large variety of chemical composition in terms of comonomers. The large variety of chitosans differing in polymer size, degree of acetylation and possible chemical modifications increases exponentially. On the other hand, the simplicity of chemical modification is one of great strengths of chitosan. The "tunable" aspect of chitosan allows its optimization to give appropriate biomaterials for therapeutic applications, in principle enabling also the optimization of its biological profile.

The counterpart of the structural variability is in the difficult characterization of its (statistical) distribution of comonomers (glucosamine and acetyl-glucosamine) and in the unsolved question of how the compositional variability and the molecular weight distribution add complexity to the identification of beneficial properties of the final product (the "bad"). However, the studies of chitosan uptake, distribution and toxicity are indeed quite few, since most commonly in its drug delivery formulations, chitosan is the carrier or a functional excipient. Nonetheless, it is also evident that the inspection of chitosan interactions with some co-solutes or other biostructures may reveal some subtle effects both in classical solution thermodynamics as well as in biological applications, such as the "interactions with endotoxins". All these undesired and often hidden effects are most appropriately classified as "ugly", because they have frustrated scientists in their research efforts.

In the following sections the title issues are presented, relating and linking the chitosan properties, whenever possible, to some basic features and to advanced solutions for specific applications. Thus, after a first section in which tribute is paid to some pioneering work, the chemistry of chitosan is explored with reference, in particular, to purity, characterization issues and chain modifications. Then, the biocompatibility, antimicrobial activity and toxicity of chitosan in relation with its solution properties are reviewed. A section is dedicated to the formulation of chitosan biomaterials, from gel to nanobeads, exploring their innovative application as active carrier nanoparticles. Finally, the toxicity issue of chitosan as a polymer and as a constructed nanomaterial is briefly referred in the conclusions.

Particular attention is devoted to critical comments on the various conflicting experimental results, highlighting the facts, the myths and the controversies. The goal is to make the best use of the ***good versatility*** of chitosan, to learn as much as possible how to manage its ***bad variability*** and to tackle in the proper way its ***ugly undesirability***.

2. The Beginning

A short historical annotation may usefully illustrate the continuous efforts dedicated to increasing the knowledge of chitosans in terms of structure-properties relations and therefore to improving and developing their applications. Another pertinent preliminary comment is also necessary on the link, established from the very beginning, between academic research directly or indirectly related to commercial applications and the scientists involved in the R&D of companies present on the market in the fields of either generic chemicals or, more specifically, polymers. The scrutiny of papers dealing with chitosan properties gives rise to a permanent feeling that while its "potential" is continuously stressed, for many years this seemingly remained wishful thinking. Several factors concur with the idea that using chitosan as a market product could be beneficial. Some of these will be reported in the specific application discussed in this paper.

It is not strange that among the first studies on chitin and chitosan is the structural determination of the chain by diffractometry. The first X-ray diffactometric investigation of chitosan was carried out by Clark and Smith in 1936 [1] on fibers prepared from lobster tendon chitin by a solid state *N*-deacetylation. A subsequent X-ray pattern was obtained and fully analyzed [2] with a tendon chitosan prepared from a crab tendon chitin by similar deacetylation reported by Clark and Smith. Therefore, as mentioned by Ogawa in his review [3] "In spite of such early finding, the pattern was analyzed in 1997 [4], 60 years later."

The relevance of crystallographic studies lies in the possibility that the chain conformation of chitosan can be affected by the chemical environment. Indeed, it has been shown that in the two polymorphs, hydrated (tendon) chitosan and anhydrous (annealed) chitosan, the crystalline structures consistently present antiparallel chains in extended two-fold helix, zigzag structure, with almost constant pitches of 1.034 and 1.043 nm, for the hydrate and anhydrous forms, respectively. This chain structure is similar to those found for chitin and cellulose. However, chitosan can easily change its solid state conformation from the extended two-fold to other structures by salt formation with acids. Although not comparable with other very flexible linear polysaccharides, such as amylose and pullulan, this moderate conformational flexibility is certainly due to the presence of free amino group on the monomer residue of chitosan; thus, the advantage of chitosan for its advanced tunable applications can be envisaged.

With regards to the more broad chemical and physico-chemical studies, very few papers were published on chitosan before 1970, some dealing with its characterization, a few on chemical reactions and derivative synthesis, and the largest number on its interaction with ionic species and the potential application in ion removal from waters. In this period, some research groups appear in the literature that dominated the later times. Between the 1970 and the 1980, some reactions (e.g., selective acylation) and properties of chitosan were assessed and many applicative properties of chitosan were explored. Among these works, it is particularly worth mentioning those on the hypocholesterolemic activity [5], the fungicidal effect [6], in addition to the reactions of chitosan to produce derivatives. A comprehensive review was offered by the First International Conference on Chitin/Chitosan held in 1977 in Boston [7], which signed the milestone for a continuous growing number of adepts in the chitin and chitosan forum. Thus, the great effort of the Muzzarelli group must be mentioned, starting with the characterization of metal chelating properties and the chromatographic uses of chitosan; as well as the contributions of the UK and Japan groups on the production and physicochemical characterization of chitosan gels.

In the following decade (1981–1990), an explosion of activities all around the world in the field of carbohydrate polymers gave rise to intensive studies in this area. In parallel, there was an increase in conferences dealing with industrial application of polysaccharides, with sessions in the very large meetings (e.g., International Carbohydrate Symposium, *etc.*) as well as some more specific thematic meetings (e.g., Grado 1981 [8] and Hoboken 1984 [9]). In the context of these meetings, several research groups emerged and the tie line for aggregation networks was marked.

In these and subsequent years, while continuous efforts were made on the preparation of well-defined samples (see, for example, [10]), many new ideas about the use of chitosan as biomaterial became explicit and other properties of chitosan and derivatives emerged in publications. The proposed uses of chitosan included the development of artificial skin, reconstruction of periodontal tissues, hemodialysis membranes, drug targeting and many other biomedical applications. The foreseen applications were always accompanied by sentences like "this novel biomolecule, biodegradable, and biocompatible, find applications in substituting or regenerating tissues" [11]. The similarity of chitosan structural characteristics with glycosamino-glycans, was interpreted as the reason for mimicking the functional behavior of the latter, while chitin was always structurally (or ancestrally?) associated with cellulose. In a slightly different direction, a derivative, *N*-carboxybutyl chitosan, was shown to display inhibitory, bactericidal, and candidacidal activities when tested against a few hundred cultures of various pathogens [12].

It is worth remembering that the first interest in the commercial applications of chitin grew in the 1930s and early 1940s, but took a backseat for many years because of competition from synthetic polymers. Large-scale production of chitin started in the mid-1970s when regulations were introduced to limit the dumping of untreated shellfish waste in coastal waters. Chitin was easily extracted from crab, lobster and prawn shells using solvents, and the production of chitin became an economical way to comply with the regulations and dispose of thousands of tons of shellfish waste. Although chitin has applications in creams and cosmetic powders and as a material for surgical stitches, the applications of chitosan expand to a multitude of areas: antibacterial lining for bandages and wound dressings, coating for seeds to boost disease resistance, agent to prevent spoilage in winemaking, in addition to its controversial use as dietary supplement (weight-loss by avoiding fat digestion). The specific area of food applications has been shortly reviewed in a 2016 paper [13], in particular for antioxidant and antimicrobial effects, which are useful in the food industry to improve food safety, quality, and shelf-life. It has been estimated that about the order of 10^{10} tons are produced annually by living organisms, although chitin and chitosan are recovered exclusively from marine sources. Applications of these polymers continue to grow and an estimated global market of chitin and chitosan derivative business was forecasted to exceed € 70 × 10^9 by the year 2015 [14]. As reported in a news agency *"Not bad for bits of leftover lobster"* [15].

Before concluding this section, one point needs to be emphasized. The chain production of a "food-waste-recycling-biopolymer" is compatible with achievement of ecological benefits and deserves further attention and effort. For chitin and chitosan, the environmentally-friendly materials produced from these renewable sources are sustainable and contribute to the creation of a circular economy, which is the current goal of an ambitious package put forward by the European Commission in December 2015 [16]. For the more specific aspects here concerned, the waste produced each year by the shellfish processing industries represents a practical challenge, since about 75% of the total weight of crustaceans (shrimps, crabs, prawns, lobsters, and krill) ends up as by-products. The current lack of acceptable waste management options creates a potentially large environmental hazard concern. All around the world, seafood wastes are thrown away at sea, burned, landfilled, or simply left out to spoil. Therefore, the extraction of chitin from shells and its use as it is or after further processing not only is a way to minimize the waste, but also to produce compounds with valuable biological properties and specialty applications, provided that all the raw material is correctly remodeled in well characterized chemicals. As recently pointed out in literature ("don't waste seafood waste") [17], the three major components of crustacean shells are calcium carbonate (20%–50%), proteins (20%–40%) and chitin (15%–40%), which could be separated by using an integrated bio-refinery with solvent-free mechano-chemical processes. Although the main objective of this work is focused on the biopolymer chitosan, it is mandatory to mention that within this sustainability view-point, several useful chemicals and primers can be obtained from chitin [18–21], giving further value to an atom efficient economy. Once again, a parallel can also be envisaged between strategies used, or to be implemented, in cellulose and chitin exploitation.

3. A critical Examination of the Physico-Chemical Properties of Chitosan Polymers

In order to emphasize, once again, the relevance of a correct understanding of the relation structure-properties in chitosan, let's just mention in these first lines that chitosan composition has become a matter of great interest in patent claiming. Documentation for these claims lies on well-defined polymer chemical characterization and this concept will be further analyzed when discussing the difficulties encountered by regulatory agencies in approving chitosan uses.

The main outcome of a critical examination of the literature results is that often (although not always!) the correlation between properties of the nanoconstructs and the structure of the macromolecular components is missed. In particular, the impression is that the advantage offered by modulating the polymer structure is turned out in the disadvantage of having final products of remarkable variability. Therefore, aiming at exploitation of chitosan and chitosan nanoparticles in modern biomedical and pharmaceutical applications, some fundamentals of physicochemical properties of this biomacromolecule cannot be overlooked. The central "dogma" is that all useful applications can *in principle* be traced back to three molecular determinants: the degree of acetyl substitution, the molecular weight and its distribution, the nature and the fraction of substituents as pendant groups. In the more general applications, the additional step to be examined is the interplay of these molecular characteristics with the supramolecular interactions that provide size, shape and surface properties in the nanoconstructed patterns [22].

3.1. Chitosan as a Copolymer: Acetylation and Substitution

The first question concerns whether it is fully correct to speak about "chitosan" or better to use the term "chitosans", making explicit the central concept that the term chitosans implies a series of copolymers that differ not only in the fraction of comonomers but also in the distributions and clength of the comonomer sequences. This aspect is well known in polymer sciences where the terms alternate-, block- and random copolymer have been introduced. It is also the case of alginate, a close polysaccharide, where the presence along the chain of the Guluronic (G) and Mannuronic (M) monomers with varying composition and sequence type, was amply recognized since long time [23], allowing to conclude, from ^{13}CNMR studies, that the relative occurrence of G-centered triads deviated significantly from those predicted by first-order Markovian statistics [24]. It is also known that in chitosan the two monomers, *N*-glucosamine and *N*-acetyl-glucosamine, display quite different solution properties, given the ionic character in acidic conditions of the first monomer and the slightly hydrophobic terminal in the other. This intrinsic difference, getting amplified for block-type sequences of acetyl substitution, may have dramatic consequences on chain conformation and aggregation, a fact that has encouraged the research on more hydrophobic substituents able to self-assembling and collapse in micelle-type lipospheres. Furthermore, derivatization of chitosan by grafting side chains, like is done in quaternization or glycosylation of amino groups, gives rise to a ter-polymer (namely, constituted by acetylated, de-acetylated and substituted monomers), worsening, therefore, the already complicated situation.

Thus, the essential issue of chitosan composition is the proper sample characterization and, first of all, the correct determination of the degree of acetyl substituents, DA, (some authors use the term DDA, degree of de-acetylation). Over the years, several papers have referred on the determination of the degree of acetylation by physical and chemical methods, although the preferred method is often simple and easy to use, not the more accurate. The long list of methods include among all, titration [25,26], IR spectroscopy [27–31], UV spectroscopy [32,33], circular dichroism (CD) [34], NMR spectroscopy [35–38] and *N*-acetyl group hydrolysis [39]. As a preliminary and important annotation, the comparison of data obtained with different techniques often show discrepancies in the DA values. Indeed, the difference in solubility of samples having DA from 1 (chitin) to 0 (fully deacetylated chitosan) affects sensibly the results and no single technique could be adopted to cover the full range, pointing out that solution methods can be used for soluble chitosans, whereas 13C CP/MAS NMR [36,37] and infrared spectroscopy (as recently summarized in a review [40]) are used for chitin and highly acetylated chitosans. Still, each method presents advantages and difficulties either in the sample preparation and/or in running the measurements. As a novel technique, the calorimetric (DSC) determination of the distinguishable decomposition of amino and acetyl groups was recently published, providing results that are independent of composition and molecular weight [41]. Thus, DSC emerged as an accurate technique to determine the degree of N-acetylation in chitin/chitosan samples, the main advantages lying on the possibility of determining the DA in the whole 0 to 1 range, without solubilization, at lower cost in relation to NMR, and more accurately than with IR. However, all techniques, except for NMR based measurements, require an accurate weighing of chitosan. Therefore, moisture needs to be eliminated carefully and the purity of the samples must be determined separately. The degree of acetylation of chitosan determined by NMR is based on the fact that the O-Ac group is easily recognizable in the spectrum of chitosan recorded at room temperature and can be integrated and normalized by the integral either of anomeric protons or of other ring protons. More important is the possibility of generalization of the NMR method to study chitosan derivatives. An interesting application is that of glycosylated samples prepared with cellobiose, maltose and lactose residues. The glycosylation introduced a pendant made by an open sugar ring C1-linked to the amino group and C4-linked to the terminal glycosyl residue. The NMR spectra show small differences for all signals of the pendant groups except for the carbon 4 and the carbon 5, which reflect the configurational change from glucose to galactose moieties. The NMR on these samples provided a DA = 0.13 and a glycosyl substitution ratio ranging from 0.43 to 0.50 for the three derivatives [42].

Not less important for chitosan varying in pendant group substitution is the knowledge of the substitution pattern of the groups (including the acetyl groups) along the chain, as it was referred as a relevant property in copolymers. This is illustrated in Figure 2, where it is schematically evidenced why not only the fraction of comonomers (e.g., DA) but also their distributions and the length of the sequences of comonomers ultimately affect the polymer behavior, *i.e.*, each copolymer species is a "specific product". The relevance of these characteristics increases with the difference in the chemical properties of the two monomer species.

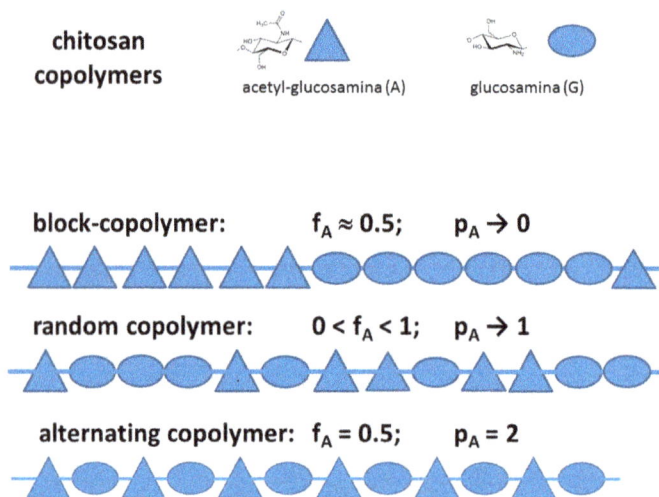

Figure 2. Schematic representation of three typical patterns of monomer distribution in a copolymer, here acetylglucosamine (**A**) and glucosamine (**B**). A perfect block-type copolymer with $f_A = 0.5$ should have a limiting p_A value $= 0$ for infinite length of blocks (top). A random distribution of the monomer (middle) is characterized a limiting p_A value $= 0.5$ for a value of f_A that can range around 0.5. The alternating copolymer, by definition, should have $f_A = 0.5$ and $p_A = 2$ (other values evidence structural irregularity). Thus, the scheme identifies limiting cases to interpret copolymer structure and underlines that nearest-neighbor effects give the three copolymer structures different physico-chemical properties.

In general, a statistical copolymer consists of macromolecules in which the sequential distribution of the monomeric units obeys known statistical laws, such as the Markovian statistics (the so-called random copolymer obey a zero-order Markovian statistics, coinciding with a Bernoulli distribution of the monomers). The only very non-random distribution of monomers is the one where monomers A and B alternate, Poly(A-alt-B), which can be clearly defined as a "new" polymer, different from Poly(A) and Poly(B) and which has rigorously fraction values $f_A = f_B = 0.5$. Another example of a copolymer distribution is that of "block copolymer" (PolyA-block-polyB), which is characterized by sequences of one monomer A interrupted by sequences of the other monomer B, the length of the sequences obeying to some n-order Markovian statistics (with $n \gg 1$). All these different copolymer sequences can be initially identified by the monomer fraction, $f_A = 1 - f_B$, and the probability of diads, f_{AA}, f_{BB}, f_{AB}. A compact mode of providing the sequence probability has been proposed for the general case of copolymers by Bovey & Mirau [43] and recently used for chitosan [44] with the definition of the parameter $p_A = f_{AB}/(2f_{AA} + f_{AB}) + f_{AB}/(2f_{BB} + f_{AB})$.

The relevance of this quantification can be traced back to the original findings of Aiba [45], who studied two samples of chitosan with similar DA but different solubility properties. This author set up the hypothesis that the samples of chitosan with DA > 0.5 prepared from highly deacetylated chitin had to be considered as random-type copolymers of *N*-acetyl-glucosamine and glucosamine units, whereas the samples with similar DA but prepared from partial deacetylation of chitin had to be considered as block-type copolymers.

In conclusion, despite some different opinions in the literature and the generally assumed random statistics of residual acetyl substitution, the accurate determination of DA and of copolymer statistics is unavoidable. NMR determination of DA has been found to be precise and accurate also for the quantification of high DA, which is usually difficult to be measured with conventional techniques like IR or titration. Additionally DA can be calculated using different combinations of peaks in order to verify that the method is consistent. Some NMR techniques described in the literature are only limited by the solubility of chitosan, which depends on the DA and the molecular weight of the polymer. The relevance of DA on the solution properties of chitosan has been mentioned and, from a more general perspective, it should be stressed that all the literature results convincingly show that chain dimension and rigidity are related not only to the degree of polymerization, but also to the degree of acetylation (with a further obvious dependence upon pH and ionic strength). This point is discussed with reference to the molecular weight determination including the methods based on the knowledge of the hydrodynamic volume, having in mind the results reached on the cognate biopolymer, *i.e.*, hyaluronan [46,47].

3.2. Chitosan Copolymers: Molecular Weight Determination and Conformation

Before reporting on molecular weight (M) determination, another subtler question arises in these data for chitosans. Not only the problem of complete solubilization (1) or even the change in the macromolecular solvation (2) affect the M measurements, but also the correct interpretation of M can be vague if the degree of substitution is not known (3). Points (1) and (2) imply solution thermodynamic issues to be reviewed, since most of the experimental methods for M determinations deal with some solution properties measurements and a dependence of the solvent goodness through the Flory interaction parameter. Point (3) is of mere analytical origin and points out on the more classical definition of macromolecular chain length in terms of degree of polymerization (DP). As an example, a monodisperse chitin sample with DP = 1000, will display a molecular weight of 203.2×10^3, while the same sample fully deacetylated, *i.e.*, the corresponding chitosan without *any* degradation, has a molecular weight of 161.2×10^3, with a difference of about 20%. Therefore, even without degradation, a continuous change in the "measured" molecular weight is expected to occur upon deacetylation of chitin. This problem is much more impressive with large substituents as it occurs in the class of glycosylated chitosans [42,44]. The experimental determination of the M and M distribution of a series of samples of glycosylated chitosan with a constant DA = 0.13 and glycosyl substitution DS 0.42–0.50, is enlightening (Table 1).

Table 1. Physico-chemical properties of chitosan and chitosan derivatives (reported in [42]). Two commercial samples of Very-Low-MW and of Low-MW and three derivatives, a trimethyl-substituted (TM-chit), a galactosyl- (Gal-chit) and a glucosyl- (Glc-chit) substituted are reported. The molecular weight of the commercial LMW sample is given to be approximately 50,000–190,000 daltons based on viscosity. TM-chit contains also substitution on O6 and O3. The sugar substituent, gal or glc, is linked to ammine via an open sugar chain, since a disaccharide is the reactant. DS (non-acetyl substitution) determined by NMR. The column "Recovery" is a parameter relevant for the data to be representative of the sample investigated.

	DS	Mw	Mn	Mw/Mn	[η]	Rh	Rg	Recovery
		kg/mol	kg/mol		dL/g	nm	nm	%
VLMW-chit	-	30	18	1.7	0.7	6.4	-	94
LMW-chit	-	128	56	2.3	2.3	15.3	-	82
TM-chit	0.71	109	46	2.3	0.7	9.7	-	80
Gal-chit	0.50	473	102	5.1	1.9	21.3	50.8	99
Glc-chit	0.42	554	96	6.4	2.4	23.8	61.7	93

Molecular weight and conformation analysis of these derivatives were investigated at UFT—Centre for Environmental Research and Sustainable Technology, Bremen, with a "triple detectors" size exclusion chromatography Viscotek system. The array, with a differential refractometer, a right angle (90°) (RALS) and a low angle (7°) light scattering detector and a four capillary, differential Wheatstone bridge viscometer, provided as final data the Molecular weight (M) and Molecular Weight Distribution (MWD) of fractionated samples, including viscosity of each fraction related to a given dilute polymer concentration [42]. The analysis of data allow to extract the weight average M_w and the parameters a and k of the Mark–Houwink–Sakurada (MHS) equation (*i.e.*, $[η] = k \, M^a$). In particular, the most evident result is that, despite their overall similarity in the values and the trend of the viscosity, an increase (about doubling) of the molecular weight is observed in the glycosylated chitosans. This effect could be interpreted as a sign of chain dimerization, whereas simply arises from the derivatization reaction by which a pendant with $M \approx 320$ is added on the chain about every two monomeric units, therefore doubling the original molecular weight in the absence of degradation. Therefore, these annotations claim for a re-examination of many published data of chitosan samples, in particular those with additional substitutions.

As far as the conformation of chitosan with low DA is concerned, it is worth reporting that SEC data with triple detectors provide the full plot of log [η] as a function of M with the interesting result that the exponent coefficient a increases from 0.5 to 1 with the molecular weight decreasing from 5×10^5 to 5×10^4, in line with the literature independent results (Figure 3) about the concurrent estimation of persistence length and mass per unit length of chitosan (see Table 3 of ref. [48]) and with the findings for the moderately stiff hyaluronan chain. These results are collected in a sort of master curve and contain the fundamental dependence of the exponent a as a function of the degree of polymerization (Figure 4).

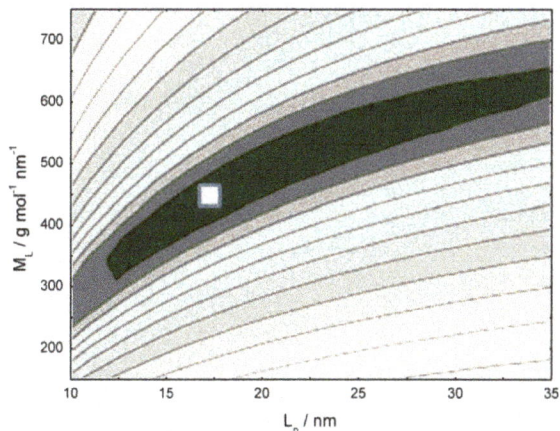

Figure 3. Descriptors of the chain conformation according to the Bushin–Bohdanecky equations. The solutions for chitosan are described in ref [48]. The diagram describes regions close to the target (dark color) and regions far from the target (pale color). The white square shows the coordinates of chitosan in the persistence length Lp (nm) *vs.* mass-per-unit length, M_L (g·mol^{-1}·nm^{-1}), plane (adapted and redrawn from ref [48]).

In conclusion, the correct polymer characterization and the robust interpretation of solution properties provide the basic knowledge for understanding the chain expansion behavior of the chitosans used for the preparation of nanoparticles, with the possibility of better tailoring the nanoparticle properties by selecting the suitable starting polymer.

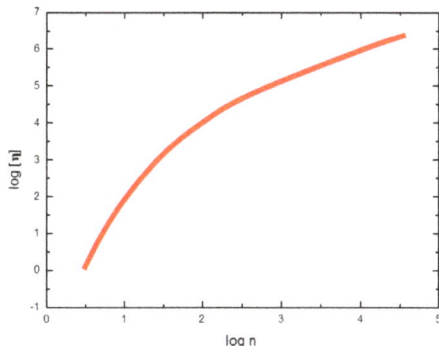

Figure 4. The dependence of intrinsic viscosity of chitosan and hyaluronan as a function of degree of polymerization (*n*) is reported as a master-plot (MHS log-log), normalized by the constant K (solvation constant). The emphasis is on the slope that increases with decreasing degree of polymerization n, *i.e.*, molecular weight M. The curve is drawn with the data of Figure 6b of ref. [47] for hyaluronan (0.5 < *n* < 4) and Figure 6.1 of ref [49] for chitosan (2 < *n* < 4.5).

Returning to the solution properties, points (1) and (2), in addition to direct methods for molecular weight determinations (e.g., those based on light scattering measurements), other indirect methods are largely used because they are simple and ready to use. Among these methods, the most popular one is the determination of the intrinsic viscosity, [η] which is log-log proportional to the molecular weight by mean of the Mark–Houwink–Sakurada (MHS) equation, [η] = k M^a. The use of the equation implies, however, that a calibration is made on monodisperse fractions of the same polymer, under the same conditions of solvent and temperature. At this stage, it is clear that the calibration constants, k and a, depend on the very chemical structure of chitosan, starting from the value of DA and possibly on its distribution, since it has been recognized that the polymer solubility is affected. The solution to this problem was afforded in the past by changing the pH and the ionic composition. Empirical correlations have been reported to adapt the calibration constants of the MHS equation to a range of DA, pH and salt concentration. The use of the empirical relations, however, opens further questions about the compensation phenomena of chain expansion and solvation. Indeed, the analysis of the data of k produced by several authors, when plotted as a function of DA, shows the absence of a clear correlation that however could arise from some important difference in the acetylation patterns. The possibility that the samples may present a range of non-statistical sequence of acetylation patterns has been very recently presented [49–51].

3.3. Chitosan Solubility

On concluding this section on the intrinsic macromolecular properties of chitosan family, it is also necessary to add some words about the common practice of derivative preparation and the consequent changes in solubility that arise when substituents are inserted as pendant groups on the chain backbone. As a rule of thumb, polysaccharide solubility is affected by the chain linkage that governs the extension and the flexibility of the polymer. This is an important factor that makes the difference between polysaccharides otherwise undistinguishable on the basis of the rule "*similia similibus solvuntur*" (similar substances are miscible). This feature is clearly shown by comparing the configurational properties of several glucans, such as cellulose, amylose, pullulan, β-glucan [52,53], where flexibility and stability of the crystalline order play the major role. Therefore, for any given polysaccharide chain, solubility can increase (or decrease) depending on whether random (or regular) substitution is achieved and whether substituents can improve or not the interactions with the solvent. These concepts are schematically summarized in Table 2, where only a few examples are given to substantiate the above-mentioned rule of thumb. Chitosan is not an exception, as the changes in the degree of acetyl substitution and the introduction of other substituents (ionic, hydrophilic, and non-polar) conform to the above scheme.

Table 2. Schematic role of structural and conformational features on solubility of polysaccharides.

Structural Features Change	Case-Polymers	Solution Behavior
chain linkage	cellulose	stiff and insoluble
	pullulan	flexible and soluble
side chains	curdlan	linear, insoluble
	scleroglucan	branched, soluble
non-sugar substituents	deacetylated gellan	gel with Ca, Mg
	native gellan (acetylated)	soluble with Ca, Mg
ionic groups (carboxyl)	Chitin	neutral, insoluble
pH > 6	hyaluronan	ionic, soluble, pH-depend.

Following these concepts, the issues of solubility of chitosan and chitosan derivatives can be focused in more detail and some issues of the past literature can be re-examined. Within the polysaccharide scientific community, the scarce water-solubility of chitosans, at neutral pH is a well-known and generally experienced fact. The solubility of chitosan(s) is an important concern representing a limit (if not an obstacle) not only to physical-chemical studies addressed to the structure-properties understandings, but also to its preparation and use as a polymer support or carrier material in biomedical applications, where neutral aqueous environment is often encountered.

It has already mentioned that the solubility of chitosan(s), *i.e.*, the entire family of partially deacetylated chitins with average degree of acetylation DA ⩽ 50%, depends on molar mass, number and distribution of acetylation sites. These structural characteristics, in turn, are all related to the source of chitin and to the means of chitin extraction and deacetylation [54–56]. Besides the exploitation of other less common sources, the main industrial source of (α)-chitin is the shell (exoskeleton) of crabs or shrimps, while that of (β)-chitin is the squid pen. More recently, chitosan from chitin extracted from cultivated edible mushrooms, like *Agaricus bisporus*, has been made commercially available. This material, prepared by Kitozyme [57] and distributed by Sigma, is claimed to be a highly pure chitosan, ideal for wound healing and hemostasis, biosurgery and ophthalmology, scaffold and cell therapy, as well as drug delivery and vaccines. Since all commercial chitosans are far from an absolute purity, it is worth mentioning that a great difference in purity of chitosan may arise on whether the commercial product is used "as it is" directly in the human applications or it is subjected to physical and chemical modifications before end-uses. As a matter of fact, most procedures to modify chitosan by introducing pendant groups, or to process it by mixing with other polymers or chelating agents, imply "*de-facto*" manipulation that may introduce new contaminants or remove original contaminants. To the best of our knowledge, an analytical report on this particular aspect is lacking in the works published, while it will possibly be required in biomedical applications in addition to safety issues. Therefore, up-to-date information about accurate analytical reports is necessary and this specific point concerns both in-house purifications and commercial products. In particular, it would be desirable that data sheets of commercial products could be integrated with the macromolecular and characterization data discussed in this paper.

Indeed, keeping in mind that the amount of the residual *N*-acetyl glucosamine (GlcNAc) and the type of distribution, play a key role in determining the water solubility of chitosans [58,59], it is also necessary to analyze some solution molecular aspects. In particular, the role played by the glucosamine residues (GlcN) is generally referred to as improving solubility in acidic medium (*i.e.*, pH $\leqslant$ 6), due to protonation of amine group. On the other hand, the acetyl groups (and so the GlcNac residues) have been viewed as more hydrophobic entities that favor chain aggregation and negatively affect the water solubility [56,60–63]. However, while these concepts seem to be clear enough, the translation of the effects on a long chain is far to be naïve. Despite decades of the active research described, a clear relationship unambiguously linking the molar mass and the acetyl content and distribution to water solubility has been still elusive. Although a poorer solubility is encountered in homogenously re-acetylated samples with high DA content with respect to commercial ones [64], the homogeneity of DA distribution is reported to be one of the factors that increases chitosan solubility in aqueous medium as long as the solution pH is kept below 6. This can be reworded by saying that, even when the acetyl groups are randomly distributed, a pH lower than 6 is a prerequisite for chitosan solubilization [61,65,66]. On the contrary, the dissolution of chitosan samples with high acetylation content remains troublesome at a molecular level especially at high molar masses [64,67,68]. These findings were originally found by Anthonsen *et al.* [69] on fractionated chitosan samples with $f_A = 0.01$ and 0.60; by using several characterization methods, a bimodal molecular weight distribution was observed in which about 5% of the sample had a very high molecular weight. The presence of supramolecular structures revealed by electron microscopy and the possibility of partially reducing this aggregated fraction by ultracentrifugation and filtration were consistent with the positive virial coefficients obtained earlier from osmotic pressure measurements. Worth mentioning is that the presence of concentration dependent aggregates are reported also in chitosan of relatively low molar mass at low DA (12%) [70,71].

At this point, it is necessary a *flashback* with some general comments, since it is not a novelty that for a given concentration and solvent the increase of a polymer molecular weight adversely affects its solubility. More specifically, polysaccharides are known, as well, as gelling or thickening agents irrespective of their having charged or neutral backbones (e.g., agarose, amylose, carrageenans, xanthan, alginate). Some of their final applicative properties linked to biological function are also often exalted by the physical form of storage samples after chemical manipulation and purification and by the procedure of solubilization. In particular, for many polysaccharides, including chitosan, these physical forms and solubilization procedures refer to solution, powder, freeze-dried, air dried and presence or addition of simple salts, heating procedure, respectively.

Solubilization processes imply that a solid form of the material has been obtained and the material "history" is involved, making evident that precipitation, gelation or freeze-drying are not equivalent. Therefore, it is clear that the question about whether the acetyl moieties favor aggregation because they impart hydrophobic characteristics to the chain, remains unclear when not controversial, until both the short range and the long range effects are properly studied. For example, although it is suggested that acetyl groups affect the chain conformation by a continuous enhancing of chain stiffness upon increasing acetyl content [72], the interpretation of the previous literature data remains somehow conflicting. It is reported for instance and in contrast with previous observations, that DA strongly influence the stiffness of the chain as evidenced by the increase in MHS exponent a and radius of gyration Rg with DA, above a certain molar mass.

Still, an apparently consistent and reliable picture of the dimensional properties of chitosan in solution can be achieved by re-analyzing some original data of viscosity and radius of gyration and properly plotting these values as function of degree of polymerization, n, (Figure 5). The controversial deductions on chitosan behavior, might likely result from disregarding that the overall chain length characteristics play the main role on viscosity and light scattering experiments, from which conformation or shape are deduced. The generic Debye relationship for an isolated chain is between the viscosity and the number of residues ($[\eta] \propto n^1$) and a direct correspondence between M and n exists only for chemically identical residues, as already stressed above. Indeed, taking into account the value n, neither the parameter a nor the exponent ν ($Rg \propto n^\nu$) seem to be much dependent on DA except for DA ~ 60%. Therefore, the fragmented picture that is given by the separated literature data [63,64,72,73] can be recomposed, as shown in the Figure 5.

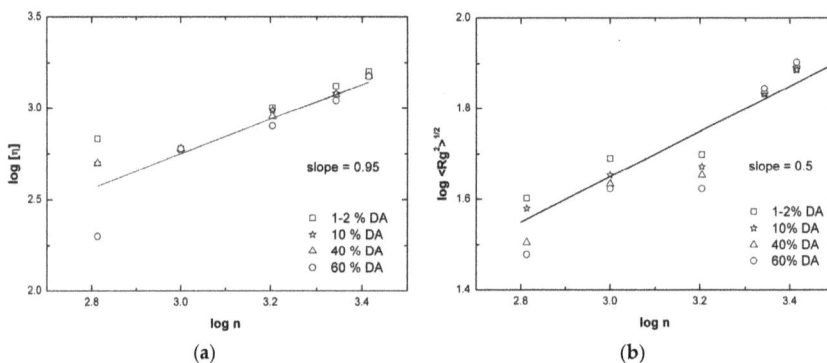

Figure 5. (**a,b**) Double logarithm plot of [η] and Rg values of chitosan samples with different DA values as a function of degree of polymerization, n. The data are recalculated from literature [63,64,72,73].

As before, here the values of $n = (M/Mru)$ are obtained from the experimental M and acetylation degree values [72,73] being Mru the average mass of repeating unit. The value of n can be identified as proportional to the chain contour length expressed in nm, being roughly 0.5 nm the average virtual bond length of the β(1,4)-D-glucose monomer (*i.e.*, the distance between two consecutive glycosidic oxygen atoms). On the other hand, the second virial coefficient is undeniably found to decrease with increasing DA and salt concentration, as expected for polyelectrolytes upon changing the effective charge density, being the effect the more evident the higher the M is [64,70].

Before concluding this section, a comment is still necessary of the mentioned changes of hydrophobicity as a function of DA. It should be clear that changes in local polarity by acetylation can scarcely affect chitosan properties as individual chains (very dilute solution). However, the influence of acetyl substitution can be more evident at high polymer concentration, where "hydrophobic" driven chitosan association can be observed. Under these circumstances, the presence of hydrophobic domains has been studied by mean of fluorescence using pyrene as hydrophobic probe, whose fluorescence spectrum is quite sensitive to the polarity of the microenvironment [71,74]. By ruling out the intrinsic acetyl contribution to the association tendency of chitosan chain, the authors assume that the formation of hydrophobic domains takes place at polymer concentrations close to and above the overlap concentration C*; for fully ionized samples with no added salt, this conclusion mainly indicates that the intermolecular character of the aggregation depends on the dimensional properties, *i.e.*, is independent from DA.

The scarce solubility of chitosan at neutral pH has been circumvented by modifying the polymer backbone with ionic or highly hydrophilic moieties. Chitosans bearing carboxylic, sulfate or *N*-alkyl groups have been synthesized for this purpose. Beside the enhanced solubility in aqueous medium, several additional features were disclosed by these modified chitosans. As a single mention, sulfate chitosan derivatives showed to be compatible with anionic polysaccharide like xanthan and the viscosity of the formed blends was found to be much greater than would have been expected from that of the individual components. Concerning the reductive alkylation route used to modify chitosan with the objective to reduce its intractability in neutral aqueous medium, the work of Yalpani and Hall [75–78] certainly deserves a specific mention; not only it showed the advantages of facility and versatility of the reaction procedure, but also it contained *in embryo* almost the entire "chitosan based biomaterials" research area whose development was still to come. It is not the scope of this review to discuss the synthetic approaches, since already reported in excellent comprehensive papers (for example, see ref [79] and references therein). However, the rationale of chitosan modification is that derivatization mainly aims at providing the following new performing properties. The first goal is to modulate physical-chemistry properties with the purpose to: (i) improve water solubility and blending ability with anionic biopolymers, enhancing and modulating rheological, hydration and compatibility properties of the new materials; and (ii) increase hydrophobic or emulsifying properties. The second goal is on the high-rated field of biological applications, with particular attention to derivatives that can impart specific biological activities to chitosan-based constructs through conjugation, and to those that can enhance antimicrobial activity. The intertwined manner with which these purposes can be found in the more recent literature might explain the continuous growth of active research on this subject. Driven by the desire of tailoring specific and complex properties, these activities led to a boost of studies on chitosan derivatives for potential applications in a very large variety of fields.

4. Biocompatibility, Antimicrobial Activity and Toxicity (Chemistry *vs.* Material)

Resuming the introductory section, since the very beginning of chitosan investigation, the properties of chitosan were interpreted in terms of a very promising technological material for a wide spectrum of applications [12,80–83]. In particular, besides all its recognized and foreseen properties, chitosan was claimed to possess the appealing feature of antimicrobial and bacteriostatic activity. Solutions, films and composites made of chitosan have been reported since the 1980s to be antimicrobial against a wide range of micro-organisms like bacteria either Gram-positive or Gram-negative, human oral and gingival pathogens [84,85] yeasts [86,87], algae [88,89] and fungi [90]. However, its effectiveness is continuously debated about whether it acts as a *bactericidal* agent, *i.e.*, capable to kill live bacteria or as a *bacteriostatic* agent, *i.e.*, able to arrest the bacterial growth without killing the microorganisms [91–93].

In a recent review, the current research in the areas of food, medical and textile industries was analyzed in order to summarize the progress in the study of antimicrobial properties of chitosan [93]. The comparison of the antimicrobial activity in relation to the structural differences in chitosan or chitosan derivative, the polymer physical form and the microorganism tested, made evident the difficulties of rationalizing the results. The conclusion to be shared is the variety of results reported by researchers even under apparently identical conditions, often due to the lack of standardization of the assay conditions. A further more general comment deals with the ambiguities in some of the studies concerning the assessment of chitosan antimicrobial potential, mainly deriving from inadequate data on the polymeric characteristics that affects this ability. Pointing at this ambiguity, in another recent analysis it has been reported that "Given the large number of proclaimed medicinal benefits of chitosan, it comes as no surprise that the literature is filled with conflicting reports about these medical potentials." [94].

With this statement out of the way, it is obvious to recognize that the mechanism of chitosan antimicrobial activity has not been completely understood so far, since it depends on many intrinsic properties of the polymer itself and extrinsic factors related to microorganisms and environmental conditions. Among a long list of factors, it is worth reporting the molecular weight values [95], degree of acetylation (DA) [96,97], pH, type of derivative, solvent composition (medium) and, in particular, presence of metal cations [98], in addition to microorganism type and its outer surface charge [99–101], growth conditions, *etc.*

In general, it can be stated that the absence of reliable experimental determination of molecular parameters makes presently impossible to pinpoint the dependence of antimicrobial activity of chitosan on M or DA. As discussed in a previous section, the reliability does not reside in an intrinsic accuracy of the determination of the molecular parameters, but rather in the poor selection of the appropriate methods or even in the absence of any determination. In order to shed light on this issue, some literature findings can be tentatively organized about the mode of action, the killing mechanism and the interaction with cell components.

Mode of action: There are many hypotheses about how chitosan could exhibit its bactericidal activity, but almost all studies underline the determinant contribution of the poly-cationic nature of chitosan. Therefore, the electrostatic interaction emerges as a fundamental feature of the killing potential, since the interaction with the negatively charged microbial surface would dramatically affect the bacterial vitality. However, in view of the variability of the microbial surfaces, different organisms expose different more or less charged molecular patterns and respond to chitosan differently. This seems to be the case of gram-negative and gram-positive bacteria, although it is still uncertain which type is more sensitive. Some studies [12,102,103] indicate that the gram positive ones are more susceptible to chitosan than other micro-organisms, because of the presence of polyanionic teichoic acids on the outer surface [104,105]. Other studies reported that also gram negative are significantly affected by chitosan, implying a role of hydrophobic interactions by the exposed lipopolysaccharides [91].

Killing mechanism: Membrane structural perturbation by chitosan results from a series of studies by using a large number of different techniques, including the measurement of membrane polarity, evaluation of minimum inhibitory concentration, electron microscopy, transcriptome and proteome analysis. Based on the evidence of Raafat *et al.* it can be argued that the molecular mechanisms of action of chitosan in inhibiting or killing bacteria is definitely a complex process, which involves a series of interrelated events, that eventually lead to micro-organisms death [105]. First, the growth-inhibitory effect is shown in this study to be dose-dependent. A permeabilization of the cell membrane to small molecules and a significant membrane depolarization occur after treatments with the polymer, with consequent loss of cell integrity, stability and functionality. Moreover, a further implication on the cell components is disclosed by the changes observed in the expression profiles of the target organism (*Staphylococcus aureus* SG511 (*S. aureus*)), especially for those genes involved in regulation of stress, autolysis and energy metabolism [93,106].

Interactions with the cell components: Different opinions emerge in literature on the possibility that chitosan enters the cytoplasm. On one side chitosan is able to penetrate the cell and to interact with nucleic acid, thus interfering with the protein synthesis [107,108]. On the other side, the molecular mass of chitosan samples used in the experiments (not oligomers) has been questioned, making dubious its uptake. Alternatively, it has been suggested that high M chitosan deposition on cell surface could lead to a blockage of nutrients reducing microbial growth [98,109]. In a growth medium, chitosan might subtract micronutrients like essential metals (Ni, Zn, Co, Fe, Mg and Cu) to bacteria, inhibiting the production of toxin and important surviving molecules [93,110]. In conclusion, the proposed bacteriostatic activity of chitosan could arise from its well-known chelating ability.

5. Applications of Bacteriostatic Activity of Chitosan and Derivatives

Chitosan has been successfully used by researchers to exploit several applications in very diverse fields, in particular pharmaceutical (drug delivery, devices, and wound dressing), cosmetic, textile and food industries, as well as in agriculture and environment (waste-water purification). Some of the most interesting applications is based on the bactericidal potential of chitosan are summarized in the Table 3 of ref [93]. A summary of antimicrobial properties of chitosan and its derivatives is given, keeping in mind the general assertion that chitosan is considered as a GRAS (Generally Recognized As Safe) compound.

Multiple constructs/models of chitosan, and its derivatives, have been evaluated to express the bactericidal potential in medical field (e.g., post-surgery wound healing, patches and bandages as antimicrobial dressing after burning, antimicrobial coating of prosthesis) [111–116]. In the emerging field of nanobiopharmaceutics, antimicrobial chitosan nanoparticles have been studied for their enhancing effect of surface-to-size value. Thus, a vast number of works report the use of chitosan in form of nanoparticles, prepared in different conditions and blend, against various microorganisms [117–121].

In food industry, maintaining the quality and extending the shelf life of food products is mandatory. One of the most promising ways for effective preservation of food from alterations is using bioactive films as packaging. The use of active and/or edible bio-film based on chitosan alone, combined or enriched with different components such as plant extracts, other natural polysaccharides and antimicrobial peptides is emerging as an answer to such problems [122–127].

Finally, the use of antibacterial agents to prevent bacteria colonization is becoming a popular procedure in textile production, especially for goods employed in medical or hygienic services or sport-wears (odor-control textile) [128]. This antimicrobial supplementation is especially needed for natural fibers, which are more vulnerable to microbial attacks. Two examples, by Gupta and Haile [129] and Ye *et al.* [130], reported that a sensitive reduction of *S. aureus* (99%) was observed in modified chitosan embedded cotton.

6. Application of Chitosan as Delivery Systems

6.1. Chitosan Microparticles

The use of chitosan in pharmaceutical technology was originally conceived as an excipient for solid dosage forms, being used as coating, film-forming, mucoadhesive, disintegrant, tablet binder and viscosity-increasing agent [131]. The first investigations on the suitability of chitosan as drug carrier date back over twenty years [132]. Since the very beginning the use of chitosan was claimed to offer numerous advantages, such as high availability in nature, relatively low toxicity, and, above all, the presence of positively charged amino groups that enable both physical and chemical cross-linking [133]. In addition, after the seminal works of Yalpani and Hall [75–78], both the amino and hydroxyl groups have been exploited for a huge number of chemical modification [134].

Although among the several formulations the initial attention was focused on chitosan microspheres [135]. The spreading of nanotechnology shifted very fast the attention towards the nanoparticles, due to the numerous advantages offered by their size (in the nanometer range) [136–138]. Nowadays the choice of formulating micro- or nanoparticles is motivated by the type of performance required. For example, for intravenous delivery, only nanoparticles can be injected since microparticles would cause obstruction of blood vessels, while in case of pulmonary delivery, would microparticles have better efficacy, since nanoparticles would be exhaled [139].

Micro- and nanoparticle production is the preliminary step for the physical generation of spherical domains, independently of the final stage of preparation. Thereafter, an appropriate gelation process is applied to stabilize the micro- or nano-domains. The most common procedures for micro-particle production are extrusion, emulsion and spray-technologies (Figure 6). The "extrusion" technique is widely used and, in the simplest case, can be performed by using a syringe with a needle. Several parameters, such as the diameter of the orifice, the flow rate, the viscosity of the solution, the distance between the hardening solution and the orifice, the polymer concentration and the temperature [140], control the size of the droplets and thus the final dimensions of the particles. The emulsion technique consists of dispersing a liquid in another one that is not miscible. Most importantly, it is possible to tune the size of the droplets by selecting the appropriate composition of the two phases, *i.e.*, the type and concentration of polymer and surfactant [141,142]. The spray-technologies are based on the atomization in heated gas (air or nitrogen) of a fluid material (solution, emulsion or suspension), which is followed by a fast removal of the solvent (usually water). The concentration of the polymer and the temperature of the system influence the final dimension.

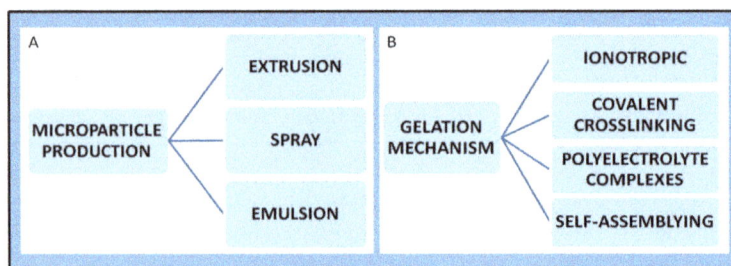

Figure 6. Graphical classification of the particles production (**A**) and mechanisms of gelation (**B**).

6.2. Chitosan Nanoparticles

Over the years the interest in the formulation of chitosan nanoparticles dramatically increased, exploring new and innovative preparation methods [143–145]. Basically, the main mechanisms for chitosan nanoparticles preparation rely on a crosslinking process (either chemical or physical [146]), formation of polyelectrolyte complexes [147] and self-assembly of hydrophobically modified chitosan [145] (Figure 6). Other possible processes include reverse micelle, desolvation, precipitation/coacervation and emulsion-droplet coalescence [144,148].

Chitosan crosslinking, such as a dropwise addition of a cross-linker to a chitosan solution, is considered an easy production technology to immobilize protein [149] or encapsulate different molecules [150,151]. Figure 7 summarizes the main characteristics of the gelation mechanisms. A classical covalent cross-linker is the glutaraldehyde; it acts as a bridge between two glucosamine units belonging to the same polymeric chain or to different chains. The amine group of the chitosan reacts with the aldehyde group of the linker leading to an imine bond via a Schiff reaction [147]. The concentration of glutaraldehyde strongly affects the textural properties of the final product [152]. The final structure results in an irreversible network with high mechanical properties, such as rigid network structures and high resistance to dissolution also in extreme pH conditions. One of the main drawback encountered is the high toxicity of the aldehydes [153]; the use of such systems is actually limited due to insufficient biocompatibility of the cross-linkers. Similar considerations can be done for epichlorohydrin [154]. A new crosslinking agent, genipin, which is found in gardenia fruit extract, is becoming an interesting alternative to glutaraldheyde [154–162].

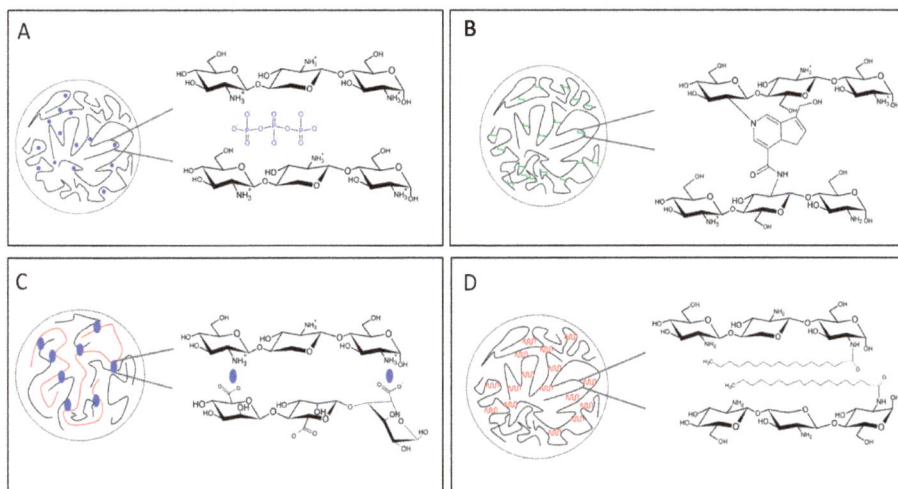

Figure 7. Gelation mechanisms for chitosan nanoparticles preparation: (**A**) ionic crosslinking; (**B**) covalent crosslinking; (**C**) polyelectrolyte complexation (PEC); and (**D**) self-assembly.

Another possibility is the use of 1-Ethyl-3-(3-dimethylaminopropyl)carbodiimide (EDC), or similar compounds, which enable the formation of an amide bond, via an intermediate hydrazide derivative, between the amino group of chitosan and the carboxyl group of different macromolecules. EDC is also known as zero length cross-linker [163]. A valuable alternative is represented by carboxylic acids (especially from natural origin) [133,164]. All the linkers mentioned above can link chitosan with different polymers, such as gelatin [165], or other compounds, such as hydroxyapatite [166]). Attempts have been addressed to perform covalent modification using dextran sulfate [167]. Alginate too is potentially able to interact covalently with chitosan [168] producing resistant hydrogel. When choosing a covalent crosslinking method, it is important to check that the cross-linker does not bind also to the drug, to avoid formation of drug–polymer conjugates.

However, such covalent derivatives of chitosan are not considered as the best choice for drug delivery due to their lack of swelling [147] and absence of pH-dependence drug release. Chitosan is indeed appealing since it is a stimulus-responsive polymer [169].

The ionic crosslinking is also very appealing due to the presence of charged amino groups on chitosan in acidic conditions that allow to crosslink the polymer by using negatively charged ions, such as tripolyphosphate (TPP), which is the most widely used (note that this name entered incorrectly in the common language and is difficult to remove; chemically speaking it is a trimer and not a poly-phosphate, and even less a "tri-poly-something"). The interaction, commonly referred as ionotropic gelation, was firstly reported by Bodmeier [170] for the preparation of beads by dropping chitosan in a TPP solution and further extensively studied and described by Calvo *et al.* [171]. The ionic gelation takes place spontaneously via electrostatic interactions between chitosan chains and TPP [172]. This methodology of nano-particles preparation allows to encapsulate several types of biomolecules, from protein [173–175], to small drugs [176,177] and DNA fragments [178], and, because of its versatility, it was extensively studied and optimized [179,180].

Since the beginning, such mild conditions of gelation, appeared immediately very attractive for the encapsulation of labile drugs, such as proteins or peptides, characterized by poor bioavailability upon oral administration [181]. Within this frame, numerous studies have been conducted for the oral delivery of insulin [182], for the pulmonary delivery [183] and also for ocular delivery [184]. In addition to drug delivery, chitosan-TPP nanoparticles have been explored for many other uses. For example Du *et al.* explored the loading of various metal ions to obtain chitosan nanoparticles characterized by antibacterial activity [117]. This study evidenced that the antibacterial activity was significantly enhanced by the metal ions load.

Another very popular mechanism for the easy formation of nanoparticles is through polyelectrolyte complexations (PEC). Chitosan, the only natural polymer with a positive charge, is an attractive polymer for its ability to complex with a wide range of negatively charged polysaccharides by electrostatic interaction, forming PEC. More specifically, the interaction involves a phase separation where the solvent is excluded from the hydrophilic colloids. This can be achieved by adding a competing hydrophilic compound, such as a salt or an alcohol [185], and it is referred as simple coacervation, while complex coacervation occurs in a solution of two oppositely charged polyelectrolytes, giving rise to two immiscible liquid phases [186]. The phase rich in colloids is the coacervate, while the other phase is the equilibrium solution [187]. The interaction is likely to occur first via the long-range Coulomb forces, which provide for the primary strong binding energy, followed by the more directional short-range hydrogen bonds. The optimum condition for complex coacervation requires a constant control of pH. The mechanical properties and permeability of the products are strongly influenced by the properties of the starting material and conditions of reaction.

Chitosan is able to form PEC with both natural and synthetic polymers and since the early work published on PECs by Fuoss and Sadek (1949) [188], different studies and reviews have been published on this topic. PEC complexes obtained using natural biopolymers with opposite charge have been formulated for peptide [189,190] and protein [191] carriers but represent also an ideal colloidal carrier for DNA delivery [192]. Another emerging field of application is the delivery of antibiotics. Tobramycin loaded nanoparticles functionalized with dornase alfa demonstrated DNA degradation and improved nanoparticles penetration, thus increasing the efficacy of tobramycin [193].

In particular, among natural polymers, chitosan can interact ionically with several polyanion [194], such as alginate where the coacervation is due to interactions between the carboxyl functions of guluronic/mannuronic units of the alginate and the positively charged amine groups of the chitosan. A slight variation of this method is known as ionotropic pre-gelation reaction, as reported by Sarmento in a study for oral delivery of insulin [195,196]. This method, firstly reported by Rajaonarivony [197], involves an interaction among alginate, calcium and poly-lysine. The reaction consists in a first step characterized by an interaction between alginate and calcium ions, due to the gelling properties of alginate for the presence of guluronic residues [196], that are able to exchange sodium ions with divalent cations such as Ca^{2+}, Sr^{2+} and Ba^{2+}, forming the characteristic egg box structures. This step in the original study was followed by the crosslinking of pre-gelled alginate with poly-lysine, but in several subsequent works chitosan substituted poly-lysine [191,195,196,198–200]. Hyaluronan, another biocompatible polymer negatively charged that can be used to form complexes with chitosan [201,202] has the ability to bind several receptors as CD44 [203]. Hyaluronan can also be used as coating for chitosan-TPP complexation [180]. Further examples of complexation are represented by the reaction of chitosan with carrageenan [204], dextran sulfate [195,205], and xanthan [206]. Studies on similar interactions using negatively charged macromolecules, such as arabic gum [207], gelatin [208] and pectin [209] have been published.

A recent review [210] reports on the advances made in this field, focusing the attention on chitosan PEC's with natural polysaccharides, such as alginate, hyaluronic acid, pectin, carrageenan, xanthan gum, gellan gum, arabic gum, and carboxymethyl cellulose, *etc.*, discussing also *in vitro* and *in vivo* data. On the other hand, synthetic polymers, such for example polyacrilic acid, have also been considered for polyeletrolytes complexation [211,212].

A sub-category of complex coacervation is represented by polyelectrolyte-colloids coacervate that has some advantage with respect to polyelectrolyte-polyelectrolyte coacervation in retaining the structure of colloids and in reducing the heterogeneity and configurational properties of the final system. In particular polymers-protein coacervates represent an interesting system for preservation of bio-functionality and are particularly important in enzyme immobilization, protein purification, antigen delivery and food stabilization [186]. Polyelectrolyte-protein coacervation [213] can occur between the amino groups of the chitosan and the reactive groups of heparin [214], BSA [186] and casein [215]. A change in the proportion between polyelectrolyte and protein can increment or suppress the coacervate.

Another appealing macromolecule is DNA, whose complexes with chitosan can be formed by simple mixing [216,217] and represent a promising non-viral vehicles for gene delivery [192,218]. A very recent review [219] reports on the promising use of chitosan as a non-viral nucleotide delivery system in spite of viral vectors that, although characterized by a high transfection efficiency, pose some safety concerns regarding immunogenicity and insertional mutagenesis. Thus, achievements towards this direction would be of great benefit for gene therapy.

Chitosan can be also modified with hydrophobic groups, resulting in grafted polymers that show a tendency to form inter- and intramolecular interactions in polar solvents by forming polymeric micelles that can be used to encapsulate hydrophobic drugs. The long polymeric chains interconnected with the hydrophobic molecules help stabilize the micelles protecting the internal drug. An abundant literature is available on this topic, investigating chitosan amphipatic behavior after the crosslinking with cholesterol [220,221], deoxycholic acid [222], stearic acid, and linoleic acid [223,224].

Finally, it is worth to stress that a full knowledge of the physicochemical properties of the chitosan is required to obtain nanoparticles with tailored characteristics. The final structure of the network is the result of the contribution of different parameters, such as molecular weight and degree of acetylation, which affect the hydrophobic interactions and the network of hydrogen bonds. Xu and Du reported a direct correlation between chitosan MW and both encapsulation efficiency and release of BSA [173]. Similarly, formulation parameters, such as the ratio between chitosan and TPP [225] and operative parameters, such as stirring time and speed, affect the properties of the nanoparticles and their yield [226]. The release profile is a consequence of the behavior of the nanoparticles in the aqueous environment; more specifically the drug release occurs basically by three main mechanisms, diffusion, swelling and erosion (Figure 8), and it is strictly dependent on the type and degree of the crosslink. In case of a covalent cross-link, the permanent network obtained prompt for a drug release, which is mainly by diffusion, and the overall release profile will depend on the cross-linking degree. Similarly, the cross-linking density and the pH of the environment, as clearly reported by Berger *et al.* [146], influence also the swelling capacity. The mechanism of erosion is instead possible in presence of an ionic cross-linker. The kinetic of swelling and erosion will affect the initial part of the release curve determining a characteristic lag-phase. A huge amount of studies have been done in the field of model equations to describe the different several release profiles [227,228].

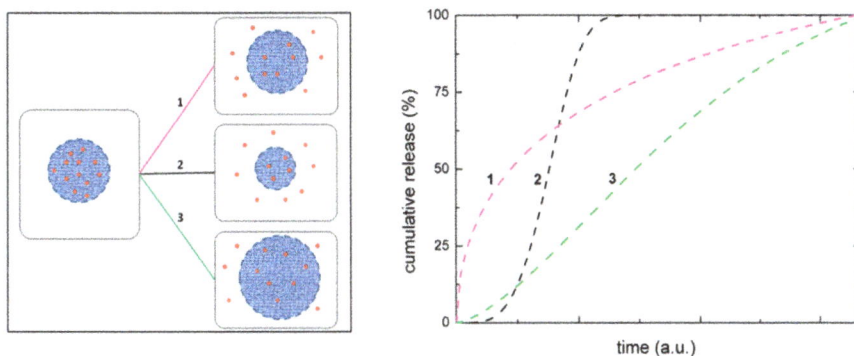

Figure 8. Example of release mechanisms (1: diffusion, 2: erosion and 3: swelling) and some related graphical trends according to mathematical models. Each mechanism is driven by the chemical structure of the network: elastic swelling and/or hydrolysis of cross-link, but also chitosan depolymerization by lysozyme action, in addition to the basic diffusion from the porous nanoconstruct.

The improvements and modifications described so far will lead further progresses in pharmaceutical applications [229]. Many papers claim the potentiality of chitosan as carrier for drug delivery. Has this promise been fulfilled? Chitosan has been recognized as safe (GRAS) and approved for dietary use in Italy, Japan and Finland and only in 2008 a specific monograph was introduced in the European Pharmacopeia and in 2011 in the US National formulary. This of course has limited its use for the development of new drug formulations. Currently, chitosan is approved for wound dressing applications and cartilage repairing formulation. A recent review [230] on the use of chitosan as an absorption enhancer in nasal drug delivery formulations describes the ChiSys[TM] technology for the delivery of peptides, proteins and small hydrophylic drugs. A chitosan-based formulation for nasal administration of morphine (Rylomine[TM]), at the moment in phase 2 (UK and EU) and phase 3 (USA) clinical trials, is expected to be released on the market in the near future.

The high level of knowledge on chitosan accumulated in the past years, both on the physico-chemical properties and on the use as drug delivery systems opens the possibility of many other applications [231].

7. Conclusive Remarks

The challenging point after reading so many articles and blending literature results with our own experience is recurrent in many lines of this article. It can be synthetized in two simple questions: How are ugly and bad tunable to good? How much we are willing to bet on a chitosan future? This final section attempts to resume the main issues that, to the best of our knowledge, should be addressed in the near future.

We would emphasize that the difficulties encountered in chitosan development as a marketed product in drug delivery and bioactive material could actually reside in the poor correlation between accurate chemical structure determination and its effective biological responses. An explicit reference can be made to references [42–44], where this issue has been recently reformulated. However, the main characteristics of chitosan to be considered, for example in the development of drug delivery systems, are not only the acetylation degree (DA), molecular weight (M), and purity, but also the relationships between internal and external parameters. Indeed, by stating that bioactivity and biocompatibility are both substrate- and host-dependent, it is clear that responses depend on investigated materials and on the experimental read-out used. In the case of chitosan, the understanding of material biocompatibility is hindered not only by the limited knowledge on the biological processes involved in material-cells interactions, but also by the poor assessment of the polymer characteristics. A recent review [222] addresses the general problem of the safety issue on chitosan uses, suggesting a more careful assessment of its safety in non-oral formulations.

A naïve answer is given by the fact that chitosan approval by the American Food and Drug Administration (FDA) is not a general definition as GRAS, but FDA and other regulatory agencies evaluate and approve materials with respect to specific applications. Not to forget that other relevant properties beyond the polymer itself are size, morphology, crystallinity, surface characteristics, degradation profile and additional products.

On the polymer side, the reported chitosan versatility and the variety of formulations add confusion to inexperienced researchers and regulatory scientists. For this reason, a thorough and systematic description of the chitosan used in a study should be either provided by the producer or be complemented by the research laboratories. Needless to say, the existence of expertise spread around the word and of large infrastructures should also be usefully utilized.

Acknowledgments: This paper wishes to David A. Brant and Professor Vittorio Crescenzi (deceased), who pioneered the experimental and theoretical studies on physicochemical properties of polysaccharides. The authors wish to acknowledge unpublished material made available by Mirko Weinhold on the characterization of chitosan derivatives. The original work has been partially carried out within the EC Project FP6 "NanoBioPharmaceutics" (NMP 026723-2) and within the project "Oral vaccine carrier for fish farming of Friuli Venezia Giulia". I.D'A. has been recipient of a grant from MIUR (Rome) during her Ph.D. studies in the School of

Mar. Drugs **2016**, *14*, 99

Nanotechnology, University of Trieste. M. Grimaldi is gratefully acknowledged for the Toc picture inspiring the article title.

Conflicts of Interest: The authors declare no conflicts of interest.

References

1. Clark, G.L.; Smith, A.F. X-ray diffraction studies. *J. Phys. Chem.* **1996**, *40*, 863–879. [CrossRef]
2. Ogawa, K. Effect of Heating an Aqueous Suspension of Chitosan on the Crystallinity and Polymorphs. *Agric. Biol. Chem.* **1991**, *55*, 2375–2379.
3. Ogawa, K.; Yui, T.; Okuyama, K. Three D structures of chitosan. *Int. J. Biol. Macromol.* **2004**, *34*, 1–8. [CrossRef] [PubMed]
4. Okuyama, K.; Noguchi, K.; Miyazawa, T.; Yui, T.; Ogawa, K. Molecular and Crystal Structure of Hydrated hitosan. *Macromolecules* **1997**, *30*, 5849–5855. [CrossRef]
5. Sugano, M.; Fujikawa, T.; Hiratsuji, Y.; Nakashima, K.; Fukuda, N.; Hasegawa, Y. A novel use of chitosan as a hypocholesterolemic agent in rats. *Am. J. Clin. Nutr.* **1980**, *33*, 787–793. [PubMed]
6. Allan, C.R.; Hadwiger, L.A. The fungicidal effect of chitosan on fungi of varying cell wall composition. *Exp. Mycol.* **1979**, *3*, 285–287. [CrossRef]
7. *Proceedings of the First International Conference on Chitin/Chitosan*; Muzzarelli, R.A.A., Pariser, E.R., Eds.; Massachusets Institute of Technology: Boston, MA, USA, 1978.
8. Special Issue: Polysaccharide Solution and Gels. *Carbohydr. Polym.* **1982**, *2*, 227–324. Available online: http://www.sciencedirect.com/science/journal/01448617/2 (accessed on 13 May 2016).
9. *New Developments in Industrial Polysaccharides*; Crescenzi, V., Dea, I.C.M., Stivala, S.S., Eds.; Gordon and Breach Science Publishers: New York, NY, USA, 1985.
10. Domard, A.; Rinaudo, M. Preparation and characterization of fully deacetylated chitosan. *Int. J. Biol. Macromol.* **1983**, *5*, 49–52. [CrossRef]
11. Chandy, T.; Sharma, C.P. Chitosan—As a biomaterial. *Biomater. Artif. cells Artif. organs* **1990**, *18*, 1–24. [CrossRef] [PubMed]
12. Muzzarelli, R.; Tarsi, R.; Filippini, O.; Giovanetti, E.; Biagini, G.; Varaldo, P.E. Antimicrobial properties of N-carboxybutyl chitosan. *Antimicrob. Agents Chemother.* **1990**, *34*, 2019–2023. [CrossRef] [PubMed]
13. Hamed, I.; Özogul, F.; Regenstein, J.M. Industrial applications of crustacean by-products (chitin, chitosan, and chitooligosaccharides): A review. *Trends Food Sci. Technol.* **2016**, *48*, 40–50. [CrossRef]
14. Chitin & Chitosan: A Global Strategic Business Report. Global Industry Analysts Inc. Available online: http://www.strategyr.com/Chitin_and_Chitosan_Market_Report.asp (accessed on 13 May 2016).
15. Stoye, E. Magnificent molecules. *Mole* **2014**, *4*. Available online: http://www.rsc.org/eic/sites/default/files/The-Mole-March-2014-corrected.pdf (accessed on 13 May 2016).
16. Stahel, W.R. The circular economy. *Nature* **2016**, *531*, 435–438. [CrossRef] [PubMed]
17. Yan, N.; Chen, X. Sustainability: Don't waste seafood waste. *Nature* **2015**, *524*, 155–157. [CrossRef] [PubMed]
18. Kerton, F.M.; Liu, Y.; Omari, K.W.; Hawboldt, K. Green chemistry and the ocean-based biorefinery. *Green Chem.* **2013**, *15*, 860. [CrossRef]
19. Chen, X.; Chew, S.L.; Kerton, F.M.; Yan, N. Direct conversion of chitin into a N-containing furan derivative. *Green Chem.* **2014**, *16*, 2204. [CrossRef]
20. Pierson, Y.; Chen, X.; Bobbink, F.D.; Zhang, J.; Yan, N. Acid-Catalyzed Chitin Liquefaction in Ethylene Glycol. *ACS Sustain. Chem. Eng.* **2014**, *2*, 2081–2089. [CrossRef]
21. Bobbink, F.D.; Zhang, J.; Pierson, Y.; Chen, X.; Yan, N. Conversion of chitin derived N-acetyl-D -glucosamine (NAG) into polyols over transition metal catalysts and hydrogen in water. *Green Chem.* **2015**, *17*, 1024–1031. [CrossRef]
22. Cesàro, A.; Bellich, B.; Borgogna, M. Nanoparticles for pharmaceutical and biomedical applications. In *Crystallography for Health and Biosciences*; Guagliardi, A., Masciocchi, N., Eds.; Insubria University Press: Como, Italy, 2012; pp. 143–161.
23. Smidsrod, O.; Haug, A.; Whittington, S.G. The molecular basis for some physical properties of polyuronides. *Acta Chem. Scand.* **1972**, *26*, 2563–2566. [CrossRef]
24. Grasdalen, H.; Larsen, B.; Smisrod, O. 13C-N.M.R. studies of monomeric composition and sequence in alginate. *Carbohydr. Res.* **1981**, *89*, 179–191. [CrossRef]

25. Terayama, H. Methods of colloids titration (a new titration between polymer ions). *J. Polym. Sci. A* **1952**, *8*, 243–253. [CrossRef]

26. Balàzs, N.; Sipos, P. Limitations of pH-potentiometric titration for the determination of the degree of deacetylation of chitosan. *Carbohydr. Res.* **2007**, *342*, 124–130. [CrossRef] [PubMed]

27. Domszy, J.G.; Roberts, G.A.F. Evaluation of infrared spectroscopic techniques for analyzing chitosan. *Macromol. Chem. Phys.* **1985**, *186*, 1671–1677. [CrossRef]

28. Miya, M.; Iwamoto, R.; Yoshikawa, S.; Mima, S. IR spectroscopic determination of CONH content in highly deacetylated chitosan. *Int. J. Biol. Macromol.* **1980**, *2*, 323–324. [CrossRef]

29. Moore, G.K.; Roberts, G.A.F. Determination of the degree of N-acetylation of chitosan. *Int. J. Biol. Macromol.* **1980**, *2*, 115–116. [CrossRef]

30. Sabnis, S.; Block, L.H. Improved infrared spectroscopic method for the analysis of degree of N-deacetylation of chitosan. *Polym. Bull.* **1997**, *39*, 67–71. [CrossRef]

31. Sannan, T.; Kurita, K.; Ogura, K.; Iwakura, Y. Studies on chitin: 7. IR spectroscopic determination of degree of deacetylation. *Polymer (Guildf).* **1978**, *19*, 458–459. [CrossRef]

32. Muraki, E.; Yaku, F.; Iyoda, J.; Kojima, H. Measurement of Degree of Deacetylation in D-Glucosamine Oligosaccharides by UV Absorption. *Biosci. Biotechnol. Biochem.* **1993**, *57*, 1929–1930. [CrossRef]

33. Tan, S.C.; Khor, E.; Tan, T.K.; Wong, S.M. The degree of deacetylation of chitosan: Advocating the first derivative UV-spectrophotometry method of determination. *Talanta* **1998**, *45*, 713–719. [CrossRef]

34. Domard, A. Determination of N-acetyl content in chitosan samples by c.d. measurements. *Int. J. Biol. Macromol.* **1987**, *9*, 333–336. [CrossRef]

35. Hirai, A.; Odani, H.; Nakajima, A. Determination of degree of deacetylation of chitosan by H NMR spectroscopy. *Polym. Bull.* **1991**, *26*, 87–94. [CrossRef]

36. Duarte, M.L.; Ferreira, M.C.; Marvao, M.R.; Rocha, J. Determination of the degree of acetylation of chitin materials by 13C CP/MAS NMR spectroscopy. *Int. J. Biol. Macromol.* **2001**, *28*, 359–363. [CrossRef]

37. Raymond, L.; Morin, F.G.; Marcessaurt, R.H. Degree of deacetylation of chitosan using conductometric titration and solid-state NMR. *Carbohydr. Res.* **1993**, *246*, 331–336. [CrossRef]

38. Vårum, K.M.; Anthonsen, M.W.; Grasdalen, H.; Smidsrod, O. Determination of the degree of N-acetylation and the distribution of N-acetyl groups in partially N-deacetylated chitins (chitosans) by high-field NMR spectroscopy. *Carbohydr. Res.* **1991**, *211*, 17–23. [CrossRef]

39. Niola, F.; Basora, N.; Chornet, E.; Vidal, P.F. A rapid method for the determination of the degree of N-acetylation of chitin-chitosan samples by acid hydrolysis and HPLC. *Carbohydr. Res.* **1993**, *238*, 1–9. [CrossRef]

40. Kasaai, M. A review of several reported procedures to determine the degree of N-acetylation for chitin and chitosan using infrared spectroscopy. *Carbohydr. Polym.* **2008**, *71*, 497–508. [CrossRef]

41. Guinesi, L.S.; Cavalheiro, É.T.G. The use of DSC curves to determine the acetylation degree of chitin/chitosan samples. *Thermochim. Acta* **2006**, *444*, 128–133. [CrossRef]

42. Rampino, A. Polysaccharide-based nanoparticles for drug delivery. Ph.D. Thesis, University of Trieste, Trieste, Italy, 2011.

43. Bovey, F.A.; Mirau, P.A. *NMR of Polymers*; Academic Press: San Diego, CA, USA, 1996.

44. Kumirska, J.; Weinhold, M.X.; Sauvageau, J.C.M.; Thöming, J.; Kaczyński, Z.; Stepnowski, P. Determination of the pattern of acetylation of low-molecular-weight chitosan used in biomedical applications. *J. Pharm. Biomed. Anal.* **2009**, *50*, 587–590. [CrossRef] [PubMed]

45. Aiba, S. Studies on chitosan: 3. Evidence for the presence of random and block copolymer structures in partially N-acetylated chitosans. *Int. J. Biol. Macromol.* **1991**, *13*, 40–44. [CrossRef]

46. Cowman, M.K.; Matsuoka, S. Experimental approaches to hyaluronan structure. *Carbohydr. Res.* **2005**, *340*, 791–809. [CrossRef] [PubMed]

47. Furlan, S.; La Penna, G.; Perico, A.; Cesàro, A. Hyaluronan chain conformation and dynamics. *Carbohydr. Res.* **2005**, *340*, 959–970. [CrossRef] [PubMed]

48. Morris, G.A.; Castile, J.; Smith, A.; Adams, G.G.; Harding, S.E. Macromolecular conformation of chitosan in dilute solution: A new global hydrodynamic approach. *Carbohydr. Polym.* **2009**, *76*, 616–621. [CrossRef]

49. Weinhold, M. Characterization of chitosan using triple detection size-exclusion chromatography and 13 C-NMR spectroscopy. Ph.D. Thesis, Universität Bremen, Bremen, Germany, 2010.

50. Weinhold, M.X.; Sauvageau, J.C.M.; Keddig, N.; Matzke, M.; Tartsch, B.; Grunwald, I.; Kübel, C.; Jastorff, B.; Thöming, J. Strategy to improve the characterization of chitosan for sustainable biomedical applications: SAR guided multi-dimensional analysis. *Green Chem.* **2009**, *11*, 498. [CrossRef]

51. Kumirska, J.; Weinhold, M.X.; Thöming, J.; Stepnowski, P. Biomedical Activity of Chitin/Chitosan Based Materials—Influence of Physicochemical Properties Apart from Molecular Weight and Degree of *N*-Acetylation. *Polymers (Basel).* **2011**, *3*, 1875–1901. [CrossRef]

52. Brant, D.A. Novel approaches to the analysis of polysaccharide structures. *Curr. Opin. Struct. Biol.* **1999**, *9*, 556–562. [CrossRef]

53. Cesàro, A.; Bellich, B.; Borgogna, M. Biophysical functionality in polysaccharides: From Lego-blocks to nano-particles. *Eur. Biophys. J.* **2012**, *41*, 379–395. [CrossRef] [PubMed]

54. Nilsen-Nygaard, J.; Strand, S.; Vårum, K.; Draget, K.; Nordgård, C. Chitosan: Gels and Interfacial Properties. *Polymers (Basel).* **2015**, *7*, 552–579. [CrossRef]

55. Younes, I.; Rinaudo, M. Chitin and chitosan preparation from marine sources. Structure, properties and applications. *Mar. Drugs* **2015**, *13*, 1133–1174. [CrossRef] [PubMed]

56. Rinaudo, M. Chitin and chitosan: Properties and applications. *Prog. Polym. Sci.* **2006**, *31*, 603–632. [CrossRef]

57. Watts, P.; Smith, A.; Hinchcliffe, M. ChiSys®as a chitosan-based delivery platform for nasal vaccination. In *Mucosal Delivery of Biopharmaceuticals: Biology, Challenges and Strategies*; das Neves, J., Sarmento, B., Eds.; Springer Science & Business Media: New York, NY, USA, 2014; pp. 499–516.

58. Ottøy, M.H.; Vårum, K.M.; Smidsrød, O. Compositional heterogeneity of heterogeneously deacetylated chitosans. *Carbohydr. Polym.* **1996**, *29*, 17–24. [CrossRef]

59. Vårum, K.M.; Ottøy, M.H.; Smidsrød, O. Water-solubility of partially *N*-acetylated chitosans as a function of pH: Effect of chemical composition and depolymerisation. *Carbohydr. Polym.* **1994**, *25*, 65–70. [CrossRef]

60. Korchagina, E.V.; Philippova, O.E. Multichain aggregates in dilute solutions of associating polyelectrolyte keeping a constant size at the increase in the chain length of individual macromolecules. *Biomacromolecules* **2010**, *11*, 3457–3466. [CrossRef] [PubMed]

61. Boucard, N.; David, L.; Rochas, C.; Montembault, A.; Viton, C.; Domard, A. Polyelectrolyte microstructure in chitosan aqueous and alcohol solutions. *Biomacromolecules* **2007**, *8*, 1209–1217. [CrossRef] [PubMed]

62. Kjøniksen, A.-L.; Iversen, C.; Nyström, B.; Nakken, T.; Palmgren, O. Light Scattering Study of Semidilute Aqueous Systems of Chitosan and Hydrophobically Modified Chitosans. *Macromolecules* **1998**, *31*, 8142–8148. [CrossRef]

63. Ottøy, M.H.; Vårum, K.M.; Christensen, B.E.; Anthonsen, M.W.; Smidsrød, O. Preparative and analytical size-exclusion chromatography of chitosans. *Carbohydr. Polym.* **1996**, *31*, 253–261. [CrossRef]

64. Berth, G.; Dautzenberg, H. The degree of acetylation of chitosans and its effect on the chain conformation in aqueous solution. *Carbohydr. Polym.* **2002**, *47*, 39–51. [CrossRef]

65. Franzén, H.; Draget, K.; Langebäck, J.; Nilsen-Nygaard, J. Characterization and Properties of Hydrogels Made from Neutral Soluble Chitosans. *Polymers (Basel).* **2015**, *7*, 373–389. [CrossRef]

66. Cho, Y.-W.; Jang, J.; Park, C.R.; Ko, S.-W. Preparation and Solubility in Acid and Water of Partially Deacetylated Chitins. *Biomacromolecules* **2000**, *1*, 609–614. [CrossRef] [PubMed]

67. Berth, G.; Dautzenberg, H.; Peter, M.G. Physico-chemical characterization of chitosans varying in degree of acetylation. *Carbohydr. Polym.* **1998**, *36*, 205–216. [CrossRef]

68. Cölfen, H.; Berth, G.; Dautzenberg, H. Hydrodynamic studies on chitosans in aqueous solution. *Carbohydr. Polym.* **2001**, *45*, 373–383. [CrossRef]

69. Anthonsen, M.W.; Vårum, K.M.; Hermansson, A.M.; Smidsrød, O.; Brant, D.A. Aggregates in acidic solutions of chitosans detected by static laser light scattering. *Carbohydr. Polym.* **1994**, *25*, 13–23. [CrossRef]

70. Buhler, E.; Rinaudo, M. Structural and Dynamical Properties of Semirigid Polyelectrolyte Solutions: A Light-Scattering Study. *Macromolecules* **2000**, *33*, 2098–2106. [CrossRef]

71. Philippova, O.E.; Volkov, E.V.; Sitnikova, N.L.; Khokhlov, A.R.; Desbrieres, J.; Rinaudo, M. Two Types of Hydrophobic Aggregates in Aqueous Solutions of Chitosan and Its Hydrophobic Derivative. *Biomacromolecules* **2001**, *2*, 483–490. [CrossRef] [PubMed]

72. Lamarque, G.; Lucas, J.-M.; Viton, C.; Domard, A. Physicochemical behavior of homogeneous series of acetylated chitosans in aqueous solution: Role of various structural parameters. *Biomacromolecules* **2005**, *6*, 131–142. [CrossRef] [PubMed]

73. Schatz, C.; Viton, C.; Delair, T.; Pichot, C.; Domard, A. Typical physicochemical behaviors of chitosan in aqueous solution. *Biomacromolecules* **2003**, *4*, 641–648. [CrossRef] [PubMed]

74. Novoa-Carballal, R.; Riguera, R.; Fernandez-Megia, E. Chitosan hydrophobic domains are favoured at low degree of acetylation and molecular weight. *Polymer (Guildf)*. **2013**, *54*, 2081–2087. [CrossRef]

75. Yalpani, M. Selective chemical modification of polysaccharides. Ph.D. Thesis, University of British Columbia, Vancouver, BC, Canada, 1980.

76. Yalpani, M. A survey of recent advances in selective chemical and enzymic polysaccharide modifications. *Tetrahedron* **1985**, *41*, 2957–3020. [CrossRef]

77. Yalpani, M.; Hall, L.D. Some chemical and analytical aspects of polysaccharide modifications. III. Formation of branched-chain, soluble chitosan derivatives. *Macromolecules* **1984**, *17*, 272–281. [CrossRef]

78. Hall, L.D.; Yalpani, M. Formation of branched-chain, soluble polysaccharides from chitosan. *J. Chem. Soc. Chem. Commun.* **1980**, 1153–1154. [CrossRef]

79. Kumar, M.N.V.R.; Muzzarelli, R.A.A.; Muzzarelli, C.; Sashiwa, H.; Domb, A.J. Chitosan chemistry and pharmaceutical perspectives. *Chem. Rev.* **2004**, *104*, 6017–6084. [CrossRef] [PubMed]

80. Qin, C.; Li, H.; Xiao, Q.; Liu, Y.; Zhu, J.; Du, Y. Water-solubility of chitosan and its antimicrobial activity. *Carbohydr. Polym.* **2006**, *63*, 367–374. [CrossRef]

81. Chen, C.-S.; Liau, W.-Y.; Tsai, G.-J. Antibacterial Effects of N-Sulfonated and N-Sulfobenzoyl Chitosan and Application to Oyster Preservation. *J. Food Prot.* **1998**, *61*, 1124–1128.

82. Kaya, M.; Asan-Ozusaglam, M.; Erdogan, S. Comparison of antimicrobial activities of newly obtained low molecular weight scorpion chitosan and medium molecular weight commercial chitosan. *J. Biosci. Bioeng.* **2016**, *121*, 678–684. [CrossRef] [PubMed]

83. Ravi Kumar, M.N. A review of chitin and chitosan applications. *React. Funct. Polym.* **2000**, *46*, 1–27. [CrossRef]

84. Choi, B.-K.; Kim, K.-Y.; Yoo, Y.-J.; Oh, S.-J.; Choi, J.-H.; Kim, C.-Y. *In vitro* antimicrobial activity of a chitooligosaccharide mixture against Actinobacillus actinomycetemcomitans and Streptococcus mutans. *Int. J. Antimicrob. Agents* **2001**, *18*, 553–557. [CrossRef]

85. İkinci, G.; Şenel, S.; Akıncıbay, H.; Kaş, S.; Erciş, S.; Wilson, C.; Hıncal, A. Effect of chitosan on a periodontal pathogen Porphyromonas gingivalis. *Int. J. Pharm.* **2002**, *235*, 121–127. [CrossRef]

86. Tsai, G.J.; Su, W.H.; Chen, H.C.; Pan, C.L. Antimicrobial activity of shrimp chitin and chitosan from different treatments and applications of fish preservation. *Fish. Sci.* **2002**, *68*, 170–177. [CrossRef]

87. Savard, T.; Beaulieu, C.; Boucher, I.; Champagne, C.P. Antimicrobial Action of Hydrolyzed Chitosan against Spoilage Yeasts and Lactic Acid Bacteria of Fermented Vegetables. *J. Food Prot.* **2002**, *65*, 828–833.

88. Cuero, R.G. Antimicrobial action of exogenous chitosan. *EXS* **1999**, *87*, 315–333. [PubMed]

89. Pelletier, E.; Bonnet, C.; Lemarchand, K. Biofouling growth in cold estuarine waters and evaluation of some chitosan and copper anti-fouling paints. *Int. J. Mol. Sci.* **2009**, *10*, 3209–3223. [CrossRef] [PubMed]

90. Martínez-Camacho, A.P.; Cortez-Rocha, M.O.; Ezquerra-Brauer, J.M.; Graciano-Verdugo, A.Z.; Rodriguez-Félix, F.; Castillo-Ortega, M.M.; Yépiz-Gómez, M.S.; Plascencia-Jatomea, M. Chitosan composite films: Thermal, structural, mechanical and antifungal properties. *Carbohydr. Polym.* **2010**, *82*, 305–315. [CrossRef]

91. Goy, R.C.; de Britto, D.; Assis, O.B.G. A review of the antimicrobial activity of chitosan. **2009**, *19*, 241–247. [CrossRef]

92. Raafat, D.; Sahl, H.-G. Chitosan and its antimicrobial potential—A critical literature survey. *Microb. Biotechnol.* **2009**, *2*, 186–201. [CrossRef] [PubMed]

93. Kong, M.; Chen, X.G.; Xing, K.; Park, H.J. Antimicrobial properties of chitosan and mode of action: A state of the art review. *Int. J. Food Microbiol.* **2010**, *144*, 51–63. [CrossRef] [PubMed]

94. Raafat, D. Chitosan as an Antimicrobial Agent: Modes of Action and Resistance Mechanisms. Ph.D. Thesis, University of Bonn, Bonn, Germany, 2008.

95. Eaton, P.; Fernandes, J.C.; Pereira, E.; Pintado, M.E.; Xavier Malcata, F. Atomic force microscopy study of the antibacterial effects of chitosans on Escherichia coli and Staphylococcus aureus. *Ultramicroscopy* **2008**, *108*, 1128–1134. [CrossRef] [PubMed]

96. Hirano, S.; Tsuchida, H.; Nagao, N. N-acetylation in chitosan and the rate of its enzymic hydrolysis. *Biomaterials* **1989**, *10*, 574–576. [CrossRef]

97. Takahashi, T.; Imai, M.; Suzuki, I.; Sawai, J. Growth inhibitory effect on bacteria of chitosan membranes regulated with deacetylation degree. *Biochem. Eng. J.* **2008**, *40*, 485–491. [CrossRef]
98. Vishu Kumar, A.B.; Varadaraj, M.C.; Gowda, L.R.; Tharanathan, R.N. Characterization of chito-oligosaccharides prepared by chitosanolysis with the aid of papain and Pronase, and their bactericidal action against Bacillus cereus and Escherichia coli. *Biochem. J.* **2005**, *391*, 167–175. [CrossRef] [PubMed]
99. Hernández-Lauzardo, A.N.; Bautista-Baños, S.; Velázquez-Del Valle, M.G.; Méndez-Montealvo, M.G.; Sánchez-Rivera, M.M.; Bello-Pérez, L.A. Antifungal effects of chitosan with different molecular weights on *in vitro* development of Rhizopus stolonifer (Ehrenb.:Fr.) Vuill. *Carbohydr. Polym.* **2008**, *73*, 541–547. [CrossRef] [PubMed]
100. Chung, Y.-C.; Su, Y.-P.; Chen, C.-C.; Jia, G.; Wang, H.-L.; Wu, J.C.G.; Lin, J.-G. Relationship between antibacterial activity of chitosan and surface characteristics of cell wall. *Acta Pharmacol. Sin.* **2004**, *25*, 932–936. [PubMed]
101. Zhong, Z.; Xing, R.; Liu, S.; Wang, L.; Cai, S.; Li, P. Synthesis of acyl thiourea derivatives of chitosan and their antimicrobial activities *in vitro*. *Carbohydr. Res.* **2008**, *343*, 566–570. [CrossRef] [PubMed]
102. Rhoades, J.; Roller, S. Antimicrobial Actions of Degraded and Native Chitosan against Spoilage Organisms in Laboratory Media and Foods. *Appl. Environ. Microbiol.* **2000**, *66*, 80–86. [CrossRef] [PubMed]
103. No, H. Antibacterial activity of chitosans and chitosan oligomers with different molecular weights. *Int. J. Food Microbiol.* **2002**, *74*, 65–72. [CrossRef]
104. Helander, I.; Nurmiaho-Lassila, E.-L.; Ahvenainen, R.; Rhoades, J.; Roller, S. Chitosan disrupts the barrier properties of the outer membrane of Gram-negative bacteria. *Int. J. Food Microbiol.* **2001**, *71*, 235–244. [CrossRef]
105. Raafat, D.; Von Bargen, K.; Haas, A.; Sahl, H.G. Insights into the mode of action of chitosan as an antibacterial compound. *Appl. Environ. Microbiol.* **2008**, *74*, 3764–3773. [CrossRef] [PubMed]
106. Zakrzewska, A.; Boorsma, A.; Brul, S.; Klaas, J.; Klis, F.M.; Hellingwerf, K.J. Transcriptional Response of Saccharomyces cerevisiae to the Plasma Membrane-Perturbing Compound Chitosan Transcriptional Response of Saccharomyces cerevisiae to the Plasma Membrane-Perturbing Compound Chitosan. *Eukaryot. Cell* **2005**, *4*, 703–715. [CrossRef] [PubMed]
107. Sudarshan, N.R.; Hoover, D.G.; Knorr, D. Antibacterial action of chitosan. *Food Biotechnol.* **1992**, *6*, 257–272. [CrossRef]
108. Sebti, I.; Martial-Gros, A.; Carnet-Pantiez, A.; Grelier, S.; Coma, V. Chitosan Polymer as Bioactive Coating and Film against Aspergillus niger Contamination. *J. Food Sci.* **2005**, *70*, 100–104. [CrossRef]
109. Tokura, S.; Ueno, K.; Miyazaki, S.; Nishi, N. Molecular weight dependent antimicrobial activity by chitosan. In *New Macromolecular Architecture and Functions*; Kamachi, M., Nakamura, A., Eds.; Springer Berlin Heidelberg: Berlin, Heidelberg, 1996.
110. Chung, Y. Effect of abiotic factors on the antibacterial activity of chitosan against waterborne pathogens. *Bioresour. Technol.* **2003**, *88*, 179–184. [CrossRef]
111. Burkatovskaya, M.; Castano, A.P.; Demidova-Rice, T.N.; Tegos, G.P.; Hamblin, M.R. Effect of chitosan acetate bandage on wound healing in infected and noninfected wounds in mice. *Wound Repair Regen.* **2008**, *16*, 425–431. [CrossRef] [PubMed]
112. Lee, S.J.; Heo, D.N.; Moon, J.-H.; Park, H.N.; Ko, W.-K.; Bae, M.S.; Lee, J.B.; Park, S.W.; Kim, E.-C.; Lee, C.H.; *et al.* Chitosan/Polyurethane Blended Fiber Sheets Containing Silver Sulfadiazine for Use as an Antimicrobial Wound Dressing. *J. Nanosci. Nanotechnol.* **2014**, *14*, 7488–7494. [CrossRef] [PubMed]
113. Tan, H.B.; Wang, F.Y.; Ding, W.; Zhang, Y.; Ding, J.; Cai, D.X.; Yu, K.F.; Yang, J.; Yang, L.; Xu, Y.Q. Fabrication and Evaluation of Porous Keratin/chitosan (KCS) Scaffolds for Effectively Accelerating Wound Healing. *Biomed. Environ. Sci.* **2015**, *28*, 178–189. [PubMed]
114. Miguel, S.P.; Ribeiro, M.P.; Brancal, H.; Coutinho, P.; Correia, I.J. Thermoresponsive chitosan-agarose hydrogel for skin regeneration. *Carbohydr. Polym.* **2014**, *111*, 366–373. [CrossRef] [PubMed]
115. Anisha, B.S.; Biswas, R.; Chennazhi, K.P.; Jayakumar, R. Chitosan-hyaluronic acid/nano silver composite sponges for drug resistant bacteria infected diabetic wounds. *Int. J. Biol. Macromol.* **2013**, *62*, 310–320. [CrossRef] [PubMed]
116. Aziz, M.A.; Cabral, J.D.; Brooks, H.J.L.; Moratti, S.C.; Hanton, L.R. Antimicrobial properties of a chitosan dextran-based hydrogel for surgical use. *Antimicrob. Agents Chemother.* **2012**, *56*, 280–287. [CrossRef] [PubMed]

117. Du, W.-L.; Niu, S.-S.; Xu, Y.-I..; Xu, Z.-R.; Fan, C.-L. Antibacterial activity of chitosan tripolyphosphate nanoparticles loaded with various metal ions. *Carbohydr. Polym.* **2009**, *75*, 385–389. [CrossRef]

118. Qi, L.; Xu, Z.; Jiang, X.; Hu, C.; Zou, X. Preparation and antibacterial activity of chitosan nanoparticles. *Carbohydr. Res.* **2004**, *339*, 2693–2700. [CrossRef] [PubMed]

119. Shi, Z.; Neoh, K.G.; Kang, E.T.; Wang, W. Antibacterial and mechanical properties of bone cement impregnated with chitosan nanoparticles. *Biomaterials* **2006**, *27*, 2440–2449. [CrossRef] [PubMed]

120. Alt, V.; Bechert, T.; Steinrücke, P.; Wagener, M.; Seidel, P.; Dingeldein, E.; Domann, E.; Schnettler, R. An *in vitro* assessment of the antibacterial properties and cytotoxicity of nanoparticulate silver bone cement. *Biomaterials* **2004**, *25*, 4383–4891. [CrossRef] [PubMed]

121. Barreras, U.S.; Méndez, F.T.; Martínez, R.E.M.; Valencia, C.S.; Rodríguez, P.R.M.; Rodríguez, J.P.L. Chitosan nanoparticles enhance the antibacterial activity of chlorhexidine in collagen membranes used for periapical guided tissue regeneration. *Mater. Sci. Eng. C. Mater. Biol. Appl.* **2016**, *58*, 1182–1187. [CrossRef] [PubMed]

122. Aider, M. Chitosan application for active bio-based films production and potential in the food industry: Review. *LWT-Food Sci. Technol.* **2010**, *43*, 837–842. [CrossRef]

123. Agulló, E.; Rodríguez, M.S.; Ramos, V.; Albertengo, L. Present and Future Role of Chitin and Chitosan in Food. *Macromol. Biosci.* **2003**, *3*, 521–530. [CrossRef]

124. Tahiri, I.; Desbiens, M.; Lacroix, C.; Kheadr, E.; Fliss, I. Growth of Carnobacterium divergens M35 and production of Divergicin M35 in snow crab by-product, a natural-grade medium. *LWT-Food Sci. Technol.* **2009**, *42*, 624–632. [CrossRef]

125. Devlieghere, F.; Vermeulen, A.; Debevere, J. Chitosan: Antimicrobial activity, interactions with food components and applicability as a coating on fruit and vegetables. *Food Microbiol.* **2004**, *21*, 703–714. [CrossRef]

126. Dutta, P.K.; Tripathi, S.; Mehrotra, G.K.; Dutta, J. Perspectives for chitosan based antimicrobial films in food applications. *Food Chem.* **2009**, *114*, 1173–1182. [CrossRef]

127. Elsabee, M.Z.; Abdou, E.S. Chitosan based edible films and coatings: A review. *Mater. Sci. Eng. C. Mater. Biol. Appl.* **2013**, *33*, 1819–1841. [CrossRef] [PubMed]

128. Kenawy, E.-R.; Worley, S.D.; Broughton, R. The chemistry and applications of antimicrobial polymers: A state-of-the-art review. *Biomacromolecules* **2007**, *8*, 1359–1384. [CrossRef] [PubMed]

129. Gupta, D.; Haile, A. Multifunctional properties of cotton fabric treated with chitosan and carboxymethyl chitosan. *Carbohydr. Polym.* **2007**, *69*, 164–171. [CrossRef]

130. Ye, W.; Leung, M.F.; Xin, J.; Kwong, T.L.; Lee, D.K.L.; Li, P. Novel core-shell particles with poly(*n*-butyl acrylate) cores and chitosan shells as an antibacterial coating for textiles. *Polymer (Guildf).* **2005**, *46*, 10538–10543. [CrossRef]

131. Illum, L. Chitosan and its use as a pharmaceutical excipient. *Pharm. Res.* **1998**, *15*, 1326–1331. [CrossRef] [PubMed]

132. Felt, O.; Buri, P.; Gurny, R. Chitosan: A unique polysaccharide for drug delivery. *Drug Dev. Ind. Pharm.* **1998**, *24*, 979–993. [CrossRef] [PubMed]

133. Rajalakshmi, R.; Indira Muzib, Y.; Aruna, U.; Vinesha, V.; Rupangada, V.; Krishna moorthy, S.B. Chitosan Nanoparticles—An Emerging Trend In Nanotechnology. *Int. J. Drug Deliv.* **2014**, *6*, 204–229.

134. Shukla, S.K.; Mishra, A.K.; Arotiba, O.A.; Mamba, B.B. Chitosan-based nanomaterials: A state-of-the-art review. *Int. J. Biol. Macromol.* **2013**, *59*, 46–58. [CrossRef] [PubMed]

135. Sinha, V.R.; Singla, A.K.; Wadhawan, S.; Kaushik, R.; Kumria, R.; Bansal, K.; Dhawan, S. Chitosan microspheres as a potential carrier for drugs. *Int. J. Pharm.* **2004**, *274*, 1–33. [CrossRef] [PubMed]

136. Liu, Z.; Jiao, Y.; Wang, Y.; Zhou, C.; Zhang, Z. Polysaccharides-based nanoparticles as drug delivery systems. *Adv. Drug Deliv. Rev.* **2008**, *60*, 1650–1662. [CrossRef] [PubMed]

137. Cheung, R.; Ng, T.; Wong, J.; Chan, W. Chitosan: An Update on Potential Biomedical and Pharmaceutical Applications. *Mar. drugs* **2015**, *13*, 5156–5186.

138. Bawarski, W.E.; Chidlowsky, E.; Bharali, D.J.; Mousa, S.A. Emerging nanopharmaceuticals. *Nanomed. Nanotech. Biol. Med.* **2008**, *4*, 273–282. [CrossRef] [PubMed]

139. Kohane, D.S. Microparticles and nanoparticles for drug delivery. *Biotechnol. Bioeng.* **2007**, *96*, 203–209. [CrossRef] [PubMed]

140. Brun-Graeppi, A.K.A.S.; Richard, C.; Bessodes, M.; Scherman, D.; Merten, O.-W. Cell microcarriers and microcapsules of stimuli-responsive polymers. *J. Control. Release* **2011**, *149*, 209–224. [CrossRef] [PubMed]

141. Anton, N.; Benoit, J.-P.; Saulnier, P. Design and production of nanoparticles formulated from nano-emulsion templates—A review. *J. Control. Release* **2008**, *128*, 185–199. [CrossRef] [PubMed]

142. Pinto Reis, C.; Neufeld, R.J.; Ribeiro, A.J.; Veiga, F. Nanoencapsulation I. Methods for preparation of drug-loaded polymeric nanoparticles. *Nanomedicine* **2006**, *2*, 8–21. [CrossRef] [PubMed]

143. Agnihotri, S.A.; Mallikarjuna, N.N.; Aminabhavi, T.M. Recent advances on chitosan-based micro- and nanoparticles in drug delivery. *J. Control. Release* **2004**, *100*, 5–28. [CrossRef] [PubMed]

144. Hamidi, M.; Azadi, A.; Rafiei, P. Hydrogel nanoparticles in drug delivery. *Adv. Drug Deliv. Rev.* **2008**, *60*, 1638–1649. [CrossRef] [PubMed]

145. Mizrahy, S.; Peer, D. Polysaccharides as building blocks for nanotherapeutics. *Chem. Soc. Rev.* **2012**, *41*, 2623. [CrossRef] [PubMed]

146. Berger, J.; Reist, M.; Mayer, J.M.; Felt, O.; Peppas, N.A.; Gurny, R. Structure and interactions in covalently and ionically crosslinked chitosan hydrogels for biomedical applications. *Eur. J. Pharm. Biopharm.* **2004**, *57*, 19–34. [CrossRef]

147. Berger, J.; Reist, M.; Mayer, J.M.; Felt, O.; Peppas, N.A.; Gurny, R. Structure and interactions in covalently and ionically crosslinked chitosan hydrogels for biomedical applications. *Eur. J. Pharm. Biopharm.* **2004**, *57*, 19–34. [CrossRef]

148. Dash, M.; Chiellini, F.; Ottenbrite, R.M.; Chiellini, E. Chitosan—A versatile semi-synthetic polymer in biomedical applications. *Prog. Polym. Sci.* **2011**, *36*, 981–1014. [CrossRef]

149. Sheu, D.-C.; Li, S.-Y.; Duan, K.-J.; Chen, C.W. Production of galactooligosaccharides by β-galactosidase immobilized on glutaraldehyde-treated chitosan beads. *Biotechnol. Tech.* **1998**, *12*, 273–276. [CrossRef]

150. Du, Y.-Z.; Ying, X.-Y.; Wang, L.; Zhai, Y.; Yuan, H.; Yu, R.-S.; Hu, F.-Q. Sustained release of ATP encapsulated in chitosan oligosaccharide nanoparticles. *Int. J. Pharm.* **2010**, *392*, 164–169. [CrossRef] [PubMed]

151. Xu, J.; Ma, L.; Liu, Y.; Xu, F.; Nie, J.; Ma, G. Design and characterization of antitumor drug paclitaxel-loaded chitosan nanoparticles by W/O emulsions. *Int. J. Biol. Macromol.* **2012**, *50*, 438–443. [CrossRef] [PubMed]

152. Monteiro, O.A.C.; Airoldi, C. Some studies of crosslinking chitosan-glutaraldehyde interaction in a homogeneous system. *Int. J. Biol. Macromol.* **1999**, *26*, 119–128. [CrossRef]

153. Leung, H.W. Ecotoxicology of glutaraldehyde: Review of environmental fate and effects studies. *Ecotoxicol. Environ. Saf.* **2001**, *49*, 26–39. [CrossRef] [PubMed]

154. Thakur, A.; Wanchoo, R.K.; Hardeep, S.K.S. Chitosan Hydrogel Beads: A Comparative Study with Glutaraldehyde, Epichlorohydrin and Genipin as Crosslinkers. *J. Polym. Mater.* **2013**, *31*, 211–223.

155. Khurma, J.R.; Rohindra, D.R.; Nand, A.V. Swelling and Thermal Characteristics of Genipin Crosslinked Chitosan and Poly(vinyl pyrrolidone) Hydrogels. *Polym. Bull.* **2005**, *54*, 195–204. [CrossRef]

156. Moura, M.J.; Figueiredo, M.M.; Gil, M.H. Rheological study of genipin cross-linked chitosan hydrogels. *Biomacromolecules* **2007**, *8*, 3823–3829. [CrossRef] [PubMed]

157. Muzzarelli, R.A.A. Genipin-crosslinked chitosan hydrogels as biomedical and pharmaceutical aids. *Carbohydr. Polym.* **2009**, *77*, 1–9. [CrossRef]

158. Karnchanajindanun, J.; Srisa-ard, M.; Baimark, Y. Genipin-cross-linked chitosan microspheres prepared by a water-in-oil emulsion solvent diffusion method for protein delivery. *Carbohydr. Polym.* **2011**, *85*, 674–680. [CrossRef]

159. Liu, Y.; Chen, W.; Kim, H.-I. pH-responsive release behavior of genipin-crosslinked chitosan/poly(ethylene glycol) hydrogels. *J. Appl. Polym. Sci.* **2012**, *125*, 290–298. [CrossRef]

160. Arteche Pujana, M.; Pérez-Álvarez, L.; Cesteros Iturbe, L.C.; Katime, I. Biodegradable chitosan nanogels crosslinked with genipin. *Carbohydr. Polym.* **2013**, *94*, 836–842. [CrossRef] [PubMed]

161. Hurst, G.A.; Novakovic, K. A facile in situ morphological characterization of smart genipin-crosslinked chitosan–poly(vinyl pyrrolidone) hydrogels. *J. Mater. Res.* **2013**, *28*, 2401–2408. [CrossRef]

162. Nwosu, C.J.; Hurst, G.A.; Novakovic, K. Hydrogels: Influence of Composition and Postsynthesis Treatment on pH Responsive Behaviour. *Adv. Mater. Sci. Eng.* **2015**, *2015*, 1–10. [CrossRef]

163. Lai, J.Y. Biocompatibility of chemically cross-linked gelatin hydrogels for ophthalmic use. *J. Mater. Sci. Mater. Med.* **2010**, *21*, 1899–1911. [CrossRef] [PubMed]

164. Bodnar, M.; Hartmann, J.F.; Borbely, J. Preparation and characterization of chitosan-based nanoparticles. *Biomacromolecules* **2005**, *6*, 2521–2527. [CrossRef] [PubMed]

165. Prata, A.S.; Grosso, C.R.F. Production of microparticles with gelatin and chitosan. *Carbohydr. Polym.* **2015**, *116*, 292–299. [CrossRef] [PubMed]

166. Frohbergh, M.E.; Katsman, A.; Botta, G.P.; Lazarovici, P.; Schauer, C.L.; Wegst, U.G.K.; Lelkes, P.I. Electrospun hydroxyapatite-containing chitosan nanofibers crosslinked with genipin for bone tissue engineering. *Biomaterials* **2012**, *33*, 9167–9178. [CrossRef] [PubMed]

167. Chaiyasan, W.; Srinivas, S.P.; Tiyaboonchai, W. Crosslinked chitosan-dextran sulfate nanoparticle for improved topical ocular drug delivery. *Mol. Vis.* **2015**, *21*, 1224–1234. [PubMed]

168. Liu, Z.; Lv, D.; Liu, S.; Gong, J.; Wang, D.; Xiong, M.; Chen, X.; Xiang, R.; Tan, X. Alginic acid-coated chitosan nanoparticles loaded with legumain DNA vaccine: Effect against breast cancer in mice. *PLoS ONE* **2013**, *8*, e60190. [CrossRef] [PubMed]

169. Koetting, M.C.; Peters, J.T.; Steichen, S.D.; Peppas, N.A. Stimulus-responsive hydrogels: Theory, modern advances, and applications. *Mater. Sci. Eng. R Reports* **2015**, *93*, 1–49. [CrossRef] [PubMed]

170. Bodmeier, R.; Chen, H.; Paeratakul, O. A novel approach to the oral delivery of micro- or nanoparticles. *Pharm. Res* **1989**, *6*, 413–417. [CrossRef] [PubMed]

171. Calvo, P.; Remunan-Lopez, C. Novel hydrophilic chitosan-polyethylene oxide nanoparticles as protein carriers. *J. Appl.* **1997**, *63*, 125–132. [CrossRef]

172. Liu, H.; Gao, C. Preparation and properties of ionically cross-linked chitosan nanoparticles. *Polym. Adv. Technol.* **2009**, *20*, 613–619. [CrossRef]

173. Xu, Y.; Du, Y. Effect of molecular structure of chitosan on protein delivery properties of chitosan nanoparticles. *Int. J. Pharm.* **2003**, *250*, 215–226. [CrossRef]

174. Fernández-Urrusuno, R.; Calvo, P.; Remuñán-López, C.; Vila-Jato, J.L.; Alonso, M.J. Enhancement of nasal absorption of insulin using chitosan nanoparticles. *Pharm. Res.* **1999**, *16*, 1576–1581. [CrossRef] [PubMed]

175. Pan, Y.; Li, Y.; Zhao, H.; Zheng, J.; Xu, H.; Wei, G.; Hao, J.; Cui, F. Bioadhesive polysaccharide in protein delivery system: Chitosan nanoparticles improve the intestinal absorption of insulin *in vivo*. *Int. J. Pharm.* **2002**, *249*, 139–147. [CrossRef]

176. da Silva, S.B.; Ferreira, D.; Pintado, M.; Sarmento, B. Chitosan-based nanoparticles for rosmarinic acid ocular delivery—In vitro tests. *Int. J. Biol. Macromol.* **2015**, *84*, 112–120. [CrossRef] [PubMed]

177. Jain, A.; Thakur, K.; Sharma, G.; Kush, P.; Jain, U.K. Fabrication, characterization and cytotoxicity studies of ionically cross-linked docetaxel loaded chitosan nanoparticles. *Carbohydr. Polym.* **2016**, *137*, 65–74. [CrossRef] [PubMed]

178. Sipoli, C.C.; Radaic, A.; Santana, N.; de Jesus, M.B.; de la Torre, L.G. Chitosan nanoparticles produced with the gradual temperature decrease technique for sustained gene delivery. *Biochem. Eng. J.* **2015**, *103*, 114–121. [CrossRef]

179. Gan, Q.; Wang, T.; Cochrane, C.; McCarron, P. Modulation of surface charge, particle size and morphological properties of chitosan-TPP nanoparticles intended for gene delivery. *Colloids Surf. B. Biointerfaces* **2005**, *44*, 65–73. [CrossRef] [PubMed]

180. Nasti, A.; Zaki, N.M.; de Leonardis, P.; Ungphaiboon, S.; Sansongsak, P.; Rimoli, M.G.; Tirelli, N. Chitosan/TPP and Chitosan/TPP-hyaluronic Acid Nanoparticles: Systematic Optimisation of the Preparative Process and Preliminary Biological Evaluation. *Pharm. Res.* **2009**, *26*, 1918–1930. [CrossRef] [PubMed]

181. des Rieux, A.; Fievez, V.; Garinot, M.; Schneider, Y.-J.; Préat, V. Nanoparticles as potential oral delivery systems of proteins and vaccines: A mechanistic approach. *J. Control. Release* **2006**, *116*, 1–27. [CrossRef] [PubMed]

182. Chaturvedi, K.; Ganguly, K.; Nadagouda, M.N.; Aminabhavi, T.M. Polymeric hydrogels for oral insulin delivery. *J. Control. Release* **2013**, *165*, 129–138. [CrossRef] [PubMed]

183. Grenha, A.; Seijo, B.; Remuñán-López, C. Microencapsulated chitosan nanoparticles for lung protein delivery. *Eur. J. Pharm. Sci.* **2005**, *25*, 427–437. [CrossRef] [PubMed]

184. De Campos, A.M.; Sánchez, A.; Alonso, M.J. Chitosan nanoparticles: A new vehicle for the improvement of the delivery of drugs to the ocular surface. Application to cyclosporin A. *Int. J. Pharm.* **2001**, *224*, 159–168. [CrossRef]

185. Peniche, C.; Argüelles-Monal, W.; Peniche, H.; Acosta, N. Chitosan: An Attractive Biocompatible Polymer for Microencapsulation. *Macromol. Biosci.* **2003**, *3*, 511–520. [CrossRef]

186. Kizilay, E.; Kayitmazer, A.B.; Dubin, P.L. Complexation and coacervation of polyelectrolytes with oppositely charged colloids. *Adv. Colloid Interface Sci.* **2011**, *167*, 24–37. [CrossRef] [PubMed]

187. Espinosa-andrews, H.; Ba, J.G.; Cruz-sosa, F.; Vernon-carter, E.J. Gum Arabic—Chitosan Complex Coacervation. *Biomacromolecules* **2007**, *8*, 1313–1318. [CrossRef] [PubMed]

188. Sadek, R.F.H. Mutual interaction of polyelectrolytes. *Science* **1949**, *110*, 552–554.

189. Rahaiee, S.; Shojaosadati, S.A.; Hashemi, M.; Moini, S.; Razavi, S.H. Improvement of crocin stability by biodegradeble nanoparticles of chitosan-alginate. *Int. J. Biol. Macromol.* **2015**, *79*, 423–432. [CrossRef] [PubMed]

190. Chen, Y.; Siddalingappa, B.; Chan, P.H.H.; Benson, H.A.E. Development of a chitosan-based nanoparticle formulation for delivery of a hydrophilic hexapeptide, dalargin. *Biopolym. Pept. Sci. Sect.* **2008**, *90*, 663–670. [CrossRef] [PubMed]

191. George, M.; Abraham, T.E. Polyionic hydrocolloids for the intestinal delivery of protein drugs: Alginate and chitosan—A review. *J. Control. Release* **2006**, *114*, 1–14. [CrossRef] [PubMed]

192. Mao, H.Q.; Roy, K.; Troung-Le, V.L.; Janes, K.A.; Lin, K.Y.; Wang, Y.; August, J.T.; Leong, K.W. Chitosan-DNA nanoparticles as gene carriers: Synthesis, characterization and transfection efficiency. *J. Control. Release* **2001**, *70*, 399–421. [CrossRef]

193. Deacon, J.; Abdelghany, S.M.; Quinn, D.J.; Schmid, D.; Megaw, J.; Donnelly, R.F.; Jones, D.S.; Kissenpfennig, A.; Elborn, J.S.; Gilmore, B.F.; *et al.* Antimicrobial efficacy of tobramycin polymeric nanoparticles for Pseudomonas aeruginosa infections in cystic fibrosis: Formulation, characterisation and functionalisation with dornase alfa (DNase). *J. Control. Release* **2015**, *198*, 55–61. [CrossRef] [PubMed]

194. Berger, J.; Reist, M.; Mayer, J.; Felt, O.; Gurny, R. Structure and interactions in chitosan hydrogels formed by complexation or aggregation for biomedical applications. *Eur. J. Pharm. Biopharm.* **2004**, *57*, 35–52. [CrossRef]

195. Sarmento, B.; Martins, S.; Ribeiro, A.; Veiga, F.; Neufeld, R.; Ferreira, D. Development and Comparison of Different Nanoparticulate Polyelectrolyte Complexes as Insulin Carriers. *Int. J. Pept. Res. Ther.* **2006**, *12*, 131–138. [CrossRef]

196. Sarmento, B.; Ribeiro, A.; Veiga, F.; Sampaio, P.; Neufeld, R.; Ferreira, D. Alginate/Chitosan Nanoparticles are Effective for Oral Insulin Delivery. *Pharm. Res.* **2007**, *24*, 2198–2206. [CrossRef] [PubMed]

197. Rajaonarivony, M.; Vauthier, C.; Couarraze, G.; Puisieux, F.; Couvreur, P. Development of a new drug carrier made from alginate. *J. Pharm. Sci.* **1993**, *82*, 912–917. [CrossRef] [PubMed]

198. Loquercio, A.; Castell-Perez, E.; Gomes, C.; Moreira, R.G. Preparation of Chitosan-Alginate Nanoparticles for Trans-cinnamaldehyde Entrapment. *J. Food Sci.* **2015**, *80*, 2305–2315. [CrossRef] [PubMed]

199. Li, P.; Dai, Y.; Zhang, J.; Wang, A.; Wei, Q. Chitosan-Alginate Nanoparticles as a Novel Drug Delivery System for Nifedipine. *Int. J. Biomed. Sci.* **2008**, *4*, 221–228. [PubMed]

200. Sarmento, B.; Ribeiro, A.J.; Veiga, F.; Ferreira, D.C.; Neufeld, R.J. Insulin-Loaded Nanoparticles are Prepared by Alginate Ionotropic Pre-Gelation Followed by Chitosan Polyelectrolyte Complexation. *J. Nanosci. Nanotechnol.* **2007**, *7*, 2833–2841. [CrossRef] [PubMed]

201. Oyarzun-Ampuero, F.A.; Brea, J.; Loza, M.I.; Torres, D.; Alonso, M.J. Chitosan-hyaluronic acid nanoparticles loaded with heparin for the treatment of asthma. *Int. J. Pharm.* **2009**, *381*, 122–129. [CrossRef] [PubMed]

202. Duceppe, N.; Tabrizian, M. Factors influencing the transfection efficiency of ultra low molecular weight chitosan/hyaluronic acid nanoparticles. *Biomaterials* **2009**, *30*, 2625–2631. [CrossRef] [PubMed]

203. Aruffo, A.; Stamenkovic, I.; Melnick, M.; Underhill, C.B.; Seed, B. CD44 is the principal cell surface receptor for hyaluronate. *Cell* **1990**, *61*, 1303–1313. [CrossRef]

204. Li, C.; Hein, S.; Wang, K. Chitosan-Carrageenan Polyelectrolyte Complex for the Delivery of Protein Drugs. *ISRN Biomater.* **2013**, *2013*, 1–6. [CrossRef]

205. Sarmento, B.; Ribeiro, A.; Veiga, F.; Ferreira, D. Development and characterization of new insulin containing polysaccharide nanoparticles. *Colloids Surf. B. Biointerfaces* **2006**, *53*, 193–202. [CrossRef] [PubMed]

206. Mounica Reddy, M.; Shanmugam, V.; Kaza, R. Design and Characterization of Insulin Nanoparticles for Oral Delivery. *Int. J. Innov. Pharm. Res.* **2012**, *3*, 238–243.

207. Avadi, M.R.; Sadeghi, A.M.M.; Mohammadpour, N.; Abedin, S.; Atyabi, F.; Dinarvand, R.; Rafiee-Tehrani, M. Preparation and characterization of insulin nanoparticles using chitosan and Arabic gum with ionic gelation method. *Nanomedicine* **2010**, *6*, 58–63. [CrossRef] [PubMed]

208. Zou, T.; Percival, S.S.; Cheng, Q.; Li, Z.; Rowe, C.A.; Gu, L. Preparation, characterization, and induction of cell apoptosis of cocoa procyanidins-gelatin-chitosan nanoparticles. *Eur. J. Pharm. Biopharm.* **2012**, *82*, 36–42. [CrossRef] [PubMed]

209. Birch, N.P.; Schiffman, J.D. Characterization of self-assembled polyelectrolyte complex nanoparticles formed from chitosan and pectin. *Langmuir* **2014**, *30*, 3441–3447. [CrossRef] [PubMed]

210. Luo, Y.; Wang, Q. Recent development of chitosan-based polyelectrolyte complexes with natural polysaccharides for drug delivery. *Int. J. Biol. Macromol.* **2014**, *64*, 353–367. [CrossRef] [PubMed]

211. Davidenko, N.; Blanco, M.D.; Peniche, C.; Becherán, L.; Guerrero, S.; Teijón, J.M. Effects of different parameters on the characteristics of chitosan-poly(acrylic acid) nanoparticles obtained by the method of coacervation. *J. Appl. Polym. Sci.* **2009**, *111*, 2362–2371. [CrossRef]

212. Hu, Y.; Jiang, X.; Ding, Y.; Ge, H.; Yuan, Y.; Yang, C. Synthesis and characterization of chitosan–poly(acrylic acid) nanoparticles. *Biomaterials* **2002**, *23*, 3193–3201. [CrossRef]

213. Cooper, C.L.; Dubin, P.L.; Kayitmazer, A.B.; Turksen, S. Polyelectrolyte–protein complexes. *Curr. Opin. Colloid Interface Sci.* **2005**, *10*, 52–78. [CrossRef]

214. Boddohi, S.; Moore, N.; Johnson, P.A.; Kipper, M.J. Polysaccharide-based polyelectrolyte complex nanoparticles from chitosan, heparin, and hyaluronan. *Biomacromolecules* **2009**, *10*, 1402–1409. [CrossRef] [PubMed]

215. Hu, B.; Ting, Y.; Zeng, X.; Huang, Q. Bioactive peptides/chitosan nanoparticles enhance cellular antioxidant activity of (−)-epigallocatechin-3-gallate. *J. Agric. Food Chem.* **2013**, *61*, 875–881. [CrossRef] [PubMed]

216. Leong, K.W.; Mao, H.-Q.; Truong-Le, V.L.; Roy, K.; Walsh, S.M.; August, J.T. DNA-polycation nanospheres as non-viral gene delivery vehicles. *J. Control. Release* **1998**, *53*, 183–193. [CrossRef]

217. Richardson, S. Potential of low molecular mass chitosan as a DNA delivery system: Biocompatibility, body distribution and ability to complex and protect DNA. *Int. J. Pharm.* **1999**, *178*, 231–243. [CrossRef]

218. Lavertu, M.; Méthot, S.; Tran-Khanh, N.; Buschmann, M.D. High efficiency gene transfer using chitosan/DNA nanoparticles with specific combinations of molecular weight and degree of deacetylation. *Biomaterials* **2006**, *27*, 4815–4824. [CrossRef] [PubMed]

219. Buschmann, M.D.; Merzouki, A.; Lavertu, M.; Thibault, M.; Jean, M.; Darras, V. Chitosans for delivery of nucleic acids. *Adv. Drug Deliv. Rev.* **2013**, *65*, 1234–1270. [CrossRef] [PubMed]

220. Yuan, X.B.; Li, H.; Yuan, Y.B. Preparation of cholesterol-modified chitosan self-aggregated nanoparticles for delivery of drugs to ocular surface. *Carbohydr. Polym.* **2006**, *65*, 337–345. [CrossRef]

221. Wang, Y.-S.; Liu, L.-R.; Jiang, Q.; Zhang, Q.-Q. Self-aggregated nanoparticles of cholesterol-modified chitosan conjugate as a novel carrier of epirubicin. *Eur. Polym. J.* **2007**, *43*, 43–51. [CrossRef]

222. Lee, K.Y.; Jo, W.H.; Kwon, I.C.; Kim, Y.; Jeong, S.Y. Structural Determination and Interior Polarity of Self-Aggregates Prepared from Deoxycholic Acid-Modified Chitosan in Water. *Macromolecules* **1998**, *31*, 378–383. [CrossRef]

223. Chen, X.G.; Lee, C.M.; Park, H.J. O/W emulsification for the self-aggregation and nanoparticle formation of linoleic acid-modified chitosan in the aqueous system. *J. Agric. Food Chem.* **2003**, *51*, 3135–3139. [CrossRef] [PubMed]

224. Du, Y.-Z.; Wang, L.; Yuan, H.; Wei, X.-H.; Hu, F.-Q. Preparation and characteristics of linoleic acid-grafted chitosan oligosaccharide micelles as a carrier for doxorubicin. *Colloids Surf. B. Biointerfaces* **2009**, *69*, 257–263. [CrossRef] [PubMed]

225. Rampino, A.; Borgogna, M.; Blasi, P.; Bellich, B.; Cesàro, A. Chitosan nanoparticles: Preparation, size evolution and stability. *Int. J. Pharm.* **2013**, *455*, 219–228. [CrossRef] [PubMed]

226. Fàbregas, A.; Miñarro, M.; García-Montoya, E.; Pérez-Lozano, P.; Carrillo, C.; Sarrate, R.; Sánchez, N.; Ticó, J.R.; Suñé-Negre, J.M. Impact of physical parameters on particle size and reaction yield when using the ionic gelation method to obtain cationic polymeric chitosan-tripolyphosphate nanoparticles. *Int. J. Pharm.* **2013**, *446*, 199–204. [CrossRef] [PubMed]

227. Korsmeyer, R.W.; Gurny, R.; Doelker, E.; Buri, P.; Peppas, N.A. Mechanisms of solute release from porous hydrophilic polymers. *Int. J. Pharm.* **1983**, *15*, 25–35. [CrossRef]

228. Siepmann, J.; Siepmann, F. Mathematical modeling of drug delivery. *Int. J. Pharm.* **2008**, *364*, 328–343. [CrossRef] [PubMed]

229. Shu, X.Z.; Zhu, K.J. A novel approach to prepare tripolyphosphate/chitosan complex beads for controlled release drug delivery. *Int. J. Pharm.* **2000**, *201*, 51–58. [CrossRef]

230. Casettari, L.; Illum, L. Chitosan in nasal delivery systems for therapeutic drugs. *J. Control. Release* **2014**, *190*, 189–200. [CrossRef] [PubMed]
231. Castro, S.P.M.; Paulín, E.G.L. Is Chitosan a New Panacea? Areas of Application. In *The Complex World of Polysacchrides*; Karunaratne, D.N., Ed.; InTech Publisher: Rijeka, Croatia, 2012; pp. 3–46.

MDPI

Review

Chitosanases from Family 46 of Glycoside Hydrolases: From Proteins to Phenotypes

Pascal Viens, Marie-Ève Lacombe-Harvey and Ryszard Brzezinski *

Département de Biologie, Faculté des Sciences, Université de Sherbrooke, 2500, boul. de l'Université, Sherbrooke, QC J1K 2R1, Canada; Pascal.Viens@USherbrooke.ca (P.V.); marie-eve.lacombe@videotron.ca (M.-E.L.-H.)

* Author to whom correspondence should be addressed; Ryszard.Brzezinski@USherbrooke.ca; Tel.: +1-819-821-7070; Fax: +1-829-821-8049.

Academic Editor: Hitoshi Sashiwa

Received: 31 August 2015; Accepted: 13 October 2015; Published: 28 October 2015

Abstract: Chitosanases, enzymes that catalyze the endo-hydrolysis of glycolytic links in chitosan, are the subject of numerous studies as biotechnological tools to generate low molecular weight chitosan (LMWC) or chitosan oligosaccharides (CHOS) from native, high molecular weight chitosan. Glycoside hydrolases belonging to family GH46 are among the best-studied chitosanases, with four crystallography-derived structures available and more than forty enzymes studied at the biochemical level. They were also subjected to numerous site-directed mutagenesis studies, unraveling the molecular mechanisms of hydrolysis. This review is focused on the taxonomic distribution of GH46 proteins, their multi-modular character, the structure-function relationships and their biological functions in the host organisms.

Keywords: chitosan; chitosanase; glycoside hydrolase; *Streptomyces*; *Bacillus*; *Microbacterium*; hydrolysis; polysaccharide; chlorovirus

1. Introduction

1.1. Why Chitosan?

The polycationic polysaccharide chitosan, a polymer of β-1,4-linked D-glucosamine residues (GlcN) with a minor and variable content of N-acetyl-D-glucosamine (GlcNAc) is increasingly attractive for science and industry. According to main bibliographical databases, the number of research articles published annually all over the world and responding to the key word "chitosan" increased from less than 2000 in 2007 to around 3000 in 2010 and slightly more than 4000 in 2012. Chitosan studies are currently in a phase of rapid growth.

The chemical and biological properties of chitosan as well as its potential applications have been exhaustively reviewed recently [1–5]. Gene therapy, drug delivery, wound repair, inhibition of microbial growth, phytoprotection or water treatment are only a few among a myriad of chitosan applications that are in advanced stages of development or commercialization [6–11]. Chitosan properties like biodegradability, lack of toxicity, solubility in mildly acidic aqueous solutions and increasing commercial availability stimulate the interest over this polymer.

1.2. Why Chitosanases?

Many properties of chitosan are influenced by its degree of polymerization. Low molecular mass chitosan displays higher solubility in water and lower viscosity. This, in turn, influences the bioactivities of chitosan. For many applications, intermediate or low molecular mass chitosan has been shown to be superior to the native polymer [12–14]. Chitosan oligosaccharides, typically formed of two to ten monomers, have, among others, well documented beneficial activities as inhibitors of tumor growth, stimulators of calcium deposition in bones, inhibitors of bacterial pathogens adhesion to animal and human tissues and elicitors of antifungal response in plants [15–17]. While many physical or chemical methods were proposed to reduce the degree of polymerization of chitosan [18], enzymatic procedures are often preferred, requiring mild conditions (pH, temperature), offering more control on the final product, and having a minimal environmental impact. As chitosan is a heterogeneous polymer of GlcN and GlcNAc, it is recognized as a substrate by chitosanases and chitinases as well [19,20]. The use of non-specific enzymes for chitosan hydrolysis has also been suggested by several groups, as reviewed in [5].

According to the sequence-based classification of glycoside hydrolases created by Henrissat [21] and developed into the CAZy database (http://www.cazy.org), enzymes with chitosanase activities belong to families 3, 5, 7, 8, 46, 75 and 80. Among these, only families 46, 75 and 80 include, so far, exclusively enzymes specific for chitosan hydrolysis. The members of the GH46 family have been characterized most extensively compared with other chitosanases.

2. GH46 Family Proteins: Phylogenetic Tree and Taxonomic Distribution

GH46 family was built around the first two chitosanase primary sequences described in the literature: the chitosanases from *Bacillus circulans* MH-K1 [22] and from *Streptomyces* sp. N174 [23]. The family itself was officially created in 1996 [24]. In contrast with several GH families populated with enzymes having many different substrate specificities [25], it became apparent, with the discovery of many new GH46 members, that this family includes exclusively enzymes specific for chitosan hydrolysis and classified as EC 3.2.1.132 in the IUBMB Enzyme Nomenclature List.

GH46 proteins are essentially present in eubacterial organisms. To analyze the phylogenetic distribution of GH46 members, we performed an alignment of a subset of 58 primary sequences, including all the sequences of biochemically and structurally studied enzymes but excluding groups of very similar sequences from closely related microorganisms (mostly originating from whole genome sequencing projects). The full-length sequences in Fasta format are shown in Figure S1. All the sequences were first analyzed for the occurrence of a signal peptide at the *N*-terminus and, when present, these segments of low sequence conservation were subtracted from the set submitted to the alignment program. Sequences were aligned with Clustal Omega at default settings [26]. The resulting alignment (Figure S2) is considered as reliable, as it shows the conservation of all the residues for which importance for chitosanase function has been demonstrated by site-directed mutagenesis or crystallography. Most secondary structures revealed by crystallography are also aligned. The alignment was used to derive the phylogenetic tree (Figure 1), which corroborates a more extensive tree based on 148 sequences, published previously [27].

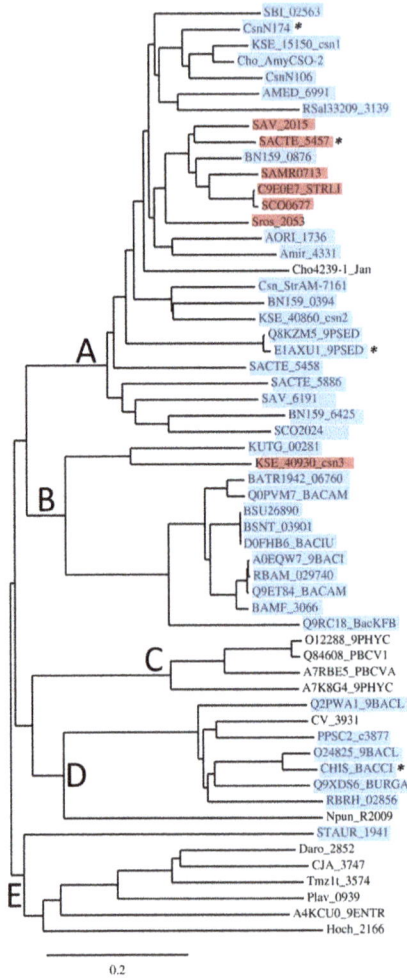

Figure 1. Phylogenetic tree of 58 primary sequences of GH46 proteins. The tree has been drawn with TreeDyn program version 198.3 [28] based on an alignment performed with Clustal Omega [26]. Asterisks (*) indicate the proteins with known 3D structure. Proteins for which a SEC-type signal peptide has been detected are highlighted in blue, while those with putative TAT-type signal peptides [29,30] are highlighted in red.

As shown in Figure 1, the proteins belonging to GH46 are essentially grouped into five clusters, named A to E.

Cluster A includes a large majority of proteins from actinobacteria, Gram-positive bacteria of which genomes are G+C-rich, with rare Gram-negative representatives, such as *Pseudomonas* sp. [31]. Cluster A represents nearly a half of all GH46 proteins listed in CAZy database. Three proteins have been crystalized and their structures were determined: the chitosanases CsnN174 from *Streptomyces* sp. N174 [32], SACTE_5457 from *Streptomyces* sp. SirexAA-E [27] and OUO1 from *Microbacterium* sp. (referred as *Pseudomonas* sp. LL2 in Protein Data Bank) [33]. Chitosanases from *Streptomyces coelicolor* A3(2) (named ScCsn46A or CsnA or SCO0677) [34,35], *Pseudomonas* sp. AO1 [31] and *Amycolatopsis* sp. CsO-2 [36] were also characterized extensively.

Cluster B is composed almost exclusively of enzymes from bacilli belonging to the low G+C branch of Gram-positive bacteria (*Firmicutes*). The chitosanase from *Bacillus subtilis* 168, encoded by locus BSU26890 is the best studied in this group [37]. Interestingly, as new genomic sequences went available in databases, a small actinobacterial sub-group emerged in Cluster B. It is represented in Figure 1 by proteins from *Kitasatospora setae* (KSE_40930_csn3) and *Kutzneria* sp. 744 (KUTG_00281). The single nucleotide or dinucleotide composition of the genes belonging to this sub-group are not significantly different from the mean composition in the entire genomes of the host organisms, indicating that recent acquisition of these genes from bacilli by lateral gene transfer is unlikely (data not shown).

Cluster C chitosanases have been found exclusively in very large, double-stranded-DNA-containing viruses, the chloroviruses, infecting some unicellular, eukaryotic green algae [38], sometimes endo-symbiotic with protozoa. Currently, more than forty GH46 sequences have been found in the genomes of these viruses, of which two have been subjected to biochemical studies [39,40].

Cluster D comprises the chitosanase from *Bacillus circulans* MH-K1, one of the most studied GH46 enzymes and the first which gene has been cloned and sequenced [22]. The 3D structure of this enzyme has also been determined [41]. Similarly to cluster B, most taxa represented inside cluster D belong to the *Firmicutes* phylum, with *Bacillus* and *Paenibacillus* as representative genera. However, the cluster D includes also proteins from Gram-negative *Betaproteobacteria*, with genera such as *Chromobacterium* or *Burkholderia*.

Cluster E groups together proteins for which no enzymatic activity has been reported so far. These are multimodular proteins often annotated as "peptidoglycan-binding proteins" in genomic databases. In most sequences, the catalytic residues and other amino acids essential for chitosanase activity seem to be present, according to sequence alignment (Figure S2). However proteins STAUR_1941 and Hoch_2166 show only limited similarity to the C-terminal half of GH46 chitosanase sequences and do not include any residues important for catalysis or substrate binding. Their relationship with GH46 chitosanases is thus doubtful. Several of these proteins possess putative peptidoglycan-binding modules, suggesting that they could be involved in cell wall metabolism.

It thus appears that GH46 chitosanases from actinobacteria are rather homogenous at the primary sequence level, being grouped essentially in one large cluster A, while chitosanases from *Firmicutes* (*Bacillus* and related genera) fall into two distinct groups, one of which (cluster B) is relatively closer to the actinobacterial proteins than the other (cluster D). This is illustrated in Figure 2, showing the percentages of identity and similarity among the sequences of the catalytic modules of the best-characterized chitosanases from each cluster.

CHIS_BACCI
cluster D

16.9
(26.4)

21.7
(32.7)

17.2
(25.7)

Bsu26890
cluster B

27.0
(38.6)

Q84608_PBCV1
cluster C

34.2
(44.0)

23.4
(36.4)

CsnN174
cluster A

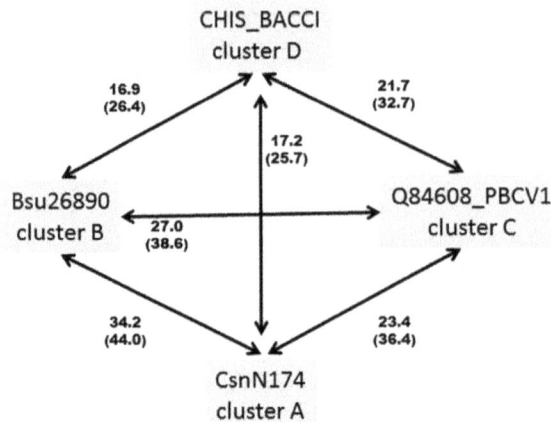

Figure 2. Percentages of identity (in bold) and similarity (into brackets) among the primary amino acid sequences of the catalytic modules of the best characterized chitosanases from clusters A–D. CHIS_BACCI: chitosanase from *Bacillus circulans* MH-K1; CsnN174: chitosanase from *Streptomyces* sp. N174; BSU26890: chitosanase from *Bacillus subtilis* 168; Q84608_PBCV1: chitosanase from Chlorella virus 1 of *Paramecium bursaria*. The sequences were aligned pairwise with Clustal Omega [26]. Identity and similarity were determined using the SIAS server (http://imed.med.ucm.es/Tools/sias.html).

Thus far, GH46 proteins have been described neither in animal nor plant organisms. Recently, two GH46 sequences, originating from the zygomycete *Lichtheimia ramosa* (formerly *Absidia idahoensis* var. *thermophila*), were added to the CaZy database as a result of a genome sequencing project [42]. These are the first two GH46 sequences of eukaryotic origin described so far. The sequences are affiliated to clusters C and D but do not fall directly in either of these clusters (not shown). The corresponding proteins have not been yet characterized biochemically. A particular trait of these proteins is their high cysteine content: ten cysteines for LRAMOSA04613 and twelve for LRAMOSA01487.

3. Multimodularity in GH46 Chitosanases

3.1. Signal Peptides and Secretion of Chitosanases

Chitosan, the target for chitosanases, is not a known constituent of prokaryotic cells. Being an extracellular target, it can be reached by chitosanases only after their secretion across the cytoplasmic membrane. Accordingly, the great majority of proteins from clusters A, B and D have well-defined signal peptides, detectable by algorithms such as Signal P [43] or PRED-TAT [29]. Indeed, many GH46 enzymes have been purified as secreted proteins from the culture supernatants, when obtained from the original producer organisms and not from recombinant *E. coli* clones [37]. Most of the signal peptides are of the SEC-type; however several twin-arginine (TAT) type signal peptides are also present. TAT signal peptides are characterized by the presence of two adjacent, highly conserved arginine residues followed by a segment less hydrophobic than the one present in SEC-type signal peptides. Most GH46 proteins having TAT-type signal peptides are grouped in a small subset of cluster A (Figure 1). The preferential secretion of the chitosanase CsnA (SCO0677) from *Streptomyces coelicolor* A3(2) by the TAT pathway has been confirmed experimentally [35]. Despite high similarity (72%) between CsnA and the chitosanase CsnN174 from *Streptomyces* sp. N174, these two chitosanases differ by their secretion preferences, being more efficiently exported by the TAT pathway and the SEC pathway, respectively [35].

Early hypotheses postulated that proteins secreted specifically through the TAT pathway must be translocated in a fully folded form due to their multimeric character or their need to bind a cytoplasmic cofactor for activity [44]. However, GH46 chitosanases do not fall into these categories. After redirection toward the "wrong" pathway (by replacement of the signal peptide), the purified CsnA chitosanase had the same specific activity as the one obtained after secretion through the native pathway, indicating that secretion through the preferred pathway was not an essential prerogative for correct folding and activity [35]. More research will be needed to understand why a given protein is more efficiently secreted through SEC or TAT pathway.

3.2. Other Modules

In addition to signal peptides, modules with other putative functions are present in 20 proteins from our subset (Figure S3). Three actinobacterial proteins, all members of Cluster A, and several others revealed by BLAST analysis (not shown) possess single (Csn_StrAM-7162; KSE_40860_csn2) or tandem (BN159_0394) "discoidin-like" domains. These segments are carbohydrate-binding modules belonging to CBM32 family. They show 50%–55% identity to modules present in the GH8 family chitosanase from *Paenibacillus* sp. IK-5, which are specific for chitosan binding [45,46] and are important for the adsorption to the chitosan component of the fungal cell wall.

Putative peptidoglycan-binding domains are present in several proteins belonging to cluster E. All the chitosanases from chloroviruses (cluster C) share an *N*-terminal domain of unknown function. This domain has no homologs in any genome sequenced so far outside the chlorovirus group. It can be then assumed that its function is closely linked to a particular trait of the biology of these viruses.

Other atypical sequence segments detected in GH46 chitosanase sequences are shown in Figure S3.

4. Structure–Function Relationships: Summary of Results from Crystallography and Site-Directed Mutagenesis

4.1. Tertiary Structure and Key Residues

The tertiary structure of GH46 chitosanases is similar to those of lysozymes (GH22, GH23, GH24) and non-processive chitinases (GH19) [47]. The CAZy database groups into "clans" the families of proteins "sharing a fold and catalytic machinery". Accordingly, GH46 proteins belong to clan GH-I together with families GH24 and GH80 [24,25]. They are mostly α-helical proteins, composed of two lobes, a major lobe and a minor lobe, separated by a substrate-binding cleft (Figure 3). Residues that function could be inferred from crystallography or site-directed mutagenesis are listed in Table 1. Secondary structures derived from crystallography are also shown in the sequence alignment in Figure S2.

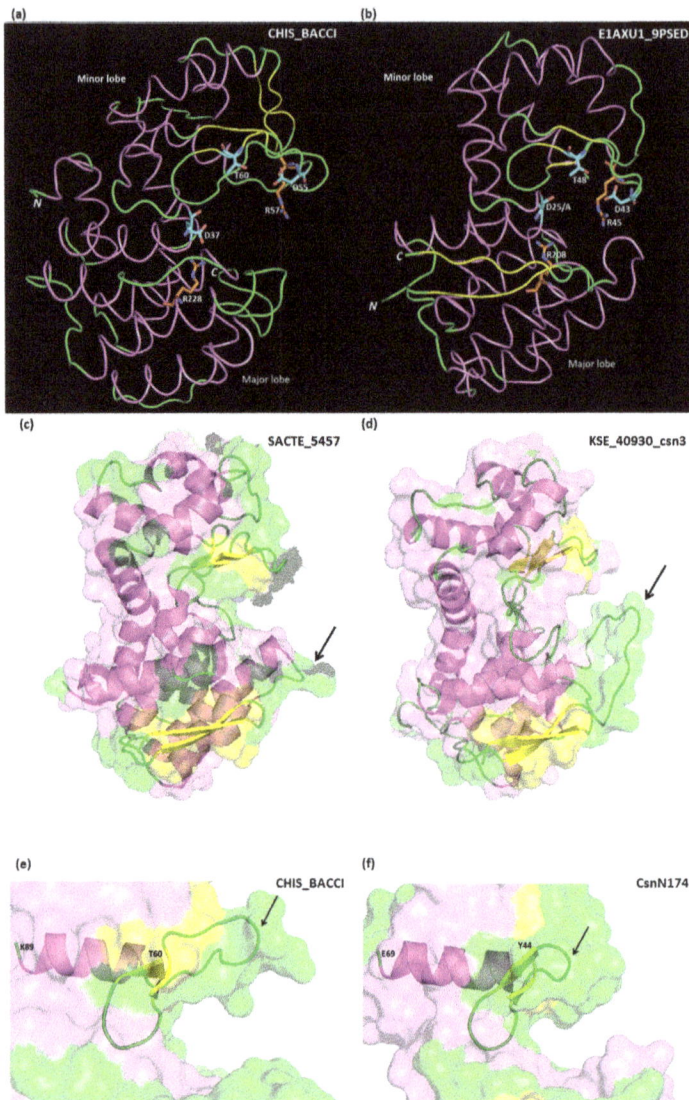

Figure 3. (**a**,**b**) comparison of tertiary structures of GH46 chitosanases from *Bacillus circulans* MH-K1 (CHIS_BACCI) and *Microbacterium* sp. OU01 (E1AXU1_9PSED). Structure drawings were derived from PDB files 1QGI and 4OLT; (**c**,**d**) major lobe loop length polymorphism shown on chitosanases from Cluster A (SACTE_5457 from *Streptomyces* sp. SirexAA-E) and B (KSE_40930_csn3 from *Kitasatospora setae*). Drawings derived from PDB file 4ILY and a homology-based model build on 3D-JIGSAW server [48]; (**e**,**f**) minor lobe loop length polymorphism shown on chitosanases from *Bacillus circulans* MH-K1 (CHIS_BACCI) and *Streptomyces* sp. N174 (CsnN174). Drawings derived from PDB files 1QGI and 1CHK. The longer loop in CHIS-BACCI allows for the accommodation of an *N*-acetyl group of the chitosan substrate at +1 subsite of chitosanase [41]. α-Helices are paint in violet, β-sheets in yellow and loops in green.

Starting from the *N*-terminus, a long helix crosses the major lobe and ends on the substrate binding cleft exposing the general acid catalytic residue, which is invariably a glutamate (Glu^{25} in *Microbacterium* sp. OU01 chitosanase; Glu^{37} in *Bacillus circulans* MH-K1 chitosanase; Figure 3a,b). Then, the amino acid chain traverses the substrate binding cleft, goes into the minor lobe and forms a series of three short β-sheets and intercalating loops, accommodating several residues essential for catalysis: the aspartate which functions as a general base (Asp^{43} and Asp^{55}, respectively; Figure 3a,b), a threonine (Thr^{48} and Thr^{60}) orientating a water molecule for nucleophilic attack and an arginine (Arg^{45} and Arg^{57}) hydrogen bound to the general base aspartate and thought to optimize its function, besides interacting with the substrate [49] (Figure 3a,b). These three residues are strictly conserved in chitosanases which enzymatic activity is experimentally confirmed, with the exception of the enzyme marked as Cho4239-1_Jan on Figure 1, originating from *Janthinobacterium* sp. 4239 [50] which lacks the aspartate and the arginine residues (Figure S2). On the other hand, the water-orientating threonine, while conserved in the vast majority of GH46 sequences, could be substituted with a serine, a similar residue but with a shorter side-chain, with only a minor decrease in activity in CsnN174 [51]. Interestingly, the chitosanase O24825_9BACL from *Paenibacillus ehimensis* has a serine in this position in its wild-type sequence (Figure S2).

Table 1. Key functional residues equivalence between chitosanases from family GH 46: summary of observations from site-directed mutagenesis and crystallography.

A. General Function Residues			
CsnN174	SACTE_5457	CHIS_BACCI	OU01
L5 Interlobe hydrophobic interaction	L57	L21	L8
E22 * [52] Catalytic general acid	E74 # [27] Catalytic general acid	E37 # [41] Catalytic general acid	E25 * [33,53] Catalytic general acid
W28 * [54] Cooperative stabilization of the protein structure via hydrophobic and carboxyl side chains interaction	W80	W43	W31
No equivalent	No equivalent	C50 # [41] Disulfide bridge with C124	No equivalent
D40 * [52] Catalytic nucleophile	D92 # [27] Catalytic nucleophile	D55 * [41] Catalytic nucleophile	D43 # [33,53] Catalytic nucleophile
R42 * [49] Electrostatic interaction with the catalytic nucleophile	R94	R57 * [55] Deprotonation of the catalytic nucleophile	R45 # [33]
T45 * [51] Water molecule positioning	T97 † [27]	T60	T48 # [33,53] Water molecule positioning
F97 † [54] Hydrophobic interaction network with W101	F149	F123	F100
No equivalent	No equivalent	C124 # [41] Disulfide bridge with C50	No equivalent
W101 * [54] Stabilization of the protein structure via hydrophobic interaction with F97	W153	I127	W104
D145 † [56] Member of ionic interaction network that stabilizes the catalytic cleft with R190 and R205	D197	D172 † [56] Member of ionic interaction network that stabilizes the catalytic cleft with R210 and R228	D148
R190 † [56] Member of ionic interaction network that stabilizes the catalytic cleft with D145 and R205	R242	R210 † [56] Member of ionic interaction network that stabilizes the catalytic cleft with D172 and R228	R193
R205 *,† [56] Member of ionic interaction network that stabilizes the catalytic cleft with D145 and R190. Also in direct interaction with the general acid catalytic residue.	R257	R228 † [56] Member of ionic interaction network that stabilizes the catalytic cleft with D172 and R210. Also in direct interaction with the general acid catalytic residue.	R208
W227 * [54] Cooperative stabilization of the protein structure via hydrophobic and carboxyl side chains interaction	W279	No equivalent	W230

Table 1. *Cont.*

B. Substrate Interaction Residues (Subsite Indicated into Brackets) §			
CsnN174	SACTE_5457 §	CHIS_BACCI	OU01
E22	E74	E37	E25 # [53] (+1)
N23	N75 † [27] (+1)	Q38	N26
S24	S76	D39	S27 *,# [33] (+2)
Q31	Q83 † [27] (+1)	Y46	Q34
K33	G85 † [27] Accommodation of an acetyl group of GlcNAc at (+1)	G48	G36
Y34 † [32] (+1)	Y86 † [27] (+1) Accommodation of an acetyl group of GlcNAc at (+1)	Y49	Y37 *,# [33] (+1)
R42 * [49] Electrostatic interaction with substrate	R94 † [27] (−2)	R57 * [55]	R45 # [33] (−2) Hydrogen bond with substrate.
T45 * [51]	T97 † [27] (−1)	T60	T48 # [33]
G46	A98 † [27] b Accommodation of an acetyl group of GlcNAc at (+1)	I61	G49
G47	G99 † [27] Accommodation of an acetyl group of GlcNAc at (+1)	G62	G50
I49 † [32] (−2)	I101 † [27] Interference with an acetyl group of GlcNAc at (−2)	F64	I52 # [33]
G50 † [32] (−1)	G102	G65	G53 # [33] (−2 and −1)
T55	T107 † [27] (−2)	H75	T58 *,# [33] (−2 and −3)
D57 * [57,58] (−2)	D109 † [27](−2)	D77	D60 # [33] (−2)
Y122 † [32] (−2)	Y174 † [27] (−2) Interference with an acetyl group of GlcNAc at (−2)	Y148 * [59] (−2)	Y125
H150	H202	N177	H153 *,# [33] (−3)
P152 † [32] (−2)	G204	A179	P155 # [33] (−3)
E197 †,* [32,58] (−1)	E249	N217	E200 *,# [33] Hydrogen bond with R45
A199	A251	Y219	A202 # [33] (+1)
H200	H252	N220	H203 # [33] (−1)
D201 †,* [32,58] (+2)	S253	K221	A204
D232	D284	T259	D235 # [33] (+3)

§ Subsite is indicated only for residues identified in crystals or models; Method of analysis: * Site-directed mutagenesis; # X-ray crystal structure; † Computational modelization; Numbering of residues in CsnN174, CHIS_BACCI and OU01 begins with the first amino acid of the mature, secreted protein. Numbering of residues in SACTE_5457 begins with the first amino acid of the immature protein; & For SACTE_5457, Takasuka *et al.* [27] adopted a subsite numbering opposite to that adopted by the authors of the other crystallographic studies. In this Table, we re-established the conventional numbering.

Typically, the small lobe is structured into five α-helices and three short β-sheets. A distinct trait of the configuration of the minor lobe in several cluster D chitosanases is the presence of a disulfide bridge (between Cys^{50} and Cys^{124} in MH-K1 chitosanase) and of two additional β-sheets (in yellow in Figure 3a). Furthermore, the loop that follows the "active-site segment" described above is much longer in MH-K1 chitosanase (residues 68–76) than in other chitosanases (Figure 3e,f). This loop reshapes significantly the substrate binding cleft compared with the structures of enzymes from cluster A [41].

Pursuing its course, the polypeptide chain forms a long bend helix at the junction between both lobes. This interlobe helix is rather rigid due to interactions between charged residues (Glu^{63}, Glu^{120} and Arg^{123} in OU01 chitosanase) and the increase in its flexibility following an E120A mutation resulted in enhanced activity toward polymeric and oligomeric chitosan substrate [53].

The major lobe is stabilized by a highly conserved network of interacting charged residues (Arg-Asp-Arg), each localized in a different helix. By site-directed mutagenesis, these residues were shown to be essential for enzyme activity [56]. The last arginine from this trio (Arg^{208} in *Microbacterium* sp. OU01 chitosanase; Arg^{228} in *Bacillus circulans* MH-K1 chitosanase; Figure 3a,b) plays an important role: it points towards the substrate binding cleft, interacting directly with the catalytic general acid glutamate and influencing its pKa. Similar networks are present in other enzyme families belonging to the lysozyme-like group [20,56].

This essential arginine is preceded by a large loop, which is much longer in proteins belonging to cluster B than in other clusters (Figure S2). As no tertiary structure is available yet for cluster B proteins, we built a model of one of them, KSE_40930_csn3. In Figure 3c,d this model is compared with the structure of the cluster A chitosanase SACTE_5457 and the discussed loop is shown by an arrow.

The C-terminal segment shows again a polymorphism among the cluster D chitosanase CHIS-BACCI, where this segment has a helical structure, and cluster A chitosanases with two β-sheets (Figure 3a,b). The C-terminus is localized in the vicinity of the N-terminus: a trait shared with most of the lysozyme-like proteins [47].

4.2. Substrate Binding and Cleavage

Crystallographic data and NMR studies showed that GH46 chitosanases act by an inverting mechanism of hydrolysis [27,32,41,60]. In chitosanase crystals obtained in the absence of substrate, the distance between the catalytic residues is larger than 9.5–10 Å, considered to be optimal for inverting glycosidases [61]: 13.8 Å, 10.9 Å and 10.3 Å for chitosanases from *Streptomyces* sp. N174, *B. circulans* MH-K1 and *Streptomyces* sp. SirexAA-E, respectively [27,32,41]. To put the catalytic residues into the right positions for substrate hydrolysis, it was suggested that the enzyme oscillates between two alternative "open" and "closed" configurations during the reaction cycle, the conformational change occurring at substrate binding and product liberation steps [32]. Following the co-crystallization of catalytically impaired chitosanase OU01 with substrate [53], the "closing" of chitosanase structure at substrate binding was further decomposed in three steps. A critical interaction with residues in subsites −2 and −1 (Asp^{60} and His^{203}, respectively) initiates the whole process of binding in OU01 chitosanase. Two more interactions with the polymeric substrate, involving distinct enzyme areas, complete the binding process [53] (Figure 4).

Compared with other glycoside hydrolases, the GH46 chitosanases have a highly electronegative substrate-binding cleft. This is due to a relatively large proportion of glutamate and aspartate among substrate binding residues (Table 1B), most of which interact with the amino groups of chitosan substrate. This abundance of acidic residues in the substrate binding cleft is thought to be responsible for the high (even if not absolute) specificity of GH46 enzymes as chitosanases and their poor recognition of chitinous, highly N-acetylated substrates [32,33].

In early structural studies, the enzyme residues potentially interacting with the chitosan substrate were identified by modeling of the mode of binding of chitosan oligosaccharides (mainly hexamers of D-glucosamine) with chitosanase. A first model build for the *Streptomyces* sp. N174 chitosanase and based on the mode of action of lysozyme [32] suggested the presence of six subsites, named A to F, with an asymmetrical cleavage of "4 + 2" type occurring between subsites D and E. Accordingly, the hydrolysis of $(GlcN)_6$ should yield dimeric and tetrameric products in equimolar proportions. However, kinetic data showed that the symmetrical "3 + 3"-type splitting is much favored over the asymmetrical one [60]. The symmetrical model was confirmed recently when Lyu *et al.* obtained a crystal of the chitosanase OU01 mutated at the general acid residue, complexed with the hexaglucosamine substrate [33]. The authors provided, for the first time, a description of the substrate-binding mechanism based on direct crystallographic observations. The substrate-interacting residues are shown in Figure 5. They are also listed in Table 1, together with the corresponding residues from other chitosanases which structures have been elucidated. Lyu *et al.* (2014) emphasize again the importance of acidic residues in the substrate binding cleft [33]. The −2 subsite is one of the most important determinants of the specificity of OU01 enzyme as a chitosanase, where the substrate interacts with two highly conserved residues: Arg^{45} and Asp^{60}. This observation confirmed previous studies by site-directed mutagenesis, which showed that mutations of corresponding residues in other chitosanases resulted in severe impairment of enzymatic activity [49,55,57]. Performing a series of mutations of residues in the substrate-binding cleft, Lyu *et al.* concluded that, "the subsites −2, −1 and +1 are probably the dominant contributors for substrate binding and essential for hydrolysis" [33].

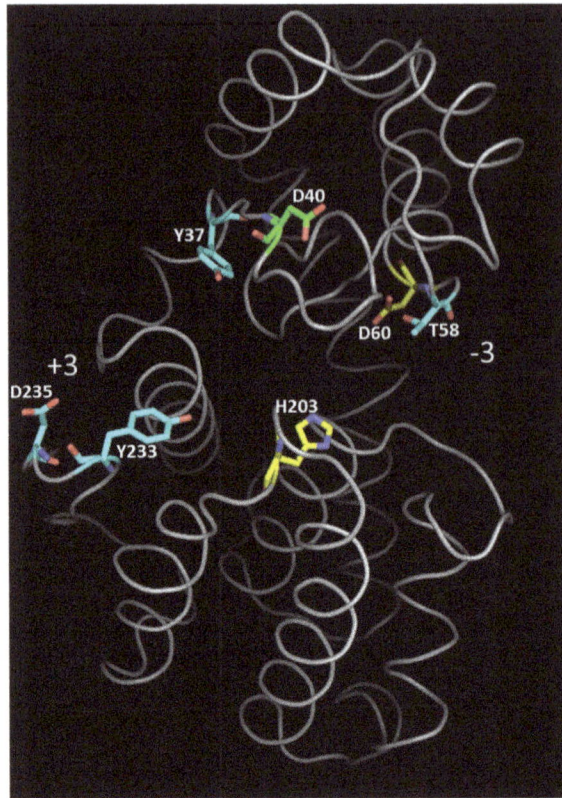

Figure 4. Tertiary structure of chitosanase from *Microbacterium* sp. OU01 with substrate-binding residues colored according to the three-step binding mechanism for polymeric substrate [53]. Yellow: residues responsible for the initial contact with substrate (step 1). Blue: residues that further stabilize the interaction with substrate (step 2). Green: residue participating in polymeric substrate binding but without effect on oligomeric substrate binding (step 3). The orientation of the substrate binding cleft between −3 and +3 subsites is also indicated. Modified from [53].

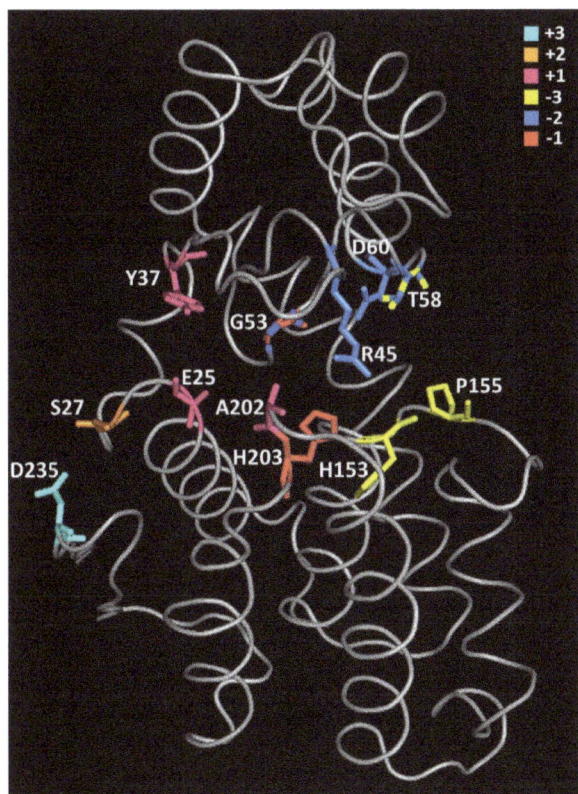

Figure 5. Tertiary structure of chitosanase from *Microbacterium* sp. OU01: distribution of substrate-binding residues among six subsites [33,53]. The colors assigned to the various subsites are shown in the upper right part of the figure. Residues painted with two colors participate simultaneously to two subsites.

Considering that chitosan polysaccharide is a mixed polymer consisting of various proportions of GlcN and GlcNAc, the concept of "cleavage specificity" has been a subject of discussions all along the chitosanase studies. Early work concentrated on the study of products obtained after extensive hydrolysis and the cleavage specificity was defined from the determination of the most frequent terminal aminosugars found in these oligosaccharidic products [62]. The aminosugars mostly found at the reducing ends corresponded to those recognized preferentially at the −1 subsite and those found preferentially at the non-reducing ends would be recognized preferentially at the +1 subsite. These studies allowed concluding that the chitosanase from *Streptomyces* sp. N174 recognized GlcN-GlcN and GlcNAc-GlcN links, while the chitosanase from *Bacillus circulans* MH-K1 had different cleavage specificity, recognizing mostly GlcN-GlcN and GlcN-GlcNAc links [60,63]. For the latter, 3D modelling showed that a loop in the minor lobe allowed to accommodate the *N*-acetyl group of GlcNAc at the +1 subsite [41] (Figure 3e).

Further studies showed, however, that this early models of cleavage specificity were oversimplified. Examination of reaction products at various stages of chitosan hydrolysis by the chitosanase ScCsn46A (SCO0677) from *Streptomyces coelicolor* A3(2) showed that this enzyme, member of Cluster A, was able to cleave at least three types of links: GlcN-GlcN, GlcNAc-GlcN and also GlcNAc-GlcNAc. The cleavage of the latter link in highly *N*-deacetylated chitosan (measured with

GlcNAc hexamer substrate) occurred more than 10^5-times slower than that of GlcN–GlcN links at initial reaction stages [34]. The authors could not conclude whether the enzyme is able to cleave the fourth type of links, *i.e.*, GlcN-GlcNAc. In fact, the lack of GlcNAc-GlcNAc and GlcNAc-GlcN dimers among final products could be explained either by the inability to cleave GlcN-GlcNAc links or by an absolute specificity for GlcN at the −2 subsite [34].

Lyu *et al.* [33] favored the second of these possibilities. Building a model of the tetrasaccharide GlcNAc-GlcN-GlcNAc-GlcNAc bound by the OU01 chitosanase from *Microbacterium* sp. (another member of Cluster A), they showed that GlcNAc units can be easily fitted into −3, −1 and +1 subsites when the GlcN unit is bound to the −2 subsite. Accordingly, OU01 has specificity for GlcN at the −2 subsite. The specificity of this chitosanase is then described as follows: (GlcN)-(GlcN/GlcNAc)-(GlcN/GlcNAc) [(−2)-(−1)-(+1)].

The main determinant of substrate specificity at −2 subsite is an aspartate residue, which interacts directly with the amino group of substrate [33,53]. This residue is strictly conserved in all GH46 members belonging to Clusters A–D. It could then be postulated that in all these four clusters, the −2 subsite has a decisive preference for GlcN binding, what could be the most characteristic trait of GH46 members, making them distinct from other chitosanases. In contrast, according to our sequence alignment (Figure S2), this aspartate does not seem to be conserved in Cluster E proteins. However, as mentioned earlier, the activity of these proteins as chitosanases was not demonstrated yet.

5. Biological Functions of GH46 Chitosanases

5.1. Metabolic Assimilation of Chitosan

Chitosan is a potentially valuable nutrient for microorganisms, being a source of carbon, nitrogen and energy. Unlike peptidoglycan in bacteria or chitin in fungi, the polysaccharide chain of chitosan in living organisms is not build by dedicated synthases [64,65]. Instead, chitosan derives from chitin through the action of chitin *N*-deacetylases, enzymes belonging to the carbohydrate esterase family CE4 [66]. Thus the distribution of chitosan in nature is limited to those organisms, which also synthesize chitin [65]. In most of these organisms, chitosan forms only a small proportion of their cell wall compared with chitin, however in zygomycetes or in the basidiomycete *Cryptococcus neoformans* the proportion of chitosan can raise to more than 50% [65]. Due to the ubiquitous presence of fungi and other chitin-containing organisms in the biosphere, chitosan can be considered as being omnipresent but not abundant.

If chitosanases are truly involved in bacterial nutrition, it would be expected that their production is induced by the presence of chitosan. Indeed, the necessity to add chitosan to the growth medium in order to obtain production of GH46 chitosanases has been confirmed in many studies [67–69]. The essential character of chitosanase for growth on chitosan was demonstrated for *Bacillus subtilis* by disrupting the *csn* gene [37].

As GH46 chitosanases are *endo*-enzymes yielding oligosaccharides (dimers and longer but no monomers) as final products, the assimilation of chitosan-derived monosaccharides (GlcN and GlcNAc) would be possible only after further degradation of oligosaccharides by *exo*-hydrolases, GlcNases and GlcNAcases, respectively. While GlcNAcases (EC 2.3.1. 52; belonging to families GH3 and GH20) are represented by almost ten thousands entries in the CaZy database, the distribution of GlcNases (belonging to sub-families of GH2 and GH9 families) is much more limited. *Streptomyces coelicolor* A3(2) in which a GlcNase could not be detected, was shown to be able to uptake the chitosan-derived oligosaccharides directly through a dedicated ABC transporter [70]. The metabolism of chitosan mediated by hydrolysis with chitosanase followed by direct uptake of oligomeric products could be shared by many microorganisms hosting chitosanases belonging to Cluster A (Figure 1), as the genes encoding ABC transporters, highly similar to that of *S. coelicolor* A3(2) are detected in almost all the genomes of streptomycetes and other phylogenetically related actinobacteria [70]. The scarcity of studies on chitosan metabolism does not allow concluding, so far, how widely such a GlcNase-less mechanism of chitosan assimilation is distributed among other taxonomic groups of organisms.

5.2. Protection against the Antimicrobial Activity of Chitosan

Chitosan as an antimicrobial has been the subject of numerous studies. It acts on Gram-positive and Gram-negative bacteria and has some antifungal activity as well. This antimicrobial effect is however dependent on the molecular weight (MW), as short-chain forms of chitosan has much lower or even undetectable antimicrobial effect. Kendra and Hadwiger [71] purified chitosan oligomers of various lengths and evaluated their antifungal activity. Monomers and dimers had no detectable antifungal effect against two strains of *Fusarium solani*. Increasing the chitosan chain length, measurable antifungal effect started with trimers and sharply increased for hexamers, heptamers and high MW chitosan. Similar studies, but with higher categories of MW were reported on antibacterial effect, using *Escherichia coli* as test organism [72,73]. The minimal inhibitory concentration (MIC) of chitosan of average MWs in the 10,500–9300 range was 0.004% against *E. coli*, while chitosan with MW in the range of 2200–4100 had no effect even at 0.5%. Additionally, using fluorescently marked chitosan, they showed that the 9300 fraction accumulated essentially in the cell wall, possibly blocking the transport of nutrients inside the cell, while the 2200 fraction penetrated into the cell allowing its further metabolism [72,73].

These observations opened the possibility that the expression of a chitosanase could render a bacterial or fungal microorganism more resistant to the antimicrobial effect of chitosan. An extracellular chitosanase would degrade chitosan into small, non-inhibitory fragments, allowing normal growth in their presence. Accordingly, wild type *Streptomyces lividans* TK24, a natural producer of an extracellular chitosanase was more resistant to chitosan than its mutant harboring a deletion of the chitosanase gene *csnA*, with MIC values of 0.2 and 0.08 g/L, respectively [35]. The metabolic activity estimated by the uptake of xylose from the culture medium was totally inhibited at 0.2 g/L of chitosan for the mutant strain, 0.3 g/L for the wild type strain but a chitosan concentration as high as 1.2 g/L was necessary to block the xylose transport in a strain expressing a recombinant GH46 chitosanase from *Kitasatospora* sp. [35].

Interestingly, recombinant *Escherichia coli* strains expressing a chitosanase that is not secreted extracellularly but remains confined to the periplasmic space are also more resistant to chitosan than wild-type strains. This effect was exploited in a mutagenesis study of the CsnN174 chitosanase originating from *Streptomyces* sp. N174 [51]. A mutated gene encoding an inactive chitosanase, with the essential Thr^{45} residue (see Table 1) replaced by a histidine (T45H mutant) has been introduced into *E. coli* and the resulting recombinant strain did not show resistance against chitosan and could not grow on chitosan medium. Then, the mutated H45 codon has been subjected to saturation mutagenesis *in vitro*, and the resulting library of recombinant *E. coli* cells was plated on chitosan medium. The colonies able to grow on chitosan medium harbored revertant genes restoring chitosanase activity. In such a way it was possible to show that, besides threonine, a serine could also successfully play the role of water-orientating residue in the active site of chitosanase [51].

5.3. Chitosanase-Aided Lysis of Algal Cell Wall as a Step in Viral Development

The presence of chitosanases in the genetic baggage of chloroviruses (or *Chlorella* viruses), giant double-stranded-DNA viruses infecting eukaryotic green algae, is consistent with the fact that many *Chlorella* strains contain a glucosamine-rich, chitin-like polymer in their cell wall [74]. Chitosanases could then participate to the lysis of host cells at various steps of the viral infection [75].

Two GH46 chitosanases present in chloroviruses were the subject of biochemical studies. Chitosanase activity was identified through the release of reducing sugars when intact particles of CVK2 virus were incubated in the presence of chitosan substrate. The presence of chitosanase was further confirmed by an in-gel staining procedure [76]. The corresponding gene, *vChta-1* was expressed in *E. coli* yielding a protein with chitosanase activity with an approximate MW of 37 kDa. However, after Western blot analysis of structural CVK2 proteins, the antibody was rather reacting with a 65-kDa protein. Transcriptional analysis of *vChta-1* gene expression, as well as further immunoblot studies, allowed concluding that the proteins with chitosanase activity are expressed from two alternative transcripts. A longer transcript, extending into a downstream ORF, encoded the larger, 65-kDa chitosanase which was incorporated into the virion and could serve during the initial attack of the host cell wall after virion attachment to the cell surface, while the smaller transcript encoded the 37-kDa enzyme which could contribute to *Chlorella* cell lysis at the final stage of infection [76].

The putative chitosanase gene from the chlorovirus PBCV-1 was expressed in *E. coli* and its activity was demonstrated by biochemical studies. During infection, the chitosanase gene *a292l* was expressed at late stages. The presence of the corresponding protein in the purified PBCV-1 virions was confirmed by immunoblotting [40]. Surprisingly, the chitosanase was not detected in a recent proteomic study of highly purified virion particles [77], nor were other lytic enzymes, such as chitinases, expected from earlier studies. This result puts into question marks the real biological function of GH46 chitosanases in chloroviruses. This function should be important, perhaps essential, as the chitosanase belongs to the core set of 155 protein families present in all the chlorovirus genomes sequenced so far [78].

5.4. Antifungal Effect

The antifungal activity of several GH46 chitosanases against fungi having partly deacetylated chitin in their cell wall is well documented. The target fungi belong mostly to zygomycetes but rare examples of chitosanase-susceptible fungi from other phyla have also been reported. The most detailed studies have been dedicated to the antifungal activity of the *Bacillus circulans* MH-K1 chitosanase [79]. The purified recombinant chitosanase inhibited the hyphal elongation of zygomycetes such as *Mucor javanicus*, *Rhizopus oryzae* and *Rhizopus stolonifer* but was without effect on the ascomycete fungi *Fusarium oxysporum* f. sp. *lycopersici racel* and *Aspergillus awamori* var. *kawachi*. After mutation of catalytic residues Glu^{37} or Asp^{55} (See Table 1) into Gln^{37} or Asn^{55}, respectively, the chitosan-binding ability was maintained but the enzymatic activity was lost. The antifungal activity was lost as well, which showed that it was due to the hydrolytic activity against chitosan and not to chitosan binding. The ability of the E37Q-mutated enzyme to bind directly to the hyphae of *Rhizopus oryzae* was confirmed using a fusion of Q37-mutated chitosanase with green fluorescent protein.

The *cho1* gene encoding the MH-K1 chitosanase has also been introduced into transgenic rice plants [80]. *cho1* gene expression was confirmed by transcriptomic and enzymological analysis. Rice blast is a disease caused by *Magnaporthe oryzae* and the transgenic plants displayed enhanced resistance against this disease [80]. This enhanced resistance phenotype is explained by the fact that *M. oryzae*, an ascomycete, performs partial *N*-deacetylation of chitin during the infection, becoming a target for chitosanase. The direct lytic action of chitosanase on the fungal cell wall is however only a part of the story. The chitosan oligomers resulting from hydrolysis potentiate the host defense response mechanisms, as exemplified by the increased release of reactive oxygen species in leaf sheaths of chitosanase-expressing rice plants [80].

Antifungal activities were also demonstrated for GH46 chitosanases from *Streptomyces* sp. N174, *Amycolatopsis* sp. CsO-2 and *Bacillus subtilis* [36,81,82]. Enzymes recognized for their antifungal activities belong to clusters A, B and C and are present in a wide variety of bacteria. It can be postulated that chitosanases, besides chitinases and chitin-binding proteins, are important players in the antagonisms between bacteria and fungi.

Acknowledgments: This work was supported by a Discovery grant from the Natural Science and Engineering Council of Canada.

Conflicts of Interest: The authors declare no conflict of interest.

References

1. Aranaz, I.; Mengíbar, M.; Harris, R.; Paños, I.; Miralles, B.; Acosta, N.; Galed, G.; Heras, A. Functional characterization of chitin and chitosan. *Curr. Chem. Biol.* **2009**, *3*, 203–230.

2. Muzzarelli, R.A.A.; Boudrant, J.; Meyer, D.; Manno, N.; DeMarchis, M.; Paoletti, M.G. Current views on fungal chitin/chitosan, human chitinases, food preservation, glucans, pectins and inulin: A tribute to Henri Braconnot, precursor of the carbohydrate polymers science, on the chitin bicentennial. *Carbohydr. Polym.* **2012**, *87*, 995–1012. [CrossRef]

3. Pillai, C.K.S.; Paul, W.; Sharma, C.P. Chitosan: Manufacture, properties and uses. In *Chitosan: Manufacture, Properties and Usage*; Davis, S.P., Ed.; Nova Publishers: Hauppauge, NY, USA, 2011; pp. 133–216.

4. Rinaudo, M. Chitin and chitosan: Properties and applications. *Prog. Polym. Sci. (Oxf.)* **2006**, *31*, 603–632. [CrossRef]

5. Zhang, J.; Xia, W.; Liu, P.; Cheng, Q.; Tahirou, T.; Gu, W.; Li, B. Chitosan modification and pharmaceutical/biomedical applications. *Mar. Drugs* **2010**, *8*, 1962–1987. [PubMed]

6. Buschmann, M.D.; Merzouki, A.; Lavertu, M.; Thibault, M.; Jean, M.; Darras, V. Chitosans for delivery of nucleic acids. *Adv. Drug Deliv. Rev.* **2013**, *65*, 1234–1270. [CrossRef] [PubMed]

7. El Hadrami, A.; Adam, L.R.; El Hadrami, I.Y.; Daayf, F. Chitosan in plant protection. *Mar. Drugs* **2010**, *8*, 968–987. [CrossRef] [PubMed]

8. Kong, M.; Chen, X.G.; Xing, K.; Park, H.J. Antimicrobial properties of chitosan and mode of action: A state of the art review. *Int. J. Food Microbiol.* **2010**, *144*, 51–63. [CrossRef] [PubMed]

9. Muzzarelli, R.A.A. Chitins and chitosans for the repair of wounded skin, nerve, cartilage and bone. *Carbohydr. Polym.* **2009**, *76*, 167–182. [CrossRef]

10. Park, J.H.; Saravanakumar, G.; Kim, K.; Kwon, I.C. Targeted delivery of low molecular drugs using chitosan and its derivatives. *Adv. Drug Deliv. Rev.* **2010**, *62*, 28–41. [CrossRef] [PubMed]

11. Wan Ngah, W.S.; Teong, L.C.; Hanafiah, M.A.K.M. Adsorption of dyes and heavy metal ions by chitosan composites: A review. *Carbohydr. Polym.* **2011**, *83*, 1446–1456. [CrossRef]

12. Blanchard, J.; Park, J.K.; Boucher, I.; Brzezinski, R. Industrial applications of chitosanases. In *Recent Advances in Marine Biotechnology; Volume 9: Biomaterials and Bioprocessing*; Fingerman, M., Nagabhushanam, R., Eds.; Science Publishers: Enfield, NH, USA, 2003; pp. 257–277.

13. Thadathil, N.; Velappan, S.P. Recent developments in chitosanase research and its biotechnological applications: A review. *Food Chem.* **2014**, *150*, 392–399. [CrossRef] [PubMed]

14. Yin, H.; Du, Y.; Zhang, J. Low molecular weight and oligomeric chitosans and their bioactivities. *Curr. Top. Med. Chem.* **2009**, *9*, 1546–1559. [CrossRef] [PubMed]

15. Aam, B.B.; Heggset, E.B.; Norberg, A.L.; Sørlie, M.; Vårum, K.M.; Eijsink, V.G.H. Production of chitooligosaccharides and their potential applications in medicine. *Mar. Drugs* **2010**, *8*, 1482–1517. [CrossRef] [PubMed]

16. Quintero-Villegas, M.I.; Aam, B.B.; Rupnow, J.; Sølie, M.; Eijsink, V.G.H.; Hutkins, R.W. Adherence inhibition of enteropathogenic *Escherichia coli* by chitooligosaccharides with specific degrees of acetylation and polymerization. *J. Agric. Food Chem.* **2013**, *61*, 2748–2754. [CrossRef] [PubMed]

17. Yin, H.; Zhao, X.; Du, Y. Oligochitosan: A plant diseases vaccine—A review. *Carbohydr. Polym.* **2010**, *82*, 1–8. [CrossRef]

18. Mourya, V.K.; Inamdar, N.N.; Choudhari, Y.M. Chitooligosaccharides: Synthesis, characterization and applications. *Polym. Sci. Ser. A* **2011**, *7*, 583–612. [CrossRef]

19. Fukamizo, T. Chitinolytic enzymes: Catalysis, substrate binding, and their application. *Curr. Protein Pept. Sci.* **2000**, *1*, 105–124. [CrossRef] [PubMed]

20. Hoell, I.A.; Vaaje-Kolstad, G.; Eijsink, V.G.H. Structure and function of enzymes acting on chitin and chitosan. *Biotechnol. Genet. Eng. Rev.* **2010**, *27*, 331–366. [CrossRef]

21. Henrissat, B. A classification of glycosyl hydrolases based on amino acid sequence similarities. *Biochem. J.* **1991**, *280*, 309–316. [CrossRef] [PubMed]

22. Ando, A.; Noguchi, K.; Yanagi, M.; Shinoyama, H.; Kagawa, Y.; Hirata, H.; Yabuki, M.; Fujii, T. Primary structure of chitosanase produced by *Bacillus circulans* MH-K1. *J. Gen. Appl. Microbiol.* **1992**, *38*, 135–144. [CrossRef]

23. Masson, J.-Y.; Denis, F.; Brzezinski, R. Primary sequence of the chitosanase from *Streptomyces* sp. N174 and comparison with other endoglycosidases. *Gene* **1994**, *140*, 103–107. [CrossRef]

24. Henrissat, B.; Bairoch, A. Updating the sequence-based classification of glycosyl hydrolases. *Biochem. J.* **1996**, *316*, 695–696. [CrossRef] [PubMed]

25. Davies, G.J.; Sinnott, M.L. Sorting the diverse: The sequence-based classifications of carbohydrate-active enzymes. *Biochem. J.* **2008**. [CrossRef]

26. Sievers, F.; Wilm, A.; Dineen, D.; Gibson, T.J.; Karplus, K.; Li, W.; Lopez, R.; McWilliam, H.; Remmert, M.; Söding, J.; *et al.* Fast, scalable generation of high-quality protein multiple sequence alignments using Clustal Omega. *Mol. Syst. Biol.* **2011**, *7*. [CrossRef] [PubMed]

27. Takasuka, T.E.; Bianchetti, C.M.; Tobimatsu, Y.; Bergeman, L.F.; Ralph, J.; Fox, B.G. Structure-guided analysis of catalytic specificity of the abundantly secreted chitosanase SACTE_5457 from *Streptomyces* sp. SirexAA-E. *Proteins* **2014**, *82*, 1245–1257. [CrossRef] [PubMed]

28. Chevenet, F.; Brun, C.; Bañuls, A.L.; Jacq, B.; Christen, R. TreeDyn: Towards dynamic graphics and annotations for analyses of trees. *BMC Bioinform.* **2006**, *7*, 439. [CrossRef] [PubMed]

29. Bagos, P.G.; Nikolaou, E.P.; Liakopoulos, T.D.; Tsirigos, K.D. Combined prediction of Tat and Sec signal peptides with hidden Markov models. *Bioinformatics* **2010**, *26*, 2811–2817. [CrossRef] [PubMed]

30. Li, H.; Jacques, P-É.; Ghinet, M.G.; Brzezinski, R.; Morosoli, R. Determining the functionality of putative Tat-dependent signal peptides in *Streptomyces coelicolor* A3(2) using two different reporter proteins. *Microbiology* **2005**, *151*, 2189–2198. [CrossRef] [PubMed]

31. Ando, A.; Saito, A.; Arai, S.; Usuda, S.; Furuno, M.; Kaneko, N.; Shida, O.; Nagata, Y. Molecular characterization of a novel family-46 chitosanase from *Pseudomonas* sp. A-01. *Biosci. Biotechnol. Biochem.* **2008**, *72*, 2074–2081. [CrossRef] [PubMed]

32. Marcotte, E.M.; Monzingo, A.F.; Ernst, S.R.; Brzezinski, R.; Robertus, J.D. X-ray structure of an anti-fungal chitosanase from *Streptomyces* N174. *Nat. Struct. Biol.* **1996**, *3*, 155–162. [CrossRef] [PubMed]

33. Lyu, Q.; Wang, S.; Xu, W.; Han, B.; Liu, W.; Jones, D.N.M.; Liu, W. Structural insights into the substrate-binding mechanism for a novel chitosanase. *Biochem. J.* **2014**, *461*, 335–345. [CrossRef] [PubMed]

34. Heggset, E.B.; Dybvik, A.I.; Hoell, I.A.; Norberg, A.L.; Sørlie, M.; Eijsink, V.G.H.; Vårum, K.M. Degradation of chitosans with a family 46 chitosanase from *Streptomyces coelicolor* A3(2). *Biomacromolecules* **2010**, *11*, 2487–2497. [CrossRef] [PubMed]

35. Ghinet, M.G.; Roy, S.; Poulin-Laprade, D.; Lacombe-Harvey, M.-È.; Morosoli, M.; Brzezinski, R. Chitosanase from *Streptomyces coelicolor* A3(2): Biochemical properties and role in protection against antibacterial effect of chitosan. *Biochem. Cell Biol.* **2010**, *88*, 907–916. [CrossRef] [PubMed]

36. Saito, A.; Ooya, T.; Miyatsuchi, D.; Fuchigami, H.; Terakado, K.; Nakayama, S.-Y.; Watanabe, T.; Nagata, Y.; Ando, A. Molecular characterization and antifungal activity of a family 46 chitosanase from *Amycolatopsis* sp. CsO-2. *FEMS Microbiol. Lett.* **2009**, *293*, 79–84. [CrossRef] [PubMed]

37. Rivas, L.A.; Parro, V.; Moreno-Paz, M.; Mellado, R.P. The *Bacillus subtilis* 168 *csn* gene encodes a chitosanase with similar properties to a *Streptomyces* enzyme. *Microbiology* **2000**, *146*, 2929–2936. [CrossRef] [PubMed]

38. Van Etten, J.L.; Meints, R.H. Giant viruses infecting algae. *Annu. Rev. Microbiol.* **1999**, *53*, 447–494. [CrossRef] [PubMed]

39. Yamada, T.; Chuchird, N.; Kawasaki, T.; Nishida, K.; Hiramatsu, S. Chlorella viruses as a source of novel enzymes. *J. Biosci. Bioeng.* **1999**, *88*, 353–361. [CrossRef]

40. Sun, L.; Adams, B.; Gurnon, J.R.; Ye, Y.; van Etten, J.L. Characterization of two chitinase genes and one chitosanase gene encoded by *Chlorella* virus PBCV-1. *Virology* **1999**, *263*, 376–387. [CrossRef] [PubMed]

41. Saito, J.; Kita, A.; Higuchi, Y.; Nagata, Y.; Ando, A.; Miki, K. Crystal structure of chitosanase from *Bacillus circulans* MH-K1 at 1.6-Å resolution and its substrate recognition mechanism. *J. Biol. Chem.* **1999**, *274*, 30818–30825. [CrossRef] [PubMed]

42. Linde, J.; Schwartze, V.; Binder, U.; Lass-Flörl, C.; Voigt, K.; Horn, F. *De novo* whole-genome sequence and genome annotation of *Lichtheimia ramosa*. *Genome Announc.* **2014**, *2*. [CrossRef]

43. Petersen, T.N.; Brunak, S.; von Heijne, G.; Nielsen, H. SignalP 4.0: Discriminating signal peptides from transmembrane regions. *Nat. Methods* **2011**, *29*, 785–786. [CrossRef] [PubMed]

44. Sargent, F. The twin-arginine transport system: Moving folded proteins across membranes. *Biochem. Soc. Trans.* **2007**, *35*, 835–847. [CrossRef] [PubMed]

45. Kimoto, H.; Akamatsu, M.; Fujii, Y.; Tatsumi, H.; Kusaoke, H.; Taketo, A. Discoidin domain of chitosanase is required for binding to the fungal cell wall. *J. Mol. Microbiol. Biotechnol.* **2010**, *18*, 14–23.

46. Shinya, S.; Ohnuma, T.; Yamashiro, R.; Kimoto, H.; Kusaoke, H.; Anbazhagan, P.; Juffer, A.H.; Fukamizo, T. The first identification of carbohydrate binding modules specific to chitosan. *J. Biol. Chem.* **2013**, *288*, 30042–30053. [CrossRef] [PubMed]

47. Monzingo, A.F.; Marcotte, E.M.; Hart, P.J.; Robertus, J.D. Chitinases, chitosanases, and lysozymes can be divided into prokaryotic and eukaryotic families sharing a conserved core. *Nat. Struct. Biol.* **1996**, *3*, 133–140. [CrossRef] [PubMed]

48. Bates, P.A.; Kelley, L.A.; MacCallum, R.M.; Sternberg, M.J.E. Enhancement of protein modelling by human intervention in applying the automatic programs 3D-JIGSAW and 3D-PSSM. *Proteins Struct. Func. Genet. Suppl.* **2001**, *5*, 39–46. [CrossRef] [PubMed]

49. Lacombe-Harvey, M.-È.; Fortin, M.; Ohnuma, T.; Fukamizo, T.; Letzel, T.; Brzezinski, R. A highly conserved arginine residue of the chitosanase from *Streptomyces* sp. N174 is involved both in catalysis and substrate binding. *BMC Biochem.* **2013**, *14*. [CrossRef] [PubMed]

50. Johnsen, M.G.; Hansen, O.C.; Stougaard, P. Isolation, characterization and heterologous expression of a novel chitosanase from *Janthinobacterium* sp. strain 4239. *Microb. Cell Fact.* **2010**, *9*. [CrossRef] [PubMed]

51. Lacombe-Harvey, M.-È.; Fukamizo, T.; Gagnon, J.; Ghinet, M.G.; Dennhart, N.; Letzel, T.; Brzezinski, R. Accessory active site residues of *Streptomyces* sp. N174 chitosanase. Variations on a common theme in the lysozyme superfamily. *FEBS J.* **2009**, *276*, 857–869. [CrossRef] [PubMed]

52. Boucher, I.; Fukamizo, T.; Honda, Y.; Willick, G.E.; Neugebauer, W.A.; Brzezinski, R. Site-directed mutagenesis of evolutionary conserved carboxylic amino acids in the chitosanase from *Streptomyces* sp. N174 reveals two residues essential for catalysis. *J. Biol. Chem.* **1995**, *270*, 31077–31082. [PubMed]

53. Lyu, Q.; Shi, Y.; Wang, S.; Yang, Y.; Han, B.; Liu, W.; Jones, D.N.M.; Liu, W. Structural and biochemical insights into the degradation mechanism of chitosan by chitosanase OU01. *Biochim. Biophys. Acta* **2015**, *1850*, 1953–1961. [CrossRef] [PubMed]

54. Honda, Y.; Fukamizo, T.; Okajima, T.; Goto, S.; Boucher, I.; Brzezinski, R. Thermal unfolding of chitosanase from *Streptomyces* sp. N174: Role of tryptophan residues in the protein structure stabilization. *Biochim. Biophys. Acta* **1999**, *1429*, 365–376. [CrossRef]

55. Ando, A.; Saito, A. Structure and function of chitosanase. In *Advances in Chitin Science: Volume XI*; Rustichelli, F., Caramella, C., Senel, S., Vårum, K.M., Eds.; The European Chitin Society: Venice, Italy, 2009; pp. 265–271.

56. Fukamizo, T.; Juffer, A.H.; Vogel, H.J.; Honda, Y.; Tremblay, H.; Boucher, I.; Neugebauer, W.A.; Brzezinski, R. Theoretical calculation of pKa reveals an important role of Arg205 in the activity and stability of *Streptomyces* sp. N174 chitosanase. *J. Biol. Chem.* **2000**, *275*, 25633–25640. [CrossRef] [PubMed]

57. Tremblay, H.; Yamaguchi, T.; Fukamizo, T.; Brzezinski, R. Mechanism of chitosanase-oligosaccharide interaction: Subsite structure of *Streptomyces* sp. N174 chitosanase and the role of Asp57 carboxylate. *J. Biochem.* **2001**, *130*, 679–686. [CrossRef] [PubMed]

58. Katsumi, T.; Lacombe-Harvey, M.-È.; Tremblay, H.; Brzezinski, R.; Fukamizo, T. Role of acidic amino acid residues in chitooligosaccharide-binding to *Streptomyces* sp. N174 chitosanase. *Biochem. Biophys. Res. Commun.* **2005**, *338*, 1839–1844. [CrossRef] [PubMed]

59. Fukamizo, T.; Amano, S.; Yamaguchi, K.; Yoshikawa, T.; Katsumi, T.; Saito, J.; Suzuki, M.; Miki, K.; Nagata, Y.; Ando, A. *Bacillus circulans* MH-K1 chitosanase: Amino acid residues responsible for substrate binding. *J. Biochem.* **2005**, *138*, 563–569. [CrossRef] [PubMed]

60. Fukamizo, T.; Honda, Y.; Goto, S.; Boucher, I.; Brzezinski, R. Reaction mechanism of chitosanase from *Streptomyces* sp. N174. *Biochem. J.* **1995**, *311*, 377–383. [CrossRef] [PubMed]

61. Rye, C.S.; Withers, S.G. Glycosidase mechanisms. *Curr. Opin. Chem. Biol.* **2000**, *4*, 573–580. [CrossRef]

62. Fukamizo, T.; Ohkawa, T.; Ikeda, Y.; Goto, S. Specificity of chitosanase from *Bacillus pumilus*. *Biochim. Biophys. Acta* **1994**, *1205*, 183–188. [CrossRef]

63. Mitsutomi, M.; Ueda, M.; Arai, M.; Ando, A.; Watanabe, T. Action patterns of microbial chitinases and chitosanases on partially N-acetylated chitosan. In *Chitin Enzymology, Volume 2*; Muzzarelli, R.A.A., Ed.; ATEC Edizioni: Grottamare, Italy, 1996; pp. 273–284.

64. Vollmer, W.; Blanot, D.; de Pedro, A.A. Peptidoglycan structure and architecture. *FEMS Microbiol. Rev.* **2008**, *32*, 149–167. [CrossRef] [PubMed]

65. Lenardon, M.D.; Munro, C.A.; Gow, N.A.R. Chitin synthesis and fungal pathogenesis. *Curr. Opin. Microbiol.* **2010**, *13*, 416–423. [CrossRef] [PubMed]

66. Zhao, Y.; Park, R.-D.; Muzzarelli, R.A.A. Chitin deacetylases: Properties and applications. *Mar. Drugs* **2010**, *8*, 24–46. [CrossRef] [PubMed]

67. Ohtakara, A. Chitosanase from *Streptomyces griseus. Method Enzymol.* **1988**, *161*, 505–510.

68. Yabuki, M.; Uchiyama, A.; Suzuki, K.; Ando, A.; Fujii, T. Purification and properties of chitosanase from *Bacillus circulans* MH-K1. *J. Gen. Appl. Microbiol.* **1988**, *34*, 255–270. [CrossRef]

69. Brzezinski, R. Uncoupling chitosanase production from chitosan. *Bioeng. Bugs* **2011**, *2*, 226–229. [CrossRef] [PubMed]

70. Viens, P.; Dubeau, M.-P.; Kimura, A.; Desaki, Y.; Shinya, T.; Shibuya, N.; Saito, A.; Brzezinski, R. Uptake of chitosan-derived D-glucosamine oligosaccharides in *Streptomyces coelicolor* A3(2). *FEMS Microbiol. Lett.* **2015**, *362*. [CrossRef] [PubMed]

71. Kendra, D.F.; Hadwiger, L.A. Characterization of the smallest chitosan oligomer that is maximally antifungal to *Fusarium solani* and elicits pisatin formation in *Pisum sativum. Exp. Mycol.* **1984**, *8*, 276–281. [CrossRef]

72. Tokura, S.; Ueno, K.; Miyazaki, S.; Nishi, N. Molecular weight dependent antimicrobial activity by chitosan. *Macromol. Symp.* **1997**, *120*, 1–9. [CrossRef]

73. Ueno, K.; Yamaguchi, T.; Sakairi, N.; Nishi, N.; Tokura, S. Antimicrobial activity by fractionated chitosan oligomers. In *Advances in Chitin Science: Volume II*; Domard, A., Roberts, G.A.F., Vårum, K.M., Eds.; Jacques André Publisher: Lyon, France, 1997; pp. 156–161.

74. Takeda, H. Cell wall composition and taxonomy of symbiotic *Chlorella* from *Paramecium* and *Acanthocystis. Phytochemistry* **1995**, *40*, 457–459. [CrossRef]

75. Sugimoto, I.; Hiramatsu, S.; Murakami, D.; Fujie, M.; Usami, S.; Yamada, T. Algal-lytic activities encoded by *Chlorella* virus CVK2. *Virology* **2000**, *277*, 119–126. [CrossRef] [PubMed]

76. Yamada, T.; Hiramatsu, S.; Songsri, P.; Fujie, M. Alternative expression of a chitosanase gene produces two different proteins in cells infected with *Chlorella* virus CVK2. *Virology* **1997**, *230*, 361–368. [CrossRef] [PubMed]

77. Dunigan, D.D.; Cerny, R.L.; Bauman, A.T.; Roach, J.C.; Lane, L.C.; Agarkova, I.V.; Wulser, K.; Yanai-Balser, G.M.; Gurnon, J.R.; Vitek, J.C.; *et al. Paramecium bursaria* chlorella virus 1 proteome reveals novel architectural and regulatory features of a giant virus. *J. Virol.* **2012**, *86*, 8821–8834. [CrossRef] [PubMed]

78. Jeanniard, A.; Dunigan, D.D.; Gurnon, J.R.; Agarkova, I.V.; Kang, M.; Vitek, J.; Duncan, G.; McClung, O.W.; Larsen, M.; Claverie, J.-M.; *et al.* Towards defining the chloroviruses: A genomic journey through a genus of large DNA viruses. *BMC Genomics* **2013**, *14*, 158. [CrossRef] [PubMed]

79. Tomita, M.; Kikuchi, A.; Kobayashi, M.; Yamaguchi, M.; Ifuku, S.; Yamashoji, S.; Ando, A.; Saito, A. Characterization of antifungal activity of the GH-46 subclass III chitosanase from *Bacillus circulans* MH-K1. *Antonie Leeuwenhoek* **2013**, *104*, 737–748. [CrossRef] [PubMed]

80. Kouzai, Y.; Mochizuki, S.; Saito, A.; Ando, A.; Minami, E.; Nishizawa, Y. Expression of a bacterial chitosanase in rice improves disease resistance to the rice blast fungus *Magnaporthe oryzae. Plant Cell Rep.* **2012**, *31*, 629–636. [CrossRef] [PubMed]

81. El Ouakfaoui, S.; Potvin, C.; Brzezinski, R.; Asselin, A. A *Streptomyces* chitosanase is active in transgenic tobacco. *Plant Cell Rep.* **1995**, *15*, 222–226. [CrossRef] [PubMed]

82. Kilani-Feki, O.; Frikha, F.; Zouari, I.; Jaoua, S. Heterologous expression and secretion of an antifungal *Bacillus subtilis* chitosanase (CSNV26) in *Escherichia coli. Bioproc. Biosyst. Eng.* **2013**, *36*, 985–992. [CrossRef] [PubMed]

marine drugs

MDPI

Article

The Effect of the Molecular Architecture on the Antioxidant Properties of Chitosan Gallate

Chunhua Wu [1,2], Liping Wang [1], Zhongxiang Fang [3], Yaqin Hu [1,*], Shiguo Chen [1], Tatsuya Sugawara [2] and Xingqian Ye [1]

[1]   College of Biosystems Engineering and Food Science, Fuli Institute of Food Science,
      Zhejiang Key Laboratory for Agro-Food Processing, Zhejiang R & D Center for Food Technology and
      Equipment, Zhejiang University, Hangzhou 310058, China; chwu0283@163.com (C.W.);
      zacamille@163.com (L.W.); chenshiguo210@163.com (S.C.); psu@zju.edu.cn (X.Y.)
[2]   Division of Applied Biosciences, Graduate School of Agriculture, Kyoto University, Kyoto 6068502, Japan;
      sugawara@kais.kyoto-u.ac.jp
[3]   Faculty of Veterinary and Agricultural Sciences, the University of Melbourne, Parkville,
      Victoria 3010, Australia; zhongxiang.fang@unimelb.edu.au
*   Correspondence: yqhu@zju.edu.cn; Tel.: +86-571-8898-2155

Academic Editors: Hitoshi Sashiwa and David Harding
Received: 13 March 2016; Accepted: 9 May 2016; Published: 13 May 2016

Abstract: To elucidate the structure–antioxidant activity relationships of chitosan gallate (CG), a series of CG derivatives with different degrees of substitution (DS's) and molecular weights (MWs) were synthesized from chitosan (CS) and gallic acid (GA) via a free radical graft reaction. A higher MW led to a lower DS of CG. The structures of CG were characterized by FT-IR and ^{1}H NMR, and results showed that GA was mainly conjugated to the C-2 and C-6 positions of the CS chain. The antioxidant activity (the DPPH radical scavenging activity and reducing power) were enhanced with an increased DS and a decreased MW of CG. A correlation between antioxidant activities and the DS and MW of CG was also established. In addition, a suitable concentration (0~250 µg/mL) of CG with different MWs (32.78~489.32 kDa) and DS's (0~92.89 mg·GAE/g CG) has no cytotoxicity. These results should provide a guideline to the application of CG derivatives in food or pharmacology industries.

Keywords: chitosan gallate; molecular architecture; antioxidant activity; grafting; gallic acid

---

1. Introduction

Free radicals closely associated with reactive oxygen species (ROS's) can cause oxidative damage to tissue and organs in biological systems, which subsequently triggers many diseases and ailments in humans (e.g. aging, cardiovascular disease, ischemic injuries, and cancer) and food deterioration [1,2]. Antioxidants obtained from natural or synthetic compounds are able to reduce or retard the rate of oxidative damage caused by ROS's in a system. With the increasing health consciousness of consumers, natural antioxidants isolated from plants, marine creatures, and microorganisms have gained great interest [3–5].

Chitosan (CS), the second most naturally abundant polysaccharide after cellulose, is a linear and natural cationic copolymer consisting of randomly distributed β-(1→4) linked N-acetyl-D-glucosamine (GlcNAc) and D-glucosamine (GlcN) units [6–8]. The unique structure of CS is produced by the deacetylation of chitin, naturally occurring biopolymers in the shells of insects, crustaceans (such as crabs and shrimp), and the cell walls of fungi [9,10]. Due to its nontoxic, biodegradable, biocompatible, and antioxidantive properties, CS and its derivatives have received wide attention as a functional biopolymer for diverse applications, such as pharmaceutical and food packaging material [11–13]. It is suggested that these functions are dependent upon not only their chemical structure (such as introducing water-soluble entities, hydrophilic moieties, bulky and hydrocarbon groups, *etc.*) but also the molecular size [4,5,14].

It was reported that CS has radical scavenging activity on the DPPH, superoxide, hydroxyl radicals, and carbon-centered free radicals [15]. However, CS is only soluble in a few dilute acid solutions, which limits its applications [6,7,16]. Furthermore, poor H-atom-donating ability of CS that leads to it has not been able to serve as a good chain breaking antioxidant [17,18]. To overcome these disadvantages, several natural antioxidative agents have been grafted to the CS chain to enhance its functional activities [7,19–21].

Gallic acid (GA) is a well-known natural phenolic acid with strong antioxidant activities extractable from plants, especially from green tea [22]. Grafting of GA to CS has already been accomplished via enzymatic grafting reactions (tyrosinase, laccase, or peroxidase) [23,24] and carbodiimide (EDC)-mediated ester reactions [20,25,26]. However, these methods are either time-consuming or contain toxic compounds that are unsuitable for use as food supplements or nutraceuticals. Compared with other conventional modifications, a H_2O_2/ascorbic acid (Vc) grafting reaction is an eco-friendly grafting procedure because toxic products are not generated, and it is possible to perform these reactions at room temperature to avoid degradation of antioxidants. Moreover, the preparation procedure is relatively simple in comparison to the above two methods [1,17,27,28]. However, it was noted that the CS MW plays a key role in designing copolymers, which has not been well elucidated in synthesized chitosan gallates (CGs) via redox pair systems in previous studies [17,28].

The molecular architecture information on the antioxidant properties of CG has never been discussed. In this study, CG derivatives with different MWs and degrees of substitution (DS's) were prepared by a H_2O_2/Vc redox pair system, and the reaction conditions were optimized. The effect of molecular architecture (MW and DS) on the antioxidant properties of CG was also investigated in order to understand the structure-activity relationships of the CG antioxidant derivatives. The potential toxicity of the derivatives was determined against HepG2 cells, which could be a guideline to the application of CG derivatives in suitable industries.

2. Results and Discussion

2.1. Preparation of CG Derivative

In this study, GA was successfully grafted onto CS chains by using a H_2O_2/Vc peroxide redox pair as radical initiators under nitrogen protection. The possible mechanism for the synthesis of CG derivatives is shown in Figure 1. The Hydroxyl radical (HO•) generated by the oxidation of Vc by H_2O_2 attacks H-atoms in R-methylene (CH_2) or NH_2 groups, hydroxyl groups (OH) of the hydroxymethylene group or of the CS chain, producing CS macro radicals [17,28]. Then, GA molecules that are in close vicinity of the reaction site become acceptors of CS macro radicals; thus, CG derivatives are formed [1,28]. Theoretically, the synthesis route is simple and does not generate toxic reaction products.

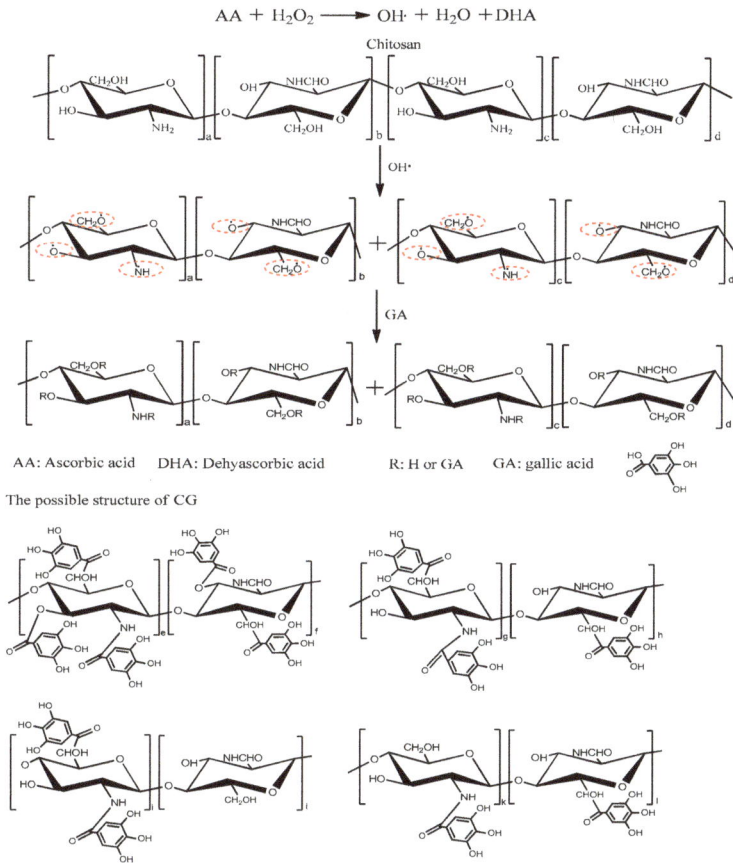

Figure 1. The proposed mechanisms for the synthesis of chitosan gallates (CGs) by free radical mediated graft copolymerization.

2.2. Effect of Reaction Conditions on the Degree of Substitution (DS)

2.2.1. Effect of the Initial Ratio of GA:CS and the MW of CS on DS

As expected, the DS values of the CG samples increased with the increase of the GA:CS ratio (Figure 2a), which could be due to an accumulation of GA monomer molecules at the close proximity of the CS backbone [28]. However, at higher molar ratios, the DS did not increase correspondingly. The GA could be saturated in the system, or the reaction could have become a dynamic chemical equilibrium. Therefore, the further increase of the molar ratio showed no influence on the DS.

It is also noted that the DS clearly increased with the decrease of CS molecule weight. It was reported that the bioactivity of CS is strongly dependent on inter- and intra-molecular hydrogen bonds [14]. LMW CS has lower hydrogen bonds than M- and HMW CS; thus, LMW CS is prone to chemical modification [16]. The maximum DS was obtained in LMW CS-GA-1, which was 112.64 ± 1.03 mg· GAE/g CG.

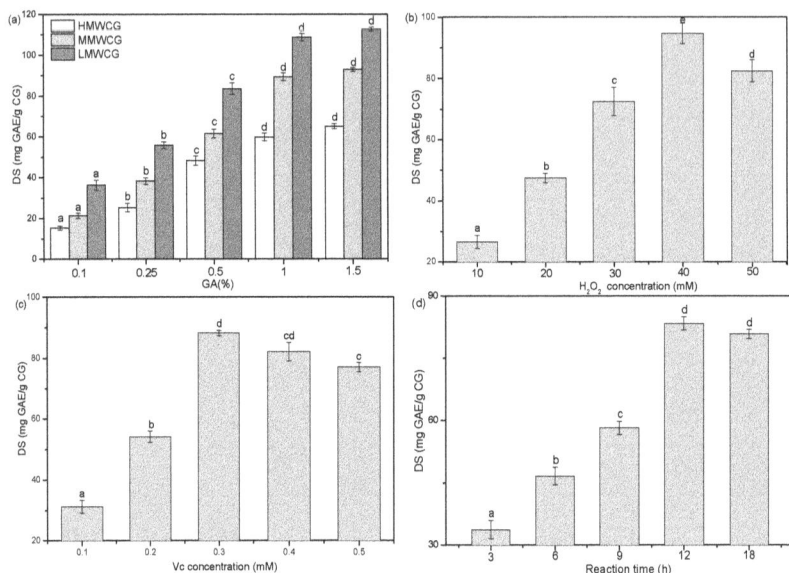

Figure 2. The effect of reaction conditions on degree of substitution (DS): (**a**) ratio of gallic acid (GA) to chitosan (CS) and molecular weight (MW) of CS on DS (t = 12 h, 20 mM H_2O_2, 0.3 mM Vc); (**b**) concentration of H_2O_2 on DS (MMWCG, t = 12 h, GA/CS = 1, 0.3 mM Vc); (**c**) concentration of Vc on DS (MMWCG, GA/CS = 1, 20 mM H_2O_2 and t = 12 h); and (**d**) reaction time on DS (MMWCG, GA/CS = 1, 20 mM H_2O_2 and 0.3 mM Vc).

2.2.2. Effect of the Concentration of H_2O_2 and Vc on DS

As presented in Figure 2b, the DS increased from 26.62 to 94.81 mg·GAE/g CG at H_2O_2 concentrations between 10 mM and 40 mM, but then decreased to 82.49 mg·GAE/g CG over 40 mM H_2O_2. A similar trend was also observed within the concentration range of ascorbic acid (0.1–0.5 mM) (Figure 2c). It is well known that the Vc is easily oxidized by H_2O_2 and generates hydroxy radicals (HO•) [17]. The enhancement of H_2O_2 or Vc concentrations in the grafting system would produce more HO• free radicals. These radicals could further react with the CS backbones to form CS macro radicals. The more CS macro radicals formed, the higher the DS obtained. However, it was also suggested that the presence of too many HO• would stop the growing grafted chain by oxidative termination or degrade the CG molecular chain for more severe reaction conditions [28,29]. Therefore, the optimal H_2O_2 and Vc concentration in this grafting system was 40 mM and 0.3 mM, respectively.

2.2.3. The Effect of Reaction Time on the DS

As shown in Figure 2d, the DS values of CG samples increased rapidly from 33.69 to 83.34 mg·GAE/g CG when the reaction time ranged from 3 to 12 h, and then decreased slightly between 12 and 18 h. Prolonging the reaction duration implied that there was more time on diffusion and absorption of the GA molecular to the active center of CS macro radicals, and more CG molecules were synthesized. The GA became saturated with CS macro radicals at 12 h, which could be the highest DS level. In addition, a longer reaction time would accelerate the degradation of CS or CG, which may be harmful to the final produce. Therefore, a further extension of reaction time did not increase the DS.

For further insight on the effect of the MW and DS on the antioxidant properties of CG, some CG derivatives were prepared according to the reaction conditions (based on the above factor analysis) in Table 1.

Table 1. Physicochemical properties of CS derivatives.

Samples	CS		Reaction Conditions	DS (mg·GAE/g CG)	CG	
	MW/kDa	PDI			MW/kDa	PDI
MWCG-1	98.67 ± 4.15	1.82 ± 0.13	GA/CS = 0.5, *t* = 15 h, 40 mM H_2O_2, 0.4 mM Vc	73.21 ± 1.60	32.78 ± 1.35	1.61 ± 0.16
MWCG-2	98.67 ± 4.15	1.82 ± 0.13	GA/CS = 1, *t* = 12 h, 20 mM H_2O_2, 0.3 mM Vc	72.57 ± 2.02	78.37 ± 2.38	1.35 ± 0.08
MWCG-3	211.59 ± 6.89	1.90 ± 0.21	GA/CS = 1, *t* = 9 h, 20 mM H_2O_2, 0.3 mM Vc	72.93 ± 2.37	183.13 ± 3.27	1.42 ± 0.13
MWCG-4	508.40 ± 5.67	1.85 ± 0.14	GA/CS = 1, *t* = 15 h, 50 mM H_2O_2, 0.5 mM Vc	73.08 ± 2.17	275.92 ± 3.25	1.61 ± 0.14
MWCG-5	508.40 ± 5.67	1.85 ± 0.14	GA/CS = 1, *t* = 12 h, 25 mM H_2O_2, 0.3 mM Vc	74.33 ± 1.49	489.32 ± 4.62	1.57 ± 0.12
CS	211.59 ± 6.89	1.90 ± 0.21	GA/CS = 0, *t* = 12 h, 20 mM H_2O_2, 0.3 mM Vc	0	182.13 ± 3.27	1.44 ± 0.07
DSCG-1	211.59 ± 6.89	1.90 ± 0.21	GA/CS = 0.1, *t* = 12 h, 20 mM H_2O_2, 0.3 mM Vc	21.37 ± 1.26	184.46 ± 1.59	1.38 ± 0.14
DSCG-2	211.59 ± 6.89	1.90 ± 0.21	GA/CS = 0.25, *t* = 12 h, 20 mM H_2O_2, 0.3 mM Vc	38.25 ± 2.03	186.13 ± 3.27	1.45 ± 0.06
DSCG-3	211.59 ± 6.89	1.90 ± 0.21	GA/CS = 0.5, *t* = 12 h, 20 mM H_2O_2, 0.3 mM Vc	61.42 ± 2.16	188.89 ± 3.83	1.50 ± 0.19
DSCG-4	211.59 ± 6.89	1.90 ± 0.21	GA/CS = 1, *t* = 12 h, 20 mM H_2O_2, 0.3 mM Vc	92.89 ± 0.93	191.52 ± 2.64	1.43 ± 0.11

MW: Molecular Weight; PDI: Polydispersity Index.

As shown in Table 1, to obtain similar DS's of CG derivatives with varied MWs, three different MWs of CS were applied in the grafting reaction, according to the above factor analysis. A longer reaction time (15 h) and higher catalyst concentration led to a more serious degradation in the MW of CG; thus, a lower MW of a CG derivative was obtained. In addition, the different DS's of CG derivatives were gained from the analysis in Section 2.2.1.

Based on the results of MWs and DS's in various reaction conditions, it could be concluded that grafting and degradation occur simultaneously during the reaction process (as shown in Table 1). The degradation of CS molecules in other chemically modified treatments was also observed [30,31]. This may be attributed to the degradation effect of oxidation or free radicals on CS molecules [32,33]. However, as the reactions were conducted under a nitrogen atmosphere, which had excluded oxygen from these reactions, a free radical degradation process could have happened in this system. As depicted in Figure 1, a HO• is not only combined with CS to form macro radicals, but is also quickly pulled off a hydrogen atom from the CS chain and combines with it to form a water molecule which degrades the CS chain as follows:

$$(GlcN)\,m - (GlcN)\,n + HO\bullet \rightarrow (GlcN)\,m + (GlcN)\,n + H_2O.$$

On the other hand, as the GA was covalently attached onto the CS backbone, the MW of the CG gradually increased with the increase of the DS, but not significantly ($p > 0.05$). The data from Table 1 also suggests that a series of CG samples has a narrow MW dispersity (lower PDI). These results demonstrated that estimating the MWs of CG products based upon the initial MW of chitosan could be misleading.

2.3. Characterization of CG

In order to confirm the chemical structure of CG, the FT-IR spectra of samples were recorded. The main characteristic peaks of CS at 3393 cm^{-1} (O–H stretch), 2899 cm^{-1} (C–H stretch), 1550 cm^{-1} (N–H bend), 1327 cm^{-1} (C–N stretch) are shown in Figure 3a. There is a weak absorption peak of amide at 1643 cm^{-1} (representing C=O groups of amide), indicating a very high deacetylation degree of CS. In addition, three strong peaks at 1030, 1076, and 1155 cm^{-1}, which were characteristic peaks of the saccharide structure, were also observed in the IR spectrum of CS [34,35]. Compared to the FT-IR spectrum of CS, the peak at 1550 cm^{-1} (N–H bending of the primary amine) of MWCG-1 was weaker, and the peak at 1638 cm^{-1} (C=O groups of amide) was enhanced, indicating that amide linkage between NH$_2$ of CS and –COOH of GA were formed. In addition, a new peak at 1732 cm^{-1} corresponding to the C=O stretching of the carbonyl group was observed in MWCG-1 samples, suggesting the formation of a ester bond between –OH (at C-3 and/or C-6) of the CS chain and –COOH of GA. Due to the steric hindrance of C-3 position of pyranose ring, the possibility of substitution of GA at C-3 was very low [17,28]. Therefore, the gallyl group of GA most likely interacted with the active hydrogen of NH$_2$ at C-2 (amide linkages) and the OH groups at C-6 position (ester linkages) of the CS chain. Similar results have been reported by Liu *et al.* (2013) [28], Cho *et al.* (2011) [1], and Spizzirri *et al.* (2010) [36].

The molecular structure of MWCG-1 was further confirmed by using ^{1}H NMR spectroscopy. As shown in Figure 3b, the CS spectrum exhibits two typical signals at δ 2.96 and 1.88 ppm due to the H-2 proton of the GlcN and N-acetyl protons of GlcNAc, respectively; the multiplet at δ 4.51, δ 3.88–3.3 ppm are attributed to H-1 and H-3 to H-6 of the CS backbone. For the ^{1}H NMR spectra of MWCG-1, it retains the characteristic signals of the parent CS; however, the chemical shifts of H-2, H-3, and H-6 of MWCG-1 were shifting to 2.86, 3.86, and 3.26 ppm, respectively, demonstrating that the substitution of GA occurred at positions C-3, C-6, or C-2. It was noted that a new signal appeared at 6.92 ppm (assigned to the phenyl protons of GA), which confirms the attachment of the phenyl group to the polymer chain. This result was consistent with that of others [1,28,36].

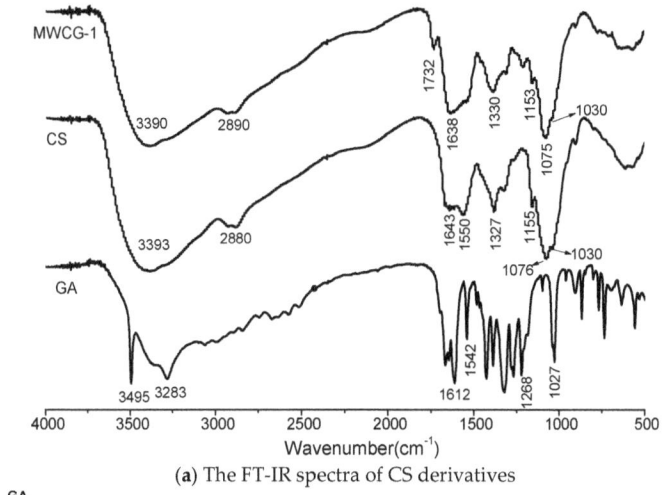

(a) The FT-IR spectra of CS derivatives

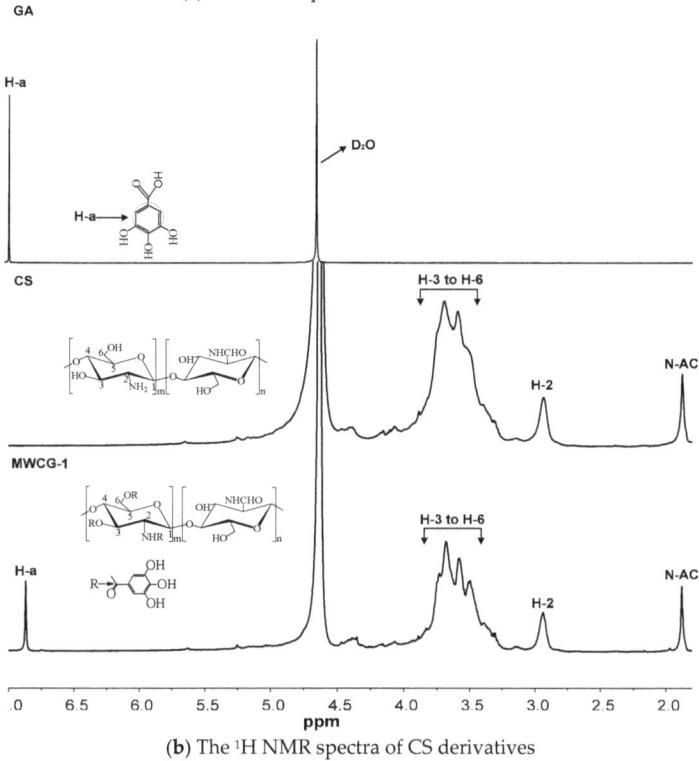

(b) The ¹H NMR spectra of CS derivatives

Figure 3. *Cont.*

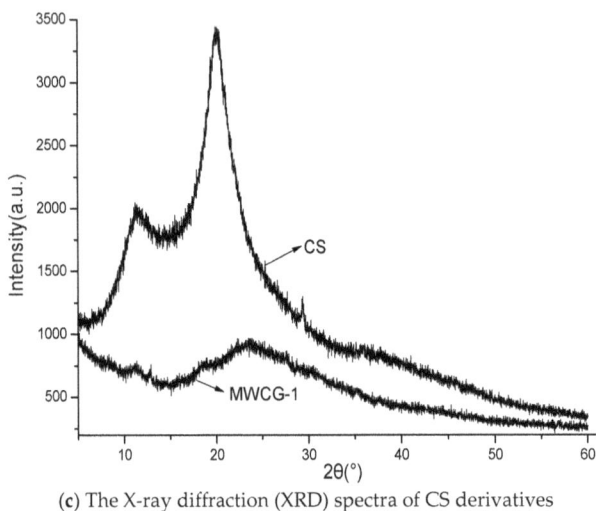

(c) The X-ray diffraction (XRD) spectra of CS derivatives

Figure 3. FT-IR spectra (**a**); ^{1}H NMR (**b**) and XRD spectra (**c**) of CS derivatives.

The crystallographic structure of MWCG-1 was determined by X-ray diffraction (XRD). As depicted in Figure 3c, two typical peaks were detected around 2θ = 10.4° and 22.1° in CS, which were assigned to crystal form I and crystal forms II, respectively. However, the XRD spectrum of MWCG-1 has much smaller peaks at around 2θ = 10.6° and 22.2°, confirming the interaction of CS with grafted GA. This result demonstrated that the incorporation of GA to the CS molecular chain reduced the crystallization of CS to some extent, suggesting that CS and GA chains were mixed well at a molecular level. This might be attributed to the fact that the intramolecular hydrogen bonding of CS had greatly decreased after grafting the GA group. As a result, the solubility of the MWCG-1 could be better than that of CS. Similar discussions for the changes of crystal structure and solubility of the CS derivatives have been found in the literature [26,28].

2.4. Antioxidant Assessments

2.4.1. Effect of Molecular Weight on Antioxidant Activity

As presented in Figure 4a, the scavenging activity of several different MW CG samples (100–2000 μg/mL) on DPPH radicals was significant and concentration-related. The scavenging rate of these CG samples increased with increasing concentration. The increase in concentration of CG resulted in the increase of total amine groups responsible for scavenging more radicals [37]. The IC50 values for MWCG-1, MWCG-2, MWCG-3, MWCG-4, and MWCG-5 were 148, 233, 305, 518, and 774 μg/mL, respectively, suggesting an inverse relationship between DPPH scavenging activity and the MW of CG (Figure 4b). The scavenging activities of LMW CG (MWCG-1 and MWCG-2) on DPPH radical were more pronounced than that of HMW CG (MWCG-4 and MWCG-5). The effect of the MW on CG scavenging activity might be attributed to the inter- and intra-molecular hydrogen bond of CS, which influences its biological activity [14]. CS has many hydrogen bonds on N2-O6 and O3-O5. HMW CG would have lower molecular mobility than the LMW CG, which would increase the possibility of inter- and intramolecular bonding among the HMW CG molecules [5]. Therefore, the chance of exposure of their amine, hydroxy, or GA groups might be restricted, which would have accounted for less radical-scavenging activity.

In terms of a reducing power test, the reducing power of CG samples correlated well with increasing concentrations and the change in reducing power for LMW CG (MWCG-1 and MWCG-2) were larger than that of HMW CG (MWCG-3 and MWCG-4) (Figure 4c). It indicates that LMW CG has a higher reducing power than that of HMW CG. Moreover, good positive correlations were observed between the reducing power and the MW of CG samples (Figure 4d), suggesting that CG with a lower MW would have relatively strong reducing power.

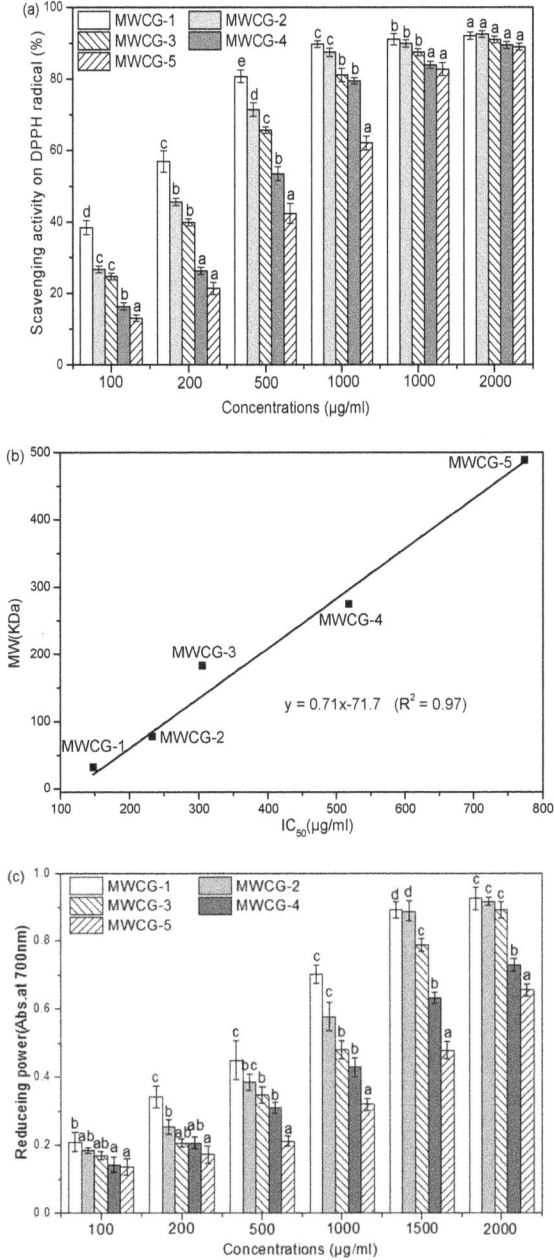

Figure 4. *Cont.*

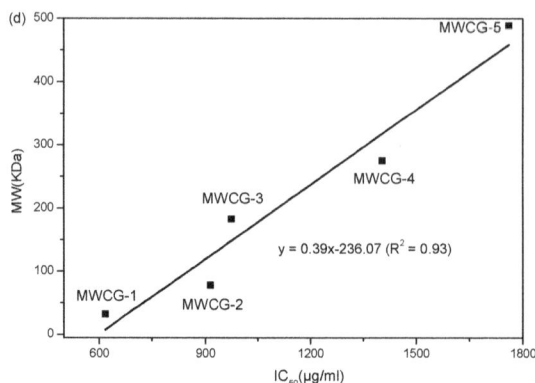

Figure 4. The effect of MW on antioxidant activity of CG samples. (**a**) Effect of MW and concentration on the DPPH scavenging activity of CG samples; (**b**) Relationship between MW and DPPH scavenging activity of CG samples; (**c**) Effect of MW and concentration on the reducing power of CG samples; (**d**) Relationship between MW and reducing power of CG samples.

2.4.2. Effect of the DS on Antioxidant Activity

As shown in Figure 5a, the DPPH scavenging activity of CS and CG samples was also a concentration-dependent manner. The scavenging rates increased with their increasing concentrations. IC_{50} of CG samples (111~945 µg/mL) were lower than that of CS samples (>2000 µg/mL). This indicated that CG may have higher activity upon the elimination of DPPH radical than the corresponding CS samples. Furthermore, good correlations were found between the radical-scavenging activity and the DS of CG samples (Figure 5b). As the DS of CG increased from 0 to 61.42 mg·GAE/g CG, the DPPH scavenging activity was enhanced significantly ($p < 0.05$). This phenomenon might be ascribed to the strong hydrogen-donating capacity of GA and -NH_2 of CG. It is well-known that the antioxidants reduce the DPPH radical to a yellow-colored compound, diphenylpicrylhydrazine, and the extent of the reaction is dependent on the hydrogen-donating ability of the antioxidants [17]. The greater the dose of GA grafted on the CG chain, the higher the hydrogen-donating capacity, and thus the faster the scavenging on the DPPH radical. In addition, a lower DS resulted in more active amino groups in the CG chain. These active amino groups could also donate hydrogen to react with the DPPH radical [4]; therefore, a non-linear correlation between radical-scavenging activity and the DS of CG were observed.

The effect of the DS on reducing power of CS and CG samples are depicted in Figure 5c. The DS of CG showed a significant effect on reducing power activity that was proportionally increased by the GA content of CG. This suggested that the capacity of CG for reducing Fe^{3+} to Fe^{2+} was closely related to GA content. The reducing properties were generally associated with the presence of reductones, which have been shown to exert antioxidant action by breaking the free radical chain through the donation of a hydrogen atom [4,38]. Thus, the increased reducing power of the CG samples might be due to the excellent hydrogen-donating ability of the GA content. In addition, the reducing power of all types of CS and CG was correlated well with their increasing concentrations (Figure 5d).

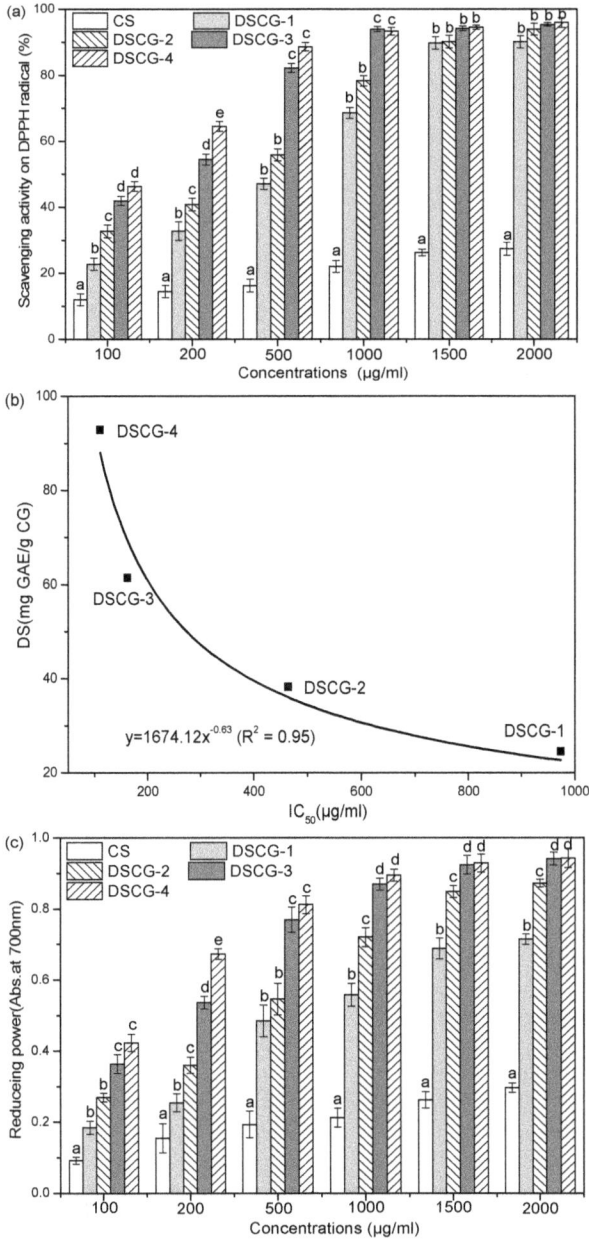

Figure 5. *Cont.*

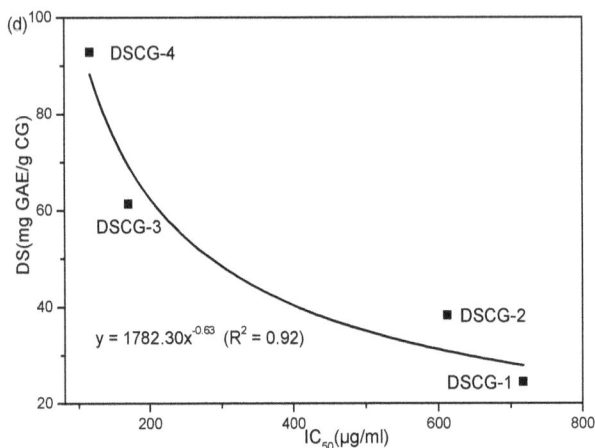

Figure 5. The effect of DS on antioxidant activity of CG samples. (**a**) Effect of DS and concentration on the DPPH scavenging activity of CG samples; (**b**) Relationship between DS and DPPH scavenging activity of CG samples; (**c**) Effect of DS and concentration on the reducing power of CG samples; (**d**) Relationship between DS and reducing power of CG samples.

Overall, the above results indicate that the antioxidant activity of the CG samples was closely related to their MW and DS and that the influence of the DS was greater than the MW.

2.5. Cytotoxicity Assessments

MTT assays were performed to test the effects of CG copolymers on the metabolic activity of cells. As shown in Figure 6, the cytotoxicity of CG derivatives was dependent on its DS and MW. With the same DS and concentration, the cytotoxicity of CGs increased with its increasing MW. The compound of MWCG-5 was particularly toxic with an IC_{50} of 275 µg/mL, whereas MWCG-1 only exhibited cytotoxicity at a high concentration (1429 µg/mL). The influence of MW on the cytotoxicity of CG derivatives could be explained by the fact that the interaction of cationic molecules with plasma membranes increases with increasing MW, due to multiple attachments to cell surfaces [39]. A similar increase in cytotoxicity with increasing MW was observed for polylysine and poly (amidoamines) [40]. Moreover, an exponential relationship between MW and IC_{50} after 24-h incubation was established (Figure 6b), which could be used to predict the cytotoxicity of different MW CG derivatives.

The effect of the DS of CG derivatives on cell viability is presented in Figure 6c. The cytotoxicity of CG derivatives had no significant difference ($p > 0.05$) with a DS value of 21.37 and 38.25 mg· GAE/g CG at concentrations between 50 µg/mL and 200 µg/mL. However, as the DS further increased, the cell viability gradually decreased. A negative correlation was also observed between the cytotoxicity and the DS of CG samples (Figure 6d). In addition, the cell viability of CG derivatives was concentration-dependent. The higher concentration of CG derivatives with stronger cytotoxic effects might be due to the changes in osmotic pressure of the polymer solutions compared to cell culture medium [41].

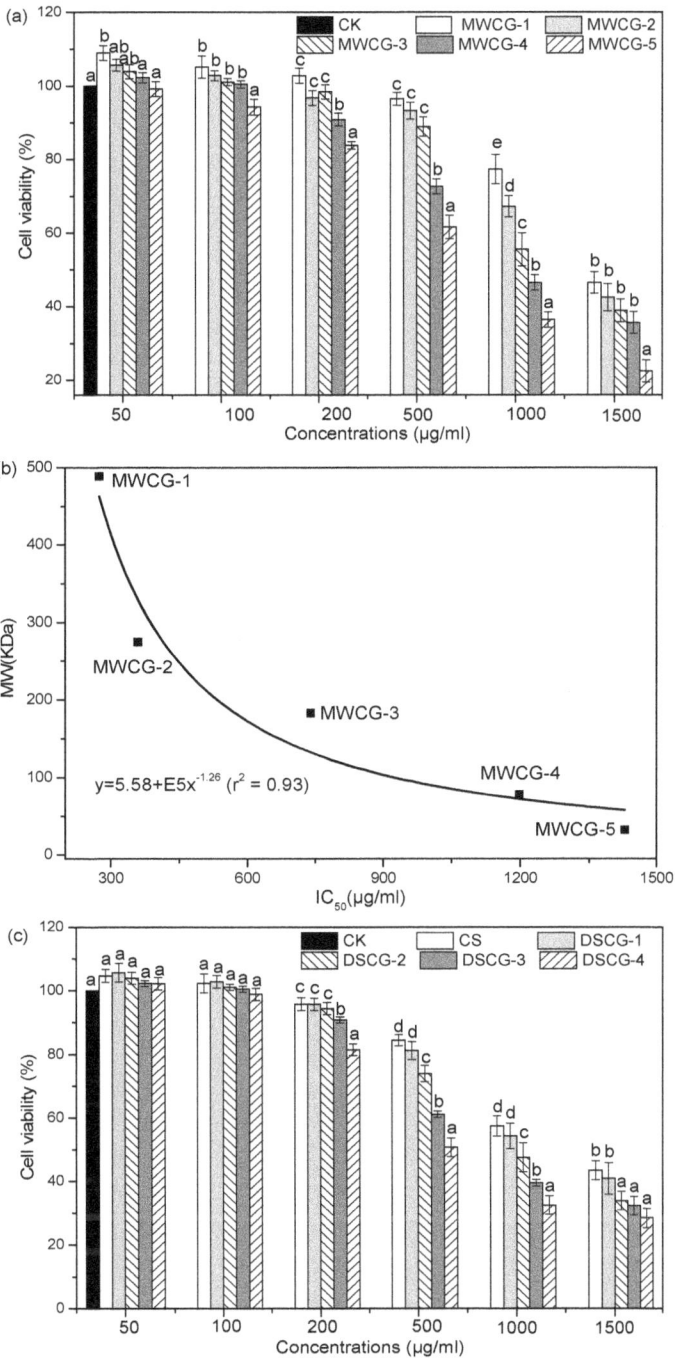

Figure 6. *Cont.*

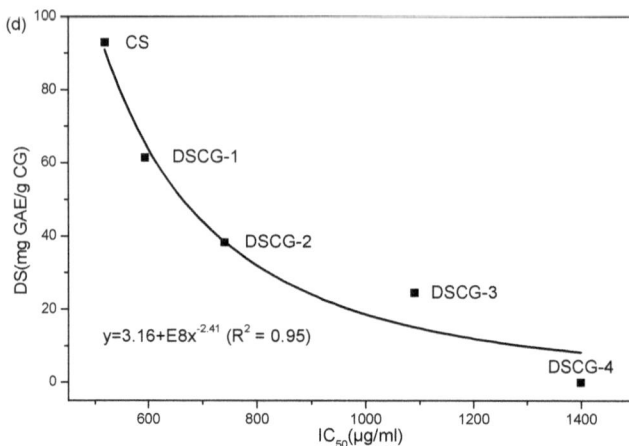

Figure 6. Cytotoxicity of the polymers at various concentrations.

3. Materials and Methods

3.1. Materials

Chitosan from shrimp with the MWs of 518.40 kDa, 211.59 kDa, and 98.67 kDa (coded as HMW-CS, MMW-CS, and LMW-CS, respectively), and a deacetylation degree of approximately 92%, was purchased from Qingdao Yunzhou Biochemistry Co. Ltd. (Qingdao, China). Chemicals of 2,2-diphenyl-1-picrylhydrazyl (DPPH), 3-(4,5-dimethylthiazol-2-yl)-2,5-diphenyltetrazolium bromide (MTT), gallic acid (GA), H_2O_2, Vc, Folin–Ciocalteu reagent, and D_2O were purchased from Sigma Chemical Co. (St. Louis, MO, USA). All other reagents were of analytical grade.

3.2. Preparation of CG Derivative

The CG derivative was prepared by using a H_2O_2/ascorbic acid redox pair under a nitrogen atmosphere according to our previous study [42]. Briefly, the CS (1 g) was dissolved in 100 mL of 1% acetic acid (v/v) in 200 mL three-necked round bottom flask. Then, a certain amount of H_2O_2 and Vc was added into the reactor, and a slow stream of oxygen-free nitrogen gas was passed for 30 min with stirring. Afterwards, GA was added to the mixture at different molar ratios of the repeating unit of CS. The reaction was allowed to proceed at different CS MWs (98.67, 211.59, and 508.40 kDa), H_2O_2 concentrations (10, 20, 30, 40, and 50 mM), Vc concentrations (0.1, 0.2, 0.3, 0.4, and 0.5 mM), ratios of CS:GA (1:0.1, 1:0.25, 1:0.5, 1:1, and 1:1.5), and times (3, 6, 9, 12, and 18 h). The reaction was stopped by letting air into the reactor and then dialyzed with distilled water using an 8–14 kDa MW cut off membrane for 72 h to remove unreacted GA. Finally, the dialyzate was lyophilized to obtain a water-soluble CG derivative.

3.3. Characterization of CG Derivatives

Structural characterization of the blank CS and CG were performed by gel permeation chromatography-multiple-angle laser light scattering (GPC-MALLS), Fourier transform infrared spectroscopy (FT-IR), proton nuclear magnetic resonance (^{1}H NMR) and X-ray diffraction (XRD) analysis. The molecular weights of the CS derivatives were analyzed by GPC with a MALLS detector (Dawn DSP, Wyatt Technology Corp., California, USA). All chitosan samples were dissolved in MQ water (5 mg/mL), filtered through a 0.22-mm syringe filter (Millipore Corp., Billerica, USA), and injected onto a TSK 3000 PWXL column. The samples were then eluted using 0.2 M ammonium acetate (pH 4.5) at a flow rate of 0.5 mL/min. The FT-IR was determined using an AVATAR 370 spectrophotometer (Thermo Nicolet Corporation, Madison, WI, USA) by scanning from 400 to 4000 cm^{-1}. ^{1}H NMR spectra were recorded at 25 °C with samples dissolved in CD$_3$COOD/D$_2$O (v/v, 1%) using a 600 MHz NMR spectrometer (Bruker Inc., Rheinstetten, Germany). The crystallographic structures of the CG derivatives were determined by a Bruker AXS D8 Advance X-ray diffractometer (Bruker Inc., Rheinstetten, Germany) using Ni-filtered Cu Kα radiation. The degree of substitution (expressed as the DS which is defined as the GA content in CG derivatives) was measured by the Folin–Ciocalteu method according to Liu *et al.* [28]. GA was used to calculate the standard curve, and the DS was expressed as milligrams of GAE per gram of the dry weight copolymer (mg· GAE/g CG).

3.4. Antioxidant Assessments

The antioxidant activities of CS and CG derivatives were evaluated using 2,2-diphenyl-1-picrylhydrazyl (DPPH) radical scavenging and reducing power assays.

The DPPH radical scavenging activity was estimated according to the previous method [20] with some modifications. Briefly, 200 µL of DPPH solution (0.4 mM DPPH in methanol) was mixed with 50 µL of samples (0.05–2 mg/mL) in a 96-well plate. The mixture was shaken vigorously and allowed to stand at room temperature for 0.5 h in the dark. Then, the absorbance of the mixture was measured at 517 nm by a microplate reader. The DPPH radical scavenging activity was calculated as followed:

$$\text{Scavenging activity (\%)} = \left[1 - \frac{A_1 - A_2}{A_0}\right] \times 100,$$

where A_0 represents the absorbance of the control (water instead of sample), A_1 represents the absorbance of the samples, and A_2 represents the absorbance of the samples only (water instead of DPPH). The IC$_{50}$ value was reported, which represents the concentration of the compounds that cause 50% inhibition of DPPH radical formation.

The reducing power was determined according to the previously described method [7] with some modifications. The reaction were carried out on 96-well plates, with each well containing a mixture of 50 µL of sample solution, sodium phosphate buffer (PBS, 0.2 M, pH 6.6), and K$_3$Fe(CN)$_6$ solution (1%, w/v), and were incubated at 50 °C for 20 min. After the addition of 50 µL of trichloroacetic acid (10%, w/v) and 30 µL of fresh FeCl$_3$ (0.1%, w/v), the absorbance was measured at 700 nm. The IC$_{50}$ value was reported, which represents the concentration of the compounds that generated 0.5 of absorbance.

3.5. Cytotoxicity Assessments

The cytotoxicity of CG derivatives was evaluated by a MTT assay using HepG2 cells according to our previous studies [43]. Cytotoxicity showing the cell viability rate was calculated by the following equation:

$$\text{Viable cell (\%)} = \frac{OD_{samples}}{OD_{control}} \times 100\%,$$

where OD$_{sample}$ and OD$_{control}$ were obtained in the presence or absence of CG derivatives, respectively.

3.6. Statistical Analysis

Analysis of variance was performed using ANOVA procedures of the IBM SSPS software (version 20.0, IBM Inc., Chicago, IL, USA). Duncan's test was used to determine the difference of means, and $p < 0.05$ was considered to be statistically significant.

4. Conclusions

In this work, a series of antioxidant copolymers based on chitosan gallate were fabricated via a free radical graft reaction in a H_2O_2/Vc redox system. The effect of the molar ratio of GA to CS, the molecular weight (MW) of CS, the concentration of H_2O_2 and Vc, and the reaction time on degrees of substitution (DS's) was investigated. The structures of CG were characterized by FT-IR, ^{1}H NMR, and XRD, which showed that GA was conjugated to the C-2 and C-6 positions of the CS chain. However, GPC analysis indicated that the grafting reaction was accompanied with a degradation of the CS molecule. The antioxidant assay indicated that the molecular architecture (various MWs and DS's) of CG samples had crucial effects on their DPPH radical scavenging activity and reducing power, and the influence of the DS on CG samples was greater than MW. In addition, the MTT assay showed no cytotoxicity for the CGs at a suitable concentration (0~250 µg/mL) with different MWs (32.78~489.32 kDa) and DS's (0~92.89 mg·GAE/g CG). The results suggested that CG derivatives with a varying molecular architecture have the potential to be used as effective antioxidants in the pharmaceutical and food industries.

Acknowledgments: The authors would like to thank the support of the Key Project of National Science and Technology Ministry of China (2012BAD38B09) and the Scientific Project of Zhejiang Province (2014C02017).

Author Contributions: All authors conceived and designed the experiments; Chunhua Wu and Liping Wang performed experiments; Chunhua Wu, Zhongxiang Fang, and Yaqin Hu analyzed the data; all authors wrote the paper.

Conflicts of Interest: The authors declare no conflict of interest. The founding sponsors had no role in the design of the study; in the collection, analyses, or interpretation of data; in the writing of the manuscript; or in the decision to publish the results.

Abbreviations

The following abbreviations are used in this manuscript:

CS	Chitosan
GA	Gallic acid
CG	Chitosan gallate
MW	Molecular weight
PDI	Polydispersity Index
HMW-CS	High molecular weight chitosan
MMW-CS	Middle molecular weight chitosan
LMW-CS	Low molecular weight chitosan
DS	Degrees of substitution
HO•	Hydroxyl radical
MTT	3-(4,5-dimethylthiazol-2-yl)-2,5-diphenyltetrazolium bromide
DPPH	2,2-diphenyl-1-picrylhydrazyl

References

1. Cho, Y.-S.; Kim, S.-K.; Ahn, C.-B.; Je, J.-Y. Preparation, characterization, and antioxidant properties of gallic acid-grafted-chitosans. *Carbohydr. Polym.* **2011**, *83*, 1617–1622. [CrossRef]
2. Apel, K.; Hirt, H. Reactive oxygen species: Metabolism, oxidative stress, and signal transduction. *Annu. Rev. Plant Biol.* **2004**, *55*, 373–399. [CrossRef] [PubMed]
3. Frei, B. *Natural Antioxidants in Human Health and Disease*; Academic Press: California, CA, USA, 2012.
4. Xing, R.; Liu, S.; Guo, Z.; Yu, H.; Wang, P.; Li, C.; Li, Z.; Li, P. Relevance of molecular weight of chitosan and its derivatives and their antioxidant activities *in vitro*. *Bioorg. Med. Chem.* **2005**, *13*, 1573–1577. [CrossRef] [PubMed]
5. Kim, K.W.; Thomas, R.L. Antioxidative activity of chitosans with varying molecular weights. *Food Chem.* **2007**, *101*, 308–313. [CrossRef]

6. Ying, G.-Q.; Xiong, W.-Y.; Wang, H.; Sun, Y.; Liu, H.-Z. Preparation, water solubility and antioxidant activity of branched-chain chitosan derivatives. *Carbohydr. Polym.* **2011**, *83*, 1787–1796. [CrossRef]

7. Liu, J.; Wen, X.Y.; Lu, J.F.; Kan, J.; Jin, C.H. Free radical mediated grafting of chitosan with caffeic and ferulic acids: Structures and antioxidant activity. *Int. J. Biol. Macromol.* **2014**, *65*, 97–106. [CrossRef] [PubMed]

8. Qinna, N.A.; Karwi, Q.G.; Al-Jbour, N.; Al-Remawi, M.A.; Alhussainy, T.M.; Al-So'ud, K.A.; Al Omari, M.M.; Badwan, A.A. Influence of molecular weight and degree of deacetylation of low molecular weight chitosan on the bioactivity of oral insulin preparations. *Mar. Drugs* **2015**, *13*, 1710–1725. [CrossRef] [PubMed]

9. Anitha, A.; Sowmya, S.; Kumar, P.T.S.; Deepthi, S.; Chennazhi, K.P.; Ehrlich, H.; Tsurkan, M.; Jayakumar, R. Chitin and chitosan in selected biomedical applications. *Prog. Polym. Sci.* **2014**, *39*, 1644–1667. [CrossRef]

10. Sahariah, P.; Gaware, V.S.; Lieder, R.; Jonsdottir, S.; Hjalmarsdottir, M.A.; Sigurjonsson, O.E.; Masson, M. The effect of substituent, degree of acetylation and positioning of the cationic charge on the antibacterial activity of quaternary chitosan derivatives. *Mar. Drugs* **2014**, *12*, 4635–4658. [CrossRef] [PubMed]

11. Van den Broek, L.A.; Knoop, R.J.; Kappen, F.H.; Boeriu, C.G. Chitosan films and blends for packaging material. *Carbohydr. Polym.* **2015**, *116*, 237–242. [CrossRef] [PubMed]

12. Chen, M.C.; Mi, F.L.; Liao, Z.X.; Hsiao, C.W.; Sonaje, K.; Chung, M.F.; Hsu, L.W.; Sung, H.W. Recent advances in chitosan-based nanoparticles for oral delivery of macromolecules. *Adv. Drug Deliv. Rev.* **2013**, *65*, 865–879. [CrossRef] [PubMed]

13. Mallick, S.; Sanpui, P.; Ghosh, S.S.; Chattopadhyay, A.; Paul, A. Synthesis, characterization and enhanced bactericidal action of a chitosan supported core–shell copper–silver nanoparticle composite. *RSC Adv.* **2015**, *5*, 12268–12276. [CrossRef]

14. Tomida, H.; Fujii, T.; Furutani, N.; Michihara, A.; Yasufuku, T.; Akasaki, K.; Maruyama, T.; Otagiri, M.; Gebicki, J.M.; Anraku, M. Antioxidant properties of some different molecular weight chitosans. *Carbohydr. Res.* **2009**, *344*, 1690–1696. [CrossRef] [PubMed]

15. Zhai, Y.; Zhou, K.; Xue, Y.; Qin, F.; Yang, L.; Yao, X. Synthesis of water-soluble chitosan-coated nanoceria with excellent antioxidant properties. *RSC Adv.* **2013**, *3*, 6833. [CrossRef]

16. Aytekin, A.O.; Morimura, S.; Kida, K. Synthesis of chitosan-caffeic acid derivatives and evaluation of their antioxidant activities. *J. Biosci. Bioeng.* **2011**, *111*, 212–216. [CrossRef] [PubMed]

17. Curcio, M.; Puoci, F.; Iemma, F.; Parisi, O.I.; Cirillo, G.; Spizzirri, U.G.; Picci, N. Covalent insertion of antioxidant molecules on chitosan by a free radical grafting procedure. *J. Agric. Food Chem.* **2009**, *57*, 5933–5938. [CrossRef] [PubMed]

18. Pasanphan, W.; Buettner, G.R.; Chirachanchai, S. Chitosan gallate as a novel potential polysaccharide antioxidant: An EPR study. *Carbohydr. Res.* **2010**, *345*, 132–140. [CrossRef] [PubMed]

19. Guo, Z.; Xing, R.; Liu, S.; Yu, H.; Wang, P.; Li, C.; Li, P. The synthesis and antioxidant activity of the Schiff bases of chitosan and carboxymethyl chitosan. *Bioorg. Med. Chem. Lett.* **2005**, *15*, 4600–4603. [CrossRef] [PubMed]

20. Xie, M.; Hu, B.; Wang, Y.; Zeng, X. Grafting of gallic acid onto chitosan enhances antioxidant activities and alters rheological properties of the copolymer. *J. Agric. Food Chem.* **2014**, *62*, 9128–9136. [CrossRef] [PubMed]

21. Lee, D.S.; Woo, J.Y.; Ahn, C.B.; Je, J.Y. Chitosan-hydroxycinnamic acid conjugates: Preparation, antioxidant and antimicrobial activity. *Food Chem.* **2014**, *148*, 97–104. [CrossRef] [PubMed]

22. Hager, A.S.; Vallons, K.J.; Arendt, E.K. Influence of gallic acid and tannic acid on the mechanical and barrier properties of wheat gluten films. *J. Agric. Food Chem.* **2012**, *60*, 6157–6163. [CrossRef] [PubMed]

23. Božič, M.; Gorgieva, S.; Kokol, V. Laccase-mediated functionalization of chitosan by caffeic and gallic acids for modulating antioxidant and antimicrobial properties. *Carbohydr. Polym.* **2012**, *87*, 2388–2398. [CrossRef]

24. Božič, M.; Štrancar, J.; Kokol, V. Laccase-initiated reaction between phenolic acids and chitosan. *React. Funct. Polym.* **2013**, *73*, 1377–1383. [CrossRef]

25. Schreiber, S.B.; Bozell, J.J.; Hayes, D.G.; Zivanovic, S. Introduction of primary antioxidant activity to chitosan for application as a multifunctional food packaging material. *Food Hydrocoll.* **2013**, *33*, 207–214. [CrossRef]

26. Pasanphan, W.; Chirachanchai, S. Conjugation of gallic acid onto chitosan: An approach for green and water-based antioxidant. *Carbohydr. Polym.* **2008**, *72*, 169–177. [CrossRef]

27. Lee, D.S.; Je, J.Y. Gallic acid-grafted-chitosan inhibits foodborne pathogens by a membrane damage mechanism. *J. Agric. Food Chem.* **2013**, *61*, 6574–6579. [CrossRef] [PubMed]

28. Liu, J.; Lu, J.F.; Kan, J.; Jin, C.H. Synthesis of chitosan-gallic acid conjugate: Structure characterization and *in vitro* anti-diabetic potential. *Int. J. Biol. Macromol.* **2013**, *62*, 321–329. [CrossRef] [PubMed]

29. Abu Naim, A.; Umar, A.; Sanagi, M.M.; Basaruddin, N. Chemical modification of chitin by grafting with polystyrene using ammonium persulfate initiator. *Carbohydr. Polym.* **2013**, *98*, 1618–1623. [CrossRef] [PubMed]

30. Cho, J.; Grant, J.; Piquette-Miller, M.; Allen, C. Synthesis and physicochemical and dynamic mechanical properties of a water-soluble chitosan derivative as a biomaterial. *Biomacromolecules* **2006**, *7*, 2845–2855. [CrossRef] [PubMed]

31. Zavaleta-Avejar, L.; Bosquez-Molina, E.; Gimeno, M.; Pérez-Orozco, J.P.; Shirai, K. Rheological and antioxidant power studies of enzymatically grafted chitosan with a hydrophobic alkyl side chain. *Food Hydrocoll.* **2014**, *39*, 113–119. [CrossRef]

32. Hsu, S.C.; Don, T.M.; Chiu, W.Y. Free radical degradation of chitosan with potassium persulfate. *Polym. Degrad. Stab.* **2002**, *75*, 73–83. [CrossRef]

33. Qin, C.Q.; Du, Y.M.; Xiao, L. Effect of hydrogen peroxide treatment on the molecular weight and structure of chitosan. *Polym. Degrad. Stab.* **2002**, *76*, 211–218. [CrossRef]

34. Bian, Y.; Gao, D.; Liu, Y.; Li, N.; Zhang, X.; Zheng, R.Y.; Wang, Q.; Luo, L.; Dai, K. Preparation and study on anti-tumor effect of chitosan-coated oleanolic acid liposomes. *RSC Adv.* **2015**, *5*, 18725–18732. [CrossRef]

35. Masoomi, M.; Tavangar, M.; Razavi, S.M.R. Preparation and investigation of mechanical and antibacterial properties of poly(ethylene terephthalate)/chitosan blend. *RSC Adv.* **2015**, *5*, 79200–79206. [CrossRef]

36. Spizzirri, U.G.; Parisi, O.I.; Iemma, F.; Cirillo, G.; Puoci, F.; Curcio, M.; Picci, N. Antioxidant–polysaccharide conjugates for food application by eco-friendly grafting procedure. *Carbohydr. Polym.* **2010**, *79*, 333–340. [CrossRef]

37. Xie, W.; Xu, P.; Liu, Q. Antioxidant activity of water-soluble chitosan derivatives. *Bioorg. Med. Chem. Lett.* **2001**, *11*, 1699–1701. [CrossRef]

38. Duh, P.-D. Antioxidant activity of burdock (*Arctium lappa* Linne): Its scavenging effect on free-radical and active oxygen. *J. Am. Oil Chem. Soc.* **1998**, *75*, 455–461. [CrossRef]

39. Kunath, K. Low-molecular-weight polyethylenimine as a non-viral vector for DNA delivery: Comparison of physicochemical properties, transfection efficiency and *in vivo* distribution with high-molecular-weight polyethylenimine. *J. Control. Release* **2003**, *89*, 113–125. [CrossRef]

40. Hill, I.R.C.; Garnett, M.C.; Bignotti, F.; Davis, S.S. *In vitro* cytotoxicity of poly(amidoamine)s: Relevance to DNA delivery. *Biochim. Biophys. Acta Gen. Subj.* **1999**, *1427*, 161–174. [CrossRef]

41. Fan, L.; Wu, H.; Cao, M.; Zhou, X.; Peng, M.; Xie, W.; Liu, S. Enzymatic synthesis of collagen peptide–carboxymethylated chitosan copolymer and its characterization. *React. Funct. Polym.* **2014**, *76*, 26–31. [CrossRef]

42. Wu, C.; Tian, J.; Li, S.; Wu, T.; Hu, Y.; Chen, S.; Sugawara, T.; Ye, X. Structural properties of films and rheology of film-forming solutions of chitosan gallate for food packaging. *Carbohydr. Polym.* **2016**, *146*, 10–19. [CrossRef] [PubMed]

43. Huang, H.; Sun, Y.; Lou, S.; Li, H.; Ye, X. *In vitro* digestion combined with cellular assay to determine the antioxidant activity in Chinese bayberry (*Myrica rubra* Sieb. et Zucc.) fruits: A comparison with traditional methods. *Food Chem.* **2014**, *146*, 363–370. [CrossRef] [PubMed]

marine drugs

MDPI

Article

Chitosan and Kappa-Carrageenan Vaginal Acyclovir Formulations for Prevention of Genital Herpes. *In Vitro* and *Ex Vivo* Evaluation

María-Pilar Sánchez-Sánchez [1], Araceli Martín-Illana [1], Roberto Ruiz-Caro [1], Paulina Bermejo [2], María-José Abad [2], Rubén Carro [2], Luis-Miguel Bedoya [2], Aitana Tamayo [3], Juan Rubio [3], Anxo Fernández-Ferreiro [4], Francisco Otero-Espinar [4] and María-Dolores Veiga [1,*]

[1] Departamento Farmacia y Tecnología Farmacéutica, Facultad de Farmacia, Universidad Complutense de Madrid, 28040-Madrid, Spain; mdelpilarss@gmail.com (M.-P.S.-S.); aracelimartin@ucm.es (A.M.-I.); rruizcar@ucm.es (R.R.-C.)

[2] Departamento Farmacología, Facultad de Farmacia, Universidad Complutense de Madrid, 28040-Madrid, Spain; naber@ucm.es (P.B.); mjabad@farm.ucm.es (M.-J.A.); rubencarrohernan@gmail.com (R.C.); lmbedoya@ucm.es (L.-M.B.)

[3] Instituto de Cerámica y Vidrio, Consejo Superior de Investigaciones Científicas, 28049-Madrid, Spain; aitanath@icv.csic.es (A.T.); jrubio@icv.csic.es (J.R.)

[4] Departamento Farmacia y Tecnología Farmacéutica, Facultad de Farmacia, Universidad de Santiago de Compostela, Campus Vida s/n, 15782 Santiago de Compostela, Spain; anxordes@gmail.com (A.F.-F.); francisco.otero@usc.es (F.O.-E.)

* Author to whom correspondence should be addressed; mdveiga@ucm.es; Tel.: +34-913-942091; Fax: +34-913-941736.

Academic Editor: Hitoshi Sashiwa
Received: 2 August 2015; Accepted: 15 September 2015; Published: 18 September 2015

Abstract: Vaginal formulations for the prevention of sexually transmitted infections are currently gaining importance in drug development. Polysaccharides, such as chitosan and carrageenan, which have good binding capacity with mucosal tissues, are now included in vaginal delivery systems. Marine polymer-based vaginal mucoadhesive solid formulations have been developed for the controlled release of acyclovir, which may prevent the sexual transmission of the herpes simplex virus. Drug release studies were carried out in two media: simulated vaginal fluid and simulated vaginal fluid/simulated seminal fluid mixture. The bioadhesive capacity and permanence time of the bioadhesion, the prepared compacts, and compacted granules were determined *ex vivo* using bovine vaginal mucosa as substrate. Swelling processes were quantified to confirm the release data. Biocompatibility was evaluated through *in vitro* cellular toxicity assays, and the results showed that acyclovir and the rest of the materials had no cytotoxicity at the maximum concentration tested. The mixture of hydroxyl-propyl-methyl-cellulose with chitosan- or kappa-carrageenan-originated mucoadhesive systems that presented a complete and sustained release of acyclovir for a period of 8–9 days in both media. Swelling data revealed the formation of optimal mixed chitosan/hydroxyl-propyl-methyl-cellulose gels which could be appropriated for the prevention of sexual transmission of HSV.

Keywords: chitosan; kappa-carrageenan; vaginal mucoadhesive formulations; acyclovir controlled release; swelling behaviour; cytotoxicity; genital herpes; *ex vivo* bioadhesion

1. Introduction

Sexually transmitted infections (STIs) are a major global cause of acute illness, infertility, long-term disability, and death, with severe medical and psychological consequences for millions of men, women and infants.

WHO/Europe advocates and assists Member States in promoting and developing human-rights-based policies and practices for STI control and prevention. According to the World Health Organization, sexually transmitted diseases (STDs) and their complications are among the top five diseases in developing countries forcing patients to seek healthcare [1]. Neonatal diseases acquired by vertical transmission are serious complications associated with significant morbidity and mortality. STDs are also the second cause of disease-related death and loss of years of good health among young women of child-bearing age (excluding HIV).

Genital herpes is one of the most common sexually transmitted infections worldwide, with a global prevalence of 536 million people infected and an annual incidence of 23.6 million new cases [2,3]. This chronic disease is caused by the Herpes simplex virus (HSV) type 2, and presents a wide variability in its clinical manifestations, ranging from asymptomatic to mild or severe signs and symptoms with potential complications. There are consequently many non-diagnosed cases of genital HSV, as many people infected with HSV are unaware of their infection [4]. Over the last thirty years, epidemiologic and molecular studies have highlighted a strong and synergistic relationship between HSV-2 and the Human Immunodeficiency Virus-1 (HIV-1), which clearly points to their capacity for co-infection [5]. It should be noted that HSV-2 infection, even without recognized lesions, is an independent risk factor for HIV infection, such that the risk of HIV acquisition is three times higher in people with HSV-2. The resulting mucosal disruption caused by genital ulcers offers an effective entry route for HIV-1. This could be prevented by high concentrations of antiviral in the genital mucosa, thereby reducing the increased susceptibility to HIV-1 infection associated with HSV-2 [6–9]. While the probability of male-to-female transmission of STDs is alarmingly high, the same is not universally true for female-to-male transmission. Current methods of preventing STDs, such as abstinence, condoms, and monogamy are frequently ineffective and out of women's control [10], making it advisable to design novel female-controlled barrier techniques, such as microbicides and female condoms [11]. Microbicides are currently emerging as a promising tool to protect women from STDs. A vaginal microbicide is any topical agent/formulation intended to prevent sexual pathogens, either by inactivating or killing cellular mechanisms, by forming a physical barrier between cells and pathogens, or by enhancing the natural protective mechanisms of the cervix and vagina. Unfortunately, many vaginal microbicide formulations may fail to elicit a protective response due to their lack of efficacy and inadequate formulation. Some of the most commonly used vaginal dosage forms include creams, gels, tablets, films, tampons, vaginal rings, and douches. Each of these formulations has specific advantages and limitations. Tablets can also be designed to perform a controlled-release of the microbicide over a prolonged period of time.

Several studies reveal that acyclovir (ACV) is a safe and effective drug for vaginal administration, and some clinical benefits have been observed in the treatment of primary or recurrent lesions from genital herpes [12,13]. Several studies on the prevention of genital herpes transmission have examined the inclusion of acyclovir as a microbicide drug in vaginal formulations such as gels [14], intravaginal rings [15,16], microporous matrices [17] or nanoparticles [18]. Vaginal bioadhesive tablets of acyclovir have been developed using different excipients such as methyl-cellulose, carboxy-methyl-cellulose, hydroxyl-propyl-cellulose, showing the dissolution results an inadequate behavior because of disintegration of tablets in the first 30 min. However, when hydroxyl-propyl-methyl-cellulose was incorporated to tablets, ACV release was prolonged during at least 8 h [19]. Gurumurthy *et al.* designed xanthan gum/Carbopol® 934P-based acyclovir vaginal tablets obtaining sustained drug release data for 12 h in simulated vaginal fluid [20].

Mucoadhesive polymers have an excellent binding capacity with mucosal tissues over a considerable period of time. Several studies have been conducted on the incorporation of tragacanth,

Carbopol®, Poloxamer 407®, pectin, sodium alginate, cellulose derivatives, and chitosan, among others, into vaginal formulations in order to increase the residence time of these formulations at the site of action [21–23].

Chitosan is a natural polysaccharide produced by the partial deacetylation of chitin, the structural element in the exoskeleton of crustaceans such as crabs and shrimps. The amino group in chitosan has a pKa value of approximately 6.5, which leads to protonation in an acidic solution with a density of charge dependent on the pH and the % deacetylation value. Chitosan has water soluble and bioadhesive properties with negatively charged surfaces, such as mucosal membranes, and can be used to transport a drug to an acidic environment where it can be degraded, thereby releasing the drug to the desired site. It is biocompatible and biodegradable and is widely used as a pharmaceutical excipient in a range of formulations such as powders, tablets, emulsions, and gels. The use of chitosan as a mucoadhesive polymer for vaginal delivery systems has been studied by several researchers [24–26].

Carrageenan is a member of the family of linear sulfated polysaccharides extracted from red edible seaweeds, which are widely used in the pharmaceutical industry for their gelling, thickening, and stabilizing properties. There are three main varieties of carrageenan, with differing degrees of sulfation. Kappa-carrageenan has one sulfate group per disaccharide, *iota*-carrageenan has two sulfates per disaccharide and *lambda* carrageenan has three. Liu *et al.* [27] developed an *in situ* kappa-carrageenan/poloxamer 407 vaginal gel with prolonged local residence. There is also evidence that carrageenan-based gel may offer some protection against HSV-2 transmission by binding with the receptors on the herpes virus, thus preventing the virus from binding to cells [28].

Cellulose derivatives have also been applied as drug delivery excipients in vaginal formulations. For instance, hydroxyl-ethyl-cellulose is a FDA-approved polymer found in a wide range of applications, because it is non-irritant and non-toxic to the vagina [29,30]. Methylcellulose has been used in the development of vaginal hydrogels due to its good biocompatibility and bioadhesion [31].

Other formulations, such as vaginal rings, gels or creams, vaginal tablets and compacts are easily manufactured, economical, stable under different environmental conditions, and easy to handle. If these solid formulations include the appropriate mucoadhesive polymer, or polymer mixture, an optimum formulation can be obtained for the *in situ* controlled release of the drug in the area where the transmission of vaginal herpes occurs.

With this background, the aim of this paper is to develop natural polymer-based vaginal mucoadhesive solid formulations for the controlled release of acyclovir, which may prevent the sexual transmission of HSV. Infection with HSV can increase the risk of infection with other pathogens, such as HIV.

2. Results and Discussion

In order to achieve optimal acyclovir controlled release mucoadhesive formulations, two types of solid systems were prepared containing 100 mg of acyclovir and natural and/or semisynthetic mucoadhesive polymers: compacts and compacted granules. Table 1 shows the composition (mg/unit) of these formulations.

Table 1. Composition (mg/unit) of prepared compacts and compacted granules.

Batch	Chitosan	*k*-Carrageenan	HPMC100	ACP	PVP	MgS	Acyclovir
CQ	225					3	100
CK		225				3	100
CH			225			3	100
CQH1	135		90			3	100
CQH2	90		135			3	100
CKH1		135	90			3	100
CKH2		90	135			3	100
CGQH1	135		90	45	27	3	100
CGQH2	90		135	45	27	3	100
CGKH1		135	90	45	27	3	100
CGKH2		90	135	45	27	3	100

2.1. Release Studies

Figure 1 shows the release profiles of acyclovir for all formulations assayed in simulated vaginal fluid (SVF) (graphs A, B, and C) and simulated vaginal fluid/simulated seminal fluid mixture (SVF/SSF) (graph D). The data from compacts prepared with polymer/acyclovir binary physical mixtures (CQ, CK, and CH) show that all the polymers control acyclovir release, but at different rates (graph A).

Kappa-carrageenan compacts (CK) produced drug release in 48 h, while samples of chitosan formulation (CQ) extend ACV release to 120 h. In both cases the swelling, disintegration, and dissolution of the formulation—the mechanisms controlling the release of A—were visually observed. Drug release was more prolonged when ACV was formulated with HPMC (CH), and the release was incomplete, as an asymptotic trend was observed in the release profile after 92 h of the test. At the conclusion of the release study it was observed that the CH compact core remained dry. The explanation is that in an aqueous medium HPMC forms a strong gel layer that prevents fluid from accessing the compact and hinders the dissolution of 100% of the dose of acyclovir. In the light of these findings, new formulations were designed combining a marine origin polymer such as chitosan or kappa-carrageenan with HPMC in order to exploit the benefits of both types of polymers. These include the swelling/dissolution—shown by the natural polymers studied—and the robustness of HPMC, which swells but will not dissolve in an aqueous medium. Figure 1 (graph B) shows the ACV release profiles obtained from the compacts formulated with the marine polymer/HPMC mixtures. As can be noted from the data, chitosan/HPMC compacts (CQH1 and CQH2) allow a complete and sustained release of ACV over a period of 168 h. Moreover, there are no appreciable differences between both profiles although they contain different chitosan/HPMC ratios. In contrast, kappa-carrageenan/HPMC compacts show total ACV release, but the time needed to achieve this depended on the kappa-carrageenan/HPMC ratio. The formulation containing the highest ratio of kappa-carrageenan (CKH1) released 100% of ACV in 120 h, whereas the CKH2 formulation required 192 h to release all of the drug. This may be the result of the ability of the kappa-carrageenan compacts to erode, as has been described by Bettini *et al.* [32] for the metoprolol/λ-carrageenan matrix. The chitosan-based compacts showed a strong interaction between chitosan and HPMC, and a more controlled release of ACV, regardless of the chitosan/HPMC ratio.

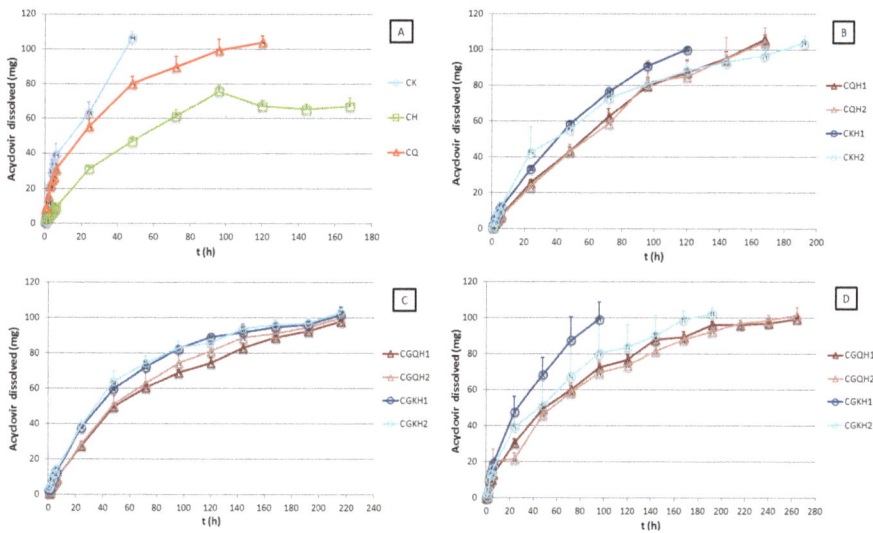

Figure 1. Acyclovir release profiles obtained from compacts and compacted granules in simulated vaginal fluid (**A**, **B**, and **C**) and simulated vaginal fluid/simulated seminal fluid mixture (**D**). All values represent the mean ± SD (*n* = 3).

In order to prolong ACV release time, compacted granules were made by granulating physical ACV/polymer mixtures. Figure 1C shows the dissolution profiles of ACV for the compacted granule formulations. In all cases total drug release was obtained in a period of nine days (216 h). Although the systems containing the chitosan/HPMC mixture showed better controlled release of ACV, the granulation process can generally be assumed to lead to unitary systems (granules) where the presence of a binder (PVP K30) ensures the ACV+HPMC+marine polymer physical bond. The subsequent granule compaction produced robust formulations which enabled the sustained release of ACV. Since these formulations present superior ACV controlled-release characteristics, they were also tested in the simulated vaginal fluid/simulated seminal fluid mixture (SVF/SSF) to assess how the nature of the medium affected the ACV release process. The results of ACV released from CGQH1, CGQH2, CGKH1, and CGKH2 in the SVF/SSF mixture are shown in Figure 2D, and indicate that formulations with chitosan (CGQH1 and CGQH2) had a higher ACV controlled release. This is because pH-dependent chitosan dissolves better in an acidic medium (SVF) than in a neutral medium (SVF/SSF). Based on the results of the release studies, it can be concluded that chitosan-based systems may be the most suitable for use in preventing the sexual transmission of genital herpes. Although controlled ACV release was one of the aims of this research, it is also essential to verify whether the formulation has mucoadhesive properties. We, therefore, determined the bioadhesion of all the systems developed.

2.2. Characterization of Bioadhesion

The bioadhesive behavior of the compacts and compacted granules containing chitosan and kappa-carrageenan and HPMC are shown in Figure 2. All formulations show higher values for bioadhesion work and maximum detachment force in vaginal mucosa than for chitosan- and HPMC-based mucoadhesive acyclovir tablets in gastric mucosa [33], and for different mucoadhesive semisolid formulations (gel-type, such as Crinone, KYJelly, zidoval, and W/S mucoadhesive emulsions) [34].

The bioadhesive behavior of the compacts and compacted granules containing chitosan, kappa-carrageenan and/or HPMC are shown in Figure 2A. As expected, all the formulations had mucoadhesive ability.

The data on maximum detachment (separation) force for all the formulations evaluated are generally more homogeneous than the corresponding data for bioadhesion work (Figure 2B). This could indicate that once the samples have adhered to the mucosa, the detachment force is similar in all the formulations regardless of their composition, although a reverse behavior can be observed compared to the results for bioadhesion work. In general, greater force was required to separate each sample of chitosan-compacted formulations from vaginal mucosa: CQ > CK, CQH1 > CKH1 and CGQH1 > CGKH1. Moreover, the comparison between the detachment forces of compacted physical mixtures and compacted granule formulations (CQH1 *vs.* CGQH1, CQH2 *vs.* CGQH2, CKH1 *vs.* CGKH1, and CKH2 *vs.* CGKH2) clearly reveals a decrease in detachment forces for compacted granules with regard to the corresponding compacted physical mixtures; when PVP (binder agent) and ADCP (structural excipient) were included in the formulations, these excipients act as impurities that prevent the polymers from performing their mucoadhesive function.

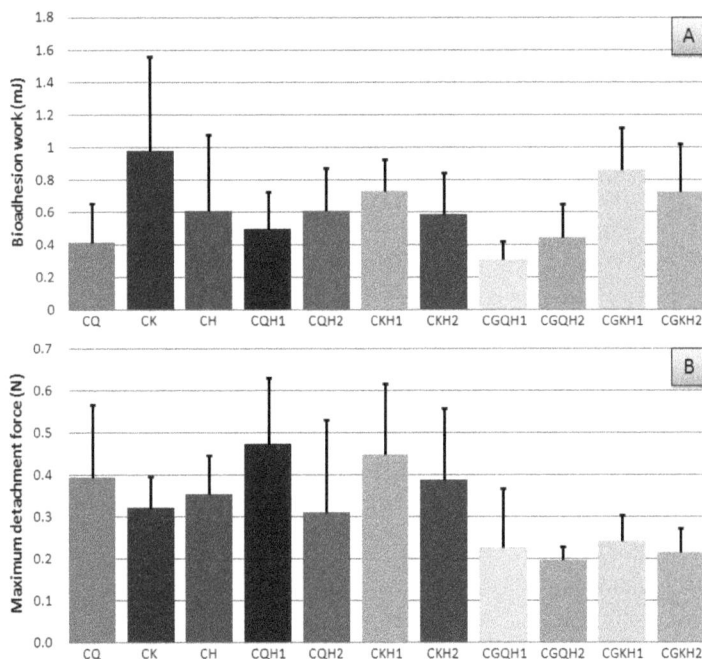

Figure 2. Vaginal bioadhesion work and maximum detachment force obtained for compacts and compacted granules (Mean Values + Standard Deviation, *n* = 6).

2.3. Bioadhesion Residence Test

Once it was verified that all the prepared formulations had mucoadhesive properties, the next step was to determine how long they remained bonded to the vaginal mucosa. Table 2 shows the periods of time this adhesion lasted when formulations were immersed in SVF and SVF/SSF. CQ and CK samples remained adhered to the vaginal mucosa for less than 30 min in the case of simulated vaginal fluid, while CH samples remained for 120 h. The mucoadhesive ability of HPMC has been extensively verified, and appears to derive from its stronger hydrogen bonding with the mucin compared to other natural polymers [35,36]. Therefore the combination of HPMC with marine polymers (chitosan or kappa-carrageenan) may produce a longer attachment according to data obtained from both natural polymers. As expected, CQH1, CQH2, CKH1, and CKH2 remained adhered to mucosa for a longer period of time (between 72 and 108 h) than CQ and CK samples. The longer bonding time of CKH1 and CKH2 samples compared to CQH1 and CQH2 is due to kappa-carrageenan's ability to form strong gels in the presence of potassium ions, which are part of the SVF composition. In other words, formulations that contain this polymer remain adhered to vaginal mucosa for longer than systems with chitosan. Chitosan is a polycationic copolymer consisting of glucosamine and *N*-acetylglucosamine units that are primarily responsible for swelling and the ensuing dissolution in contact with acidic fluid, and also have OH and NH_2 groups that are considered essential for mucoadhesion [37].

Table 2. Bioadhesion residence time of acyclovir compacts and compacted granule formulations in simulated vaginal fluid (SVF) and simulated vaginal fluid/simulated seminal fluid (SVF/SSF) mixture.

Sample	Bioadhesion Residence Time in SVF	Bioadhesion Residence Time in SVF/SSF
CQ	<30 min	—
CK	<30 min	—
CH	120 h	—
CQH1	72 h	—
CQH2	72 h	—
CKH1	96 h	—
CKH2	108 h	—
CGQH1	72 h	72 h
CGQH2	72 h	96 h
CGKH1	72 h	72 h
CGKH2	72 h	72 h

The incorporation of a structural agent—anhydrous calcium hydrogen phosphate (ACP)—and a binder—polyvinyl pyrrolidone (PVP)—into the polymer blend to prepare wet granules produces compacted granules that remained adhered to the vaginal mucosa for the same period of time (72 h), regardless of the formulation composition or the pH of the medium (SFV or SVF/SSF mixture). This indicates that formulations obtained by combining HPMC with a natural polymer are reliable enough to remain adhered to the vaginal site for the time required to achieve ACV release, and consequently prevent the sexual transmission of genital herpes. If formulations were entirely composed of one polymer (chitosan, kappa-carrageenan or HPMC) they would have inappropriate mucoadhesion values, which may be deficient in the case of chitosan or kappa-carrageenan and excessive for HPMC.

2.4. Swelling Behavior

Swelling processes were visually detected during release studies and mucoadhesion tests; however, it is very important to quantify this process, as the swelling curves can help explain the release results. Swelling ratio (SR) values obtained from all formulations determined in SVF are shown in Figure 3. Each positive SR value indicates that at a given time the swollen matrix weight was higher than that of the dry system weight ($t = 0$). Conversely, each negative SR value shows that the weight of the swollen system was lower than the weight of the dry system ($t = 0$). The swelling data for CQ and CK formulations reveal that both polymers are erodible in acidic media, while CH has the highest swelling profile. When HPMC is combined with chitosan or carrageenan, the formulations showed intermediate swelling, which would be more acceptable in terms of patient comfort. An analysis of all the formulations made with polymer mixtures showed that their swelling behavior was conditioned by two factors: the nature of the natural polymer (chitosan or kappa-carrageenan), and the type of formulation (compact or compacted granules). Compact formulations with chitosan—CQH1 and CQH2—displayed lower swelling values than the kappa-carrageenan-based samples (CKH1 and CKH2). The explanation is that the presence of chitosan prevents HPMC's own swelling pattern, as it produces a mixed HPMC/chitosan gel. CKH1 and CKH2 had very similar swelling profiles for the first 48 h of the test, with a subsequent decrease in weight in inverse proportion to their HPMC content (CKH1 weight loss is greater than CKH2 weight loss), as kappa-carrageenan is unable to interact with HPMC. CKH1 and CKH2 swelling curves therefore have a similar shape to the CH swelling curve. The areas under the swelling curve are related to the amount of HPMC in each formulation.

The swelling behavior of all compacted granule formulations is very similar, as they all showed swelling peaks between 350% and 510%, although the chitosan-based formulations CGQH1 and CGQH2 had a lower maximum swelling (350% and 452%, respectively). ACP plays a key role in the swelling process in these formulations. In formulations with kappa-carrageenan (CGKH1 and CGKH2), ACP creates a structure that prevents the free swelling of HPMC. In formulations with chitosan, the ACP structure prevents the formation of the HPMC-chitosan mixed gel. In other words, ACP acts by modulating the swelling of the polymers, resulting in systems with a similar swelling ratio which are very different from those obtained from formulations without this carrier.

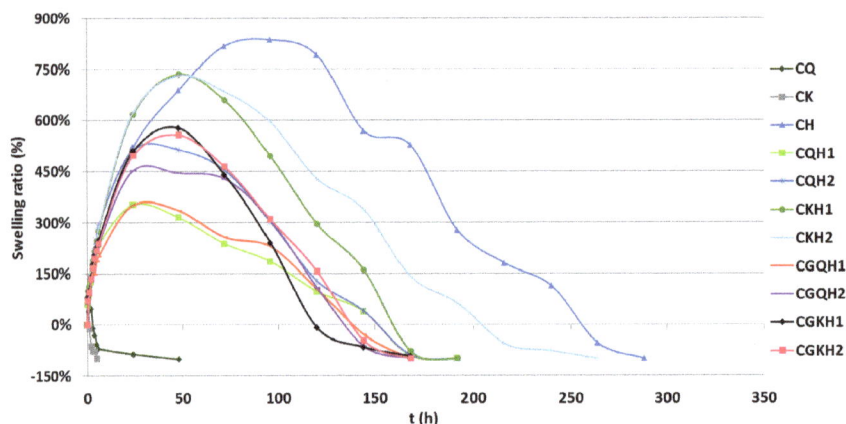

Figure 3. Swelling profiles obtained from compacts and compacted granules in simulated vaginal fluid.

2.5. Cell Toxicity

The biocompatibility of the different formulations was evaluated through *in vitro* cellular toxicity assays. The toxicity of the components was studied by incubation at room temperature for 48 h at different concentrations before the assay to ensure that any potential toxic component of the materials would be present in the dilution. The cell culture was then treated with a suspension of the different dilutions. All the components were tested at a maximum concentration of 1000 µg/mL, except for chitosan, which was tested at 250 µg/mL due to solubility problems.

Experiments were performed on a lymphoblastic cell line (MT-2) to evaluate toxicity on the immune cells present in vaginal or uterine mucosae and in the uterus epithelial cell line (HEC-1A) to assess potential damage to mucosae integrity. CC_{50} were calculated for the drug or empty material when possible.

As shown in Table 3 and Figure 4, no toxicity was detected at the concentrations tested except in the case of ACP, which showed mild toxicity in both cell lines, with a CC_{50} value of around 1000 µg/mL. Acyclovir and all other materials showed no cytotoxicity at the maximum concentration tested, so CC_{50} values were not calculated.

Table 3. CC$_{50}$ values of acyclovir, chitosan, kappa-carrageenan, magnesium stearate, HPMC-K100M, ACP and PVP K30 in the cytotoxic assay in both MT-2 and HEC-1A cell lines. CC$_{50}$: cytotoxic concentration 50%.

Drug	Cell Line	CC$_{50}$
Acyclovir	MT-2	>500 µM
	HEC-1A	>500 µM
Chitosan	MT-2	>250 µg/mL
	HEC-1A	>250 µg/mL
Kappa-carrageenan	MT-2	>1000 µg/mL
	HEC-1A	>1000 µg/mL
Magnesium stearate	MT-2	>1000 µg/mL
	HEC-1A	>1000 µg/mL
HPMC-K100M	MT-2	>1000 µg/mL
	HEC-1A	>1000 µg/mL
ACP	MT-2	≈1000 µg/mL
	HEC-1A	≈1000 µg/mL
PVP	MT-2	>1000 µg/mL
	HEC-1A	>1000 µg/mL

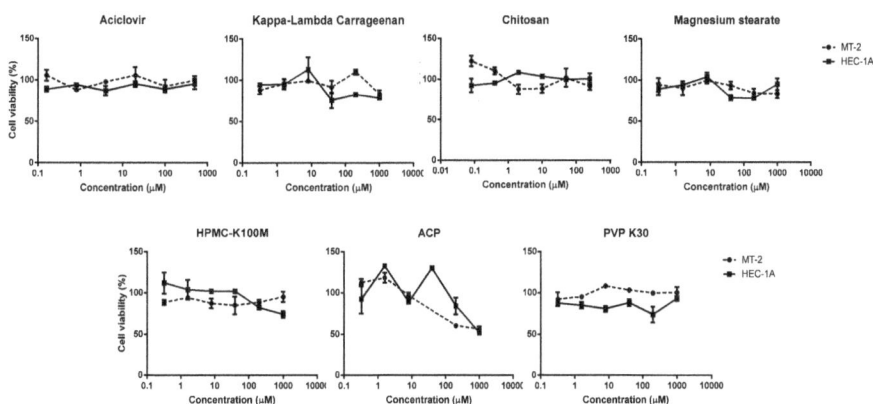

Figure 4. Graphic representation of the cytotoxic evaluation of acyclovir, chitosan, kappa-carrageenan, magnesium stearate, HPMC-K100M, ACP, and PVP K30. Cytotoxicity was measured in MT-2 cells and HEC-1A cells for 48 h. Cell viability is expressed as percentage of living cells compared to a non-treated control (100%).

3. Experimental Section

3.1. Materials

Acyclovir was obtained from Lab. Reig Jofré (Toledo, Spain). Chitosan with a deacetylation degree of 97% (CH, Lot: 8826900003) was supplied by Nessler (Madrid, Spain). Kappa-carrageenan (*k*; batch 088K0178) was provided by Sigma-Aldrich (St. Louis, MO, USA). Hydroxyl-propyl-methyl-cellulose Methocel® K 100 M (HPMC; lot: DT352711) was a gift from Colorcon Ltd (Kent, UK). Anhydrous calcium hydrogen phosphate (ACP; batch: K93487944416) was provided by Merck (Darmstadt, Germany). Magnesium stearate PRS-CODEX (MGSt; batch: 85269 ALP) was purchased from Panreac (Barcelona, Spain). Kollidon® 30 (PVP K30; batch: 98-0820) was supplied by

BASF (Ludwingshafen, Germany). All other reagents in this study were of analytical grade and used without further purification. Demineralized water was used in all cases.

3.2. Preparation of Compacts and Compacted Granules

Two types of solid formulations, compacts and compacted granules, were prepared containing natural and/or semisynthetic mucoadhesive polymers, and 100 mg of acyclovir. This selected drug amount is based in a common application of a marketed vaginal formulation (acyclovir topical cream 5% *w/w*). The compacts (CQ, CK, CH, CQH1, CQH2, CKH1, and CKH2) were prepared from physical mixtures of the components, while the compacted granules (CGQH1, CGQH2, CGKH1, and CGKH2) were made from the corresponding granules in a pre-compacted stage. Granules were produced by adding a PVP K30 ethanol solution (binder agent) to a physical mixture of ACV, chitosan or kappa-carrageenan, HPMC, and ACP. ACP acts as structural agent. The wet mass was then passed through a 0.5 mm mesh and dried at 40 °C for 12 h. Magnesium stearate was added to the granules before compaction. Both types of systems were compacted according to the IR spectroscopy technique. A stainless steel disk was placed in a die assembly and the sample (physical mixture or granules) was added. A stainless steel disk was inserted in the cylinder bore on top of the sample. A piston was then placed inside the cavity, and the sample was pressed with a force of 5 t for 4 min. Finally, piston and disks were dismantled from the die and the acyclovir compact was removed and stored in a desiccator until use.

3.3. Release Study

All samples were placed in 100 mL DURAN laboratory borosilicate glass bottles containing 80 mL of release medium and stored in a shaking water bath (37 °C, 15 opm). Aliquots of 5 mL were withdrawn daily and filtered through Millipore® filters (0.45 μm). The medium was replaced with an equal volume of release medium equilibrated to temperature. Acyclovir concentrations were quantified by UV spectroscopy at a wavelength of 251 nm in a Shimadzu® UV-1700 spectrophotometer (Kyoto, Japan). The assay was made in triplicate in each case. In order to ensure that the formulations act as effective drug delivery carriers even in the presence of semen, the acyclovir release studies were carried out in two dissolution media: SVF (pH 4.2) and a SVF/SSF mixture (pH 7.6). In the first three hours of assay, the release medium was simulated vaginal fluid which was removed from each bottle and substituted by the SVF/SSF 1:4 *v/v* mixture until the end of the assay. Both media were prepared according to the methodology described in the bibliography [38,39].

3.4. Characterization of Bioadhesion

The bioadhesive capacity of the compacts and compacted granules was determined by measuring the maximum detachment force and the bioadhesion work using a TA-XT Plus texture analyzer (TA Instruments, Newcastle, UK), and bovine vaginal mucosa as substrate. Samples of bovine vaginal mucosa were obtained from a local slaughterhouse (Compostelana de Carnes S.L., Santiago de Compostela, Spain) immediately after slaughter. Each formulation was fixed to the upper support and the vaginal mucosa samples to the lower support using a cyanoacrylate adhesive. The mucosa was wetted with 0.5 mL of warm simulated vaginal fluid, and the system was equilibrated and maintained at 37 °C for 1 min. Finally a compression/extension stage was applied with an initial contact force of 0.5 N for 60 s to ensure close contact between the substrate and the sample, followed by an extension phase at a rate of 1 mm/s until the total separation of the components. The results were recorded as force *versus* elongation. Bioadhesion work was calculated as the area under the force-elongation curve. All the formulations were evaluated six-fold.

3.5. Bioadhesion Residence Test

A complementary *ex vivo* mucoadhesion test was conducted based on the rotating cylinder technique [26,40] to evaluate the permanence of the bioadhesion of the mucoadhesive formulations to the vaginal mucosa. A sample of freshly excised vaginal bovine mucosa obtained from a local slaughterhouse was fixed with acrylic adhesive to a stainless steel prism with a width of 2 cm anda height of 3 cm. In order to provide a homogeneous force of adhesion between the vaginal mucosa and the solid formulation, each sample was placed over the vaginal mucosa and pressed with a weight of 500 g for 30 s. The prism was then placed in a dissolution USP apparatus 1 (the basket was substituted by the prism) containing 500 mL of SVF at 37 ± 0.1 °C. The prism was fully immersed and agitated at 30 rpm. The time elapsed until the complete detachment, dissolution and/or erosion of the dosage forms was visually assessed. To determine whether the presence of semen influences the time the formulation remains attached to the vaginal mucosa, the bioadhesion remanence test was conducted under the same conditions but changing the nature of the medium. The samples were maintained for three hours in SVF and the medium was then exchanged for the SVF/SSF 1:4 *v/v* mixture [26]. Two replicates of each bioadhesion residence test were carried out in all cases.

3.6. Swelling Characterization

The influence of composition on the behavior of compacts and compacted granules was studied by determining the degree of swelling in simulated vaginal fluid according to a method described by Ruiz-Caro and Veiga [41]. Before the samples were exposed to the SFV they were glued to metallic discs with a diameter of 30 mm. The discs were then placed in beakers containing 100 mL of SFV. These beakers were placed in a thermostated water shaking bath (Selecta® UNITRONIC320 OR, Barcelona, Spain) with an experimental temperature of 37 ± 0.1 °C and a shaking rate of 15 U/min. At specific time intervals, the samples were extracted from the test medium and blotted with filter paper to absorb the excess liquid on the sample surface before the weight evaluation. The swelling ratio (SR) of each sample was calculated according to the following equation: SR = ((Ts − Td)/Td) × 100, where Ts and Td were the weights of the swollen and dried samples respectively [42]. All swelling tests were performed in triplicate.

3.7. Cell and Cytotoxicity

Two human cell lines were used: a lymphoblastic cell line, MT-2 [43] and a uterus/endometrium epithelial cell line, HEC-1-A [44] (kindly provided by María Angeles Muñoz, Hospital Gregorio Marañón, Madrid). Both cells were grown and propagated in RPMI 1640 medium supplemented with 10% (*v/v*) foetal bovine serum, 2 mM L-glutamine and 50 µg/mL streptomycin at 37 °C with a humidified atmosphere of 5% CO_2. The HEC-1-A cells were detached by removing the medium and rinsing the flask for 10 min with 1 to 2 mL of Trypsin 0.25%—EDTA 0.03% solution. The medium was replaced every third day after cell centrifugation at 1000 rpm for five minutes.

Cell toxicity was measured by the widely-used MTT (3-(4, 5-dimethylthylthiazol-2-yl)-2,5-diphenyltetrazolium bromide) method. Cells were incubated in 96-well plates at a density of 10×10^4 cells per well in the case of MT-2, and 2×10^4 in the case of HEC-1-A in complete medium. To assess the cytotoxic effects of acyclovir and the excipients used in the different formulations, cells were exposed to fresh medium containing various concentrations of compounds, or with the same concentration of vehicle to dissolved compound (DMSO) for 48 h as controls in triplicate. The particles were suspended in water following a standard method [45]. After 48 h of incubation, 20 µL of MTT solution (7.5 mg/mL) was added to the plate and incubated for another two hours for MT-2 cells and three hours for HEC-1-A cells. The supernatant was then carefully removed and 100 µL of dimethyl sulfoxide (DMSO) was added to each well. Absorbance at a wavelength of 550 nm was measured in a Labtech LT-4000 microplate spectrophotometer. Values of cytotoxic concentrations 50 (CC_{50}) were calculated using GraphPad Prism Software (non-linear regression, log inhibitor *versus* response). The results of the MTT assay represent the average of at least three individual experiments.

Mar. Drugs **2015**, *13*, 5976–5992

4. Conclusions

From the results presented above it can be concluded that the combination of a polymer of marine origin—chitosan or kappa-carrageenan—with a semisynthetic polymer—hydroxyl-propyl-methyl-cellulose—in solid compacted formulations achieves the complete and sustained release of acyclovir over a period of between 220 and 260 h. These formulations can also adhere to the vaginal mucosa and remain adhered for 72 h. The swelling data of the formulations are influenced by the nature and quantity of the polymers they contain. Thus hydroxyl-propyl-methyl-cellulose has a higher degree of swelling than chitosan and kappa-carrageenan, which have lower swelling values and additionally are erodible in an acidic medium. The presence of anhydrous calcium hydrogen phosphate as a structural agent allows the modulation of the polymers' swelling properties, reduces the swelling behavior of the hydroxyl-propyl-methyl-cellulose and prevents the formation of a mixed HPMC/chitosan gel. Of all the formulations developed, those containing chitosan are more indicated for preventing the sexual transmission of genital herpes, as added to their beneficial characteristics of mucoadhesion and sustained release of ACV, they showed a preventive effect of chitosan against cell damage.

Acknowledgments: This work was supported by the Spanish Ministry of Economy and Competitiveness (MAT2012-34552).

Author Contributions: Roberto Ruiz-Caro, Paulina Bermejo, María-José Abad, Luis-Miguel Bedoya, Aitana Tamayo, Juan Rubio, Francisco Otero-Espinar and María-Dolores Veiga designed and planned the experiments. María-Pilar Sánchez-Sánchez, Araceli Martín-Illana, Rubén Carro and Anxo Fernández-Ferreiro conducted the experiments. All authors contributed to the preparation of the manuscript. María-Dolores Veiga is a senior author and project leader.

Conflicts of Interest: The authors declare no conflict of interest.

References

1. Sexually Transmitted Infections. Available online: http://www.who.int/topics/sexually_transmitted_infection (accessed on 19 February 2015).
2. Antoine, T.E.; Mishra, Y.K.; Trigilio, J.; Tiwari, V.; Adelung, R.; Shukla, D. Prophylactic, therapeutic and neutralizing effects of zinc oxide tetrapod structures against herpes simplex virus type-2 infection. *Antivir. Res.* **2012**, *96*, 363–375. [PubMed]
3. Cherpes, T.L.; Matthews, D.B.; Maryak, A. Neonatal herpes simplex virus infection. *Clin. Obstet. Gynecol.* **2012**, *55*, 938–944. [CrossRef] [PubMed]
4. Garland, S.M.; Steben, M. Genital herpes. *Best Pract. Res. Clin. Obstet. Gynaecol.* **2014**, *28*, 1098–1110. [CrossRef] [PubMed]
5. Tan, D.H.; Murphy, K.; Shah, P.; Walmsley, S.L. Herpes simplex virus type 2 and HIV disease progression: A systematic review of observational studies. *BMC Infect. Dis.* **2013**, *13*, 502. [CrossRef] [PubMed]
6. Freeman, E.E.; Weiss, H.A.; Glynn, J.R.; Cross, P.L.; Whitworth, J.A.; Hayes, R.J. Herpes simplex virus 2 infection increases HIV acquisition in men and women: Systematic review and meta-analysis of longitudinal studies. *AIDS* **2006**, *20*, 73–83. [CrossRef] [PubMed]
7. Corey, L.; Wald, A.; Celum, C.L.; Quinn, T.C. The effects of herpes simplex virus-2 on HIV-1 acquisition and transmission: A review of two overlapping epidemics. *J. Acquir. Immune Defic. Syndr.* **2004**, *35*, 435–445. [CrossRef] [PubMed]
8. Hen, M.; Heng, S.; Allen, S. Co-infection and synergy of human immunodeficiency virus-1 and herpes simplex virus-1. *Lancet* **1994**, *343*, 255–258.
9. Moriuchi, M.; Moriuchi, H.; Williams, R.; Straus, S.E. Herpes simplex virus infection induces replication of human immunodeficiency virus type 1. *Virology* **2000**, *278*, 534–540. [CrossRef] [PubMed]
10. Podaralla, S.; Alt, C.; Shankar, G.N. Formulation development and evaluation of innovative two-polymer (SR-2P) bioadhesive vaginal gel. *AAPS PharmSciTech* **2014**, *15*, 928–938. [CrossRef] [PubMed]
11. Garg, S.; Tambwekar, K.R.; Vermani, K.; Kandarapu, R.; Garg, A.; Waller, D.P.; Zaneveld, L.J.D. Development pharmaceutics of microbicide formulations. Part II: Formulation, evaluation, and challenges. *AIDS Patient Care STDS* **2003**, *17*, 377–399. [CrossRef] [PubMed]

12. Corey, L.; Benedetti, J.K.; Critchlow, C.W.; Remington, M.R.; Winter, C.A.; Fahnlander, A.L.; Smith, K.; Salter, D.L.; Keeney, R.E.; Davis, L.G.; *et al.* Double-blind controlled trial of topical acyclovir in genital herpes simplex virus infections. *Am. J. Med.* **1982**, *73*, 326–334. [CrossRef]

13. Corey, L.; Nahmias, A.J.; Guinan, M.E.; Benedetti, J.K.; Critchlow, C.W.; Holmes, K.K. A trial of topical acyclovir in genital herpes simplex virus infections. *N. Engl. J. Med.* **1982**, *306*, 1313–1319. [CrossRef] [PubMed]

14. Shankar, G.N.; Alt, C. Prophylactic treatment with a novel bioadhesive gel formulation containing acyclovir and tenofovir protects from HSV-2 infection. *J. Antimicrob. Chemother.* **2014**, *69*, 3282–3293. [CrossRef]

15. Moss, J.A.; Malone, A.M.; Smith, T.J.; Kennedy, S.; Kopin, E.; Nguyen, C.; Gilman, J.; Butkyavichene, I.; Vincent, K.L.; Motamedi, M.; *et al.* Simultaneous delivery of tenofovir and acyclovir via an intravaginal ring. *Antimicrob. Agents Chemother.* **2012**, *56*, 875–882. [CrossRef] [PubMed]

16. Baum, M.M.; Butkyavichene, I.; Gilman, J.; Kennedy, S.; Kopin, E.; Malone, A.M.; Nguyen, C.; Smith, T.J.; Friend, D.R.; Clark, M.R.; *et al.* An intravaginal ring for the simultaneous delivery of multiple drugs. *J. Pharm. Sci.* **2012**, *101*, 2833–2843. [CrossRef] [PubMed]

17. Asvadi, N.H.; Dang, N.T.T.; Davis-Poynter, N.; Coombes, A.G.A. Evaluation of microporous polycaprolactone matrices for controlled delivery of antiviral microbicides to the female genital tract. *J. Mater. Sci. Mater. Med.* **2013**, *24*, 2719–2727. [CrossRef] [PubMed]

18. Ensign, L.M.; Tang, B.C.; Wang, Y.Y.; Tse, T.A.; Hoen, T.; Cone, R.; Hanes, J. Mucus-penetrating nanoparticles for vaginal drug delivery protect against herpes simplex virus. *Sci. Transl. Med.* **2012**, *4*, 138ra79. [CrossRef] [PubMed]

19. Genç, L.; Oğuzlar, C.; Güler, E. Studies on vaginal bioadhesive tablets of acyclovir. *Pharmazie* **2000**, *55*, 297–299.

20. Gurumurthy, V.; Deveswaran, R.; Bharath, S.; Basavaraj, B.V.; Madhavan, V. Design and optimization of bioadhesive vaginal tablets of acyclovir. *Ind. J. Pharm. Educ. Res.* **2013**, *47*, 140–147.

21. Andersen, T.; Vanic, Z.; Flaten, G.E.; Mattsson, S.; Tho, I.; Skalko-Basnet, N. Pectosomes and chitosomes as delivery systems for metronidazole: The one-pot preparation method. *Pharmaceutics* **2013**, *5*, 445–456. [CrossRef] [PubMed]

22. Berginc, K.; Suljakovic, S.; Skalko-Basnet, N.; Kristl, A. Mucoadhesive liposomes as new formulation for vaginal delivery of curcumin. *Eur. J. Pharm. Biopharm.* **2014**, *87*, 40–46. [CrossRef] [PubMed]

23. Li, C.; Han, C.; Zhu, Y.; Lu, W.; Li, Q.; Liu, Y. *In vivo* evaluation of an in-situ hydrogel system for vaginal administration. *Pharmazie* **2014**, *69*, 458–460. [PubMed]

24. Gavini, E.; Sanna, V.; Juliano, C.; Bonferoni, M.C.; Giunchedi, P. Mucoadhesive vaginal tablets as veterinary delivery system for the controlled release of an antimicrobial drug, acriflavine. *AAPS PharmSciTech* **2002**, *3*, 32–38. [CrossRef]

25. Perioli, L.; Ambrogi, V.; Pagano, C.; Scuota, S.; Rossi, C. FG90 chitosan as a new polymer for metronidazole mucoadhesive tablets for vaginal administration. *Int. J. Pharm.* **2009**, *377*, 120–127. [CrossRef] [PubMed]

26. Kast, C.E.; Valenta, C.; Leopold, M.; Bernkop-Schnürch, A. Design and *in vitro* evaluation of a novel bioadhesive vaginal drug delivery system for clotrimazole. *J. Control. Release* **2002**, *81*, 347–354. [CrossRef]

27. Liu, Y.; Zhu, Y.; Wei, G.; Lu, W. Effect of carrageenan on poloxamer-based *in situ* gel for vaginal use: Improved *in vitro* and *in vivo* sustained-release properties. *Eur. J. Pharm. Sci.* **2009**, *37*, 306–312. [CrossRef] [PubMed]

28. Fernández-Romero, J.A.; Abraham, C.J.; Rodriguez, A.; Kizima, L.; Jean-Pierre, N.; Menon, R.; Begay, O.; Seidor, S.; Ford, B.E.; Gil, P.I.; *et al.* Zinc acetate/carrageenan gels exhibit potent activity *in vivo* against high-dose herpes simplex virus 2 vaginal and rectal challenge. *Antimicrob. Agents Chemother.* **2012**, *56*, 358–368. [CrossRef] [PubMed]

29. Mahalingam, A.; Smith, E.; Fabian, J.; Damian, FR.; Peters, J.J.; Clark, M.R.; Friend, D.R.; Katz, D.F.; Kiser, P.F. Design of a semisolid vaginal microbicide gel by relating composition to properties and performance. *Pharm. Res.* **2010**, *27*, 2478–2491. [CrossRef] [PubMed]

30. Yang, S.; Chen, Y.; Gu, K.; Dash, A.; Sayre, C.L.; Davies, N.M.; Ho, E.A. Novel intravaginalnanomedicine for the targeted delivery of saquinavir to CD4+ immune cells. *Int. J. Nanomed.* **2013**, *8*, 2847–2858.

31. Li, N.; Yu, M.; Deng, L.; Yang, J.; Dong, A. Thermosensitive hydrogel of hydrophobically-modified methylcellulose for intravaginal drug delivery. *J. Mater. Sci. Mater. Med.* **2012**, *23*, 1913–1919. [CrossRef] [PubMed]

32. Bettini, R.; Bonferoni, M.C.; Colombo, P.; Zanelotti, L.; Caramella, C. Drug release kinetics and front movement in matrix tablets containing diltiazem or metoprolol/λ-carrageenan complexes. *Biomed. Res. Int.* **2014**, *2014*, 671532. [CrossRef] [PubMed]

33. Ruiz-Caro, R.; Gago-Guillán, M.; Otero-Espinar, F.J.; Veiga, M.D. Mucoadhesive tablets for controlled release of Acyclovir. *Chem. Pharm. Bull.* **2012**, *60*, 1249–1257. [CrossRef]

34. Campaña-Seoane, M.J.; Otero-Espinar, F.J. Nuevas emulsiones mucoadhesivas para la liberación controlada de progesterona. In Proceedings of the X Congreso de la Sociedad Española de Farmacia Industrial y Galénica, Madrid, Spain, 2–4 February 2011; pp. 141–142.

35. Mankala, S.K.; Korla, A.C.; Gade, S. Development and evaluation of aceclofenac loaded mucoadhesive microcapsules. *J. Adv. Pharm. Technol. Res.* **2011**, *2*, 245–254. [CrossRef] [PubMed]

36. Tuğcu-Demiröz, F.; Acartürk, F.; Erdoğan, D. Development of long-acting bioadhesive vaginal gels of oxybutynin: Formulation, *in vitro* and *in vivo* evaluations. *Int. J. Pharm.* **2013**, *457*, 25–39. [CrossRef] [PubMed]

37. Valenta, C. The use of mucoadhesive polymers in vaginal drug delivery. *Adv. Drug Deliv. Rev.* **2005**, *57*, 1692–1712. [CrossRef] [PubMed]

38. Owen, D.H.; Katz, D.F. A vaginal fluid simulant. *Contraception* **1999**, *59*, 91–95. [PubMed]

39. Owen, D.H.; Katz, D.F. A review of the physical and chemical properties of human semen and the formulation of a semen simulant. *J. Androl.* **2005**, *26*, 459–469. [CrossRef] [PubMed]

40. Bernkop-Schnürch, A.; Steininger, S. Synthesis and characterization of mucoadhesive thiolated polymers. *Int. J. Pharm.* **2000**, *194*, 239–247. [CrossRef]

41. Ruiz-Caro, R.; Veiga, M.D. Characterization and dissolution study of chitosan freeze-dried systems for drug controlled release. *Molecules* **2009**, *14*, 4370–4386. [CrossRef] [PubMed]

42. Haupt, S.; Zioni, T.; Gati, I.; Kleinstern, J.; Rubinstein, A. Luminal delivery and dosing considerations of local celecoxib administration to colorectal cancer. *Eur. J. Pharm. Sci.* **2006**, *28*, 204–211. [CrossRef] [PubMed]

43. Harada, S.; Koyanagi, Y.; Yamamoto, N. Infection of HTLV-III/LAV in HTLV-I-carrying cells MT-2 and MT-4 and application in a plaque assay. *Science* **1985**, *229*, 563–566. [CrossRef] [PubMed]

44. Kuramoto, H. Studies of the growth and cytogenetic properties of human endometrial adenocarcinoma in culture and its development into an established line. *Acta Obstet. Gynaecol. Jpn.* **1972**, *19*, 47–58. [PubMed]

45. Krug, H.F. *Quality Handbook: Standard Procedures for Nanoparticle Testing*; Nanommune Deliverable XYZ (EMPA): Zurich, Switzerland, 2011; pp. 130–137.

marine drugs

MDPI

Article

Physical Stability Studies of Semi-Solid Formulations from Natural Compounds Loaded with Chitosan Microspheres

Niuris Acosta [1,*]**, Elisa Sánchez** [1]**, Laura Calderón** [1]**, Manuel Cordoba-Diaz** [2]**, Damián Cordoba-Diaz** [2]**, Senne Dom** [2] **and Ángeles Heras** [1,*]

[1] Department of Physical Chemistry II, Faculty of Pharmacy, Institute of Biofunctional Studies, Complutense University, Paseo Juan XXIII n° 1, Madrid 28040, Spain; elisas01@ucm.es (E.S.); lcaldero@ucm.es (L.C.)

[2] Department of Pharmacy and Pharmaceutical Technology, Faculty of Pharmacy, Institute of Industrial Pharmacy, Complutense University, Plaza Severo Ochoa s/n., Madrid 28040, Spain; mcordoba@ucm.es (M.C.-D.); damianco@ucm.es (D.C.-D.); sene_dom@hotmail.com (S.D.)

* Authors to whom correspondence should be addressed; facosta@ucm.es (N.A.); aheras@ucm.es (A.H.); Tel.: +34-913943284 (N.A. & A.H.).

Academic Editor: Hitoshi Sashiwa
Received: 5 August 2015; Accepted: 9 September 2015; Published: 16 September 2015

Abstract: A chitosan-based hydrophilic system containing an olive leaf extract was designed and its antioxidant capacity was evaluated. Encapsulation of olive leaf extract in chitosan microspheres was carried out by a spray-drying process. The particles obtained with this technique were found to be spherical and had a positive surface charge, which is an indicator of mucoadhesiveness. FTIR and X-ray diffraction results showed that there are not specific interactions of polyphenolic compounds in olive leaf extract with the chitosan matrix. Stability and release studies of chitosan microspheres loaded with olive leaf extract before and after the incorporation into a moisturizer base were performed. The resulting data showed that the developed formulations were stable up to three months. The encapsulation efficiency was around 44% and the release properties of polyphenols from the microspheres were found to be pH dependent. At pH 7.4, polyphenols release was complete after 6 h; whereas the amount of polyphenols released was 40% after the same time at pH 5.5.

Keywords: chitosan; spray-drying; semi-solid formulation; antioxidant encapsulation

1. Introduction

Chitosan is a linear biopolyaminosaccharide obtained by alkaline deacetylation of chitin, which is the second most abundant polysaccharide, surpassed only by cellulose [1]. Chitosan possesses free amino and hydroxyl groups, which facilitate its cross-linking reaction to form chitosan cross-linked microspheres [2]. The use of chitosan for the encapsulation of active components has attracted interest in recent years due to its mucoadhesiveness, non-toxicity, biocompatibility, and biodegradability. Antibacterial and moisturizing properties and other beneficial features of chitosan, especially in emulsions and topical gels for application in biomedicine and cosmetics, have been extensively described [3,4]. The cationic character of chitosan, along with the presence of reactive functional groups, provides particular possibilities for utilization in controlled-release technologies [5]. Numerous varieties of chitosan can be obtained due to the different chitin sources and to differences in deacetylation conditions. The presence of free amino groups is responsible for the interaction of chitosan with biological systems, and the distribution of deacetylated groups along the chitosan molecule may regulate these interactions. The wide variety of products that can be obtained as a result of the chemical modification of chitosan can enhance its already valuable properties. A

Mar. Drugs **2015**, *13*, 5901–5919

wise selection of the suitable type of chitosan can lead to the development of customized delivery systems. Chitosan-based modified release systems may prolong the duration of active agent activity, improving its efficiency and reducing side effects under desired conditions, through the modification of its pharmacokinetic characteristics. It has been found that chitosan has a slight positive charge, which makes it soluble in acid or neutral solutions depending on the pH and degree of deacetylation. This feature also provides bioadhesive properties because chitosan can ionically link to mucosal surfaces. Due to this physical property, chitosan facilitates the transport of polar active ingredients through the epithelial surfaces. The benefits of encapsulating active agents in a polymer matrix include their protection from the surrounding medium or processing conditions and the possibility of controlling release [6]. The combination of chitosan, a natural polymer, with other compounds such as antioxidants generates a new system joining the properties of both components, which improves the stability of the antioxidants and controls their pharmacokinetic properties [7].

The olive leaf extract (*Olea europea* L.) is rich in phenolic compounds, with strong antioxidant activity. The most abundant phenolic compounds available in the olive leaf are tyrosol and hidroxytyrosol. Hydroxytyrosol is known for its capacity to stop oxidative stress and neutralize free radicals [8]. Due to its beneficial properties and abundance, olive leaf extract has been chosen as a source of antioxidant compounds for our study.

The popularity of natural ingredients in cosmetics is increasing in the market to satisfy the needs of consumers. The possibility of allergies and skin irritations due to synthetic preservatives (such as parabens and stabilizers) has not yet been fully tested for long-term consequences on the health of consumers [9]. Some vegetable oils containing essential fatty acids (EFAs) have proven to be of great use in the formulation of cosmetics, either as a component of high potential activity or as precursors for the synthesis of novel compounds. It is well described that EFAs are easily integrated into the hydro-lipid layer structures of the skin and they provide nourishing, moisturizing, and protective properties. Apart from their moisturizing, smoothing, and anti-inflammatory effects, these products reduce skin aging with their antioxidant and stabilizing action on the cellular membranes [10].

The human skin is an important target for drug delivery due to the inherent advantages of the transdermal route in the treatment of some pathologies. Many formulations have been designed for transdermal delivery of different substances using different strategies, in spite of the fact that one of the most important functions of the skin is as a barrier against xenobiotic agents [11].

In this study, different topical formulations including essential oils were designed in order to examine its stability over time under different conditions and to study their beneficial effects. The work described here was complemented with the use of chitosan microparticles loaded with olive leaf extracts to examine its influence on the organoleptic characteristics of the emulsions and to enhance their antioxidant effects over the skin in terms of the release of antioxidant compounds.

The main aim of this study is focused in the incorporation of chitosan microparticles loaded with olive leaf extracts in different topical formulations for cosmetic use. The obtained microspheres were previously characterized in terms of morphology, zeta potential, particle size, drug-polymer interactions, encapsulation efficiency, and *in vitro* release profile. Physicochemical and stability studies of the different formulations obtained and the release profiles of the active ingredient were also carried out over time.

2. Results and Discussion

2.1. Physico-Chemical Characterization of Chitosan Batches

The average molecular weight and deacetylation degree of chitosan are two of the main characteristics that affect the physico-chemical and biological properties of chitosan and its derivatives [12]; a structural analysis and determination of this parameter became necessary in order to establish a relationship to the functional properties. The molecular weight average was determined by gel permeation chromatography, obtaining a value of 84,500 Da; this value indicates the presence of relatively long polymer chains that generate potential crossovers that can provoke a decrease in water solubility [13]. The degree of deacetylation was determined by nuclear magnetic resonance (^{1}H-NMR), obtaining an average value of 84%, which denotes the presence of a high percentage of NH_2 protons with respect to the acetamides groups. It is well known that the molecular weight is related to the acetylation degree. In this way, an increase of the molecular weight is associated to an increase of the acetylation degree. Chitosan has a deacetylation degree with high value, which indicates a high solubility of the chitosan in an acidic medium due to the protonation of its amino groups.

2.2. Antioxidant Activity and Total Amount of Polyphenols in the Olive Leaf Extract

The values of the antioxidant activity of the olive leaf extract were obtained by comparing the absorbance change at 595 nm in test reaction mixtures with those containing ferrous ions at a certain known concentration using a calibration curve. The antioxidant activity of the olive leaf extract selected for our study was 73.34 ± 0.7 mM (mean ± SD of six determinations).

The Folin-Ciocalteu method [14] was used to measure the total amount of polyphenols presents in the olive leaf extract. Absorbance of the developed coloration was measured at 765 nm against a blank sample. Gallic acid was used as the standard and the results were expressed as g/L of gallic acid equivalents. The value of the total amount of polyphenols present in our olive leaf extract was 9.28 ± 0.03 g/L.

These results correlate quite well to those obtained by other authors [15] in terms of total phenol values that were congruent with FRAP and Folin assays: antioxidant values with R^2 values of 0.932 and 0.736, respectively. The results suggested that the olive leaf extract selected for our study constitutes a natural alternative to synthetic antioxidants due to the high amount of antioxidants present within its composition, being a promising natural alternative for the prevention of oxidative damage.

2.3. Physico-Chemical Characterization of Chitosan/OLE Microspheres

Chitosan microparticles were prepared by spray-drying, a technique commonly used in pharmacy to produce a dry powder from a liquid phase [16,17]. The yield of the spray-drying process was around 70%. The morphology of the resulting chitosan microspheres loaded with olive leaf extract is shown in Figure 1A,B. All prepared microparticles were spherical and showed a smooth surface without any indentations or irregularities. Particle size analysis revealed a highly dispersed distribution, although it could be observed that the vast majority of the studied particles showed a diameter below 5 μm.

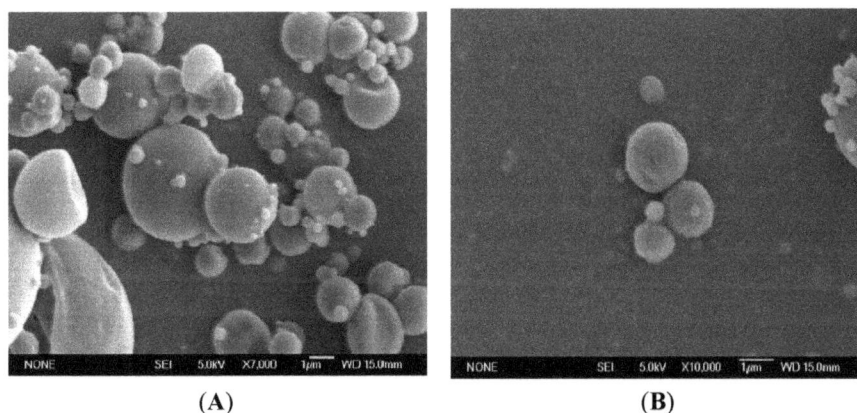

(A)   (B)

Figure 1. SEM micrographs of chitosan microspheres loaded with olive leaf extract. (**A**) Microspheres at 7000 magnification and (**B**) microspheres at 10000 magnification.

The total polyphenol content in the obtained chitosan microspheres was measured and the encapsulation efficiency was near 44%. Therefore, it was proven that our chitosan microspheres maintained the stability of the polyphenols over time and constitute a good vehicle for the encapsulation of these compounds, which corroborates the findings of other authors for this kind of system e.g., [5], who also observed that the encapsulation of olive leaf extract by spray-drying did not lead to the inactivation of the polyphenolic compounds.

The chitosan FTIR spectra shown in Figure 2 exhibited distinctive absorption bands at 1658 cm^{-1} (Amide I), 1595 cm^{-1} (-NH$_2$ bending) and 1320 cm^{-1} (Amide III). The absorption bands at 1154 cm^{-1} (anti-symmetric stretching of the C-O-C bridge), 1080, and 1030 cm^{-1} (skeletal vibrations involving the C-O stretching) are characteristic of its saccharide structure [18,19]. It can be also observed that in the spectrum of chitosan microspheres the amine band shifted to 1560 cm^{-1}. This shift has been previously reported for chitosan acetate films [20]. The intensity of the band at 1560 cm^{-1} was increased by the contribution of the asymmetric COO$^-$ stretching vibration, resulting from the carboxylate groups of the acetate ion. The symmetric COO$^-$ stretching band can be observed at 1406 cm^{-1}.

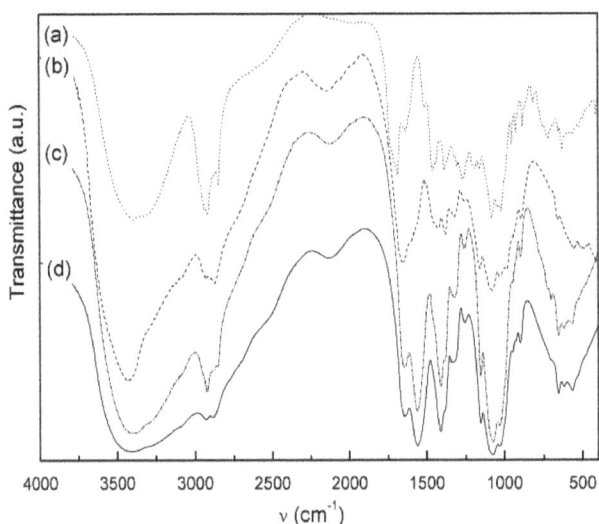

Figure 2. FTIR of olive leaf extract (**a**); chitosan (**b**); olive leaf extract loaded microspheres (**c**); chitosan microspheres (**d**).

The IR spectrum of the chitosan microspheres loaded with olive leaf extract was found to be very similar to the spectrum obtained from the unloaded microspheres. The main difference between them lies in the presence of a stronger C-H stretching absorption band at 2919 cm^{-1} in the loaded microspheres. This band was also observed in the spectrum of the olive leaf extract shown in Figure 2.

The above results corroborate the presence of olive leaf extract in the microspheres. However, from the analysis of IR spectra no specific interactions between the extract and chitosan were evident.

In order to ascertain the possible interaction of polyphenols and the polymer, some experiments were performed. XRD plays a prominent role in the characterization of polymeric particles because it is able to provide structural information on the dispersed particles [21]. The X-ray diffractions of chitosan, empty microspheres, olive leaf extract, and OLE/chitosan microspheres are shown in Figure 3. The diffractogram from the chitosan sample showed a weak peak at 2θ of 10° and a more intense peak at 2θ of 20°, caused by diffraction from (020) and (110) planes of the crystalline lattice with inter-planar distances of 0.88 nm and 0.45 nm [22]. The diffractogram of the chitosan sample indicated that the crystalline phase coexists with a considerable amount of amorphous phase [23]. Comparing the diffractograms of chitosan and chitosan microspheres, it can be seen that the crystallinity decreases with the formation of chitosan microspheres. On the other hand, olive leaf diffractograms showed various peaks not present in the diffractogram of OLE/chitosan microspheres, which indicated that the olive leaf extract is molecularly dispersed within the polymer matrix. It was proven that the incorporation of the olive leaf extract into the chitosan microspheres did not affect the amorphous structure of the particles, which agrees with findings reported by Yenilmez *et al.* [24].

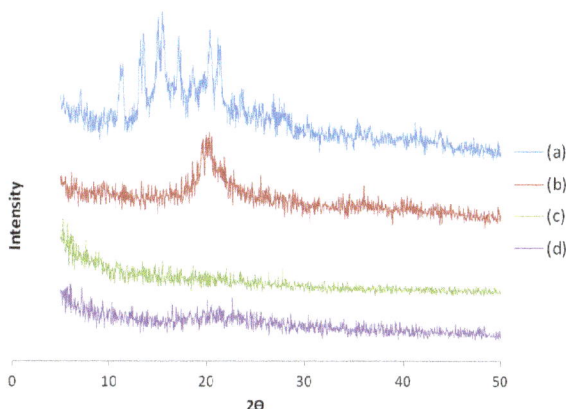

Figure 3. XRD of olive leaf extract (**a**); chitosan (**b**); olive leaf extract loaded microspheres (**c**); and chitosan microspheres (**d**).

The surface charge of the microspheres was measured to evaluate the stability of the suspended particles. The positive surface charge of the chitosan microspheres loaded with olive leaf extract promotes mucoadhesion and hinders aggregation, which results in particles with a good stability. The zeta potential of microspheres was +47.58 ± 1.48 mV (mean ± SD of three determinations). These delivery systems have mucoadhesive potential and absorption enhancement properties [25].

2.4. In Vitro Release Profile of Olive Leaf Extract Loaded Microspheres

The objective of the encapsulation of antioxidants is their release into the skin to prevent damage caused by free radicals and to help reversing the signs of aging. The extract of olive leaves is rich in polyphenols and these have a high antioxidant power.

Although the pH of the skin is 5.5, the pH of cosmetic formulations can range between 5.5 and 7 [26]. The release studies of chitosan microspheres loaded with olive leaf extract were performed in two buffers with different pH values (pH 5.5 and 7.4). Release profiles at pH 5.5 and 7.4 are shown in Figure 4.

The release profiles of chitosan microspheres loaded with olive leaf extract were investigated *in vitro* at 37 °C. As can be seen in Figure 4, a complete release of polyphenols was achieved after 6 h at pH 7.4, whereas at pH 5.5 the maximum release was near 35% of the initial amount of polyphenols. It can be seen that the differences of the release rate at different pH values were clear after the first 2 h. It was found that the release rate at a pH of 7.4 doubled the one obtained at a pH value of 5.5. According to the figures, it can be concluded that the polyphenols' release was pH dependent [27]. The low solubility of chitosan with a smaller number of acetyl groups is caused by difficulties in attaching protons to amino groups that are close to protonated amino groups, due to electrostatic repulsive forces. The diffusion of the polyphenols was enhanced at high pH due to the deprotonation of chitosan, resulting in an improved release. So we can conclude with these results that an effective release was obtained at a pH of 7.4. These results also confirm the antioxidant profile release, corroborating the previous findings of other authors [18] who observed that the encapsulation of OLE by spray-drying did not lead to an inactivation of the polyphenolic compounds.

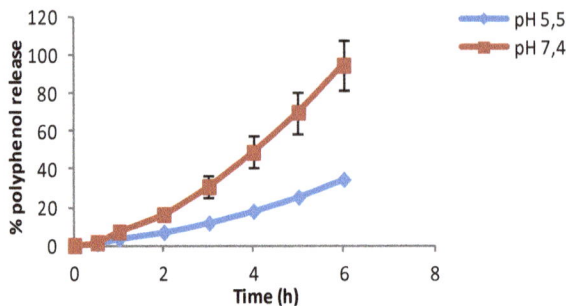

Figure 4. *In vitro* release profile of polyphenols of chitosan microspheres loaded with olive leaf extract in buffers pH 5.5 and pH 7.4.

2.5. Physico-Chemical Characterization of Semi-Solid Formulations

A physicochemical characterization of the three semi-solid moisturizing formulations was performed before carrying out the *in vitro* release assays, in order to measure their extensibility, penetrometry, and pH. The pH was around 5 in formulation B and C. This pH value was conditioned mainly by the essential oils present in their composition. Emulsion A had a higher pH value, caused by a greater amount of water in its composition. The extensibility coefficient turned out to be 2.000 mm^2 for formulations B and C. In formulation A, the extensibility was 2.376 mm^2 due to the higher amount of water, resulting in a reduction of the consistency. Formulations B and C showed penetrometry values of 3.300 g·cm/s^2. This parameter resulted to be only 2.600 g·cm/s^2 for formulation A. No significant data differences among the different formulations were observed in extensibility and penetrometry tests.

Under normal conditions, it is accepted that skin pH ranges between 5 and 6. Hence, a topical formulation should have a pH near that range to avoid irritability and problems of tolerance. For this reason, pH was one of the parameters studied to evaluate the physical stability of our emulsions. It was found that the pH values of all formulations ranged between 5 and 6 depending on storage time. The stability studies were performed comparing, among other parameters, the controlled pH variations along different times. The results showed no significant differences between the studied formulations.

All the formulations were also tested from an organoleptic point of view due to the big importance of these properties for consumers, especially in cosmetics. The physical appearance of the emulsion formulations kept at 25 °C was examined visually at different times. In this connection, the formulations prepared were found to be stable in structure, without any change in physical appearance, and were found to keep their clear, transparent, and uniform appearance. No phase separation was visually observed in formulations after stored for three months at the same conditions.

Viscosity data and rheograms are shown in Table 1 and Figure 5A–C. Typical flow curves for semisolid formulations with or without microspheres as a function of ascending-descending shear rate sweep are shown in Figure 5. In all experiments, shear stress increased to a maximum value at shear rate values near 25 s^{-1}, then dropped off dramatically. These results suggest that structural disorganization occurs within the moisturizer base [28]. Viscosity values decreased in all experiments when the shear rate increased; this behavior is defined as pseudoplasticity or shears thinning, and it is indicative of structural breakdown. As can be seen in Figure 5, different behavior can be observed comparing the upward and downward curves of the rheograms, due to a delay in the reorganization of the internal structure of the semi-solid formulation. This phenomenon is known as thixotropy and can be associated with a decrease in viscosity due to a structural disorder after a continuous increase of the shear rate for a certain period of time [29].

Table 1. Viscosity data of the different formulations with and without microspheres.

Formulation	Viscosity V (Pa·s)	Thixotropy T (Pa)
A	3.146	8.65
A + MP	2.674	5.64
B	6.763	46.27
B + MP	5.426	37.62
C	4.404	47.58
C + MP	2.595	9.17

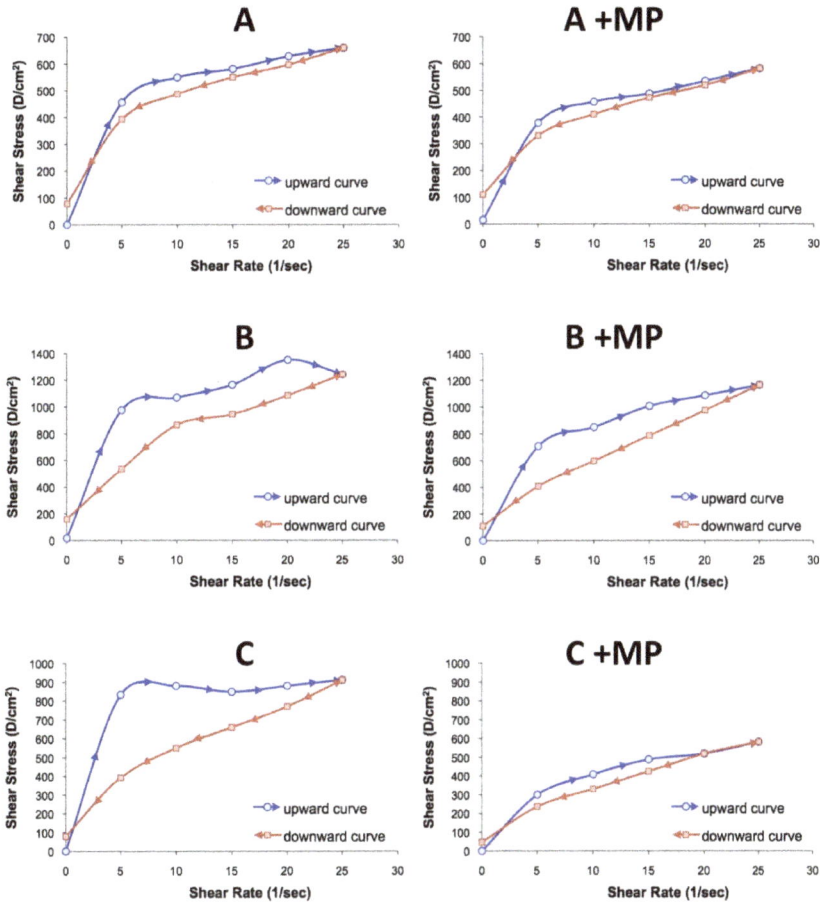

Figure 5. Rheological behavior of the semi-solid formulations with increasing shear rate. (**A–C**) Formulations without microspheres and A + MP, B + MP, and C + MP formulations with microspheres.

It was observed that the incorporation of chitosan microspheres into the semisolid formulations provoked significant changes over the rheological behavior. Formulations with a pseudoplastic flow produce a coherent film covering the skin surface, and this is important for a better antioxidant protection of the skin surface [30]. The results suggest that the incorporation of the microspheres into the semi-solid formulation produces changes in the internal structure of the formulation, modifying the apparent viscosity and reducing the thixotropy. These findings suggest that the inclusion of the microspheres loaded with olive leaf extract within the oil phase can modify the semi-solid structure.

Thixotropy is desirable in topical formulations because it helps to maintain the suspending components' stability; moreover it can influence the active substances' release to the skin due to the structural disarrangement of the system, where the active substances' diffusion is facilitated. Although the thixotropy and viscosity values decreased for formulations with added microspheres, the rheograms obtained showed no instability signals. These results showed that these formulations could be considered stable comparing the rheological properties.

Likewise, a total bacteria recount was performed just after the elaboration of the formulations to test the possible errors in the manipulation, obtaining a negative result.

2.6. In Vitro Release Profile of Semi-Solid Formulations

The results in Figure 6 showed that the release of antioxidants from chitosan microspheres loaded with olive leaf extract after 8 h ranged from 60% to 80% at time 0 (just after elaboration of formulations) depending on the formulation. The results indicated that the release of the polyphenols of the olive leaf from chitosan microspheres varied with the time. Moisturizer bases A + MP released 60% of the total polyphenols after 8 h, whereas formulations B + MP and C + MP released around 80% of polyphenols at 8 h; these two moisturizers are rich in vegetable oils. Moisturizers with more vegetable oils in their composition had a higher release just after elaboration.

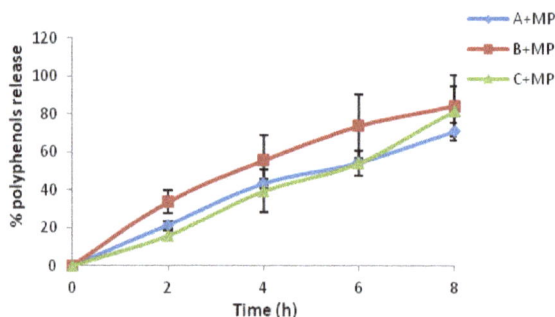

Figure 6. *In vitro* release profile of three different moisturizer bases with chitosan microspheres loaded with olive leaf extract at time 0.

After 3 months of storage at 25 °C, formulation C + MP showed the highest percentage of polyphenols release (Figure 7). This might be due to the fact that the composition of this cream is rich in fatty oils and the emulsion is less stable, which can probably lead to a slight breakdown of the emulsion, or to the fact that fatty oils interfered with the measurements. The olive leaf extract release increases proportionally over time.

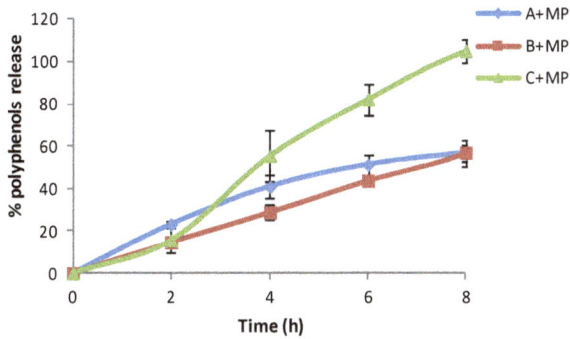

Figure 7. *In vitro* release profile of three different moisturizer bases with chitosan microspheres loaded with olive leaf extract stored for three months at 25 °C.

After 3 months of storage at 4 °C, formulation A + MP showed the higher percentage of polyphenol release (Figure 8). This might be due to the preservation of the formulation in low temperatures. The olive leaf extract release increases proportionally over time.

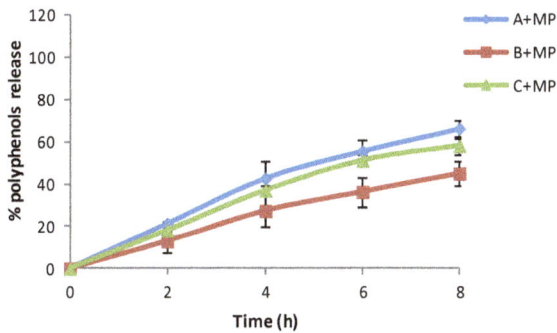

Figure 8. *In vitro* release profile of three different moisturizer bases with chitosan microspheres loaded with olive leaf extract stored for three months at 4 °C.

An individual analysis of each formulation at different storage temperatures evidenced some particularities.

Figure 9 shows the *in vitro* release profile of Formulation A + MP stored at three temperatures for different lengths of time. In this formulation the percentage of *in vitro* release is maintained for long periods of time. The microspheres and the composition of the formulation are stable after three months at different storage temperatures.

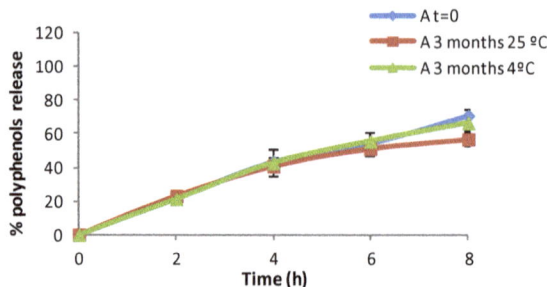

Figure 9. *In vitro* release profile of Formulation A + MP.

Figure 10 shows that at the initial time of the elaboration, the *in vitro* release of microspheres in formulation B + MP is more effective than that obtained after three months. Here, the temperature affects the physical stability of the formulation. It was found that the physical stability was improved by refrigerating the formulation (storage at 4 °C). Formulation C + MP in Figure 11 showed a higher release at high temperatures. This could be due to the breaking of the phases of the emulsion, resulting in the detection of polyphenols in the spectrophotometer. At time 0 (just after elaboration) the percentage of *in vitro* release is higher than after three months stored at 4 °C. This result is remarkable as the release was expected to remain more stable after storage in the fridge than after storage at room temperature. After comparing the *in vitro* release at different temperatures, we concluded that storage at 4 °C is the best method for maintaining the stability of the present formulations.

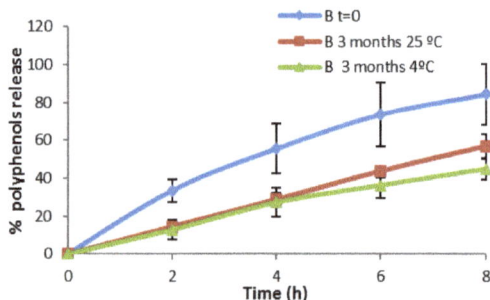

Figure 10. *In vitro* release profile of Formulation B + MP.

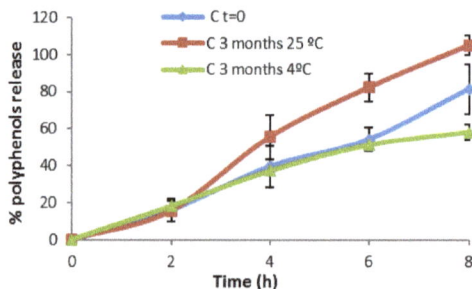

Figure 11. *In vitro* release profile of Formulation C + MP.

The encapsulation of olive leaf extract into chitosan microspheres aimed to protect the polyphenols from degradation and to control their release to obtain a longer lasting effect. It is clear that formulations A + MP and B + MP appeared to have the most promising results if we compare the release profiles of the different formulations.

A comparison of the different permeability values of the formulations is represented. The permeability coefficients for formulation A + MP and B + MP follow the same pattern: at time 0 the permeability coefficients are the highest (0.0073 cm/h and 0.0044 cm/h, respectively); these values decreased after 3 months of storage at 25 °C (0.0034 cm/h and 0.0022 cm/h, respectively). This decrease was also found under storage conditions of 4 °C (0.0043 cm/h and 0.0038 cm/h, respectively). This decrease is bigger for formulation A. After 3 months at 4 °C permeability, constants are higher than after 3 months at 25 °C, which is normal as storage in the fridge is considered to be better for the stability of the formulation. In contrast, Formulation C + MP showed different results: compared to time 0 (0.0065 cm/h), the permeability coefficient increased slightly after 3 months at 25 °C (0.0067 cm/h) and decreased after 3 months at 4 °C (0.0055 cm/h), which can be associated with a degradation process. The permeability coefficients obtained in these experiments gave an estimation of the release of polyphenols from the vehicle into the surface structures of the skin, providing an idea of the permeability of polyphenols through the skin.

Chitosan microspheres loaded with olive leaf extracts enhance the antioxidant effect of the polyphenols present in the moisturizer base. The chitosan microspheres have greater facility to diffuse into the medium and the release of the polyphenols was improved, enhancing the desired effect. Both the polyphenols present in vegetable oils and the polyphenols from the microspheres loaded with olive leaf extracts add value to the moisturizer base by providing antioxidant activity; this plays an important role in skin damage caused by oxidizing agents that are good regenerators and effective for reconstructing skin.

3. Experimental Section

Chitosan was obtained from InFiQuS S.L. The degree of deacetylation was 84% and the molecular weight (Mw) was 84.5 KDa. Olive leaf extract was provided by the University of Cordoba (Research group FQM-227, Spain). Formulation ingredients (calendula oil, jojoba oil, olivem® 1000, glycerol stearate, xanthan gum, candy dye, eucalyptus water, sodium stearoyl lactylate, beeswax, shea butter, roses water, avocado oil, and chlorophyll dye) were purchased from Aromazone and were of pharmaceutical quality. All other reagents were commercially available and used as received.

3.1. Physico-Chemical Characterization of Chitosan

3.1.1. Determination of Molecular Weight by Size Exclusion Chromatography

SEC-HPLC was performed in a Waters 625 LC System pump equipped with an Ultrahydrogel (Waters, i.d = 7.8 mm l = 300 mm) column thermostated at 35 °C. Waters 2414 differential refractometer and evaporative light scattering (ELS Waters 2424, Milford, MA, USA) were connected online. A 0.15 M ammonium acetate/0.2 M acetic acid buffer (pH 4.5) was used as eluent. Analytical samples were dissolved in the buffer solution (0.05%, w/v) and filtered through a 0.45 µm pore size membrane (Millipore Corporation, Beverly, MA, USA) before injection of aliquots of 20 µL. The flow rate was 0.6 mL/min. The Mw of the different fractions were obtained from the SEC profiles by extrapolation in a calibration curve using different known Mw chitosans as standards.

3.1.2. Determination of Acetylation Degree by Nuclear Magnetic Resonance (^{1}H-NMR)

Structural characterization of chitosan samples was carried out using nuclear magnetic resonance spectroscopy (^{1}H-NMR). Samples were dissolved in a DCl/D2O 1% (w/v) mixture and placed in 5 mm NMR tubes. TSP (sodium 3-(trimethylsilyl) propionate-d4) was used as an internal standard. The measurements in the experiments were performed on an AMX 500 spectrometer (Bruker Ettinglen,

Germany). The acquisition conditions were: 2s number of scans 128 for ^{1}H-NMR experiments and a frequency of 125.77 MHz.

3.2. Antioxidant Activity and Total Amount of Polyphenols in the Olive Leaf Extract

Antioxidant activity was determined by the ferric reducing/antioxidant power (FRAP) assay at 595 nm. The method is based on the reduction of a ferric 2, 4, 6-tripyridyl-s-triazine complex (Fe^{3+}-TPTZ) to the ferrous form (Fe^{2+}-TPTZ) [31,32] using trolox (a water-soluble vitamin E analogue) as the standard.

Total polyphenols content was spectrophotometrically quantified at 750 nm using the Folin–Ciocalteu reagent [14] and gallic acid as the standard. GBC UV/Vis 920 spectrophotometer (GBC Scientific Equipment PTY LTD, Melbourne, Victoria, Australia) was used in both methods.

3.3. Preparation of Chitosan Microparticles Loaded with Olive Leaf Extract

Microspheres were prepared by the spray-drying technique [33]. Briefly, an accurately weighed amount of chitosan was dissolved in 0.2% acetic acid aqueous solution to obtain a concentration of 0.83% w/v. Olive leavf extract (OLE) (10% w/v) mixed with ethanol (96%) was added dropwise to the acidic chitosan solution, under mild agitation. The resulting solution was spray-dried using a mini spray-dryer (Büchi Mini Spray-Dryer B-290, Switzerland) with an inlet temperature of 160 °C and an outlet temperature of 80 °C. The effect of the encapsulating system on the active compound stability and its release profiles was analyzed.

3.4. Physico-Chemical Characterization of the Chitosan Microparticles

3.4.1. Estimation of the Encapsulation Efficiency of Polyphenolic Compounds

Five milligrams of chitosan microspheres loaded with olive leaf extracts were incubated with 1 N hydrochloric acid solution for 24 h. The amount of encapsulated olive leaf extract was estimated spectrophotometrically by the Folin-Ciocalteu method, expressed as a loading percent in terms of the amount of gallic acid in 100 g of microspheres.

3.4.2. Evaluation of the Microparticles' Morphology by Scanning Electron Microscopy

The shape and surface of the obtained microparticles were studied by scanning electron microscopy (SEM). The samples were examined using a scanning electron microscope JEOL JSM-6400 (JEOL, Tokyo, Japan). Microparticle surfaces were sputter-coated with a thin layer of gold for SEM visualization. The SEM images were taken by applying an electron-accelerating voltage of 15 kV.

3.4.3. Drug-Polymer Interactions

Infrared Spectroscopy FT-IR

The FT-IR spectra were obtained using a Magna-IR 750 (Nicolet) spectrometer with a resolution of 4 cm^{-1} (50 scans, resolution 4.000, wavenumber range 4000–500 cm^{-1}). X-ray diffraction patterns were obtained using an X-ray diffractometer (PHILIPS X0PERT SW) equipped with a copper anode. The samples were scanned continuously from 0° to 50° (2θ) at 45 kV and 40 mA.

3.4.4. Zeta Potential Measurements

The Z potential value of the chitosan microspheres in aqueous solution was measured using a Nano-Zetasizer system (Malvern Instruments, Herrenberg, Germany). Every measurement was carried out in three serial measurements. Electrophoretic mobility, Z-potential, average size, and polydispersity index were obtained for each batch of microparticles.

3.4.5. *In Vitro* Release Profile of Olive Leaf Extract Encapsulated in Microspheres

The release of olive leaf extracts expressed as the total amount of polyphenols under predetermined conditions was investigated. Microspheres (50 mg) (Mw cut off 12,000 Da, Sigma-Aldrich, Madrid, Spain) were suspended in 5 mL of PBS solutions (Sigma-Aldrich, Madrid, Spain) with different pH values (pH 5.5 and pH 7.4), inside a cellulose dialysis bag. The dialysis bag was immersed in 50 mL of phosphate buffered saline (PBS) release medium with different pH values (pH 5.5 and pH 7.4) at 37 °C and 100 rpm (Rotabit horizontal shaker, Selecta, Barcelona, Spain). One-milliliter samples were taken at specific time intervals and replaced with fresh medium. The release of the polyphenols was quantified using the Folin-Ciocalteu method. All the experiments were carried out in triplicate.

3.5. Design of Semi-Solid Formulations Containing Chitosan Microspheres

Microspheres were incorporated into three different base formulation prepared by o/w and w/o emulsions. The water phase and oil phase were heated separately up to 60 °C until homogenization of all the components. Microspheres were added into the oil phase at a concentration of 0.2% (w/w). Finally, the water phase was mixed with the oil phase and vigorously stirred together. For control purposes, a moisturizer sample without added microspheres was also prepared. Samples with and without microspheres were analyzed. Compositions of the semi-solid formulations are shown in Table 2.

Table 2. Compositions of the semi-solid formulations.

Formulation	Oil Phase Ingredients	Water Phase Ingredients
A(o/w emulsion)	0.8% (w/w) calendula oil 2.5% (w/w) jojoba oil 13% (w/w) olive oil 1.6% (w/w) olivem® 1000 (emulsifier) 2.5% (w/w) glycerol stearate (emollient) 0.35% (w/w) xanthan gum (thickener) 0.25% (w/w) candy dye 0. 2% (w/w) OLE/chitosan microspheres	52.4% (w/w) purified water 26.4% (w/w) eucalyptus water
B (o/w emulsion)	6.5% (w/w) olive oil 3.7% (w/w) jojoba oil 5.5% (w/w) calendula oil 4.5% (w/w) sodium stearoyl lactylate 0.9% (w/w) beeswax 8% (w/w) glycerol stearate (emollient) 1.7% (w/w) shea butter 0. 2% (w/w) OLE/chitosan microspheres	46% (w/w) purified water 23% (w/w) roses water
C (w/o emulsion)	9.7% (w/w) jojoba oil 29% (w/w) avocado oil 5.5% (w/w) beeswax 3.2% (w/w) sodium stearoyl lactylate (emulsifier) 4.3% (w/w) shea butter 0.10% (w/w) chlorophyll dye 0. 2% (w/w) OLE/chitosan microspheres	32% (w/w) purified water 16% (w/w) eucalyptus water

Physico-Chemical Characterization of Semi-Solid Formulations

The organoleptic properties including general appearance, consistence, suffusing onto the skin, absorption, homogeneity of the formulation, absence of phase separation, and creaming and oily sensation on the skin were evaluated. Furthermore, certain changes in the product emerging in short-term storage such as color and odor were observed.

Rheological characterization of all the developed semi-solid formulation was carried out. Flow properties were studied using a previously calibrated Brookfield rotational viscosimeter (model HB, Brookfield Engineering Laboratories, Stoughton, MA, USA) equipped with a cone-plate spindle (CP52). The rheograms were recorded at different shear rates increasing from 0 to 25 s^{-1} for the upward curves, and then decreasing from 25 to 0 s^{-1} for the downward curves. All the rheograms were recorded within a total time of 240 s, in a thermostated cell at 25 °C. Data were collected and processed using the Rheocalc 32 software (Brookfield Eng., Middleborough, MA, USA). Rheological behavior was studied for each formulation and a viscosity value (in Pa·s) was recorded at a fixed rotational speed of 10 rpm for comparison.

Extensibility is defined as the area occupied by a given amount of sample to be subjected to a standard pressure between two glass plates. Extensibility tests were performed following the methods described by Campaña-Seoane [34]. Briefly, an extensometer consisting of two glass slides of 5 cm × 5 cm was used. One gram of the sample was deposited on the very center of the base plate and the second plate was disposed onto the sample. After a 3 min equilibrium time, the area of the formulation spread between the two plates was measured. The samples were compressed to uniform thickness by placing different weights (50, 100, 200, and 250 g) for 30 s. After each time, the area of the sample was determined. All experiments were conducted in triplicate and at room temperature.

Penetrometry experiments were carried out to determine the consistency of the formulations, using an Analis penetrometer equipped with a penetrating cone (Analis Instruments, Namir, Belgium), following the method described by European Pharmacopeia [35].

Microbiological quality was also determined for all the formulations. For the total bacteria recount, 0.5 g of each formulation was dissolved in 4.5 mL of peptone water. Serial dilutions were performed and seeded in Agar medium after incubation for two days at 37 °C. For the recount of coliform enterobacteria, 0.5 g of the formulations was dissolved in 4.5 mL of peptone water. After serial dilutions, the samples were seeded in Violet Red Bile Glucose (VRBG) medium after incubation for 24 h at 37 °C.

3.6. In Vitro Release Profile of Semi-Solid Formulations

A previously validated [36] PermeGear ILC-07 automated system (PermeGear, Riesgelsville, PA, USA) was used to perform the *in vitro* release studies. Briefly, the equipment incorporated seven in-line flow-through diffusion cells, made of Kel F, in which the donor and receptor chambers and the diffusion membrane placed over a support with a hole of 1 cm in diameter (having a diffusional area of 0.785 cm^2) were clamped by threaded rods with adjustable locking nuts. Cellulose acetate dialysis membrane (average cutoff 12.000 Da) was selected as the diffusion membrane throughout all the experiments. All cells were placed in a cell warmer connected with a Haake-DC10 circulating bath (Gebruder Haake, Karlsruhe, Germany) to permit the system to be brought up to 37 °C. The receptor fluid consisting of phosphate buffered saline solutions (PBS pH 7.4 Sigma-Aldrich, Madrid, Spain) was pumped at a flow rate of 2 mL/h and collected in the receptor tubes in an Isco R$^{©}$ Retriever IV fraction collector (Isco, Lincoln, NE, USA).

The evolution of the concentration of olive leaf extracts into the receiver chamber (Crec) of a flow-through diffusion cell *versus* time (*t*) was determined using the following equation:

$$\text{Vrec} \times d\text{Crec}/dt = J \times A - \text{Frec} \times \text{Crec},$$

where, Vrec is the volume of the receiver chamber, J the flux of drug through the membrane, A the diffusion area, and Frec the flow rate of receptor fluid. In this way, the term dCrec/dt could be easily estimated from the concentration *versus* time raw data.

The amounts of antioxidant components released from the different formulations containing olive leaf extracts were determined spectrophotometrically by the Folin–Ciocalteu method. All data were expressed as mean ± confidence interval. A *p*-value < 0.05 was considered to be statistically significant using the *t*-test between the two means for the unpaired data. Data analyses were conducted with SPSS software, version 14.0 (SPSS Science, Chicago, IL, USA).

4. Conclusions

Chitosan microspheres are adequate vehicles for the encapsulation of natural antioxidants from olive leaf extract. The methodology of the incorporation of the microspheres comprises its addition to the oil phase, guaranteeing a good dispersion. According to the *in vitro* release studies, chitosan microspheres loaded with olive leaf extract showed a progressive release. Likewise, it has been found that the incorporation of these microspheres in cosmetic moisturizers was effective because an adequate release of the active antioxidant ingredients from the microspheres was obtained.

The antioxidant effect was maintained in the long term in the developed formulations. The physico-chemical characteristics of the formulations were not negatively affected by the incorporation of the microspheres into the formulation. The antioxidant capacity of olive leaf extract was maintained in the long term, thus opening promising new perspectives for the application of chitosan microcapsules loaded with olive leaf extract in the cosmetic industry.

It was demonstrated that this microencapsulated system containing olive leaf extract and chitosan in semi-solid formulation is valid for applications in dermatology and cosmetics. We therefore propose a new system based on chitosan and olive leaf extracts with potential applications in the cosmetic dermatology and cosmetic industries.

Acknowledgments: This work was supported by a project (MAT 2014) from the Spanish Ministry of Economy and Competitiveness (Spain).

Author Contributions: Niuris Acosta and Ángeles Heras wrote the manuscript. Elisa Sánchez, Laura Calderón, Manuel Córdoba-Diaz, Damian Córdoba-Díaz, Senne Dom contributed with the article with a critical revision. All authors read and approved the final manuscript.

Conflicts of Interest: The authors declare no conflict of interest.

References

1. Sailaja, A.K.; Awareshwar, P.R. Chitosan nanoparticles as a drug delivery system. *Res. J. Pharm. Biol. Chem. Sci.* **2010**, *1*, 474–478.
2. Gupta, K.C.; Jabrail, F.H. Glutaraldehyde and glyoxal cross-linked chitosan microspheres for controlled delivery of centchroman. *Carbohydr. Res.* **2006**, *341*, 744–756. [CrossRef] [PubMed]
3. Caiqin, Q.; Huirong, L.; Qi, X.; Yi, L.; Juncheng, Z.; Yumin, D. Water-solubility of chitosan and its antimicrobial activity. *Carbohydr. Polym.* **2006**, *63*, 367–374.
4. He, W.; Guo, X.; Zhang, M. Transdermal permeation enhancement of *N*-trimethyl chitosan for testosterone. *Int. J. Pharm.* **2008**, *356*, 82–87. [CrossRef] [PubMed]
5. Kosaraju, S.L.; D'ath, L.; Lawrence, A. Preparation and characterization of chitosan microspheres for antioxidant delivery. *Carbohydr. Polym.* **2006**, *64*, 163–167. [CrossRef]
6. Belscak-Cvitanovic, A.; Stojanovic, R.; Manojlovic, V.; Komes, D.; Cindric, I.J.; Nedovic, V.; Bugarski, B. Encapsulation of polyphenolic antioxidants from medicinal plant extracts in alginate-chitosan system enhanced with ascorbic acid by electrostatic extrusion. *Food Res. Int.* **2011**, *44*, 1094–1101. [CrossRef]
7. Harris, R.; Lecumberri, E.; Mateos-Aparicio, I.; Mengíbar, M.; Heras, A. Chitosan nanoparticles and microspheres for encapsulation of natural antioxidants extracted from Ilex paraguariensis. *Carbohydr. Polym.* **2011**, *83*, 803–806. [CrossRef]
8. Chimi, H.; Cillard, J.; Cillard, P.; Rahmani, M. Peroxyl and hydroxyl radical scavenging activity of some natural phenolic antioxidants. *J. Am. Oil Chem. Soc.* **1991**, *68*, 307–312. [CrossRef]
9. Glampedaki, P.; Deuth, V. Stability studies of cosmetic emulsions prepare from natural products such as wine, grape seed oil and mastic resin. *J. Cosmet. Sci.* **2006**, *57*, 205–214.
10. Badiu, D.; Luque, R.; Rajendram, R. Effect of Olive Oil on the Skin. In *Olives and Olive Oil in Health and Disease Prevention*; Victor, R., Ed.; Preedy and Ronald Ross Watson: Constanza, Romania, 2010.

11. Desai, P.; Patlolla, R.R.; Singh, M. Interaction of nanoparticles and cell-penetrating peptides with skin for transdermal drug delivery. *Mol. Membr. Biol.* **2010**, *27*, 247–259. [CrossRef] [PubMed]

12. Aranaz, I.; Mengibar, M.; Harris, R.; Miralles, B.; Acosta, N.; Calderon, L.; Sanchez, A.; Heras, A. Role of Physicochemical Properties of Chitin and Chitosan on their Functionality. *Curr. Chem. Biol.* **2014**, *8*, 27–42. [CrossRef]

13. Aranaz, I.; Mengíbar, M.; Heras, A. Functional Characterization of Chitin and Chitosan. *Curr. Chem. Biol.* **2009**, *3*, 203–230.

14. Montreau, F.R. On the analysis of total phenolic compounds in wines by the Folin-Ciocalteu method. *Connaiss. Vigne Vin* **1972**, *24*, 397–340.

15. Sifaoui, I. Activity of olive leaf extracts against the promastigote stage of *Leishmania* species and their correlation with the antioxidant activity. *Exp. Parasitol.* **2014**, *141*, 106–111. [CrossRef] [PubMed]

16. Stulzer, H.K.; Tagliari, M.P.; Parize, A.L.; Silva, M.A.S.; Laranjeira, M.C.M. Evaluation of cross-linked chitosan microparticles containing acyclovir obtained by spray-drying. *Mater. Sci. Eng.* **2009**, *29*, 387–392. [CrossRef]

17. Fang, Z.; Bhandair, B. Encapsulation of polyphenols—A Review. *Trends Food Sci. Technol.* **2010**, *21*, 510–523. [CrossRef]

18. Argüelles-Monal, W.; Peniche-Covas, C. Study of the Interpolyelectrolyte Reaction between Chitosan and Carboxymethyl Cellulose. *Die Makromol. Chem. Rapid Commun.* **1988**, *10*, 693–697. [CrossRef]

19. Bocourt, M.; Argüelles, W.; Cauich, J.V.; Bada, N.; Peniche, C. Interpenetrated chitosan-poly (acrylic acid-co acrylamide) hydrogels. Synthesis, characterization and sustained protein release studies. *Mater. Sci. Appl.* **2011**, *2*, 509–520.

20. Osman, Z.; Arof, A.K. FTIR studies of chitosan acetate based polymer electrolytes. *Electrochim. Acta* **2003**, *48*, 993–999. [CrossRef]

21. Bunjes, H.; Unruh, T. Characterization of lipid nanoparticles by differential scanning calorimetry, X-ray and neutron scattering. *Adv. Drug Deliv. Rev.* **2007**, *59*, 379–402. [CrossRef] [PubMed]

22. Ioelovich, M. Crystallinity and Hydrophility of Chitin and Chitosan. *J. Chem.* **2014**, *3*, 7–14.

23. Hidalgo, C.; Fernández, M.; Nieto, O.M.; Paneque, A.A.; Fernández, G.; Lópiz, J.C.L. Estudio de quitosanos cubanos derivados de la quitina de la langosta. *Rev. Iberoam. Polím.* **2009**, *10*, 11–27.

24. Yenilmez, E.; Başaran, E.; Yazan, Y. Release characteristics of vitamin E incorporated chitosan microspheres and *in vitro-in vivo* evaluation for topical application. *Carbohydr. Polym.* **2011**, *8*, 807–811. [CrossRef]

25. Bernkop-Schnürch, A. Mucoadhesive polymers: Strategies, achievements and future challenges. *Adv. Drug Deliv. Rev.* **2005**, *57*, 1553–1555. [CrossRef] [PubMed]

26. Rosen, R.M. *Delivery System Handbook for Personal Care and Cosmetic Products: Technology, Applications and Formulations*; William Andrew: Norwich, NY, USA, 2005.

27. Popa, M.; Aelenei, N.; Popa, V.I.; Andrei, D. Study of the interactions between polyphenolic compounds and chitosan. *React. Funct. Polym.* **2000**, *42*, 35–43. [CrossRef]

28. Ramírez-Moreno, E.; Córdoba-Díaz, M.; de Cortes Sánchez-Mata, M.; Marqués, C.; Isabel Goñi, I. The addition of cladodes (*Opuntia ficus indica* L. Miller) to instant maize flour improves physicochemical and nutritional properties of maize tortillas. *LWT—Food Sci. Technol.* **2015**, *62*, 675–681. [CrossRef]

29. Prudencio, I.D.; Prudencio, E.S.; Gauche, C.; Barreto, P.; Bordignon-Luiz, M.T. Flow properties of petit Suisse cheeses use of cheese whey as a partial milk substitute. *Ital. J. Food Sci.* **2008**, *20*, 169–179.

30. Di Mambro, V.M. Assays of physical stability and antioxidant activity of a topical formulation added with different plant extracts. *J. Pharm. Biomed.* **2005**, *37*, 287–295. [CrossRef] [PubMed]

31. Benzie, I.F.F.; Strain, J.J. Ferric Reducing/Antioxidant Power Assay: Direct Measure of Total Antioxidant Activity of Biological Fluids and Modified Version for Simultaneous Measurement of Total Antioxidant Power and Ascorbic Acid Concentration. In *Methods in Enzymology*; Packer, L., Ed.; Academic Press: Waltham, MA, USA, 1999; pp. 15–27.

32. Pulido, R.; Bravo, L.; Saura-Calixto, F. Antioxidant activity of dietary polyphenols as determined by a modified ferric reducing/antioxidant power assay. *J. Agric. Food Chem.* **2000**, *48*, 3396–3402. [CrossRef] [PubMed]

33. Harris, R.; Paños, I.; Acosta, N.; Heras, A. Preparation and characterization of chitosan microspheres for controlled release of tramadol. *J. Control. Release* **2008**, *132*, 76–77. [CrossRef]

34. Campaña-Seoane, P.A. Bioadhesive emulsions for control release of progesterone resistant to vaginal fluids clearance. *Int. J. Pharm.* **2014**, *477*, 495–505. [CrossRef] [PubMed]

35. Measurement of Consistency by Penetrometry, 267. In *European Pharmacopoeia 7.0*; European Directorate for Quality Medicines & HealthCare: Strasbourg, France, 2008.

36. Córdoba-Díaz, M.; Nova, M.; Elorza, B.; Córdoba-Díaz, D.; Chantres, J.R.; Córdoba-Borrego, M. Validation protocol of an automated in-line flow-through diffusion equipment for *in vitro* permeation studies. *J. Control. Release* **2000**, *69*, 357–367. [CrossRef]

![marine drugs logo] *marine drugs*

MDPI

Review

Chitosan-Based Multifunctional Platforms for Local Delivery of Therapeutics

Seong-Chul Hong [†], Seung-Yup Yoo [†], Hyeongmin Kim and Jaehwi Lee *

College of Pharmacy, Chung-Ang University, Seoul 06974, Korea; shotgun30@naver.com (S.-C.H.);
marin4906@naver.com (S.-Y.Y.); hm.kim8905@gmail.com (H.K.)
* Correspondence: jaehwi@cau.ac.kr; Tel.: +82-02-820-5606
† These authors contributed equally to this work.

Academic Editors: Hitoshi Sashiwa and David Harding
Received: 26 January 2017; Accepted: 24 February 2017; Published: 1 March 2017

Abstract: Chitosan has been widely used as a key biomaterial for the development of drug delivery systems intended to be administered via oral and parenteral routes. In particular, chitosan-based microparticles are the most frequently employed delivery system, along with specialized systems such as hydrogels, nanoparticles and thin films. Based on the progress made in chitosan-based drug delivery systems, the usefulness of chitosan has further expanded to anti-cancer chemoembolization, tissue engineering, and stem cell research. For instance, chitosan has been used to develop embolic materials designed to efficiently occlude the blood vessels by which the oxygen and nutrients are supplied. Indeed, it has been reported to be a promising embolic material. For better anti-cancer effect, embolic materials that can locally release anti-cancer drugs were proposed. In addition, a complex of radioactive materials and chitosan to be locally injected into the liver has been investigated as an efficient therapeutic tool for hepatocellular carcinoma. In line with this, a number of attempts have been explored to use chitosan-based carriers for the delivery of various agents, especially to the site of interest. Thus, in this work, studies where chitosan-based drug delivery systems have successfully been used for local delivery will be presented along with future perspectives.

Keywords: chitosan; local delivery; anti-cancer drugs; medical device

1. Introduction

Oral and intravascular routes are most commonly used for administration of drugs due to their advantages such as patient convenience and fast onset of action. However, drugs administered through these routes are distributed to the whole body via the systemic blood circulation, and thereby only a small portion of the drugs reaches the organs or tissues of interest while most of the drugs are diffused to unwanted areas of the body. This feature of the oral and intravascular administration results in low efficacy of drugs in the intended site and side effects of the drugs in the unwanted tissues [1–3]. For this reason, local drug delivery has attracted great attention due to its distinct advantages such as high concentration and efficacy of drugs at the site of interest in the body, decreased side effects of the drugs in the rest of the organs or tissues, and reduced dosing frequency and fluctuation in circulating drug levels [4]. To facilitate local drug delivery, various drug delivery systems of natural or synthetic polymers have been exploited such as particulates at a micro-/nano-scale, hydrogels, and films [5]. The most crucial step for designing such drug delivery systems is to select a polymer with appropriate physico-chemical properties, biocompatibility, biodegradability, and biological characteristics because the polymer largely determines the fate of the local drug delivery systems after administration.

Among diverse biomaterials and synthetic polymers explored as a fundamental material for preparing pharmaceutical formulations for local drug delivery, chitosan, a copolymer consisting of β-(1→4)-linked D-glucosamine and N-acetyl-D-glucosamine, has been extensively investigated for the purpose. Chitosan is generally obtained by deacetylation of chitin, which is the primary component of the exoskeleton of many living organisms, in particular marine crustaceans such as shrimps and crabs. By the deacetylation process, chitosan develops primary amine groups in its chemical structure and thereby it is positively charged in diluted acidic aqueous solutions, which makes chitosan distinguishable from other biomaterials [6]. Due to this feature, mechanical properties of chitosan-based drug delivery systems can be controlled by forming a polyelectrolyte complex with other anionic compounds [7–9]. In addition, chitosan can be simply fabricated to various morphologies including micro-/nano-spheres, fibers, gels, and films [10–12]. The biopolymer can also be conjugated with other molecules through its reactive amino groups of D-glucosamine residues, offering great possibilities to modify the chitosan-based formulations [13–15]. Because it is a biomaterial, chitosan exhibits good biocompatibility and biodegradability that can be tuned by varying its molecular weight and deacetylation degree [16,17].

Beyond numerous studies that only focused on investigating the feasibility of chitosan as a main material for fabricating drug delivery systems for simple local drug delivery, innovative chitosan-based platforms with multi-functions have recently been emerging. With a variety of approaches for conferring different functions of chitosan, the chitosan-based drug delivery systems have demonstrated versatile properties, thereby overcoming major challenges posed in diverse research fields. In particular, in the field of cancer therapy and tissue regeneration, much literature has recently reported remarkable achievements using the chitosan-based multifunctional systems (Table 1). For further studies to advance the chitosan-based systems, deep understanding of approaches for designing the drug delivery systems of chitosan with different functions that have been done to date is indispensable. Thus, in this review, representative chitosan-based multifunctional platforms for local drug delivery will be introduced, mainly focusing on their applications for cancer therapy and tissue regeneration.

2. Conventional Chitosan-Based Local Drug Delivery

2.1. Mucosal Surface

Most human organs are covered by a slippery material, namely mucus. The mucus is mainly composed of glycoproteins called mucins, in which a large portion of sialic acid residues take place. Due to the abundant sialic acid residues, the mucus layer shows a negatively charged surface [18], which provides a favorable environment for chitosan to be electronically adhered (Figure 1). Many researchers have used such a phenomenon to deliver the drug locally to the mucus-covered organs [19–21].

To effectively deliver drugs into the eye, one must overcome lacrimal elimination in order to provide sufficient time for drugs to penetrate [22]. Pavan et al. [23] produced ultra-small chitosan nanoparticles (US-CNPs) for the encapsulation of bovine lactoferrin to treat pesticide-induced ocular toxicity. The synthesized CNPs were small in size, ranging about 30–50 nm. The biodistribution results demonstrated that a large portion of US-CNPs were presented on the mucus layer compared to large CNPs or poly(lactide-*co*-glycolide) (PLGA)-based NPs. In addition, they were able to penetrate into the rat eye even at low concentration (100 µg/mL) compared to conventional CNPs. The interaction between chitosan and the mucus layer resulted in prolonged contact time on the eye, which might have assisted the penetration [24].

Table 1. Various therapeutic uses and applications of chitosan-based platforms.

Therapeutic Use	Application	Key Findings	Reference
Embolization	Deformable chitosan microspheres	• Highly spherical and porous chitosan microspheres were formed • Deformable microspheres were able to pass through the microcatheter.	[25]
	Adriamycin-loaded alginate-chitosan microcapsule	• Drug rapidly released in acidic condition. • Successful occlusion of rabbit renal artery. • Observed synergistic effect of embolization and anti-cancer drug.	[26]
	Doxorubicin (DX)-loaded chitosan microsphere	• Microspheres were designed and evaluated under different conditions. • Observed synergistic effect of embolization and anti-cancer drug.	[27]
	Superparamagnetic iron oxides (SPIOs) loaded chitosan microsphere	• Deformable microspheres were able to pass through the microcatheter. • The released SPIOs from the microsphere were detectable via magnetic resonance imaging (MRI) allowing to monitor the embolization outcome of the patient.	[28]
Theragnosis	DX-loaded ZnO folate-chitosan quantum dot	• Folate allows receptor-specific targeting of the anticancer drug. • Long-term fluorescence stability of ZnO allows in vivo visualization.	[29]
	Cy5.5-labled paclitaxel-loaded chitosan nanoparticle	• The anti-cancer drug was selectively delivered to tumor tissue by enhanced permeation and retention effect. • The Cy5.5 dye in the tumor tissue was detectable by near-infrared fluorescence detection.	[30]
	Chitosan-based DX-loaded magnetic nanoparticle	• The DX and nanoparticles were released in a pH-dependent manner. • Under acidic condition, the tumor tissue was detectable by MRI.	[31]
Tissue engineering	Chitosan hyaluronic acid (HA) hydrogel	• Addition of HA showed tighter networks, smaller pore size, increased stability. • HA provides a suitable environment for chondrocytes culture.	[32]
	Alginate-*O*-carboxymethyl chitosan hydrogel	• The modified chitosan was able to enhance the adhesion, differentiation and survival of adipose-derived stem cells on the scaffold.	[33]
	Arginine-glycine-aspartate (RGD)-conjugated chitosan scaffold	• Applied specific method to fabricate the RGD-conjugated, crosslinked chitosan scaffold. • Mesenchymal stem cells were well adhered, differentiated and survived on the scaffold.	[34]
Wound healing	Chitosan-pectin-TiO_2 nanodressing	• Addition of titanium dioxide improved mechanical strength. • The nanodressing showed good anti-microbial and blood-compatibility along with significant wound healing and closure rate.	[35]
	Human epidermal growth factor (EGF)-loaded chitosan film	• EGF showed significant vascular healing effect compared to conventional formulation.	[36]
	Neurotensin (NT)-loaded chitosan dressing	• Bioactive NT enhanced the healing effect on diabetic wounds. • Chitosan dressing was able to modulate immune response along with sustained release.	[37]

Researchers are interested in the fact that nasal administration is an effective delivery route to the central nervous system. It is found that direct diffusion of drugs through olfactory epithelium allows to circumvent the blood–brain barrier [38]. Matthias et al. [39] developed siRNA conjugated CNPs for nose-to-brain delivery to treat glioblastoma. The electric interaction of positively charged chitosan and negatively charged siRNAs successfully encapsulated siRNAs in the nanoparticles and protected them from RNases. The administered CNPs strongly adhered to the nasal mucosa and the siRNAs were detectable up to 8 h after administration, compared to naked siRNAs which showed only mild adherence. In addition, encapsulated siRNAs were effectively transported to glioma cells through nasal cavity. These results were the outcome of the combined effect of chitosan. The mucoadhesive property of chitosan allowed the CNPs to overcome mucosal clearance in the nasal cavity. Afterwards, tight junction opening of the polymer [40] made it easier to transport CNPs through olfactory epithelium. These features make CNPs a promising nose-to-brain local drug delivery system.

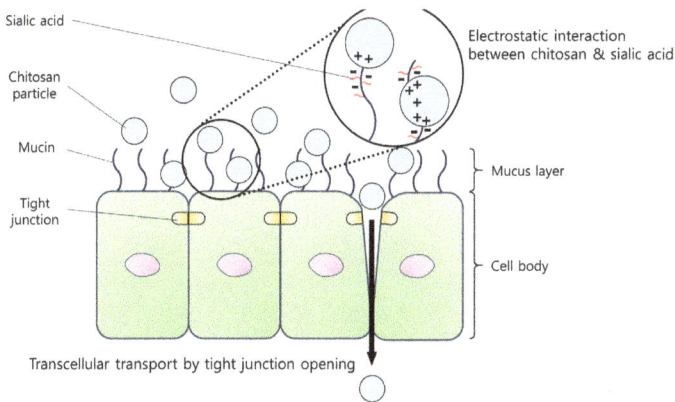

Figure 1. A diagrammatic presentation of the electrostatic interaction between chitosan particles and the mucus layer. Positively charged chitosan particles are prone to interact with the negatively charged sialic acid, resulting in enhanced adherence. In addition, a tight junction opening could be induced by chitosan particles, promoting local delivery of drugs.

Local drug delivery to the oral cavity using chitosan was mainly studied to treat oral candidiasis. To treat such disease, Seda et al. [41] developed chitosan-coated nanoparticles (CS-NPs) containing fluconazole (FLZ) for local treatment. The team performed an in vitro release study as well as an ex vivo diffusion test, and in vivo anti-fungal observations using rabbits as the animal model. Not surprisingly, the CS-NPs were able to interact with mucins, prolong the release of the drug, and successful recovery from candidiasis was seen in rabbit oral cavities. An interesting result from the ex vivo Franz diffusion cell test using cow buccal mucosa was observed. No drug was found from the receptor phase samples, but rather 20% of FLZ was found in the mucosa. This indicates that CS-NPs resided locally to reach effective salivary concentration of FLZ, and did not penetrate through the buccal membrane for systemic absorption.

2.2. Skin Surface

In dermal delivery, the use of chitosan exhibits many benefits. Even though the outermost skin layer is not covered with mucus, the slight negative charge of mammalian skin offers a suitable environment for chitosan to be adhered [42]. Interestingly, it was studied that positively charged particles are less prone to deeper skin penetration, thus showing less systemic exposure, which is also another benefit of using chitosan for local delivery [43].

In recent years, lecithin-chitosan nanoparticles (LC-CNPs) have been largely studied for dermal delivery. Lecithin, a natural lipid mixture of phospholipids, has negative charge which makes it prone to interact with positively charged chitosan [44]. Qi et al. [45] studied LC-CNPs as the dermal delivery system for quercetin, an anti-oxidant. Both in vitro and in vivo study of permeation studies showed significant drug accumulation, especially in the epidermis, compared to quercetin propylene glycol solution. Ipek et al. [46] and Taner et al. [47] produced LC-CNPs and further incorporated them into chitosan gel for more suitable formulation. These particles are both in agreement that the chitosan gel formulation had fair compatibility with skin pH, sufficient rheological and mechanical properties for topical administration. In addition, when these LC-CNPs were incorporated into chitosan gel, they showed better efficacy compared to commercial cream, even though the drug amount of LC-CNPs was 10 times less than the latter. Again, the LC-CNPs of the two particles also showed higher drug accumulation in the epidermis. The fact that LC-CNPs did not penetrate further into deep dermis is in alignment with the benefit of positively charged chitosan, again emphasizing the advantage of using chitosan for the skin delivery system.

3. Diversity of Chitosan-Based Local Drug Delivery

3.1. Cancer Therapy

3.1.1. Chitosan-Based Platforms for Embolic Therapy

Anti-cancer drugs for chemotherapy have been continually studied for decades. Although they are largely developed up to date, some fundamental problems of anti-cancer drugs are still being mentioned. The main drawback of these drugs is severe side effects due to their systemic administration and low specificity to the tumor tissues [48]. Thus, there have been unmet needs for targeted local delivery. Selective embolization is an attractive therapy to these needs because it uses a typical feature of cancer cells, angiogenesis [49]. By occluding arteries generated by cancer cells, embolic materials can induce starvation of cancer cells (Figure 2). Until now, many types of embolic materials, such as polyvinyl alcohol (PVA) [50] or PLGA [51] were used. Despite its potency of cheapness as well as biocompatibility, non-toxicity, and biodegradability [52], chitosan has been disregarded as a candidate due to its brittleness when formulated into microspheres.

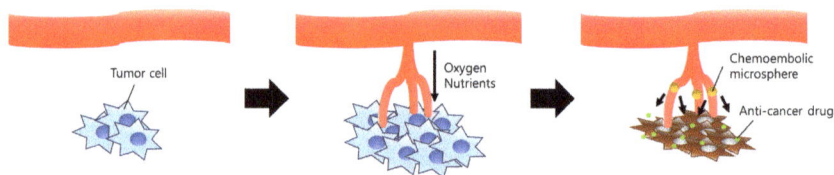

Figure 2. Schematic diagram of the therapeutic mechanism of chemoembolic microspheres. Tumor cells rapidly proliferate due to the angiogenesis. Oxygen and nutrients are supplied via newly generated arteries. Chemoembolic microspheres efficiently occlude vessels which were formed by angiogenesis, leading to starvation of the tumor tissue. Drugs released from the microspheres exhibit a synergistic anti-cancer effect on the tumor tissue.

In this background, Kang et al. [25] focused on preparing deformable chitosan microspheres (CMs) for embolization by ionotropic gelation with polyethylene glycol (PEG). PEG was used as a porogen and was extracted later to form porous CMs, which made them overcome their rigidity. Stable pore structures were observed in SEM analysis and water retention increased proportionally with the amount of PEG. Although blended shapes were observed when passing through the narrow end of the catheter, they recovered their original form after administration. These results indicated that porous CMs could be a great candidate for embolization material which would be able to overcome clogging problems in catheters [53]. A later study continued by Park et al. [27] investigated CMs as a multifunctional embolic platform. Park and his team formulated doxorubicin (DX)-loaded CMs to present the synergistic effect of embolization and sustained-release of the anti-cancer drug, that is, chemoembolization. Tripolyphosphates were used to load DX more efficiently by neutralizing the positive charge of chitosan and to enhance the sustained-release profile. The DX-loaded CMs were spherical in shape, deformable, and homogeneous in size, satisfying the essential factors of efficient chemoembolization [54]. An in vitro drug release test showed an initial burst, followed by sustained release of DX. These features resulted in an efficient anti-cancer effect in the in vivo test. Tumor size was decreased by about one-third of the initial size in CMs treated groups, whereas other groups treated with normal saline, DX solution, and blank chitosan microspheres showed increased tumor size. Kim et al. [55] conducted another study to improve conventional CMs for chemoembolization.

They noticed that the drug loading prior to clinical use was time consuming and the drug release was instable. The team planned to overcome these defects by loading drug encapsulated liposome in CMs. It was found that liposomal CMs presented a stable sustained-release profile of DX, while liposomal DX without CMs showed a burst release of DX. Combining with previous studies, these results indicate that liposomal CMs could be used for sustained-release chemoembolization.

Another consideration of CMs for embolization is monitoring them. Although CMs can be delivered selectively to tumor arteries, the help of X-ray angiography is almost inevitable [56]. Since excessive exposure to X-rays is harmful [57], patients would be safer if physicians could observe their changes visually with reduced usage of them. Kang et al. [58] used superparamagnetic iron oxide nanoparticles (SPIOs), nanocrystals of iron oxides coated with hydrophilic polymers, to achieve such an objective. SPIOs are originally used as a contrast agent for magnetic resonance imaging (MRI). Therefore, they hypothesized that SPIOs loaded CMs could be traceable via MRI. Spherical CMs with narrow size distribution were prepared by emulsion and the cross-linking method. The amount of loaded SPIOs was larger than the detectable amount of an early report [59], indicating possibility for detectability. It was also revealed that highly-cross linked CMs had a lower degree of swelling, leading to secure entrapment of SPIOs. The study was continued by Chung et al. [28] to strengthen the feasibility for detectable embolization. SPIO-CMs were prepared by ionotropic gelation including pores that made them deformable. SPIOs were loaded with minimal loss, showing 94% of loading efficiency. The microspheres passed through the catheter without any damage, confirming their deformability. In vivo MR tracing in renal arteries of rabbit showed long-term occlusion of target blood vessels for 18 weeks. The effect of embolization was further supported by anatomical observation and Prussian blue staining. The embolized kidney showed decreased size 8 weeks after treatment. Prussian blue staining indicated SPIO-CMs located in the main and segmental renal arteries, enhancing evidence of selective embolization.

3.1.2. Chitosan-Based Platforms for Theragnosis

Diagnosis is an important aspect for cancer therapy, because accurate diagnosis can give information regarding whether a treatment will be effective to the individual patient [60]. Molecular imaging is an evolving area of diagnosis due to its non-invasive feature and real-time monitoring [61]. Thus, molecular imaging was clinically used with diverse imaging modalities, by loading contrast agents to the molecular carrier. Nanoparticles have been the focus for clinical diagnosis of cancer via molecular imaging since they can be efficiently accumulated to tumor tissues by the enhanced permeation and retention (EPR) effect [62]. Besides, there have been numerous studies to encapsulate anti-cancer drugs into nanoparticles for cancer therapy. By integrating these two strategies, a new concept called 'theragnosis' has come to the fore. The concept of theragnosis is attractive because of its possibility of personalized therapy [63]. Chitosan is a bio-friendly material that is ideal for nanoparticle-based theragnosis, since it can be easily formulated into nanoparticles making it suitable for the EPR effect [64,65]. It is also degradable in human bodies after its diagnostic and drug delivering role. Kim et al. [30] investigated the feasibility of chitosan nanoparticles (CNPs) for theragnosis. CNPs were formulated to encapsulate paclitaxel (PTX) and conjugated with Cy5.5, a near-infrared fluorescent (NIRF) dye. When intravenously injected, CNPs were traceable in real-time by in vivo NIRF imaging, showing a tumor-selective signal proportional to tumor size. The therapeutic effect also could be observed by NIRF imaging. Tumor growth rates were investigated and CNPs proved their feasibility for cancer therapy, showing reduced tumor volume compared to normal saline groups.

Theragnosis can be achieved not only by optical imaging but also by other techniques such as MRIs. An additional study conducted by Lim et al. [31] traced the CNPs by MRI. The research team also attempted to enhance the selectivity of CNPs. They focused on a feature of cancer cells that exhibit abnormally high acidic environment, which makes them distinguishable from normal cells [66]. Thus, pH-sensitive formulation was suggested to be a potent drug delivery system for cancer therapy. pH-sensitive CNPs were prepared by adding maleoyl groups on the chitosan backbone.

Moreover, magnetic nanocrystals (MNCs) were encapsulated into CNPs, making them detectable by MRI. CNPs showed increased particle size in pH 5.5 due to the hydrolysis of maleoyl groups. In accordance with hydrolysis, 90% of DX was released within 24 h, while only 20% did at higher pH conditions (pH 7.4 and 9.8). When intravenously injected in mice, MR contrast effects were seen in tumor tissues by the EPR effect and 80% of DX was released from CNPs because of acidic condition, as expected. CNPs also reduced the tumor growth rate, compared to naive DX and normal saline.

Although diverse imaging techniques for diagnosis have been used clinically, limitations of conventional techniques still act as an obstacle for accurate diagnosis. Positron emission tomography (PET) is a good example. PET has the advantage that it can provide long-term image and quantitative information [67,68]. Utilizing these advantages, PET is applied to diagnose many types of cancer such as lung, colorectal, and breast cancers [69–71]. However, there is a limitation; it was difficult to detect specific molecular activity with PET, reported by Lee et al. [72]. To overcome this defect, Lee and his colleagues investigated the application of various imaging techniques in one formulation simultaneously. CNPs were radiolabeled with Cu^{64} for PET imaging and conjugated with special moieties which can be activated by matrix metalloproteinase (MMP) for NIRF imaging. Since MMPs are overexpressed in tumor cells [73,74], they could be molecular targets for theragnosis. Results from NIRF imaging showed rapid response, showing a plateau 6 h after intravenous injection. In contrast, PET images presented their peak response 24 h after injection and it was maintained over 48 h. Both images showed accumulation of CNPs in tumor tissues. Multimodal CNPs showed integrated advantages of two methods; quantitative information and molecular accuracy. These results indicate that multimodal CNPs could be promising theranosis agents.

3.1.3. Chitosan-Based Platforms for Cancer Radiotherapy

Radiotherapy is a concept using radiation for therapeutic purposes. Beta-emitting materials have been considered as useful anti-cancer therapeutics because of their cytotoxic activity [75]. Although radiotherapy was expected as a novel therapy for cancers, there were also disadvantages that limit their clinical usage. Kim et al. [76] emphasized safety issues. They claimed that because the distribution of the radioisotope is dependent on tumor vasculature, leakage can occur if vascular-disturbing incidents happen. Therefore, the need for local isotope delivery has risen and studies about the biodistribution of the isotope were conducted. A biodistribution study of the holmium-166–chitosan complex conducted by Yuka et al. indicated that it could be a potential candidate for tumor radiotherapy [77]. 166-Holmium (^{166}Ho) possesses an appropriate half-life of 26.8 h and a high beta energy of 1.85 MeV which make it suitable for the therapeutic radioisotope. Meanwhile, chitosan has the physicochemical feature that it is soluble in acidic conditions but insoluble in neutral or basic conditions. In their reports, it was also suggested that chitosan, a biocompatible material, can form chelate with ^{166}Ho. It means that acidic solutions can be prepared for stable ^{166}Ho–chitosan injection. Once injected into human organs, chitosan forms a solid structure, establishing a radioactive pharmaceutical device. In the results, the radioactivity of ^{166}Ho–chitosan was localized in the administration site when intrahepatically and intratumorally injected, while formulation with ^{166}Ho alone spread to the whole body via the vascular system. A step further, Lee et al. [78] researched the feasibility of the ^{166}Ho–chitosan complex as a therapeutic agent for hepatocellular carcinoma. The ^{166}Ho–chitosan complex showed coagulation necrosis in vivo because of the radiation effect of ^{166}Ho, when injected directly to tumor. The area of necrosis was proportional to the dosage and the required dosage for total necrosis was dependent on the size of the tumor. It was also found that ^{166}Ho–chitosan accumulated within the tumor, showing no significant leakage in neighboring organs. Furthermore, ^{166}Ho–chitosan formulation for hepatocarcinoma successfully finished its phase IIb clinical trial [76]. Forty patients participated in this clinical trial and they were treated with ^{166}Ho–chitosan formulation which is designed to be injected intratumorally. A total of 77.5% of patients experienced complete tumor necrosis. The ^{166}Ho-chitsoan complex proved its effect clinically and later, was approved by Korean Food and Drug Administration (KFDA) for hepatocarninoma (Milican, DongWha Pharmaceutical Co., Seoul, Korea).

3.2. Tissue Regeneration

3.2.1. Chitosan-Based Platforms for Enhanced Cell Adhesion Properties

Tissue engineering is a promising method for tissue regeneration. By providing an artificial extracellular matrix (ECM) platform, cells can adhere and grow in the platform. The non-immunogenicity, low allergenicity, biodegradability, formability, and chemical variability of chitosan allow us to produce various types of platforms for diverse tissue engineering. Injectable hydrogels have been widely used as scaffolds and have several advantages for tissue engineering; minimal invasive procedure, physical flexibility, and easiness of incorporating cells. However, one critical limitation is the lack of adhesion in some cells to scaffolds. It is necessary to increase the cell adhesion rate in order to successfully load the desired cell and regenerate the tissue. Dhanya et al. [33] made an attempt to prepare a scaffold for adipose tissue engineering by improving conventional alginate hydrogels. Although alginate hydrogels have been widely used as a biomaterial for tissue regeneration, lack of cell adhesion was the limitation for its clinical usage. *O*-carboxymethyl chitosan (O-CMC) was adopted to solve this problem, due to its good water solubility, high viscosity, and high hydrodynamic volume. The alginate/O-CMC/nano-fibrin (AOF) scaffold was found to have high viscous moduli when measured by the rheometer, proving it as a useful scaffold for cell incorporation. In addition, AOF showed a relatively rapid and higher swelling ratio than PVA due to abundant hydrophilic groups which can help the scaffold to be hydrated. This demonstrates that the AOF can obtain a sufficient area/volume ratio and porosity which allows cells to be infiltrated into scaffolds more efficiently. Coinciding with the results, adipose-derived stem cells were well attached, proliferated, and differentiated in culture. To increase cell adhesion, another useful method is to conjugate the cell-binding sequence arginine-glycine-aspartate (RGD) to chitosan. However, one of the drawbacks of using RGD-chitosan is the difficulty of conjugating RGD uniformly and simultaneously crosslinking the scaffold. To overcome this drawback, Tsai and his team previously developed a specific method to fabricate a RGD-conjugated, crosslinked chitosan scaffold for bone tissue engineering (Figure 3), described elsewhere [79]. Using their method, the team evaluated the suitability of RGD-chitosan scaffolds for bone tissue engineering using mesenchymal stem cells (MSCs) [34]. By visually observing the attachment of MSCs, RGD-chitosan film displayed elongated and spread morphology compared to normal chitosan film, on which only few cells were visible. Further quantification of cell numbers also confirmed that the number of attached MSCs on RGD-chitosan film was approximately twice the number observed on normal chitosan. This result is in alignment with RGD-chitosan scaffold results, which also showed a significant increase in MSCs attachment. Moreover, MSCs seeded in the RGD-chitosan scaffold were able to successfully differentiate, showing the highest alkaline phosphatase activity and calcium deposition. These results indicate that RGD-chitosan scaffolds can thus act as a favorable environment for the MSCs to adhere.

A chitosan-based platform can also be used in a more delicate area. One of the good examples is a cardiac patch for a congenital heart defect, a defect in the contractile myocardial tissue. Although there are surgical patches on the market to treat this type of defects, there have been problems such as inabilities to grow and regenerate cells, and limited durability. Incorporating stem cells into the chitosan scaffold can be a solution to these drawbacks. Pok et al. [80] prepared a gelatin-chitosan hydrogel-based multilayer cardiac patch. They adopted polycaprolactone (PCL) to attain high patch tensile strength, and gelatin to consolidate cell attachments. Chitosan was the main component of the patch, because it is biodegradable and can be fabricated into a porous structure for cell migration easily by freeze-drying the gel solutions. Multilayer scaffolds—50:50 (gelatin:chitosan)—showed a narrow pore size distribution and moderate porosity, suggesting an effective mixture ratio. Samples with such a ratio demonstrated their potential by presenting high cell attachment and viability.

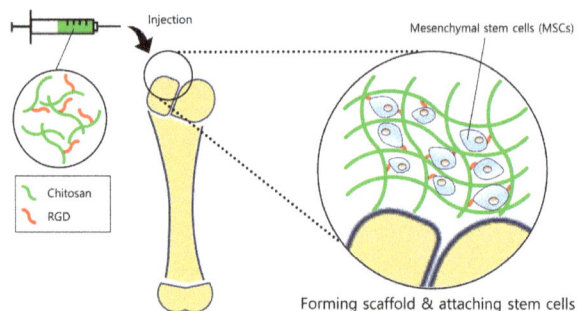

Figure 3. Schematic diagram of tissue regeneration. Chitosan forms a scaffold similar to extracellular matrix (ECMs), providing a suitable environment for cells to be adhered. In order to enhance the attachment of stem cells, RGD moieties were linked with the chitosan backbone. Stem cells then could proliferate on the scaffold more efficiently, ultimately regenerating the bone or cartilage tissue.

3.2.2. Chitosan-Based Platforms for Enhanced Wound Healing

Wound healing products are another medical platform of chitosan. Chitosan was found to influence all steps of wound healing [81] by regulating the immune system. It can attract neutrophils and macrophages towards the wound, controlling fibroplasia and reepithelization [82] (Figure 4). Anti-microbial and hemostatic effects also make chitosan a potential wound healing material [83]. In addition, chitosan has biocompatibility, biodegradability, bioadhesiveness, and low toxicity, which makes it more a powerful candidate for wound healing. Therefore, by combining chitosan with other materials which assist the healing process, the multi-functional effect could be obtained, thereby having a synergistic effect in wound healing.

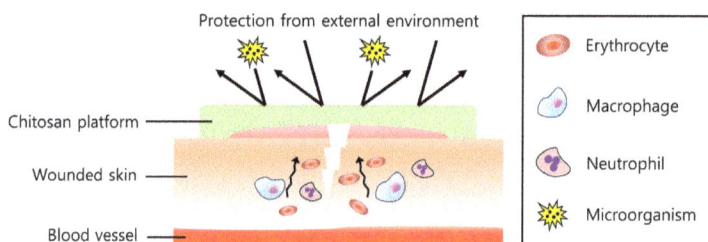

Figure 4. Effect of the chitosan platform on wound healing. Chitosan attracts immune cells such as macrophages and neutrophils, regulating inflammatory responses to recover the wound quickly. Erythrocytes and platelets (not described in the figure) also agglomerate at the wound site, showing hemostasis. In addition, a harmful external environment and microorganisms can be efficiently isolated by the chitosan platform.

In this background, Lee et al. [36] investigated the hemostatic and vascular healing effect of chitosan film in combination with epidermal growth factor (EGF). Compared to a formulation on the market, both single chitosan film and EGF-chitosan film showed similar in vivo hemostatic activity. However, in a histological study, EGF-chitosan film showed a significantly improved vascular healing effect. The researchers proposed that impregnating EGF to chitosan film does not interrupt the action of each other, but rather they have a synergistic effect of vascular healing and hemostasis. Proving its effectiveness, applications of chitosan have been extended to treat diverse types of wounds such as diabetic wounds [37]. Moura et al. fabricated foams with chitosan derivatives to treat diabetic wounds. Chitosan foams were prepared using a chitosan derivative, 5-methyl pyrrolidinone chitosan

(MPC), to cover the wound site and help regeneration. A bioactive neuropeptide neurotensin (NT) [84] was added in order to improve wound healing. Significantly decreasing the wound size, the result suggested a major wound healing impact of NT-loaded MPC foams in diabetic animals, not to mention the sustained release of NT due to MPC. In addition, the combined formulation could modulate the immune response, which is also important in the healing process.

3.3. Chitosan-Based Platforms for Miscellaneous Therapeutic Purposes

Chitosan, in combination with various materials—newly attempted multiplatforms—has been developed to improve therapeutic effects. Recently, Behl et al. [85] developed a contact lens containing chitosan nanoparticles (CNPs) as an alternative to eye drops. The negatively charged dexamethasone sodium phosphate (DXP) could be successfully incorporated into positively charged chitosan by ionic interaction to form stable chitosan nanoparticles (DXP-CNPs). These nanoparticles were then mixed with polymer material and molded to form a small lens. The CNPs mixed lens was found to have an average transmittance of 95%–98%, with almost no visual disturbance and no irritation to the eye expected, giving the advantage of using chitosan in combination with such formulation. The mucoadhesive CNPs can thus adhere to the mucus layer of the eye, offering prolonged residence time and resistance to lacrimal elimination [86]. Also, the release rate of DXP from the lens was nearly three-fold slower than the DXP-CNPs itself, demonstrating the combined effect of the lens which acts as an additional obstacle that slows down the drug release [87]. The authors calculated that compared to conventional DXP eye drops, an increase of up to 72% in bioavailability could be reached with this formulation. Another formulation, such as a hydrogel, can also be useful to increase the retention time and bioavailability of the drug. Cho et al. [88] previously developed a glycol chitosan-based thermogel for biomedical applications. The authors chemically cross-linked hexanoyl glycol chitosan (HGC) by exposing it to UV light. In this study, enhanced thermosensitivity and low sol-gel transition temperature of the polymer was observed even at low concentrations (3–5 wt. %) of chitosan compared to typical thermogelling polymers [89]. Again, a more recent study by Cho et al. [90] created a HGC-based thermosensitive gel with the combination of the anti-glaucoma drug brimonide tartarate (BRT). The BRT-HGC showed a promising result in which the viscosity was increased at body temperature and prolonged intraocular pressure decrease was observed, compared to the conventional eye drop. The advantage of the low concentration requirement of chitosan along with body temperature-activated rheology proved chitosan as the cost-effective polymer for its use. As a new multiplatform for dermal use, Tu et al. [91] combined the use of *N*-trimethyl chitosan nanoparticles (TMCNPs) with polypropylene electret to promote the dermal delivery of protein drugs. Electret is a dielectric material that can hold onto its electric charge for a long time. The research team hypothesized and verified the fact that with the use of both positively charged electret and TMCNPs, improved penetration of TMCNPs into the epidermis and dermis could be observed. They proposed a powerful electrostatic repulsive force between TMCNPs and electret as the mechanism for successful dermal delivery. In addition, the tight junction opening of chitosan also acts as a delivery enhancing effect [92]. If the degree of repulsive force could be adjusted just enough to deliver CNPs only to the upper skin, this combined formulation could be a promising method for effectively delivering drugs without systemic absorption.

4. Conclusions and Future Perspectives

Under the increasing demand of local drug delivery, amongst various polymers, chitosan, a bio-friendly and easily obtainable polymer, has been widely exploited for the purpose. Chitosan with its derivatives plays a predominant role in medical and pharmaceutical sciences. Its biocompatibility, biodegradability, and non-toxicity promote the use of chitosan as a useful drug carrier to the human body. As documented in the previous sections, chitosan also has unique properties that coincide with its specific use. For example, the mucoadhesive property of chitosan benefits the targeted delivery to the mucus-covered organs and induces prolonged residence time of drugs. The tight junction opening property of the polymer allows the drugs to penetrate more easily through the organs, thereby increasing the accumulation of the drugs at the target of interest. In addition, the chemical structure of chitosan is easily modified due to the reactive primary amines on the chitosan backbone. To sum up, these unique properties have given chitosan the privilege of being frequently used as a fundamental material for fabricating local drug delivery systems.

Innovative chitosan-based multifunctional platforms are continuously under development as versatile characteristics of chitosan are being understood. However, there are still some unsolved problems that must be taken into account when applying chitosan formulations. When chitosan is combined with other polymers or synthesized into diverse derivatives, the safety issue must be considered as the principal factor. Many more chitosan mixtures and derivatives are being explored to develop novel multifunctional platforms for local drug delivery. In response to such drastic development, the lack of toxicity studies is the primary assignment that needs to be fulfilled. The majority of toxicity studies on the platforms are limited to in vitro tests. Thus, performing additional in vivo or ex vivo studies will help to ensure its safety. Furthermore, clinical studies on human bodies should be carefully conducted based on these results. Along with its safety, the stability of the platform should also be taken into account. Parameters such as degradation time or elimination rate should be carefully controlled in order to validate the consistent performance of the platforms. Because many researchers have concentrated on proving the efficacy of new formulations, from now on, researchers should focus their work on establishing safety and stability.

Herein, we have gathered recent articles about the chitosan-based multifunctional platform regarding the purpose of its use. Chitosan may be one of the well-explored and widely used polymers as a platform, yet there is more to be studied. We hope that this review will give an insight into designing better chitosan platforms and ultimately, further help to extend the boundaries of the use of chitosan-based multifunctional platforms.

Acknowledgments: This work was supported by the National Research Foundation of Korea (NRF) grant funded by the Korea government (MSIP) (No. 2015R1A5A1008958). This research was also supported by the Chung-Ang University Graduate Research Scholarship in 2017.

Conflicts of Interest: The authors declare no conflict of interest.

References

1. Davila, G.W.; Daugherty, C.A.; Sanders, S.W.; Transdermal Oxybutynin Study Group. A short-term multicenter, randomized double-blind dose titration study of the efficacy and anticholinergic side effects of transdermal compared to immediate release oral oxybutynin treatment of patients with urge urinary incontinence. *J. Urol.* **2001**, *166*, 140–145. [CrossRef]
2. Pires, A.; Fortuna, A.; Alves, G.; Falcao, A. Intranasal drug delivery: How, why and what for? *J. Pharm. Pharm. Sci.* **2009**, *12*, 288–311. [CrossRef] [PubMed]
3. Gaudana, R.; Ananthula, H.K.; Parenky, A.; Mitra, A.K. Ocular drug delivery. *AAPS J.* **2010**, *12*, 348–360. [CrossRef] [PubMed]
4. Weiser, J.R.; Saltzman, W.M. Controlled release for local delivery of drugs: Barriers and models. *J. Control. Release* **2014**, *190*, 664–673. [CrossRef] [PubMed]
5. Uhrich, K.E.; Cannizzaro, S.M.; Langer, R.S.; Shakesheff, K.M. Polymeric systems for controlled drug release. *Chem. Rev.* **1999**, *99*, 3181–3198. [CrossRef] [PubMed]
6. Giotra, P.; Singh, S.K. Chitosan: An emanating polymeric carrier for drug delivery. In *Handbook of Polymers for Pharmaceutical Technologies: Biodegradable Polymers*; Thakur, V.K., Thakur, M.K., Eds.; Wiley: Hoboken, NJ, USA, 2015; Volume 3, pp. 33–39.
7. Avadi, M.R.; Sadeghi, A.M.M.; Mohammadpour, N.; Abedin, S.; Atyabi, F.; Dinarvand, R.; Rafiee-Tehrani, M. Preparation and characterization of insulin nanoparticles using chitosan and Arabic gum with ionic gelation method. *Nanomed. Nanotechnol. Biol. Med.* **2010**, *6*, 58–63. [CrossRef] [PubMed]
8. Fan, W.; Yan, W.; Xu, Z.S.; Ni, H. Formation mechanism of monodisperse, low molecular weight chitosan nanoparticles by ionic gelation technique. *Colloid Surf. B Biointerfaces* **2012**, *90*, 21–27. [CrossRef] [PubMed]

9. Shu, X.Z.; Zhu, K.J. Chitosan/gelatin microspheres prepared by modified emulsification and ionotropic gelation. *J. Microencapsul.* **2001**, *18*, 237–245. [PubMed]

10. Ko, J.A.; Park, H.J.; Hwang, S.J.; Park, J.B.; Lee, J.S. Preparation and characterization of chitosan microparticles intended for controlled drug delivery. *Int. J. Pharm.* **2002**, *249*, 165–174. [CrossRef]

11. Azuma, K.; Ifuku, S.; Osaki, T.; Okamoto, Y.; Minami, S. Preparation and biomedical applications of chitin and chitosan nanofibers. *J. Biomed. Nanotechnol.* **2014**, *10*, 2891–2920. [CrossRef] [PubMed]

12. Croisier, F.; Jerome, C. Chitosan-based biomaterials for tissue engineering. *Eur. Polym. J.* **2013**, *49*, 780–792. [CrossRef]

13. Sashiwa, H.; Aiba, S.I. Chemically modified chitin and chitosan as biomaterials. *Prog. Polym. Sci.* **2004**, *29*, 887–908. [CrossRef]

14. Aranaz, I.; Harris, R.; Heras, A. Chitosan amphiphilic derivatives. Chemistry and applications. *Curr. Org. Chem.* **2010**, *14*, 308–330. [CrossRef]

15. Thakur, V.K.; Thakur, M.K. Recent advances in graft copolymerization and applications of chitosan: A review. *ACS Sustain. Chem. Eng.* **2014**, *2*, 2637–2652. [CrossRef]

16. Huang, M.; Khor, E.; Lim, L.Y. Uptake and cytotoxicity of chitosan molecules and nanoparticles: Effects of molecular weight and degree of deacetylation. *Pharm. Res.* **2004**, *21*, 344–353. [CrossRef] [PubMed]

17. Hsu, S.H.; Whu, S.W.; Tsai, C.L.; Wu, Y.H.; Chen, H.W.; Hsieh, K.H. Chitosan as scaffold materials: Effects of molecular weight and degree of deacetylation. *J. Polym. Res.* **2004**, *11*, 141–147. [CrossRef]

18. Alonso, M.J.; Sanchez, A. The potential of chitosan in ocular drug delivery. *J. Pharm. Pharmacol.* **2003**, *55*, 1451–1463. [CrossRef] [PubMed]

19. Hermans, K.; van den Plas, D.; Kerimova, S.; Carleer, R.; Adriaensens, P.; Weyenberg, W.; Ludwig, A. Development and characterization of mucoadhesive chitosan films for ophthalmic delivery of cyclosporine A. *Int. J. Pharm.* **2014**, *472*, 10–19. [CrossRef] [PubMed]

20. Luo, Y.C.; Teng, Z.; Li, Y.; Wang, Q. Solid lipid nanoparticles for oral drug delivery: Chitosan coating improves stability, controlled delivery, mucoadhesion and cellular uptake. *Carbohydr. Polym.* **2015**, *122*, 221–229. [CrossRef] [PubMed]

21. Mouez, M.A.; Zaki, N.M.; Mansour, S.; Geneidi, A.S. Bioavailability enhancement of verapamil HCL via intranasal chitosan microspheres. *Eur. J. Pharm. Sci.* **2014**, *51*, 59–66. [CrossRef] [PubMed]

22. Seyfoddin, A.; Shaw, J.; Al-Kassas, R. Solid lipid nanoparticles for ocular drug delivery. *Drug Deliv.* **2010**, *17*, 467–489. [CrossRef] [PubMed]

23. Sunkireddy, P.; Kanwar, R.K.; Ram, J.; Kanwar, J.R. Ultra-small algal chitosan ocular nanoparticles with iron-binding milk protein prevents the toxic effects of carbendazim pesticide. *Nanomedicine* **2016**, *11*, 495–511. [CrossRef] [PubMed]

24. Di Colo, G.; Zambito, Y.; Burgalassi, S.; Nardini, I.; Saettone, M.F. Effect of chitosan and of N-carboxymethylchitosan on intraocular penetration of topically applied ofloxacin. *Int. J. Pharm.* **2004**, *273*, 37–44. [CrossRef] [PubMed]

25. Kang, M.J.; Park, J.M.; Choi, W.S.; Lee, J.; Kwak, B.K.; Lee, J. Highly spherical and deformable chitosan microspheres for arterial embolization. *Chem. Pharm. Bull.* **2010**, *58*, 288–292. [CrossRef] [PubMed]

26. Li, S.; Wang, X.T.; Zhang, X.B.; Yang, R.J.; Zhang, H.Z.; Zhu, L.Z.; Hou, X.P. Studies on alginate-chitosan microcapsules and renal arterial embolization in rabbits. *J. Control. Release* **2002**, *84*, 87–98. [CrossRef]

27. Park, J.M.; Lee, S.Y.; Lee, G.H.; Chung, E.Y.; Chang, K.M.; Kwak, B.K.; Kuh, H.J.; Lee, J. Design and characterisation of doxorubicin-releasing chitosan microspheres for anti-cancer chemoembolisation. *J. Microencapsul.* **2012**, *29*, 695–705. [CrossRef] [PubMed]

28. Chung, E.Y.; Kim, H.M.; Lee, G.H.; Kwak, B.K.; Jung, J.S.; Kuh, H.J.; Lee, J. Design of deformable chitosan microspheres loaded with superparamagnetic iron oxide nanoparticles for embolotherapy detectable by magnetic resonance imaging. *Carbohydr. Polym.* **2012**, *90*, 1725–1731. [CrossRef] [PubMed]

29. Yuan, Q.; Hein, S.; Misra, R.D.K. New generation of chitosan-encapsulated ZNO quantum dots loaded with drug: Synthesis, characterization and in vitro drug delivery response. *Acta Biomater.* **2010**, *6*, 2732–2739. [CrossRef] [PubMed]

30. Kim, K.; Kim, J.H.; Park, H.; Kim, Y.S.; Park, K.; Nam, H.; Lee, S.; Park, J.H.; Park, R.W.; Kim, I.S.; et al. Tumor-homing multifunctional nanoparticles for cancer theragnosis: Simultaneous diagnosis, drug delivery, and therapeutic monitoring. *J. Control. Release* **2010**, *146*, 219–227. [CrossRef] [PubMed]

31. Lim, E.K.; Sajomsang, W.; Choi, Y.; Jang, E.; Lee, H.; Kang, B.; Kim, E.; Haam, S.; Suh, J.S.; Chung, S.J.; et al. Chitosan-based intelligent theragnosis nanocomposites enable pH-sensitive drug release with MR-guided imaging for cancer therapy. *Nanoscale Res. Lett.* **2013**, *8*, 12. [CrossRef] [PubMed]

32. Park, H.; Choi, B.; Hu, J.L.; Lee, M. Injectable chitosan hyaluronic acid hydrogels for cartilage tissue engineering. *Acta Biomater.* **2013**, *9*, 4779–4786. [CrossRef] [PubMed]

33. Jaikumar, D.; Sajesh, K.M.; Soumya, S.; Nimal, T.R.; Chennazhi, K.P.; Nair, S.V.; Jayakumar, R. Injectable alginate-O-carboxymethyl chitosan/nano fibrin composite hydrogels for adipose tissue engineering. *Int. J. Biol. Macromol.* **2015**, *74*, 318–326. [CrossRef] [PubMed]

34. Tsai, W.B.; Chen, Y.R.; Li, W.T.; Lai, J.Y.; Liu, H.L. RGD-conjugated UV-crosslinked chitosan scaffolds inoculated with mesenchymal stem cells for bone tissue engineering. *Carbohydr. Polym.* **2012**, *89*, 379–387. [CrossRef] [PubMed]

35. Archana, D.; Dutta, J.; Dutta, P.K. Evaluation of chitosan nano dressing for wound healing: Characterization, in vitro and in vivo studies. *Int. J. Biol. Macromol.* **2013**, *57*, 193–203. [CrossRef] [PubMed]

36. Lee, S.; Jung, I.; Yu, S.; Hong, J.P. Effect of recombinant human epidermal growth factor impregnated chitosan film on hemostasis and healing of blood vessels. *Arch. Plast. Surg.* **2014**, *41*, 466–471. [CrossRef] [PubMed]

37. Moura, L.I.F.; Dias, A.M.A.; Leal, E.C.; Carvalho, L.; de Sousa, H.C.; Carvalho, E. Chitosan-based dressings loaded with neurotensin-an efficient strategy to improve early diabetic wound healing. *Acta Biomater.* **2014**, *10*, 843–857. [CrossRef] [PubMed]

38. Illum, L. Transport of drugs from the nasal cavity to the central nervous system. *Eur. J. Pharm. Sci.* **2000**, *11*, 1–18. [CrossRef]

39. Van Woensel, M.; Wauthoz, N.; Rosiere, R.; Mathieu, V.; Kiss, R.; Lefranc, F.; Steelant, B.; Dilissen, E.; Van Gool, S.W.; Mathivet, T.; et al. Development of siRNA-loaded chitosan nanoparticles targeting Galectin-1 for the treatment of glioblastoma multiforme via intranasal administration. *J. Control. Release* **2016**, *227*, 71–81. [CrossRef] [PubMed]

40. Casettari, L.; Illum, L. Chitosan in nasal delivery systems for therapeutic drugs. *J. Control. Release* **2014**, *190*, 189–200. [CrossRef] [PubMed]

41. Rencber, S.; Karavana, S.Y.; Yilmaz, F.F.; Erac, B.; Nenni, M.; Ozbal, S.; Pekcetin, C.; Gurer-Orhan, H.; Hosgor-Limoncu, M.; Guneri, P.; et al. Development, characterization, and in vivo assessment of mucoadhesive nanoparticles containing fluconazole for the local treatment of oral candidiasis. *Int. J. Nanomed.* **2016**, *11*, 2641–2653. [CrossRef] [PubMed]

42. Wu, X.; Landfester, K.; Musyanovych, A.; Guy, R.H. Disposition of charged nanoparticles after their topical application to the skin. *Skin Pharmacol. Physiol.* **2010**, *23*, 117–123. [CrossRef] [PubMed]

43. Yang, Y.; Sunoqrot, S.; Stowell, C.; Ji, J.L.; Lee, C.W.; Kim, J.W.; Khan, S.A.; Hong, S. Effect of size, surface charge, and hydrophobicity of poly(amidoamine) dendrimers on their skin penetration. *Biomacromolecules* **2012**, *13*, 2154–2162. [CrossRef] [PubMed]

44. Sonvico, F.; Cagnani, A.; Rossi, A.; Motta, S.; Di Bari, M.T.; Cavatorta, F.; Alonso, M.J.; Deriu, A.; Colombo, P. Formation of self-organized nanoparticles by lecithin/chitosan ionic interaction. *Int. J. Pharm.* **2006**, *324*, 67–73. [CrossRef] [PubMed]

45. Tan, Q.; Liu, W.D.; Guo, C.Y.; Zhai, G.X. Preparation and evaluation of quercetin-loaded lecithin-chitosan nanoparticles for topical delivery. *Int. J. Nanomed.* **2011**, *6*, 1621–1630.

46. Ozcan, I.; Azizoglu, E.; Senyigit, T.; Ozyazici, M.; Ozer, O. Enhanced dermal delivery of diflucortolone valerate using lecithin/chitosan nanoparticles: In Vitro and in vivo evaluations. *Int. J. Nanomed.* **2013**, *8*, 461–475. [CrossRef] [PubMed]

47. Senyigit, T.; Sonvico, F.; Rossi, A.; Tekmen, I.; Santi, P.; Colombo, P.; Nicoli, S.; Ozer, O. In vivo assessment of clobetasol propionate-loaded lecithin-chitosan nanoparticles for skin delivery. *Int. J. Mol. Sci.* **2016**, *18*. [CrossRef] [PubMed]

48. Park, J.H.; Saravanakumar, G.; Kim, K.; Kwon, I.C. Targeted delivery of low molecular drugs using chitosan and its derivatives. *Adv. Drug Deliv. Rev.* **2010**, *62*, 28–41. [CrossRef] [PubMed]

49. Nishida, N.; Yano, H.; Nishida, T.; Kamura, T.; Kojiro, M. Angiogenesis in cancer. *Vasc. Health Risk Manag.* **2006**, *2*, 213–219. [CrossRef] [PubMed]

50. Lee, D.H.; Yoon, H.K.; Song, H.Y.; Kim, G.C.; Hwang, J.C.; Sung, K.B. Embolization of severe arterioportal shunts in the patients with hepatocellular carcinoma: Safety and influence on patient survival. *J. Korean Radiol. Soc.* **1999**, *41*, 1117–1125. [CrossRef]

51. Grandfils, C.; Flandroy, P.; Jerome, R. Control of the biodegradation rate of poly(DL-lactide) microparticles intended as chemoembolization materials. *J. Control. Release* **1996**, *38*, 109–122. [CrossRef]

52. Cheung, R.C.F.; Ng, T.B.; Wong, J.H.; Chan, W.Y. Chitosan: An update on potential biomedical and pharmaceutical applications. *Mar. Drugs* **2015**, *13*, 5156–5186. [CrossRef] [PubMed]

53. Kubo, M.; Kuwayama, N.; Hirashima, Y.; Takaku, A.; Ogawa, T.; Endo, S. Hydroxyapatite ceramics as a particulate embolic material: Report of the physical properties of the hydroxyapatite particles and the animal study. *Am. J. Neuroradiol.* **2003**, *24*, 1540–1544. [PubMed]

54. Bendszus, M.; Klein, R.; Burger, R.; Warmuth-Metz, M.; Hofmann, E.; Solymosi, L. Efficacy of trisacryl gelatin microspheres versus polyvinyl alcohol particles in the preoperative embolization of meningiomas. *Am. J. Neuroradiol.* **2000**, *21*, 255–261. [PubMed]

55. Kim, H.M.; Lee, G.H.; Kuh, H.J.; Kwak, B.K.; Lee, J. Liposomal doxorubicin-loaded chitosan microspheres capable of controlling release of doxorubicin for anti-cancer chemoembolization: In vitro characteristics. *J. Drug Deliv. Sci. Technol.* **2013**, *23*, 283–286. [CrossRef]

56. Barnett, B.P.; Kraitchman, D.L.; Lauzon, C.; Magee, C.A.; Walczak, P.; Gilson, W.D.; Arepally, A.; Bulte, J.W.M. Radiopaque alginate microcapsules for X-ray visualization and immunoprotection of cellular therapeutics. *Mol. Pharm.* **2006**, *3*, 531–538. [CrossRef] [PubMed]

57. Balter, S.; Hopewell, J.W.; Miller, D.L.; Wagner, L.K.; Zelefsky, M.J. Fluoroscopically guided interventional procedures: A review of radiation effects on patients' skin and hair. *Radiology* **2010**, *254*, 326–341. [CrossRef] [PubMed]

58. Kang, M.J.; Oh, I.Y.; Choi, B.C.; Kwak, B.K.; Lee, J.; Choi, Y.W. Development of superparamagnetic iron oxide nanoparticles (SPIOs)-embedded chitosan microspheres for magnetic resonance (MR)-traceable embolotherapy. *Biomol. Ther.* **2009**, *17*, 98–103. [CrossRef]

59. Lee, H.S.; Kim, E.H.; Shao, H.P.; Kwak, B.K. Synthesis of SPIO-chitosan microspheres for MRI-detectable embolotherapy. *J. Magn. Magn. Mater.* **2005**, *293*, 102–105. [CrossRef]

60. Roberts, P.J.; Stinchcombe, T.E.; Der, C.J.; Socinski, M.A. Personalized medicine in non-small-cell lung cancer: Is KRAs a useful marker in selecting patients for epidermal growth factor receptor-targeted therapy? *J. Clin. Oncol.* **2010**, *28*, 4769–4777. [CrossRef] [PubMed]

61. Na, J.H.; Koo, H.; Lee, S.; Min, K.H.; Park, K.; Yoo, H.; Lee, S.H.; Park, J.H.; Kwon, I.C.; Jeong, S.Y.; et al. Real-time and non-invasive optical imaging of tumor-targeting glycol chitosan nanoparticles in various tumor models. *Biomaterials* **2011**, *32*, 5252–5261. [CrossRef] [PubMed]

62. Matsumura, Y.; Maeda, H. A new concept for macromolecular therapeutics in cancer chemotherapy: Mechanism of tumoritropic accumulation of proteins and the antitumor agent smancs. *Cancer Res.* **1986**, *46*, 6387–6392. [PubMed]

63. Baetke, S.C.; Lammers, T.; Kiessling, F. Applications of nanoparticles for diagnosis and therapy of cancer. *Br. J. Radiol.* **2015**, *88*, 12. [CrossRef] [PubMed]

64. Son, Y.J.; Jang, J.S.; Cho, Y.W.; Chung, H.; Park, R.W.; Kwon, I.C.; Kim, I.S.; Park, J.Y.; Seo, S.B.; Park, C.R.; et al. Biodistribution and anti-tumor efficacy of doxorubicin loaded glycol-chitosan nanoaggregates by EPR effect. *J. Control. Release* **2003**, *91*, 135–145. [CrossRef]

65. Bisht, S.; Maitra, A. Dextran-doxorubicin/chitosan nanoparticles for solid tumor therapy. *Wiley Interdiscip. Rev. Nanomed. Nanobiotechnol.* **2009**, *1*, 415–425. [CrossRef] [PubMed]

66. Kim, J.K.; Garripelli, V.K.; Jeong, U.H.; Park, J.S.; Repka, M.A.; Jo, S. Novel pH-sensitive polyacetal-based block copolymers for controlled drug delivery. *Int. J. Pharm.* **2010**, *401*, 79–86. [CrossRef] [PubMed]

67. Hong, H.; Zhang, Y.; Sun, J.T.; Cai, W.B. Molecular imaging and therapy of cancer with radiolabeled nanoparticles. *Nano Today* **2009**, *4*, 399–413. [CrossRef] [PubMed]

68. Liu, Y.J.; Welch, M.J. Nanoparticles labeled with positron emitting nuclides: Advantages, methods, and applications. *Bioconjug. Chem.* **2012**, *23*, 671–682. [CrossRef] [PubMed]

69. Quon, A.; Gambhir, S.S. FDG-pet and beyond: Molecular breast cancer imaging. *J. Clin. Oncol.* **2005**, *23*, 1664–1673. [CrossRef] [PubMed]

70. Gambhir, S.S. Molecular imaging of cancer with positron emission tomography. *Nat. Rev. Cancer* **2002**, *2*, 683–693. [CrossRef] [PubMed]

71. Liu, S. Bifunctional coupling agents for radiolabeling of biornolecules and target-specific delivery of metallic radionuclides. *Adv. Drug Deliv. Rev.* **2008**, *60*, 1347–1370. [CrossRef] [PubMed]

72. Lee, S.; Kang, S.W.; Ryu, J.H.; Na, J.H.; Lee, D.E.; Han, S.J.; Kang, C.M.; Choe, Y.S.; Lee, K.C.; Leary, J.F.; et al. Tumor-homing glycol chitosan-based optical/pet dual imaging nanoprobe for cancer diagnosis. *Bioconjug. Chem.* **2014**, *25*, 601–610. [CrossRef] [PubMed]

73. Egeblad, M.; Werb, Z. New functions for the matrix metalloproteinases in cancer progression. *Nat. Rev. Cancer* **2002**, *2*, 161–174. [CrossRef] [PubMed]

74. Na, J.H.; Lee, S.Y.; Lee, S.; Koo, H.; Min, K.H.; Jeong, S.Y.; Yuk, S.H.; Kim, K.; Kwon, I.C. Effect of the stability and deformability of self-assembled glycol chitosan nanoparticles on tumor-targeting efficiency. *J. Control. Release* **2012**, *163*, 2–9. [CrossRef] [PubMed]

75. Hamoudeh, M.; Kamleh, M.A.; Diab, R.; Fessi, H. Radionuclides delivery systems for nuclear imaging and radiotherapy of cancer. *Adv. Drug Deliv. Rev.* **2008**, *60*, 1329–1346. [CrossRef] [PubMed]

76. Kim, J.K.; Han, K.H.; Lee, J.T.; Paik, Y.H.; Ahn, S.H.; Lee, J.D.; Lee, K.S.; Chon, C.Y.; Moon, Y.M. Long-term clinical outcome of phase IIB clinical trial of percutaneous injection with holmium-166/chitosan complex (Milican) for the treatment of small hepatocellular carcinoma. *Clin. Cancer Res.* **2006**, *12*, 543–548. [CrossRef] [PubMed]

77. Suzuki, Y.S.; Momose, Y.; Higashi, N.; Shigematsu, A.; Park, K.B.; Kim, Y.M.; Kim, J.R.; Ryu, J.R. Biodistribution and kinetics of holmium-166-chitosan complex (DW-166HC) in rats and mice. *J. Nucl. Med.* **1998**, *39*, 2161–2166. [PubMed]

78. Lee, Y.H. Effect of holmium-166 injection into hepatocellular carcinomas (SK-HEP1) heterotransplanted in mice. *J. Korean Radiol. Soc.* **1998**, *38*, 83–92. [CrossRef]

79. Tsai, W.B.; Chen, Y.R.; Liu, H.L.; Lai, J.Y. Fabrication of UV-crosslinked chitosan scaffolds with conjugation of RGD peptides for bone tissue engineering. *Carbohydr. Polym.* **2011**, *85*, 129–137. [CrossRef]

80. Pok, S.; Myers, J.D.; Madihally, S.V.; Jacot, J.G. A multilayered scaffold of a chitosan and gelatin hydrogel supported by a PCL core for cardiac tissue engineering. *Acta Biomater.* **2013**, *9*, 5630–5642. [CrossRef] [PubMed]

81. Howling, G.I.; Dettmar, P.W.; Goddard, P.A.; Hampson, F.C.; Dornish, M.; Wood, E.J. The effect of chitin and chitosan on the proliferation of human skin fibroblasts and keratinocytes in vitro. *Biomaterials* **2001**, *22*, 2959–2966. [CrossRef]

82. Ishihara, M.; Nakanishi, K.; Ono, K.; Sato, M.; Kikuchi, M.; Saito, Y.; Yura, H.; Matsui, T.; Hattori, H.; Uenoyama, M.; et al. Photocrosslinkable chitosan as a dressing for wound occlusion and accelerator in healing process. *Biomaterials* **2002**, *23*, 833–840. [CrossRef]

83. Patrulea, V.; Ostafe, V.; Borchard, G.; Jordan, O. Chitosan as a starting material for wound healing applications. *Eur. J. Pharm. Biopharm.* **2015**, *97*, 417–426. [CrossRef] [PubMed]

84. Brun, P.; Mastrotto, C.; Beggiao, E.; Stefani, A.; Barzon, L.; Sturniolo, G.C.; Palu, G.; Castagliuolo, I. Neuropeptide neurotensin stimulates intestinal wound healing following chronic intestinal inflammation. *Am. J. Physiol. Gastrointest. Liver Physiol.* **2005**, *288*, G621–G629. [CrossRef] [PubMed]

85. Behl, G.; Iqbal, J.; O'Reilly, N.J.; McLoughlin, P.; Fitzhenry, L. Synthesis and characterization of poly(2-hydroxyethylmethacrylate) contact lenses containing chitosan nanoparticles as an ocular delivery system for dexamethasone sodium phosphate. *Pharm. Res.* **2016**, *33*, 1638–1648. [CrossRef] [PubMed]

86. Lehr, C.M.; Bouwstra, J.A.; Schacht, E.H.; Junginger, H.E. Invitro evaluation of mucoadhesive properties of chitosan and some other natural polymers. *Int. J. Pharm.* **1992**, *78*, 43–48. [CrossRef]

87. Kapoor, Y.; Chauhan, A. Ophthalmic delivery of cyclosporine A from Brij-97 microemulsion and surfactant-laden p-HEMA hydrogels. *Int. J. Pharm.* **2008**, *361*, 222–229. [CrossRef] [PubMed]

88. Cho, I.S.; Cho, M.O.; Li, Z.; Nurunnabi, M.; Park, S.Y.; Kang, S.W.; Huh, K.M. Synthesis and characterization of a new photo-crosslinkable glycol chitosan thermogel for biomedical applications. *Carbohydr. Polym.* **2016**, *144*, 59–67. [CrossRef] [PubMed]

89. Jeong, B.; Kim, S.W.; Bae, Y.H. Thermosensitive sol-gel reversible hydrogels. *Adv. Drug Deliv. Rev.* **2002**, *54*, 37–51. [CrossRef]

90. Cho, I.S.; Park, C.G.; Huh, B.K.; Cho, M.O.; Khatun, Z.; Li, Z.Z.; Kang, S.W.; Bin Choy, Y.; Huh, K.M. Thermosensitive hexanoyl glycol chitosan-based ocular delivery system for glaucoma therapy. *Acta Biomater.* **2016**, *39*, 124–132. [CrossRef] [PubMed]

91. Tu, Y.; Wang, X.X.; Lu, Y.; Zhang, H.; Yu, Y.; Chen, Y.; Liu, J.J.; Sun, Z.G.; Cui, L.L.; Gao, J.; et al. Promotion of the transdermal delivery of protein drugs by *N*-trimethyl chitosan nanoparticles combined with polypropylene electret. *Int. J. Nanomed.* **2016**, *11*, 5549–5561. [CrossRef] [PubMed]

92. Yeh, T.H.; Hsu, L.W.; Tseng, M.T.; Lee, P.L.; Sonjae, K.; Ho, Y.C.; Sung, H.W. Mechanism and consequence of chitosan-mediated reversible epithelial tight junction opening. *Biomaterials* **2011**, *32*, 6164–6173. [CrossRef] [PubMed]

marine drugs

MDPI

Article

Effect of Chitosan Properties on Immunoreactivity

Sruthi Ravindranathan, Bhanu prasanth Koppolu, Sean G. Smith and David A. Zaharoff *

Department of Biomedical Engineering, University of Arkansas, Fayetteville, AR 72701, USA;
ravindra@uark.edu (S.R.); bkoppolu@uark.edu (B.K.); sgs004@uark.edu (S.G.S.)
* Correspondence: zaharoff@uark.edu; Tel.: +1-479-575-2005

Academic Editors: Hitoshi Sashiwa and David Harding
Received: 24 February 2016; Accepted: 2 May 2016; Published: 11 May 2016

Abstract: Chitosan is a widely investigated biopolymer in drug and gene delivery, tissue engineering and vaccine development. However, the immune response to chitosan is not clearly understood due to contradicting results in literature regarding its immunoreactivity. Thus, in this study, we analyzed effects of various biochemical properties, namely degree of deacetylation (DDA), viscosity/polymer length and endotoxin levels, on immune responses by antigen presenting cells (APCs). Chitosan solutions from various sources were treated with mouse and human APCs (macrophages and/or dendritic cells) and the amount of tumor necrosis factor-α (TNF-α) released by the cells was used as an indicator of immunoreactivity. Our results indicate that only endotoxin content and not DDA or viscosity influenced chitosan-induced immune responses. Our data also indicate that low endotoxin chitosan (<0.01 EU/mg) ranging from 20 to 600 cP and 80% to 97% DDA is essentially inert. This study emphasizes the need for more complete characterization and purification of chitosan in preclinical studies in order for this valuable biomaterial to achieve widespread clinical application.

Keywords: chitosan; immunoreactivity; endotoxin; immune response

1. Introduction

Chitosan, or β-(1-4)-linked D-glucosamine and *N*-acetyl-D-glucosamine, is widely explored for use in numerous biomedical applications. A recent PubMed search indicated that, in 2015, more than 2000 biomedical-related publications cited chitosan as a keyword. To put this in perspective, there are more studies investigating chitosan than the ubiquitous biomaterials poly (lactic-co-glycolic acid) (PLGA), poly(L-lactic acid) (PLA) and polycaprolactone (PCL) combined.

Chitosan's popularity is a result of its versatility, availability and biocompatibility. Chitosan is frequently used in the development of novel drug delivery systems and controlled release platforms [1–7]. It is soluble in mildly acidic, aqueous solutions and thus, is easily formulated with variety of biopharmaceuticals from small, organic cytotoxic drugs to large, labile proteins [8–11]. In addition, due to its polycationic charge in solutions, chitosan interacts efficiently with polyanionic nucleic acids to form effective non-viral gene delivery complexes [12–14]. Chitosan's polycationic charge also allows it to function in wound healing and anti-microbial applications as well [15–20]. Additionally, this positive charge enables chitosan to loosen epithelial gap junctions, making it a promising candidate in mucosal delivery applications [21–26]. Chitosan and chitosan derivatives in the form of hydrogels, sponges and films are also under investigation for use as tissue engineering scaffolds [27–31].

Manipulation of chitosan's biophysical properties by changing its chemical composition further contributes to its versatility. For instance, both the molecular weight and degree of deacetylation (DDA) of chitosan can be modified via simple chemical treatments [32,33]. Controlling these parameters directly affects solubility, viscosity, charge and bioadhesion. In addition, chitosan's accessible amine and hydroxyl functional groups allow for facile conjugation of a variety of side chain moieties for limitless customization [34].

Despite the wealth of literature describing a plethora of potential biomedical uses, data are scarce and often conflicting, with regard to the nature and strength of immune responses induced by chitosan following parenteral injection or implantation. For example, in a study evaluating the biocompatibility of chitosan in mice, chitosan scaffolds were found to induce a typical acute inflammatory response marked by mild neutrophilic infiltration which dissipated over time [35]. By 12 weeks after implantation, the number of neutrophils at the implantation site was significantly reduced [35]. Similarly, in another study, chitosan hydrogel injected subcutaneously and intraperitoneally in rats induced a lower inflammatory response than the response against Vicryl® (Ethicon), a PLGA absorbable surgical suture [36]. Tissue surrounding the chitosan hydrogel was found to result in a typical wound healing response without the development of hemorrhage or necrosis. In contrast, a separate study found that chitosan could activate macrophages to secrete nitric oxide, leading to long term damage of surrounding tissues [37]. Transmission electron microscopy images of chitosan and the surrounding tissue 14 days after subcutaneous implantation in male Wistar rats, showed increased fibroblast proliferation and accumulation of collagen fibrils potentially indicating chronic inflammation and fibrosis.

Given that APCs, such as dendritic cells (DCs) and macrophages, are key regulators of immunity, understanding the effects of chitosan on these cells may shed some light on chitosan's immunoreactivity. Unfortunately, there is no consensus regarding chitosan's impact on APC function. For instance, it was determined that chitosan activates macrophages to produce monokines, such as colony stimulating factor and IL-1 [38,39]. In another study, chitosan-based microspheres increased the cytolytic activity of peritoneal macrophages following intraperitoneal injection [40]. Similarly, other studies showed that chitosan oligosaccharides induced the robust production of pro-inflammatory cytokines, such as TNF-α and IL-1β, by macrophages [41,42]. A contrasting study found no significant increase in the production of these cytokines when DCs were treated with chitosan [43].

Even if one assumes that chitosan is capable of activating macrophages and DCs, the mechanism of activation is not well understood. In a particular study, it was determined that oligochitosan with a polymerization degree (PD) of 7–16 upregulated the expression of major histocompatibility complex class II (MHCII) and CD86 on murine splenic CD11c$^+$ DCs, increased the production of tumor necrosis factor-α (TNF-α) and resulted in the proliferation of CD4$^+$ T cells. Toll-like receptor 4 (TLR4) on splenic DCs were found to play an important role since silencing the receptor reduced the expression of MHCII, CD86 and TNF-α. Additionally, these effects were observed only upon treatment with oligochitosan of PD 7–16 and not with PD of 3–7 [44]. In contrast, another study found that oligochitosan of PD 3–10 increased the production of TNF-α by mouse macrophages. Unlike the TLR4 study, mannose receptors were found to play a key role in chitosan-induced activation [45]. Another type of receptor, NOD-like receptor family pyrin domain containing 3 (NLRP3), was found to be activated by chitosan and induce the release of IL-1β [46].

Thus, it is evident from the literature that there are a number of discrepancies related to chitosan's immunoreactivity. Chitosan is viewed by some as an inert biomaterial that induces no more than a mild, transient foreign body reaction. To others, chitosan induces a specific, inflammatory response initiated by direct molecular recognition. It is possible that both sides are correct. Chitosan is a diverse class of molecules manufactured from a variety of raw materials. Different chitosans may induce different immune responses. Moreover, some researchers also suspect most, if not all, chitosans to be contaminated with varying levels of endotoxin which would impact immune responses significantly [47]. Thus, there is a compelling need for a systematic study to resolve large, contradictory discrepancies and to gain a better understanding of the immunoreactivity of chitosan. In particular, the issue of endotoxin contamination must be addressed for chitosan to achieve widespread clinical application. For now, chitosan's clinical use is limited to topical applications, e.g., hemostatic bandages, despite its considerable potential in other arenas.

The goal of this study was to assess the effect of chitosan's modifiable parameters, *i.e.*, molecular weight and DDA, as well as endotoxin contamination on the function of critical APC populations *in vitro*. Findings from this study should be of interest to any investigator developing chitosan-based implants and injectables. This study also makes the case for more detailed testing and reporting of biochemical parameters when using chitosan in biomedical applications.

2. Results

2.1. Chitosan Induced Cytokine Production

Mouse macrophages treated with chitosan from different sources secreted an enormous range of TNF-α (Figure 1). Among the six chitosan sources, the concentration of TNF-α was highest in supernatant collected from cells treated with chitosan from MP Biomedicals (5553.1 $\pm$ 373.7 pg/mL) and lowest with chitosan from Acros Organics (397.0 $\pm$ 27.1 pg/mL). Chitosan from MP Biomedicals induced a similar response as pure LPS (6606.3 $\pm$ 416.6 pg/mL). Untreated cells produced less than 200 pg/mL TNF-α. All chitosan treatments induced TNF-α levels that were significantly different from one another and higher than the untreated control group ($p < 0.05$ via ANOVA and Dunnett's post test).

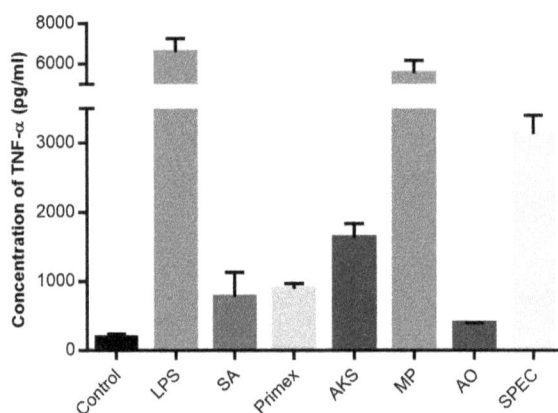

Figure 1. Immune response to commercially available chitosan. TNF-α released by Raw 264.7 macrophages upon 24-h incubation with 0.1 µg/mL LPS, 0.1 mg/mL of chitosan solutions from six different manufacturers; Sigma-Aldrich (SA), Primex, AK Scientific (AKS), MP Biomedicals (MP), Acros Organics (AO) and Spectrum (SPEC) or the cell medium alone as negative control. Data presented are mean with SEM from three independent measurements.

2.2. Characterization of Commercially Available Chitosan

Viscosity, DDA and endotoxin content in each of the six chitosan samples were measured and tabulated. As seen in Table 1, each chitosan was found to have different levels of endotoxin contamination and also greatly varied with respect to viscosity and DDA. Chitosan from Spectrum chemicals was found to be highly contaminated with endotoxin levels averaging 3.45 $\pm$ 0.04 EU/mg, whereas chitosan from Primex contained only 0.22 $\pm$ 0.06 EU/mg. The DDA ranged from 74% to 98% while viscosities of 1% chitosan solutions varied from 13 to 265 cP among the six different chitosan samples respectively.

Table 1. Comparison of properties of commercially available chitosan. The amount of endotoxin contamination, degree of deacetylation (DDA) and viscosity for each of the six different commercially available chitosan considered in this study were measured as described. Data presented are mean $\pm$ standard deviation from three independent measurements.

Chitosan	Molecular Weight Provided by Manufacturer (kDa)	Amount of Endotoxin (EU/mg)	DDA (%)	Viscosity (cP)
Sigma-Aldrich	50–190	1.48 ± 0.05	74 ± 0.1	265 ± 5
Primex	Not provided	0.22 ± 0.06	80 ± 0.3	95 ± 25
AK Scientific	Not provided	1.44 ± 0.03	96 ± 0.4	65 ± 20
MP Biomedicals	Not provided	2.27 ± 0.03	88 ± 0.9	19 ± 6
Acros Organics	100–300	1.09 ± 0.05	98 ± 0.4	58 ± 16
Spectrum chemicals	Not provided	3.45 ± 0.04	92 ± 0.4	13 ± 5

2.3. Effect of DDA on Cytokine Release

Based on the data shown in Table 1, it was not possible to correlate the effects of any of the three properties measured with TNF-α production (Figure 1). Therefore, in order to isolate the effect of DDA on cytokine release, custom purified chitosan of two different DDAs, 80% and 97%, but of same viscosity (95 ± 25 cP) were obtained from UABC. This purified chitosan contained undetectable (<0.01 EU/mg) levels of endotoxin. Upon incubation with mouse macrophages, no significant difference was found in the TNF-α released between the two chitosan samples (Figure 2). The amount of TNF-α released in both cases was similar to media alone control group, 193.7 ± 67.5 pg/mL ($p > 0.05$ via ANOVA and Dunnett's post test).

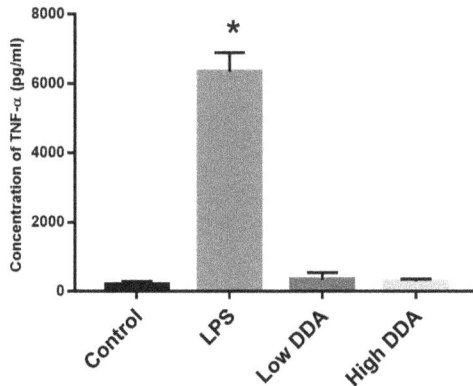

Figure 2. Effect of DDA on mouse macrophages. TNF-α released by mouse macrophages upon 24-h incubation with low endotoxin (<0.01 EU/mg) chitosan of 80% (Low) and 97% (High) DDA from UABC. LPS (0.1 µg/mL) and media alone served as positive and negative controls, respectively. Data presented are mean plus SEM from three independent measurements. Amount of TNF-α released upon treatment with LPS was significantly different with $p < 0.05$ and is represented with (*).

2.4. Effect of Viscosity on Cytokine Release

To isolate the effect of viscosity or polymer chain length, RAW 264.7 macrophages were treated with chitosans from both Primex and UABC with three different viscosity ranges, <20 cP, 20–200 cP and 200–600 cP, at a single DDA of 80%. Significant differences in TNF-α production between the different viscosity samples from the Primex chitosan was observed (Figure 3). However, when purified chitosan from UABC was used, no significant difference was found in the TNF-α release between samples of different viscosities ($p > 0.05$ via ANOVA). Additionally, for UABC chitosans, the amount of TNF-α release was indistinguishable from the untreated control group, 128 ± 17.5 pg/mL ($p > 0.05$ via ANOVA and Dunnett's post test).

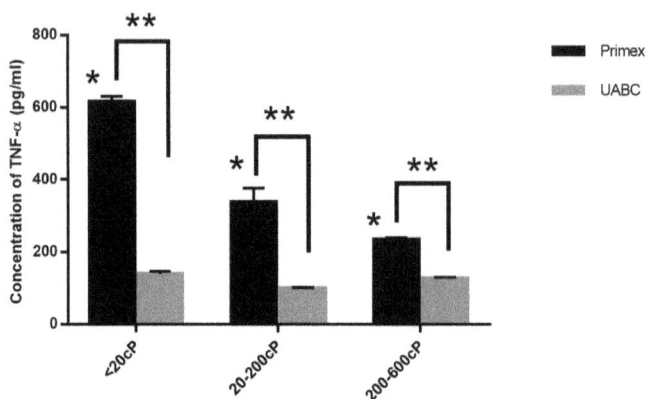

Figure 3. Effect of endotoxin and viscosity on immune response of chitosan. TNF-α released by RAW 264.7 macrophages upon 24 h incubation with chitosans with viscosities of <20 cP, 20–200 cP and 200–600 cP from Primex and UABC. Data presented are mean plus SEM from three independent measurements. * $p < 0.05$ for Primex chitosan of viscosities <20 cP, 20–200 cP and 200–600 cP. ** $p < 0.05$ for Primex *vs.* UABC.

2.5. Effect of Endotoxin Contamination on Cytokine Release

To determine if endotoxin content is directly responsible for immunoreactivity, UABC chitosan spiked in with varying levels of endotoxin (0.5 and 1 EU) were exposed to RAW 264.7 cells. Both spiked samples elicited significantly higher levels of TNF-α when compared to the untreated control (Figure 4). In fact, chitosan spiked with 1 EU released higher levels of TNF-α (376 ± 38 pg/mL) when compared to chitosan with 0.5 EU (177 ± 5 pg/mL). Here, the amount of TNF released by the cells upon treatment with media alone or LPS were 39 ± 5 pg/mL and $12,700 \pm 251$ pg/mL respectively.

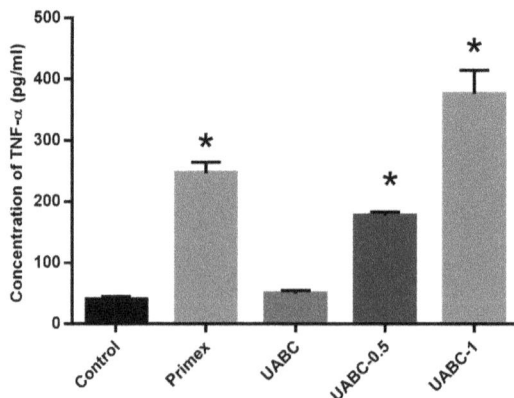

Figure 4. Effect of endotoxin content on immune response of chitosan. TNF-α released by mouse macrophages upon 24-h incubation with Primex, UABC and UABC chitosan spiked with 0.5 or 1 EU (UABC−0.5/UABC 1). Cells treated with media alone served as negative control. Data presented are mean plus SEM from three independent measurements. * $p < 0.05$ *vs.* control via ANOVA and Tukey's *post hoc* test.

2.6. Effect of Endotoxin Contamination on Mouse Dendritic Cells

To verify that differences in chitosan-induced immunoreactivity were not an artifact of RAW 264.7 macrophages, chitosans from both Primex and UABC were exposed to BMDCs isolated from healthy mouse femurs. Chitosans from Primex and UABC were similar in viscosity, 20–200 cP, and DDA, 80%, however, endotoxin levels were 0.22 ± 0.06 EU/mg and <0.01 EU/mg, respectively. Chitosan from Primex induced significantly higher levels of TNF-α compared to purified chitosan from UABC -29.0 ± 6.7 pg/mL *vs.* 14.9 ± 1.8 pg/mL (Figure 5). Once again the response to UABC chitosan was similar to untreated controls ($p > 0.05$ via Tukey's post test). LPS-induced TNF-α secretion by BMDCs was orders of magnitude higher at 2597.2 ± 76.2 pg/mL (data not shown on graph).

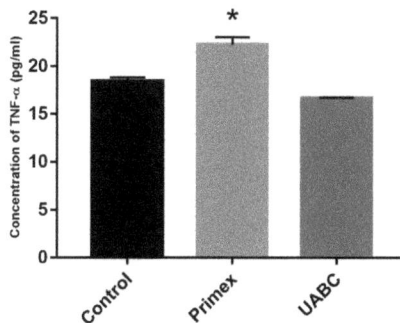

Figure 5. Effect of endotoxin on BMDCs. TNF-α released by BMDCs upon 24 h incubation with chitosan from Primex and UABC. Data presented are mean plus SEM from three independent measurements. Media alone served as negative control. * $p < 0.05$ *vs.* control and UABC via Tukey's *post hoc* analysis.

2.7. Effect of Endotoxin Contamination on Human Macrophage Cell Line

To confirm that differences in chitosan-induced immunoreactivity were not restricted to murine cells, human macrophages differentiated from THP1 cells were treated with chitosan (80% DDA and 20–200 cP viscosity) from Primex and UABC as above (Figure 6). The amount of TNF-α produced by THP1 cells exposed to Primex chitosan, 101.1 ± 2.6 pg/mL, was significantly higher than TNF-α following exposure to the UABC chitosan, 34.7 ± 2.8 pg/mL ($p < 0.05$ via Tukey's post test). With media alone and LPS, THP1 cells produced 4.3 ± 0.4 pg/mL and 4522.7 ± 149.9 pg/mL TNF-α, respectively.

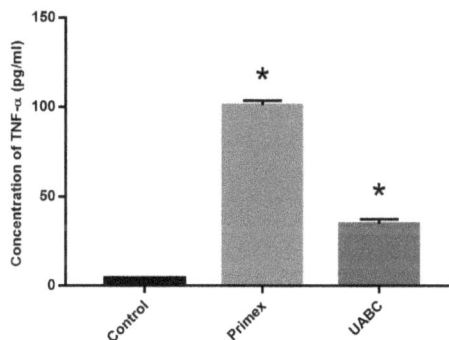

Figure 6. Effect of endotoxin contamination of chitosan on human macrophage cell line. Human macrophages were differentiated from THP1 cells and treated with chitosan from Primex and UABC or media alone (control). Data presented are mean plus SEM from three independent measurements. * $p < 0.05$ *vs.* control.

3. Discussion

Chitosan is a promising, versatile biopolymer with numerous potential clinical applications. However, there is no clear consensus regarding the immunoreactivity of chitosan. This is partly due to a lack of standardization in chitosan-based studies. In a recent review, it was correctly noted that most studies involving chitosan fail to fully disclose important characteristics such as the molecular weight, DDA, source and purity of chitosan [48]. As a result, it is difficult to compare results between groups which use chitosan from different sources. Our characterization studies showed that there were indeed large differences in chitosans from different manufacturers in terms of viscosity, DDA and endotoxin contamination (Table 1).

Because chitosan is a natural biopolymer produced primarily via multiple cycles of alkaline and acidic treatments of crustacean exoskeletons, some variation in physico-chemical properties can be expected. Different treatment conditions inevitably lead to differences in molecular weights/viscosities and DDAs. As a result, chitosan from a single source can vary from batch-to-batch.

To understand the effects of chitosan's properties on immunoreactivity, our studies focused on cellular immune responses to chitosan. By focusing on the cell level, we are able to isolate chitosan-specific immunoreactivity from more complex processes, such as foreign body reactions and inflammatory responses that occur following *in vivo* administrations. In particular, professional APCs such as macrophages and dendritic cells are key mediators of immunity. Thus, understanding how these APCs respond to different chitosans is essential for helping resolve discrepancies in the literature and supporting clinical translation of chitosan-based applications.

TNF-α was used as an indicator of immunoreactivity since it is an important early signaling protein and a highly sensitive indicator of immune responsiveness. Our preliminary studies which also measured IL-1β and IL-6 levels in response to chitosan exposure, demonstrated that these cytokines were not as sensitive as TNF-α (data not shown).

After the initial characterization (Table 1) and TNF-α release (Figure 1) studies, it was clear that (1) immune responses were highly variable; and (2) it was not possible to isolate the specific effects of viscosity, DDA and endotoxin contamination using commercially available chitosans. In addition, despite endotoxins having a major effect on chitosan immunoreactivity, it was not possible to correlate endotoxin levels in the different chitosans with TNF-α release. This is due to the fact that the activity of endotoxins vary widely with the intrinsic activity of its lipid associated protein [49]. Because these chitosans were derived from different sources, the types of endotoxin were likely to be different. Nevertheless, in anticipation that endotoxin contamination would be a significant factor in driving the immune response, we obtained, from the UABC, purified chitosan that contained undetectable levels of endotoxin. This purified chitosan allowed us to uncouple the effects of molecular weight, DDA and endotoxin contamination.

By treating mouse macrophages with chitosans of different viscosities and DDAs but with the same amount of endotoxin (<0.01 EU/mg), we found that neither DDA (Figure 2) nor viscosity/molecular weight (Figure 3) had any effect on the immune response. Additionally, by spiking in known amounts of endotoxin into UABC chitosan, we confirmed that the immunoreactivity is directly affected by amount of endotoxin contamination in chitosan.

Regarding the effect of viscosity/molecular weight, it was previously shown that higher molecular weight oligochitosans, PD 7–16, were more immunoreactive, in terms of upregulation of antigen presenting and co-stimulatory molecules as well as TNF-α production, than lower molecular weight oligochitosans, PD 3–7 [26]. We estimate that even our smallest chitosan had a PD of about 20–40 which is much higher than the oligochitosans of previous studies. As a result, it is possible that viscosity/molecular weight does not impact immunoreactivity above a threshold chain length. It is also possible that oligochitosans used in previous studies contained varying amounts of endotoxin which, as we have shown above, can lead to significant differences in immunoreactivity. Endotoxin levels were not reported in the oligochitosan study [26]. Nevertheless, planned studies will explore smaller, purified oligochitosans to determine if we find the same effect of molecular weight with smaller chitosans.

Regarding the effect of DDA, it was previously shown that chitosan scaffolds of 85% DDA led to greater mononuclear cell infiltration as compared to 96% DDA [50]. Furthermore, the density of infiltrates increased over a 2-week period for the lower DDA scaffold. While our study found no effect of DDA, we looked at a simple system aimed at understanding cellular responses and not *in vivo* foreign body response. It is possible that immune responses may change *in vivo* when more cells and processes are involved. This is the subject of future studies.

It is important to reiterate that chitosan-induced responses were not restricted to a single cell line or species. Similar results were obtained with RAW 264.7 murine macrophages and murine BMDCs. Since mice are generally less sensitive to endotoxins than humans, it was also useful to observe that human macrophages differentiated from THP1 similarly responded to endotoxin contamination in chitosan by producing higher levels of TNF-α (Figure 4). While it is not surprising that higher endotoxin levels lead to higher pro-inflammatory cytokine production, these data emphasize the need for chitosan-based preclinical studies to provide a complete biochemical characterization, including endotoxin levels, of any chitosans used.

4. Materials and Methods

4.1. Reagents

Chitosan was obtained from Sigma-Aldrich (St. Louis, MO, USA), Acros Organics (Bridgewater, NJ, USA), AK Scientific (Union City, CA, USA), MP Biomedicals (Solon, Ohio, OH, USA), Spectrum Chemical Mfg Corp. (New Brunswick, NJ, USA) and Primex (Siglufjordor, Iceland). Purified, low endotoxin chitosan (<0.01 EU/mg) was obtained from the University of Arkansas Biologics Center (UABC) (Fayetteville, AR, USA). Sodium hydroxide was purchased from Thermo Fisher Scientific Inc. (Waltham, MA, USA). Hydrochloric acid and acetic acid were purchased from Sigma-Aldrich (St. Louis, MO, USA). Unless otherwise specified, chitosan solutions were prepared by dissolving chitosan in 0.1 M HCl and adjusted to pH 6.0 using NaOH.

Cell culture media components including Dulbecco's modified Eagle's medium (DMEM), Roswell Park Memorial Institute 1640 (RPMI-1640), fetal bovine serum (FBS), L-glutamine and penicillin-streptomycin solution were purchased from Hyclone Laboratories (Logan, UT, USA). Ammonium-chloride-potassium (ACK) buffer used in the process of isolating bone marrow derived dendritic cells from mouse was purchased from Lonza (Allendale, NJ, USA). Recombinant murine granulocyte-macrophage colony stimulating factor (rmGM-CSF) was purchased from Peprotech (Rocky Hill, NJ, USA). Lipopolysaccharide (LPS) from Salmonella enterica serotype enteritidis was purchased from Sigma-Aldrich.

4.2. Laboratory Animals

Female C57BL/6 J mice were purchased from The Jackson Laboratory (Bar Harbor, ME, USA). Mice were housed in microisolator cages and used at 8–12 weeks of age. All experimental procedures were approved by the Institutional Animal Care and Use Committee at the University of Arkansas. Animal care was in compliance with The Guide for Care and Use of Laboratory Animals (National Research Council).

4.3. Cell Culture

RAW 264.7 murine macrophages and THP1 human monocyte leukemia cells were purchased from American Type Culture Collection (Manassas, VA, USA). RAW 264.7 macrophages were cultured in Dulbecco's Modified Eagle Medium (DMEM) with 10% fetal bovine serum (FBS) and 1% penicillin/streptomycin. THP1 cells were differentiated into macrophages and cultured in RPMI-1640 with 10% FBS as described previously [51]. Bone marrow derived dendritic cells were isolated from C57BL/6 mice, as described elsewhere [52]. All cultures were maintained in a humidified CO_2 incubator at 37 °C and 5% CO_2.

4.4. Determination of Relative Viscosity, DDA and Endotoxin Content

Relative viscosity was used as a surrogate for polymer chain length or molecular weight. A 1% (w/v) solution of each chitosan was prepared in 1% (v/v) acetic acid (Sigma-Aldrich) and the viscosity was measured using an LV DVIII rheometer from Brookfield Engineering Laboratories (Middleboro, MA, USA). Readings were taken at 25 °C using a CP-40 spindle. The DDA of the chitosan solutions were determined, as described previously [32]. Recombinant Factor C assay (rFC) assay from Lonza was used to measure the level of endotoxin in chitosan solutions following the procedure described by the manufacturer.

4.5. Measurement of Immune Responses

RAW 264.7 macrophages were exposed to 0.1 mg/mL of each chitosan solution for 24 h. The amount of TNF-α in cell culture supernatants was quantified via enzyme linked immunosorbent assay (ELISA) kits from eBiosciences (San Diego, CA, USA) following the protocol provided by the manufacturer. Cells treated with 0.1 µg/mL LPS or media alone served as positive and negative controls, respectively. Endotoxin standard (*E.coli*, O55:B5) provided in the rFC assay kit was spiked into UABC chitosan to confirm its effect on immunoreactivity.

4.6. Statistical Analysis

All experiments were performed in triplicate. Data are reported as mean ± standard error of the mean (SEM) or standard deviation. Student's *t*-test was used to compare two groups of interest. For more than two groups, analysis of variance (ANOVA) was performed followed by Tukey's or Dunnett's post test analyses. Statistical differences were accepted at the $p < 0.05$ level.

5. Conclusions

Despite its remarkable potential for use in a range of medical applications, discrepancies in chitosan's immunoreactivity have limited its clinical use. By exploring the direct effects of chitosan on immune responses at the cellular level, we are able to isolate chitosan-specific immunoreactivity from more complex processes such as foreign body reactions and inflammatory responses that occur following injections *in vivo*. Initial characterization studies demonstrated that commercially available chitosans, in solution, induce varying degrees of immune responses. Using purified, low endotoxin chitosan, it was determined that viscosity/molecular weight and DDA within the ranges 20–600 CP and 80%–97%, respectively, have no impact on chitosan's immunoreactivity. Similar results were found using murine and human macrophages as well as murine dendritic cells. In retrospect, large differences in immune responses, both in our initial characterization of commercially available chitosan as well as results found in the literature, may be explained by highly variable endotoxin levels. It should be noted that because immunoreactivity was assessed solely based on TNF-α production, to fully appreciate the immunoreactivity of chitosan, *in vitro* studies characterizing other inflammatory cytokines, such as IL-1 and IL-6, as well as *in vivo* studies using purified chitosans must be performed. Ultimately, endotoxin contamination is expected to remain as the chief factor influencing immunoreactivity. Therefore, we suggest that endotoxin levels be explicitly reported in all future biomedical studies involving chitosan. It should also be noted that chitosan solutions were evaluated in this study. While we expect to find similar low reactivity to chitosan in solid forms, such as particles, scaffolds, films, *etc.*, no conclusions can be drawn without additional experimentation.

Author Contributions: S.R. and D.A.Z. were responsible for study concept and design. S.R., B.K. and S.G.S. were responsible for acquisition of data. S.R., B.K., S.G.S. and D.A.Z. were responsible for analysis and interpretation of data. S.R. and D.A.Z. were responsible for manuscript preparation.

Conflicts of Interest: The authors declare no conflict of interest.

References

1. Agnihotri, S.A.; Mallikarjuna, N.N.; Aminabhavi, T.M. Recent advances on chitosan-based micro-and nanoparticles in drug delivery. *J Control. Release* **2004**, *100*, 5–28. [CrossRef] [PubMed]

2. Al Rubeaan, K.; Rafiullah, M.; Jayavanth, S. Oral insulin delivery systems using chitosan-based formulation: A review. *Expert Opin. Drug Deliv.* **2016**, *13*, 223–237. [CrossRef] [PubMed]

3. Dang, Q.; Liu, C.; Wang, Y.; Yan, J.; Wan, H.; Fan, B. Characterization and biocompatibility of injectable microspheres-loaded hydrogel for methotrexate delivery. *Carbohydr. Polym.* **2016**, *136*, 516–526. [CrossRef] [PubMed]

4. Park, J.H.; Saravanakumar, G.; Kim, K.; Kwon, I.C. Targeted delivery of low molecular drugs using chitosan and its derivatives. *Adv. Drug Deliv. Rev.* **2010**, *62*, 28–41. [CrossRef] [PubMed]

5. Risbud, M.V.; Hardikar, A.A.; Bhat, S.V.; Bhonde, R.R. pH-sensitive freeze-dried chitosan—Polyvinyl pyrrolidone hydrogels as controlled release system for antibiotic delivery. *J. Control. Release* **2000**, *68*, 23–30. [CrossRef]

6. Ruel-Gariepy, E.; Leclair, G.; Hildgen, P.; Gupta, A.; Leroux, J. Thermosensitive chitosan-based hydrogel containing liposomes for the delivery of hydrophilic molecules. *J. Control. Release* **2002**, *82*, 373–383. [CrossRef]

7. Vo, J.L.; Yang, L.; Kurtz, S.L.; Smith, S.G.; Koppolu, B.P.; Ravindranathan, S.; Zaharoff, D.A. Neoadjuvant immunotherapy with chitosan and interleukin-12 to control breast cancer metastasis. *Oncolmmunology* **2014**, *3*, e968001. [CrossRef] [PubMed]

8. Gan, Q.; Wang, T. Chitosan nanoparticle as protein delivery carrier—Systematic examination of fabrication conditions for efficient loading and release. *Colloids Surf. B Biointerfaces* **2007**, *59*, 24–34. [CrossRef] [PubMed]

9. Koppolu, B.; Zaharoff, D.A. The effect of antigen encapsulation in chitosan particles on uptake, activation and presentation by antigen presenting cells. *Biomaterials* **2013**, *34*, 2359–2369. [CrossRef] [PubMed]

10. Mehrotra, A.; Nagarwal, R.C.; Pandit, J.K. Lomustine loaded chitosan nanoparticles: Characterization and *in vitro* cytotoxicity on human lung cancer cell line L132. *Chem. Pharm. Bull.* **2011**, *59*, 315–320. [CrossRef] [PubMed]

11. Koppolu, B.; Smith, S.G.; Ravindranathan, S.; Jayanthi, S.; Kumar, T.K.S.; Zaharoff, D.A. Controlling chitosan-based encapsulation for protein and vaccine delivery. *Biomaterials* **2014**, *35*, 4382–4389. [CrossRef] [PubMed]

12. Bao, H.; Pan, Y.; Ping, Y.; Sahoo, N.G.; Wu, T.; Li, L.; Li, J.; Gan, L.H. Chitosan-functionalized graphene oxide as a nanocarrier for drug and gene delivery. *Small* **2011**, *7*, 1569–1578. [CrossRef] [PubMed]

13. Lu, H.; Dai, Y.; Lv, L.; Zhao, H. Chitosan-graft-polyethylenimine/DNA nanoparticles as novel non-viral gene delivery vectors targeting osteoarthritis. *PLoS ONE* **2014**, *9*, e84703. [CrossRef] [PubMed]

14. Raftery, R.; O'Brien, F.J.; Cryan, S. Chitosan for gene delivery and orthopedic tissue engineering applications. *Molecules* **2013**, *18*, 5611–5647. [CrossRef] [PubMed]

15. Charernsriwilaiwat, N.; Opanasopit, P.; Rojanarata, T.; Ngawhirunpat, T. Lysozyme-loaded, electrospun chitosan-based nanofiber mats for wound healing. *Int. J. Pharm.* **2012**, *427*, 379–384. [CrossRef] [PubMed]

16. Dai, T.; Tanaka, M.; Huang, Y.; Hamblin, M.R. Chitosan preparations for wounds and burns: Antimicrobial and wound-healing effects. *Expert Rev. Anti-Infect. Ther.* **2011**, *9*, 857–879. [CrossRef] [PubMed]

17. Kweon, D.; Song, S.; Park, Y. Preparation of water-soluble chitosan/heparin complex and its application as wound healing accelerator. *Biomaterials* **2003**, *24*, 1595–1601. [CrossRef]

18. Lih, E.; Lee, J.S.; Park, K.M.; Park, K.D. Rapidly curable chitosan—PEG hydrogels as tissue adhesives for hemostasis and wound healing. *Acta Biomater.* **2012**, *8*, 3261–3269. [CrossRef] [PubMed]

19. Moura, L.I.; Dias, A.M.; Leal, E.C.; Carvalho, L.; de Sousa, H.C.; Carvalho, E. Chitosan-based dressings loaded with neurotensin—An efficient strategy to improve early diabetic wound healing. *Acta Biomater.* **2014**, *10*, 843–857. [CrossRef] [PubMed]

20. Wang, B.; Liu, X.; Ji, Y.; Ren, K.; Ji, J. Fast and long-acting antibacterial properties of chitosan-Ag/polyvinylpyrrolidone nanocomposite films. *Carbohydr. Polym.* **2012**, *90*, 8–15. [CrossRef] [PubMed]

21. Casettari, L.; Illum, L. Chitosan in nasal delivery systems for therapeutic drugs. *J. Control. Release* **2014**, *190*, 189–200. [CrossRef] [PubMed]

22. Kumar, M.; Behera, A.K.; Lockey, R.F.; Zhang, J.; Bhullar, G.; de la Cruz, C.P.; Chen, L.C.; Leong, K.W.; Huang, S.K.; Mohapatra, S.S. Intranasal gene transfer by chitosan-DNA nanospheres protects BALB/c mice against acute respiratory syncytial virus infection. *Hum. Gene Ther.* **2002**, *13*, 1415–1425. [CrossRef] [PubMed]

23. Luo, Y.; Wang, Q. Recent development of chitosan-based polyelectrolyte complexes with natural polysaccharides for drug delivery. *Int. J. Biol. Macromol.* **2014**, *64*, 353–367. [CrossRef] [PubMed]
24. Smith, S.G.; Koppolu, B.; Ravindranathan, S.; Kurtz, S.L.; Yang, L.; Katz, M.D.; Zaharoff, D.A. Intravesical chitosan/interleukin-12 immunotherapy induces tumor-specific systemic immunity against murine bladder cancer. *Cancer Immunol. Immunother.* **2015**, *64*, 689–696. [CrossRef] [PubMed]
25. Ven der Lubben, I.M.; Verhoef, J.C.; Borchard, G.; Junginger, H.E. Chitosan and its derivatives in mucosal drug and vaccine delivery. *Eur. J. Pharm. Sci.* **2001**, *14*, 201–207. [CrossRef]
26. Yao, W.; Peng, Y.; Du, M.; Luo, J.; Zong, L. Preventative vaccine-loaded mannosylated chitosan nanoparticles intended for nasal mucosal delivery enhance immune responses and potent tumor immunity. *Mol. Pharm.* **2013**, *10*, 2904–2914. [CrossRef] [PubMed]
27. Bhardwaj, N.; Kundu, S.C. Chondrogenic differentiation of rat MSCs on porous scaffolds of silk fibroin/chitosan blends. *Biomaterials* **2012**, *33*, 2848–2857. [CrossRef] [PubMed]
28. Jana, S.; Florczyk, S.J.; Leung, M.; Zhang, M. High-strength pristine porous chitosan scaffolds for tissue engineering. *J. Mater. Chem.* **2012**, *22*, 6291–6299. [CrossRef]
29. Levengood, S.K.L.; Zhang, M. Chitosan-based scaffolds for bone tissue engineering. *J. Mater. Chem. B* **2014**, *2*, 3161–3184. [CrossRef] [PubMed]
30. Mukhopadhyay, P.; Sarkar, K.; Bhattacharya, S.; Bhattacharyya, A.; Mishra, R.; Kundu, P. pH sensitive *N*-succinyl chitosan grafted polyacrylamide hydrogel for oral insulin delivery. *Carbohydr. Polym.* **2014**, *112*, 627–637. [CrossRef] [PubMed]
31. Nettles, D.L.; Elder, S.H.; Gilbert, J.A. Potential use of chitosan as a cell scaffold material for cartilage tissue engineering. *Tissue Eng.* **2002**, *8*, 1009–1016. [CrossRef] [PubMed]
32. Yuan, Y.; Chesnutt, B.M.; Haggard, W.O.; Bumgardner, J.D. Deacetylation of chitosan: Material characterization and *in vitro* evaluation via albumin adsorption and pre-osteoblastic cell cultures. *Materials* **2011**, *4*, 1399–1416. [CrossRef]
33. Mao, S.; Shuai, X.; Unger, F.; Simon, M.; Bi, D.; Kissel, T. The depolymerization of chitosan: Effects on physicochemical and biological properties. *Int. J. Pharm.* **2004**, *281*, 45–54. [CrossRef] [PubMed]
34. Riva, R.; Ragelle, H.; des Rieux, A.; Duhem, N.; Jérôme, C.; Préat, V. Chitosan and chitosan derivatives in drug delivery and tissue engineering. In *Chitosan for Biomaterials II*; Springer: Berlin Heidelberg, Germany, 2011; pp. 19–44.
35. Vande Vord, P.J.; Matthew, H.W.T.; DeSilva, S.P.; Mayton, L.; Wu, B.; Wooley, P.H. Evaluation of the biocompatibility of a chitosan scaffold in mice. *J. Biomed. Mater. Res.* **2002**, *59*, 585–590. [CrossRef] [PubMed]
36. Azab, A.K.; Doviner, V.; Orkin, B.; Kleinstern, J.; Srebnik, M.; Nissan, A.; Rubinstein, A. Biocompatibility evaluation of crosslinked chitosan hydrogels after subcutaneous and intraperitoneal implantation in the rat. *J. Biomed. Mater. Res. A* **2007**, *83*, 414–422. [CrossRef] [PubMed]
37. Peluso, G.; Petillo, O.; Ranieri, M.; Santin, M.; Ambrosic, L.; Calabro, D.; Avallone, B.; Balsamo, G. Chitosan mediated stimulation of macrophage function. *Biomaterials* **1994**, *15*, 1215–1220. [CrossRef]
38. Suzuki, K.; Okawa, Y.; Hashimoto, K.; Suzuki, S.; Suzuki, M. Protecting effect of chitin and chitosan on experimentally induced murine candidiasis. *Microbiol. Immunol.* **1984**, *28*, 903–912. [CrossRef] [PubMed]
39. Nishimura, K.; Ishihara, C.; Ukei, S.; Tokura, S.; Azuma, I. Stimulation of cytokine production in mice using deacetylated chitin. *Vaccine* **1986**, *4*, 151–156. [CrossRef]
40. Nishimura, K.; Nishimura, S.; Seo, H.; Nishi, N.; Tokura, S.; Azuma, I. Effect of multiporous microspheres derived from chitin and partially deacetylated chitin on the activation of mouse peritoneal macrophages. *Vaccine* **1987**, *5*, 136–140. [CrossRef]
41. Feng, J.; Zhao, L.; Yu, Q. Receptor-mediated stimulatory effect of oligochitosan in macrophages. *Biochem. Biophys. Res. Commun.* **2004**, *317*, 414–420. [CrossRef] [PubMed]
42. Chen, C.; Wang, Y.; Liu, C.; Wang, J. The effect of water-soluble chitosan on macrophage activation and the attenuation of mite allergen-induced airway inflammation. *Biomaterials* **2008**, *29*, 2173–2182. [CrossRef] [PubMed]
43. Villiers, C.; Chevallet, M.; Diemer, H.; Couderc, R.; Freitas, H.; Van Dorsselaer, A.; Marche, P.N.; Rabilloud, T. From secretome analysis to immunology: Chitosan induces major alterations in the activation of dendritic cells via a TLR4-dependent mechanism. *Mol. Cell Proteom.* **2009**, *8*, 1252–1264. [CrossRef] [PubMed]

44. Dang, Y.; Li, S.; Wang, W.; Wang, S.; Zou, M.; Guo, Y.; Fan, J.; Du, Y.; Zhang, J. The effects of chitosan oligosaccharide on the activation of murine spleen CD11c dendritic cells via Toll-like receptor 4. *Carbohydr. Polym.* **2011**, *83*, 1075–1081. [CrossRef]

45. Han, Y.; Zhao, L.; Yu, Z.; Feng, J.; Yu, Q. Role of mannose receptor in oligochitosan-mediated stimulation of macrophage function. *Int. Immunopharmacol.* **2005**, *5*, 1533–1542. [CrossRef] [PubMed]

46. Bueter, C.L.; Lee, C.K.; Rathinam, V.A.; Healy, G.J.; Taron, C.H.; Specht, C.A.; Levitz, S.M. Chitosan but not chitin activates the inflammasome by a mechanism dependent upon phagocytosis. *J. Biol. Chem.* **2011**, *286*, 35447–35455. [CrossRef] [PubMed]

47. Lieder, R.; Gaware, V.S.; Thormodsson, F.; Einarsson, J.M.; Ng, C.H.; Gislason, J.; Masson, M.; Petersen, P.H.; Sigurjonsson, O.E. Endotoxins affect bioactivity of chitosan derivatives in cultures of bone marrow-derived human mesenchymal stem cells. *Acta Biomater.* **2013**, *9*, 4771–4778. [CrossRef] [PubMed]

48. Vasiliev, Y.M. Chitosan-based vaccine adjuvants: Incomplete characterization complicates preclinical and clinical evaluation. *Expert Rev. Vaccines* **2014**, *14*, 1–17. [CrossRef] [PubMed]

49. Morrison, D.C.; Ulevitch, R.J. The effects of bacterial endotoxins on host mediation systems. A review. *Am. J. Pathol.* **1978**, *93*, 526–618. [PubMed]

50. Barbosa, J.N.; Amaral, I.F.; Aguas, A.P.; Barbosa, M.A. Evaluation of the effect of the degree of acetylation on the inflammatory response to 3D porous chitosan scaffolds. *J. Biomed. Mater. Res. A* **2010**, *93*, 20–28. [CrossRef] [PubMed]

51. Schwende, H.; Fitzke, E.; Ambs, P.; Dieter, P. Differences in the state of differentiation of THP-1 cells induced by phorbol ester and 1,25-dihydroxyvitamin D3. *J. Leukoc. Biol.* **1996**, *59*, 555–561. [PubMed]

52. Inaba, K.; Inaba, M.; Romani, N.; Aya, H.; Deguchi, M.; Ikehara, S.; Muramatsu, S.; Steinman, R.M. Generation of large numbers of dendritic cells from mouse bone marrow cultures supplemented with granulocyte/macrophage colony-stimulating factor. *J. Exp. Med.* **1992**, *176*, 1693–1702. [CrossRef] [PubMed]

marine drugs

MDPI

Article

Resonance Rayleigh Scattering Spectra of an Ion-Association Complex of Naphthol Green B–Chitosan System and Its Application in the Highly Sensitive Determination of Chitosan

Weiai Zhang [1], Caijuan Ma [1], Zhengquan Su [1,2,*] and Yan Bai [1,*]

[1] School of Public Health, Guangdong Pharmaceutical University, Guangzhou 510310, China; zhangweiai0629@126.com (W.Z.); mcjane2013@163.com (C.M.)
[2] Key Research Center of Liver Regulation for Hyperlipidemia SATCM/Class III Laboratory of Metabolism SATCM, Guangdong TCM Key Laboratory for Metabolic Diseases, Guangdong Pharmaceutical University, Guangzhou 510006, China
* Correspondence: suzhq@scnu.edu.cn (Z.S.); angell_bai@163.com (Y.B.); Tel.: +86-20-3935-2067 (Z.S.); +86-20-3405-5161 (Y.B.); Fax: +86-20-3935-2067 (Z.S.); +86-20-3405-5355 (Y.B.)

Academic Editors: Hitoshi Sashiwa and David Harding
Received: 29 October 2015; Accepted: 29 March 2016; Published: 18 April 2016

Abstract: This work describes a highly-sensitive and accurate approach for the determination of chitosan (CTS) using Naphthol Green B (NGB) as a probe in the Resonance Rayleigh scattering (RRS) method. The interaction between CTS and NGB leads to notable enhancement of RRS, and the enhancement is proportional to the concentration of CTS over a certain range. Under optimum conditions, the calibration curve of ΔI against CTS concentration was $\Delta I = 1860.5c + 86.125$ (c, $\mu g/mL$), $R^2 = 0.9999$, and the linear range and detection limit (DL) were 0.01–5.5 $\mu g/mL$ and 8.87 ng/mL. Moreover, the effect of the molecular weight of CTS on the accurate quantification of CTS was studied. The experimental data were analyzed through linear regression analysis using SPSS20.0, and the molecular weight was found to have no statistical significance. This method has been applied to assay two CTS samples and obtained good recovery and reproducibility.

Keywords: chitosan; Naphthol Green B; Resonance Rayleigh scattering spectra

1. Introduction

Among biopolymers, chitosan (CTS), which is produced from the deacetylation of natural chitin, has seen increased use due to the presence of amino groups on the polymer backbone that make it a natural cationic polymer [1].With the extensive application of CTS in different fields [2], especially in the application of reducing weight and drug delivery system. It is very important to study the accurate and sensitive quantification of CTS for quantity monitoring [3–5]. In recent years, the main methods for the determination of CTS were spectrophotometric methods [6–8] and HPLC methods [9–11]. Spectrophotometric methods have the advantages of simplicity and low cost, but they are not sensitive enough [12]. HPLC methods have the advantages of sensitivity and accuracy, but HPLC cannot directly determine CTS without hydrolysis. Thus far, 100% hydrolysis efficiency of CTS still cannot be achieved, which affects the accuracy of the determination of CTS and is the main imperfection in the use of HPLC to assay CTS. Therefore, it remains worthwhile to develop a highly sensitive, convenient, and rapid method for determining CTS.

Resonance Rayleigh scattering (RRS) is a special elastic scattering which is produced when the wavelength of Rayleigh scattering is close to the molecular absorption band [13,14]. It provides useful information concerning molecular structure, form, size, state of combination, charge distribution, and other factors [15]. RRS is a highly-sensitive analytical technique for the determination of certain inorganic [16–18] and organic substances [19–21].

Naphthol Green B (NGB) (Figure 1) is a complexometric indicator with three SO_3^- groups and a naphthalene structure, leading to excellent water solubility and good stability. In this assay, it is the first time that NGB was proposed as a highly-sensitive probe for the determination of CTS. The experimental results have showed that both CTS and NGB produce very weak RRS signals. However, when the two agents react by virtue of electrostatic interaction to form an ion-association complex, the RRS intensity could be enhanced greatly. In this paper, the reaction principle, UV-VIS spectral, and RRS spectral characteristics, optimum reaction conditions, and analytical properties have been studied. It is worthwhile to mention in this context that the effect of the molecular weight of CTS was investigated. The experimental data analyzed through linear regression analysis has shown that there is no statistical significance on the molecular weight.

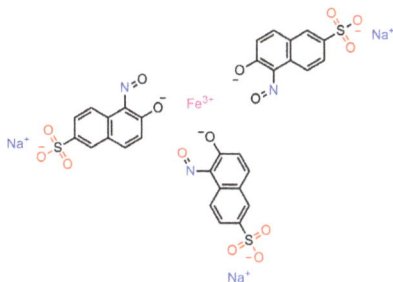

Figure 1. The structure of NGB.

Therefore, CTS could be accurately quantified by this method even if the molecular weight of sample CTS is different from that of CTS standard. A highly sensitive method has been established and applied to the determination of complicated CTS capsules.

2. Results and Discussion

2.1. Mechanism

In this assay, the RRS method is used to test the change of CTS. The working principle of our sensing system is schematically represented in Figure 2. First, in an acidic solution, CTS becomes a positively-charged macromolecule as the $-NH_2$ of CTS is protonated to $-NH_3^+$ [22]. NGB takes a negative charge on the surface [23].With positively-charged CTS in solution, so that it is very easy to form an ion-association complex, the scattering intensity is enhanced. In addition, when the CTS-NGB complex is placed in a 75 °C water bath for 3 min, the color of the CTS-NGB complex changes to yellow from green. When the solution is green, the CTS-NGB complex solution has greater molecular absorption above 600 nm. After heating and the solution turns yellow, the CTS-NGB complex solution has greater molecular absorption at 300–500 nm, almost overlapping with the scattering wavelength of the solution. Thus, the resonance between the absorption and the scattering is formed. As a result, the RRS intensity is greatly enhanced.

2.2. UV-VIS Absorption Spectra

Figure 3 compares the unheated and heated UV-VIS spectral characteristics of the CTS-NGB system and shows that the unheated and heated CTS have almost the same UV-VIS spectral characteristics. In contrast to the CTS solution, the heated NGB solution has very different UV-VIS spectral characteristics from that of unheated NGB solution. The unheated NGB solution exhibits the maximum absorption peaks at $\lambda_1 = 264$ nm, $\lambda_2 = 364$ nm, and $\lambda_3 = 716$ nm, whereas the heated NGB solution exhibits the maximum absorption peaks at $\lambda_1 = 264$ nm and $\lambda_2 = 364$ nm. Thus, the heated NGB solution undergoes a color change from green to yellow which is visible to the naked eyes. These results all confirmed that the spatial structure and chromophoric group of NGB might be changed in the process of heating. The UV-VIS spectral characteristics of CTS-NGB complex solution unheated and heated are similar to that of the pure NGB solution, but after heating, the absorbance of CTS- NGB complex solution is slightly higher than that of pure NGB solution.

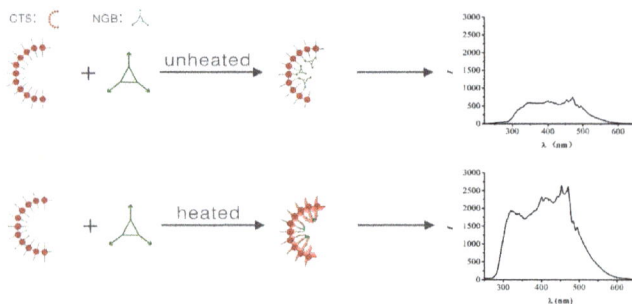

Figure 2. Schematic diagram of CTS-NGB system.

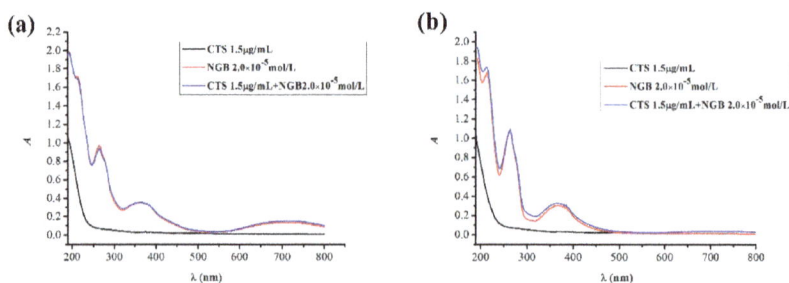

Figure 3. Comparison of the unheated and heated UV-VIS spectral characteristics of CTS-NGB system. (a) The UV-VIS absorption spectral characteristics for the unheated CTS-NGB system, and (b) the UV-VIS absorption spectral characteristics for the heated CTS-NGB system.

2.3. RRS Spectra

Figure 4 depicts the RRS spectra of the CTS-NGB system and shows that the RRS intensities of CTS and NGB solution were individually weak under the measurement conditions. When CTS reacted with NGB to form an anion-association complex, the RRS was enhanced remarkably, and a new spectrum appeared. The maximum RRS peak was at $\lambda = 470$ nm, and the enhancement of RRS intensities was proportional to the concentration of CTS. Therefore, the new method of monitoring CTS could be established.

2.4. Optimum Experimental Conditions

In the experimental conditions optimization process, we chose a medium molecular weight chitosan as standard.

2.4.1. Effects of Buffer Solution

Three buffer solutions, namely, HAc-NaAc buffer solution, glycine-HCl buffer solution, and B-R buffer solution, were used to investigate the influence of acidity on ΔI and the linear relationship of the standard curve in the CTS-NGB system (Figure 5). The results showed that B-R buffer solution was the most suitable reaction medium and it was, therefore, selected for further study.

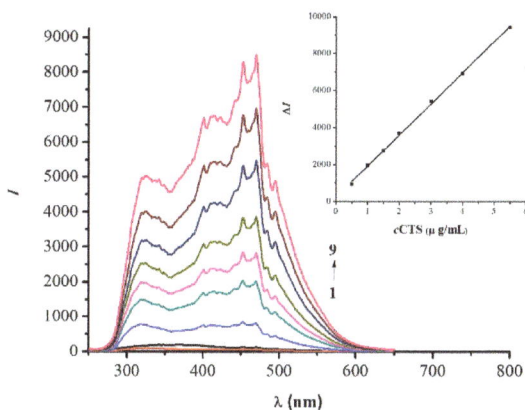

Figure 4. RRS spectra of the CTS-NGB dye systems.1. CTS 2.0 µg/mL; 2. NGB2.0 × 10^{-5} mol/L; 3–9. CTS (0.5, 1.0, 1.5, 2.0, 3.0, 4.0, and 5.0 µg/mL) –NGB (2.0 × 10^{-5} mol/L) complex.

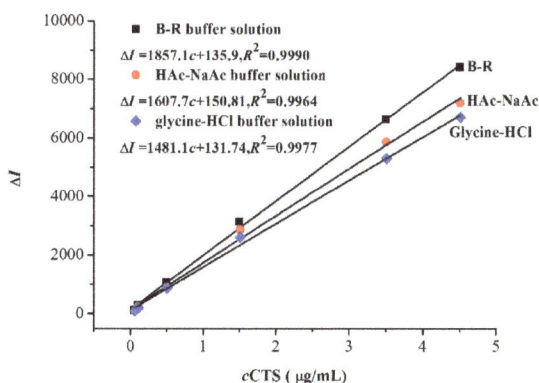

Figure 5. Effects of buffer solution. (■) B-R buffer solution; (●) HAc-NaAc buffer solution; (♦) Glycine-HCl buffer solution; The pH of three buffer solutions is 2.0; t = 75 °C.

2.4.2. Effects of pH and Amount of B-R Buffer Solution

When B-R buffer solution was used as the reaction medium, the scattering intensity (ΔI) of CTS-NGB complex solution was the strongest in the pH range 1.9–2.5. When the pH value was outside this range, ΔI decreased. The pH 2.0 buffer solution was chosen as the reaction medium for the following experiments, and the most suitable amount of buffer solution was 1.5 mL for the reaction system.

2.4.3. Effect of Concentration of Naphthol Green B

By increasing the amount of NGB in the experimental solution, the RRS intensity of this system was increased first and then decreased (the concentration of CTS was 5.5 µg/mL). The results indicated that the optimum concentration range was (1.0–3.0) × 10^{-5} mol/L. When the concentration of NGB in the solution was lower than 1.0 × 10^{-5} mol/L, the RRS intensity of the solution was decreased because of the CTS in the solution not completely reacted. When the concentration of NGB was higher than 3.0×10^{-5} mol/L, ΔI decreased. ΔI was the strongest at concentration of 2.0×10^{-5} mol/L, so this concentration was chosen as the optimum concentration for the reaction system.

2.4.4. Effect of Reaction Temperature

The effect of reaction temperature on the RRS intensity was examined. Figure 6 displays the results at room temperature (23.5 °C), 30 °C, 40 °C, 50 °C, 60 °C, 70 °C, 75 °C, 80 °C, 90 °C and 100 °C, which shows that temperatures had a great influence on the RRS intensity. When the temperature was 70–80 °C, the RRS intensity was stronger. However, when the temperature was higher than 80 °C, the RRS intensity decreased significantly. Therefore, 75 °C was selected as the optimum reaction temperature for the CTS-NGB system.

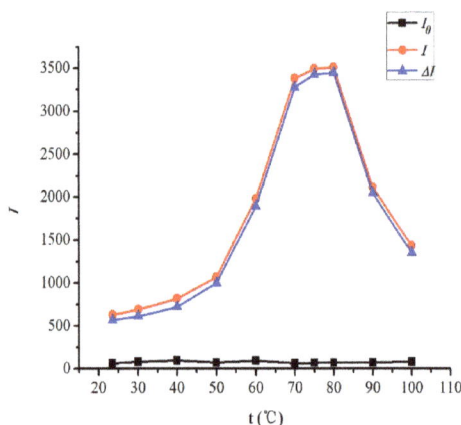

Figure 6. Effects of reaction temperature. CTS 2.0 μg/mL, NGB 2.0×10^{-5} mol/L, B-R pH = 2.0, 1.5 mL.

2.4.5. Effects of Reaction Time and Stability

Under the optimum experimental conditions, the reaction time in a 75 °C water bath and stability at room temperature were studied. The CTS-NGB system reacted completely in 3 min and could remain stable for 4.0 h at room temperature. These results indicated that this system reacted quickly at 75 °C and exhibited good stability.

2.4.6. Effect of Addition Sequence

Under the optimum experimental conditions, three addition sequences of reagents were tested: first, CTS-B-R buffer solution-NGB; second, CTS-NGB-B-R buffer solution; and third, NGB-B-R buffer solution-CTS. The experimental results of the three sets clearly showed that the RRS intensity of the third addition sequence was the highest. The following experiments all use NGB-B-R buffer solution-CTS as the addition sequence.

2.4.7. Effect of Ionic Strength

The effect of ionic strength on the CTS-NGB system was tested by varying the concentration of NaCl. As shown in Figure 7, when the concentration of NaCl was controlled below 0.03 mol/L, the determination results of the CTS-NGB system were relatively stable. When the concentration of NaCl was increased, the RRS intensity decreased. The reason was that large amounts of Na^+ cations and Cl^- anions in solution would be combined with CTS and the anion dye to form ion complexes, opposing the binding of CTS with the anion dye.

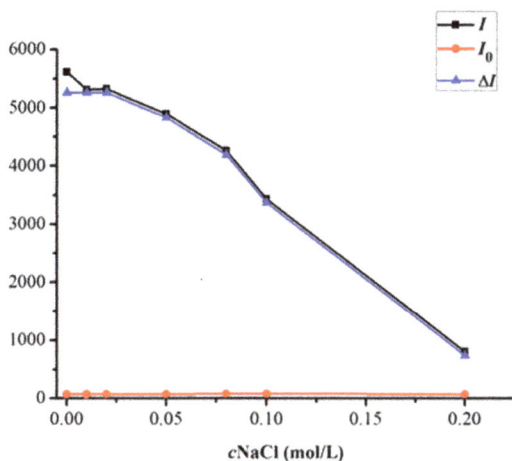

Figure 7. Effect of ionic strength on the RRS intensities of the CTS-NGB system. CTS 3.0 µg/mL, NGB 2.0×10^{-5} mol/L, B-R pH = 2.0 1.5 mL, t = 75 °C

2.5. Analytical Application

2.5.1. Calibration Curves

Under the optimum conditions, CTS of variant concentrations reacted with NGB to form ion-association complexes, and ΔI was measured after 3 min reaction in a 75 °C water bath. A calibration curve of ΔI against CTS concentration over a certain range was constructed: $\Delta I = 1860.5c + 86.125$ (c, µg/mL).The correlation coefficient R^2 = 0.9999, and the linear range and the detection limit(DL) were 0.01–5.5 µg/mL and 8.87 ng/mL, respectively. This work and other methods for the determination of CTS are compared in Table 1, which shows that this method exhibits higher sensitivity and will be a valuable tool for the determination of CTS.

Table 1. Comparison of sensitivities for determination of CTS between present method and some other methods.

Methods	Linear Range	Detection Limits	Investigated the Effect of the Molecular Weight of CTS	References
UV-VIS spectra	10–80 µg/mL	.	N	[6]
UV-VIS spectra	4.55–30.30 mg/L	1.41 mg/mL	Y	[24]
Cathodic Stripping Voltammetry	5.0×10^{-7}–1.7×10^{-5} g/mL	5.0×10^{-7} g/mL	N	[25]
RRS	0.042–3.0 µg/mL	1.2 µg/mL	N	[26]
SOS	0.016–3.0µg/mL	4.9 µg/mL	N	[27]
FDS	0.005–3.0 µg/mL	1.6 µg/mL	N	[27]
RRS	0.10–20.0 µg/mL	29 ng/mL	N	[27]
HPLC	(2–20 mg/mL) Glucosamine	.	N	[11]
RRS	0.01–5.5 µg/mL	8.87 ng/mL	Y	This paper

RRS: Resonance Rayleigh scattering, SOS: second-order scattering, FDS: frequency doubling scattering, HPLC: High-performance liquid chromatographic, N: No, Y: Yes.

2.5.2. Effect of the Molecular Weight of CTS

Three kinds of chitosan with different molecular weights (low, medium, high) were selected as standards, and a series of concentrations (0.1, 0.5, 1.5, 3.5, and 4.5 µg/mL) were prepared for each solution. The reagent blank was prepared at the same time, and all the solutions were determined under the optimum experimental conditions. A series of calibration curves of ΔI against the concentrations of CTS were constructed. Finally, the results were analyzed by linear regression analysis using the statistical product SPSS20.0 (IBM Company, Armonk, NY, USA) to determine the effect of the molecular weight of CTS. The results showed that p = 0.224 > 0.05, which suggested that the effect of molecular weight on the determination results has no statistical significance. Thus, the molecular weight does not interfere with the determination of CTS, even though there are significant differences in molecular weight between the CTS sample and the CTS standard.

2.5.3. Effect of the Degree of Deacetylation

When the degree of deacetylation of chitosan is greater than 85%, its biological activity and solubility are better. Therefore, the degree of deacetylation of chitosan applied to the health food was more than 85%, generally.

Two kinds of chitosan with different deacetylation degree (the molecular weight of CTS is 60CPS and the degree of deacetylation are 85% and 90%) were selected as standards. Two calibration curves were constructed as in 2.5.2. The linear regression analysis results exhibited $p > 0.05$, which showed that the effects of different degree of deacetylation of chitosan on the determination results were not statistically significant.

2.5.4. Effect of Foreign Substances

CTS capsule is composed of chitosan and a small amount of other auxiliary components. Under the optimum conditions, the effects of certain common foreign substances on the determination of CTS capsules were detected, and the results are given in Table 2, which shows that when the concentration of CTS was 3.0 μg/mL, some metal ions such as Fe^{3+}, Ca^{2+}, Mg^{2+}, Cu^{2+}, and Zn^{2+} were only tolerated in small amounts. Other additives, such as glucose, glycine, L-lysine, and L-tryptophan, even if there are large amounts, will not affect the accurate determination of chitosan, In order to eliminate the interference of metal ions, we chose tartaric acid and EDTA as masking agents. Firstly, we studied the effect of tartaric acid or EDTA on the dose. The results are shown in Table 3. In addition, we compared one set having tartaric acid or EDTA with another set not containing the two masking agents by t-test. The result of t-test was $p > 0.05$, which showed that the addition of tartaric acid or EDTA had no effect on the determination results.

Table 2. Effects of foreign substances (c_{CTS}: 3.0 μg/mL).

Foreign Substance	Amount Tolerated (μg/mL)	RE (%)	Foreign Substance	Amount Tolerated (μg/mL)	RE (%)
Glucose	9600.00	−3.59	Fe^{3+}	0.30	−4.76
β-Cyclodextrin	120.00	3.33	Mg^{2+}	0.40	−3.67
Soluble starch	100.00	−5.51	Ca^{2+}	0.40	4.64
VitaminC	85.00	−4.39	K^+	10.00	3.33
Citricacid	37.50	−5.05	Na^+	5.00	−3.38
Glycine	1600.00	4.69	Cu^{2+}	0.12	3.84
L-lysine	900.00	−5.38	Zn^{2+}	0.80	−5.07
L-tryptophan	250.00	3.42	NH_4^+	441.85	−5.02
L-leucine	135.00	−4.01			

Table 3. The results of anti-interference experiment (c_{CTS}: 3.0 μg/mL).

Cation	The Concentration of Cation (μg/mL)	Masking Agent	The Concentration of Masking Agent	RE (%)
Fe^{3+}	3.0	EDTA	0.001 mol/L	−4.22
Mg^{2+}	2.0	Tartaric acid	0.08%	4.78
Ca^{2+}	3.0	Tartaric acid	0.08%	2.88
Cu^{2+}	3.0	Tartaric acid	0.02%	1.31
Zn^{2+}	3.0	EDTA	0.001 mol/L	−3.04

For unknown samples, it is difficult to know what interfering substances are in the sample. However, we know that the presence of metal ions will increase the conductivity of the sample solution. Therefore, we added metal ions Fe^{3+}, Ca^{2+}, Mg^{2+}, Cu^{2+}, and Zn^{2+} (amount tolerated in

Table 2), respectively, in the chitosan standard solution with a concentration of 3.0 µg/mL and detected the electrical conductivity. The results were 433.7 ± 25.54 µs/cm, 431.7 ± 16.56 µs/cm, 444.7 ± 5.507 µs/cm, 427.7 ± 6.351 µs/cm, and 535.3 ± 9.451 µs/cm.

We also detected the electrical conductivity of two CTS sample solutions (Olevy chitosan capsules, AiDeLan chitosan capsules), the results were 490.0 ± 5.196 µs/cm and 490.7 ± 8.083 µs/cm, which showed that there might be some interfering substances in the samples. However, when the concentration of EDTA in the sample solutions was 0.001 mol/L, the conductivity of the solutions was smaller than 200 µs/cm. At this point, the relative error of the measurement results can be controlled less than 5%.

2.5.5. Precision and Recovery

The method was applied for the determination of CTS in health products, specifically Olevy and AiDeLan CTS capsules. The CTS capsules were weighed accurately to obtain 0.4 g and dissolved in a 100-mL volumetric flask with 0.5 mol/L HAc for 36 h. Then, 10.00 mL was obtained after high speed centrifugation as a sample stock solution. Next, 2.50 mL stock solution was added to a 100-mL volumetric flask, diluted to the mark with high-purity water and thoroughly mixed to produce a working solution.

The following operations were similar to the general procedure with 0.5 mL working solution used as a determination sample (with EDTA added as a masking agent). The corresponding results were calculated according to the calibration graphs of CTS, and the results are listed in Table 4. Moreover, the recoveries of CTS in this method were investigated, and the results are listed in Table 5. The recoveries of the Olevy capsule and AiDeLan capsule were 103.2%–104.6% and 102.7%–103.5%, respectively, and the relative standard deviations were 0.71% and 0.42%, respectively. Thus, this method exhibited good recovery and reproducibility.

Table 4. Results of the determination of CTS capsules.

Sample	Olevy (mg/g)	AiDeLan (mg/g)
1	920.2	871.5
2	938.5	871.1
3	930.1	878.4
4	927.2	876.7
5	922.4	877.2
6	925.6	871.8
Average (mg/g)	927.3	874.5
RSD (%)	0.70	0.38

Table 5. The results of recoveries.

Sample	Found Value (µg/mL)	Added(µg/mL)	Found Value (µg/mL) (n=6)	Recoveries (%)	RSD (%)
Olevy	0.4658	0.4000	0.8830	104.3	
		1.000	1.512	104.6	0.71
		1.500	2.014	103.2	
AiDeLan	0.4498	0.4000	0.8605	102.7	
		1.000	1.485	103.5	0.42
		1.500	2.001	103.4	

3. Experimental

3.1. Materials and Reagents

Three types of chitosan (CTS, low molecular weight: ≤200 mPa.S, medium molecular weight: 200–400 mPa.S, high molecular weight: 400–1000 mPa.S; 1% in 1% acetic acid, 20 °C; Sigma, St. Louis, MO, USA). The stock solution of 400.0 μg/mL chitosan was prepared by mixing a suitable chitosan and 0.5 mol/L HAc solution, and a working solution of 10.0 μg/mL chitosan was prepared for use in the experiment. A working solution of Naphthol Green B (NGB, 2.0×10^{-4} mol/L, Tokyo Chemical Industry Co. Ltd., Tokyo, Japan) was prepared and kept at 4°C. Britton-Robinson (B-R) buffer solutions with different pH values were prepared by combining the mixed acid (consisting of 2.71 mL of H_3PO_4, 2.36 mL of HAC and 2.47 g of H_3BO_3)/L with 0.2 mol/L NaOH in different proportions, and the pH values were adjusted using a pH meter. Olevy chitosan capsules (Weihai South Gulf Biological Co. Ltd., Weihai, China) and AiDeLan chitosan capsules (Shanghai TongJi Biological Co. Ltd., Shanghai, China) were used. All reagents were of analytical grade without further purification, and high-purity water was used throughout.

A Hitachi F-2500 spectrofluorophotometer (Hitachi Ltd., Tokyo, Japan) was used for recording RRS spectra and measuring RRS intensity. A UV-3010 spectrophotometer (Hitachi Ltd., Tokyo, Japan) was used to record the absorption spectra and measure the absorbance. A PHS-3C pH meter (Shanghai Scientific Instruments Company, Shanghai, China) was used to measure the pH values of the solutions, and a CP124C electronic analytical balance (Ohaus Instrument Co. Ltd., Shanghai, China) was used in this experiment.

3.2. Procedures

Appropriate amounts of NGB, B-R buffer solution, and CTS were added to a 10 mL cuvette. Then, each cuvette was diluted with water to a final volume of 10.0 mL. The solution was mixed and set in a 75 °C thermostat water bath for 3 min, followed by rapid cooling to room temperature. The RRS spectra were recorded by scanning synchronously with the same excitation and emission wavelengths and measuring RRS intensity by time scan pattern. The slit ($\lambda_{ex} = \lambda_{em}$) was 5 nm/5 nm, and the PMT was 400 V. The RRS intensity I was measured for the reaction product and I_0 for the reagent blank at the maximum RRS wavelength, $\Delta I = I - I_0$.

4. Conclusions

In conclusion, a novel and innovative methodology was developed to quantify CTS and successfully applied to the determination of CTS samples. The main advantages of this assay are that it is simple, sensitive, accurate, and rapid. Vitally, the molecular weight and the deacetylation degree of CTS exhibit no interference with its accurate quantification. The detection limit (DL) of the method is good, and the RSD of the method is better. In contrast to HPLC methods, this method does not require sophisticated pretreatment processes.

Acknowledgments: The authors gratefully acknowledge financial support for this study by grants from the National Natural Science Foundation of China (No. 81173107) and the Science and Technology Planning Project of Guangdong, China (No. 2013B021100018).

Author Contributions: Conceived and designed the experiments: Yan Bai, Zhengquan Su, Weiai Zhang. Performed the experiments: Weiai Zhang, Caijuan Ma. Analyzed the data: Weiai Zhang, Yan Bai, Caijuan Ma. Wrote the paper: Weiai Zhang, Yan bai.

Conflicts of Interest: The authors declare no conflict of interest.

References

1. Kadib, A.E. Chitosan as a Sustainable Organocatalyst: A Concise Overview. *ChemSusChem* **2015**, *8*, 217–244. [CrossRef] [PubMed]

2. Paños, I.; Acosta, N.; Heras, A. New drug delivery systems based on chitosan. *Curr. Drug Discov. Technol.* **2008**, *5*, 333–341. [CrossRef] [PubMed]
3. Su, Z.Q.; Wu, S.H.; Zhang, H.L.; Feng, Y.F. Development and validation of an improved Bradford method for determination of insulin from chitosan nanoparticulate systems. *Pharm. Biol.* **2010**, *48*, 966–973. [CrossRef] [PubMed]
4. Zhang, H.L.; Wu, S.H.; Tao, Y.; Zang, L.Q.; Su, Z.Q. Preparation and Characterization of Water-Soluble Chitosan Nanoparticles as Protein Delivery Syste. *J. Nanomater.* **2010**, *2010*, 1–5.
5. Hsieh, Y.L.; Yao, H.T.; Cheng, R.S.; Chiang, M.T. Chitosan Reduces Plasma Adipocytokines and Lipid Accumulation in Liver and Adipose Tissues and Ameliorates Insulin Resistance in Diabetic Rats. *J. Med. Food* **2012**, *15*, 453–460. [CrossRef] [PubMed]
6. Wischke, C.; Borchert, H. Increased sensitivity of chitosan determination by a dye binding method. *Carbohydr. Res.* **2006**, *341*, 2978–2979. [CrossRef] [PubMed]
7. Mendelovits, A.; Prat, T.; Gonen, Y.; Rytwo, G. Improved Colorimetric Determination of Chitosan Concentrations by Dye Binding. *Appl. Spectrosc.* **2012**, *66*, 979–982. [CrossRef] [PubMed]
8. Qian, Z.G.; Pan, S.K.; Xia, Z.Q.; Wu, S.J. A new assay of chitosan in the presence of protein by hydrolysis with commercial α-amylase. *Eur. Food Res. Technol.* **2011**, *233*, 717–719. [CrossRef]
9. Zhu, X.L.; Cai, J.B.; Yang, J.; Su, Q.D. Determination of glucosamine in impure chitin samples by high-performance liquid chromatography. *Carbohydr. Res.* **2005**, *340*, 1732–1738. [CrossRef] [PubMed]
10. El-Saharty, Y.S.; Bary, A.A. High-performance liquid chromatographic determination of neutraceuticals, glucosamine sulphate and chitosan, in raw materials and dosage forms. *Anal. Chim. Acta* **2002**, *462*, 125–131. [CrossRef]
11. Li, B.; Zhang, J.L.; Bu, F.; Xia, W.S. Determination of chitosan with a modified acid hydrolysis and HPLC method. *Carbohydr. Res.* **2013**, *366*, 50–54. [CrossRef] [PubMed]
12. Abou-Shoer, M. A Simple Colorimetric Method for the Evaluation of Chitosan. *Am. J. Anal. Chem.* **2010**, *1*, 91–94. [CrossRef]
13. Xu, D.P.; Liu, S.P.; Liu, Z.F.; Hu, X.L. Determination of verapamil hydrochloride with 12-tungstophosphoric acid by resonance Rayleigh scattering method coupled to flow injection system. *Anal. Chim. Acta* **2007**, *588*, 10–15. [CrossRef] [PubMed]
14. Bi, S.Y.; Wang, Y.; Wang, T.J.; Pang, B.; Zhao, T.T. The analytical application and spetral investigation of DNA-CPB-emodin and sensitive determination of DNA by resonance Rayleigh light scattering technique. *Spectrochim. Acta Part A: Mol. Biomol. Spectrosc.* **2013**, *101*, 233–238. [CrossRef] [PubMed]
15. Gao, Z.F.; Song, W.W.; Luo, H.Q.; Li, N.B. Detection of mercury ions (II) based on non-cross-linking aggregation of double-stranded DNA modified gold nanoparticles by resonance Rayleigh scattering method. *Biosensors Bioelectron.* **2015**, *65*, 360–365. [CrossRef] [PubMed]
16. Wen, G.Q.; Yang, D.; Jiang, Z.L. A new resonance Rayleigh scattering spectral method for determination of O3 with victoria blue B. *Spectrochim. Acta Part A: Mol. Biomol. Spectrosc.* **2014**, *117*, 170–174. [CrossRef] [PubMed]
17. Yang, Q.L.; Lu, Q.M.; Liu, Z.F.; Liu, S.P.; Chen, G.C.; Duan, H.; Song, D.; Wang, J.; Liu, J. Resonance Rayleigh scattering spectra of ion-association nanoparticles of [Co(4-[(5-Chloro-2-pyridyl) azo]-1, 3-diaminobenzene)$_2$]$^{2+}$-sodium dodecyl benzene sulfonate system and its analytical application. *Anal. Chim. Acta* **2009**, *632*, 115–121. [CrossRef] [PubMed]
18. Long, X.F.; Bi, S.P.; Ni, H.Y.; Tao, X.C.; Gan, N. Resonance Rayleigh scattering determination of trace amounts of Al in natural waters and biological samples based on the formation of an Al(III)–morin–surfactant complex. *Anal. Chim. Acta* **2004**, *501*, 89–97. [CrossRef]
19. Cui, Z.P.; Hu, X.L.; Liu, S.P.; Liu, Z.F. A dual-wavelength overlapping resonance Rayleigh scattering method for the determination of chondroitin sulfate with nile blue sulfate. *Spectrochim. Acta Part A: Mol. Biomol. Spectrosc.* **2011**, *83*, 1–7. [CrossRef] [PubMed]
20. Shi, Y.; Li, C.Y.; Liu, S.P.; Liu, Z.F.; Yang, J.D.; Zhu, J.H.; Qiao, M.; Duan, R.L.; Hu, X.L. A novel method for detecting allura red based on triple-wavelength overlapping resonance Rayleigh scattering. *RSC Adv.* **2014**, *4*, 37100. [CrossRef]
21. Parham, H.; Saeed, S. Resonance Rayleigh scattering method for determination of ethion using silver nanoparticles as probe. *Talanta* **2015**, *131*, 570–576. [CrossRef] [PubMed]

Mar. Drugs **2016**, *14*, 71

22. Ye, K.; Felimban, R.; Traianedes, K.; Moulton, S.E.; Wallace, G.G.; Chung, J.; Quigley, A.; Choong, P.F.M.; Myers, D.E. Chondrogenesis of Infrapatellar Fat Pad Derived Adipose Stem Cells in 3D Printed Chitosan Scaffold. *PLoS ONE* **2014**, *9*, e102638.

23. Gu, B.; Zhong, H.; Li, X.M.; Wang, Y.Z.; Ding, B.C.; Cheng, Z.P.; Zhang, L.L.; Li, S.P.; Yao, C. Sensitve Determination of Proteins With Naphthol Green B by Resonance Light Scattering Technique. *J. Appl. Spectrosc.* **2013**, *80*, 486–491. [CrossRef]

24. Beatriz, M.; Marian, M.; Ruth, H.; Angeles, H. Suitability of a colorimetric method for the selective determination of chitosan in dietary supplements. *Food Chem.* **2011**, *126*, 1836–1839.

25. Lu, G.H.; Wang, L.R.; Wang, R.X.; Zeng, Y.; Huang, X. Determination of Chitosan by Cathodic Stripping Voltammetry. *Anal. Sci.* **2006**, *22*, 575–578. [CrossRef] [PubMed]

26. Wang, Y.W.; Li, N.B.; Luo, H.Q. Resonance Rayleigh scattering method for the determination of chitosan with some anionic surfactants. *Luminescence* **2008**, *23*, 126–131. [CrossRef] [PubMed]

27. Peng, J.J.; Liu, S.P.; Wang, L.; Liu, Z.W.; He, Y.Q. Study on the interaction between CdSe quantum dots and chitosan by scattering spectra. *J. ColloidInterface Sci.* **2009**, *338*, 578–583. [CrossRef] [PubMed]

![marine drugs logo](marine drugs)

MDPI

Article

Conversion of Squid Pen to Homogentisic Acid via *Paenibacillus* sp. TKU036 and the Antioxidant and Anti-Inflammatory Activities of Homogentisic Acid

San-Lang Wang [1,2,*], Hsin-Ting Li [2], Li-Jie Zhang [3], Zhi-Hu Lin [3] and Yao-Haur Kuo [3,4,*]

1   Life Science Development Center, Tamkang University, No. 151, Yingchuan Rd., Tamsui,
    New Taipei City 25137, Taiwan
2   Department of Chemistry, Tamkang University, New Taipei City 25137, Taiwan; cindy810924@yahoo.com.tw
3   Division of Chinese Materia Medica Development, National Research Institute of Chinese Medicine,
    Taipei 11221, Taiwan; lijiezhang@hotmail.com (L.-J.Z.); tiger77749@gmail.com (Z.-H.L.)
4   Graduate Institute of Integrated Medicine, College of Chinese Medicine, China Medical University,
    Taichung 40402, Taiwan
*   Correspondence: sabulo@mail.tku.edu.tw (S.-L.W.); kuoyh@nricm.edu.tw (Y.-H.K.);
    Tel.: +886-2-2621-5656 (S.-L.W.); +886-2-2820-1999 (Y.-H.K.);
    Fax: +886-2-2620-9924 (S.-L.W.); +886-2-2823-6150 (Y.-H.K.)

Academic Editors: Hitoshi Sashiwa and David Harding
Received: 14 September 2016; Accepted: 7 October 2016; Published: 12 October 2016

Abstract: The culture supernatant of *Paenibacillus* sp. TKU036, a bacterium isolated from Taiwanese soils, showed high antioxidant activity (85%) when cultured in a squid pen powder (SPP)-containing medium at 37 °C for three days. Homogentisic acid (2,5-dihydroxyphenylacetic acid, HGA) was isolated and found to be the major antioxidant in the culture supernatant of the SPP-containing medium fermented by *Paenibacillus* sp. TKU036. Tryptophan was also present in the culture supernatant. The results of high-performance liquid chromatography (HPLC) fingerprinting showed that HGA and tryptophan were produced via fermentation but did not pre-exist in the unfermented SPP-containing medium. Neither HGA nor tryptophan was found in the culture supernatants obtained from the fermentation of nutrient broth or other chitinous material, i.e., medium containing shrimp head powder, by *Paenibacillus* sp. TKU036. The production of HGA via microorganisms has rarely been reported. In this study, we found that squid pen was a potential carbon and nitrogen source for *Paenibacillus* sp. Tryptophan (105 mg/L) and HGA (60 mg/L) were recovered from the culture supernatant. The isolated HGA was found to have higher antioxidant activity (IC_{50} = 6.9 µg/mL) than α-tocopherol (IC_{50} = 17.6 µg/mL). The anti-inflammatory activity of the isolated HGA (IC_{50} = 10.14 µg/mL) was lower than that of quercetin (IC_{50} = 1.14 µg/mL). As a result, squid pen, a fishery processing byproduct, is a valuable material for the production of tryptophan and the antioxidant and anti-inflammatory HGA via microbial conversion.

Keywords: squid pen; chitin; homogentisic acid; tryptophan; *Paenibacillus*; antioxidant; anti-inflammatory

1. Introduction

Chitin is one of the most abundant biopolymers in the world, and these natural polymers have versatile properties, such as biocompatibility and non-toxicity. Among the natural chitinous resources, fishery processings (shrimp shells, crab shells, and squid pens) have the highest chitin content. Conventionally, chitin is obtained from shrimp shells, crab shells, and squid pens using a strong alkali or an inorganic acid for deproteinization or demineralization, respectively [1]. However, these chemical processes have several drawbacks, such as the creation of pollutant alkali or acid liquid.

Furthermore, the unutilized bioresource of the deproteinized liquid is reduced due to the presence of an alkali [1].

Among the chitin-containing fishery processings, squid pens contain the highest ratio of protein (approximately 70%) [2]. For recycling squid pens in order to produce additional highly value-added products other than chitin or chitosan, we investigated the reutilization of this fishery processings via microbial conversion in order to produce enzymes [2–5], exopolysaccharides [6,7], chitooligomers [3], antioxidants [8,9], insecticidal materials [10], and biosorbents [11,12].

Many strains of *Paenibacillus* have been reported to use squid pen powder (SPP) as the sole carbon and nitrogen (C/N) source. Recently, we isolated strains of *Paenibacillus* species that converted squid pen to exopolysaccharides [7], chitosanase [3], and chitooligomers [3].

Microbial fermentation can result in the production of some antioxidants, such as ellagic acid produced by *Aspergillus niger* [13], gallic acid produced by *Bacillus sphaericus* [14], ferulic and acid produced by *Saccharomyces cerevisiae* [15]. In this study, we screened antioxidant-producing bacteria from Taiwanese soils by using squid pen as the sole C/N source. A potential bacterial strain, TKU036, was isolated and identified as *Paenibacillus* sp. The optimized culture conditions for antioxidant production via *Paenibacillus* sp. TKU036 was studied.

Here, the antioxidant compound produced in the culture supernatant of *Paenibacillus* sp. TKU036 was isolated and identified as HGA. HGA has shown to have antioxidant and anti-inflammatory activities [16,17]. In this study, the antioxidant and anti-inflammatory activities of the isolated HGA were investigated and compared with those activities of other well-known antioxidant (α–tocopherol) and anti-inflammatory compound (quercetin).

2. Results and Discussion

2.1. Screening and Identification of Strain TKU036

Over 350 bacterial strains isolated from the soils of Northern Taiwan were cultivated at 37 °C in a medium containing 1% squid pen powder (SPP). Among these strains, strain TKU036 exhibited the strongest antioxidant activity and was chosen for more intensive examination. Based on morphological and biochemical studies, as well as 16S rDNA sequences [7], this strain was confirmed to be *Paenibacillus* sp. Analytical profile index (API) identification was further used to identify the species name [7]; however, no match was found. Therefore, the TKU036 strain was identified as *Paenibacillus* sp. and was used for further investigation.

2.2. Comparing the Non-Exopolysaccharide Antioxidants Produced by Paenibacillus Species

Many strains of *Paenibacillus*, such as *P. mucilaginosus* TKU032 [7], *Paenibacillus* sp. TKU023 [18], and *P. macerans* TKU029 [19], have been reported as the sole C/N source for the production of exopolysaccharides (EPOs) using SPP, and some of these EPOs showed antioxidant activity. In this study, EPOs were also found in the culture broth of SPP-containing medium fermented by *Paenibacillus* sp. TKU036 (data not shown). To investigate whether the antioxidant activity was from the EPOs, the EPO-containing culture supernatant of strain TKU036 underwent ethanol precipitation (final concentration of 70%, v/v) to remove the EPOs. The obtained EPO-deficient culture supernatants were than lyophilized to remove the ethanol and were used for analyzing antioxidant activity. The EPO-deficient culture supernatant of *Paenibacillus* sp. TKU036 showed high antioxidant activity (85%). The culture conditions for the production of antioxidants and the isolation of the non-EPO antioxidants were studied subsequently.

2.3. Culture Conditions for Antioxidant Production

Different concentrations (0.5%, 1.0%, and 1.5% w/v) of squid pen powder (SPP), shrimp head powder (SHP), and cicada shell powder (CSP) were used as the sole C/N source for the production of antioxidant by *Paenibacillus* sp. TKU036. The effects of the medium volume, the medium pH, and the culture temperature on antioxidant activity were also examined. Commercial nutrient broth (NB) medium, which does not contained chitin, was used for comparison. The result showed that the highest antioxidant activity (85%) was obtained by fermentation within the 0.5% SPP-containing medium (100 mL medium in 250 mL Erlenmeyer flask) at 37 °C in a reciprocal shaker at 150 rpm for three days.

The results of this study are remarkably different from those of other reports, such as studies of Bacillus subtilis using red bean [20], *Aspergillus awamori* and *Aspergillus oryzae* using soybean [21], *Aspergillus usami* using sesamin [22], *A. awamori* using black bean [23], and *Monascus pilosus* using potato dextrose broth [24] as a C/N source for antioxidant production.

A novel antioxidant (serraticin) with antitumor activity was isolated from the culture supernatant of SPP-containing medium fermented by *Serratia ureilytica* TKU013 [8]. *S. ureilytica* TKU013 used SPP (1.5%) for the production of the antioxidant, but the maximal antioxidant activity was 82% after four days of fermentation [25]. In this study, *Paenibacillus* sp. TKU036 used a cheaper C/N source of 0.5% SPP and produced a higher antioxidant activity (85%) in a shorter time (three days). The studied culture condition was then used for antioxidant production.

2.4. Isolation of Antioxidant Compounds

As described in the Materials and Methods section below, the 95% ethanol extract was separated into 14 fractions by column chromatography. All the fractions were evaluated for antioxidant activity using a scavenging 2,2-diphenyl-1-picrylhydrazyl (DPPH) radical test. As shown in Figure 1, Fraction 4 had the highest antioxidant activity.

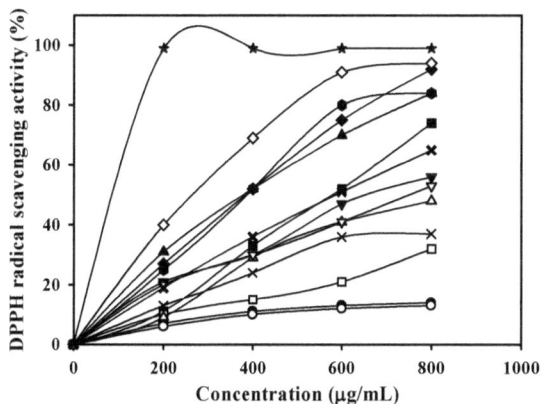

Figure 1. DPPH radical scavenging activity of the 14 fractions eluted by different concentrations of methanol. —●—, Fraction 1 (1.1414 g, eluted with 0% methanol); —○—, Fraction 2 (0.4464 g, eluted with 5% methanol); —●—, Fraction 3 (0.1765 g, eluted with 10% methanol); —★—, Fraction 4 (0.1698 g, eluted with 15% methanol); —■—, Fraction 5 (0.1253 g, eluted with 20% methanol); —□—, Fraction 6 (0.0864 g, eluted with 25% methanol); —◆—, Fraction 7 (0.0599 g, eluted with 30% methanol); —◇—, Fraction 8 (0.0482 g, eluted with 35% methanol); —▲—, Fraction 9 (0.0363 g, eluted with 40% methanol); —△—, Fraction 10 (0.0275 g, eluted with 45% methanol); —▼—, Fraction 11 (0.0263 g, eluted with 50% methanol); —▽—, Fraction 12 (0.0321 g, eluted with 55% methanol); —✕—, Fraction 13 (0.0324 g, eluted with 60% methanol); and —×—, Fraction 14 (0.0424 g, eluted with 100% methanol).

At a concentration of 200 µg/mL, Fraction 4 showed approximately 99% antioxidant activity. Fraction 4 was further purified with a preparative HPLC column. In total, five sub-fractions (4-1, 4-2, 4-3, 4-4, and 4-5) (Figure 2) were obtained, and the antioxidant activity of these fractions were analyzed (Figure 3). As shown in Figure 3, Fraction 4-4 showed the highest antioxidant activity (IC$_{50}$ of

6.9 µg/mL) compared with those of Fraction 4-5 (62.8 µg/mL, which showed a little antioxidant activity due to containing minor 4-4) and the other three fractions, as well as the antioxidant activity of the positive control, α-tocopherol (17.6 µg/mL). The results showed Fraction 4-4 contained a potential antioxidant that was valuable for further identification.

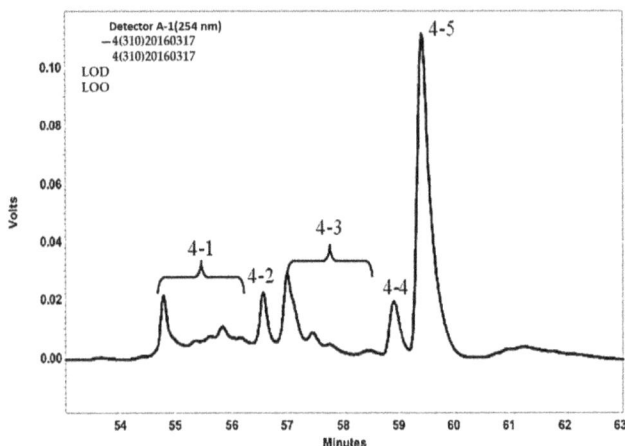

Figure 2. The HPLC profile of Fraction 4 (13% acetonitrile, 254 nm).

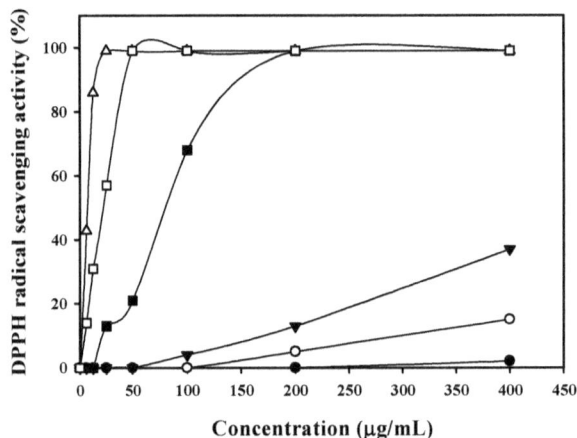

Figure 3. DPPH radical scavenging activities of 4-1 to 4-5. —●—, 4-1; —○—, 4-2; —▼—, 4-3; —△—, 4-4; —■—, 4-5; and —□—, α-tocopherol.

2.5. Identification of HGA and Tryptophan by NMR

The chemical structures of the isolated compounds were elucidated using detailed spectroscopic analyses, including 1D (^{1}H NMR, ^{13}C NMR) and 2D NMR experiments (^{1}H–^{1}H COSY, HSQC, and HMBC), together with the spectroscopic comparisons of previously reported compounds. Fractions 4-4 and 4-5 were shown to contain HGA [26] and tryptophan [27], respectively (Figure 4).

HGA (4-4) was obtained as a white amorphous powder. ^{1}H NMR data (400 MHz, MeOH-d_4, δ_H ppm): 6.62 (d, J = 8.4 Hz, H-3), 6.59 (d, J = 2.8 Hz, H-6), 6.53 (dd, J = 8.4, 2.8 Hz, H-4), and 3.50 (s, 2H, H-7). ^{13}C NMR data (100 MHz, MeOH-d_4, δ_C ppm): 177.0 (C-8), 151.6 (C-5), 150.4 (C-2), 124.3 (C-1), 119.0 (C-6), 117.4 (C-3), 116.0 (C-4), and 37.7 (C-7).

Tryptophan was obtained as a white amorphous powder. ^{1}H NMR data (600 MHz, MeOH-d_4, δ_H ppm): 7.69 (d, *J* = 7.8 Hz, H-3), 7.35 (d, *J* = 7.8 Hz, H-6), 7.19 (s, H-8), 7.11 (td, *J* = 7.8, 0.6 Hz, H-5), 7.04 (td, *J* = 7.8, 0.6 Hz, H-4), 3.86 (dd, *J* = 9.6, 4.2 Hz, H-10), 3.51 (dd, *J* = 15.6, 9.6 Hz, H-9), 3.15 (dd, *J* = 15.6, 4.2 Hz, H-9). ^{13}C NMR data (150 MHz, MeOH-d_4, δ_C ppm): 174.5 (C-11), 138.4 (C-1), 128.5 (C-2), 125.2 (C-8), 122.7 (C-5), 120.1 (C-4), 119.3 (C-3), 112.4 (C-6), 109.5 (C-7), 56.7 (C-10), and 28.5 (C-9).

HGA is an important intermediate in the metabolism of phenylalanine and tyrosine [28]. The production HGA via microorganisms has only been shown in a few reports, such as *Aspergillus niger* (using phenyl acetic acid as a C/N source) [29], *Vibrio cholerae* (using marine broth with 4 mM tyrosine as a C/N source) [30], and *Yarrowia lipolytica* (using tyrosine as a C/N source) [28]. Tryptophan is widely used in human food and medicine, as well as in animal feed. The production of tryptophan has been reported in two typical bacteria strains: *Escherichia coli* FB-04 (using glucose, yeast, tryptone, and citric acid as C/N sources) [31] and *Corynebacterium glutamicum* KY9218 (using sucrose, corn steep liquor, tyrosine, phenylalanine etc. as C/N sources) [32]. In this study, *Paenibacillus* sp. TKU036 cultured with SPP, a seafood processing, was used for the production of HGA and tryptophan and may have potential for further investigation.

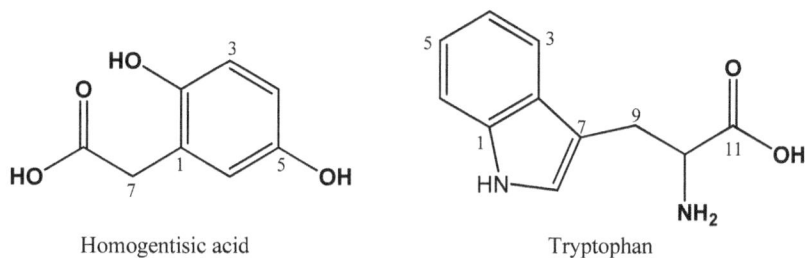

Figure 4. The chemical structures of HGA (**left**) and L-tryptophan (**right**).

2.6. The Effect of HGA on Cytotoxicity and Anti-Inflammation

Nitric oxide (NO) is recognized as a key pro-inflammatory mediator that is involved in certain inflammatory disorders, including chronic hepatitis, pulmonary fibrosis, and rheumatoid arthritis [33]. In our previous study [3], we discovered chitosan oligomers with a low degree of polymerization that showed both antioxidant and anti-inflammatory activity. In this study, the anti-inflammatory activity of Fraction 4-4 (HGA) was estimated using an in vitro model: LPS-stimulated RAW 264.7 cells. The inhibition of LPS-stimulated NO secretion was due to anti-inflammatory activity. First, to examine the potential cell cytotoxicity induced by Fraction 4-4, a MTT assay was conducted. When RAW 264.7 macrophages were treated with Fraction 4-4 at concentrations of 0, 5, 10, 20, and 40 µg/mL, along with 1 µg/mL LPS, the resulting viabilities of RAW 264.7 cells were recorded and are summarized in Figure 5. The results of a statistical analysis indicated that treatment with Fraction 4-4 (40 µg/mL) had no noticeable toxic effect on cell growth when compared with the cell growth of the 0.05% DMSO treated group (100%). Fraction 4-4 (40 µg/mL) was capable of inhibiting NO production by 77.79% in LPS-stimulated cells. The IC_{50} value of Fraction 4-4, representing the anti-inflammatory effect, was 10.14 µg/mL (Figure 5). A similar result was found when purchased HGA was used. Quercetin is a potent dietary antioxidant that also displays anti-inflammatory activity [34]. Thus, the anti-inflammatory activity (IC_{50}) of quercetin was investigated. These results indicated that HGA exhibited an acceptable anti-inflammatory activity (IC_{50} = 10.14 µg/mL) compared with the anti-inflammatory activity of quercetin (IC_{50} = 1.14 µg/mL).

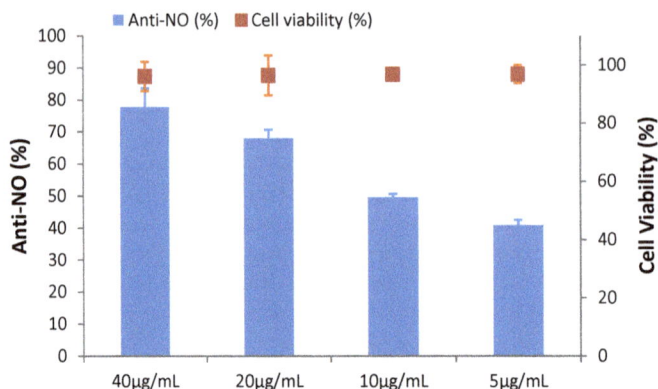

Figure 5. NO inhibitory activities of HGA isolated from culture supernatant of *Paenibacillus* sp. TKU036 in the SPP-containing medium. Cell lines: Murine RAW 264.7 monocyte/macrophage cells. The cells were treated with LPS (1 μg/mL) or in combination with the tested agents (40, 20, 10, and 5 μg/mL) for 24 h.

2.7. Confirmation of HGA and Tryptophan Produced from SPP by Fermentation

The 95% ethanol extracts were extracted from the culture fermented supernatant of *Paenibacillus* sp. TKU036 and were then compared with the extract of unfermented medium via HPLC analysis. The results found that HGA and tryptophan appeared in the fermentation supernatant at *Rt* 12.75 min and at *Rt* 16.25 min, respectively, revealing that the two components did not pre-exist in the SPP-containing medium (data not shown).

To confirm whether HGA and tryptophan were also produced using other chitin-containing materials as the sole C/N source, the 0.5% SHP-containing medium was also studied. Furthermore, nutrient broth (NB), a commercial medium for bacteria cultivation, was also tested. As shown in Figure 6, after fermentation by *Paenibacillus* sp. TKU036 for three days, no HGA and tryptophan were detected from the ethanol extract of the culture supernatant. To the best of our knowledge, there have been no reports of materials harmful to humans produced by *Paenibacillus* species [35]. The transformation of squid pen to functional foods of HGA and tryptophan via *Paenibacillus* sp. strain TKU036 may have the potential to be intensively investigated.

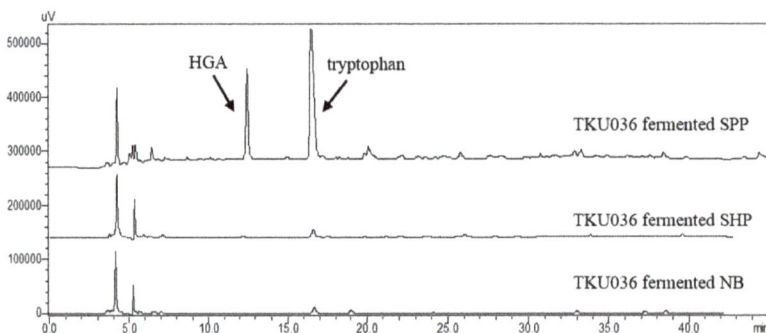

Figure 6. The HPLC fingerprints of the ethanol extracts from the *Paenibacillus* sp. TKU036 fermented SPP-containing medium, the *Paenibacillus* sp. TKU036 fermented SHP-containing medium, and *Paenibacillus* sp. TKU036 fermented nutrient broth (NB).

3. Materials and Methods

3.1. Materials

Squid pens were obtained from Shin-Ma Frozen Food Co. (I-Lan, Taiwan). Shrimp head power (SHP) was obtained from Fwu-Sow Industry. (Taichun, Taiwan). Cicada shells were collected at the Tamsui Campus of Tamkang University (New Taipei, Taiwan). HGA, tryptophan, and 2,2-diphenyl-1-picrylhydrazyl (DPPH) were purchased from Sigma-Aldrich (St. Louis, MO, USA). Nutrient broth was obtained from Difco. Octadecylsilane (ODS) gel was purchased from Merck (Darmstadt, Germany).

3.2. Antioxidant Activity Assay

The antioxidant samples (1.2 mL) were mixed with 0.3 mL of a methanolic solution containing 0.75 mM DPPH radicals. The mixture was vigorously shaken and incubated for 30 min in the dark, and the absorbance was then measured at 517 nm against a blank [3]. The scavenging ability was calculated as described in our previous paper [3].

3.3. Screening of Antioxidant-Producing Strain

The bacteria were isolated from soil samples that were collected at different locations in Northern Taiwan. They were cultivated in a medium containing squid pen powder (SPP) (pH 7.2) supplemented with 0.05% $MgSO_4 \cdot 7H_2O$ and 0.1% K_2HPO_4 to screen for antioxidant activity. The strains were cultivated in a 250 mL Erlenmeyer flask that contained 50 mL of medium at 37 °C in a reciprocal shaker at 150 rpm for 1–2 days. The supernatants obtained via centrifugation were used for the estimation of antioxidant activity using the protocol described in our previous paper [7]. Strain TKU036, which showed the highest activity, was selected for further study.

3.4. Extraction and Isolation of HGA and Tryptophan

The culture supernatant (2 L) of *Paenibacillus* sp. TKU036 was lyophilized (5.824 g) and extracted via ultrasonication in 300 mL of 95% ethanol at 60 °C in triplicate. The extract was concentrated under reduced pressure. The obtained ethanol extract (2.4905 g) was dissolved in H_2O, and it was loaded onto an open ODS column and eluted with 0%–60% MeOH in H_2O (to maintain the acetic acid concentration at 0.4%), resulting in 14 fractions (Fractions 1 to 14). The tryptophan was found in Fractions 5 (0.1253 g) and 6 (0.0864 g). Fraction 4 was further purified via preparative HPLC, equipped with a 250 mm × 20 mm i.d. preparative Cosmosil 5C18-AR-II column (Nacalai Tesque, Kyoto, Japan) and a UV detector at 254 nm, and it was eluted with 13% ACN in H_2O (0.4% acetic acid), yielding 5 fractions (Fractions 4-1 to 4-5). HGA (0.012 g) and tryptophan (0.008 g) were obtained from Fractions 4-4 and 4-5, respectively. In total, HGA (0.012 g) and tryptophan (0.21 g) were produced from 2 L of the culture supernatant (5.824 g) by employing the method described above. The recovery of HGA (60 mg/L) in this study was comparable to that of HGA in strawberry tree honey, which has been reported to contain HGA at a concentration of 414 mg/kL [19].

3.5. The Analysis of HGA and Tryptophan by HPLC:

The chromatographic separation was carried out on an Agilent HC-C18 (5 μm, 250 mm × 4.6 mm i.d., Agilent Technologies, Tokyo, Japan). The binary gradient elution system consisted of 0.4% acetic acid aq. (A) and 0.4% acetic acid in acetonitrile (B), and the HPLC profile separation was achieved using the following gradient: 0–20 min, 5%–15% B; 20–35 min, 15%–35% B; and 35–40 min, 35%–100% B. The UV detection wavelength was 292 nm. The column was kept at room temperature. The flow rate was 0.8 mL/min, and the injection volume was 10 μL.

4. Conclusions

To efficiently reutilize seafood processings via microbial transformation, squid pen was used as the sole C/N source for screening antioxidant-producing bacteria from Taiwanese soils. The culture supernatant of strain TKU036 produced potential antioxidant activity and was identified as *Paenibacillus* species. The antioxidant compound in the culture supernatant was identified as HGA, which showed anti-inflammatory effects as well. Tryptophan was also identified in the culture supernatant. Neither HGA nor tryptophan was found in the unfermented SPP-containing medium or in other chitinous materials (shrimp head powder)-containing medium. The results showed that squid pen is a promising material for the production of antioxidants and anti-inflammatories by *Paenibacillus* sp. TKU036.

HGA showed higher antioxidant activity ($IC_{50} = 6.9$ μg/mL) than α-tocopherol ($IC_{50} = 17.6$ μg/mL). The anti-inflammation activity of HGA ($IC_{50} = 10.14$ μg/mL) was lower than that of quercetin ($IC_{50} = 1.14$ μg/mL). HGA has been reported as the most abundant phenolic compound in strawberry tree honey (414 mg/kg) [36]. The recovery of 60 mg of HGA per liter of culture supernatant, compared with that of strawberry tree honey, seems to have potential for HGA production.

Acknowledgments: This work was supported in part by a grant from the Ministry of Science and Technology, Taiwan (MOST 105-2313-B-032-001, MOST 104-2811-B-032-001, MOST 104-2320-B-077-006-MY3, and MOST 101-2320-B-077-006-MY3).

Author Contributions: S.-L.W. conceived and designed the experiment; H.-T.L. and Z.-H.L. performed the experiments; S.-L.W., Y.-H.K. and L.-J.Z. analyzed the data; S.-L.W. wrote the paper.

Conflicts of Interest: The authors declare no conflict of interest.

References

1. Wang, S.L.; Liang, T.W. Microbial reclamation of squid pen and shrimp shell. *Res. Chem. Intermed.* **2016**. [CrossRef]
2. Wang, S.L. Microbial reclamation of squid pen. *Biocatal. Agric. Biotechnol.* **2012**, *1*, 177–180. [CrossRef]
3. Liang, T.W.; Chen, W.T.; Lin, Z.H.; Kuo, Y.H.; Nguyen, A.D.; Pan, P.S.; Wang, S.L. An amphiprotic novel chitosanase from *Bacillus mycoides* and its application in the production of chitooligomers with their antioxidant and anti-inflammatory evaluation. *Int. J. Mol. Sci.* **2016**, *17*, 1302. [CrossRef] [PubMed]
4. Nguyen, A.D.; Huang, C.C.; Liang, T.W.; Nguyen, V.B.; Pan, P.S.; Wang, S.L. Production and purification of a fungal chitosanase and chitooligomers from *Penicillium janthinellum* D4 and discovery of the enzyme activators. *Carbohydr. Polym.* **2014**, *108*, 331–337. [CrossRef] [PubMed]
5. Wang, S.L.; Liang, T.W.; Yen, Y.H. Bioconversion of chitin-containing wastes for the production of enzymes and bioactive materials. *Carbohydr. Polym.* **2011**, *84*, 732–742. [CrossRef]
6. Liang, T.W.; Wang, S.L. Recent advances in exopolysaccharides from *Paenibacillus* spp.: Production, isolation, structure, and bioactivities. *Mar. Drugs* **2015**, *13*, 1847–1863. [CrossRef] [PubMed]
7. Liang, T.W.; Tseng, S.C.; Wang, S.L. Production and characterization of antioxidant properties of exopolysaccharides from *Paenibacillus mucilaginosus* TKU032. *Mar. Drugs* **2016**, *14*, 40–51. [CrossRef] [PubMed]
8. Kuo, Y.H.; Liang, T.W.; Liu, K.C.; Hsu, Y.W.; Hsu, H.C.; Wang, S.L. Isolation and identification of a novel antioxidant with antitumor activity from *Serratia ureilytica* using squid pen as fermentation substrate. *Mar. Biotechnol.* **2011**, *13*, 451–461. [CrossRef] [PubMed]
9. Kuo, Y.H.; Hsu, H.C.; Chen, Y.C.; Liang, T.W.; Wang, S.L. A novel compound with antioxidant activity produced by *Serratia ureilytica* TKU013. *J. Agric. Food Chem.* **2012**, *60*, 9043–9047. [CrossRef] [PubMed]
10. Liang, T.W.; Chen, C.H.; Wang, S.L. Production of insecticidal materials from *Pseudomonas tamsuii*. *Res. Chem. Intermed.* **2015**, *41*, 7965–7971. [CrossRef]
11. Wang, S.L.; Chen, S.Y.; Yen, Y.H.; Liang, T.W. Utilization of chitinous materials in pigment adsorption. *Food Chem.* **2012**, *135*, 1134–1140. [CrossRef] [PubMed]
12. Liang, T.W.; Lo, B.C.; Wang, S.L. Chitinolytic bacteria-assisted conversion of squid pen and its effect on dyes and pigments adsorption. *Mar. Drugs* **2015**, *13*, 4576–4593. [CrossRef] [PubMed]

13. Sepúlveda, L.; Aguilera-Carbó, A.; Ascacio-Valdés, J.A.; Rodríguez-Herrera, R.; Martínez-Hernández, J.L.; Aguilar, C.N. Optimization of ellagic acid accumulation by *Aspergillus niger* GH1 in solid state culture using pomegranate shell powder as a support. *Process Biochem.* **2012**, *47*, 2199–2203. [CrossRef]

14. Raghuwanshi, S.; Dutt, K.; Gupta, P.; Misra, S.; Saxena, R.K. *Bacillus sphaericus*: The highest bacterial tannase producer with potential for gallic acid synthesis. *J. Biosci. Bioeng.* **2011**, *111*, 635–640. [CrossRef] [PubMed]

15. Lambert, F.; Zucca, J.; Ness, F.; Aigle, M. Production of ferulic acid and coniferyl alcohol by conversion of eugenol using a recombinant strain of *Saccharomyces cerevisiae*. *Flavour Fragr. J.* **2014**, *29*, 14–21. [CrossRef]

16. Kang, K.A.; Chae, S.; Lee, K.H.; Zhang, R.; Jung, M.S.; You, H.J.; Kim, J.S.; Hyun, J.W. Antioxidant effect of homogentisic acid on hydrogen peroxide induced oxidative stress in human lung fibroblast cells. *Biotechnol. Bioprocess. Eng.* **2005**, *10*, 556–563. [CrossRef]

17. Chen, P.; Wang, Y.; Chen, L.; Jiang, W.; Niu, Y.; Shao, Q.; Gao, L.; Zhao, Q.; Yan, L.; Wang, S. Comparison of the anti-inflammatory active constituents and hepatotoxic pyrrolizidine alkaloids in two *Senecio* plants and their preparations by LC-UV and LC-MS. *J. Pharm. Biomed. Anal.* **2015**, *115*, 260–271. [CrossRef] [PubMed]

18. Wang, C.L.; Huang, T.H.; Liang, T.W.; Wang, S.L. Production and characterization of exopolysaccharides and antioxidant from *Paenibacillus* sp. TKU023. *New Biotechnol.* **2011**, *28*, 559–565. [CrossRef] [PubMed]

19. Liang, T.W.; Wu, C.C.; Cheng, W.T.; Chen, Y.C.; Wang, C.L.; Wang, I.L.; Wang, S.L. Exopolysaccharides and antimicrobial biosurfactants produced by *Paenibacillus macerans* TKU029. *Appl. Biochem. Biotechnol.* **2014**, *172*, 933–950. [CrossRef] [PubMed]

20. Chung, Y.C.; Chang, C.T.; Chao, W.W.; Lin, C.F.; Chou, S.T. Antioxidative activity and safety of the 50% ethanolic extract from red bean fermented by *Bacillus subtilis* IMR-NK1. *J. Agric. Food Chem.* **2002**, *50*, 2454–2458. [CrossRef] [PubMed]

21. Dyah, H.W.; Joe, A.V.; Severino, S.P. Mathematical modeling of the development of antioxidant activity in soybeans fermented with *Aspergillus oryzae* and *Aspergillus awamori* in the solid state. *J. Agric. Food Chem.* **2009**, *57*, 540–544.

22. Miyake, Y.; Fukumoto, S.; Okada, M.; Sakaida, K.; Nakamura, Y.; Osawa, T. Antioxidative catechol lignans converted from sesamin and sesaminol triglucoside by culturing with *Aspergillus*. *J. Agric. Food Chem.* **2005**, *53*, 22–27. [CrossRef] [PubMed]

23. Lee, I.H.; Hung, Y.H.; Chou, C.C. Total phenolic and anthocyanin contents, as well as antioxidant activity, of black been koji fermented by *Aspergillus awamori* under different culture conditions. *Food Chem.* **2007**, *104*, 936–942. [CrossRef]

24. Kuo, C.F.; Hou, M.H.; Wang, T.S.; Chyau, C.C.; Chen, Y.Y. Enhanced antioxidant activity of *Monascus pilosus* fermented products by addition of ginger to the medium. *Food Chem.* **2009**, *116*, 915–922. [CrossRef]

25. Wang, S.L.; Lin, C.L.; Liang, T.W.; Liu, K.C.; Kuo, Y.H. Conversion of squid pen by *Serratia ureilytica* for the production of enzymes and antioxidants. *Bioresour. Technol.* **2009**, *100*, 316–323. [CrossRef] [PubMed]

26. Cabras, P.; Angioni, A.; Tuberoso, C.; Floris, I.; Reniero, F.; Guillou, C.; Ghelli, S. Homogentisic acid: A phenolic acid as a marker of strawberry-tree (*Arbutus unedo*) honey. *J. Agric. Food Chem.* **1999**, *47*, 4064–4067. [CrossRef] [PubMed]

27. Yan, X.; Suzuki, M.; Ohnishi-Kameyama, M.; Sada, Y.; Nakanishi, T.; Nagata, T. Extraction and identification of antioxidants in the roots of yacon (*Smallanthus sonchifolius*). *J. Agric. Food Chem.* **1999**, *47*, 4711–4713. [CrossRef] [PubMed]

28. Carreira, A.; Ferreira, L.M.; Loureiro, V. Brown pigments produced by *Yarrowia lipolytica* result from extracellular accumulation of homogentisic acid. *Appl. Environ. Microbiol.* **2001**, *67*, 3463–3468. [CrossRef] [PubMed]

29. Kluyver, A.J.; van Zijp, J.C.M. The production of homogentisic acid out of phenylacetic acid by *Aspergillus niger*. *Antonie Van Leeuwenhoek* **1951**, *17*, 315–324. [CrossRef] [PubMed]

30. Kotob, S.I.; Coon, S.L.; Quintero, E.J.; Weiner, R.M. Homogentisic acid is the primary precursor of melanin synthesis in *Vibrio cholerae*, a hyphomonas strain, and *Shewanella colwelliana*. *Appl. Environ. Microbiol.* **1995**, *61*, 1620–1622. [PubMed]

31. Liu, L.; Duan, X.; Wu, J. Modulating the direction of carbon flow in *Escherichia coli* to improve L-tryptophan production by inactivating the global regulator FruR. *J. Biotechnol.* **2016**, *231*, 141–148. [CrossRef] [PubMed]

32. Ikeda, M.; Katsumata, R. Hyperproduction of tryptophan by *Corynebacterium glutamicum* with the modified pentose phosphate pathway. *Appl. Environ. Microbiol.* **1999**, *65*, 2497–2502. [PubMed]

33. Chung, M.J.; Park, J.K.; Park, Y.I. Anti-inflammatory effects of low-molecular weight chitosan oligosaccharides in IgE-antigen complex-stimulated RBL-2H3 cells and asthma model mice. *Int. Immunopharmcol.* **2012**, *12*, 453–459. [CrossRef] [PubMed]

34. Boots, A.W.; Drent, M.; de Boer, V.C.; Bast, A.; Haenen, G.R. Quercetin reduces markers of oxidative stress and inflammation in sarcoidosis. *Clin. Nutr.* **2011**, *30*, 506–512. [CrossRef] [PubMed]

35. Guo, Y.; Huang, E.; Yuan, C.; Zhang, L.; Yousef, A.E. Isolation of a *Paenibacillus* sp. strain and structural elucidation of its broad-spectrum lipopeptide antibiotic. *Appl. Environ. Microbiol.* **2012**, *78*, 3156–3165. [CrossRef] [PubMed]

36. Rosa, A.; Tuberoso, C.I.G.; Atzeri, A.; Melis, M.P.; Bifulco, E.; Dessì, M.A. Antioxidant profile of strawberry tree honey and its marker homogentisic acid in several models of oxidative stress. *Food Chem.* **2011**, *129*, 1045–1053. [CrossRef] [PubMed]

marine drugs

MDPI

Article

Influence of Chitosan Swelling Behaviour on Controlled Release of Tenofovir from Mucoadhesive Vaginal Systems for Prevention of Sexual Transmission of HIV

Fernando Notario-Pérez [1], Araceli Martín-Illana [1], Raúl Cazorla-Luna [1], Roberto Ruiz-Caro [1], Luis-Miguel Bedoya [2], Aitana Tamayo [3], Juan Rubio [3] and María-Dolores Veiga [1,*]

[1] Departamento Farmacia y Tecnología Farmacéutica, Facultad de Farmacia, Universidad Complutense de Madrid, 28040 Madrid, Spain; fnotar01@ucm.es (F.N.-P.); aracelimartin@ucm.es (A.M.-I.); racazorl@ucm.es (R.C.-L.); rruizcar@ucm.es (R.R.-C.)
[2] Departamento Farmacología, Facultad de Farmacia, Universidad Complutense de Madrid, 28040 Madrid, Spain; lmbedoya@ucm.es
[3] Instituto de Cerámica y Vidrio, Consejo Superior de Investigaciones Científicas, 28049 Madrid, Spain; aitanath@icv.csic.es (A.T.); jrubio@icv.csic.es (J.R.)
* Correspondence: mdveiga@ucm.es; Tel.: +34-913-942-091; Fax: +34-913-941-736

Academic Editors: David Harding and Hitoshi Sashiwa
Received: 30 September 2016; Accepted: 16 February 2017; Published: 21 February 2017

Abstract: The main challenges facing efforts to prevent the transmission of human immunodeficiency virus (HIV) are the lack of access to sexual education services and sexual violence against young women and girls. Vaginal formulations for the prevention of sexually transmitted infections are currently gaining importance in drug development. Vaginal mucoadhesive tablets can be developed by including natural polymers that have good binding capacity with mucosal tissues, such as chitosan or guar gum, semisynthetic polymers such as hydroxypropylmethyl cellulose, or synthetic polymers such as Eudragit® RS. This paper assesses the potential of chitosan for the development of sustained-release vaginal tablets of Tenofovir and compares it with different polymers. The parameters assessed were the permanence time of the bioadhesion—determined ex vivo using bovine vaginal mucosa as substrate—the drug release profiles from the formulation to the medium (simulated vaginal fluid), and swelling profiles in the same medium. Chitosan can be said to allow the manufacture of tablets that remain adhered to the vaginal mucosa and release the drug in a sustained way, with low toxicity and moderate swelling that ensures the comfort of the patient and may be useful for the prevention of sexual transmission of HIV.

Keywords: Human Immunodeficiency Virus; Acquired Immunodeficiency Syndrome; chitosan; mucoadhesive vaginal tablets; Tenofovir; controlled release; ex vivo bioadhesion; swelling behaviour; swelling witness microstructure

1. Introduction

Acquired Immunodeficiency Syndrome (AIDS) continues to be one of the main public health problems around the world, especially in countries with the fewest resources. It is estimated that 36.7 million people are currently living with HIV [1]. The latest available data indicate that significant progress has been made over the last decade [2–4], and yet HIV continues to highlight the world's inequalities. The main challenges in preventing HIV transmission are the lack of access to sexual education services, and sexual violence against young women and girls [5]. It is, therefore, necessary to have methods such as microbicides that are controlled by women themselves in order to prevent transmission, so they no longer depend on men to prevent the acquisition of the virus.

Tenofovir (TFV) is a drug that acts by blocking reverse transcriptase activity in HIV infection. It is currently being investigated for its potential microbicidal effect against HIV [6,7]. TFV microbicide formulations have had proven antiviral efficacy in animal models and are currently in phase III clinical trials. Recent studies have demonstrated that TFV vaginal administration has no significant cytotoxicity in women and that TFV has no toxicity for vaginal mucosa at concentrations commonly used as a microbicide [8,9]. Numerous reports have assessed and confirmed the effectiveness of TFV vaginal formulations. A wide range of dosage forms containing this drug have been evaluated, including gels [10,11], films [12,13], and intravaginal rings [14–17].

Solid formulations have the advantage of high dose accuracy and long-term stability, as compared to semi-solid systems. The polymers used in these formulations must, therefore, be able to adhere to the vaginal mucosa and modulate drug release from the dosage form. The term "adhesion" describes the ability of certain macromolecules to adhere to the body's tissues; when this occurs in mucosa it is known as mucoadhesion. Although any material can adhere to the mucosa thanks to its viscous nature, there can be no real bioadhesion without an interrelation between some specific chemical groups in the polymers and biological tissues, or without establishing an interpenetration of chains. The dosage forms that bind to mucous membranes are described as mucoadhesive, as their purpose is to remain fixed at the point where the release and/or absorption of the drug occurs by prolonging its residence time [18,19].

All bioadhesive systems owe their properties to the inclusion of one or more types of polymeric molecules which, under appropriate conditions, establish interactions with the biological surface. One of these polymers is chitosan (CH), a natural, biocompatible, biodegradable, bioadhesive, and water-soluble polymer that degrades in acidic medium. It is obtained from the deacetylation of chitin, one of the most abundant polysaccharides in nature, as it is the structural element in the exoskeleton of crustaceans, such as crabs and shrimps. The amino and hydroxyl groups allow the adhesion to mucous through hydrogen bonds, and are protonated in an acid medium, which improves adhesion to negatively charged surfaces such as mucous. This polymer has been widely applied in the development of different pharmaceutical dosage forms for vaginal administration such as gels and tablets [20–23].

Possibly the most widely studied polymer for the development of such formulations is hydroxypropylmethyl cellulose (HPMC), a cellulose ether with methyl and hydroxypropyl groups used for the controlled release of drugs in hydrophilic matrix systems [24]. It is a FDA-approved polymer found in a wide range of applications, and was initially used in vaginal formulations as an excipient in the manufacture of films [25,26] and gels [27], although its use in vaginal administration tablets [20,28–30] has recently become more widespread. Another very similar polymer to HPMC and CH is guar gum (GG), which is also soluble in water, where it produces a viscous gel. GG is a biocompatible and biodegradable polysaccharide obtained from the seeds of *Cyamopsis tetragonoloba* used in the pharmaceutical industry as a binder or a disintegrant in tablets, and there are also several references to its use in the development of vaginal dosage forms. GG is sometimes present in bioadhesive vaginal gels, and has also been combined with HPMC to develop a bioadhesive vaginal tablet formulation [31,32]. All of the above-mentioned polymers are hydrophilic, and since the purpose of these formulations is their dissolution in the vaginal environment, it is also worth mentioning hydrophobic polymers such as Eudragit RS PO® (ERS). This is a copolymer of ethyl acrylate, methyl methacrylate, and a low content of methacrylic acid ester with quaternary ammonium groups. The ammonium groups are present as salts and render the polymers permeable. It is insoluble in aqueous medium and has low permeability and pH-independent swelling [33]. Its inclusion in pharmaceutical forms of vaginal administration to date is much scarcer than for the polymers described above, and it is mainly used in nanocapsules, microspheres, and microparticles [21,34,35].

With this background, the aim of this study is to assess the potential of chitosan to develop sustained release mucoadhesive tablets of TFV, where the drug release from these systems depends on the properties of each polymer. These properties are also analysed in other natural, semisynthetic, and synthetic polymers in order to assess the advantages offered by CH in the development of these formulations.

2. Results and Discussion

2.1. Swelling Tests

Figure 1 shows the swelling ratio (*SR*) profiles of the different batches studied. The maximum swelling ratio (SR_{max}) is included for each swelling curve. The curves in Figure 1 show that swelling and erosion processes are present in most cases. Hydrophilic polymers such as CH, HPMC, and GG swell when in contact with an aqueous medium as opposed to disintegrating. These batches increase in size due to the relaxation of the polymer chains. A temperature of 37 °C causes a decrease in the vitreous transition temperature, forming an area where the polymers change from a crystalline to a rubbery state (known as the gel layer) [36]. It is, thus, possible to distinguish a first stage for the GG and HPMC batches in which the swelling process takes precedence until the SR_{max} (96 h) is reached, followed by the erosion of the formulations. HPMC and GG are significant for having a high SR; this is higher in the case of GG, which also takes much longer to dissolve completely. GG is well known for its high water-absorbent capacity, which is the reason it is used in the development of superabsorbent hydrogels [37].

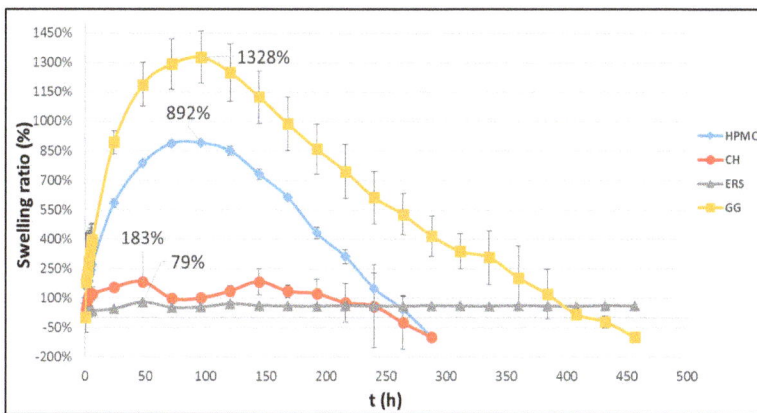

Figure 1. Swelling profile of each batch in simulated vaginal fluid (SVF). Data on the maximum swelling ratio (SR_{max}) are indicated.

In contrast, CH undergoes moderate but sustained swelling and acquires an aqueous volume of 183% of its weight, corroborating the results of Chen et al. [38]. This is because CH has an unusual swelling process, and after about 48 h the pressure from the gel causes the breakdown of its core (Figure 2). From the time when the fracture occurs, the core portion which the gel prevented from swelling is exposed to the aqueous medium and absorbs water, causing a new increase in SR values (Figure 1). Finally, ERS is not a water-soluble polymer and hardly absorbs water from the medium—although the porous matrix adsorbs a small amount—and, thus, remains undissolved throughout the test (19 days). This renders it inadequate, as once the drug has been released, the compact would need to be removed. In view of the results, the batch with CH would be the most comfortable, since it undergoes moderate swelling and complete erosion. This factor, the comfort of women, is crucial for the adherence to the use of the formulation. In this respect there is no problem since studies show that vaginal tablets are the solid dosage form preferred by women for intravaginal administration [39]. In addition, the small size of the compacts developed (2.2–2.3 mm in height) makes them even more comfortable.

Figure 2. Chitosan compact swelling pattern. First the compact has a given shape (**A**), although the upper and lower layers swell in the presence of SFV, exerting pressure on the core (**B**) until finally this pressure causes the compact to break (**C**).

2.2. Release Study

A drug's release rate from a dosage form can be influenced by different phenomena, ranging from drug dissolution to water absorption, polymer swelling, and the dissolution and diffusion of the drug through the polymer network [40]. Figure 3 shows the TFV release profiles corresponding to the prepared batches. The release data shows that HPMC and CH are the polymers that best control the drug release from the compact. One interesting result is that in the first 48 h these formulations released lower drug amounts than those containing ERS or GG, owing to the characteristics of ERS and GG. ERS is an insoluble polymer with pH-independent swelling and low permeability that is unable to gel in aqueous medium and, thus, barely controls the release of the drug [33]. Although GG produces the gel layer with the highest *SR* (Figure 1), it has very little consistency and the drug diffuses rapidly through it, so it does not represent a delayed release mechanism [41]. However, when HPMC and CH compacts are introduced into simulated vaginal fluid (SVF) the outer layers form a strong consistency gel, as reported by other authors, which controls TFV release [20]. This result would ensure women were protected against the transmission of HIV for at least 3–4 days (90%–95% TFV released). The inhibitory concentration 50 (IC_{50}) of TFV has been found to be between 1.08–1.22 μM depending on the HIV strain used [42]. Thus, using these compacts IC_{50} is reached in a few minutes after administration.

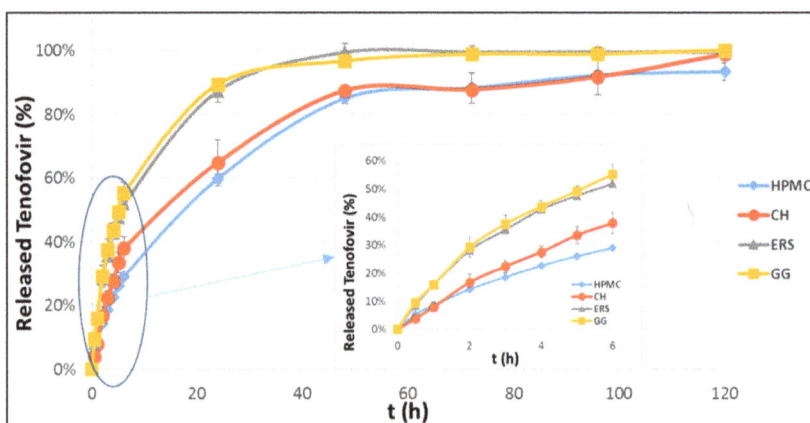

Figure 3. Tenofovir release profiles obtained from different batches in SVF.

In order to investigate the kinetics of TFV release from these formulations, mathematical model dependent methods (zero-order, first-order, Higuchi, Korsmeyer-Peppas, Hixson-Crowell, Hopfenberg and Weibull) were used to fit the experimental results. After analysing each batch, the models found to have the best fit to the curves in Figure 3 (high correlation coefficients R^2) were Korsmeyer-Peppas and Higuchi and Weibull [43,44].

According to Korsmeyer-Peppas, the drug release as a function of time follows Equation (1):

$$M_t/M_\infty = Kt^n \tag{1}$$

Which can also be expressed as Equation (2):

$$Ln(M_t/M_\infty) = Ln(K_{KP}) + nLn(t) \tag{2}$$

where M_t/M_∞ is the fraction of drug released at time t, K_{KP} is a constant incorporating the structural and geometric characteristics of the compact, and n is the release exponent [44]. In the case of cylindrical compacts, depending on the n value the drug release could follow a pure diffusion process ($n \leq 0.45$), an anomalous transport with simultaneous diffusion and structural modification of the polymer matrix ($0.45 < n < 0.89$), transport case II ($n = 0.89$) or transport Supercase II ($n > 0.89$). Both Case II and Supercase II involve the structural modification of the polymer matrix [43,45].

A good fit to the Higuchi model indicates that the drug diffuses through pores in the polymer matrix (this process is equivalent to Korsmeyer–Peppas for $n \leq 0.45$). The Higuchi equation for fitting the curves of Figure 3 is in this case Equation (3), where Q is the amount of drug released in time t and K_H is the Higuchi dissolution constant:

$$Q = K_H t^{1/2} \tag{3}$$

Finally, the curves in Figure 3 were also evaluated by the Weibull model, with the following mathematical equation:

$$M = M_0 \left[1 - e^{-\frac{(t-t_{lag})^b}{a}} \right] \tag{4}$$

where M is the dissolved drug, M_0 is the total amount of drug in the compact and t_{lag} is the lag time, a is a scale parameter that describes the dependence on time, and b describes the shape of the dissolution curve [43].

In our case, where there is no lag time (see curves in Figure 3), and because the drug release profiles have an exponential shape, b is equal to 1. If we take the constant $K_W = 1/a$, then Equation (4) can be summarised as Equation (5):

$$\ln\left(1 - \frac{M}{M_0}\right) = -K_W t \tag{5}$$

Figure 4 shows the corresponding fit of the experimental data to these drug release models, and Table 1 shows the n, K_{KP}, K_H, and K_W kinetic constants for these three models.

Table 1. TFV release kinetics from HPMC, CH, ERS, and GG batches.

Batch	Korsmeyer-Peppas			Higuchi		Weibull	
	K_{KP}	n	R^2	K_H	R^2	K_W	R^2
HPMC	0.088	0.63	0.9899	0.124	0.9980	0.036	0.9931
CH	0.077	0.92	0.9926	0.130	0.9815	0.040	0.9839
ERS	0.152	0.73	0.9887	0.148	0.9453	0.098	0.9859
GG	0.161	0.71	0.9929	0.145	0.9231	0.068	0.9756

According to the data in Table 1, the batches with HPMC and CH have a good fit with the Higuchi kinetic (Figure 4A). However, the Higuchi kinetic can only be applied when the swelling and dissolution of the matrix are negligible [43], so it cannot be used to explain the drug release behaviour from these batches. All of the batches tested have a good fit to the Korsmeyer–Peppas kinetic (Figure 4B). The analysis of n values for different batches reveals that those prepared with HPMC, ERS and GG have similar values of close to 0.7, and that in the case of CH n is higher and close to 1. It can, therefore, be said that the TFV release from HPMC, ERS, and GG corresponds to an anomalous (non-Fickian) transport, while for CH—whose n value is over 0.89—it follows a Supercase II release, a mechanism that implies an extreme drug transport [45]. During polymer swelling the breakage of the compact occurs because the upper and lower layers of the compact swell to form a gel, causing a compressive stress on the core that prevents axial swelling. As the gel continues pressing on the core, the internal compact pressure increases until the core breaks (Figure 2) [46].

The n values for HPMC, ERS, and GG fall in the range 0.45–0.89, indicating that the drug release is governed by simultaneous structural modification and diffusion through the polymer matrix processes. In the HPMC and GG batches, the polymer swells at the same time as the drug diffuses through the gel formed. As has been shown in the swelling test, these two polymers form a long-lasting gel and the rearrangement of chains occurs slowly; the simultaneous diffusion is the process that causes the time-dependent anomalous effect [46].

ERS captures very little water, which rules out polymer swelling as a possible explanation. When the drug release from the ERS batch is fit to the Weibull model, the constant K_W is much higher than for the other polymers—including GG (Figure 4C)—although the differences between GG and ERS in the other models are insignificant (Figure 4A,B). The Kw values in the Weibull equation represent the drug release rate constant. HPMC and chitosan have similar K_W values that are much lower than the other two formulations, signalling the greater control over the release of TFV from the compacts made with these polymers. In contrast, GG and ERS have much higher K_W values, since the drug also diffuses from these compacts at a greater rate. The Weibull model is used to analyse the release profile of matrix-type drug delivery, and this is the mechanism of TFV release in ERS. This is because ERS is a permeable polymer that allows water into the compact, followed by the dissolution of the drug in the medium, and finally the diffusion of TFV through the polymer.

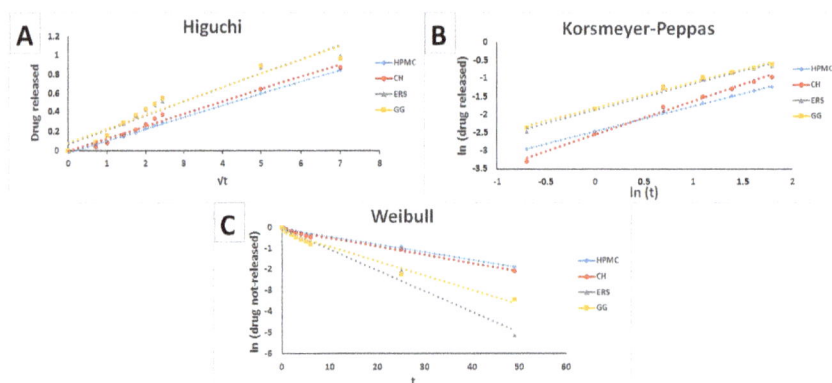

Figure 4. Fit of TFV release from batches HPMC, CH, ERS, and GG to the Higuchi (**A**) Korsmeyer-Peppas (**B**), and Weibull (**C**) models.

2.3. Microstructure of Witnesses. FE-SEM, and Hg Porosimetry

It is well known that water is removed during freeze-drying of a hydrated polymer system, and the space that was originally occupied by the solvent is transformed into pores, generating a porous structure similar to a sponge [47]. As can be seen from the micrographs in Figure 5, the witness microstructures vary considerably depending on the polymer type. Figure 5A, corresponding to the swelling witness of HPMC, has a channelled microstructure formed when water enters the polymer during swelling. These channels allow the compacts to maintain their shape while the drug diffuses slowly between them. This perfectly homogeneous microstructure is maintained because HPMC swelling occurs progressively; the outer layers become swollen but the core remains unswollen until the outer gel erodes and water reaches the core [48]. This is observed in our release studies, thus, water mobility plays a role in controlling drug release.

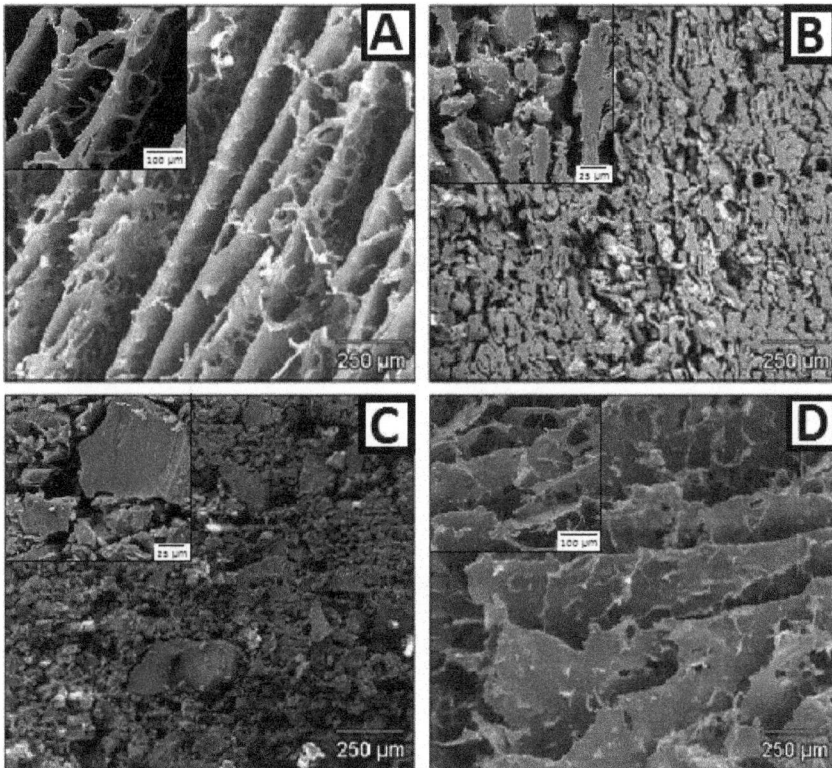

Figure 5. Electron microscopy micrographs of swollen witnesses of HPMC (**A**); chitosan (**B**); Eudragit® RS PO (**C**); and guar gum (**D**).

The micrograph of the CH witness (Figure 5B) shows a sponge-like microstructure with numerous pores in the polymer through which the SVF circulates, albeit with difficulty. This result explains the controlled release of TFV from CH in spite of its moderate swelling capacity. In contrast, no defined microstructure is observed for the ERS witness (Figure 5C); the formulation has a grainy microstructure with different-sized particles, but is unable to swell, which explains the failure of this formulation to control TFV delivery. Finally, the micrograph of the GG witness (Figure 5D) shows a perfectly microstructured formulation where the polymer is arranged in parallel sheets with the absorbed water between them. This microstructure explains why GG formulations swell the most and remain swollen the longest, as there is a high capacity for very effectively retaining water between these sheets. However, although the water cannot escape, the drug is able to diffuse through the polymer sheets, so their ability to retain the drug is minimal.

The above porous microstructures have been characterized by Hg porosimetry. Figure 6 shows the corresponding pore size distributions (PSD).

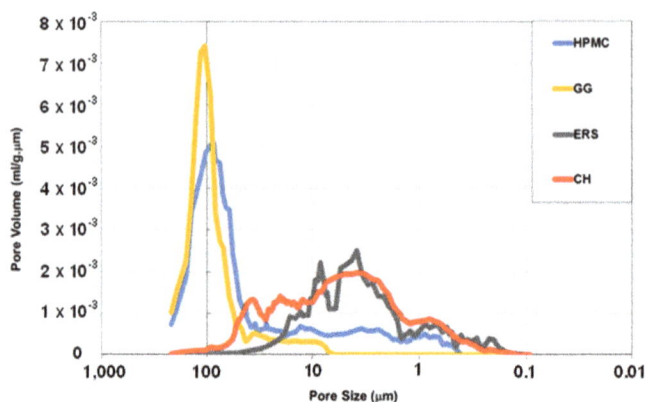

Figure 6. Pore distributions of HPMC, CH, ERS, and GG witnesses.

According to Figure 6, two types of PSD can be described: one has a narrow PSD where most of the pores are close to 100 μm, while the other has a wide PSD with pores of between 100 and 0.1 μm. Both PSD types are unambiguously associated to the swelling behaviours in Figure 1 and the FSEM micrographs in Figure 5. The HPMC and GG batches have high swelling characteristics with well-defined channelled microstructures and produce a narrow PSD with high pore sizes. In contrast, ERS and CH, with minimal swelling properties and grainy microstructures, produce a wide PSD with small pore sizes. There are two interesting observations: the first is that while the PSD of the GG batch also has a small number of pores with sizes between 50 and 10 mm, the PSD of HPMC also has small pores with sizes between 50 and 0.5 μm; and the other is that the PSD of CH has pores between 10 and 100 μm, while the PSD of ERS is below 50 μm. This points to the conclusion that as the compact has higher swelling properties, the corresponding PSD must have a high pore size.

Table 2 contains a summary of the results obtained from these PSD, and shows that mean pore size (Dp) values are related to SR values, as mentioned earlier. Hence, the higher the swelling capacity, the higher the Dp values. However, pore volumes (Vp) are more closely related to the total number of pores present in the witness, so the lowest Vp correspond to the ERS batch containing the smallest pore size. The CH witness has a higher Vp value than ERS due to its larger pore size, as indicated by the Dp. It is followed by GG, with a high Vp value but lower than HPMC, although it has a higher Dp. Finally, the highest Vp corresponds to HPMC. HPMC's higher Vp compared to GG is due to the small pores of between 50 and 0.5 μm in HPMC, but not in GG. In contrast, pore area (Sp) values show the opposite pattern; namely the higher the Dp, the lower the Sp. This is because pore area increases as pore volume decreases. As may be expected, porosity (P) values are related to Dp and Vp values, as porosity increases with both pore size and pore volume. Finally, bulk density (ρ_B) values are related to Vp and P values, as ρ_B corresponds to a sample where pores and material are measured as a whole. However, apparent density (ρ_A) corresponds to the sample with no pores over 0.1 μm, i.e., a dense sample, and these values are characteristic of the chemical sample composition.

Table 2. Pore volume (*Vp*), pore area (*Sp*), mean pore size (*Dp*), bulk and apparent densities (ρ_B, ρ_A), and porosity (*P*) of HPMC, CH, ERS, and GG witnesses.

Witness	Vp (cm$^3 \cdot$g^{-1})	Sp (m$^2 \cdot$g^{-1})	Dp (μm)	ρ_B (cm$^3 \cdot$g^{-1})	ρ_A (cm$^3 \cdot$g^{-1})	P (%)
HPMC	5.97	0.36	91.89	0.14	0.90	84
CH	1.74	0.43	28.74	0.38	1.19	67
ERS	0.35	3.59	9.16	0.77	1.06	27
GG	5.89	0.25	106.08	0.14	0.97	85

These results show that *P*, *Vp*, *Dp*, and ρ_B are related to the SR_{max} of the corresponding batches (Figure 1), but are not clearly related to the release profiles (Figure 3) or release kinetics (Table 1). Thus, the following relationship (with $R^2 = 0.992$) has been found between *Dp* and SR_{max}:

$$Dp = 35.4 \cdot Ln(SR_{max}) + 13.4 \tag{6}$$

This equation indicates that for a polymer with a very low swelling capacity the release of the TFV drug in aqueous medium causes pores of around 13 μm. In our case the ERS polymer presented pores with a mean size of 9 μm, close to the value obtained by this equation.

2.4. Evaluation of Mucoadhesion

An analysis of the mucoadhesion results (Figure 7) reveals that HPMC, ERS, and GG formulations remain attached to the mucosa for extended periods of time, even after all the TFV has been released. In contrast, the CH formulation shows a good initial adhesion to vaginal mucosa, and a residence time of about 48 h. This agrees with the results of other studies, highlighting the lower mucoadhesive ability of CH compared to cellulose derivatives [49]. This seems to be because the bonding to the mucosa by positively charged groups, as in the case of chitosan, is less durable than bonding through hydrogen bonds, which is typical of HPMC and GG.

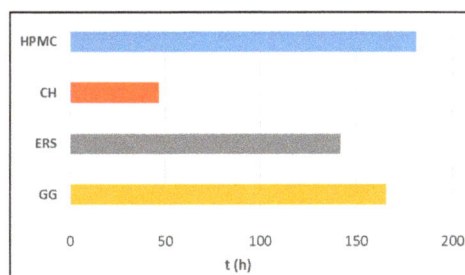

Figure 7. Mucoadhesion residence time of each batch in SVF.

The other three polymers show longer adhesion times to the mucosa of over 140 h in all cases. The formulation that remains attached for longest is HPMC, followed by GG and ERS. HPMC has been studied in depth, and this research corroborates its high mucoadhesive potential, caused by hydrogen bonding effects [50]. Although the mucoadhesive properties of GG have been poorly studied, another work shows its mucoadhesive strength is similar to HPMC [51]. Lastly, the most surprising results derive from the evaluation of the mucoadhesion of ERS, which in the previous literature is not classified as a mucoadhesive polymer. Our study shows that it has substantial mucoadhesive properties, and confirms a previous study comparing ERS with materials typically regarded in the literature as being good adhesives [52]. The adhesion to mucosa may be due to the presence of the quaternary ammonium group, which is protonated and may bind to negative charges in mucosa. Although these good mucoadhesion results highlight the high binding ability of the polymers, shorter times are required in this case—similar to the time for CH—since the TFV release occurs over a shorter period and there is no therapeutic justification for retaining the formulation adhered to the patient's vaginal mucosa after the drug has been completely released. In addition to discomfort, it could induce the rejection of the formulation.

2.5. Cell Toxicity

The biocompatibility of the formulations was evaluated through an in vitro cellular toxicity assay. All of the components of the different formulations were incubated at 37 °C in a 5% CO_2 atmosphere for five days before the assay to ensure that any potential toxic component would be present in the suspension. MT-2, a lymphoblastoid cell line, and HEC-1A, a uterus-derived cell line, were seeded and treated with different dilutions of the suspensions. All of the components were tested at a maximum concentration of 1000 µg/mL in base-5 serial dilutions. Experiments were performed on MT-2 cells to evaluate toxicity on the immune cells present in vaginal or uterine mucosae, and also on the uterus epithelial cell line (HEC-1A) to assess any potential damage to the integrity of the mucosae. Cytotoxic concentration 50 (CC_{50}) were calculated when possible.

As shown in Table 3 and Figure 8, no toxicity was detected at the concentrations tested for any of the compounds. Interestingly, Tenofovir did not show cytotoxicity even at the highest tested concentration of 1000 µg/mL (around 3.3 mM).

Table 3. CC_{50} values of TFV, GG, CH, ERS, and HPMC obtained from the cytotoxicity assay in both MT-2 and HEC-1A cell lines. CC_{50}: cytotoxic concentration 50%.

Evaluated Substance	Cell Line	CC50
TFV	MT-2	>1000 µg/mL
	HEC-1A	>1000 µg/mL
GG	MT-2	>1000 µg/mL
	HEC-1A	>1000 µg/mL
CH	MT-2	>1000 µg/mL
	HEC-1A	>1000 µg/mL
ERS	MT-2	>1000 µg/mL
	HEC-1A	>1000 µg/mL
HPMC	MT-2	>1000 µg/mL
	HEC-1A	>1000 µg/mL

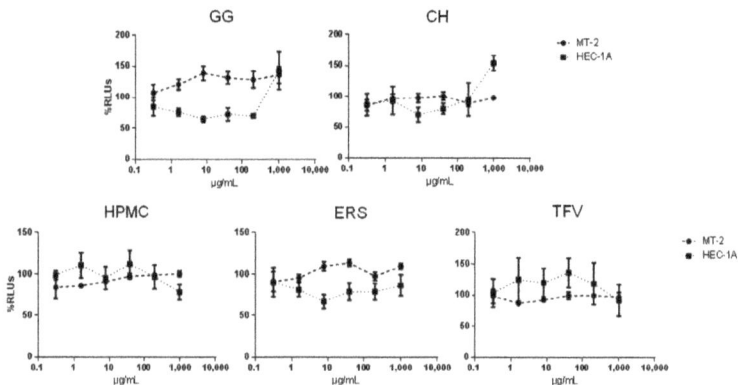

Figure 8. Cytotoxic evaluation of TFV, GG, CH, ERS, and HPMC measured in MT-2 cells and HEC-1A cells.

3. Experimental Section

3.1. Materials and Preparation of the Compact

Tenofovir (TFV, lot: FT104801401, MW: 287.21 g/mol) was supplied by Carbosynth Limited (Berkshire, UK). Chitosan, with 97% deacetylation and a viscosity of 92 mpa·s (CH, lot: 8826900003), was provided by Nessler (Madrid, Spain). The molecular weight, 10^5 g/mol, was estimated by viscometric measurements. Hydroxypropylmethylcellulose—Methocel® K 100 M (HPMC; lot: DT352711, MW: 72×10^4 g/mol) was kindly supplied by Colorcon Ltd. (Kent, UK). Eudragit RS® (ERS; lot: G120238035, MW: 407.932 g/mol) was supplied by Evonik (Essen, Germany). Guar gum (GG; lot: SLBH5231V, MW: 22×10^4 g/mol) was acquired from Sigma-Aldrich (Saint Louis, MO, USA). Magnesium stearate PRS-CODEX (MgSt; lot: 85269 ALP) was acquired from Panreac (Barcelona, Spain). All other reagents used in this study were of analytical grade and used without further purification. Demineralized water was used in all cases.

Four batches of compacts were prepared from physical mixtures of the corresponding polymer (HPMC, CH, ERS, GG), TFV and MgSt. In all cases, each compact contained 290, 30, and 3 mg of polymer, TFV, and MgSt respectively.

In all cases the compacts were prepared with a press similar to the one used for preparing solid samples for analysis by IR spectroscopy. A stainless-steel disc was placed in the die, and the physical mixture of the components was placed on top. A second stainless steel disc was then placed on top. Five tons of constant pressure was applied using a punch for four minutes. Finally, the piston and the discs were removed, and the compact was stored in a desiccator until its subsequent evaluation. The manufactured compacts were cylindrical in shape and measured 13 mm in diameter and 2.2–2.3 mm in height.

3.2. Methods

3.2.1. Swelling Tests

The swelling pattern of the different batches in SVF were analysed using the method described by Ruiz-Caro et al. [53]. Each analysis was tested in triplicate. Swelling tests were carried out in a shaking water bath at 37 °C and 15 opm. In order to maintain the contact between the compact and the medium, and for more convenient handling of the samples during the test, each compact was previously fixed to a stainless steel disc, 3 cm in diameter, with a cyanoacrylate adhesive. At given time intervals (every hour during the first six hours and once a day for the remainder of the analysis), the discs were removed from the medium, placed on filter paper to eliminate the liquid excess and weighed. SR was calculated according to Equation (7):

$$SR = \left(\frac{C_s - C_d}{C_d} \right) \cdot 100 \tag{7}$$

where C_s and C_d correspond to the swollen and dry compact weights, respectively.

3.2.2. Release Study

The method described by Sánchez-Sánchez et al. [20] was used concurrently with the swelling test to assess the release behaviour of TFV in each batch. Each sample was inserted in a borosilicate glass bottle containing 80 mL of the SVF [54] and placed in a shaking water bath (Selecta® UNITRONIC320 OR, Barcelona, Spain) at 37 °C and 15 opm. Every day at given times, 5 mL samples were removed and filtered. The medium was replaced with the same volume of SVF at the same temperature. TFV release concentrations were quantified by UV spectroscopy at a wavelength of 260 nm in a Shimadzu® UV-1700 spectrophotometer (Kyoto, Japan). The test was performed in triplicate in each case. Studies have been conducted on the solubility of TFV in SFV at room temperature and the results are 4 mg/mL.

Therefore, this ensures that the release study is performed under sink conditions and the release of drug is not conditioned by its solubility but by the release system.

The drug release experimental data were fitted to different model-dependent methods (zero order, first order, Higuchi, Korsmeyer-Peppas, Hixson-Crowell, Hopfenberg, and Weibull models) to investigate the kinetics of drug release from the various batches [43].

3.2.3. Swelling Witnesses

In order to characterize the microstructure acquired by the compacts when introduced in the SVF, swelling witnesses were prepared by determining the time each batch reached the maximum SR. This was done by attaching two compacts from each batch to the same stainless steel discs used for the swelling test. The discs were treated in the same way as in the swelling test. They were immersed in a beaker with SVF which was then placed in the shaking water bath (Selecta®UNITRONIC320 OR, Barcelona, Spain) (37 °C and 15 opm). The compacts were left under these conditions until they reached the maximum SR. Each compact was then extracted from the medium and lyophilized, and then stored in a desiccator until analysis. Witness microstructures were analysed by electron microscopy using a field emission scanning electron microscope (FE-SEM, Hitachi 4700, Tokyo, Japan) at an accelerating voltage of 15 V. Pore size distributions (PSD) were determined by mercury porosimetry using an Autopore II 9215 (Micromeritics Corp., Norcross, GA, USA). The corresponding pore volumes (Vp), pore areas (Sp), mean pore sizes (Dp), bulk and apparent densities (ρ_B, ρ_A), and porosities (P) of the samples were calculated from these PSD, assuming cylindrical pore shapes in all cases.

3.2.4. Assessment of Mucoadhesion

A new ex vivo mucoadhesion test was used to determine how long the compact remained adhered to the vaginal mucosa. A sample of freshly excised veal vaginal mucosa (obtained from a local slaughterhouse) was fixed with a cyanoacrylate adhesive to an 8.5 cm × 5 cm stainless steel plate (SSP). Each compact was then adhered to the mucosa, applying a given pressure (500 g for 30 s). The SSP was placed at an angle of 60° inside a beaker containing 150 mL of SVF, and this system was inserted in the shaking water bath (Selecta® UNITRONIC320 OR, Barcelona, Spain) at 37 °C and 15 opm. The bioadhesion time of each batch was assessed by visual observation of the samples. All batches were tested in duplicate.

3.2.5. Cytotoxicity Assessment

Two human cell lines were used: a lymphoblastic cell line, MT-2 [55] and a uterus/endometrium epithelial cell line, HEC-1-A (kindly provided by M. A. Muñoz, Hospital Gregorio Marañón, Madrid, Spain). Both cells were grown in RPMI 1640 medium supplemented with 10% (v/v) foetal bovine serum, 2 mM L-glutamine and 50 µg/mL streptomycin at 37 °C with a humidified atmosphere of 5% CO_2. HEC-1-A cells were detached by treatment with a trypsin 0.25% and EDTA 0.03% solution. Cell cultures were split twice a week.

Cell toxicity was measured by the CellTiter Glo viability assay (Promega). Briefly, GG, CH, ERS, HPMC, and TFV were suspended in water at a concentration of 10 mg/mL and left in culture (5% CO_2 and 37 °C) for five days [56]. Cells were then seeded in 96 microwell plates at a density of 1×10^5 cells per well in the case of MT-2, and 2×10^4 in the case of HEC-1-A, in complete RPMI medium, and treated with fresh medium containing different concentrations of suspensions (1000, 200, 40, 8, 1.6, and 0.32 µg/mL), or with the same concentration of vehicle (water). After 48 h of incubation, cell viability was measured following the manufacturer's instructions (CellTiter Glo viability assay), and RLUs were obtained in a luminometer. Data were normalized using RLUs obtained from cells treated with vehicle (100%) as a reference. Values of CC_{50} were calculated using GraphPad Prism Software (non-linear regression, log inhibitor versus response).

4. Conclusions

Vaginal compacts can be a very useful tool for the prevention of HIV transmission from men to women. Adherence to the use of microbicides has been one of the main drawbacks in demonstrating the efficacy of the formulations studied to date. In contrast, these compacts would decrease the frequency of administration by achieving sustained release of the drug over several days, resulting in a greater adherence to the treatment.

From the results obtained it can be concluded that there are two polymers (CH and HPMC) with the potential to achieve sustained and complete release of TFV from vaginal mucoadhesive compacts. However, the good bioadhesive properties of CH, which allow the formulation to remain attached to the vaginal mucosa only until all of the drug has been released, its moderate SR, which ensures more comfort for the patient than the other polymers tested, and its low cytotoxicity warrants that CH compacts containing TFV have proven to be a suitable formulation for the prevention of sexual transmission of HIV.

Acknowledgments: This work was supported by the Spanish Ministry of Economy and Competitiveness (MAT2012-34552).

Author Contributions: Roberto Ruiz-Caro, Luis-Miguel Bedoya, Aitana Tamayo, Juan Rubio and María-Dolores Veiga designed and planned the experiments. Fernando Notario-Pérez, Araceli Martín-Illana and Raúl Cazorla-Luna conducted the experiments. All authors contributed to the preparation of the manuscript. María-Dolores Veiga is the senior author and project leader.

Conflicts of Interest: The authors declare no conflict of interest.

References

1. World Health Organization. HIV/AIDS. Available online: http://www.who.int/topics/hiv_aids/en/ (accessed on 9 September 2015).
2. World Health Organization. Global HIV Prevalence Has Levelled off. Available online: http://www.who.int/mediacentre/news/releases/2007/pr61/en/ (accessed on 9 September 2015).
3. UNAIDS. AIDS by the Numbers. Available online: http://www.unaids.org/sites/default/files/media_asset/20131120_AIDSbynumbers_A5brochure_en_2.pdf (accessed on 9 September 2015).
4. UNAIDS. AIDS by the Numbers 2015. Available online: http://www.unaids.org/sites/default/files/media_asset/AIDS_by_the_numbers_2015_en.pdf (accessed on 9 September 2015).
5. UNAIDS. The Gap Report. Available online: http://www.unaids.org/sites/default/files/media_asset/UNAIDS_Gap_report_en.pdf (accessed on 9 September 2015).
6. McConville, C.; Friend, D.R.; Clark, M.R.; Malcolm, K. Preformulation and development of a once-daily sustained-release tenofovir vaginal tablet containing a single excipient. *J. Pharm. Sci.* **2013**, *102*, 1859–1868. [CrossRef] [PubMed]
7. Viread®, European Medicines Agency (EMA). Available online: http://www.ema.europa.eu/docs/en_GB/document_library/EPAR_-_Product_Information/human/000419/WC500051737.pdf (accessed on 9 September 2015).
8. Mayer, K.H.; Maslankowski, L.A.; Gai, F.; El-Sadr, W.M.; Justman, J.; Kwiecien, A.; Mâsse, B.; Eshleman, S.H.; Hendrix, C.; Morrow, K.; et al. Safety and tolerability of tenofovir vaginal gel in abstinent and sexually active HIV-infected and uninfected women. *AIDS* **2006**, *20*, 543–551. [CrossRef] [PubMed]
9. Johnson, T.J.; Gupta, K.M.; Fabian, J.; Albright, T.H.; Kiser, P.F. Segmented polyurethane intravaginal rings for the sustained combined delivery of antiretroviral agents dapivirine and tenofovir. *Eur. J. Pharm. Sci.* **2010**, *39*, 203–212. [CrossRef] [PubMed]
10. Abdool Karim, Q.; Abdool Karim, S.S.; Frohlich, J.A.; Grobler, A.C.; Baxter, C.; Mansoor, L.E.; Kharsany, A.B.; Sibeko, S.; Mlisana, K.P.; Omar, Z.; et al. Effectiveness and safety of tenofovir gel, an antiretroviral microbicide, for the prevention of HIV infection in women. *Science* **2010**, *329*, 1168–1174. [CrossRef] [PubMed]
11. Celum, C.; Baeten, J.M. Tenofovir-based pre-exposure prophylaxis for HIV prevention: Evolving evidence. *Curr. Opin. Infect. Dis.* **2012**, *25*, 51–57. [CrossRef] [PubMed]

12. Akil, A.; Agashe, H.; Dezzutti, C.S.; Moncla, B.J.; Hillier, S.L.; Devlin, B.; Shi, Y.; Uranker, K.; Rohan, L.C. Formulation and characterization of polymeric films containing combinations of antiretrovirals (ARVs) for HIV prevention. *Pharm. Res.* **2015**, *32*, 458–468. [CrossRef] [PubMed]

13. Akil, A.; Devlin, B.; Cost, M.; Rohan, L.C. Increased Dapivirine tissue accumulation through vaginal film codelivery of dapivirine and Tenofovir. *Mol. Pharm.* **2014**, *11*, 1533–1541. [CrossRef] [PubMed]

14. Baum, M.M.; Butkyavichene, I.; Churchman, S.A.; Lopez, G.; Miller, C.S.; Smith, T.J.; Moss, J.A. An intravaginal ring for the sustained delivery of tenofovir disoproxil fumarate. *Int. J. Pharm.* **2015**, *495*, 579–587. [CrossRef] [PubMed]

15. Smith, J.M.; Srinivasan, P.; Teller, R.S.; Lo, Y.; Dinh, C.T.; Kiser, P.F.; Herold, B.C. Tenofovir disoproxil fumarate intravaginal ring protects high-dose depot medroxyprogesterone acetate-treated macaques from multiple SHIV exposures. *J. Acquir. Immune Defic. Syndr.* **2015**, *68*, 1–5. [CrossRef] [PubMed]

16. Moss, J.A.; Malone, A.M.; Smith, T.J.; Kennedy, S.; Nguyen, C.; Vincent, K.L.; Motamedi, M.; Baum, M.M. Pharmacokinetics of a Multipurpose Pod-Intravaginal Ring Simultaneously Delivering Five Drugs in an Ovine Model. *Antimicrob. Agents Chemother.* **2013**, *57*, 3994–3997. [CrossRef] [PubMed]

17. Smith, J.M.; Rastogi, R.; Teller, R.S.; Srinivasan, P.; Mesquita, P.M.; Nagaraja, U.; McNicholl, J.M.; Hendry, R.M.; Dinh, C.T.; Martin, A.; et al. Intravaginal ring eluting tenofovir disoproxil fumarate completely protects macaques from multiple vaginal simian-HIV challenges. *Proc. Natl. Acad. Sci. USA* **2013**, *110*, 16145–16150. [CrossRef] [PubMed]

18. Valenta, C. The use of mucoadhesive polymers in vaginal delivery. *Adv. Drug Deliv. Rev.* **2005**, *57*, 1692–1712. [CrossRef] [PubMed]

19. Rodriguez, I.C.; Cerezo, A.; Salem, I.I. Bioadhesive delivery systems. *ARS Pharm.* **2000**, *41*, 115–128.

20. Sánchez-Sánchez, M.P.; Martín-Illana, A.; Ruiz-Caro, R.; Bermejo, P.; Abad, M.J.; Carro, R.; Bedoya, L.M.; Tamayo, A.; Rubio, J.; Fernández-Ferreiro, A.; et al. Chitosan and Kappa-Carrageenan Vaginal Acyclovir Formulations for Prevention of Genital Herpes. In Vitro and Ex Vivo Evaluation. *Mar. Drugs* **2015**, *13*, 5976–5992. [CrossRef] [PubMed]

21. Frank, L.A.; Sandri, G.; D'Autilia, F.; Contri, R.V.; Bonferoni, M.C.; Caramella, C.; Frank, A.G.; Pohlmann, A.R.; Guterres, S.S. Chitosan gel containing polymeric nanocapsules: A new formulation for vaginal drug delivery. *Int. J. Nanomed.* **2014**, *9*, 3151–3161.

22. Senyiğit, Z.A.; Karavana, S.Y.; Eraç, B.; Gürsel, O.; Limoncu, M.H.; Baloğlu, E. Evaluation of chitosan based vaginal bioadhesive gel formulations for antifungal drugs. *Acta Pharm.* **2014**, *64*, 139–156. [CrossRef] [PubMed]

23. Szymańska, E.; Winnicka, K.; Amelian, A.; Cwalina, U. Vaginal chitosan tablets with clotrimazole-design and evaluation of mucoadhesive properties using porcine vaginal mucosa, mucin and gelatine. *Chem. Pharm. Bull.* **2014**, *62*, 160–167. [CrossRef] [PubMed]

24. Sánchez, R.; Damas, R.; Domínguez, P.; Cerezo, P.; Salcedo, I.; Aguzzi, C.; Viseras, C. Uso de la HidroxiPropilMetilCelulosa (HPMC) en la Liberación Modificada de Fármacos. Available online: http://innovacion.gob.sv/inventa/attachments/article/8566/HPMC%20en%20f%C3%A1rmacos.pdf (accessed on 9 September 2015).

25. Ghosal, K.; Hazra, B.T.; Bhowmik, B.B.; Thomas, S. Formulation Development, Physicochemical Characterization and In Vitro-in vivo drug release of vaginal Films. *Curr. HIV Res.* **2016**, *14*, 295–306. [CrossRef] [PubMed]

26. Grammen, C.; Van den Mooter, G.; Appeltans, B.; Michiels, J.; Crucitti, T.; Ariën, K.K.; Augustyns, K.; Augustijns, P.; Brouwers, J. Development and characterization of a solid dispersion film for the vaginal application of the anti-HIV microbicide UAMC01398. *Int. J. Pharm.* **2014**, *475*, 238–244. [CrossRef] [PubMed]

27. Tuğcu-Demiröz, F.; Acartürk, F.; Özkul, A. Preparation and characterization of bioadhesive controlled-release gels of cidofovir for vaginal delivery. *J. Biomater. Sci. Polym. Ed.* **2015**, *26*, 1237–1255. [CrossRef] [PubMed]

28. Wang, L.; Tang, X. A novel ketoconazole bioadhesive effervescent tablet for vaginal delivery: Design, in vitro and 'in vivo' evaluation. *Int. J. Pharm.* **2008**, *350*, 181–187. [CrossRef] [PubMed]

29. Karasulu, H.Y.; Hilmioğlu, S.; Metin, D.Y.; Güneri, T. Efficacy of a new ketoconazole bioadhesive vaginal tablet on Candida albicans. *Farmaco* **2004**, *59*, 163–167. [CrossRef] [PubMed]

30. Hiorth, M.; Nilsen, S.; Tho, I. Bioadhesive mini-tablets for vaginal drug delivery. *Pharmaceutics* **2014**, *6*, 494–511. [CrossRef] [PubMed]

31. Baloğlu, E.; Ozyazici, M.; Yaprak Hizarcioğlu, S.; Senyiğit, T.; Ozyurt, D.; Pekçetin, C. Bioadhesive controlled release systems of ornidazole for vaginal delivery. *Pharm. Dev. Technol.* **2006**, *11*, 477–484. [CrossRef] [PubMed]

32. Ahmad, F.J.; Alam, M.A.; Khan, Z.I.; Khar, R.K.; Ali, M. Development and in vitro evaluation of an acid buffering bioadhesive vaginal gel for mixed vaginal infections. *Acta Pharm.* **2008**, *58*, 407–419. [CrossRef] [PubMed]

33. Eudragit®RS PO, Evonik Industries. Available online: http://eudragit.evonik.com/product/eudragit/en/products-services/eudragit-products/sustained-release-formulations/rs-po/pages/default.aspx (accessed on 9 September 2015).

34. Gupta, N.V.; Natasha, S.; Getyala, A.; Bhat, R.S. Bioadhesive vaginal tablets containing spray dried microspheres loaded with clotrimazole for treatment of vaginal candidiasis. *Acta Pharm.* **2013**, *63*, 359–372. [CrossRef] [PubMed]

35. Parodi, B.; Russo, E.; Caviglioli, G.; Baldassari, S.; Gaglianone, N.; Schito, A.M.; Cafaggi, S. A chitosan lactate/poloxamer 407-based matrix containing Eudragit RS microparticles for vaginal delivery of econazole: Design and in vitro evaluation. *Drug Dev. Ind. Pharm.* **2013**, *39*, 1911–1920. [CrossRef] [PubMed]

36. Maderuelo, C.; Zarzuelo, A.; Lanao, J.M. Critical factors in the release of drugs from sustained release hydrophilic matrices. *J. Control. Release* **2011**, *154*, 2–19. [CrossRef] [PubMed]

37. Chandrika, K.P.; Singh, A.; Rathore, A.; Kumar, A. Novel cross linked guar gum-g-poly(acrylate) porous superabsorbent hydrogels: Characterization and swelling behaviour in different environments. *Carbohydr. Polym.* **2016**, *149*, 175–185. [CrossRef] [PubMed]

38. Chen, Y.C.; Ho, H.O.; Chiu, C.C.; Sheu, M.T. Development and characterization of a gastroretentive dosage form composed of chitosan and hydroxyethyl cellulose for alendronate. *Drug Des. Dev. Ther.* **2013**, *8*, 67–78.

39. Palmeira-de-Oliveira, R.; Duarte, P.; Palmeira-de-Oliveira, A.; das Neves, J.; Amaral, M.H.; Breitenfeld, L.; Martinez-de-Oliveira, J. Women's experiences, preferences and perceptions regarding vaginal products: Results from a cross-sectional web-based survey in Portugal. *Eur. J. Contracept. Reprod. Health Care* **2015**, *20*, 259–271. [CrossRef] [PubMed]

40. Siepmann, J.; Siepmann, F. Mathematical modeling of drug dissolution. *Int. J. Pharm.* **2013**, *453*, 12–24. [CrossRef] [PubMed]

41. Zakaria, A.S.; Afifi, S.A.; Elkhodairy, K.A. Newly Developed Topical Cefotaxime Sodium Hydrogels: Antibacterial Activity and In Vivo Evaluation. *Biomed. Res. Int.* **2016**, *2016*, 6525163. [CrossRef] [PubMed]

42. Musumeci, G.; Bon, I.; Lembo, D.; Cagno, V.; Re, M.C.; Signoretto, C.; Diani, E.; Lopalco, L.; Pastori, C.; Martin, L.; et al. M48U1 and Tenofovir combination synergistically inhibits HIV infection in activated PBMCs and human cervicovaginal histocultures. *Sci. Rep.* **2017**, *7*, 41018. [CrossRef] [PubMed]

43. Dash, S.; Murthy, P.N.; Nath, L.; Chowdhury, P. Kinetic modeling on drug release from controlled drug delivery systems. *Acta Pol. Pharm.* **2010**, *67*, 217–223. [PubMed]

44. Mamani, P.L.; Ruiz-Caro, R.; Veiga, M.D. Matrix tablets: The effect of hydroxypropyl methylcellulose/anhydrous dibasic calcium phosphate ratio on the release rate of a water-soluble drug through the gastrointestinal tract I. In vitro tests. *AAPS PharmSciTech* **2012**, *13*, 1073–1083. [CrossRef] [PubMed]

45. Costa, P.; Sousa Lobo, J.M. Modeling and comparison of dissolution profiles. *Eur. J. Pharm. Sci.* **2001**, *13*, 123–133. [CrossRef]

46. Bruschi, M.L. *Strategies to Modify the Drug Release from Pharmaceutical Systems*; Elsevier: Amsterdam, The Netherlands, 2015.

47. Hazzah, H.A.; Farid, R.M.; Nasra, M.M.; El-Massik, M.A.; Abdallah, O.Y. Lyophilized sponges loaded with curcumin solid lipid nanoparticles for buccal delivery: Development and characterization. *Int. J. Pharm.* **2015**, *492*, 248–257. [CrossRef] [PubMed]

48. Kulinowski, P.; Dorożyński, P.; Młynarczyk, A.; Węglarz, W.P. Magnetic resonance imaging and image analysis for assessment of HPMC matrix tablets structural evolution in USP Apparatus 4. *Pharm. Res.* **2011**, *28*, 1065–1073. [CrossRef] [PubMed]

49. Agarwal, S.; Murthy, R.S. Effect of Different Polymer Concentration on Drug Release Rate and Physicochemical Properties of Mucoadhesive Gastroretentive Tablets. *Indian J. Pharm. Sci.* **2015**, *77*, 705–714. [CrossRef] [PubMed]

50. Odeniyi, M.A.; Khan, N.H.; Peh, K.K. Release and mucoadhesion properties of diclofenac matrix tablets from natural and synthetic polymer blends. *Acta Pol. Pharm.* **2015**, *72*, 559–567. [PubMed]

51. Tasdighi, E.; Jafari Azar, Z.; Mortazavi, S.A. Development and In Vitro Evaluation of a Contraceptive Vagino-Adhesive Propranolol Hydrochloride Gel. *Iran. J. Pharm. Res.* **2012**, *11*, 13–26. [PubMed]

52. Morales, J.O.; Su, R.; McConville, J.T. The influence of recrystallized caffeine on water-swellable polymethacrylate mucoadhesive buccal films. *AAPS PharmSciTech* **2013**, *14*, 475–484. [CrossRef] [PubMed]

53. Ruiz-Caro, R.; Veiga-Ochoa, M.D. Characterization and dissolution study of chitosan freeze-dried systems for drug controlled release. *Molecules* **2009**, *14*, 4370–4386. [CrossRef] [PubMed]

54. Owen, D.H.; Katz, D.F. A vaginal fluid simulant. *Contraception* **1999**, *59*, 91–95. [CrossRef]

55. Harada, S.; Koyanagi, Y.; Yamamoto, N. Infection of HTLV-III/LAV in HTLV-I-carrying cells MT-2 and MT-4 and application in a plaque assay. *Science* **1985**, *229*, 563–566. [CrossRef] [PubMed]

56. Krug, H.F. *Quality Handbook: Standard Procedures for Nanoparticle Testing*; Nanommune Deliverable XYZ (EMPA): Zurich, Switzerland, 2011; pp. 130–137.

marine drugs

MDPI

Review

Multifaceted Applications of Chitosan in Cancer Drug Delivery and Therapy

Anish Babu [1,2] and Rajagopal Ramesh [1,2,3,]*

[1] Department of Pathology, University of Oklahoma Health Sciences Center, Oklahoma City, OK 73104, USA;
anish-babu@ouhsc.edu
[2] Stephenson Cancer Center, University of Oklahoma Health Sciences Center, Oklahoma City, OK 73104, USA
[3] Graduate Program in Biomedical Sciences, University of Oklahoma Health Sciences Center,
Oklahoma City, OK 73104, USA
* Correspondence: rajagopal-ramesh@ouhsc.edu; Tel.: +1-(405)-271-6101

Academic Editor: Hitoshi Sashiwa
Received: 27 January 2017; Accepted: 20 March 2017; Published: 27 March 2017

Abstract: Chitosan is a versatile polysaccharide of biological origin. Due to the biocompatible and biodegradable nature of chitosan, it is intensively utilized in biomedical applications in scaffold engineering as an absorption enhancer, and for bioactive and controlled drug release. In cancer therapy, chitosan has multifaceted applications, such as assisting in gene delivery and chemotherapeutic delivery, and as an immunoadjuvant for vaccines. The present review highlights the recent applications of chitosan and chitosan derivatives in cancer therapy.

Keywords: chitosan; gene delivery; drug delivery; adjuvant; cancer; nanoparticle

---

1. Introduction

Marine products have been in the forefront of natural materials used in therapeutic applications against human diseases [1]. The marine biopolymer chitin, which is isolated from crustaceans and is the second most abundant polymer in nature, has recently received increased attention for healthcare applications [2]. Chitin transforms into chitosan by partial deacetylation under strong alkaline conditions. Chitosan is composed of $(1 \rightarrow 4)$-2-acetamido-2-deoxy-β-D-glucan (*N*-acetyl D-glucosamine) and $(1 \rightarrow 4)$-2-amino-2-deoxy-β-D-glucan (D-glucosamine) units, and has long been used in health care materials, such as nasal absorption enhancers of peptide drugs [3], and in 3D or 2D scaffold preparations for wound healing [4]. In addition, chitosan has been widely used as a promising non-viral delivery vector for biomacromolecules and low molecular weight drugs [5]. Figure 1 shows the major applications of chitosan in healthcare and cancer therapy. Chitosan exhibits high biocompatibility and biodegradability, attractive properties for the development of a safe and active drug delivery tool [5,6]. Chitosan is cationic in nature and its solubility in water is poor but it is soluble in low pH solutions. Different derivatives of chitosan have been made to overcome this limitation for controlled drug delivery purposes [6,7]. The cationic charge of chitosan has been utilized for ionic gelation methods using materials with strong anionic charge for nanoparticle preparation [8]. Additionally, this cationic nature has been harnessed for electrostatic interaction with nucleic acids, and chitosan has been used as a gene delivery carrier for cancer therapy. Another important application of chitosan as a potential immune-adjuvant for cancer vaccines has been realized recently [9]. The present review summarizes some of the significant recent developments in which chitosan is used as nanoparticle carrier for gene therapeutics, chemotherapeutic drugs, and in immune-adjuvant therapy for cancer.

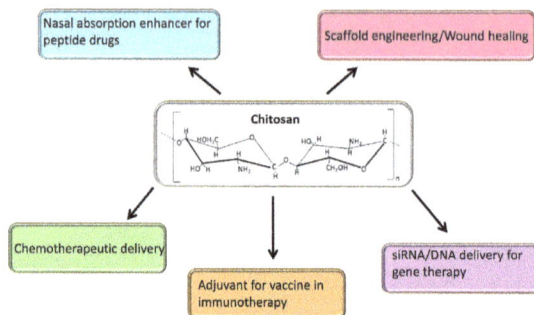

Figure 1. Diagram showing the various applications of chitosan in healthcare and cancer therapy. Abbreviations: siRNA (small interfering siRNA).

2. Chitosan as Gene Delivery Vehicle for Cancer Therapy

In gene therapy, a gene of interest that has been implicated in cancer pathology is altered or manipulated by delivering exogenous nucleic acid material into the tumor cells or the milieu [10]. However, when nucleic acid therapeutics are systemically administered, they encounter many hurdles across their circulation to reach the target tissue in the human body that might reduce their therapeutic potential. These hurdles include short plasma half-life due to enzymatic degradation and rapid bio-clearance of nucleic acid therapeutics from the circulation. In vivo circulating nucleic acid therapeutics also face cellular entry limitations, such as charge-based repulsion from cell membranes and poor endosomal escape. To overcome these barriers, delivery vehicles are required for nucleic acids [11,12].

Viral or non-viral vectors are major systems used for gene delivery applications. Viral vectors are excellent transfection agents, however mutagen and carcinogen properties of many viral vectors limits their use in cancer gene therapy [13]. As an alternative to viral-vector nanotechnology, non-viral vectors have made remarkable advances in recent years [14]. Non-viral vectors for gene delivery include liposomes, polymer-based carriers and nanoparticles of various kinds. Among them, liposome is a highly-investigated gene carrier because of its high transfection efficiency and ease of preparation. However, poor encapsulation efficiency, short shelf-life, non-specific toxicity and low in vivo stability are major limitations of liposomes [15–17]. Though PEGylation (PEG: poly-ethylene glycol) appears to improve the circulation time of liposomes, the accelerated blood clearance (ABC) phenomenon resulting from repeated liposome administration enhanced its bio-clearance from the body [18]. As alternative gene carriers, cationic polymers were extensively used for gene delivery due to their improved transfection efficiency, high gene encapsulation and in vivo stability [19]. The presence of numerous free amine groups in cationic polymers such as polyethylene imine (PEI), chitosan (CS), poly-l-lysine (PLL) and polyamidoamine (PAMAM) effectively condense oligonucleotides or DNA. The high cationic charge density in these polymers allow for enhanced intracellular trafficking of nanoparticles via endosomal disruption, however this imparts undesired cellular toxicity [20]. Interestingly, chitosan is an exception in that it exhibits no apparent toxicity as gene delivery vehicle. Chitosan has excellent physicochemical properties that appear to be favorable for nucleic acid delivery by overcoming the systemic barriers of gene delivery. Chitosan readily forms complexes, microspheres or nanoparticles upon electrostatic interaction with nuclei acids [21]. Because of these promising characteristics, chitosan has been increasingly studied as a gene delivery system in cancer therapy.

2.1. Influencing Factors in Chitosan-Based Gene Delivery

The charge-to-charge ratio between chitosan and DNA is a critical factor for successful electrostatic binding of DNA or siRNA to chitosan [21]. The nitrogen-to phosphate (N/P) ratio, i.e., the ratio between the positively charged nitrogen of free amines in chitosan to negatively charged phosphates in nucleic acids potentially influences the efficiency of the chitosan polymer to condense and protect the DNA or siRNA. Reports suggests that a low N/P ratio would affect the stability of chitosan-DNA complexes, whereas extremely high N/P ratios results in low transfection efficiency [22,23].

Molecular weight and de-acetylation degrees are other important factors that determine the stability of chitosan-nucleic acid complexes. The molecular weight of chitosan has a broad range from low molecular weight (LMW, <100 KDa) to medium molecular weight (MMW, <300 KDa) and high molecular weight (HMW, >300 KDa). The ability of chitosan to transfect functionally active siRNA into cells is strongly dependent on the molecular weight of chitosan, among other factors [24]. Huang et al. (2005) reported that a 213-KDa chitosan formulation showed superior uptake and transfection efficiency with plasmid DNA (pDNA) in cancer cells compared with a LMW chitosan formulation [25]. In their report, the LMW chitosan was less capable of condensing and protecting DNA, and could not retain DNA upon dilution [25]. However, there are reports of superior cell uptake and gene delivery using LMW chitosan in cancer cells compared to higher molecular weight chitosan used in the formulations [23,26]. The enhanced cell uptake and transfection efficiency can be attributed to relatively low binding of LMW chitosan to DNA, allowing for easy dissociation compared to higher molecular weight chitosan. Reports also suggest that LMW chitosan formed small nanoparticles and showed significant transfection efficiency for DNA or siRNA polyplexes, when chitosan nanoparticles were modified with targeting ligands [27,28] or with other polymers [29]. When complexed with 25 and 50 kDa of chitosan, siRNA showed <220-nm size and good gene silencing effect in HeLa cells [30]. When LMW chitosan conjugated with another cationic polymer protamine was used for gene delivery at physiological pH, the transfection efficiency and gene expression in host cells were significantly improved [29]. This complex reportedly had low toxicity both in vitro and in vivo. These studies indicate that molecular weight has a great influence in chitosan's biological and physicochemical properties. However, literature lacks a unanimous opinion about appropriate chitosan molecular weight to be chosen for the best possible transfection efficiency. Nevertheless, the above studies suggest that the chitosan used for gene transfection should have intermediate degree of stability and appropriate molecular weight, which appears to be better achieved with LMW chitosan, that provides a balance between DNA protection ability and intracellular release [21,31].

Along with molecular weight, the degree of deacetylation in chitosan is important in determining its efficiency for stable complex formation with DNA or siRNA. Most of the chitosan used in gene delivery applications has a high degree of deacetylation (DDA), since high DDA corresponds to more free amines and increased positive charge for efficient DNA binding. Moreover, studies suggest that the binding efficiency of nucleic acids decreased with decreasing DDA in chitosan, resulting in incomplete complex formation [23]. This is because chitosan with low DDA has fewer primary amine groups that are freely available for electrostatic interaction with negatively charged nucleic acids. Based on reports, it is recommended that DDA % should be above 65% in chitosan for efficient complex formation with DNA [32]. The particle size also has been shown to decrease upon increased DDA % of chitosan when complexed with DNA [25].

The pH of the transfection medium must also be considered, since the protonation of amine groups in chitosan requires an acidic pH range (5.0–6.0). This low acidic pH increases the DNA binding efficiency of chitosan and thereby enhances transfection efficiency [27]. The presence of serum is an important factor that determines the stability of a cationic gene delivery system [27]. Interestingly, compared with other cationic polymers, chitosan-based DNA transfection is improved in the presence of serum [33]. Sato et al. (2001) studied the transfection efficiency of pDNA/chitosan complexes in the presence of serum (0%–50%). Their data showed that the 20% serum conditions resulted in highest transfection efficiency, whereas 50% serum in the medium produced the lowest transfection efficiency [26].

The difference in size and charge density of DNA and siRNA influence the complex formation with chitosan of same length. Due to the bigger size and good charge density of plasmid DNA, it has the ability for strong electrostatic binding compared to small-sized siRNA (19–25 bp) [21]. Taken together, to develop a successful gene delivery system using chitosan, all of the above-mentioned parameters should be considered carefully. Apart from the physicochemical characteristics of chitosan-gene complexes, the cell type may influence the transfection efficiency. Therefore, when handling a difficult-to-transfect cell-line, the chitosan-gene complexes should be tailor-made. This includes the use of chitosan derivatives and the addition of other polymers that favor the best possible transfection efficiency in the specific cell line.

2.2. Formulation Methods

There are two commonly used formulation methods for chitosan-based gene delivery systems: (a) simple complexation and (b) ionic gelation (Figure 2). Simple complexation between the chitosan polymer and siRNA or DNA involves electrostatic interactions between cationic chitosan and anionic nucleic acids [34–36]. At proper N/P ratios, chitosan forms complexes in micro- or nano-sized particles. Formation of smaller particles requires optimization of molecular weight, DDA, pH, and sometimes external force, like mild stirring, for proper condensation and complex formation. Ionic gelation is a common method to prepare crosslinked nanoparticles. In ionic gelation, the siRNA or DNA is entrapped, rather than fully depending on electrostatic interactions [8,37]. Crosslinkers that are oppositely and strongly negatively charged to chitosans, such as tripolyphosphate, thiamine pyrophosphate, and hyaluronic acid, are used in the ionic gelation process. While crosslinking enhances the stability of the nanoparticles, it slows down the release of entrapped nucleic acids. Nevertheless, this crosslinking strategy may be useful in gene delivery that requires slow and sustained release of nucleic acids over time.

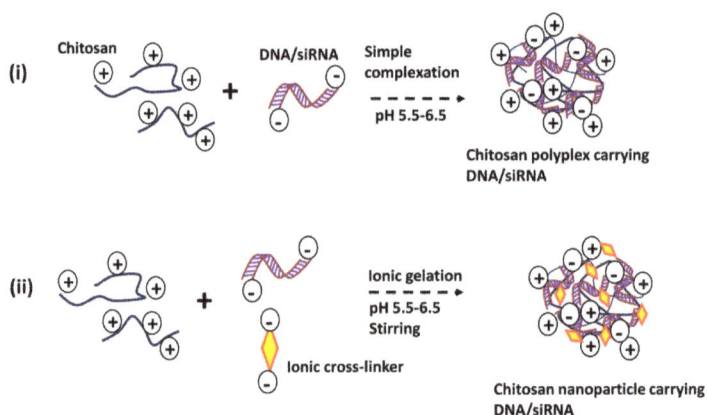

Figure 2. Common preparation methods of chitosan nanocarrier for DNA/siRNA delivery. (**a**) simple complexation; (**b**) ionic gelation.

2.3. Chitosan Derivatives in Gene Delivery

Poor water solubility in physiological pH is a limitation of chitosan in gene delivery applications. Since chitosan requires the protonation of its free amines for effective complexation with siRNA or DNA, which is possible only at acidic pH, transfection in physiological pH may result in early dissociation of siRNA into the medium without achieving effective cellular transfection [38]. Another issue is the slow release of nucleic acid materials from chitosan, possibly affecting transfection efficiency. Therefore, chitosan derivatives were synthesized by chemical modifications of chitosan structure or by grafting polymers with distinct properties to overcome water insolubility and poor gene delivery efficiency. Quaternization is a commonly used method to modify chitosan by alkylation of tertiary amines by different methods for improved gene transfection [39,40]. A recent example used a quaternary ammonium salt crystal called *N*-2-hydroxypropyl trimethyl ammonium chloride chitosan (HACC) for gene delivery in human cells [41]. The HACC particles were not cytotoxic, and HACC/pDNA complexes showed comparable transfection efficiency to liposome/pDNA complexes, indicative of their potential as a novel tool for gene delivery. Chitosan hydroxybenzotriazole (chitosan-HOBT) is another derivative of chitosan known for its safe and efficient siRNA delivery capacity [42]. Chitosan-HOBT could condense siRNA, formed stable complexes, and exhibited good gene silencing efficiency. However, its full gene therapeutic potential in cancer cells is yet to be realized. Another study used dendronized chitosan derivative prepared by modification of 6-azido-6-deoxy-chitosan with propargyl focal point poly(amidoamine) dendron [43]. Compared with PEI non-viral vector, these novel dendronized chitosan/DNA complexes showed enhanced gene transfection efficiency in human kidney and nasopharyngeal carcinoma cells. In a different derivatization approach, hybrid-type chitosan (MixNCH) was synthesized using 2-chloroethylamine hydrochloride and N, *N*-dimethyl-2-chloroethylamine hydrochloride, for gene delivery to cancer cells [44]. MixNCH nanoparticles showed good physicochemical characteristics for gene delivery, transfected HepG2 cancer cells, and effectively inhibited cell proliferation.

Trimethyl chitosan (TMC) is one of the intensively studied quarternized derivatives of chitosan in gene delivery applications [45–49]. One of the important advantages of trimethylation is that chitosan's solubility can be increased in physiological pH. Compared to chitosan polyplexes, trimethyl chitosan nanoparticles strongly reduces the aggregation tendency and pH dependency of nucleic acid complexation [45]. Studies in NIH/3T3 (mouse embryonic fibroblasts) cells showed a huge increase in transfection efficiency of pDNA using TMC nanoparticles compared to chitosan polyplexes [45]. Moreover, the same study showed that PEG grafting onto TMC enhanced the particle stability, decreased particle size in physiological pH and reduced the toxicity showed by unmodified TMC. Finally, PEG-TMC nanoparticles enhanced the transfection efficiency of pDNA by 10-fold compared to unmodified TMC.

Conjugating targeting moieties to TMC enhanced the gene delivery efficiency according to a report by Zheng et al. [46]. The TMC nanoparticles efficiently condensed pDNA and the presence of folate on its surface allowed its target-specific delivery of pDNA in SKOV3 (human ovarian adenocarcinoma) and KB (HeLa contaminant, carcinoma) cells which overexpresses folate receptor. The drug carrying ability of TMC nanoparticles has also been harnessed in drug gene co-delivery towards cancer cells. In a recent study, the triple negative MDA-MB-231 (human breast adenocarcinoma) cell line was successfully transfected by high mobility group protein 2 (HMGA-2) siRNA with simultaneous delivery of chemotherapeutic doxorubicin using a TMC nanoparticle system [49]. The anti-cancer effect of doxorubicin has been enhanced by conjunctional delivery of siRNA that silenced HMGA-2 gene expression. All these studies point towards the importance of various factors such as particle size, stability, toxicity, targeting ability and the modifications required for TMC-based gene delivery systems to achieve successful gene transfection.

2.4. PEG Modification of Chitosan in Gene Delivery

To make chitosan more water soluble and enhance its blood-circulation time, conjugation of poly-ethylene glycol (PEG) polymer with chitosan is a common approach [50]. The PEGylated nanoparticle is generally known as "stealth" nanoparticle. PEG is a neutral polymer that increases the hydrophilicity of chitosan and delays the reticulo-endothelial system clearance while in the circulation. This improves the chances of the chitosan-gene delivery system to passively accumulate in tumor areas by enhanced permeation and retention effect (EPR) in a time-dependent manner. However, the EPR effect applies to only those nanoparticles with particle sizes less than 200 nm in most cases. Moreover, PEGylation may reduce the charge-based affinity of cationic chitosan towards net negatively charged cell membranes and affect the cellular delivery of gene therapeutics. This issue has been addressed by attaching targeting ligands or stimuli responsive polymers to nanoparticles for receptor target delivery of nucleic acids to tumor cells. While PEG improves the circulation half-life of nanoparticles, conjugation of targeting ligands enhances cell-specific delivery of gene therapeutics [30] (Figure 3). PEG also serves as the linker molecule for nanoparticle modification with targeting ligands. Chan et al. (2007) developed a chitosan gene delivery system with PEG-folate modification for targeted delivery to folic acid receptor-overexpressing tumor cells [51]. This chitosan nanoparticle system carrying DNA not only improved the water solubility upon PEG addition, but also showed low cytotoxicity towards normal HEK 293 (Human embryonic kidney cells 293) cells. A recent study demonstrated the use of transferrin (Tf)-functionalized chitosan nanoparticles, where PEG was used to conjugate Tf onto chitosan [52]. Thus, PEG modification is an important step in designing water-soluble, long-circulating, and target-specific nanoparticles.

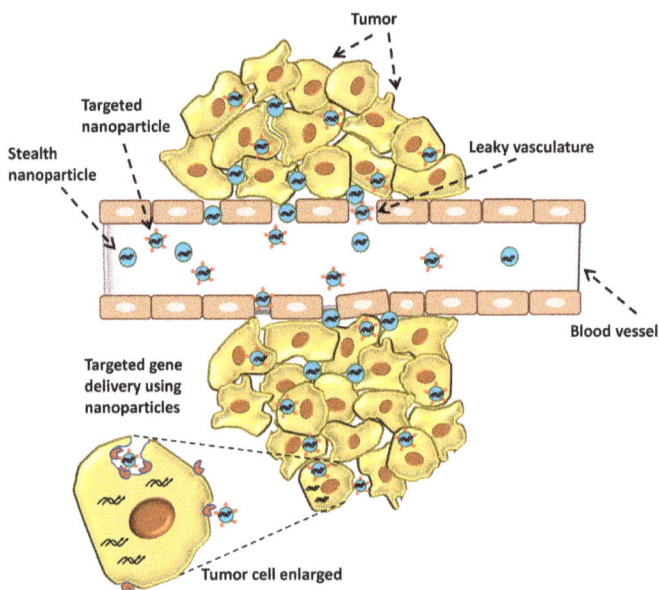

Figure 3. Gene delivery to tumors using PEGylated (stealth) nanoparticles or by using receptor targeted nanoparticles. Nanoparticle (stealth or targeted) enter tumor area via leaky vasculature, while targeted nanoparticles specifically enter tumor cells via receptor mediated pathway (see enlarged portion of the figure). Gene therapeutics are then released into the cytoplasm escaping from the endo-lysosomes. PEG: poly-ethylene glycol.

Altogether, chitosan is a promising gene delivery system for in vitro and in vivo applications, however requires several formulation parameters to be optimized. Structure modification or incorporation of other polymers is an effective way to enhance the potential of chitosan by improving the in vivo stability, target specificity and desirable intracellular release of gene therapeutics. Some recent examples of chitosan-based gene delivery an application are described in the Table 1.

Table 1. Recent examples of gene delivery systems based on chitosan for cancer therapy. PEG: poly-ethylene glycol.

Chitosan or Chitosan-Associated Nanoparticles	Gene Material/ Molecular Target	Cancer/Cell Type	Special Features of the Study/Formulation	Reference
Low molecular weight (LMW) chitosan/2-acrylamido-2-methylpropane sulphonic acid	Model pDNA/Luc (plasmid DNA/Luciferase)	A549 (lung adenocarcinoma), HeLa (cervical carcinoma) and HepG2 (hepatocellular carcinoma)	• Incorporation of 2-acrylamido-2-methylpropane sulphonic acid made chitosan water soluble. • Higher transfection efficiency in cancer cells and mouse model.	[7]
Alginic acid-coated chitosan nanoparticles	Legumain pDNA	Murine 4T1 (mouse mammary tumor cell line)	• Used as oral delivery system for DNA vaccine. • Legumain pDNA delivery improved autoimmune response to breast cancer in mice.	[53]
Glycol-chitosan nanoparticles	MDR1 (Multi drug resistant 1)-siRNA	MCF-7 (Human breast adenocarcinoma; Adriamycin resistant, ADR)	• Nanoparticles accumulated in MCF7/ADR tumors and downregulated P-gp expression. • Chemo-siRNA combination therapy significantly inhibited tumor growth without systemic toxicity in mice.	[54]
Polyethylene glycol-chitosan	Survivin-siRNA	Murine 4T1 (mouse mammary tumor cell line)	• The PEG–Chitosan nanoparticles carrying siRNA were efficiently taken up by cancer cells and induced antitumor activity in xenografts.	[55]
Biotinylated chitosan-graft-polyethyleneimine	antiEGFR (Epidermal growth factor receptor)-siRNA	Hela (cervical carcinoma), OVCAR-3 (Human ovarian adenocarcinoma)	• The biotinylated chitosan-graft-polyethyleneimine was less cytotoxic than polyethyleneimine. • Efficient cell uptake and epidermal growth factor siRNA delivery was possible in cancer cells.	[56]
Folate-targeted chitosan polymeric nanoparticles	METHFR (Methylenetetrahydrofolate Reductase) shRNA (coloaded with 5-FU)	SGC-7901 (Human gastric carcinoma)	• Folate-targeted chitosan polymeric nanoparticles (CPNs) could reverse drug-resistant SGC-7901 cells by co-delivery of METHFR shRNA and 5-fluorouracil (5-FU). • Folate-targeted CPN system showed significantly enhanced therapeutic efficacy compared to non-targeted CPN.	[57]
Polyethyleneimine/poly(allylamine)-citraconic anhydride/ gold nanoparticle (PEI/PAH-Cit/AuNP)- chitosan nanoparticle	MDR1 (Multi drug resistant 1) siRNA	MCF-7 (Human breast adenocarcinoma; drug-resistant)	• Gold nanoparticle reduced and stabilized by chitosan was coated by charge-reversible polymer PAH-cit and PEI by layer-by-layer deposition. • This charge-reversible core/shell nanosystem were effective in protecting, cell uptake and endosomal escape of siRNA; facilitated safe siRNA delivery and gene silencing in cancer cells.	[58]
Chitosan	Plasmid IL-12 (Interleukin-12)	WEHI-164 (Human fibrosarcoma)	• Chitosan formed polyplex with IL-12 plasmid. • Treatment with IL-12 resulted in significant tumor regression in mouse fibrosarcoma model.	[59]
Chitosan/Polylactic-acid nanoparticle	Plasmid Beta-5/siP62 (P62 or Sequestosome 1 siRNA)	2008S, 2008/C13 (Human ovarian carcinoma; drug-resistant)	• Chitosan-coated polylactic acid nanoparticles were co-loaded with siRNA/pDNA and chemotherapeutic. • Drug resistant ovarian cancer cells were sensitized to cisplatin by simultaneous delivery P62 siRNA, Proteasome beta-5 plasmid and cisplatin.	[60]

3. Chitosan Nanoparticles in Chemotherapeutic Delivery

Nano-drug delivery systems using chitosan offer many advantages. These systems minimize drug clearance in the circulation, control release of drug, reduce drug cytotoxicity, and increase therapeutic index. Moreover, the biodegradability and biocompatibility have made chitosan a suitable material for chemo-drug delivery in cancer therapy. Chitosan is mucoadhesive, and its cationic nature allows for enhanced affinity towards mucous membrane, thereby assisting trans-mucosal drug delivery. These properties of chitosan would be useful in intra-nasal and intrapulmonary delivery of chemotherapeutics for cancers especially of the nasopharyngeal and lung tissues.

3.1. Delivery of Hydrophilic Chemotherapeutics

Chitosan nanoparticles can be used to deliver both hydrophilic drugs [61,62], and hydrophobic drugs [63,64]. The presence of many free amine groups can be easily functionalized for conjugation of chemotherapeutic drugs. For example, in a recent study, water-soluble drug doxorubicin (DOX) was conjugated to chitosan using a succinic anhydride spacer [62]. The succinic anhydride could react with the amine of DOX and functionalize to become carboxylic. This carboxylic acid of DOX was then conjugated with chitosan's free amine groups using carbodiimide chemistry. The chitosan-DOX was then self-assembled to form nanoparticles in aqueous solution under stirring at room temperature. However, the introduction of more DOX reduced the conjugation efficiency to chitosan. The Her2+ (human epidermal growth factor receptor 2+) targeting monoclonal antibody, trastuzumab was also conjugated to chitosan-DOX nanoparticles via thiolation of lysine residues (by reacting with primary amines) and subsequent linking of the resulted thiols to chitosan. The trastuzumab conjugated chitosan-DOX nanoparticles showed target specificity towards Her2+ cancer cells, resulting in enhanced uptake compared to chitosan-DOX and free drug. Also, trastuzumab conjugated chitosan-DOX nanoparticles could efficiently discriminate between Her2+ and Her2– cells, demonstrating its potential for active targeted drug delivery.

In another strategy, a chitosan-pluronic micelle was designed and fabricated for the encapsulation of water-soluble DOX [65]. They grafted Pluronic® F127 polymer into chitosan and fabricated a co-polymer micelle that can encapsulate DOX with high drug loading capacity with a particle size of 50 nm. The chitosan-pluronic micelle carrying DOX (DOX-NP) showed better in vitro therapeutic activity than free DOX in MCF7 breast cancer cell lines.

3.2. Delivery of Hydrophobic Chemotherapeutics

For the delivery of poorly water-soluble drugs, chitosan derivatives have been synthesized with suitable characteristics that can support hydrophobic drugs. Paclitaxel, a hydrophobic chemotherapeutic, showed enhanced activity when encapsulated in a glyceryl monooleate-chitosan core-shell nanoparticle prepared using an emulsification-evaporation technique [66]. Strikingly, a 1000-fold reduction in paclitaxel IC_{50} (Inhibitory Concentration 50) was observed with this core-shell nanosystem in MDA-MB-231 human breast cancer cells. This huge reduction in IC_{50} value would reduce the cytotoxicity of paclitaxel towards normal cells. In a different study, Kim et al. (2006) introduced an amphiphilic derivative of chitosan for paclitaxel delivery [63]. They combined glycol chitosan and 5β-cholanic acid to produce nanoparticles (Glycol chitosan hydrophobically modified with 5beta-cholanic acid or HGC nanoparticles). The drug loading achieved for paclitaxel was 80% in HGC nanoparticles. The cytotoxicity of HGC nanoparticles were negligible compared to conventional Cremophor EL formulation used for paclitaxel administration. Further, when administered in mice tumor model, the tumor regression ability of paclitaxel delivered using HGC nanoparticles was comparable to Cremophor EL at 20 mg/kg dose, whereas a higher concentration of paclitaxel (50 mg/kg) in HGC nanoparticles caused complete regression of tumors in four out of six treated mice. Their study clearly indicated a superior anticancer effect of HGC nanoparticle formulation paclitaxel compared to Cremophor EL formulation. Later, the same group studied cisplatin (CDDP) loaded-HGC nanoparticles for their physicochemical properties intended for anti-cancer therapy [67]. CDDP, a low water soluble drug (up to 1 mg/mL) was encapsulated in hydrophobic cores of HGC nanoparticles and showed sustained drug release. In vivo delivery of CDDP-HGC nanoparticles accumulated in solid tumors in a mouse model via the EPR effect. Finally, they showed promising antitumor efficiency of CDDP-HGC nanoparticles in tumor-bearing mice.

Chitosan-copolymer nanoparticles are also used to encapsulate hydrophobic anti-cancer drug 5-flurouracil (5-FU), as reported by Rajan et al. [68]. They prepared a hyaluronidase-5-fluoruracil (5-FU)-loaded chitosan-PEG-gelatin polymer nanocomposite using the ionic gelation technique. A short-time incubation (3–12 h) of hyaluronidase-5-fluoruracil (5-FU)-loaded chitosan formulations showed less toxicity than chemotherapeutic 5-FU. Hyaluronic acid conjugation with biopolymers imparted targeting capability for the drug delivery vehicle towards cancer cells. The physicochemical characteristics such as particle size, homogenous distribution, morphology, drug loading capacity and low toxicity of these chitosan-based nanocomposite formulations are promising for the drug delivery system in anti-cancer studies.

Recently, Cavalli et al. (2014) formulated chitosan nanospheres with 5-fluorouracil using a combination of coacervation and emulsion droplet coalescence methods [69]. The resulting 5-FU-loaded chitosan nanospheres were not only able to reduce the proliferation of HT29 (Human colorectal adenocarcinoma) and PC-3 (Human prostate cancer-3) tumor cell lines in a time- and concentration-dependent manner but also inhibited their adhesion to human umbilical vein endothelial cells (HUVEC). These examples suggest that chitosan-based nanoparticles have the potential to deliver a wide range of drugs with different physicochemical properties. Table 2 shows some recent examples of chitosan or chitosan-based nanoparticles in chemotherapeutic drugs of hydrophilic, hydrophobic or amphiphilic properties for cancer therapy.

Table 2. Examples of chemotherapeutic delivery using chitosan or chitosan based nanoparticles.

Solubility Property	Chemotherapeutic	Nanoparticle	Special Features/Application	Cancer Model/Cell Lines	Reference
Hydrophilic	Doxorubicin	Chitosan diacetate and chitosan triacetate nanoparticles	• Sustained release of anticancer drugs • Increased oral bioavailability of doxorubicin in animal model	MCF-7 and Caco-II tumor cell lines	[70]
		Cholesterol-modified glycol chitosan (CHGC) self-aggregated nanoparticles	• High drug loading (9.36%) and enhanced drug release in low pH range • Prolonged circulation in plasma	S180 murine cancer	[71]
		Self-assembled chitosan-doxorubin conjugate (CS-DOX) nanoparticles	• Trastuzumab decoration enhanced the uptake of CS-DOX nanoparticles in Her2+ cancer cells compared with nontargeted CS-DOX nanoparticles	MCF7 (breast cancer) and SKOV3 (ovarian cancer) cell lines	[62]
		CD44 targeted-doxorubicin-encapsulated polymeric nanoparticle surface decorated with chitosan	• Drug release in acidic tumor environment • Nanoparticle delivery Increased cytotoxicity to cancer-stem cells by six times compared to free doxorubicin	3D mammary tumor spheroids	[72]
	Taxanes	Paclitaxel-loaded chitosan nanoparticles	• Nanoparticle exhibited sustained release pattern of paclitaxel • Low hemolytic toxicity observed for nanoparticles compared to free drug • Nanoparticle demonstrated enhanced antitumor activity in vitro compared to naïve drug	MDA-MB-231 breast cancer cell lines	[73]
Hydrophobic		Ionically cross-linked docetaxel loaded chitosan nanoparticles	• Nanoparticles exhibited 78%–92% drug encapsulation efficiency • Nanoparticle delivery enhanced cytotoxicity of docetaxel compared to free drug	MDA-MB-231 breast cancer cell lines	[74]
		Paclitaxel-loaded N-octyl-O-sulfate chitosan micelles	• N-octyl-O-sulfate chitosan inhibited p-glycoprotein overcoming multi-drug resistance • Paclitaxel- N-octyl-O-sulfate chitosan micelles showed superior blood persistence, tumor accumulation, and therapeutic efficacy in tumor bearing mice	Human hepatocellular liver carcinoma (HepG2) cells and the multidrug resistance HepG2 (HepG2-P) cells	[75]

Table 2. *Cont.*

Solubility Property	Chemotherapeutic	Nanoparticle	Special Features/Application	Cancer Model/Cell Lines	Reference
Sparingly-water soluble	Platinum drugs	Folic acid-conjugated chitosan-coated poly(d-l-lactide-co-glycolide) (PLGA) nanoparticles (FPCC)	• Presence of protective chitosan layer controlled the overall release rate of carboplatin • FPCC displayed higher cell uptake and reduced IC_{50} (Inhibitory concentration 50) values of carboplatin compared to non-targeted nanoparticles	Hela cervical cancer cells	[76]
		Cisplatin-loaded cholanic acid-modified glycol chitosan nanoparticles	• Drug loading was 80% • Cisplatin-loaded nanoparticles showed prolonged blood circulation and accumulated in tumor by utilizing enhanced permeation and retention effect (EPR) effect • Nanoparticles delivery showed higher anti-tumor efficacy and lower toxicity compared to free cisplatin	MDA-MB231 human breast tumor	[67]
		Cisplatin loaded-chitosan-nanolayered solid lipid nanoparticles (CChSLN)	• Nanoparticle exhibited excellent biocompatibility • IC_{50} value of cisplatin was lowered by CChSLN delivery • CChSLN enhanced apoptosis in cancer cells compared to free cisplatin	HeLa cervical carcinoma	[77]

3.3. Targeted Delivery of Chemotherapeutics Using Chitosan-Based Nanoparticles

Conjugation of tumor-specific ligands onto chitosan nanoparticles has been developed for active targeting [78]. Many surface receptors specifically overexpressed in cancer cells are exploited for receptor-targeted delivery of chemotherapeutics using chitosan nanoparticles. Specific interaction between targeting ligands in nanoparticles and cell surface receptors results in receptor-mediated endocytosis nanoparticles. In cells, the internalized drug-loaded chitosan nanoparticles escape from endo-lysosomal compartments and accumulate in cytoplasm, where the nanoparticles release the drug payload over time. Transferrin, epidermal growth factor receptor (EGFR), folate receptor, CD44 (known as HCAM or homing cell adhesion molecule) receptor, integrins, and low density lipoprotein receptors are commonly exploited for targeted drug delivery in cancer cells [79]. The expression levels of these receptors in each cancer type varies; therefore, it is important to know the cell type and receptor expression levels before formulating targeted drug delivery systems. When conjugated with drug via pH-cleavable bonds, chitosan nanoparticles undergo dissociation of the assembly within the acidic pH of endo-lysosomes and release the drug into the cytoplasm [80]. Figure 4 shows the diagrammatic representation of stimuli responsive drug delivery of chitosan-based nanoparticles with acid-cleavable bonds conferred by a pH-sensitive linker.

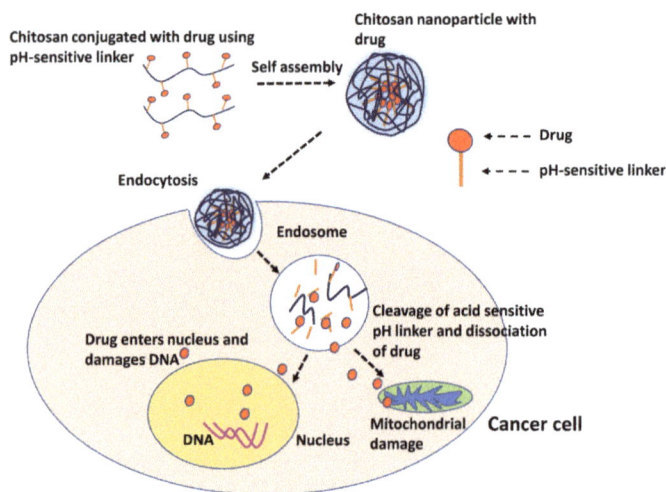

Figure 4. Acid responsive drug delivery using chitosan nanoparticles. Chitosan is linked to drug molecules with a pH-sensitive linker. After endocytotic uptake of nanoparticles, the pH-sensitive linker is dissolved (bond breakage) in the acidic pH of the endosomes, resulting in the release of conjugated drug into the cytoplasm. The drug is then transported to the nucleus or mitochondria and causes DNA damage and apoptosis.

4. Chitosan in Cancer Immunotherapy

Vaccines require adjuvants for enhancing the immune response. Aluminum hydroxide, lipopolysaccharide derivative monophosphoryl lipid A, antimicrobial peptide, and TLR9 (Toll like receptor 9) combinations were among the adjuvants commonly used with vaccines. However, due to possible side effects, scientists worldwide are in search of safe and potential adjuvants for vaccine development, especially in cancer therapy.

Polysaccharides from plant, animal, and fungal sources have emerged as possible adjuvants for cancer vaccines [81]. Among these, chitosan has the potential to become an ideal vaccine adjuvant due to its safety, biocompatibility, cationic nature, and its ability to be used as an antigen carrier [82]. For more than two decades, the immunostimulatory activity of chitosan has been known. However, its potential as a safe and non-toxic adjuvant in cancer vaccine development has only recently been realized [9,83]. Recent studies explored the adjuvant properties of chitosan in vaccines against cancer and infectious diseases [9,83–85]. The bioadhesive property of chitosan aids in its cell-uptake, leading to strong systemic and mucosal immune responses.

The striking feature of chitosan is that it can enhance both humoral and cell-mediated immune responses [86]. Chitosan showed comparable potency to incomplete Freund's adjuvant, and showed immune activity superior to that of the traditional immunoadjuvant, aluminum hydroxide (Imject Alum) [87]. Chitosan retains the peptide antigen in the administration site for a longer time, allowing antigen to be presented for efficient immune activity. Zaharoff et al. (2007) reported that more than 60% of antigen is retained in the subcutaneous site of injection, even after 7 days [87]. This strategy may reduce the booster doses of vaccine to be used for enhanced immune response.

The mechanism of immune-adjuvant activity of chitosan has recently been elucidated. Chitosan induces immune activity via the NLRP3 (NLR Family Pyrin Domain Containing 3) inflammasome in phagocytic cells and promotes IL-1β (Interleukin 1β) secretion [88]. It is also reported that chitosan induces mitochondrial DNA-mediated cGAS-STING (Cyclic GMP-AMP synthase-Stimulator of Interferon Genes) pathway activation, resulting in the secretion of IFN (Interferon) type I. IFN type I in turn stimulates the maturation of dendritic cells, resulting in antigen presentation, followed by a Th1 (Type 1 T helper) immune response [89]. Chitosan is also known to elicit a balanced Th1/Th2 immune response [90].

A simplified schematic of chitosan's adjuvant activity when delivering cancer vaccine is depicted in Figure 5. Zaharoff et al. (2010) reported the use of chitosan as adjuvant for IL-12 therapy in colorectal (MC32a) and pancreatic (Panc02) solid tumors in mice [85]. Upon intratumoral injection, chitosan prolonged the retention of IL-12 in the injection site and resulted in tumor regression in more than 80% of mice. The resultant systemic tumor immunity was able to prevent tumor recurrence. As a result of chitosan/IL-12 therapy, CD8$^+$ (Cluster of differentiation 8+) cells and NK (Natural killer) cells were revealed as the predominant immune cells involved in the regression of aggressive murine tumors. The same group also demonstrated the efficacy of chitosan/IL-12 adjuvant therapy in superficial bladder cancer treatment [91].

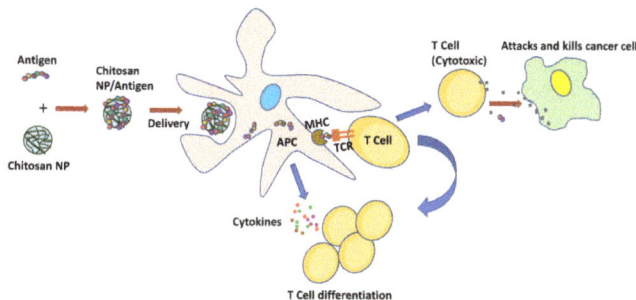

Figure 5. Chitosan nanoparticles act as carriers and enhance the immunostimulatory activity of protein antigen for its presentation by antigen-presenting cells (APC). The stimulated cytotoxic T cells attack and kill cancer cells, whereas cytokines released by APC activate T cell differentiation and expansion. MHC: major histocompatibility complex; TCR: T cell receptor; Chitosan-NP: Chitosan-nanoparticle

In a different study, Heffernan and colleagues (2011) explored the chitosan/IL-12 adjuvant system in stimulating protein vaccine immune responses [92]. Protein-based vaccines have potential for cancer immunotherapy; however, their poor immunostimulatory effect is a limitation. The immunoadjuvant consisted of a viscous chitosan solution and Il-12 cytokine; when injected along with ovalbumin (OVA; model protein antigen), this treatment elicited increased antigen-specific CD4$^+$ and CD8$^+$ T-cell responses. Further, the chitosan/IL-2 adjuvant system enhanced IgG2a and IgG2b (Imunoglobulin G2a and b) antibody responses to OVA. Another study reported that chitosan nanoparticles enhanced the Th1 and Th2 immune responses induced by OVA in mice [90]. Chitosan nanoparticles improved not only Th1 (IL-2 and IFN-γ) and Th2 (Il-10) cytokine levels but also increased the killing activity of NK cells. Therefore, chitosan may be a safe and promising immune-adjuvant for cancer vaccine, by promoting both humoral and cellular immune responses.

Since chitosan comprise of a large group of glucosamine polymers, its proper standardization is warranted, although challenging, for the development of a successful vaccine adjuvant. Key characteristics, such as chitosan's molecular weight, degree of deacetylation, viscosity, and endotoxin levels [93], should be considered when testing chitosan for adjuvant applications. It would be helpful to refer Vasiliev's (2015) step-by-step approaches in the proper evaluation and standardization of chitosan for use as vaccine-adjuvants [94].

5. Conclusions

Chitosan, the natural biodegradable and non-toxic polymer, holds promise as a suitable material for biomedical applications. There are multifaceted applications of chitosan in cancer therapy, including gene delivery, chemotherapeutic delivery, and immunotherapy. Although chitosan-based drug delivery systems and gene delivery vectors are not yet approved by the FDA (Food and Drug Administration), great progress in cancer therapy research is being made. Physico-chemical characteristics, such as its cationic nature, molecular weight, DDA, and pH of transfection medium are major factors that influence the gene delivery efficacy of chitosan nanoparticles. The genetic material, i.e., siRNA or DNA, and cell type also contribute to the efficiency of transfection using chitosan vectors.

However, chitosan's low water solubility is a major limitation for gene and drug delivery applications. To improve the water solubility, new functional groups or addition of neutral polymers like PEG have been commonly employed. PEG addition also has the advantages of prolonged in vivo circulation and reduced bio-clearance of chitosan nanoparticles. Alone, chitosan has difficulty encapsulating hydrophilic drugs; therefore, conjugation strategies are employed to achieve high drug loading. Derivatization of chitosan with hydrophobic molecules or polymers has enhanced the ability of chitosan to encapsulate hydrophobic drugs.

Targeting of ligands or antibodies is frequently used to improve the target specificity of chitosan in gene or drug delivery applications for cancer. Intracellular delivery of therapeutics can be improved by modification of chitosan with stimuli-responsive polymers or moieties. Apart from these, the immune-adjuvant properties of chitosan are highly promising. Chitosan is known to induce both humoral and cellular immune responses and enhance the immune-stimulatory activity of cancer vaccines. However, the choice of chitosan polymer for immunotherapy is still a challenge, since chitosan is a generically used name for all forms of de-acetylated chitins with versatile properties. Importantly, the molecular weight, DDA, and endotoxin levels of chitosan should be considered in immunoadjuvant applications of chitosan. Worldwide, researchers are engaged in the development of cancer vaccines. It is hoped that chitosan's promising characteristics as an immunoadjuvant will be advantageous for its future application in cancer vaccines. Overall, chitosan's multifaceted characteristics show the potential of this marine biopolymer in cancer therapy applications.

Acknowledgments: The study was supported in part by a grant received from the National Institutes of Health (NIH), R01 CA167516, and by funds received from the Stephenson Cancer Center Seed Grant, Presbyterian Health Foundation Seed Grant, Presbyterian Health Foundation Bridge Grant, and Jim and Christy Everest Endowed Chair in Cancer Developmental Therapeutics, at the University of Oklahoma Health Sciences Center. The authors thank Kathy Kyler at the office of the Vice President of Research, OUHSC, for editorial assistance. Rajagopal Ramesh is an Oklahoma TSET Research Scholar and holds the Jim and Christy Everest Endowed Chair in Cancer Developmental Therapeutics.

Author Contributions: Anish Babu and Rajagopal Ramesh equally contributed in the writing, and editing of the manuscript.

Conflicts of Interest: The authors declare no conflict of interest.

References

1. Venugopal, V. *Marine Products for Healthcare: Functional and Bioactive Nutraceutical Compounds from the Ocean, Functional Foods and Nutraceuticals Series*; CRC Press, Taylor and Francis Group: Boca Raton, FL, USA, 2008; pp. 185–508.
2. Jayakumar, R.; Menon, D.; Manzoor, K.; Nair, S.V.; Tamura, H. Biomedical applications of chitin and chitosan based nanomaterials—A short review. *Carbohydr. Polym.* **2010**, *82*, 227–232. [CrossRef]
3. Tengamnuay, P.; Sahamethapat, A.; Sailasuta, A.; Mitra, A.K. Chitosans as nasal absorption enhancers of peptides: Comparison between free amine chitosans and soluble salts. *Int. J. Pharm.* **2000**, *197*, 53–67. [CrossRef]
4. Croisier, F.; Jérôme, C. Chitosan-based biomaterials for tissue engineering. *Eur. Polym. J.* **2013**, *49*, 780–792. [CrossRef]
5. Saikia, C.; Gogoi, P.; Maji, T.K. Chitosan: A promising biopolymer in drug delivery applications. *J. Mol. Genet. Med.* **2015**, S4:006. [CrossRef]
6. Lee, M.; Nah, J.W.; Kwon, Y.; Koh, J.J.; Ko, K.S.; Kim, S.W. Water-soluble and low molecular weight chitosan-based plasmid DNA delivery. *Pharm. Res.* **2001**, *18*, 427–431. [CrossRef] [PubMed]
7. Kumar, S.; Garg, P.; Pandey, S.; Kumari, M.; Hoon, S.; Jang, K.J.; Kapavarapu, R.; Choung, P.H.; Sobrala, A.J.; Chung, J.H. Enhanced chitosan—DNA interaction by 2-acrylamido-2-methylpropane coupling for an efficient transfection in cancer cells. *J. Mater. Chem. B* **2015**, *3*, 3465–3475. [CrossRef]
8. Csaba, N.; Köping-Höggård, M.; Alonso, M.J. Ionically crosslinked chitosan/tripolyphosphate nanoparticles for oligonucleotide and plasmid DNA delivery. *Int. J. Pharm.* **2009**, *382*, 205–214. [CrossRef] [PubMed]
9. Highton, A.J.; Girardin, A.; Bell, G.M.; Hook, S.M.; Kemp, R.A. Chitosan gel vaccine protects against tumour growth in an intracaecal mouse model of cancer by modulating systemic immune responses. *BMC Immunol.* **2016**, *17*, 39. [CrossRef] [PubMed]
10. Cross, D.; Burmester, J.K. Gene therapy for cancer treatment: Past, present and future. *Clin. Med. Res.* **2006**, *4*, 218–227. [CrossRef] [PubMed]
11. Wang, J.; Lu, Z.; Wientjes, M.G.; Au, J.L.S. Delivery of siRNA therapeutics: Barriers and carriers. *AAPS J.* **2010**, *12*, 492–503. [CrossRef] [PubMed]
12. Gottfried, L.F.; Dean, D.A. Extracellular and intracellular barriers to non-viral gene transfer. In *Novel Gene Therapy Approaches*; Wei, M., Good, D., Eds.; InTech: Rijeka, Croatia, 2013.
13. Nayerossadat, N.; Maedeh, T.; Ali, P.A. Viral and nonviral delivery systems for gene delivery. *Adv. Biomed. Res.* **2012**, *1*, 27. [CrossRef] [PubMed]
14. Yin, H.; Kanasty, R.L.; Eltoukhy, A.A.; Vegas, A.J.; Dorkin, J.R.; Anderson, D.G. Non-viral vectors for gene-based therapy. *Nat. Rev. Genet.* **2014**, *15*, 541–555. [CrossRef] [PubMed]
15. Kedmi, R.; Ben-Arie, N.; Peer, D. The systemic toxicity of positively charged lipid nanoparticles and the role of Toll-like receptor 4 in immune activation. *Biomaterials* **2010**, *31*, 6867–6875. [CrossRef] [PubMed]
16. Tao, W.; Mao, X.; Davide, J.P.; Ng, B.; Cai, M.; Burke, P.A.; Sachs, A.B.; Sepp-Lorenzino, L. Mechanistically probing lipid-siRNA nanoparticle-associated toxicities identifies Jak inhibitors effective in mitigating multifaceted toxic responses. *Mol. Ther.* **2011**, *19*, 567–575. [CrossRef] [PubMed]
17. Whitehead, K.A.; Langer, R.; Anderson, D.G. Knocking down barriers: Advances in siRNA delivery. *Nat. Rev. Drug Discov.* **2009**, *8*, 129–138. [CrossRef] [PubMed]
18. Ishida, T.; Harada, M.; Wang, X.Y.; Ichihara, M.; Irimura, K.; Kiwada, H. Accelerated blood clearance of PEGylated liposomes following preceding liposome injection: Effects of lipid dose and PEG surface-density and chain length of the first-dose liposomes. *J. Control. Release* **2005**, *105*, 305–317. [CrossRef] [PubMed]
19. Cun, D.; Jensen, L.B.; Nielsen, H.M.; Moghimi, M.; Foged, C. Polymeric nanocarriers for siRNA delivery: Challenges and future prospects. *J. Biomed. Nanotechnol.* **2008**, *4*, 258–275. [CrossRef]
20. Singha, K.; Namgung, R.; Kim, W.J. Polymers in small-interfering RNA delivery. *Nucleic Acid Ther.* **2011**, *21*, 133–147. [CrossRef] [PubMed]
21. Mao, S.; Sun, W.; Kissel, T. Chitosan-based formulations for delivery of DNA and siRNA. *Adv. Drug Deliv. Rev.* **2010**, *62*, 12–27. [CrossRef] [PubMed]
22. Alameh, M.; Dejesus, D.; Jean, M.; Darras, V.; Thibault, M.; Lavertu, M.; Buschmann, M.D.; Merzouki, A. Low molecular weight chitosan nanoparticulate system at low N:P ratio for nontoxic polynucleotide delivery. *Int. J. Nanomed.* **2012**, *7*, 1399–414.

23. Lavertu, M.; Méthot, S.; Tran-Khanh, N.; Buschmann, M.D. High efficiency gene transfer using chitosan/DNA nanoparticles with specific combinations of molecular weight and degree of deacetylation. *Biomaterials* **2006**, *27*, 4815–4824. [CrossRef] [PubMed]

24. Techaarpornkul, S.; Wongkupasert, S.; Opanasopit, P.; Apirakaramwong, A.; Nunthanid, J.; Ruktanonchai, U. Chitosan-mediated siRNA delivery In Vitro: Effect of polymer molecular weight, concentration and salt forms. *AAPS PharmSciTech* **2010**, *11*, 64–72. [CrossRef] [PubMed]

25. Huang, M.; Fong, C.W.; Khor, E.; Lim, L.Y. Transfection efficiency of chitosan vectors: Effect of polymer molecular weight and degree of deacetylation. *J. Control. Release* **2005**, *106*, 391–406. [CrossRef] [PubMed]

26. Sato, T.; Ishii, T.; Okahata, Y. In Vitro gene delivery mediated by chitosan. Effect of pH, serum, and molecular mass of chitosan on the transfection efficiency. *Biomaterials* **2001**, *22*, 2075–2080. [CrossRef]

27. Nimesh, S.; Thibault, M.M.; Lavertu, M.; Buschmann, M.D. Enhanced gene delivery mediated by low molecular weight chitosan/DNA complexes: Effect of pH and serum. *Mol. Biotechnol.* **2010**, *46*, 182–196. [CrossRef] [PubMed]

28. Agirre, M.; Zarate, J.; Ojeda, E.; Puras, G.; Desbrieres, J.; Pedraz, J.L. Low Molecular Weight Chitosan (LMWC)-based Polyplexes for pDNA Delivery: From Bench to Bedside. *Polymers* **2014**, *6*, 1727–1755. [CrossRef]

29. Patil, S.; Bhatt, P.; Lalani, R.; Amrutiya, J.; Vhora, I.; Kolte, A.; Misra, A. Low molecular weight chitosan–protamine conjugate for siRNA delivery with enhanced stability and transfection efficiency. *RSC Adv.* **2016**, *6*, 110951–110963. [CrossRef]

30. Fernandes, J.C.; Qiu, X.; Winnik, F.M.; Benderdour, M.; Zhang, X.; Dai, K.; Shi, Q. Low molecular weight chitosan conjugated with folate for siRNA delivery In Vitro: Optimization studies. *Int. J. Nanomed.* **2012**, *7*, 5833–5845.

31. Köping-Höggård, M.; Varum, K.M.; Issa, M.; Danielsen, S.; Christensen, B.E.; Stokke, B.T.; Artursson, P. Improved chitosan-mediated gene delivery based on easily dissociated chitosan polyplexes of highly defined chitosan oligomers. *Gene Ther.* **2004**, *11*, 1441–1452. [CrossRef] [PubMed]

32. Köping-Höggård, M.; Tubulekas, I.; Guan, H.; Edwards, K.; Nilsson, M.; Vårum, K.M.; Artursson, P. Chitosan as a nonviral gene delivery system. Structure-property relationships and characteristics compared with polyethylenimine In Vitro and after lung administration In Vivo. *Gene Ther.* **2001**, *8*, 1108–1121.

33. Erbacher, P.; Zou, S.; Bettinger, T.; Steffan, A.M.; Remy, J.S. Chitosan-based vector/DNA complexes for gene delivery: Biophysical characteristics and transfection ability. *Pharm. Res.* **1998**, *15*, 1332–1339. [CrossRef] [PubMed]

34. Amaduzzi, F.; Bomboi, F.; Bonincontro, A.; Bordi, F.; Casciardi, S.; Chronopoulou, L.; Diociaiuti, M.; Mura, F.; Palocci, C.; Sennato, S. Chitosan-DNA complexes: Charge inversion and DNA condensation. *Colloids Surf. B Biointerfaces* **2014**, *114*, 1–10. [CrossRef] [PubMed]

35. Bravo-Anaya, L.M.; Soltero, J.F.; Rinaudo, M. DNA/chitosan electrostatic complex. *Int. J. Biol. Macromol.* **2016**, *88*, 345–353. [CrossRef] [PubMed]

36. Liu, X.; Howard, K.A.; Dong, M.; Andersen, M.Ø.; Rahbek, U.L.; Johnsen, M.G.; Hansen, O.C.; Besenbacher, F.; Kjems, J. The influence of polymeric properties on chitosan/siRNA nanoparticle formulation and gene silencing. *Biomaterials* **2007**, *28*, 1280–1288. [CrossRef] [PubMed]

37. Ragelle, H.; Vanvarenberg, K.; Vandermeulen, G.; Préat, V. Chitosan nanoparticles for siRNA delivery In Vitro. *Methods Mol. Biol.* **2016**, *1364*, 143–150.

38. Ishii, T.; Okahata, Y.; Sato, T. Mechanism of cell transfection with plasmid/chitosan complexes. *Biochim. Biophys. Acta* **2001**, 51–64. [CrossRef]

39. Ouchi, T.; Murata, J.; Ohya, Y. Gene delivery by quaternary chitosan with antennary galactose residues. In *Polysaccharide Applications*; ACS Symposium Series; American Chemical Society: Washington, DC, USA, 1999; Volume 737, pp. 15–23, Chapter 2.

40. Wei, W.; Lv, P.P.; Chen, X.M.; Yue, Z.G.; Fu, Q.; Liu, S.Y.; Yue, H.; Ma, G.H. Codelivery of mTERT siRNA and paclitaxel by chitosan-based nanoparticles promoted synergistic tumor suppression. *Biomaterials* **2013**, *34*, 3912–3923. [CrossRef] [PubMed]

41. Li, G.F.; Wang, J.C.; Feng, X.M.; Liu, Z.D.; Jiang, C.Y.; Yang, J.D. Preparation and testing of quaternized chitosan nanoparticles as gene delivery vehicles. *Appl. Biochem. Biotechnol.* **2015**, *175*, 3244–3257. [CrossRef] [PubMed]

42. Opanasopit, P.; Techaarpornkul, S.; Rojanarata, T.; Ngawhirunpat, T.; Ruktanonchai, U. Nucleic acid delivery with chitosan hydroxybenzotriazole. *Oligonucleotides* **2010**, *20*, 127–136. [CrossRef] [PubMed]

43. Deng, J.; Zhou, Y.; Xu, B.; Mai, K.; Deng, Y.; Zhang, L.M. Dendronized chitosan derivative as a biocompatible gene delivery carrier. *Biomacromolecules* **2011**, *12*, 642–649. [CrossRef] [PubMed]

44. Zhong, J.; Huang, H.L.; Li, J.; Qian, F.C.; Li, L.Q.; Niu, P.P.; Dai, L.C. Development of hybrid-type modified chitosan derivative nanoparticles for the intracellular delivery of midkine-siRNA in hepatocellular carcinoma cells. *Hepatobiliary Pancreat. Dis. Int.* **2015**, *14*, 82–89. [CrossRef]

45. Germershaus, O.; Mao, S.; Sitterberg, J.; Bakowsky, U.; Kissel, T. Gene delivery using chitosan, trimethyl chitosan or polyethylenglycol-graft-trimethyl chitosan block copolymers: Establishment of structure–activity relationships in vitro. *J. Control. Release* **2008**, *125*, 145–154. [CrossRef] [PubMed]

46. Zheng, Y.; Cai, Z.; Song, X.; Yu, B.; Bi, Y.; Chen, Q.; Zhao, D.; Xu, J.; Hou, S. Receptor mediated gene delivery by folate conjugated N-trimethyl chitosan In Vitro. *Int. J. Pharm.* **2009**, *382*, 262–269. [CrossRef] [PubMed]

47. Eivazy, P.; Atyabi, F.; Jadidi-Niaragh, F.; Aghebati Maleki, L.; Miahipour, A.; Abdolalizadeh, J.; Yousefi, M. The impact of the codelivery of drug-siRNA by trimethyl chitosan nanoparticles on the efficacy of chemotherapy for metastatic breast cancer cell line (MDA-MB-231). *Artif. Cells Nanomed. Biotechnol.* **2016**, 1–8. [CrossRef] [PubMed]

48. Gao, Y.; Wang, Z.Y.; Zhang, J.; Zhang, Y.; Huo, H.; Wang, T.; Jiang, T.; Wang, S. RVG-peptide-linked trimethylated chitosan for delivery of siRNA to the brain. *Biomacromolecules* **2014**, *15*, 1010–1018. [CrossRef] [PubMed]

49. Zheng, H.; Tang, C.; Yin, C. Exploring advantages/disadvantages and improvements in overcoming gene delivery barriers of amino acid modified trimethylated chitosan. *Pharm. Res.* **2015**, *32*, 2038–2050. [CrossRef] [PubMed]

50. Zhang, Y.; Chen, J.; Zhang, Y.; Pan, Y.; Zhao, J.; Ren, L.; Liao, M.; Hu, Z.; Kong, L.; Wang, J. A novel PEGylation of chitosan nanoparticles for gene delivery. *Biotechnol. Appl. Biochem.* **2007**, *46*, 197–204. [PubMed]

51. Chan, P.; Kurisawa, M.; Chung, J.E.; Yang, Y.Y. Synthesis and characterization of chitosan-g-poly(ethylene glycol)-folate as a non-viral carrier for tumor-targeted gene delivery. *Biomaterials* **2007**, *28*, 540–549. [CrossRef] [PubMed]

52. Nag, M.; Gajbhiye, V.; Kesharwani, P.; Jain, N.K. Transferrin functionalized chitosan-PEG nanoparticles for targeted delivery of paclitaxel to cancer cells. *Colloids Surf. B Biointerfaces* **2016**, *148*, 363–370. [CrossRef] [PubMed]

53. Liu, Z.; Lv, D.; Liu, S.; Gong, J.; Wang, D.; Xiong, M.; Chen, X.; Xiang, R.; Tan, X. Alginic acid-coated chitosan nanoparticles loaded with legumain DNA vaccine: Effect against breast cancer in mice. *PLoS ONE* **2013**, *8*, e60190. [CrossRef] [PubMed]

54. Yhee, J.Y.; Song, S.; Lee, S.J.; Park, S.G.; Kim, K.S.; Kim, M.G.; Son, S.; Koo, H.; Kwon, I.C.; Jeong, J.H.; et al. Cancer-targeted MDR-1 siRNA delivery using self-cross-linked glycol chitosan nanoparticles to overcome drug resistance. *J. Control. Release* **2015**, *198*, 1–9. [CrossRef] [PubMed]

55. Sun, P.; Huang, W.; Jin, M.; Wang, Q.; Fan, B.; Kang, L.; Gao, Z. Chitosan-based nanoparticles for survivin targeted siRNA delivery in breast tumor therapy and preventing its metastasis. *Int. J. Nanomed.* **2016**, *11*, 4931–4945. [CrossRef] [PubMed]

56. Darvishi, M.H.; Nomani, A.; Amini, M.; Shokrgozar, M.A.; Dinarvand, R. Novel biotinylated chitosan-graft-polyethyleneimine copolymer as a targeted non-viral vector for anti-EGF receptor siRNA delivery in cancer cells. *Int. J. Pharm.* **2013**, *456*, 408–416. [CrossRef] [PubMed]

57. Xin, L.; Fan, J.C.; Le, Y.G.; Zeng, F.; Cheng, H.; Hu, X.Y.; Cao, J.Q. Construction of METHFR shRNA/5-fluorouracil co-loaded folate-targeted chitosan polymeric nanoparticles and its anti-carcinoma effect on gastric cells growth. *J. Nanopart. Res.* **2016**, *18*, 105. [CrossRef]

58. Han, L.; Zhao, J.; Zhang, X.; Cao, W.; Hu, X.; Zou, G.; Duan, X.; Liang, X.J. Enhanced siRNA delivery and silencing gold-chitosan nanosystem with surface charge-reversal polymer assembly and good biocompatibility. *ACS Nano* **2012**, *6*, 7340–7351. [CrossRef] [PubMed]

59. Soofiyani, S.R.; Hallaj-Nezhadi, S.; Lotfipour, F.; Hosseini, A.M.; Baradaran, B. Gene therapy based on interleukin-12 loaded chitosan nanoparticles in a mouse model of fibrosarcoma. *Iran. J. Basic Med. Sci.* **2016**, *11*, 1238–1244.

60. Babu, A.; Wang, Q.; Muralidharan, R.; Shanker, M.; Munshi, A.; Ramesh, R. Chitosan coated polylactic acid nanoparticle-mediated combinatorial delivery of cisplatin and siRNA/Plasmid DNA chemosensitizes cisplatin-resistant human ovarian cancer cells. *Mol. Pharm.* **2014**, *11*, 2720–2733. [CrossRef] [PubMed]

61. Jeong, Y.; Jin, S.G.; Kim, I.Y.; Pei, J.; Wen, M.; Jung, T.Y.; Moon, K.S.; Jung, S. Doxorubicin-incorporated nanoparticles composed of poly(ethylene glycol)-grafted carboxymethyl chitosan and antitumor activity against glioma cells in vitro. *Colloids Surf. B Biointerfaces* **2010**, *79*, 149–155. [CrossRef] [PubMed]

62. Yousefpour, P.; Atyabi, F.; Vasheghani-Farahani, E.; Mousavi, A.A.M.; Dinarvand, R. Targeted delivery of doxorubicin-utilizing chitosan nanoparticles surface-functionalized with anti-Her2 trastuzumab. *Int. J. Nanomed.* **2011**, *6*, 1977–1990.

63. Kim, J.H.; Kim, Y.S.; Kim, S.; Park, J.H.; Kim, K.; Choi, K.; Chung, H.; Jeong, S.Y.; Park, R.W.; Kim, I.S.; et al. Hydrophobically modified glycol chitosan nanoparticles as carriers for paclitaxel. *J. Control. Release* **2006**, *111*, 228–234. [CrossRef] [PubMed]

64. Li, F.; Li, J.; Wen, X.; Zhou, S.; Tonga, X.; Sua, P.; Lia, H.; Shib, D. Anti-tumor activity of paclitaxel-loaded chitosan nanoparticles: An in vitro study. *Mater. Sci. Eng. C* **2009**, *29*, 2392–2397. [CrossRef]

65. Naruphontjirakul, P.; Viravaidya-Pasuwat, K. Development of Doxorubicin—Core Shell Chitosan Nanoparticles to Treat Cancer. In *Proceedings of the International Conference on Biomedical Engineering and Technology*; IACSIT Press: Singapore, 2011; Volume 11, pp. 90–94.

66. Trickler, W.J.; Nagvekar, A.A.; Dash, A.K. A novel nanoparticle formulation for sustained paclitaxel delivery. *AAPS PharmSciTech* **2008**, *9*, 86–93. [CrossRef] [PubMed]

67. Kim, J.H.; Kim, Y.S.; Park, K.; Lee, S.; Nam, H.Y.; Min, K.H.; Jo, H.G.; Park, J.H.; Choi, K.; Jeong, S.Y.; et al. Antitumor efficacy of cisplatin-loaded glycol chitosan nanoparticles in tumor-bearing mice. *J. Control. Release* **2008**, *127*, 41–49. [CrossRef] [PubMed]

68. Rajan, M.; Raj, V.; Al-Arfaj, A.A.; Murugan, A.M. Hyaluronidase enzyme core-5-fluorouracil-loaded chitosan-PEG-gelatin polymer nanocomposites as targeted and controlled drug delivery vehicles. *Int. J. Pharm.* **2013**, *453*, 514–522. [CrossRef] [PubMed]

69. Cavalli, R.; Leone, F.; Minelli, R.; Fantozzi, R.; Dianzani, C. New chitosan nanospheres for the delivery of 5-fluorouracil: Preparation, characterization and in vitro studies. *Curr. Drug Deliv.* **2014**, *11*, 270–278. [CrossRef] [PubMed]

70. Khdair, A.; Hamad, I.; Alkhatib, H.; Bustanji, Y.; Mohammad, M.; Tayem, R.; Aiedeh, K. Modified-chitosan nanoparticles: Novel drug delivery systems improve oral bioavailability of doxorubicin. *Eur. J. Pharm. Sci.* **2016**, *93*, 38–44. [CrossRef] [PubMed]

71. Yu, J.M.; Li, Y.J.; Qiu, L.Y.; Jin, Y. Polymeric nanoparticles of cholesterol-modified glycol chitosan for doxorubicin delivery: Preparation and In Vitro and In Vivo characterization. *J. Pharm. Pharmacol.* **2009**, *61*, 713–719. [CrossRef] [PubMed]

72. Rao, W.; Wang, H.; Han, J.; Zhao, S.; Dumbleton, J.; Agarwal, P.; Zhang, W.; Zhao, G.; Yu, J.; Zynger, D.L.; et al. Chitosan-decorated doxorubicin-encapsulated nanoparticle targets and eliminates tumor reinitiating cancer stem-like cells. *ACS Nano* **2015**, *9*, 5725–5740. [CrossRef] [PubMed]

73. Gupta, U.; Sharma, S.; Khan, I.; Gothwal, A.; Sharma, A.K.; Singh, Y.; Chourasia, M.K.; Kumar, V. Enhanced apoptotic and anticancer potential of paclitaxel loaded biodegradable nanoparticles based on chitosan. *Int. J. Biol. Macromol.* **2017**, *98*, 810–819. [CrossRef] [PubMed]

74. Jain, A.; Thakur, K.; Sharma, G.; Kush, P.; Jain, U.K. Fabrication, characterization and cytotoxicity studies of ionically cross-linked docetaxel loaded chitosan nanoparticles. *Carbohydr. Polym.* **2016**, *137*, 65–74. [CrossRef] [PubMed]

75. Jin, X.; Mo, R.; Ding, Y.; Zheng, W.; Zhang, C. Paclitaxel-loaded N-octyl-O-sulfate chitosan micelles for superior cancer therapeutic efficacy and overcoming drug resistance. *Mol. Pharm.* **2014**, *11*, 145–157. [CrossRef] [PubMed]

76. Jing, J.; Zuo, P.; Wang, Y.-L. Enhanced antiproliferative effect of carboplatin in cervical cancer cells utilizing folate-grafted polymeric nanoparticles. *Nanoscale Res. Lett.* **2015**, *10*, 453–461.

77. Wang, J.Y.; Wang, Y.; Meng, X. Chitosan nanolayered cisplatin-loaded lipid nanoparticles for enhanced anticancer efficacy in cervical cancer. *Nanoscale Res. Lett.* **2016**, *11*, 524–532. [CrossRef] [PubMed]

78. Ghaz-Jahanian, M.A.; Abbaspour-Aghdam, F.; Anarjan, N.; Berenjian, A.; Jafarizadeh-Malmiri, H. Application of chitosan-based nanocarriers in tumor-targeted drug delivery. *Mol. Biotechnol.* **2015**, *57*, 201. [CrossRef] [PubMed]

79. Xu, S.; Olenyuk, B.Z.; Okamoto, C.T.; Hamm-Alvarez, S.F. Targeting receptor-mediated endocytotic pathways with nanoparticles: Rationale and advances. *Adv. Drug Deliv. Rev.* **2013**, *65*, 121–138. [CrossRef] [PubMed]

80. Chen, C.; Zhou, J.L.; Han, X.; Song, F.; Wang, X.L.; Wang, Y.Z. A prodrug strategy based on chitosan for efficient intracellular anticancer drug delivery. *Nanotechnology* **2014**, *25*, 255101. [CrossRef] [PubMed]

81. Petrovsky, N.; Cooper, P.D. Carbohydrate-based immune adjuvants. *Expert Rev. Vaccines* **2011**, *10*, 523–537. [CrossRef] [PubMed]

82. Tokura, S.; Tamura, H.; Azuma, I. Immunological aspects of chitin and chitin derivatives administered to animals. *EXS* **1999**, *87*, 279–292. [PubMed]

83. Zhao, K.; Chen, G.; Shi, X.-M.; Gao, T.-T.; Li, W.; Zhao, Y.; Zhang, F.Q.; Wu, J.; Cui, X.; Wang, Y.F. Preparation and efficacy of a live newcastle disease virus vaccine encapsulated in chitosan nanoparticles. *PLoS ONE* **2012**, *7*, e53314. [CrossRef] [PubMed]

84. Doavi, T.; Mousavi, S.L.; Kamali, M.; Amani, J.; Ramandi, M.F. Chitosan-based intranasal vaccine against escherichia coli O157:H7. *Iran. Biomed. J.* **2016**, *20*, 97–108. [PubMed]

85. Zaharoff, D.A.; Hance, K.W.; Rogers, C.J.; Schlom, J.; Greiner, J. Intratumoral immunotherapy of established solid tumors with Chitosan/IL-12. *J. Immunother.* **2010**, *33*, 697–705. [CrossRef] [PubMed]

86. Arca, H.C.; Günbeyaz, M.; Şenel, S. Chitosan-based systems for the delivery of vaccine antigens. *Expert Rev. Vaccines* **2009**, *8*, 937–953. [CrossRef] [PubMed]

87. Zaharoff, D.A.; Rogers, C.J.; Hance, K.W.; Schlom, J.; Greiner, J.W. Chitosan solution enhances both humoral and cell-mediated immune responses to subcutaneous vaccination. *Vaccine* **2007**, *25*, 2085–2094. [CrossRef] [PubMed]

88. Bueter, C.L.; Lee, C.K.; Rathinam, V.A.; Healy, G.J.; Taron, C.H.; Specht, C.A.; Levitz, S.M. Chitosan but not chitin activates the inflammasome by a mechanism dependent upon phagocytosis. *J. Biol. Chem.* **2011**, *286*, 35447–35455. [CrossRef] [PubMed]

89. Carroll, E.C.; Jin, L.; Mori, A.; Muñoz-Wolf, N.; Oleszycka, E.; Moran, H.B.T.; Mansouri, S.; McEntee, C.P.; Lambe, E.; Agger, E.M.; et al. The vaccine adjuvant chitosan promotes cellular immunity via DNA sensor cGAS-STING-dependent induction of Type I interferons. *Immunity* **2016**, *44*, 597–608. [CrossRef] [PubMed]

90. Wen, Z.S.; Xu, Y.L.; Zou, X.T.; Xu, Z.R. Chitosan nanoparticles act as an adjuvant to promote both Th1 and Th2 immune responses induced by ovalbumin in mice. *Mar. Drugs* **2011**, *9*, 1038–1055. [CrossRef] [PubMed]

91. Zaharoff, D.A.; Hoffman, B.S.; Hooper, H.B.; Benjamin, C.J.; Khurana, K.K.; Hance, K.W.; Rogers, C.J.; Pinto, P.A.; Schlom, J.; Greiner, J.W. Intravesical immunotherapy of superficial bladder cancer with Chitosan/Interleukin-12. *Cancer Res.* **2009**, *69*, 6192–6199. [CrossRef] [PubMed]

92. Heffernan, M.J.; Zaharoff, D.A.; Fallon, J.K.; Schlom, J.; Greiner, J.W. In vivo efficacy of a chitosan/IL-12 adjuvant system for protein-based vaccines. *Biomaterials* **2011**, *32*, 926–932. [CrossRef] [PubMed]

93. Zaharoff, D.A.; Rogers, C.J.; Hance, K.W.; Schlom, J.; Greiner, J.W. Chitosan solution enhances the immunoadjuvant properties of GM-CSF. *Vaccine* **2007**, *25*, 8673–8686. [CrossRef] [PubMed]

94. Vasiliev, Y.M. Chitosan-based vaccine adjuvants: Incomplete characterization complicates preclinical and clinical evaluation. *Expert Rev. Vaccines* **2015**, *14*, 37–53. [CrossRef] [PubMed]

marine drugs

MDPI

Article

Short-Chain Chitin Oligomers: Promoters of Plant Growth

Alexander J. Winkler [1,2,†], Jose Alfonso Dominguez-Nuñez [1], Inmaculada Aranaz [3],
César Poza-Carrión [4], Katrina Ramonell [5], Shauna Somerville [6,‡] and Marta Berrocal-Lobo [1,7,*]

[1] Department of Systems and Natural Resources, MONTES (School of Forest Engineering and Natural Environment), Universidad Politécnica de Madrid, Ciudad Universitaria s/n, 28040 Madrid, Spain; alexander.winkler@mpimet.mpg.de (A.J.W.); josealfonso.dominguez@upm.es (J.A.D.-N.)

[2] Department for Wood Biology, Centre for Wood Science and Technology, Universität Hamburg, Leuschnerstr. 91d, D-2103 Hamburg, Germany

[3] Departamento de Físico-Química, Instituto de Estudios Bifuncionales, Facultad de Farmacia, Universidad Complutense, Paseo Juan XXIII, 1, 28040 Madrid, Spain; iaranaz@hotmail.com

[4] Centro Nacional de Biotecnología, Calle Darwin, 3, 28049 Madrid, Spain; cpoza@cnb.csic.es

[5] Department of Biological Sciences, P.O. Box 870344, University of Alabama, Tuscaloosa, AL 35487, USA; kramonel@bama.ua.edu

[6] Plant Biology, Carnegie Institution of Science, 260 Panama St., Stanford, CA 94305, USA; ssomerville@berkeley.edu

[7] Centro de Biotecnología y Genómica de Plantas, Instituto Nacional de Investigación y Tecnología Agraria y Alimentaria (INIA), Campus Montegancedo UPM, Universidad Politécnica de Madrid (UPM), 28223 Pozuelo de Alarcón (Madrid), Spain

* Correspondence: m.berrocal@upm.es; Tel.: +34-91-3366-408

† Current Address: Max-Planck-Institute for Meteorology, Bundesstrasse 53, 20146 Hamburg, Germany.

‡ Current Address: Plant and Microbial Biology Department, Energy Biosciences Building, University of California-Berkeley, 2151 Berkeley Way, Berkeley, CA 94720, USA.

Academic Editors: Hitoshi Sashiwa and David Harding
Received: 20 December 2016; Accepted: 6 February 2017; Published: 15 February 2017

Abstract: Chitin is the second most abundant biopolymer in nature after cellulose, and it forms an integral part of insect exoskeletons, crustacean shells, krill and the cell walls of fungal spores, where it is present as a high-molecular-weight molecule. In this study, we showed that a chitin oligosaccharide of lower molecular weight (tetramer) induced genes in *Arabidopsis* that are principally related to vegetative growth, development and carbon and nitrogen metabolism. Based on plant responses to this chitin tetramer, a low-molecular-weight chitin mix (CHL) enriched to 92% with dimers (2mer), trimers (3mer) and tetramers (4mer) was produced for potential use in biotechnological processes. Compared with untreated plants, CHL-treated plants had increased in vitro fresh weight (10%), radicle length (25%) and total carbon and nitrogen content (6% and 8%, respectively). Our data show that low-molecular-weight forms of chitin might play a role in nature as bio-stimulators of plant growth, and they are also a known direct source of carbon and nitrogen for soil biomass. The biochemical properties of the CHL mix might make it useful as a non-contaminating bio-stimulant of plant growth and a soil restorer for greenhouses and fields.

Keywords: chitin oligosaccharides; bio-stimulator; fertilizer; soil biomass; biodiversity; soil health, soil biomass, bio-diversity

---

1. Introduction

Chitin is the second most abundant carbohydrate (after cellulose) in the biosphere. It is a nitrogen-containing polysaccharide that is generally composed of *N*-acetyl-D-glucosamine (GlcNAc,

A) and D-glucosamine (GlcN, D) monomers bound by beta-1,4 linkages. Chitin is the major structural component of fungal cell walls and spores, crustacean shells, insect exoskeletons, mollusks and some protozoa. In the soil, chitin comes principally from insects and fungi. Millions of tons of chitin are discharged onto the sea floor annually as "marine snow" by copepod (planktonic crustaceans). The decomposition of chitin is very significant in the natural soil ecosystem, and it removes tons of chitin that accumulate every year from dead insects and later used by soil biomass [1]. The release of organically-bound nitrogen and carbon from chitin is an important factor that should be taken into account when investigating carbon and nitrogen cycling in ecosystems. Chitin is also the principal source of carbon and nitrogen for chitinolytic organisms, which are largely marine and soil bacteria belonging to the genera of the *Proteobacteria*, *Bacteroidetes*, *Actinobacteria* and *Firmicutes*, as well as soil fungi [2–4].

These chitin-decomposing organisms contain the metabolic machinery necessary to detect, modify and transport small chitin oligosaccharides (generally from two–four monomers long), and incorporate them directly into their glycolytic and nitrogen metabolic pathways as glucose (carbon) and ammonia, respectively [5–7].

Chitin is primarily converted by organisms to the more soluble biopolymer, chitosan, via a modification catalyzed by deacetylase enzymes that recognize a sequence of four GlcNAc units, one of which undergoes de-acetylation. Chitosan is not only a recognized antibacterial biopolymer [4], but it is also a source of nutrients for insects, bacteria and fungi living in the soil [8]. The biochemical properties of chitosan make it particularly useful for biomedical applications, such as wound dressing, weight loss agent, blood cholesterol control, surgical sutures, cataract surgery or periodontal disease treatment, and chitinolytic enzymes from bacteria and fungi are useful to pharmacological enterprises as a source of antifungal agents [9].

The exposure of plants to chitosan results in the activation of defense response genes associated with biotic stress [10], and chitosan is normally used in agriculture as an agent to induce innate plant protection [11]. In addition, the literature describes the use of chitosan to stimulate various plant growth parameters in potatoes (*Solanum tuberosum*), tomatoes (*Solanum lycopersicum*), orchids (*Orchidaceae*), grape vines (*Vitis vinifera*) or pines (*Pinaceae*) [11,12]. The stimulating effect of chitin or chitosan on plant growth has traditionally been attributed to the positive effects on soil biomass and on the association of symbiotic organisms with plants, more than to a direct effect on plant growth itself [4,13]. Additionally, high-molecular-weight chitin has been shown to produce an increase in eukaryotic and prokaryotic microflora when it is used in the soil as a source of nutrients [13–16].

In parallel, chitin, the principal component of fungal spores, is a well-characterized elicitor of plant responses, as it can activate the plant innate immune response by inducing the expression of genes related principally to a biotic stress in response to phytopathogenic fungi [17,18]. Plants can recognize high-molecular-weight chitin via specific receptors, which have been characterized in several plant species, including rice (*Oryza sativa*) [19], *Arabidopsis* (*Arabidopsis thaliana*) [20–22] and *Medicago* (*Medicago truncatula*) [23]. The chitin receptors implicated in plant defense, such as CERK1 and Lyk 5, can bind a chitin 8mer with higher affinity than smaller fragments [21,22,24,25]. Previously, we showed that a high-molecular-weight chitin mix (CHH) derived from crab shells and a purified chitin 8mer oligosaccharide induced a similar suite of genes (related to the defense response) in *Arabidopsis* seedlings [17,18]. Some of these genes were essential for a successful plant defense response against phytopathogenic fungi [26]. In soybean (*Glycine max*), compared with small oligomers of chitin, oligomers of chitin and chitosan that were larger than four monomers produced an increase in the production of phenolic compounds mediated by phenylalanine ammonia lyase [27].

Additionally, chitosan heptamers can target the chromatin within the plant nucleus, altering the chromatin conformation, which has been shown to change the expression and gene activation of the plant cell [28,29].

In contrast, short-chain oligomers of chitin (chitooligosaccharides, COs) have been found to be associated with non-stress-related plant responses. It was shown that while the foliar application of CHH decreased the net photosynthetic rate of maize (*Zea mays*) and soybean, this effect was not observed in plants treated with a chitin 5mer [30]. In addition, COs shorter than 6–7mers induced a lower production of reactive oxygen species in plants, compared to larger chitin oligomers [31]. Moreover, unlike CHH, COs were unable to induce the defense marker gene, mitogen-activated protein kinase 3 (*MAPK3*) [17] and to activate the defense-associated MAPK cascade [32,33]. More recently, it has been suggested that COs play an important role in the initiation of legume-*Rhizobium* symbiosis [34,35], and in the activation of the initial stages of root colonization by arbuscular mycorrhizal fungi [36]. Short forms of chitin oligosaccharides bound to lipids, which are known as lipochitooligosaccharides (LCOs), and nodulation (Nod) factors, were found to be secreted by rhizobial bacteria and mycorrhizal fungi during the establishment of symbiotic interactions, and they have been shown to increase the early plant growth of both soybean and maize [37,38].

In this study, we showed that treating plants with a chitin 4mer activated a transcriptional response in genes that were principally related to plant development and nitrogen and carbon metabolism. Our analysis revealed that this response differs significantly from previously-described plant defense-related response activated by a chitin 8mer or CHH [18] and by high-molecular-weight chitosan [10]. Additionally, we produced a CHL that was enriched with 2mer–4mers (92.3%) that produced a direct effect in vitro, activating plant growth and producing an increase in total nitrogen and carbon content compared with controls.

Our results point to a mechanism whereby naturally occurring low-molecular-weight forms of chitin might contribute not only to nutrient allocation in soil microorganisms, but also to the stimulation of plant development. These compounds could potentially be used as natural bio-stimulants of plant growth and soil restorers in agriculture.

2. Results

2.1. Analysis of Arabidopsis Transcriptional Response to the Chitin 4mer

In order to better understand the different types of plant responses induced by chitin in nature, we treated *Arabidopsis* seedlings with a highly purified chitin oligomer, specifically chito-fourmer (4mer), and we compared these results with our previous results associated with a chito-octamer (8mer) and a high-molecular-weight chitin mix (CHH) derived from crab shells [32]. The genomic response of the plants was then determined using Affymetrix full-genome microarrays (ATH1).

Of the approximately 14,373 genes on the arrays, 5435 genes showed altered expression according to the Statistical Analysis of Microarrays program (SAM) (Figure 1A). In order to narrow down the list, genes that were induced or repressed by 1.5-fold were identified and grouped using Venn diagrams (Figure 1B). A set of 191 genes was induced in all three treatments (Figure 1(B1)), while a set of 91 genes was repressed in all three treatments (Figure 1(B2)). The chitin 4mer induced 71 genes specifically, and the CHH and 8mer treatments (Nmer) induced a much larger set of genes (1598). Interestingly, after treatment with the chitin 4mer, more genes that responded specifically to the 4mer treatment (rather than the Nmer treatments) were significantly repressed (325) than induced (71). Roughly equivalent numbers of genes were uniquely repressed by the chitin 4mer and CHH (339 and 329, respectively), while 239 genes were repressed by treatment with the chitin 8mer (and a total of 802 genes were repressed by the Nmer treatments; Figure 1(B2)).

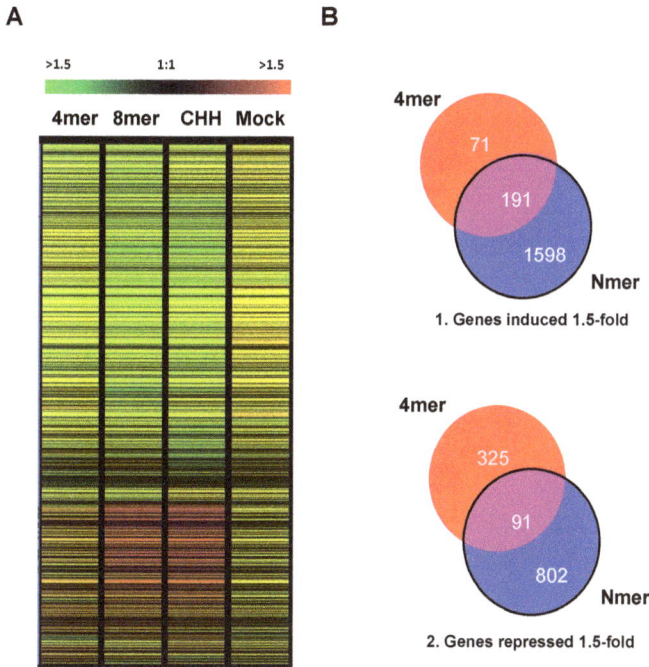

Figure 1. Microarray analysis of Arabidopsis after chitin treatments. (**A**) Hierarchical cluster of the ratio values of the genes that responded to different sizes of chitin, 4mer, 8mer and high-molecular-weight chitin mix (CHH). Each gene is represented by a single row, and each column represents an individual treatment. Red represents upregulated genes, green downregulated genes and black genes with no change (the signals are relative to the control treatment, which was water). (**B**) Venn diagrams of hierarchical clustering results; (**B1**) Venn diagram of genes showing a ≥1.5-fold increase in expression after 4mer treatment and 8mer or high-molecular-weight chitin (CHH) treatments (i.e., Nmer treatments); (**B2**) Venn diagram of genes showing a ≥1.5-fold decrease in expression after 4mer treatment and Nmer treatments. Two-way analysis of variance (ANOVA) was used for clustering. The genotypes and treatment groups were analyzed using a p-value of 0.5, with p-value > 0.5 = not significant and p-value < 0.5 = significant. Three array data replicates were used for the analysis.

In order to determine the enrichment of functional categories or overrepresented genes that responded to the chitin 4mer, genes induced greater than a 1.5-fold by the 4mer and not by other treatments were classified using three different in silico tools; the Bar Toronto Classification Super Viewer tool (http://bar.utoronto.ca/; Supplementary Materials Figure S1), the agriGO Tool [39], (http://bioinfo.cau.edu.cn/agriGO/) (Figure S2) and the PageMan tool from MapMan Software [40], (Figure S3). These analyses showed that genes and functional categories induced by the chitin 4mer belonged principally to the categories of developmental processes, cell organization, biogenesis, multicellular organismal development, membrane transport and primary amino acid metabolism (Figures S1 and S3). Among the genes repressed by the 4mer, genes related to biotic stress responses were overrepresented, and the proportion of these biotic stress-related genes (among all of the repressed genes) was higher than the proportion of induced biotic stress-related genes by the 4mer (Figure S3). These results are consistent with a non-biotic stress plant molecular response to the chitin 4mer, more closely related to the promotion of plant development and completely different from the previously-identified response to Nmer treatment, related to activation of plant defense and innate immunity [32].

2.2. Analysis of Chitin Mix Enriched with Low-Molecular-Weight Chitin Oligosaccharides

Taking into account previous works [41–46], we produced a low-molecular-weight chitin mix (CHL) using combined thermal treatment and sonication on a high-molecular-weight chitin mix (CHH, see the Material and Methods Section). In order to characterize this CHL mix, several analyses were performed. First, to determine the composition of the CHL mix, the sample was analyzed using matrix-assisted laser desorption/ionization time-of-flight (MALDI-TOF) mass spectrometry. The mix of chitin fragments obtained by MALDI-TOF mass spectrometry is shown in Figure 2A. The MALDI-TOF spectrum of the chitin oligomers revealed that the sample was composed of chito-oligomers that were 100% acetylated with a degree of polymerization (DP) ranging from 2 to 6 (A2 to A6 at Table 1). The relative ion intensity of each signal can reflect the quantification of the products (Table 1, [47]). The theoretical molecular weight (TMW) and obtained molecular weight sizes (OMW) of the chitin fragments in the CHL mix and an estimation of the composition of the CHL mix are given in Table 1. The most abundant oligomers were A2, A3 and A4 (34.5%, 35.6% and 22.2%, respectively), then A5 and A6 at 6.48% and 1.20%, respectively, with no oligomers with a DP higher than six (Table 1). This 100% acetylated CHL mix was used for further experiments on plants.

Secondly, the effect of the thermal treatment and sonication on the samples was explored using X-ray crystallography (XRD). As seen in Figure 2B, both non-treated (CHH) and thermal-treated chitin (TCH) showed XRD patterns with strong reflections at 9.2° and 19.2° and minor reflections at 12.6°, 22.9°, and 26.2°. The results showed that the non-treated chitin (CHH) had a crystallinity index (CrI) of 73%, and this value was reduced to 67% after the thermal treatment as expected (thermally-treated chitin (TCH) in Figure 2B). Additionally, after the thermal treatment, plus sonication, (sonicated chitin after thermal treatment, TSCH sample in Figure 2B), the sample had a much lower CrI value (around 50%), and the intensity of the diffraction was less intense than that of CHH. Both the crystallinity pattern and CrI were in good agreement with those previously reported for α-chitin [48].

Finally, in order to confirm the presence or absence of larger molecular-weight chitin oligosaccharides in the CHL sample after the thermal treatment plus sonication, a proton magnetic spectroscopy (^{1}H-NMR) analysis was performed. The ^{1}H-NMR profile (Figure 2C) shows that the characteristic signals of acetyl protons and the ring protons were around 2.6 ppm and 3.6–4.5 ppm, respectively. The H-1 of internal acetylated units resonated at 4.91 ppm, while the characteristic resonances in the anomeric region of the acetylated α- and β-anomers were 5.43 and 5.05 ppm, respectively. Signals corresponding to N-acetylglucosamine residues at 5.07 (H-1), 5.65 (H-1 reducing end, α), 5.21 (H-1 reducing end, β), 3.44 (H-2), 3.57 (H-1 reducing end, α) and 3.32 (H-1 reducing end, β) were detected [49]. Based on this spectrum, a degree of polymerization (DP) of five was estimated.

These results confirmed that thermal treatment plus sonication induced the opening and breaking of the CHH biopolymer, forming the CHL mix enriched with A2, A3 and A4 fragments (a total of 92.3%, see Table 1), with no contamination with higher molecular-weight chitin oligosaccharides with more than six monomers.

Table 1. Estimated composition of CHL mix. A comparison between Theoretical Molecular Weights MNa$^+$ (TMW) and Obtained Molecular Weights (OMW) is shown. The corresponding intensity and percentage of each oligosaccharide in the mix CHL obtained by MALDI-TOF shown at Figure 2A is also indicated, Amer: number of N-acetylglucosamine oligosaccharides.

Amer	TMW (*m/z*)	OMW (*m/z*)	Intensity	%
A2	446.85	447.16	1166	34.53
A3	650.09	650.24	1203	35.63
A4	853.28	853.31	750	22.21
A5	1056.33	1056.39	219	6.48
A6	1259.56	1259.47	38	1.12

(A)

(B)

(C)

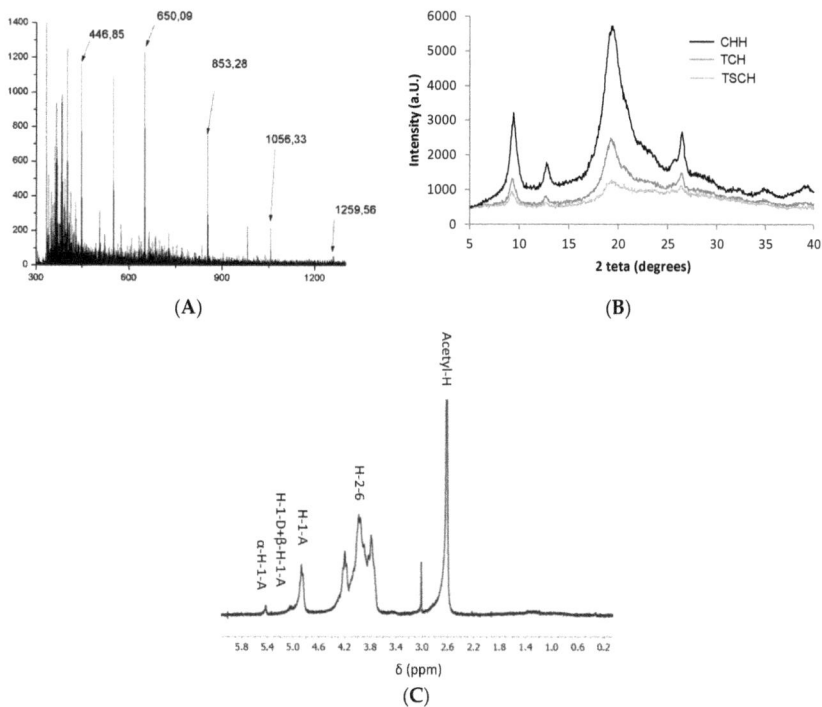

Figure 2. CHL sample characterization. (**A**) Matrix-assisted laser desorption/ionization time-of-flight (MALDI-TOF) Ultraflex profile of CHL mix in a 2,5-dihydroxybenzoic acid (DHB) matrix. The obtained molecular weights (OMW) are marked with arrows; (**B**) X-ray diffraction (XRD) pattern of the following chitin mixes: untreated chitin (CHH), thermally-treated chitin (TCH) and sonicated chitin after thermal treatment (TSCH). The intensity is in arbitrary units (a.u.), and 2θ degrees represents the diffraction angles; (**C**) Proton nuclear magnetic spectroscopy (1H-NMR) spectrum (300 MHz) of CHL in concentrated deuterium chloride (DCL) at room temperature. The signals of *N*-acetylglucosamine residues are marked.

2.3. Analysis of the Vegetative Growth of Chitin-Treated Arabidopsis

As the microarray analysis of genes that responded to the chitin 4mer suggested that the 4mer mostly induced genes related to plant development and nitrogen and carbon metabolism, a study was designed to test whether small chitin fragments might produce a direct effect on plant growth. For these experiments, plants were treated with CHL under in vitro conditions, to control the composition of the growth medium (thereby avoiding the confounding factors that might occur in a greenhouse experiment). *Arabidopsis* seedlings were grown for 21 days under these in vitro conditions with a low concentration of nitrogen in the medium (see the Materials and Methods), in either the presence or absence of CHH [18,26] and CHL.

After 20 days, there was an increase in the radicle length of 6% in the CHL group treatment, and 11.5% in the CHH group treatment was observed in plants after 20 days, compared with the control (Figure 3A,C). The total plant fresh weight increased by 10% in the CHL group treatment compared with the controls, although the total plant fresh weight decreased by 10% in the CHH group (Figure 3B).

A

B

C

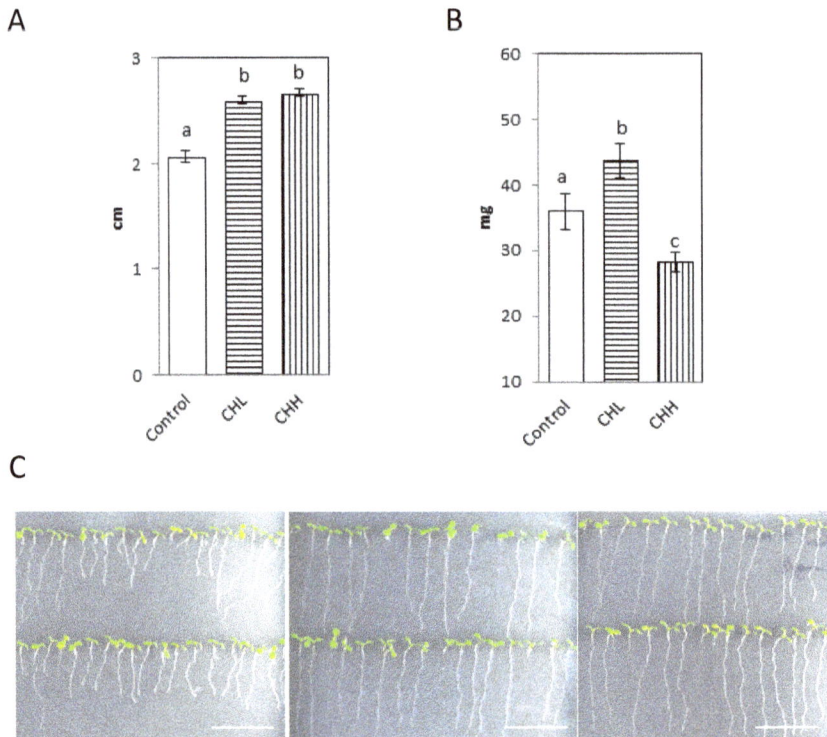

Figure 3. Chitin increased plant growth in vitro. (**A**) The radicle length of controls and plants treated with low-molecular-weight chitin mix (CHL) or high molecular weight chitin mix (CHH); (**B**) Fresh weight of controls and plants treated with CHL or CHH. The plants were grown for 20 days; (**C**) Representative plates of control seedlings (left), seedlings treated with CHL (center) and treated with CHH (right) after seven days. The experiments were performed at least three times with similar results. The data were analyzed using one-way analysis of variance (ANOVA) and the Statgraphics program Centurion XVI.II. Different letters indicate significant (*p*-value < 0.05) differences between treatment groups, according to Duncan's test. Bars: 2 cm.

2.4. Analysis of the Total Content of Nitrogen and Carbon Content of Chitin-Treated Arabidopsis

To determine whether chitin acted as a source of nutrients for the plants, total nitrogen and carbon were quantified in the CHL- and CHH-treated plants. An increase in plant total nitrogen content was observed in all treatments groups after 10 days, with increases of 5% in the CHH group and 8% in the CHL group compared to the controls (Figure 4A,B). A similar effect on total carbon content was observed in the CHL- and CHH-treated plants after 10 days, relative to the controls (Figure 4C), with no significant differences observed between the two treatment groups (Figure 4D).

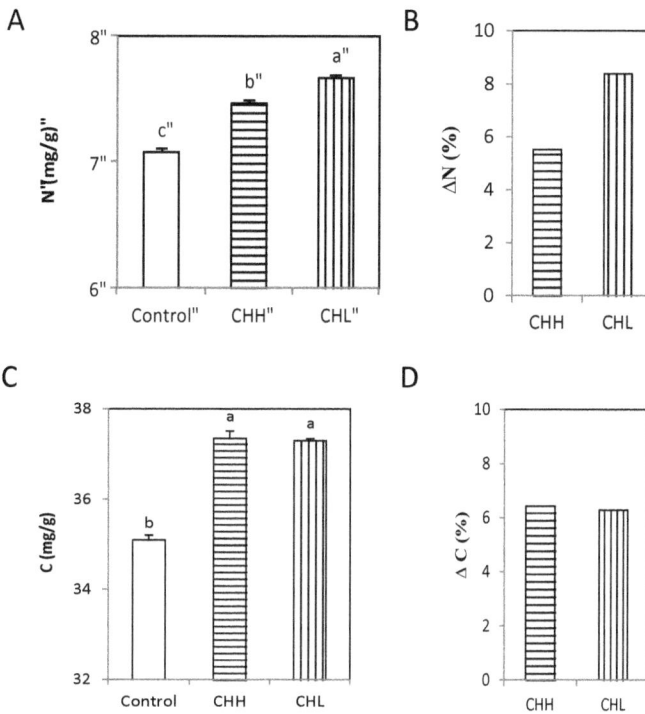

Figure 4. Chitin induces an increase of the total nitrogen and carbon content of *Arabidopsis* plants in vitro. (**A**) Total nitrogen content in controls (white bars) and plants treated with the low-molecular-weight chitin mix (CHL) or high-molecular-weight chitin mix (CHH); (**B**) Percentage of increase (ΔN) in total nitrogen content in plants treated with CHL and CHH relative to the controls; (**C**) Total carbon content in controls and plants treated with CHL and CHH; (**D**) Percentage increase (ΔC) in total carbon content in plants treated with CHL and CHH relative to the controls. Different letters indicate significant (p-value < 0.05) differences between treatment groups according to Duncan's test. The measures were taken after 10 days of growth. The experiments were performed at least three times with similar results.

2.5. Analysis of the Vegetative Growth of Chitin-Treated Poplar Explants

In order to determine the effect of chitin on plant growth in other species, poplar explants of a *Populus trichocarpa* clone were grown under similar in vitro conditions for 45–70 days, following the same growth parameters as those in the *Arabidopsis* experiments (see the Materials and Methods). An increase in root length (up to 5%) and shoot length (up to 28%) was observed in the CHL-treated poplar explants relative to the controls (Figure 5A,B).

It was not possible, however, to experimentally determine the narrow range of exchange of the total nitrogen, carbon content and fresh weight of the treated poplar explants compared with the controls because of the high variability found between the plants in the measurements on these parameters. Additionally, because the poplar experiments allowed us to follow the plant growth for longer periods than the *Arabidopsis* experiments (as the *Arabidopsis* experiments always involved younger plants, before *Arabidopsis* transitioned to flowering), we explored longer-term poplar growth in the presence of CHL and CHH. However, leaf chlorosis and cell death were observed after 70 days (growth parameters for these experiments were not measured). These symptoms were not observed in the presence of CHL (Supplementary Figure S5).

A
B

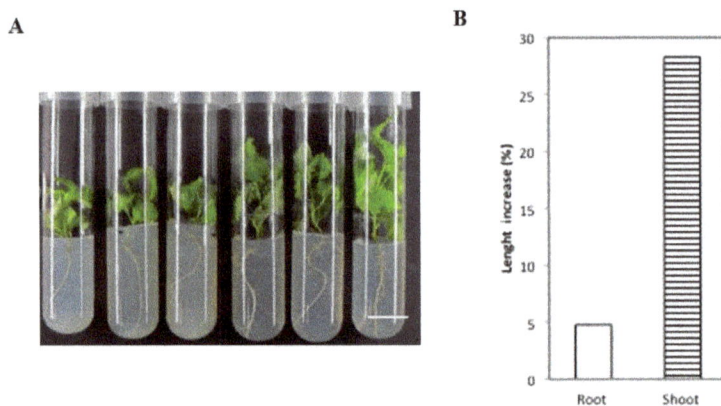

Figure 5. CHL produces an increase in shoot and radicle length in poplar explants. (**A**) Representative triplicates of poplar explants in the absence of CHL (in the three tubes on the left) or in the presence of CHL (in the three tubes of the right); (**B**) Increase in radicle and shoot length of explants grown in the presence of 100 µg/mL CHL compared to controls. The photos were taken after 45 days of growth. Bars: 2 cm.

3. Discussion

It has been extensively documented in the literature that high-molecular-weight chitin induces plant defense-related responses, protein phosphorylation and reactive oxygen species production at the molecular level [35]. In this study, we examined the expression profiles of *Arabidopsis* seedlings treated with a chitin 4mer and compared this response to plants treated with an 8mer and a high-molecular-weight chitin mix (CHH) using full-genome Affymetrix microarrays. Previously, we showed that treatment with a chitin 8mer elicited a host defense-related cellular response in *Arabidopsis* similar to that induced by CHH in *Arabidopsis* [18,26]. The induction of specific defense-related genes after treating *Arabidopsis* with CHH, has also been documented by other groups experimenting with *Arabidopsis* [21,50,51], rice and soybean [19,52–54]. Our results showed that chitin 4mer treatment induced an expression pattern that was distinct from the pattern observed after 8mer or CHH treatments. Interestingly, unique subsets of genes responded to each chitin treatment, suggesting that different chitin oligomer lengths activate different signaling pathways in *Arabidopsis*. While there was significant overlap in the expression patterns elicited by the 8mer and CHH, the gene expression patterns elicited by the 4mer were completely different. In silico analysis of the genes that were uniquely induced or repressed by the 4mer showed that the gene families that were overrepresented in the set of genes induced were not present in the set of repressed genes. These genes were related to lipid and nitrogen metabolism, nutrient physiology signaling and the transport of ammonium, sugars and potassium. In contrast, defense- and stress-related genes were overrepresented in the set of repressed genes, while a similar set of genes (i.e., stress response genes) was completely absent from the set of induced genes.

Analysis performed using the agriGO tool (http://bioinfo.cau.edu.cn/agriGO/) showed that the genes induced by the 4mer belonged principally to functional categories of developmental processes, cell organization, biogenesis, membrane transport and primary amino acid metabolism. A large number of genes related to biotic stress responses were overrepresented in the set of the genes that were repressed by the 4mer.

The 4mer induced several previously characterized development-related genes. These genes included *IPT5*, which has been implicated in stem cell initiation and meristem formation [55], *PTL* involved in auxin signaling [56], *EXPA22*, which has been implicated in cell elongation [57], *JAZ7*, which is involved in secondary growth [58], and ANAC101, which is involved in xylogenesis [59]. Other known genes that are known to be related to nutrient transport were also induced specifically by the 4mer, such as the sucrose transporter *AtSUC 7*, the Golgi sugar transporter *GONST5* [60,61], the *IRT1* iron transporter [62] and the amino acid transporter *AtPUP9* [63] (see selected genes in Table S1). In addition, several previously-characterized genes were associated with lower (but still upregulated) translational activation levels. These genes included genes related to xylogenesis [59,64], vascular patterning [65,66], early seedling development [67], cellular differentiation [68] and shoot development [69]. In parallel, genes that are known to be involved in inhibiting the activation of the jasmonic acid-mediated defense pathway in the absence of pathogens were also specifically induced by the 4mer, such as genes in the JAZ family [70].

However, there was no 4mer-induced upregulation of marker genes related to the defense response, such as pathogenesis-related protein 1, *PR1* (AT2G14610), defensin *PDF1.2* (AT5G44420), basic chitinase *PR3* (AT3G12500) and *MAPK3* (At3g45640). The transcriptional level of *MAPK3*, after 4mer treatment, was confirmed using qRT-PCR (see the Supplementary Materials). In addition, several unknown genes that belong to categories related to plant development were induced by the chitin 4mer. These changes in gene expression after 4mer treatment concur with the hypothesis that the exposure of a plant to low-molecular-weight forms of chitin might results in a molecular adaptation response and bio-stimulation of plant growth, rather than a stress response.

Small LCOs have recently been shown to produce an increase in root growth in maize [38]. We found that the transcriptional plant response to the chitin 4mer overlapped with the transcriptional response observed in maize in this previous study, especially regarding the over-represented gene families related to nutrient and ion transport, embryogenesis, secondary metabolism and gluconeogenesis. It should be highlighted that, according to the results of the study on maize, short-chain chitin oligomers and lipo-chitin oligomers may have overlapping roles in plant growth promotion and, perhaps, in plant-symbiont interactions. Other studies have shown that a chitin 4mer can serve as the backbone of Nod factors during the interaction between the symbiotic bacteria, *Rhizobia*, and the roots of legumes [34,35,71,72]. Additionally, short-chain chitin oligomers can trigger nuclear calcium spikes, a cellular event that also occurs also during Nod and mycorrhizal (Myc) factor-mediated symbiotic signaling [36]. The induction of symbiotic signaling by the arbuscular mycorrhizal fungal-produced LCOs and COs has been observed in both legumes and rice. In parallel, the detection of these LCOs by grasses and other non-legumes that act as hosts for arbuscular mycorrhizal fungi is potentially controlled by Myc receptors. Recently, it has been found that the intraradical colonization by arbuscular mycorrhizal fungi triggers the induction of a new LysM-type LCO receptor, LYS11 [73]. In addition, in *Medicago truncatula*, a high-affinity LCO-binding protein (LYR3) interacted with a key symbiotic receptor (LYK3) [74]. The detection of LCO during the establishments of legume-Rhizobium symbiosis is controlled by a LysM receptor-like kinase known as nodulation factor perception (NFP) in *M. truncatula* [75]. In rice, OsCERK1 was found to regulate both chitin-triggered immunity and symbiosis with arbuscular mycorrhizal fungi [76]. LCOs can modulate plant host immunity to enable endosymbiosis [77], and the ability to detect LCOs might have evolved from plant innate immunity signaling [35].

Additionally, chitosan oligomers can target the chromatin within the plant nucleus, altering the chromatin conformation and gene expression of the plant cell [28,29]. Although chitosan has a high affinity for DNA and chitin does not, a still unknown effect of chitin oligomers in the plant cell should not be discarded. All of these data indicate that the role of small COs in nature related to plant growth still needs more biochemical clarification.

The results from the in vitro experiments of plants grown in the presence of CHL and CHH might be in line with the hypothesis that these oligosaccharides might have a bio-stimulating effect on plant growth. This hypothesis is further supported by the transcriptional data, since the genes that responded differentially were related to ammonium, amino acid and glutamate metabolism, the hexosamine biosynthetic pathway and nutrient transport. However, additional experiments are necessary to determine the validity of this hypothesis. It is still unclear whether the observed effects on development are only a consequence of the observed transcriptional activation.

A previous study indicated that a highly purified chitin 5mer was able to increase root length in *Arabidopsis*, although the mechanism behind this effect remains unknown [78]. The increase in root length produced by treatment with CHL in this study was similar to that observed in the study on the chitin 5mer (i.e., around a 25% increase compared with controls). The 5mer comprised only 6.48% of the CHL, and there was a higher percentage (a total of 92.37%) of 2mer (dimer), 3mer (trimer) and 4mers (tetramer), so it is plausible that the principal effect observed in both studies was a consequence of a response specifically to fragments of five or fewer monomers.

We also observed that poplar explants were able to grow in medium containing only CHH as a source of carbon, although leaf chlorosis and cell death were observed after long periods of growth under in vitro conditions. Interestingly, these symptoms were not observed in the presence of CHL. The stress-response symptoms were also not observed in in vitro experiments involving *Arabidopsis* grown in the presence of CHH (but not in the presence of CHL). This may be because our experiments and analysis in *Arabidopsis* experiments were always performed using younger plants (in comparison with the poplar assays) well before *Arabidopsis* transitioned to flowering.

It is reasonable to hypothesize that plants may be able to degrade CHH, thereby releasing smaller fragments of chitin that could promote plant development (which may have happened in the case of treatments with CHL) or be used as a nutrient source for both the plant itself and the soil microorganisms present in the rhizosphere. This process would require a prior stress-response activation to induce chitinases production, thereby putatively leading to a negative trade-off concerning plant development. However, more detailed experiments are needed to determine whether this is the case.

It should be noted that previous studies have shown that plants can respond to CHH with increased expressions of innate immunity- and defense-related genes with induction ratios of more than 200-fold compared with controls [18,26]. However, the plant transcriptional changes produced by the chitin 4mer in the present study were much lower, with no inductions above eight-fold compared with the controls, and the genes induced by the 4mer were related principally to plant development, with a suppression of the genes involved in biotic stress. These results are in line with a molecular mechanism of adaptation response (rather than a transcriptional response to stress in order to aid survival), and the results suggest that the CHL described in this study might be useful as a bio-fertilizer, or, at a minimum, a bio-stimulator of plant development when combined with other nitrogen, phosphate and magnesium (NPK) fertilizers currently used in agriculture.

It is also important to determine whether plants possess the molecular machinery needed to transport COs and use them as a direct source of carbon and nitrogen, similar to chitinolytic microorganisms [5]. It is obvious that the energetic cost associated with the production of chitinases and for the subsequent transport and assimilation of chitin fragments by a plant would be higher than the energetic cost required to transport and assimilate small oligosaccharides. This fitness cost might explain why the increase in total nitrogen content observed in the CHL treatment-plants was higher than in the CHH-treated plants. Analogously, based on the microarray data, the stress response induced by the chitin 8mer and CHH may be associated with the cost of the synthesis of chitinases, which was not observed after the 4mer treatment. This fact might explain the stress that CHH treatment produced on poplar explants after longer periods of growth, which was not observed in CHL-treated plants. Our experiments showed that several chitinases were highly induced by the 8mer and CHH, but not at all by the 4mer. However, at the time points tested, no significant difference was observed in total carbon content or radicle length between the 4mer and 8mer/CHH treatments, although there were increases in comparison to the controls. The increase in growth seen in the CHH group might be attributable to the degradation of CHH oligosaccharides by the plant, and the energetic costs associated with this process might make CHH a less efficient bio-stimulator compared with CHL.

Although the molecular mechanism behind increased plant growth in the presence of CHL is still unknown, plants exhibit well-known chitinolytic activity in response to phytopathogens [79], sharing similarities with chitinoclastic organisms. Chitinoclastic activity has been described previously in several microorganisms, including *Vibrio furnissii* [5], *Vibrio carchariae* [80], *Amycolatopsis orientalis* or *Kitasatospora* sp. [81], *Serratia marcescens* [82], *Escherichia coli* [83] and *Streptomyces coelicolor* [84]. The chitinoclastic cascade in *Vibrio* has been characterized [6] and found to involve 10 genes that are implicated in chitin catabolism. *Vibrio* is able to incorporate the chitin-derived glucose into the glycolytic pathway and the amino groups into amino acids [5,85]. The presence of orthologous genes related to a chitinolytic pathway in plants and the induction by the chitin 4mer, but not by 8mer or CHH of *Arabidopsis* genes related to amino acid, sugar and ammonium transport (Figure S1), indicate that plants might utilize the amino groups derived from chitin as a source of nitrogen. However, this response might only involve transcriptional stimulation of plant growth rather than the stimulation of a chitinoclastic pathway. Along the same lines, the 4mer slightly induced enzymes that are related to glutamine and glutamate synthesis, a mechanism that is known for its roles in the induction of cell growth and as a nutrient-responsive signaling pathway [86].

As commercially available purified small chitin 4mers and chitosan fragments or mixes are extremely expensive to be used in the high volumes, in this work, we developed a less expensive method to obtain a chitin mix, enriched with low-molecular-weight oligosaccharides. This method could be used in the future in greenhouses and fields. The mainstream process for obtaining chitin- or chitosan-derived oligomers involves acid hydrolysis, several deproteinization steps (depending on the natural origin of the chitin), deacetylation and purification by high-performance liquid chromatography, which can lead to the production of oligomers with very high levels of purity. In addition, if chitin oligomers are required, chitosan oligomers need to be re-acetylated [41,42]. As we were interested in producing a mixture of chitin fragments with as high a degree of acetylation as possible and a low molecular weight, we optimized a protocol based on previous work that indicated that chitin biopolymers can be fragmented using sonication. Previous studies indicated that sonicated chitin is more susceptible to chitinase activity than the non-treated biopolymer [39,40]. Sonication of chitosan also decreased its molecular weight and crystallinity grade [43]. In addition, we observed that other studies showed that heating chitin during the depolymerization process made it more accessible to the action of chitinase [44]. Both processes had no effect on the functional groups found in the polymers. Our newly-developed protocol is a quick method that allows CHL to be produced at a low cost, which means that CHL could potentially be used in agriculture and industry, something that might be impossible with previous biochemical methods used to obtain small forms of chitin oligosaccharides.

Our present and previous results highlight the high variability of chitin-derived compounds in nature, making the study of plant responses to chitin very complex. Additional experiments are currently in progress. These experiments aim to determine the activity of CHL on soil, greenhouse and field conditions and the molecular mechanisms that allow plants to respond differentially to different chitin oligosaccharides in natural conditions, in order to determine the biotechnological potential of the CHL mix for promoting plant growth in the field and to be used as a bio-stimulator of plant development.

4. Materials and Methods

4.1. Plant Growth and Chitin Treatments

For the microarray analysis, *Arabidopsis thaliana* (*Col-0* ecotype, obtained from Arabidopsis Biological Resource Center stock) seeds were treated according to a previously-reported procedure [26], with small modifications. The seeds were surface sterilized and grown in liquid Murashige and Skoog culture medium at a density of approximately 500 seeds (10 mg) per 125-mL flask. The flasks with the seeds were incubated at 4 °C for 6 days and then placed in a shaking incubator at 150 rpm for 2 weeks under constant illumination (125 µmol·m^{-2}·s^{-1}) at 23 °C. After 14 days, the seedlings were treated with purified chitin oligo-4mer (4mer, Seikagaku Corporation, Tokyo, Japan) at a final concentration of 100 µg/mL. After 30 min of treatment, the seedlings were harvested, flash-frozen in liquid N$_2$ and stored at −80 °C until analysis.

For the in vitro *Arabidopsis* experiments, *Arabidopsis thaliana* (*Col-0* ecotype, obtained from ABRC stock) seedlings were surface sterilized (30% bleach and 0.01%, sodium dodecyl sulfate for 20 min), stratified (i.e., cold treated) for 2 days at 4 °C and placed in small square petri dishes containing 70 mL of 1/2× Murashige and Skoog (MS) Basal Salt Mixture (2.28 g/L, modification 1B micro and 1/2 macro elements including vitamins, # M0233.0050, Duchefa Biochemie, Haarleem, The Netherlands) plus 2% sucrose at pH 5.8. The plates were then transferred into a growth chamber. The medium contained half of the nitrogen present in standard media (825 mg/L, 10.3 mM).

For in vitro poplar experiments, the hybrid poplar *Populus tremula* × *P. alba INRA clone 717 1B4* was used; explants of 3–4 cm obtained from 60-day-old plants were transferred into glass tubes with 15 mL of the MS medium described above and placed into an in vitro chamber. The *Arabidopsis* seedlings were grown for 21 days and poplar explants for 70 days at 60% humidity (*v/v*), temperatures of 24 °C during the day and 22 °C during the night, with a 14-h light/10-h dark photoperiod and a light intensity of 150 µE·m^{-2} per s for all experiments.

In order to obtain the CHL mix for the CHL treatments, ultrapure chitin from shrimp shells (Cat#C9752, acetylation degree higher than 95% Sigma-Aldrich, St. Louis, MO, USA) was finely ground using a grinder and a porcelain mortar to obtain a homogeneous mixture of chitin. This was then suspended in ddH$_2$O (1–4 g/L, Milli-Q purity grade) and autoclaved for 20 min at 121 °C. Once the solution was cooled to room temperature, the mix was subjected to sonication (50 Hz) for 15 min at a temperature lower than 25 °C. The suspension was then stored at 4 °C until further use or lyophilized for further analysis.

4.2. Oligomer Characterization

For CHL characterization, a known amount of the filtrate of the CHL solution was initially analyzed, and no soluble oligomers were detected in the liquid. The sample was then freezing dried.

XRD patterns were obtained using a Bruker D8 Advance diffractometer with CuKa radiation (step size, 0.05, counting time, 3.5 s). Sample crystallinity (CI) was determined using the following equation previously described [87]: CI (%) = $[(I_{110} - I_{am})/I_{110}] \times 100$, where I_{110} (arbitrary units) is the maximum intensity of the (110) peak at around 2θ = 19° and I_{am} (arbitrary units) is the amorphous diffraction at 2θ = 12.6°.

^{1}H-NMR spectra were recorded using a Varian spectrometer at 300 MHz (spectral width = 8000 Hz, number of transients = 128, block size = 4, recycle delay = 5 s). The mean polymerization degree (DP$_n$) and the fraction of acetylated units (FA) were calculated using a previously described method by [47].

For the ^{1}H-NMR measurements, a sample was dissolved at 4 °C in concentrated deuterium chloride (Sigma-Aldrich) at a concentration of 15 mg/mL. DPn was calculated based on the integrated area associated with all of the H1 protons divided by the integrated area associated with the reducing end protons. The mass spectra were recorded using a Bruker Ultraflex (Bruker Daltonik, Bremen, Germany) with MALDI-TOF/TOF equipment in positive-ion mode. For ionization, 2,5-dihydroxybenzoic acid was used as the matrix. The oligomers were soaked in a mixture of 1:1 water: methanol and mixed with the matrix prior to the analysis.

4.3. Microscopy and Photography Techniques

A stereomicroscope (MZ9, Leica Microsystems, Leica, Deerfield, IL, USA) with a charge-coupled device (CCD) camera (DC 280, Leica Microsystems) was used to obtain photos of the *Arabidopsis* seedlings growing on the plates. Image processing was performed using the ImageJ Software [88].

4.4. Nitrogen and Carbon Content Analysis

For each plate, 12 seedlings per treatment group were collected and grouped as a single sample. The aerial parts of each plant were separated from the roots, cleaned and dried in an oven at 65–70 °C for at least 48 h. The dried tissues were finely ground (to sizes less than 150 μm) in a porcelain mortar. The concentrations of nitrogen and carbon were determined using a mass elemental analyzer for macro-samples (LECO CHN-600, Leco Corp. St. Joseph, MI, USA) according to the manufacturer's instructions.

4.5. Data Analysis of Growth Parameters

All statistical analyses were performed using StatGraphics Centurion XVI.II (StatPoint Technologies, Inc., Warrenton, VA, USA). A one-way analysis of variance (ANOVA) and Duncan's mean comparison test were performed for the all experiments or *t*-test with a significance level of 0.05%. In the case of non-homogeneous variance, a nonparametric Kruskal–Wallis test was applied.

4.6. Microarray Preparation, Hybridization and Data Extraction

Total RNA samples were processed according to the manufacturer's protocols with the following modifications (Affymetrix GeneChip Expression Analysis Technical Manual, Affymetrix, Inc., Santa Clara, CA, USA). Single-stranded, then double-stranded cDNA was synthesized from the polyA$^+$ mRNA present in the isolated total RNA (20 μg of total RNA was used as the starting material in each sample reaction) using the SuperScript Double-Stranded cDNA Synthesis Kit (Invitrogen Corp., Carlsbad, CA, USA) and custom poly (T)-nucleotide primers that contained a sequence recognized by T$_7$ RNA polymerase. The resulting double-stranded cDNA was used as a template to generate biotin-tagged cRNA from an in vitro transcription reaction, using the Bio-Array High-Yield RNA Transcript Labeling Kit (Enzo Diagnostics, Inc., Farmingdale, NY, USA). In accordance with the prescribed protocols, 20 μg of the resulting biotin-tagged cRNA were fragmented to strands of less than 100 bases in length following prescribed protocols (Affymetrix GeneChip Expression Analysis Technical Manual, Affymetrix, Inc., Santa Clara, CA, USA). The 20 μg fragmented target cRNA (20 μg) were hybridized at 45 °C with rotation speed of 60 rpm for 16 h (Affymetrix GeneChip Hybridization Oven 640) with the probe sets present on an Affymetrix ATH1 GeneChip array. The GeneChip arrays were washed and then stained (with streptavidin-phycoerythrin) on an Affymetrix Fluidics Station 400, followed by scanning on a Hewlett-Packard GeneArray scanner (Hewlett-Packard, Palo Alto, CA, USA). Three independent biological replicates were performed for each sample. Image analysis and pixel intensity were quantified using MicroArray Suite 5.0 software (Affymetrix, Inc., Santa Clara, CA, USA). Text files were then generated and exported to TM4 Microarray Software swift Version MeV 4.9 (http://www.tm4.org/index.html) for normalization and further analysis.

The data discussed in this publication have been deposited in the U.S. National Center for Biotechnology Information's Gene Expression Omnibus (GEO) database [89] (see online resources below).

4.7. Microarray Data Analysis

The data were filtered and analyzed using the Statistical Analysis of Microarrays (SAM) program [90]. Genes identified as significant using SAM were exported and clustered using GeneSpring 6.0 (Figure 2A). Text files containing raw data were imported to TMEV [91] and were normalized as follows. First, values below 0.01 were set to 0.01, and then, each measurement was divided by the measurement at the 50th percentile of all of the measurements in the sample. The specific samples were then normalized to one another: three replicates of each treatment were normalized against the median of the control samples (water treatment). Each measurement for each gene in the specific samples was divided by the median of that gene's measurements in the corresponding control samples. The raw data on all of the genes were then extracted and analyzed for significance using the Statistical

Analysis of Microarrays (SAM) software program [90]. Genes determined to be statistically significant were listed, and the resulting information was imported into TMEV for further analysis. The genes were grouped by their biological function according to their gene ontology (GO) annotation using the Bar Toronto Super Viewer tool (Figure S1), MapMan tool (Figure S2; [40]) and agriGO toolkit (Figure S3; [39]).

4.8. Quantitative Reverse Transcription-PCR Analysis for Microarray Data Validation

Total RNA was isolated from the frozen plant tissues using TRIzol Reagent (Invitrogen®, Carlsbad, CA, USA) according to the manufacturer's protocol. The RNA samples were treated with RQ1 DNase (Promega, Madison, WI, USA). Trace amounts of genomic DNA were removed by digestion with Turbo DNA-*free*™ (Ambion, Austin, TX, USA). First-strand cDNA synthesis was primed using an oligo $(dT)_{15}$ anchor primer, and cDNA was synthesized using a First-Strand Synthesis Kit (Amersham-Pharmacia, Rainham, UK) according to the manufacturer's protocol. An aliquot of 1.5 µL of the first-strand synthesis reaction was used as the template for PCR amplification. To ensure that the sequence amplified was specific, a nested PCR was performed using 1 µL of a 1:50 dilution of the products synthesized in the first PCR reaction as a template. The RT-PCR, PCR and nested PCR program consisted of: 3 min at 96 °C, 40 cycles of 30 s at 94 °C, 30 s at 65 °C and 1 min at 72 °C. The final extension step consisted of 7 min at 72 °C. The amplified PCR fragments were visualized using 1.5% agarose gels.

The qRT-PCR experiments were performed using a SYBR® Green qPCR kit (Finnzymes, Espoo, Finland) with reactions at a final volume of 20 µL per well and using the cycle protocol recommended by the manufacturer. The samples were run in a DNA Engine Opticon® 2 System instrument with the PTC-200 DNA Engine Cycler and CFD-3220 Opticon™ 2 Detector (Bio-Rad, Hercules, CA, USA). Gene-specific primers were designed using the Primer Express 2.0 program (Applied Biosystems, Foster City, CA, USA), and minimal self-hybridization and dimer formation of primers were determined using the Oligo 6.0 program (Molecular Biology Insights, West Cascade, CO, USA). Primers with annealing temperatures of 62 °C–65 °C that amplified products with lengths of about 300 bp were selected and then verified for specificity using Basic Local Alignment Search Tool (BLAST) searches. The efficiency of amplification for each gene was calculated as recommended by the manufacturer (Bio-Rad, Hercules, CA, USA). The following gene specific primers were used for quantitative RT-PCR: β-*ACTIN* (At3g18780): 5′-GTGATGAAGCACAATCCAAG-3′ (forward) and 5′-GAACAAGACTTCTGGGCAT-3′ (reverse); *MAPK3* (At3g45640): 5′-ATGAACACCGGCGGTGGCC-3′ (forward) and 5′-GGCATTCACGGGGCTGCTG-3′ (reverse); *ATSUC9* (At5g06170) 5′-AGCCGTTGGTTTCTTCGT-3′ (forward) and 5′-CTAATCACTCCAATAACAAG-3′ (reverse); *ATJAZ7* (At2g34600) 5′-CGGATCCTCCAACAATCC-3′ (forward) and 5′-GACAATTGGATTATTATG-3′ (reverse); *ATECP31* (At3g22500) 5′-GTCGAAGCACCTGATGTAGC-3′ (forward) and 5-GAGCAATGACGTTGGTACC-3′ (reverse); *ATEXPA22* (At5g39270) 5′-GTCGAAGCACCTGATGTAGC-3′ (forward) and 5′-CCACAAGCTCCCTGTTGAG-3′ (reverse). Data acquisition was performed using the Opticon Monitor Analysis software (Version 2.01), and changes in the transcript levels were determined using the $2^{-\Delta\Delta CT}$ method [92]. Data points were compared using *t*-tests. Three independent biological replicates were used in each experiment.

4.9. Bioinformatics Analysis

Additional information about gene expression and data analysis tools was obtained from the following webpages: http://affymetrix.arabidopsis.info/narrays/, https://www.genevestigator.ethz.ch/, http://aramemnon.botanik.uni-koeln.de/, http://www.us.expasy.com/tools/, http://www.ncbi.nlm.nih.gov/ and http://www.ebi.ac.uk/Tools.

5. Conclusions

Using genomic approaches, we have identified several plant genes that respond to low-molecular-weight chitin-derived oligosaccharides. We identified a genomic and physiological link between transcriptional activation and plant development and a concomitant increase in plant growth and nutrient content in vitro. In addition, this study demonstrates the power of microarray data to identify potential transcriptionally activated metabolic networks in order to characterize novel signaling pathways. Our results concur with those of previous studies that show that, in nature, plants might activate a developmental response if small chitin fragments are present in the rhizosphere. Additional work is in progress to determine the exact molecular pathways that allow these compounds to stimulate plant development and their role in the natural environment.

Supplementary Materials: The following are available online at www.mdpi.com/1660-3397/15/2/40/s1. Figure S1: In silico data analysis of genes induced by the 4mer, Figure S2: Functional classification of genes specifically induced by the 4mer, Figure S3: Overrepresentation analysis of genes differentially induced or repressed by the 4mer, Figure S4: Verification of microarray results for selected genes responding differentially to 4mer measured by qRT-PCR, Figure S5: Poplar growth under different conditions into the medium, Table S1: Data represent the corresponding array signal values on selected genes analyzed by qRT-PCR (Figure S4), Table S2: Selected known development related genes specifically induced by the 4mer more than 1.5 related to controls. Online Resources: U.S. National Center for Biotechnology Information's Gene Expression Omnibus (GEO) database. Accession Number: GSE83858 (https://www.ncbi.nlm.nih.gov/geo/query/acc.cgi?acc=GSE83858).

Acknowledgments: Funding was provided in part by grants to Shauna Somerville from the Carnegie Institution of Science and the National Science Foundation, USA (#0114783). Alexander J. Winkler was funded by the Erasmus Program at Universidad Politécnica de Madrid (UPM). We acknowledge the kind contributions, at different stages of this project, of Fernando García Arenal, Pablo González Melendi and Mark Wilkinson from CBGP (UPM-INIA), Luis Díaz Balteiro and Carlos Calderón from MONTES, Carmen Muñoz from E.T.S.I. Forestales (UPM), Norma García from UPM and Elisabeth Magel from Hamburg University.

Author Contributions: Marta Berrocal-Lobo and Shauna Somerville conceived of and designed the experiments. Alexander J. Winkler and Jose Alfonso Domínguez-Núñez and Shauna Somerville contributed reagents/materials/analysis tools, and Inmaculada Aranaz performed experiments. César Poza-Carrión, Inmaculada Aranaz, Katrina Ramonell and Marta Berrocal-Lobo analyzed the data. Marta Berrocal-Lobo wrote the paper.

Conflicts of Interest: The authors declare no conflict of interest.

1. Suresh, P.V. Biodegradation of shrimp processing bio-waste and concomitant production of chitinase enzyme and *N*-acetyl-D-glucosamine by marine bacteria: Production and process optimization. *World J. Microbiol. Biotechnol.* **2012**, *28*, 2945–2962. [CrossRef] [PubMed]

2. Keyhani Nemat, O. Saul Roseman Physiological aspects of chitin catabolism in marine bacteria. *Biochim. Biophys. Acta* **1999**, *1473*, 108–122. [CrossRef]

3. Donderski, W.; Swiontek Brzezinska, M. The Utilization of *N*-acetyloglucosamine and Chitin as Sources of Carbon and Nitrogen by Planktonic and Benthic Bacteria in Lake Jeziorak. *Pol. J. Environ. Stud.* **2003**, *6*, 685–692.

4. Khoushab, F.; Yamabhai, M. Chitin research revisited. *Mar. Drugs* **2010**, *8*, 1988–2012. [CrossRef] [PubMed]

5. Bassler, B.L.; Gibbons, P.J.; Yu, C.; Roseman, S. Chitin utilization by marine bacteria. Chemotaxis to chitin oligosaccharides by *Vibrio furnissii*. *J. Biol. Chem.* **1991**, *266*, 24268–24275. [PubMed]

6. Li, X.; Roseman, S. The chitinolytic cascade in Vibrios is regulated by chitin oligosaccharides and a two-component chitin catabolic sensorkinase. *Proc. Natl. Acad. Sci. USA* **2004**, *101*, 627–631. [CrossRef] [PubMed]

7. Killiny, N.; Prado, S.S.; Almeida, R.P. Chitin Utilization by the Insect-Transmitted Bacterium *Xylella fastidiosa*. *Appl. Environ. Microbiol.* **2010**, *76*, 6134. [CrossRef] [PubMed]

8. Zhao, Y.; Park, R.D.; Muzzarelli, R.A. Chitin deacetylases: Properties and applications. *Mar. Drugs* **2010**, *8*, 24–46. [CrossRef] [PubMed]

9. Kandra, P.; Challa, M.M.; Jyothi, H.K. Efficient use of shrimp waste: Present and future trends. *Appl. Microbiol. Biotechnol.* **2012**, *93*, 17–29. [CrossRef] [PubMed]

10. Povero, G.; Loreti, E.; Pucciariello, C.; Santaniello, A.; Di Tommaso, D.; Di Tommaso, G.; Kapetis, D.; Zolezzi, F.; Piaggesi, A.; Perata, P. Transcript profiling of chitosan-treated Arabidopsis seedlings. *J. Plant Res.* **2011**, *124*, 619–629. [CrossRef] [PubMed]

11. El Hadrami, A.; Adam, L.R.; El Hadrami, I.; Daayf, F. Chitosan in Plant Protection. *Mar. Drugs* **2010**, *8*, 968–987. [CrossRef] [PubMed]

12. Lárez-Velásquez, C. Algunas potencialidades de la quitina y el quitosano para usos relacionados con la Agricultura en Latinoamérica. *Rev. UDO Agríc.* **2008**, *8*, 1–22.

13. Ramírez, M.A.; Rodríguez, A.T.; Alfonso, L.; Peniche, C. Chitin and its derivatives as biopolymers with potential agricultural applications. *Biotechnol. Apl.* **2010**, *27*, 270–276.

14. Manucharova, N.A.; Vlasenko, A.N.; Men'ko, E.V.; Zvyagintsev, D.G. Specificity of the chitinolytic microbial complex of soils incubated at different temperatures. *Microbiology* **2011**, *80*, 205–215. [CrossRef]

15. Wongkaew, P.; Homkratoke, T. Enhancement of soil microbial metabolic activity in tomato field plots by chitin application. *Asian J. Food Agro Ind.* **2009**, *2*, S325–S335.

16. Kielak, A.M.; Cretoiu, M.S.; Semenov, A.V.; Sorensen, S.J.; van Elsas, J.D. Bacterial chitinolytic communities respond to chitin and pH alteration in soil. *Appl. Environ. Microbiol.* **2013**, *79*, 263–272. [CrossRef] [PubMed]

17. Zhang, B.; Ramonell, K.; Somerville, S.; Stacey, G. Characterization of early, chitin-induced gene expression in Arabidopsis. *Mol. Plant Microbe Interact.* **2002**, *15*, 963–970. [CrossRef] [PubMed]

18. Ramonell, K.; Berrocal-Lobo, M.; Koh, S.; Wan, J.; Edwards, H.; Stacey, G.; Somerville, S. Loss-of-function mutations in chitin responsive genes show increased susceptibility to the powdery mildew pathogen Erysiphe cichoracearum. *Plant Physiol.* **2005**, *138*, 1027–1036. [CrossRef] [PubMed]

19. Shimizu, T.; Nakano, T.; Takamizawa, D.; Desaki, Y.; Ishii-Minami, N.; Nishizawa, Y.; Minami, E.; Okada, K.; Yamane, H.; Kaku, H.; et al. Two LysM receptor molecules, CEBiP and OsCERK1, cooperatively regulate chitin elicitor signaling in rice. *Plant J.* **2010**, *64*, 204–214. [CrossRef] [PubMed]

20. Miya, A.; Albert, P.; Shinya, T.; Desaki, Y.; Ichimura, K.; Shirasu, K.; Narusaka, Y.; Kawakami, N.; Kaku, H.; Shibuya, N. CERK1, a LysM receptor kinase, is essential for chitin elicitor signaling in Arabidopsis. *Proc. Natl. Acad. Sci. USA* **2007**, *104*, 19613–19618. [CrossRef] [PubMed]

21. Wan, J.; Zhang, X.C.; Neece, D.; Ramonell, K.M.; Clough, S.; Kim, S.Y.; Stacey, M.G.; Stacey, G. A LysM receptor-like kinase plays a critical role in chitin signaling and fungal resistance in Arabidopsis. *Plant Cell* **2008**, *20*, 471–481. [CrossRef] [PubMed]

22. Liu, T.; Liu, Z.; Song, C.; Hu, Y.; Han, Z.; She, J.; Fan, F.; Wang, J.; Jin, C.; Chang, J.; et al. Chitin-induced dimerization activates a plant immune receptor. *Science* **2012**, *336*, 1160–1164. [CrossRef] [PubMed]

23. Pietraszewska-Bogiel, A.; Lefebvre, B.; Koini, M.A.; Klaus-Heisen, D.; Takken, F.L.; Geurts, R.; Cullimore, J.V.; Gadella, T.W. Interaction of *Medicago truncatula* Lysin Motif Receptor-Like Kinases, NFP and LYK3, Produced in *Nicotiana benthamiana* Induces Defence-Like Responses. *PLoS ONE* **2013**, *8*, e65055. [CrossRef] [PubMed]

24. Cao, Y.; Liang, Y.; Tanaka, K.; Nguyen, C.T.; Jedrzejczak, R.P.; Joachimiak, A.; Stacey, G. The kinase LYK5 is a major chitin receptor in Arabidopsis and forms a chitin-induced complex with related kinase CERK1. *eLife* **2014**, *3*, e03766. [CrossRef] [PubMed]

25. Le, M.H.; Cao, Y.; Zhang, X.C.; Stacey, G. LIK1, a CERK1-interacting kinase, regulates plant immune responses in Arabidopsis. *PLoS ONE* **2014**, *9*, e102245. [CrossRef] [PubMed]

26. Berrocal-Lobo, M.; Stone, S.; Yang, X.; Antico, J.; Callis, J.; Ramonell, K.M.; Somerville, S. ATL9, a RING zinc finger protein with E3 ubiquitin ligase activity implicated in chitin- and NADPH oxidase-mediated defense responses. *PLoS ONE* **2010**, *5*, e14426. [CrossRef] [PubMed]

27. Khan, W.; Prithiviraj, B.; Smith, D.L. Chitosan and chitin oligomers increase phenylalanine ammonia-lyase and tyrosine ammonia-lyase activities in soybean leaves. *J. Plant Physiol.* **2003**, *160*, 859–863. [CrossRef] [PubMed]

28. Hadwiger, L.A. Multiple effects of chitosan on plant systems: Solid science or hype. *Plant Sci.* **2013**, *208*, 42–49. [CrossRef] [PubMed]

29. Hadwiger, L.A. Anatomy of a nonhost disease resistance response of pea to Fusarium solani: PR gene elicitation via DNase, chitosan and chromatin alterations. *Front. Plant Sci.* **2015**, *6*, 373. [CrossRef] [PubMed]

30. Khan, W.; Prithiviraj, B.; Smith, D.L. Effect of Foliar Application of Chitin and Chitosan Oligosaccharides on Photosynthesis of Maize and Soybean. *Photosynthetica* **2002**, *40*, 621–624. [CrossRef]

31. Day, R.B.; Okada, M.; Ito, Y.; Tsukada, K.; Zaghouani, H.; Shibuya, N.; Stacey, G. Binding site for chitin oligosaccharides in the soy-bean plasma membrane. *Plant Physiol.* **2001**, *126*, 1162–1173. [CrossRef] [PubMed]

32. Ramonell, K.M.; Zhang, B.; Ewing, R.M.; Chen, Y.; Xu, D.; Stacey, G.; Somerville, S. Microarray analysis of chitin elicitation in *Arabidopsis thaliana*. *Mol. Plant Pathol.* **2002**, *3*, 301–311. [CrossRef] [PubMed]

33. Wan, J.; Zhang, S.; Stacey, G. Activation of a mitogen-activated protein kinase pathway in Arabidopsis by chitin. *Mol. Plant Pathol.* **2004**, *5*, 125–135. [CrossRef] [PubMed]

34. Hamel, L.P.; Beaudoin, N. Chitooligosaccharide sensing and downstream signaling: Contrasted outcomes in pathogenic and beneficial plant–microbe interactions. *Planta* **2010**, *232*, 787–806. [CrossRef] [PubMed]

35. Liang, Y.; Tóth, K.; Cao, Y.; Tanaka, K.; Espinoza, C.; Stacey, G. Lipochitooligosaccharide recognition: An ancient story. *New Phytol.* **2014**, *204*, 289–296. [CrossRef] [PubMed]

36. Genre, A.; Chabaud, M.; Balzergue, C.; Puech-Pagès, V.; Novero, M.; Rey, T.; Fournier, J.; Rochange, S.; Bécard, G.; Bonfante, P.; et al. Short-chain chitin oligomers from arbuscular mycorrhizal fungi trigger nuclear Ca^{2+} spiking in *Medicago truncatula* roots and their production is enhanced by strigolactone. *New Phytol.* **2013**, *198*, 190–202. [CrossRef] [PubMed]

37. Souleimanov, A.; Prithiviraj, B.; Smith, D.L. The major Nod factor of *Bradyrhizobium japonicum* promotes early growth of soybean and corn. *J. Exp. Bot.* **2002**, *53*, 1929–1934. [CrossRef] [PubMed]

38. Tanaka, K.; Cho, S.H.; Lee, H.; Pham, A.Q.; Batek, J.M.; Cui, S.; Qiu, J.; Khan, S.M.; Joshi, T.; Zhang, Z.J.; et al. Effect of lipo-chitooligosaccharide on early growth of C4 grass seedlings. *J. Exp. Bot.* **2015**, *66*, 5727–5738. [CrossRef] [PubMed]

39. Du, Z.; Zhou, X.; Ling, Y.; Zhang, Z.; Su, Z. agriGO: A GO analysis toolkit for the agricultural community. *Nucleic. Acids Res.* **2010**, *38*, W64–W70. [CrossRef] [PubMed]

40. Usadel, B.; Nagel, A.; Steinhauser, D.; Gibon, Y.; Blasing, O.E.; Redestig, H.; Sreenivasulu, N.; Krall, L.; Hannah, M.A.; Poree, F.; et al. PageMan: An interactive ontology tool to generate, display, and annotate overview graphs for profiling experiments. *BMC Bioinform.* **2006**, *7*, 535. [CrossRef] [PubMed]

41. Machová, E.; Kvapilová, K.; Kogan, G.; Sandula, J. Effect of ultrasonic treatment on the molecular weight of carboxymethylated chitin-glucan complex from *Aspergillus niger*. *Ultrason. Sonochem.* **1999**, *5*, 169–172. [CrossRef]

42. Mislovicová, D.; Masárová, J.; Bendzálová, K.; Soltés, L.; Machová, E. Sonication of chitin-glucan, preparation of water-soluble fractions and characterization by HPLC. *Ultrason. Sonochem.* **2000**, *7*, 63–68. [CrossRef]

43. Kurita, K. Controled funcionalization of polysacharide chitin. *Prog. Polym. Sci.* **2001**, *26*, 1921–1971. [CrossRef]

44. Aranaz, I.; Mengíbar, M.; Harris, R.; Paños, I.; Miralles, B.; Acosta, N.; Galed, G.; Heras, A. Functional Characterization of Chitin and Chitosan. *Curr. Chem. Biol.* **2009**, *3*, 203–230. [CrossRef]

45. Azra, Y.; Linggar, S.; Emma, S.; Anita, R. The Effect of Sonication on the Characteristic of Chitosan. In Proceedings of the International Conference on Chemical and Material Engineering, Semarang, Indonesia, 12–13 September 2012.

46. Villa-Lerma, G.; González-Márquez, H.; Gimeno, M.; López-Luna, A.; Bárzana, E.; Shirai, K. Ultrasonication and steam-explosion as chitin pretreatments for chitin oligosaccharide production by chitinases of *Lecanicillium lecanii*. *Bioresour. Technol.* **2013**, *146*, 794–798. [CrossRef] [PubMed]

47. Trombotto, S.; Ladavière, C.; Delolme, F.; Domard, A. Chemical Preparation and Structural Characterization of a Homogeneous Series of Chitin/Chitosan Oligomers. *Biomacromolecules* **2008**, *9*, 1731–1738. [CrossRef] [PubMed]

48. Kumirska, J.; Czerwicka, M.; Kaczyński, Z.; Bychowska, A.; Brzozowski, K.; Thöming, J.; Piotr Stepnowski, P. Application of Spectroscopic Methods for Structural Analysis of Chitin and Chitosan. *Mar. Drugs* **2010**, *8*, 1567–1636. [CrossRef] [PubMed]

49. Einbu, A.; Vårum, K.M. Characterization of Chitin and Its Hydrolysis to GlcNAc and GlcN. *Biomacromolecules* **2008**, *9*, 1870–1875. [CrossRef] [PubMed]

50. Libault, M.; Wan, J.; Czechowski, T.; Udvardi, M.; Stacey, G. Identification of 118 Arabidopsis transcription factor and 30 ubiquitin-ligase genes responding to chitin, a plant-defense elicitor. *Mol. Plant Microbe Interact.* **2007**, *20*, 900–911. [CrossRef] [PubMed]

51. Son, G.H.; Wan, J.; Kim, H.J.; Nguyen, X.C.; Chung, W.S.; Hong, J.C.; Stacey, G. Ethylene-responsive element-binding factor 5, ERF5, is involved in chitin-induced innate immunity response. *Mol. Plant Microbe Interact.* **2012**, *25*, 48–60. [CrossRef] [PubMed]

52. Minami, E.; Kouchi, H.; Carlson, R.W.; Cohn, J.R.; Kolli, V.K.; Day, R.B.; Ogawa, T.; Stacey, G. Cooperative action of lipo-chitin nodulation signals on the induction of the early nodulin, ENOD2, in soybean roots. *Mol. Plant Microbe Interact.* **1996**, *9*, 574–583. [CrossRef] [PubMed]

53. Stacey, G.; Shibuya, N. Chitin recognition in rice and legumes. *Plant Soil* **1997**, *194*, 161–169. [CrossRef]

54. Wang, N.; Khan, W.; Smith, D.L. Changes in soybean global gene expression after application of lipo-chitooligosaccharide from *Bradyrhizobium japonicum* under sub-optimal temperature. *PLoS ONE* **2012**, *7*, e31571. [CrossRef] [PubMed]

55. Cheng, Z.J.; Wang, L.; Sun, W.; Zhang, Y.; Zhou, C.; Su, Y.H.; Li, W.; Sun, T.T.; Zhao, X.Y.; Li, X.G.; et al. Pattern of auxin and cytokinin responses for shoot meristem induction results from the regulation of cytokinin biosynthesis by AUXIN RESPONSE FACTOR3. *Plant Physiol.* **2013**, *161*, 240–251. [CrossRef] [PubMed]

56. Lampugnani, E.R.; Kilinc, A.; Smyth, D.R. Auxin controls petal initiation in Arabidopsis. *Development* **2013**, *140*, 185–194. [CrossRef] [PubMed]

57. Irshad, M.; Canut, H.; Borderies, G.; Pont-Lezica, R.; Jamet, E. A new picture of cell wall protein dynamics in elongating cells of *Arabidopsis thaliana*: Confirmed actors and newcomers. *BMC Plant Biol.* **2008**, *16*, 94. [CrossRef] [PubMed]

58. Sehr, E.M.; Agusti, J.; Lehner, R.; Farmer, E.E.; Schwarz, M.; Greb, T. Analysis of secondary growth in the Arabidopsis shoot reveals a positive role of jasmonate signalling in cambium formation. *Plant J.* **2010**, *63*, 811–822. [CrossRef] [PubMed]

59. Yamaguchi, M.; Mitsuda, N.; Ohtani, M.; Ohme-Takagi, M.; Kato, K.; Demura, T. Vascular-related nac-domain7 directly regulates the expression of a broad range of genes for xylem vessel formation. *Plant J.* **2011**, *66*, 579–590. [CrossRef] [PubMed]

60. Handford, M.G.; Sicilia, F.; Brandizzi, F.; Chung, J.H.; Dupree, P. *Arabidopsis thaliana* expresses multiple Golgi-localised nucleotide-sugar transporters related to GONST1. *Mol. Genet. Genom.* **2004**, *272*, 397–410. [CrossRef] [PubMed]

61. Sauer, N.; Ludwig, A.; Knoblauch, A.; Rothe, P.; Gahrtz, M.; Klebl, F. AtSUC8 and AtSUC9 encode functional sucrose transporters, but the closely related AtSUC6 and AtSUC7 genes encode aberrant proteins in different Arabidopsis ecotypes. *Plant J.* **2004**, *40*, 120–130. [CrossRef] [PubMed]

62. Boonyaves, K.; Gruissem, W.; Bhullar, N.K. NOD promoter-controlled AtIRT1 expression functions synergistically with NAS and FERRITIN genes to increase iron in rice grains. *Plant Mol. Biol.* **2015**, *90*, 207–215. [CrossRef] [PubMed]

63. Gillissen, B.; Bürkle, L.; André, B.; Kühn, C.; Rentsch, D.; Brandl, B.; Frommer, W.B. A New Family of High-Affinity Transporters for Adenine, Cytosine, and Purine Derivatives in Arabidopsis. *Plant Cell* **2000**, *12*, 291–300. [CrossRef] [PubMed]

64. Chen, Z.H.; Jenkins, G.I.; Nimmo, H.G. pH and carbon supply control the expression of phosphoenolpyruvate carboxylase kinase genes in *Arabidopsis thaliana*. *Plant Cell Environ.* **2008**, *31*, 1844–1850. [CrossRef] [PubMed]

65. Della Rovere, F.; Fattorini, L.; D'Angeli, S.; Veloccia, A.; Del Duca, S.; Cai, G.; Falasca, G.; Altamura, M.M. Arabidopsis SHR and SCR transcription factors and AUX1 auxin influx carrier control the switch between adventitious rooting and xylogenesis in planta and in in vitro cultured thin cell layers. *Ann. Bot.* **2015**, *115*, 617–628. [CrossRef] [PubMed]

66. Fàbregas, N.; Formosa-Jordan, P.; Confraria, A.; Siligato, R.; Alonso, J.M.; Swarup, R.; Bennett, M.J.; Mähönen, A.P.; Caño-Delgado, A.I.; Ibañes, M. Auxin influx carriers control vascular patterning and xylem differentiation in *Arabidopsis thaliana*. *PLoS Genet.* **2015**, *11*, e1005183. [CrossRef] [PubMed]

67. Hoyos, M.E.; Palmieri, L.; Wertin, T.; Arrigoni, R.; Polacco, J.C.; Palmieri, F. Identification of a mitochondrial transporter for basic amino acids in *Arabidopsis thaliana* by functional reconstitution into liposomes and complementation in yeast. *Plant J.* **2003**, *33*, 1027–1035. [CrossRef] [PubMed]

68. Sterken, R.; Kiekens, R.; Boruc, J.; Zhang, F.; Vercauteren, A.; Vercauteren, I.; De Smet, L.; Dhondt, S.; Inzé, D.; De Veylder, L.; et al. Combined linkage and association mapping reveals CYCD5;1 as a quantitative trait gene for endoreduplication in Arabidopsis. *Proc. Natl. Acad. Sci. USA* **2012**, *109*, 4678–4683. [CrossRef] [PubMed]

69. Johnson, K.L.; Kibble, N.A.; Bacic, A.; Schultz, C.J. A fasciclin-like arabinogalactan-protein (FLA) mutant of *Arabidopsis thaliana*, fla1, shows defects in shoot regeneration. *PLoS ONE* **2011**, *6*, e25154. [CrossRef] [PubMed]

70. Thatcher, L.F.; Cevik, V.; Grant, M.; Zhai, B.; Jones, J.D.; Manners, J.M.; Kazan, K. Characterization of a JAZ7 activation-tagged Arabidopsis mutant with increased susceptibility to the fungal pathogen *Fusarium oxysporum*. *J. Exp. Bot.* **2016**, *67*, 2367–2386. [CrossRef] [PubMed]

71. Spaink, H.P.; Sheeley, D.M.; Van Brussel, A.A.N.; Glushka, J.; York, W.S.; Tak, T.; Geiger, O.; Kennedy, E.P.; Reinhold, V.N.; Lugtenberg, B.J.J. A novel highly unsaturated fatty acid moi-ety of lipo-oligosaccharide signals determines host-specificity of *Rhi-zobium*. *Nature* **1991**, *354*, 125–130. [CrossRef] [PubMed]

72. Spaink, H.P. Root nodulation and infection factors produced by rhizobial bacteria. *Annu. Rev. Microbiol.* **2000**, *54*, 257–288. [CrossRef] [PubMed]

73. Rasmussen, S.R.; Füchtbauer, W.; Novero, M.; Volpe, V.; Malkov, N.; Genre, A.; Bonfante, P.; Stougaard, J.; Radutoiu, S. Intraradical colonization by arbuscular mycorrhizal fungi triggers induction of a lipochitooligosaccharide receptor. *Sci. Rep.* **2016**, *20*, 29733. [CrossRef] [PubMed]

74. Fliegmann, J.; Jauneau, A.; Pichereaux, C.; Rosenberg, C.; Gasciolli, V.; Timmers, A.C.; Burlet-Schiltz, O.; Cullimore, J.; Bono, J.J. LYR3, a high-affinity LCO-binding protein of *Medicago truncatula*, interacts with LYK3, a key symbiotic receptor. *FEBS Lett.* **2016**, *590*, 1477–1487. [CrossRef] [PubMed]

75. Gough, C.; Jacquet, C. Nod factor perception protein carries weight in biotic interactions. *Trends Plant Sci.* **2013**, *18*, 566–574. [CrossRef] [PubMed]

76. Miyata, K.; Kozaki, T.; Kouzai, Y.; Ozawa, K.; Ishii, K.; Asamizu, E.; Okabe, Y.; Umehara, Y.; Miyamoto, A.; Kobae, Y.; et al. The bifunctional plant receptor, OsCERK1, regulates both chitin-triggered immunity and arbuscular mycorrhizal symbiosis in rice. *Plant Cell Physiol.* **2014**, *55*, 1864–1872. [CrossRef] [PubMed]

77. Limpens, E.; van Zeijl, A.; Geurts, R. Lipochitooligosaccharides modulate plant host immunity to enable endosymbioses. *Annu. Rev. Phytopathol.* **2015**, *53*, 311–334. [CrossRef] [PubMed]

78. Khan, W.; Costa, C.; Souleimanov, A.; Prithiviraj, B.; Smith, D.L. Response of *Arabidopsis thaliana* roots to lipo-chitooligosaccharide from *Bradyrhizobium japonicum* and other chitin-like compounds. *Plant Growth Regul.* **2011**, *63*, 243–249. [CrossRef]

79. Jashni, M.K.; Dols, I.H.; Iida, Y.; Boeren, S.; Beenen, H.G.; Mehrabi, R.; Collemare, J.; de Wit, P.J. Synergistic Action of a Metalloprotease and a Serine Protease from *Fusarium. oxysporum* f. sp. *lycopersici* Cleaves Chitin-Binding Tomato Chitinases, Reduces Their Antifungal Activity, and Enhances Fungal Virulence. *Mol. Plant Microbe Interact.* **2015**, *28*, 996–1008. [CrossRef] [PubMed]

80. Pantoom, S.; Songsiriritthigul, C.; Suginta, W. The effects of the surface-exposed residues on the binding and hydrolytic activities of *Vibrio carchariae* chitinase A. *BMC Biochem.* **2008**, *9*, 2. [CrossRef] [PubMed]

81. Zitouni, M.; Fortin, M.; Scheerle, R.K.; Letzel, T.; Matteau, D.; Rodrigue, S.; Brzezinski, R. Biochemical and molecular characterization of a thermostable chitosanase produced by the strain *Paenibacillus.* sp. 1794 newly isolated from compost. *Appl. Microbiol. Biotechnol.* **2013**, *97*, 5801–5813. [CrossRef] [PubMed]

82. Gómez Ramírez, M.; Rojas Avelizapa, L.I.; Rojas Avelizapa, N.G.; Cruz Camarillo, R. Colloidal chitin stained with Remazol Brilliant Blue R, a useful substrate to select chitinolytic microorganisms and to evaluate chitinases. *J. Microbiol. Methods* **2004**, *56*, 213–219. [CrossRef] [PubMed]

83. Shen, C.R.; Chen, Y.S.; Yang, C.J.; Chen, J.K.; Liu, C.L. Colloid chitin azure is a dispersible, low-cost substrate for chitinase measurements in a sensitive, fast, reproducible assay. *J. Biomol. Screen.* **2010**, *15*, 213–217. [CrossRef] [PubMed]

84. Colson, S.; van Wezel, G.P.; Craig, M.; Noens, E.E.; Nothaft, H.; Mommaas, A.M.; Titgemeyer, F.; Joris, B.; Rigali, S. The chitobiose-binding protein, DasA, acts as a link between chitin utilization and morphogenesis in Streptomyces coelicolor. *Microbiology* **2008**, *154*, 373–382. [CrossRef] [PubMed]

85. Meibom, K.L.; Li, X.B.; Nielsen, A.T.; Wu, C.Y.; Roseman, S.; Schoolnik, G.K. The Vibrio cholerae chitin utilization program. *Proc. Natl. Acad. Sci. USA* **2005**, *101*, 2524–2529. [CrossRef]

86. Love, D.C.; Krause, M.W.; Hanover, J.A. O-GlcNAc cycling: Emerging roles in development and epigenetics. *Semin. Cell Dev. Biol.* **2010**, *21*, 646–654. [CrossRef] [PubMed]

87. Focher, B.; Beltranme, P.L.; Naggi, A.; Torri, G. Alkaline Ndeacetylation of chitin enhanced by flash treatments: Reaction kinetics and structure modifications. *Carbohydr. Polym.* **1990**, *12*, 405–418. [CrossRef]

88. Schneider, C.A.; Rasband, W.S.; Eliceiri, K.W. NIH Image to ImageJ: 25 years of image analysis. *Nat. Methods* **2012**, *9*, 671–675. [CrossRef] [PubMed]

89. Edgar, R.; Domrachev, M.; Lash, A.E. Gene Expression Omnibus: NCBI gene expression and hybridization array data repository. *Nucleic Acids Res.* **2002**, *30*, 207–210. [CrossRef] [PubMed]

90. Tusher, V.G.; Tibshirani, R.; Chu, G. Significance analysis of microarrays applied to the ionizing radiation response. *Proc. Natl. Acad. Sci. USA* **2001**, *98*, 5116–5121. [CrossRef] [PubMed]

91. Saeed, A.I.; Sharov, V.; White, J.; Li, J.; Liang, W.; Bhagabati, N.; Braisted, J.; Klapa, M.; Currier, T.; Thiagarajan, M.; et al. TM4: A free, open-source system for microarray data management and analysis. *Biotechniques* **2003**, *34*, 374–378. [PubMed]

92. Livak, K.J.; Schmittgen, T.D. Analysis of relative gene expression data using real-time quantitative PCR and the 2(T)(-Delta Delta C) method. *Methods* **2001**, *25*, 402–408. [CrossRef] [PubMed]

![marine drugs logo] *marine drugs*

MDPI

Article

Fabrication of Gelatin-Based Electrospun Composite Fibers for Anti-Bacterial Properties and Protein Adsorption

Ya Gao [1], Yingbo Wang [1,*], Yimin Wang [1] and Wenguo Cui [2,*]

1   College of Chemical Engineering, Xinjiang Normal University, Urumqi 830054, China;
     gaoya1965810837@sina.com (Y.G.); wangyimin73@sina.com (Y.W.)
2   Department of Orthopedics, The First Affiliated Hospital of Soochow University, Orthopedic Institute,
     Soochow University, 708 Renmin Road, Suzhou 215006, China
*   Correspondence: ybwang20002575@163.com (Y.W.); wgcui@suda.edu.cn (W.C.);
     Tel.: +86-991-433-3279 (Y.W.); +86-512-6778-1420 (W.C.)

Academic Editor: Hitoshi Sashiwa
Received: 5 September 2016; Accepted: 17 October 2016; Published: 21 October 2016

Abstract: A major goal of biomimetics is the development of chemical compositions and structures that simulate the extracellular matrix. In this study, gelatin-based electrospun composite fibrous membranes were prepared by electrospinning to generate bone scaffold materials. The gelatin-based multicomponent composite fibers were fabricated using co-electrospinning, and the composite fibers of chitosan (CS), gelatin (Gel), hydroxyapatite (HA), and graphene oxide (GO) were successfully fabricated for multi-function characteristics of biomimetic scaffolds. The effect of component concentration on composite fiber morphology, antibacterial properties, and protein adsorption were investigated. Composite fibers exhibited effective antibacterial activity against *Staphylococcus aureus* and *Escherichia coli*. The study observed that the composite fibers have higher adsorption capacities of bovine serum albumin (BSA) at pH 5.32–6.00 than at pH 3.90–4.50 or 7.35. The protein adsorption on the surface of the composite fiber increased as the initial BSA concentration increased. The surface of the composite reached adsorption equilibrium at 20 min. These results have specific applications for the development of bone scaffold materials, and broad implications in the field of tissue engineering.

Keywords: electrospinning; composite fibers; antibacterial properties; protein adsorption

---

1. Introduction

Human bones are composed of organic and inorganic components, particularly hydroxyapatite (HA) and collagen. Composite materials that mimic the bone matrix have important clinical applications. HA exhibits excellent biocompatibility and biodegradability and, therefore, is a high-profile artificial bone material. However, its insufficient flexural and compressive strength and high brittleness limit its medical applications [1]. Gelatin (Gel) is a modified collagen product and a natural polymer; it is structurally similar to collagen in extracellular matrices [2–4]. These composite materials can adapt well to the internal environment of humans [5]. Gel is rich in amino and carboxyl hydrophilic groups [6] and is beneficial for nutrients and oxygen infiltration. Chitosan (CS) possesses good biocompatibility, antimicrobial properties [7], and has the potential for various chemical modifications and combinations to obtain specific properties [8]. Accordingly, it has a wide range of applications in tissue engineering [9].

Graphene oxide (GO) is a derivative of graphene. Carboxylic, epoxy, and hydroxyl groups, in addition to many other highly active response groups, facilitate the combination of GO with other substances to form new composite materials. GO is widely used for biomedical applications. Owing to its high-strength mechanical properties, GO can be used in medical implants, as a filler, or as reinforcement material in tissue engineering scaffolds [10]. GO also has excellent antibacterial properties [11] and can be used for external wound healing to prevent infection [12]. GO-based composite nanofibers have been successfully prepared by electrospinning. Lu et al. [13] fabricated reduced graphene oxide (RGO)/CS/polyvinyl alcohol nanofiber scaffolds for wound healing and observed that RGO is beneficial for cellular attachment and growth. Andreia et al. [14] developed a nanocomposite comprised of GO sheets with silver nanoparticles (GO-Ag), and found that it can inhibit the growth of microbial adherent cells, thus preventing biofilm formation; however, the sudden release of silver was observed. Isis et al. [15] prepared GO/polyvinyl carbazole nanocomposites using electrochemical technology, and these had stronger antimicrobial effects than unmodified GO. According to these previous results, GO and RGO are beneficial for cell adhesion and growth. GO possesses antibacterial activity and can combine with polymers to form composites with favorable antimicrobial properties.

RGO and GO interact with the phospholipid bilayer of cells to form a stable structure [16], and the large specific surface area can promote cell adsorption [17], enhance cell adhesion, and induce proliferation via extracellular matrix protein adsorption [18]. In addition, the non-biodegradable RGO and GO can be discharged through lysosomes and enter the cell via phagocytes [19].

In this study, we demonstrate the design and facile fabrication of Gel/CS/HA/GO and Gel/CS/HA/RGO composite fibers using co-electrospinning, as shown in Scheme 1. With the development of materials science, biological materials have progressed from those that are passively adapted to the biological environment to those that are purposefully designed, with respect to material composition and microstructure, to confer specific functions [20]. Fibers generated by electrospinning are similar to the extracellular matrix with respect to structural morphology. The multipath structure in the Scheme 1 is similar to the collagen fiber structure of the extracellular matrix, which can be used for cellular attachment and growth [21]. At the same time, the large specific surface-area of the nanofibrous scaffold enhances its protein absorption ability, which is vital for cell anchoring [22]. Inorganic HA in bone tissue promotes bone remodeling via cell signaling to regulate osteoblast formation. This process is associated with an extracellular matrix protein adsorption effect. HA adsorption on these specific proteins can promote osteoblast proliferation, differentiation, and adhesion [23]. Accordingly, the adsorption of proteins has a very important effect on osteoblasts; it is, therefore, necessary to prepare a composite fiber with a similar composition to that of bone tissue that has good antibacterial properties and protein adsorption performance for use as a bone scaffold material. In this study, for the first time, the advantages of four materials: HA, Gel, CS, and GO, were combined to prepare composite nanofibers using electrospinning technology. The antimicrobial properties and the protein adsorption performance of these nanofibers were evaluated.

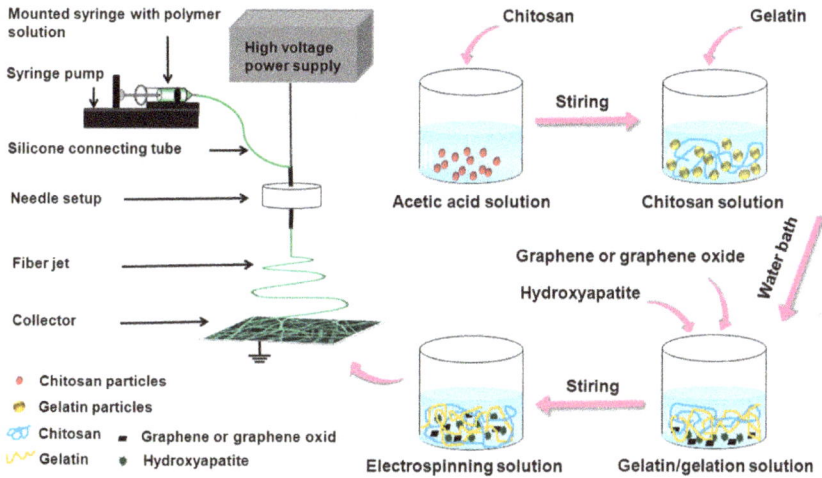

Scheme 1. Schematic illustration of the fabrication process of Gel/CS/HA/GO composite fibers.

2. Results

2.1. Influence of Gel Concentration on Fibers

Gel was the main component in the composite fiber in this experiment. The Gel concentration was the main factor determining the morphology of the composite fiber. Figure 1 shows that low Gel concentrations (i.e., ≤5 wt. %) resulted in insufficient surface tension, and the phenomenon of "electrospray" was observed (Figure 1a). As the Gel concentration increased (10 wt. %), surface tension gradually increased, and fiber formation was observed. There were some fiber junctions and bead structures in the fibers (Figure 1b). When the Gel concentration reached 15–20 wt. %, uniform fibers were obtained (Figure 1c,d). However, further increases in the Gel concentration (25 wt. %) led to increased surface tension. The electric force could not overcome the surface tension, resulting in the formation of larger fibers. This also caused slow solvent evaporation, which led to the formation of fiber junctions and beads (Figure 1e). The Gel concentration was further increased to 30 wt. %; however, the electrospinning liquid viscosity was too high and solvent evaporation was too slow, which led to fiber junctions and flat fibers (Figure 1f). These results indicate that a Gel concentration of 15–20 wt. % for electrospinning fibers is optimal.

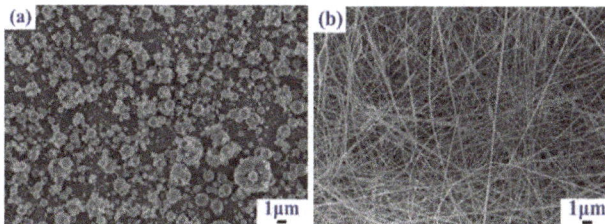

Figure 1. *Cont.*

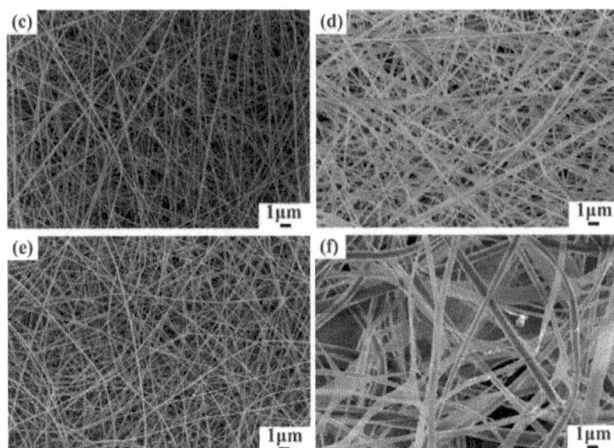

Figure 1. Scanning electron microscope (SEM) micrographs of fibers prepared with a series of Gel concentration; (**a**) 5 wt. %; (**b**) 10 wt. %; (**c**) 15 wt. %; (**d**) 20 wt. %; (**e**) 25 wt. %; (**f**) 30 wt. %.

2.2. Influence of CS Concentration on Fibers

CS possesses natural antibacterial activity. Figure 2 shows scanning electron microscope (SEM) photographs of fibers after the addition of chitosan for Gel concentrations of 15 wt. % and 20 wt. %. When the Gel concentration was 15 wt. % and the CS concentration was low, electrospinning was easier and electrospun fibers with relatively smooth surfaces were obtained (Figure 2a). As the CS concentration increased, the electrospinning liquid surface tension increased, electrospunfibers exhibited uniformity, and fiber diameters increased (Figure 2b). When the Gel concentration was 20 wt. %, the addition of CS resulted in fine fibers. Junctions were detected in some fibers for low CS concentrations (Figure 2c). As the CS concentration increased, beads were observed during the process of electrospinning (Figure 2d). This may be attributed to the presence of hydrogen bonds in the CS molecules, which increase rigidity and can enhance the mechanical properties of the material. However, increasing the surface tension of the electrospinning liquid also makes electrospinning more difficult, which limits the CS concentration. For a Gel concentration of 15 wt. %, a CS concentration of 1 wt. % was optimal.

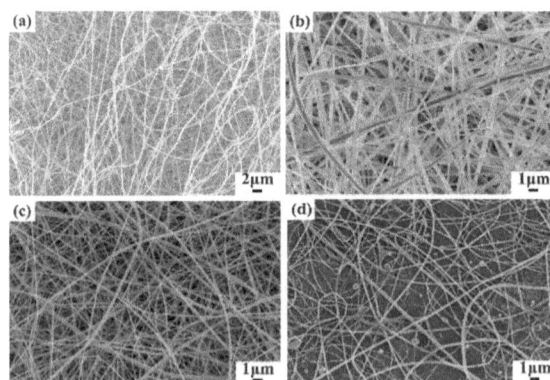

Figure 2. SEM micrographs of fibers prepared with a series of CS concentration; (**a**) 15 wt. % Gel, 0.75 wt. % CS; (**b**) 15 wt. % Gel, 1 wt. % CS; (**c**) 20 wt. % Gel, 0.75 wt. % CS; (**d**) 20 wt. % Gel, 1 wt. % CS.

2.3. Influence of HA Concentration and Particle Size on Fibers

The major inorganic constituent of human bones is HA. Therefore, HA particles were combined with Gel (Gel concentration, 15 wt. %) and CS (CS concentration, 1 wt. %) to enhance fiber biocompatibility. In addition, CS and HA tend to interact with each other, through hydrogen bonds between $-NH_2$ and $-OH$ as well as chelation between $-NH_2$ and Ca^{2+} [24], to make more uniform fibers. We evaluated HA particles to determine what effect concentration (2 wt. %, 5 wt. %, and 8 wt. %) and particle size (12 µm and 60 nm) has on fiber morphology.

For the HA (12 µm, 2 wt. %) electrospinning solution, the solvent content was relatively high and not completely volatile. This resulted in fiber junctions (Figure 3a). As the HA concentration increased, the number of junctions in the fibers decreased, and gradually more uniform and smooth fibers were produced (Figure 3b,c). HA particles at 12 µm exhibit corrosion in acidic environments, which can lead to HA grain refinement. As shown in Figure 4, grain refinement of HA in acidic conditions was observed for an average particle size of 149 ± 29 nm. Therefore, HA particles can be completely enclosed by the fiber due to the 1 µm fiber diameter.

For HA particles of 60 nm, nanoparticle aggregation resulted in more junctions in the fibers than those observed for HA particles of 12 µm (Figure 3d,e). As the HA concentration was increased to 8 wt. %, the number of junctions decreased (Figure 3f), but the electrospinning solution exhibited HA sedimentation.

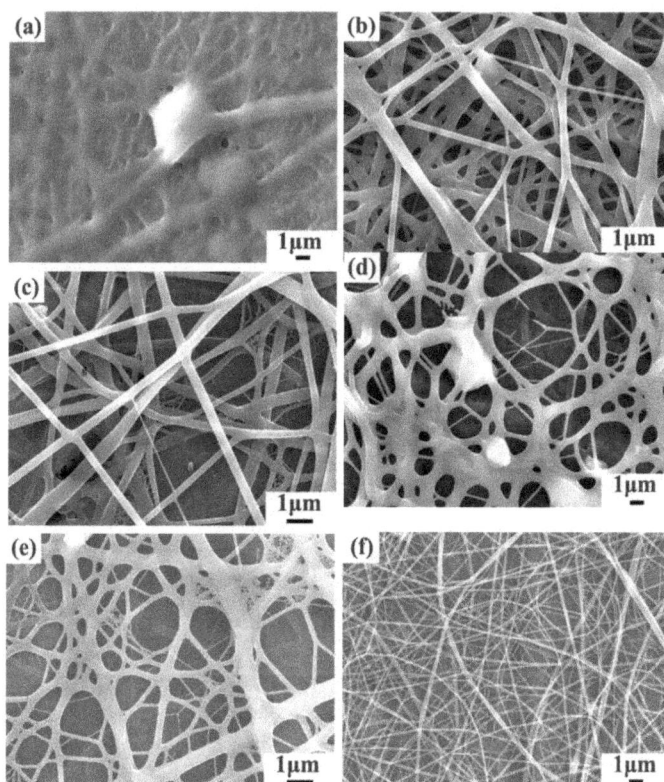

Figure 3. SEM micrographs of fibers prepared with different HA concentration and HA particle size; (**a**) 2 wt. % 12 µm; (**b**) 5 wt. % 12 µm; (**c**) 8 wt. % 12 µm; (**d**) 2 wt. % 60 nm; (**e**) 5 wt. % 60 nm; (**f**) 8 wt. % 60 nm.

Figure 4. SEM micrographs of HA with a size of 12 μm, before (**a**) and after (**b**) dissolution.

Based on these results, the particle size of HA had a significant effect on composite fiber morphology. When HA particles were too small (on the order of nanometers), an aggregation effect was observed. This resulted in a non-uniform electrospinning solution and uneven fibers. When the HA particle size was increased to the micrometer scale, this aggregation effect disappeared. The electrospinning solution was uniform, and the composite fiber morphology improved. In summary, composite fibers can be optimized using a HA particle size of 12 μm and HA concentration of 5 wt. %.

2.4. Influence of the GO Concentration on Fibers

The goal of this study was to develop a human bone material with antimicrobial activity. CS possesses weak antimicrobial properties, while GO has excellent antibacterial properties. Accordingly, GO was added to the electrospinning solution with 15 wt. % Gel, 1 wt. % CS, and 5 wt. % HA (12 μm). Figure 5a,b show SEM images of uniform, composite fibers formed from the aforementioned ternary compound electrospinning solution after the addition of 2 wt. % GO and 2 wt. % RGO, respectively. However, if the concentrations of GO and RGO were to be increased in the electrospinning solution, solvent evaporation would cause blockage of the needles during electrospinning. Figure 5 shows that the composite fibers with RGO exhibit bonding (Figure 5b), the GO composite fibers are good (Figure 5a), and the inorganic phase was completely covered in the fiber, as seen by transmission electron microscopy (Figure 5c).

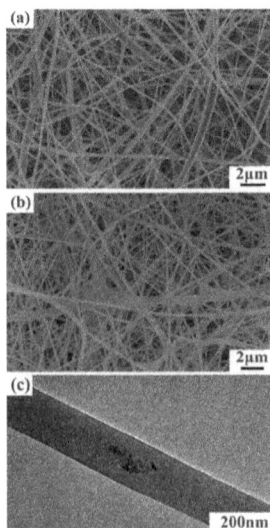

Figure 5. SEM micrographs of fibers with GO and RGO; (**a**) 2 wt. % GO; (**b**) 2 wt. % RGO; and TEM micrographs of the composite fiber (**c**).

Figure 6 shows the Fourier transform infrared (FTIR) spectrum of Gel/CS/HA/GO composite fibers. The peaks observed at 562 cm^{-1} correspond to the stretching vibration bands of P–O from PO_4^{3-} in HA. The peaks appearing at 2887, 1650, and 1544 cm^{-1} are attributed to methylene (–CH_2), C=O in the amide group (amide I band), and the NH-bending vibration in the amide group from chitosan, respectively. The characteristic IR band for amide III is at 1244 cm^{-1}, and carboxylate from Gel is at 1449 cm^{-1}. The existence of a chemical bond between the carboxylate group in Gel and the Ca^{2+} ion in HA that binds the particulate-reinforced composite together was confirmed by the characteristic IR band at 1334 cm^{-1}. FTIR studies show that there was strong interaction between HA, CS, Gel, and GO networks in the composite fibers. Redshifting of the characteristic amine (1660 cm^{-1}) and C=O (1592 cm^{-1}) bands in chitosan were caused by electrostatic interactions between –NH_3^+ and PO_4^{3-}, as well as between C=O– and Ca^{2+} in Gel/CS/HA/GO composite fibers. In summary, the structure of the four species has not changed, which means the composition of the composite fibers has not changed.

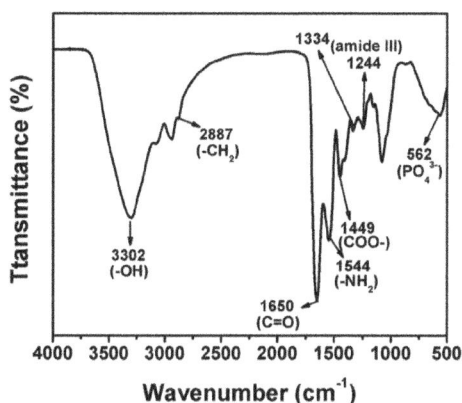

Figure 6. FTIR spectra of Gel/CS/HA/GO composite fibers.

2.5. Test of Antibacterial Properties

Based on the above experiments, a Gel concentration of 15 wt. %, CS concentration of 1 wt. %, HA (12 μm) concentration of 5 wt. %, and GO concentration of 2 wt. % were optimal conditions. Gel/CA/HA/GO composite fibers were prepared using these concentrations to evaluate antimicrobial properties.

To investigate the antibacterial effects of the composite fibers against *Escherichia coli* (*E. coli*) and *Staphylococcus albus* (*S. aureus*), the spread plate method was used for a qualitative antibacterial analysis. After the bacteria were cultivated for 24 h, the Gel/CS/HA/GO surface did not show *E. coli* colonies (Figure 7c), and Gel/CS/HA/RGO had a few colonies (Figure 7b). The surface of a blank sample of Gel/CS/HA (Figure 7a) had some colonies that were stacked to form a lawn. For antimicrobial properties against *S. aureus*, the Gel/CS/HA/RGO and Gel/CS/HA surfaces had more colonies than the other surfaces, and showed no significant macroscopic differences between samples (Figure 7d,e), while the Gel/CS/HA/GO surface exhibited significantly fewer colonies (Figure 7f). These results indicated that and Gel/CS/HA/GO had stronger antibacterial ability against *E. coli* and *S. aureus*. Meanwhile, the Gel/CS/HA/GO is far more effective against Gram-negative bacteria than against Gram-positive. These results may result from the distinct structure of the cell wall between Gram-positive and Gram-negative bacteria. Gram-positive bacteria contain a thick peptidoglycan layer (20–80 nm) on the outside of the cell wall, and lack an outer membrane. In contrast, the cell wall of Gram-negative bacteria is composed of a thin peptidoglycan layer (7–8 nm) with an additional outer membrane. The thick peptidoglycan of Gram-positive bacteria is a meshlike polymer consisting of amino acids and sugars and may also include other components, such as teichoic and lipoteichoic acids. Therefore, the peptidoglycan layer protects against antibacterial agents, such as antibiotics, toxins, chemicals, and degradative enzymes.

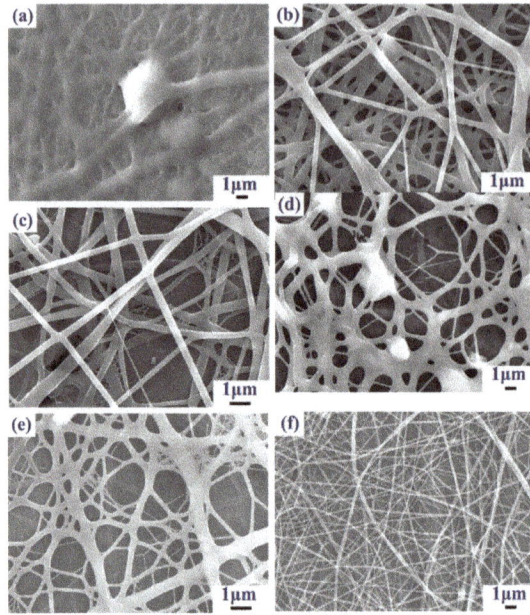

Figure 7. The composite fibers antibacterial test results; (**a**) and (**d**) Gel/CS/HA; (**b**) and (**e**) Gel/CS/HA/RGO; (**c**) and (**f**) Gel/CS/HA/GO; (**a–c**) *Escherichia coli* and (**b–d**) *Staphylococcus albus*.

A quantitative analysis was performed to evaluate the antibacterial properties of the composites, and the results are summarized in Figure 8. For Gel/CS/HA/GO, the *E. coli* antibacterial rate was 100% and for Gel/CS/HA/RGO, the antibacterial rate was only 38.6%. The Gel/CS/HA/GO *S. aureus* antibacterial rate was 73.2%, while Gel/CS/HA/RGO exhibited a weaker antibacterial effect, i.e., 3.4%. The results of the quantitative and qualitative analyses demonstrated that Gel/CS/HA/GO has strong antibacterial properties against *E. coli*, and has good antibacterial properties against *S. aureus*.

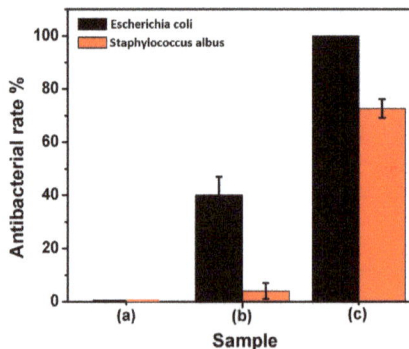

Figure 8. The composite fibers' antibacterial rate of *Escherichia coli* (black histograms) and *Staphylococcus albus* (red histograms); (**a**) Gel/CS/HA; (**b**) Gel/CS/HA/RGO; and (**c**) Gel/CS/HA/GO.

2.6. Adsorption Performance of BSA

After the addition of GO, the composite fiber morphology was more uniform and excellent antibacterial activity was observed. Protein adsorption has a substantial effect on osteoblasts. The effects of GO on the BSA adsorption ability of the composite fibers were examined. Specifically, the effects of pH and initial concentration of the BSA solution on the protein adsorption of the composite fiber were examined.

The pH of the BSA solution had a significant effect on composite fiber protein adsorption. For various pH values, variation in the BSA concentration versus time is shown in Figure 9a,b. For Gel/CS/HA and Gel/CS/HA/GO, the adsorption capacity all increased sharply during the first 20 min, then levelled-off gently, and finally reached equilibrium. The adsorption ability of the composite fiber increased as the solution pH increased. When a pH value of 3.90 was applied, protein adsorption on the surface of all composite fibers was minimal. Adsorbed protein for composite fibers was maximal for a pH value of 5.32–6.00. From the two figures we can see that the Gel/CS/HA/GO composite balance concentration is greater than that of Gel/CS/HA. Thus, we mainly investigated the effect of initial BSA concentration on the Gel/CS/HA/GO composite fibers' protein adsorption.

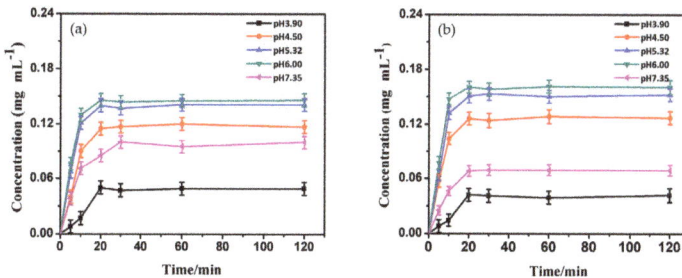

Figure 9. Protein adsorption curve of composite fibers immersed in the BSA solution with different pH values: (**a**) Gel/CS/HA; and (**b**) Gel/CS/HA/GO.

The initial BSA concentration also had a significant effect on composite fiber protein adsorption. Variation in BSA concentration over time was observed when the composite fibers were immersed in BSA solutions with a pH value of 7.35. The results are summarized in Figure 10. The saturated protein adsorption on the surface of the composite fiber increased as the initial BSA concentration increased. When the initial concentration was constant, the surface of the composite reached the adsorption equilibrium in 20 min.

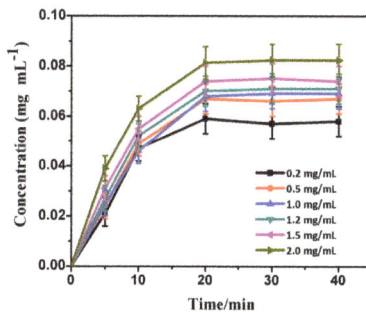

Figure 10. Protein adsorption curve of the Gel/CS/HA/GO composite fibers immersed in the BSA solution with different concentrations.

3. Discussion

Gel/CS/HA/GO composite nanofibers were prepared using electrospinning to evaluate their antibacterial properties. Recent studies have emphasized [25] the strong influence of substance concentration in electrospinning liquids on fiber morphology. Therefore, the effects of the composition on fiber morphology and antibacterial properties were investigated.

According to a recent report [26,27], the conductivity of the electrospinning liquid is mainly determined by the ionized salt type, polymer type, and concentration. Some ionized substances added to the electrospinning solution do not change the electrically-neutral property of the electrospinning solution. However, decomposition into positive and negative ions can obviously change the electric charge density of the electrospinning solution, improve electrical conductivity, and affect the morphology and diameter of fibers. In our study, as the HA concentration increased, the dissolution of HA increased in an acidic solution, and the inorganic ion concentration increased in the electrospinning solution, thereby enhancing the conductivity of the electrospinning solution, resulting in thinner fibers. Essentially, natural polymers, including polyelectrolytes, like $-NH_2$, generates $-NH_3^+$ when CS is added to an acidic solution. This increases the electrospinning solution conductivity. A hybrid electrospinning solution consisting of HA and CS, N and Ca^{2+} can be hybridized by the protonation of CS molecules [24], resulting in complex formation, but this does not change the charge density of the electrospinning solution. Therefore, the reaction between CS and the inorganic ions does not influence spinnability. However, various CS and inorganic ion concentrations affect the conductivity of the electrospinning solution and, thus, influence the morphology and diameter of electrospun fibers. Gel shows a positively-charged point below isoelectric, which is similar to CS because the isoelectric point of Gel is between 6.00 and 8.00. When Gel is added to an acidic solution it is positively charged, which increases the electrospinning solution conductivity. A hybrid electrospinning solution that consists of HA and Gel, N and Ca^{2+} can be hybridized in the protonation of Gel molecules, resulting in complex formation.

In this experiment, for a lower CS content in composite fibers, the antimicrobial properties were mainly attributed to GO and RGO. According to recent reports, both GO and RGO have shown good antimicrobial properties, and the antibacterial mechanism is a result of oxidative stress and cell membrane damage. Oxidative stress in target cells is caused by the generation of reactive oxygen species. Antioxidant enzymes in the cell can be used to reduce and eliminate reactive oxygen species. If homeostasis is not achieved, cellular macromolecules, such as proteins, DNA, and lipids, can be damaged [28]. Cell membrane damage via physical interactions with sharp-edged graphene is another possible antibacterial mechanism [29]. Feng et al. [30] found that *E. coli* can interact directly with GO to induce the loss of bacterial membrane integrity and glutathione oxidation, suggesting that GO antimicrobial action contributes to both membrane disruption and oxidative stress. Additionally, the antibacterial mechanism of grapheme-based materials largely depends on the surface of nanomaterials; when GR is well dispersed in the composite material, its antibacterial effect is stronger. In this experiment, GO had good hydrophilicity, which promotes dispersion in a composite fiber. Accordingly, evenly dispersed GO has shown strong antibacterial activity [31].

Additionally, the effects of different pH values on the properties of protein adsorption have been investigated. When a pH value of 3.90 was applied, protein adsorption on the surface of composite fibers was minimal. Since the pH was lower than the BSA isoelectric point (4.90), BSA had a positive charge, and the gelatin isoelectric point was 6.00–8.00. In other words, when the pH value ≤ 4.90, gelatin also has a positive charge, and electrostatic repulsion interactions between the surface of composite fibers and BSA decreases the adsorption capacity of the surface of the composite fibers. When the pH exceeds the BSA isoelectric point (4.90), BSA has a negative charge. When $4.90 \leq pH \leq 6.00$, gelatin has a positive charge, and the electrostatic attraction between BSA and Gel is beneficial for protein adsorption; accordingly, adsorbed protein for composite fibers was maximal for a pH value of 5.32–6.00. pH values in the range of 5.32–7.35 characterize the normal physiological environment of the human body. In an alkaline environment, BSA and Gel are negatively charged and repel each other, which is not conducive to protein adsorption. Meanwhile, when investigating the effect of initial concentration on protein adsorption, the results demonstrated that Gel/CS/HA/GO composite fibers have good protein adsorption performance. Furthermore, these scaffolds and their effect in supporting stem cells for bone regeneration will be reported in the future.

4. Experimental Section

4.1. Materials

Acetic acid (CH_3COOH) and 30% hydrogen peroxide (H_2O_2) were produced by Tianjin Yong Sheng Chemical Co., Ltd. (Tianjin, China). CS (low molecular weight) with a degree of deacetylation of about 91% was obtained from Sigma (St. Louis, MO, USA). Gel and sulfuric acid (H_2SO_4) were supplied by Beijing Chemical Factory (Beijing, China). HA, with average particle sizes of 12 μm and 60 nm, was obtained from Shanghai Blue Reagent, Co., Ltd. (Shanghai, China). Graphite was obtained from Shanghai Mountain Pu Chemical Co., Ltd. (Shanghai, China). Potassium permanganate ($KMnO_4$) was obtained from Xian Bo Station Always Sells On Commission (Shanxi, China). Hydrazine hydrate ($N_2H_4 \cdot H_2O$) was produced by Luoyang Chemical Reagents (Luoyang, China). Bovine serum albumin (BSA) was supplied by Shanghai Blue Technology Development Co., Ltd. (Shanghai, China).

The following instruments were used in the study: a TL01 Electrostatic Spinning Machine (Shenzhen Tong Li Wei Technology Co., Ltd., Shenzhen, China), a scanning electron microscope (SEM, Carl Zeiss, LEO-1430 vp; Oberkochen, Germany), a transmission electron microscope (TEM, JEOL JEM-2100-f; Tokyo, Japan), a Fourier transform infrared spectrometer (FTIR, BRUKER VERTEX70; Brook, Germany), and an ultraviolet spectrophotometer (U3310; Hitachi, Tokyo, Japan).

4.2. Preparation of Graphene Oxide and Reduction of Graphene Oxide Using the Hummers Method

Graphite (5.0 g) was added to a 500 mL distillation bottle with concentrated H_2SO_4 (115 mL). $KMnO_4$ (25 g) was added to the distillation bottle very slowly with vigorous stirring for 2 h at 0 °C. The mixture was then transferred to an oil bath with lateral flow agitation at 35 °C overnight. Water was then added slowly to an oil bath at 90 °C, followed by heating for 30 min (the solution color immediately turned from black to chocolate brown). At this time, H_2O_2 (15 mL) was added. The solution was stirred for 30 min, filtered, and washed several times with deionized water and hydrochloric acid (HCl) solution until the pH was nearly neutral, and then dried overnight using a drying oven at 60 °C to obtain GO.

Aqueous GO was added to a flask that was connected with condenser pipe backflow devices. Hydrazine hydrate (2 mL) was added. Agitation backflow was performed for 24 h at 100 °C in oil bath conditions. The product was washed with ethanol and distilled water three times; it was then dried to a constant weight in a vacuum at 60 °C to obtain RGO.

4.3. Configuration of Electrospinning Solution

CS (1 wt. %) was added to 20% (v/v) acetic acid (20 mL) and stirred for 12 h. Gelatin (15 wt. %) was then added and dissolved at 60 °C in a water bath. HA particles (5 wt. %), powdery GO (2 wt. %), or RGO (2 wt. %) were added, the temperature was maintained, and the solutions were stirred for 30 min, resulting in the uniform dispersion of HA and GO in the solution. Configurations with particular concentrations of electrospinning liquid were obtained.

4.4. Preparation of Electrospun Fibers

A dry plastic syringe was filled with the solution and connected to a blunt-end stainless steel needle (#6). The syringe was fixed to a syringe pump. A stainless steel plate covered with aluminum foil (20 × 13 cm) was used as a collector and grounded. The fixed distance from the syringe needle tip to the aluminum foil was 15 cm. The syringe pump rate was adjusted, and a uniform flow rate of 2 mL·h^{-1} was used. The high voltage power supply was opened and the electrospinning voltage was adjusted to prepare electrospinning fibers. The laboratory temperature was 25 °C and the relative humidity was 45%–55%. The electrospunfibers were dried for three days under a vacuum at room temperature to remove residual solvents.

4.5. Test of Antibacterial Properties

A Gram-negative species (*Escherichia coli*) and a Gram-positive species (*Staphylococcus aureus*) were used to examine antibacterial properties. Qualitative and quantitative evaluations of the antibacterial activity of the Gel/GO/CS/HA composite fibers were performed. The spread plate method was used for qualitative analysis and the film adhering method was used for quantitative analysis. The antibacterial experiment included three groups: Gel/CS/HA, Gel/CS/HA/RGO, and Gel/CS/HA/GO. All experiments were repeated at least three times.

For the qualitative analysis, *E. coli* and *S. aureus* were plated on LB (Luria-Bertani) culture medium. The bacteria were aerobically cultivated at 37 °C for 24 h. Adequate amounts of bacteria were applied using inoculation loops, and were incubated in liquid medium for 24 h to obtain bacterial solutions. Secondly, the two bacteria were adjusted to a concentration of 1×10^8 cell mL^{-1} (10 mL) using phosphate-buffered saline (PBS), and the cultures were incubated at 37 °C with shaking at 200 rpm for 12 h. After shaking well, the 100 μL coated tablet was removed and incubated for 24 h at 37 °C in a constant temperature incubator. Plates were examined for bacterial growth and images were obtained.

For the quantitative analysis, the antimicrobial rate of samples was evaluated using the sample surface adhesion method for the two types of bacteria plated on the LB culture medium. The bacteria were cultivated at 37 °C for 24 h. This process was repeated three times to obtain pure colonies. One bacterium was inoculated in liquid medium at 37 °C with shaking at 220 rpm for 12 h. The two bacteria were configured to 3.0×10^7 cell mL^{-1} using PBS and diluted 10^3 times. The measured sample was loaded onto a Petri dish and sterile water was added to the bottom of the dish to prevent evaporation. Next, 50 μL drops were added to the surface and cultured for 12 h at 37 °C. The bacterial fluid was added to PBS (500 μL), mixed well, and 50 μL was plated. After cultivation for 12 h, colony-forming units were obtained. The antibacterial activity of the samples was estimated by calculating the antibacterial rate of samples based on the following formula. Gel/CS/HA was used as the control group, and three parallel experiments were conducted in order to obtain the average rate.

$$\text{Antibacterial rate} = (\text{colonies of control group} - \text{colonies of experimental group}) \times 100\%/\text{colonies of control group}$$

4.6. Protein Adsorption

BSA was prepared with phosphate buffer solution (PBS, pH 7.40), and a 10 times concentration of PBS solution formula as shown in Table 1. The samples with dimensions of 1 cm × 1 cm were incubated in 10 mL of the BSA solution in a centrifuge tube at room temperature. The change in the concentrations of BSA in the solution was determined by measuring the absorbance at 278 nm using an ultraviolet spectrophotometer.

Table 1. Ten times concentration of phosphate buffer solution.

Reagent	NaCl	KCl	KH$_2$PO$_4$	Na$_2$HPO$_4$
Concentration (g·L^{-1})	80.0	2.0	2.4	14.4

4.7. Statistical Analyses

All data are presented as means ± standard deviation. Statistical analysis was carried out using a one-sample *t*-test (assuming unequal variance). The difference between two sets of data was considered statistically significant when $p < 0.05$.

5. Conclusions

In summary, we demonstrated that quaternary Gel/CS/HA/GO composite fibers could be prepared by a simple electrospinning method. The powdery HA and GO were uniformly dispersed into Gel and CS matrix. Furthermore, the Gel/CS/HA/GO composite fibers displayed better antibacterial effects against *E. coli* and *S. aureus* compared with Gel/CS/HA composite fibers. This study observed that the composite fibers have a good adsorption capacity of BSA at the normal physiological environment of the human body. The protein adsorption on the surface of the composite fiber increased as the initial BSA concentration increased. At a certain initial concentration, the surface of the composite can reach the adsorption equilibrium, namely, the maximum adsorption value. These scaffolds and their effect in supporting stem cells for bone regeneration will be reported in the future.

Acknowledgments: This work was supported by The Natural Science Foundation of Xinjiang Uygur Autonomous Region (2015211A037). We thank Wasim Kapadia (University of Waterloo) for helpful language revision.

Author Contributions: Ya Gao and Yingbo Wang designed the experiments; Ya Gao and Yimin Wang performed the experiments; Ya Gao, Yingbo Wang and Wenguo Cui analyzed the data; Ya Gao, Yingbo Wang and Wenguo Cui wrote the manuscript.

Conflicts of Interest: The authors declare no conflict of interest.

References

1. Lao, L.; Wang, Y.; Zhu, Y.; Zhang, Y.; Gao, C. Poly(lactide-co-glycolide)/hydroxyapatite nanofibrous scaffolds fabricated by electrospinning for bone tissue engineering. *J. Mater. Sci. Mater. Med.* **2011**, *22*, 1873–1884. [CrossRef] [PubMed]
2. Hofman, K.; Tucker, N.; Stanger, J.; Staiger, M.; Marshall, S.; Hall, B. Effects of the molecular format of collagen on characteristics of electrospun fibres. *J. Mater. Sci.* **2012**, *47*, 1148–1155. [CrossRef]
3. Jaiswal, A.K.; Chhabra, H.; Soni, V.P.; Bellare, J.R. Enhanced mechanical strength and biocompatibility of electrospun polycaprolactone-gelatin scaffold with surface deposited nano-hydroxyapatite. *Mater. Sci. Eng. C* **2013**, *33*, 2376–2385. [CrossRef] [PubMed]
4. Sajkiewicz, P.; Kołbuk, D. Electrospinning of gelatin for tissue engineering—Molecular conformation as one of the overlooked problems. *J. Biomater. Sci. Polym. Ed.* **2014**, *25*, 2009–2022. [CrossRef] [PubMed]
5. Wang, H.; Feng, Y.; Fang, Z.; Xiao, R.; Yuan, W.; Khan, M. Fabrication and characterization of electrospun gelatin-heparin nanofibers as vascular tissue engineering. *Macromol. Res.* **2013**, *21*, 860–869. [CrossRef]
6. Meng, Z.X.; Li, H.F.; Sun, Z.Z.; Zheng, W.; Zheng, Y.F. Fabrication of mineralized electrospun PLGA and PLGA/gelatin nanofibers and their potential in bone tissue engineering. *Mater. Sci. Eng. C* **2013**, *33*, 699–706. [CrossRef] [PubMed]
7. Saravanan, S.; Nethala, S.; Pattnaik, S.; Tripathi, A.; Moorthi, A.; Selvamurugan, N. Preparation, characterization and antimicrobial activity of a bio-composite scaffold containing chitosan/nano-hydroxyapatite/nano-silver for bone tissue engineering. *Int. J. Biol. Macromol.* **2011**, *49*, 188–193. [CrossRef] [PubMed]
8. Okuyama, K.; Noguchi, K.; Hanafusa, Y.; Osawa, K.; Ogawa, K. Structural study of anhydrous tendon chitosan obtained via chitosan/acetic acid complex. *Int. J. Biol. Macromol.* **1999**, *26*, 285–293. [CrossRef]
9. Jayakumar, R.; Prabaharan, M.; Nair, S.V.; Tokura, S.; Tamura, H.; Selvamurugan, N. Novel carboxymethyl derivatives of chitin and chitosan materials and their biomedical applications. *Prog. Mater. Sci.* **2010**, *55*, 675–709. [CrossRef]
10. Mu, Q.; Su, G.; Li, L.; Gilbertson, B.O.; Yu, L.H.; Zhang, Q.; Sun, Y.-P.; Yan, B. Size-Dependent Cell Uptake of Protein-Coated Graphene Oxide Nanosheets. *ACS Appl. Mater. Interfaces* **2012**, *4*, 2259–2266. [CrossRef] [PubMed]
11. Artiles, M.S.; Rout, C.S.; Fisher, T.S. Graphene-based hybrid materials and devices for biosensing. *Adv. Drug Deliv. Rev.* **2011**, *63*, 1352–1360. [CrossRef] [PubMed]
12. Wang, G.; Qian, F.; Saltikov, C.W.; Jiao, Y.; Li, Y. Microbial reduction of graphene oxide by *Shewanella*. *Nano Res.* **2011**, *4*, 563–570. [CrossRef]

13. Lu, B.; Li, T.; Zhao, H.; Li, X.; Gao, C.; Zhang, S.; Xie, E. Graphene-based composite materials beneficial to wound healing. *Nanoscale* **2012**, *4*, 2978–2982. [CrossRef] [PubMed]
14. Faria, A.F.D.; Martinez, D.S.T.; Meira, S.M.M.; de Moraes, A.C.M.; Brandelli, A.; Filho, A.G.S.; Alves, O.L. Anti-adhesion and antibacterial activity of silver nanoparticles supported on graphene oxide sheets. *Colloids Surf. B Biointerfaces* **2014**, *113*, 115–124. [CrossRef] [PubMed]
15. Carpio, I.E.M.; Santos, C.M.; Wei, X.; Rodrigues, D.F. Toxicity of a polymer-graphene oxide composite against bacterial planktonic cells, biofilms, and mammalian cells. *Nanoscale* **2012**, *4*, 4746–4756. [CrossRef] [PubMed]
16. Titov, A.V.; Král, P.; Pearson, R. Sandwiched Graphene–Membrane Superstructures. *ACS Nano* **2010**, *4*, 229–234. [CrossRef] [PubMed]
17. Goenka, S.; Sant, V.; Sant, S. Graphene-based nanomaterials for drug delivery and tissue engineering. *J. Control. Release* **2014**, *173*, 75–88. [CrossRef] [PubMed]
18. Shi, X.; Chang, H.; Chen, S.; Lai, C.; Khademhosseini, A.; Wu, H. Regulating Cellular Behavior on Few-Layer Reduced Graphene Oxide Films with Well-Controlled Reduction States. *Adv. Funct. Mater.* **2012**, *22*, 751–759. [CrossRef]
19. Seabra, A.B.; Paula, A.J.; de Lima, R.; Alves, O.L.; Durán, N. Nanotoxicity of Graphene and Graphene Oxide. *Chem. Res. Toxicol.* **2014**, *27*, 159–168. [CrossRef] [PubMed]
20. Shirkhanzadeh, M. Direct formation of nanophase hydroxyapatite on cathodically polarized electrodes. *J. Mater. Sci. Mater. Med.* **1998**, *9*, 67–72. [CrossRef] [PubMed]
21. Mohamed, K.R.; Mostafa, A.A. Preparation and bioactivity evaluation of hydroxyapatite-titania/chitosan-gelatin polymeric biocomposites. *Mater. Sci. Eng. C* **2008**, *28*, 1087–1099. [CrossRef]
22. Lee, J.Y.; Chung, W.-J.; Kim, G. A mechanically improved virus-based hybrid scaffold for bone tissue regeneration. *RSC Adv.* **2016**, *6*, 55022–55032. [CrossRef]
23. Lin, K.; Xia, L.; Gan, J.; Zhang, Z.; Chen, H.; Jiang, X.; Chang, J. Tailoring the Nanostructured Surfaces of Hydroxyapatite Bioceramics to Promote Protein Adsorption, Osteoblast Growth, and Osteogenic Differentiation. *ACS Appl. Mater. Interfaces* **2013**, *5*, 8008–8017. [CrossRef] [PubMed]
24. Pang, X.; Zhitomirsky, I. Electrodeposition of composite hydroxyapatite–chitosan films. *Mater. Chem. Phys.* **2005**, *94*, 245–251. [CrossRef]
25. Cui, W.; Li, X.; Zhou, S.; Weng, J. Investigation on process parameters of electrospinning system through orthogonal experimental design. *J. Appl. Polym. Sci.* **2007**, *103*, 3105–3112. [CrossRef]
26. Lu, X.; Zhou, J.; Zhao, Y.; Qiu, Y.; Li, J. Room Temperature Ionic Liquid Based Polystyrene Nanofibers with Superhydrophobicity and Conductivity Produced by Electrospinning. *Chem. Mater.* **2008**, *20*, 3420–3424. [CrossRef]
27. Cheng, W.; Yu, Q.; Qiu, Z.; Yan, Y. Effects of different ionic liquids on the electrospinning of a polyacrylonitrile polymer solution. *J. Appl. Polym. Sci.* **2013**, *130*, 2359–2368. [CrossRef]
28. Sanchez, V.C.; Jachak, A.; Hurt, R.H.; Kane, A.B. Biological Interactions of Graphene-Family Nanomaterials: An Interdisciplinary Review. *Chem. Res. Toxicol.* **2012**, *25*, 15–34. [CrossRef] [PubMed]
29. Akhavan, O.; Ghaderi, E. Toxicity of Graphene and Graphene Oxide Nanowalls Against Bacteria. *ACS Nano* **2010**, *4*, 5731–5736. [CrossRef] [PubMed]
30. Feng, L.; Liu, Z. Graphene in biomedicine: Opportunities and challenges. *Nanomedicine* **2011**, *6*, 317–324. [CrossRef] [PubMed]
31. Kuila, T.; Bose, S.; Khanra, P.; Mishra, A.K.; Kim, N.H.; Lee, J.H. Recent advances in graphene-based biosensors. *Biosens. Bioelectron.* **2011**, *26*, 4637–4648. [CrossRef] [PubMed]

marine drugs

MDPI

Article

Chitosan-Based Nanomedicine to Fight Genital *Candida* Infections: Chitosomes

Toril Andersen [1], Ekaterina Mishchenko [2], Gøril Eide Flaten [1], Johanna U. Ericson Sollid [2], Sofia Mattsson [3], Ingunn Tho [4] and Nataša Škalko-Basnet [1,*]

[1] Drug Transport and Delivery Research Group, Department of Pharmacy, Faculty of Health Sciences, University of Tromsø The Arctic University of Norway, 9037 Tromsø, Norway; toril_andersen@hotmail.com (T.A.); goril.flaten@uit.no (G.E.F.)
[2] Host Microbe Interactions Research Group, Department of Medical Biology, Faculty of Health Sciences, University of Tromsø The Arctic University of Norway, 9037 Tromsø, Norway; ekaterina.mishchenko@uit.no (E.M.); johanna.e.sollid@uit.no (J.U.E.S.)
[3] Department of Pharmacology and Clinical Neuroscience, Division of Clinical Pharmacology, Umeå University, SE-90187 Umeå, Sweden; sofia.mattsson@pharm.umu.se
[4] Personalized Dosage Form Design Research Group, School of Pharmacy, Faculty of Mathematics and Natural Sciences, University of Oslo, 0316 Oslo, Norway; ingunn.tho@farmasi.uio.no
* Correspondence: natasa.skalko-basnet@uit.no; Tel.: +47-776-46640; Fax: +47-776-46151

Academic Editors: Hitoshi Sashiwa and David Harding
Received: 27 January 2017; Accepted: 1 March 2017; Published: 4 March 2017

Abstract: Vaginal infections are associated with high recurrence, which is often due to a lack of efficient treatment of complex vaginal infections comprised of several types of pathogens, especially fungi and bacteria. Chitosan, a mucoadhesive polymer with known antifungal effect, could offer a great improvement in vaginal therapy; the chitosan-based nanosystem could both provide antifungal effects and simultaneously deliver antibacterial drugs. We prepared chitosan-containing liposomes, chitosomes, where chitosan is both embedded in liposomes and surface-available as a coating layer. For antimicrobial activity, we entrapped metronidazole as a model drug. To prove that mucoadhesivness alone is not sufficient for successful delivery, we used Carbopol-containing liposomes as a control. All vesicles were characterized for their size, zeta potential, entrapment efficiency, and in vitro drug release. Chitosan-containing liposomes were able to assure the prolonged release of metronidazole. Their antifungal activity was evaluated in a *C. albicans* model; chitosan-containing liposomes exhibited a potent ability to inhibit the growth of *C. albicans*. The presence of chitosan was crucial for the system's antifungal activity. The antifungal efficacy of chitosomes combined with antibacterial potential of the entrapped metronidazole could offer improved efficacy in the treatment of mixed/complex vaginal infections.

Keywords: chitosan; liposomes; drug delivery; vaginal therapy; *Candida albicans*; metronidazole

1. Introduction

Vaginal infections are extremely prevalent and are reported to be one of the most common reasons for women to seek professional healthcare. Surveys estimate that more than 70% of adult women have experienced vaginal problems and have used vaginal products to treat infections [1]. The most common among vaginal infections are bacterial vaginosis, trichomoniasis, and vulvovaginal candidiasis [2–4]. In addition to causing discomfort and affecting the quality of life of infected individuals, these infections are also associated with an increased risk of related complications. There is an increased susceptibility to other pathogens such as human immunodeficiency virus (HIV), *Herpes simplex* virus, Neisseria *gonorrhoeae*, and *Chlamydia trachomatis*. Among pregnant women, there is a risk of preterm labor and birth and late fetal loss [5,6]. Current treatment regimens for these vaginal infections are considered to be insufficient due to a high degree of recurrence and limited patient compliance. The emergence of antimicrobial resistance to chemotherapeutics and the management of recurrent infections puts strong emphasis on the need for more efficient local therapies. Local treatment of vaginal infections has become favored over oral drug administration because, if properly applied, it can assure a higher local drug concentration and reduce the incidence of drug interaction, as well as interference with the gastrointestinal tract [1]. In past years, more attention has been given to the importance of biofilms and the treatment failure due to the inability of the selected drug to disrupt biofilms leading to bacterial survival and the recurrence of infection [7]. In spite of the increased interest to address the need for superior local treatment, most of the conventional dosage forms applied to the vagina are still suffering from poor distribution and retention at the vaginal site [8]. Moreover, the causes of vaginal infections might be often both of bacterial and fungal origin [9]. We aimed to develop an advanced mucoadhesive delivery system based on the synergistic antimicrobial action of a commonly prescribed drug to treat bacterial vaginosis, metronidazole [10], incorporated in chitosan-based mucoadhesive vesicles expected to exhibit antifungal activity on their own.

Chitosan is a natural polysaccharide obtained by the deacetylation of chitin. It is biocompatible, biodegradable, and non-toxic and thereby an interesting substance as a pharmaceutical excipient [11,12]. Chitosan has demonstrated excellent mucoadhesive properties and is therefore a good candidate for incorporation in dosage forms and delivery systems targeting the vaginal site; chitosan is expected to assure a prolonged vaginal residence time of the drug dosage form/delivery system [13–15]. In addition to its wide use in conventional dosage forms, chitosan has also been extensively studied as a constituent in nanosystems, such as chitosan-based nanoparticles or nanoemulsions, or as a coating material for liposomes [16–22]. We were particularly interested in the fact that chitosan exhibits potent antimicrobial, particularly antifungal, effects [23–25]. Chitosan was able to disrupt the biofilm integrity and exhibited antifungal activity against *Candida* species [26] or bactericidal effects against *Pseudomonas aeruginosa*. Importantly, the biofilm disruption was not depended on the simulated pH conditions; it was slightly more potent at the less acidic pH [27].

To prove the superiority of chitosan-based delivery systems in antimicrobial vaginal therapy, we used the mucoadhesive polymer Carbopol as a control. Carbopol is widely used in pharmaceutical applications as a constituent in different mucoadhesive drug delivery systems [8], including the ophthalmic, nasal, buccal, topical, intestinal, and vaginal therapy sites [28,29], due to its accessibility, stability, and safety. In vaginal drug delivery, Carbopol is often studied as a constituent in bioadhesive gels [30].

To prepare novel mucoadhesive formulations with either chitosan or Carbopol, we applied our own one-pot preparation method for polymer-containing liposomes [14,31]. This preparation method results in a polymer coating at the surface of the liposomes as well as a proportion of the polymer coating being inside the aqueous compartments of the liposomes and ensures sufficient drug loading. To prove the antifungal activity of chitosan and the ability of chitosan-containing liposomes to successfully carry antimicrobial drugs, we prepared chitosan-containing liposomes loaded with metronidazole, a known antimicrobial drug without intrinsic antifungal potential against *Candida albicans*. The activity was compared to the control, Carbopol-containing liposomes, as well as plain liposomes, both loaded with metronidazole, at the pH conditions mimicking the infected vagina (neutral pH) [4].

2. Results and Discussion

Chitosan-based nanomedicine for the localized therapy of vaginal infections should be characterized by optimal vesicular size, surface properties, and drug load, as well as the ability to assure the prolonged release of the incorporated drug within the vaginal cavity.

2.1. Characterization of Vesicles

The size of non-sonicated chitosan-containing vesicles is known to be greater than 1 micron, exhibiting a high polydispersity (expressed as high polydispersity index, PI) [31]. To reduce the polydispersity and vesicle size, the parameters that affect the retention time, the release of incorporated drug, and consequently its efficacy, sonication was applied and its duration optimized to prepare vesicles of a suitable size (100–300 nm), considering targeted vaginal administration [22]. Smaller more uniformly sized vesicles are expected to have a better distribution in the vaginal cavity and to penetrate deeper into the mucosal layer [32]. Deeper penetration will improve the residence time as the vesicles could penetrate through the rapidly clearing upper layer of the mucus, thereby improving the treatment efficacy [33]. The prepared vesicles fitted to a bimodal size distribution (NICOMP) and the volume-weighted percentages of particles in each population are presented in Table 1. All the vesicles were initially sonicated for 1 min. As previously reported, the chitosan-containing liposomes were smaller than the plain liposomes after the size reduction [14,31]. The Carbopol-containing liposomes seemed to resist size reduction to a greater extent than the other types of liposomes. After 1 min of sonication, the size of the Carbopol-containing liposomes was only reduced to 500 nm, as opposed to around 200 nm for both the chitosan-containing and the plain liposomes (Table 1). The Carbopol-containing liposomes were therefore subjected to another 1 min of sonication and the size was determined again. The size of the vesicles was this time reduced to 400 nm and remained larger than other vesicles. It seems that Carbopol, at the concentrations used, has a stabilizing effect on the polymer-containing liposomal structure; the same stabilizing effect was not observed for chitosan.

Table 1. Size distribution of liposomes. All values represent the mean size and are volume-weighted bimodal distribution ($n = 3$).

Type of Liposomes (Sonication Time)	Peak 1 *		Peak 2 *		PI
	Size (nm)	%	Size (nm)	%	
Chitosan-containing (1 min)	44.9	35	188.7	59	0.357
Carbopol-containing (1 min)	90.9	15	508.6	83	0.456
Carbopol-containing (2 min)	72.0	15	401.6	85	0.517
Plain (1 min)	41.4	13	224.5	86	0.368

* The values are presented as NICOMP distribution, which provided the best fit for the measured data (Fit error <1.5; residual error <10).

The surface properties of the vesicles were characterized through the zeta potential of the different formulations and are presented in Table 2. Plain liposomes exhibited a neutral zeta potential as was expected for phosphatidylcholine (PC) liposomes, which is a neutral lipid. The chitosan-containing liposomes exhibited a positive zeta potential due to the presence of positively charged chitosan on the surface, in agreement with our previous work. However, the values were slightly higher than previously reported [14], which can be attributed to the batch to batch differences in chitosan properties. The differences were not significant. The Carbopol-containing liposomes of two different sizes had a negative zeta potential, although not pronounced. This could be a reflection of both the amount of Carbopol at the surface of the liposomes and the degree of ionization of the carboxylic acids of Carbopol. Carbopols have a pKa of 6 [34] and the pH of the solution was 4.9; therefore the polymer would not be fully ionized. An additional explanation could also be that some of the Carbopol was embedded within the liposomal bilayers, as we have previously reported for chitosan in chitosomes [14]. This hypothesis might be supported by the published data on Carbopol-coated liposomes (post-coating of formed liposomes), wherein no Carbopol could be found embedded within liposomal bilayers [20]; the authors reported a zeta potential of −10.5 mV at the concentration of the coating solution similar to that used in current study. The differences in the expected and measured zeta potential for Carbopol-coated and our Carbopol-containing liposomes confirm that the preparation method rather than the type of polymer used determines the polymer distribution within/onto the vesicles.

Table 2. Zeta potential of liposomes. All values represent the mean ± SD (*n* = 3).

Type of Liposomes (Sonication Time)	Zeta Potential (mV)
Chitosan-containing (1 min)	10.6 ± 1.3
Carbopol-containing (1 min)	-4.2 ± 0.4
Carbopol-containing (2 min)	-2.3 ± 0.5
Plain (1 min)	-0.5 ± 0.7

2.2. Metronidazole Entrapment Efficiency

A key aspect of successful treatment of local vaginal infections is assuring that a sufficient amount of the drug is readily available at the vaginal site for a sufficient period of time [35]. Localized metronidazole therapy is a safe alternative to oral metronidazole treatment; vaginal metronidazole gel exhibited a 96% reduction in total systemic exposure [36], which is of great importance in the therapy of pregnant patients.

Figure 1 represents the entrapment efficiency of the different liposomal formulations, expressed as the amount of metronidazole per mg of liposomal lipid, normalized after the accurately determined amount of lipid in each formulation. Smaller and larger Carbopol-containing liposomes are presented as Carbopol-containing liposomes sonicated for one and two minutes, respectively.

As previously shown [31], the plain liposomes entrapped the lowest amount of metronidazole at about 6 µg/mg lipid, which is significantly less than the polymer-containing formulations ($p < 0.001$). The chitosan-containing liposomes and both of the Carbopol-containing liposomes were characterized by the entrapment efficiency of 11–12 µg/mg lipid and were not significantly different from each other. This indicates that the presence of polymer, rather than the type of polymer used, determines the entrapment efficacy. Chitosan-containing liposomes in the current study (1 min sonication) had a higher entrapment efficacy than in our previous work [31]. The finding can be attributed to a change/optimization in the sonication method; in the previous study [31] the vesicles were not diluted prior to sonication. The sonication of the more concentrated suspensions could lead to the higher loss of the entrapped drug, resulting in a lower final drug load.

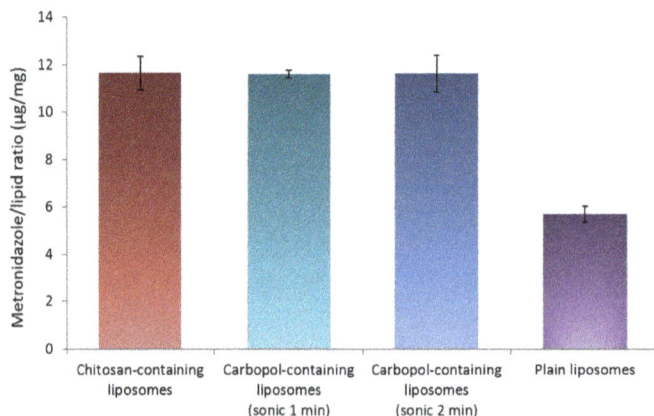

Figure 1. Entrapment efficiency of metronidazole in chitosan-containing liposomes, Carbopol-containing liposomes, and plain liposomes. All liposomes were sonicated for one minute, unless stated differently. All values represent the mean ± SD (*n* = 3).

2.3. In Vitro Release Studies

Nanomedicine has great potential in the development of superior delivery systems to treat vaginal diseases locally [37]. Metronidazole is an antimicrobial with a high potential for local administration against anaerobic bacteria and protozoa pathogens [38]; its importance in the local treatment of vaginal infections is decades long and remains attractive.

Figure 2 shows the cumulative release of metronidazole from the different types of vesicles containing metronidazole; metronidazole in solution served as a control. The testing was performed at neutral pH, mimicking the conditions of an infected vagina. Based on our previous work with chitosan-coated vesicles, we expected that the mucoadhesiveness of the chitosomes would not be affected by the pH [39]. All the polymer-containing liposomes sonicated for one minute exhibited a sustained release of metronidazole compared to the control solution. The Carbopol-containing liposomes, sonicated for longer time (2 min), exhibited a release that did not differ significantly from the control (metronidazole in solution), failing to assure a sustained drug release. The explanation of why Carbopol-containing vesicles of different sizes exhibit different drug release profiles might be that the drug was more loosely associated with smaller Carbopol-containing vesicles and was released faster from these than from larger vesicles of the same type (sonicated for 1 min). Figure 1 indicates that both smaller and larger Carbopol-containing vesicles entrapped a similar amount of metronidazole; the difference was in their size and surface charge (Table 2). It seems that Carbopol as a polymer can protect a vesicular structure from forced size reduction to a certain degree; when less Carbopol becomes surface available after longer sonication (closer to neutral zeta potential, Table 2), the vesicles become leakier. The phenomenon needs to be further evaluated, and, at this stage, we can only postulate on the reasons behind the observations and the difference in packing within vesicle bilayers between chitosomes and Carbopol-containing vesicles. However, the Carbopol-containing vesicles only served as a control in the antimicrobial testing; we did not invest time in further exploration of this interesting observation.

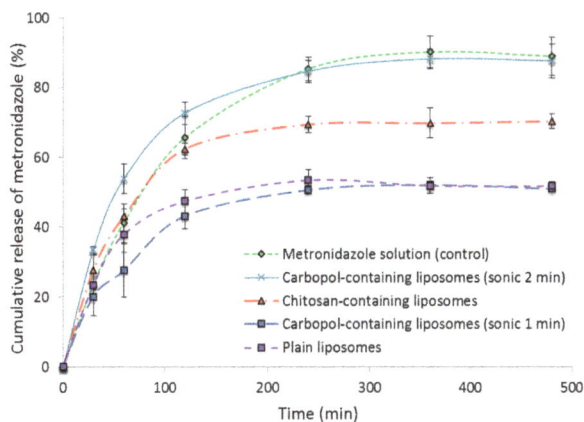

Figure 2. Cumulative release of metronidazole from different types of liposomes at pH 7. All liposomes were sonicated for one minute unless stated differently. The values represent the mean $\pm$ SD ($n = 3$).

The Carbopol-containing liposomes, sonicated for 1 min, sustained the delivery of metronidazole the most, similarly to plain liposomes (Figure 2). The finding could be attributed to their size and multilamelarity as compared to smaller vesicles which were probably oligolamellar. Chitosan-containing liposomes release drug in a more sustainable manner. We expected the drug to be completely released from the vesicles without reaching a plateau, as in the case of the metronidazole solution. Considering that the equilibrium reached between the drug associated with the vesicles in the donor chamber and the released drug in the acceptor chamber can slow the concentration gradient and contribute to the plateau and that different vesicles tend to accumulate on the membrane surface to a different extent, we cannot exclude the interaction between the membrane and the vesicles. However, considering the normal vaginal clearance and expected residence time within vagina [40], the drug release patterns for all three liposomal formulations are acceptable.

2.4. Antifungal Activity

Chitosan has been previously confirmed to exhibit high activity against *C. albicans* and to act on the prevention and disruption of biofilms of *C. albicans* [25,41–43]. It has also been reported that *C. albicans* biofilms are resistant to commonly used antifungal agents such as fluconazole [44,45]; therefore we expected our chitosan-containing liposomes to act on *C. albicans* inhibition and express potent antifungal activity. This would enable us to apply a synergistic approach in the therapy of vaginal infections, namely a carrier system with antifungal potential bearing a drug with antibacterial potential. This synergistic approach would combat the mixed infections and possibly act on antimicrobial resistance. Moreover, empty, drug free, chitosan-containing liposomes could be applied as a safe and effective antifungal treatment, which avoids exposure to drugs, which is particularly important in the treatment of pregnant patients [40].

As a first step we wanted to prove that chitosan-containing vesicles prevent the growth of *C. albicans* and that the presence of the drug within liposomes does not interfere with the activity. Therefore, the antifungal activity of the polymer-containing liposomes was tested by challenging the growth of *C. albicans* in the presence of the liposomal formulations during a 24 h-period. The chitosan-containing liposomes, both loaded with metronidazole as well as empty liposomes, exhibited a potent ability to inhibit the growth of *C. albicans* (Table 3). The lowest concentration of antifungal activity varied between 0.22 and 0.11 mg/mL of chitosan. The content of metronidazole in the drug-containing samples, 18 to 36 µg/mL, for chitosan- and Carbopol-containing vesicles respectively, did neither significantly alter nor improve the *C. albicans* inhibition compared to the non-drug containing (empty) liposomes.

Table 3. Antifungal activity of different formulations.

Formulation	*C. albicans* Inhibition
	Chitosan (mg/mL)
Chitosan-containing (MTZ)	0.11–0.22
Chitosan-containing (no drug)	0.11–0.22
Carbopol-containing (MTZ)	No inhibition
Plain (MTZ)	No inhibition
Metronidazole in solution (control)	No inhibition

MTZ: Metronidazole.

For the Carbopol-containing liposomes and plain liposomes, as well as the metronidazole control solution, no inhibition of *C. albicans* growth was observed. This is expected since metronidazole has no known activity against *C. albicans*. The results are highly encouraging considering our hypothesis. Vaginal infections can be complicated and result in infection with multiple pathogens, both of bacterial and fungal origin, which may occur either simultaneously or successively, leading to a high degree of recurrence associated with mixed infections [9,46]. The addition of an antibacterial agent, such as metronidazole, to a drug delivery system that is potently able to inhibit fungal growth would be greatly beneficial for the overall treatment and prevention of recurrence. In addition, chitosan has demonstrated the ability to disrupt bacterial biofilms involved in bacterial vaginosis [27]. This confirms chitosan as a superior component in our vaginal drug delivery system since chitosan is expected to disrupt the biofilm allowing the vesicle-associated antibacterial agent to reach the bacteria that would otherwise be shielded by the biofilm. Importantly, chitosan has also been reported to act against *Trihomonas vaginalis* infections, further strengthening its potential in antimicrobial vaginal therapy [47].

Metronidazole is a good choice among antibacterial agents since many vaginal infections are susceptible to metronidazole [4]. In the treatment of metronidazole-sensitive vaginal infections, such as bacterial vaginosis and trichomoniasis, there is an additional advantage. While many antimicrobial agents have a damaging effect on the natural microflora, metronidazole has been demonstrated not to affect the *Lactobacillus* strains in the vagina. This leaves the flora able to maintain the beneficial pH in the vaginal environment, protecting against further infection from opportunistic pathogens [48].

Considering chitosan's safety profile, mucoadhesiveness, and potent antimicrobial potential, chitosan can be considered a crucial constituent of novel nanosystems for vaginal administration. Moreover, its wound healing activities could assist in the treatment of vaginal lesions, which might be the pre-condition of vaginal infections.

3. Materials and Methods

3.1. Materials

Soy phosphatidylcholine (SPC; Lipoid S100, Lipoid GmbH, Ludwigshafen, Germany) was a generous gift from Lipoid GmbH. Chitosan (77% degree of deacetylation (DD), Fiske-SubbaRow reducer reagent, metronidazole, methanol, *n*-propanol, phosphorus standard, and Triton X solution were purchased from Sigma Aldrich Inc. (St. Luis, MO, USA). Ammonium molybdate and peroxide were purchased from Merck KGaA (Darmstadt, Germany), while sulfuric acid was purchased from May and Baker LTD (Dagenham, UK). Potato dextrose broth was purchased from Difco (BD, Franklin Lakes, NJ, USA) All other chemicals used in the experiments were of analytical grade. Carbopol®974P NF was a product from Lubrizol, Billingham, UK.

3.2. Preparation of Vesicles

Polymer-containing liposomes were prepared by the one-pot method previously described by Andersen et al. [31] (Appendix A Figure A1). In brief, SPC (200 mg) for the drug-free liposomes and SPC (200 mg) and 20 mg of metronidazole for metronidazole-containing liposomes were dissolved in methanol. The solvent was evaporated using a rotoevaporator system (Büchi rotavapor R-124 with vacuum controller B-721, Büchi Vac V-500, Büchi Labortechnik, Flawil, Switzerland) under a vacuum at 45 °C. The resulting lipid film was redispersed in 100 μL of *n*-propanol with a micro syringe pipette (Hamilton Company, Bonaduz, Switzerland). The dispersion was injected via a needle into 2 mL of aqueous solution of chitosan (0.17%, *w/w*, in 0.1% acetic acid), or Carbopol (0.10%, *w/w* in distilled water), and stirred for 2 h at room temperature. The dispersions were left in a refrigerator (4–8 °C) overnight prior to vesicle size reduction and characterization.

Plain (polymer-free) liposomes were prepared under the same conditions using the same lipid and metronidazole ratio to prepare the film. The film was subsequently re-dispersed (100 μL *n*-propanol), injected into distilled water, and stirred on a magnetic stirrer for 2 h. The dispersions were left in a refrigerator (4–8 °C) overnight prior to vesicle size reduction and characterization.

3.3. Size Reduction of Vesicles

The polymer-containing and plain (non-coated) liposomes were reduced to the desired size by sonication using a Sonics High Ultrasonic Processor (Sigma Aldrich Chemie GmbH, Steinheim, Germany). Prior to sonication, the samples were diluted to a suitable volume (5 mL) with distilled water. The duration of sonication was adjusted to achieve the desired size range of the vesicles (1 min, except for Carbopol-containing liposomes, which were also sonicated for 2×1 min). An ice bath was used to prevent heating of the samples.

3.4. Entrapment of Metronidazole

To remove unentrapped metronidazole from the polymer-containing, as well as the plain vesicles, the vesicles were dialyzed (Mw cut off: 12,000–14,000 Daltons; Medicell International Ltd., London, UK) against distilled water for 4 h at room temperature. The volume was adjusted to assure the sink conditions.

The amount of metronidazole entrapped in the vesicles was determined by UV-spectrophotometry (Agilent Technologies, Santa Clara, CA, USA). The samples were dissolved in methanol and metronidazole concentrations measured at 311 nm. The standard curve of metronidazole in methanol was prepared using the concentrations in the range of 2 to 20 µg/mL (R^2 = 0.9999).

3.5. Lipid Content

The content of phosphatidylcholine (PC) in the vesicles was measured using the modified Bartlett method [49]. In brief, the samples were diluted by distilled water to appropriate concentration and an aliquot (1 mL) was mixed with 0.5 mL of 10 N H_2SO_4 and heated at 160 °C for a minimum of 3 h. After cooling, two drops of H_2O_2 were added, and the mixture was heated at 160 °C for 1.5 h. Then ammonium molybdate (4.6 mL; 0.22% w/v) and 0.2 mL of Fiske-SubbaRow reducer reagent were added after cooling, the samples were vortexed, and the mixture was heated for 7 min at 100 °C. All samples were analyzed by UV spectrophotometry at 830 nm. The phosphorus standard solution was used to prepare a standard curve in appropriate concentrations.

3.6. Particle Size Analysis

The size distributions of the sonicated vesicles were measured by photon correlations spectroscopy using a Submicron Particle-sizer (Model 370, Nicomp, Santa Barbara, CA, USA), according to the method described earlier [14]. Briefly, the samples were diluted with filtered distilled water (0.2 µm Millipore filters) to provide the appropriate count intensity (approx. 250–350 kHz) and measured in three parallels (runtime 10 min at 23 °C). Both Gaussian and Nicomp algorithms were fitted to the experimental data to find the distribution that best describes the particle population. The volume-weighted distribution was used to determine the mean diameter and polydispersity index (PI) of all samples.

3.7. Determination of Zeta Potential

The zeta potential of all vesicles was measured on a Malvern Zetasizer Nano ZS (Malvern Instruments Ltd., Oxford, UK). The instrument was calibrated throughout the measurements using the Malvern zeta potential transfer standard (−42 ± 4.2 mV). The samples were diluted in filtered water until an appropriate count rate was achieved and measured in a disposable folded capillary cell. All measurements were performed at 23 °C [22]. The results represent an average of at least three independent measurements.

3.8. In Vitro Release Studies

Release studies were performed using Franz diffusion cells (PermeGear, Hellertown, PA, USA), with the heating circulator (Julabo Labortechnik F12-ED, Seelback, Germany) maintaining the temperature at 37 °C and a neutral pH to mimic the pH of the bacterial vaginosis infected vagina [4]. Cells with 12 mL volume acceptor chambers and a diffusion area of 1.77 cm^2 were used [14]. Polyamide membranes (0.2 µm pore size; Sartorius polyamide membrane; Sartorius AG, Göttingen, Germany) were used. The formulations were added to the donor compartment with a volume of 600 µL. The acceptor chambers were filled with distilled water and kept at 37 °C. Samples (500 µL) from the acceptor chamber were taken at 30, 60, 120, 240, 360, and 480 min and replaced with fresh medium. Both the sampling port and the donor chamber were covered with quadruple layers of para-film to prevent evaporation. Quantification of the released model substance was determined by spectrophotometry, wherein the aqueous solution of metronidazole was measured at 319 nm. The standard curve of metronidazole in distilled water was prepared using concentrations in the range of 2 to 20 µg/mL (R^2 = 0.9988). All experiments were carried out in triplicate.

3.9. Antifungal Activity Testing

Antifungal activity was tested according to the method described by Sperstad et al. [50] with modifications, using the yeast strain *C. albicans* (ATCC 10231). Fungal cells were suspended in the potato dextrose broth (Difco, Sparks, NV, USA) with 2% glucose; the cell concentration was determined and adjusted after counting in a Bürker chamber. Aliquots (50 µL) of fungal cells (final concentration approx. 2×10^5 cells/mL) were inoculated in a 96-well microtitre plate (Nunc[TM], Roskilde, Denmark) along with 50 µL of vesicular formulation dissolved in Milli-Q water. The vesicular formulations were diluted in a two-fold sequence and tested at final concentrations ranging from 2.3 to 300 µg/mL. Cultures were grown in a dark chamber, without shaking, for 24 h at 37 °C. The growth inhibition was determined by plating aliquots of the samples on potato dextrose agar plates with 2% glucose (Figure 3) and was defined as the concentration at which no growth was observed after 24 h incubation at 37 °. In addition, the negative controls containing neither *C. albicans* nor vesicles and the controls containing only vesicles were also tested for growth.

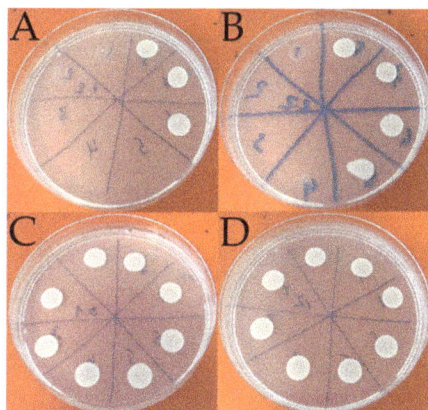

Figure 3. Representative photographs of the *C. albicans* growth on agar plates in the presence of different liposomal formulations; (**A**) Chitosan-containing liposomes (no drug); (**B**) Chitosan-containing liposomes (MTZ); (**C**) Carbopol-containing liposomes (MTZ); (**D**) plain liposomes (MTZ). Each sector contains an aliquot from a test well with a dilution of a formulation; aliquots were inoculated anticlockwise, i.e., from the highest concentration in sector 1 to the lowest concentration in sector 8. White spots represent the 'lawn' growth of *C. albicans*. The inhibition of *C. albicans* growth is indicated by the absence of white spots.

3.10. Statistical Analysis

Students' *t*-test was used for comparison of two means. A significance level of $p < 0.05$ was considered acceptable.

4. Conclusions

We developed chitosomes, chitosan-containing liposomes, of optimal size and sufficient metronidazole load for the localized therapy of mixed vaginal infections and have proven in vitro that this novel nanosystem can act on *C. albicans*. The chitosan-containing liposomes inhibited the growth of *C. albicans* independently of the presence of the entrapped model antibacterial drug, metronidazole, whereas the Carbopol-containing vesicles failed to prevent *C. albicans* growth. We have also shown that chitosomes provide a sustained release of the model drug, making it suitable for the administration of antibacterial drugs assuring the synergistic treatment of vaginal infections caused by several pathogens.

Acknowledgments: The authors are grateful to Lipoid GmbH, Ludwigshafen, Germany for continuous support in providing the lipids. Hans-Matti Blecke and Elizaveta Igumnova are acknowledged for their suggestions regarding the design of the antifungal assay. The publication charges for this article have been funded by a grant from the publication fund of University of Tromsø, The Arctic University of Norway.

Author Contributions: T.A., J.U.E.S., G.E.F., I.T., S.M. and N.Š.-B. designed and planned the experiments. T.A. and E.M. conducted the experiments. All authors contributed to the manuscript preparation. N.Š.-B. is the senior author and project leader.

Conflicts of Interest: The authors declare no conflict of interest.

Appendix A

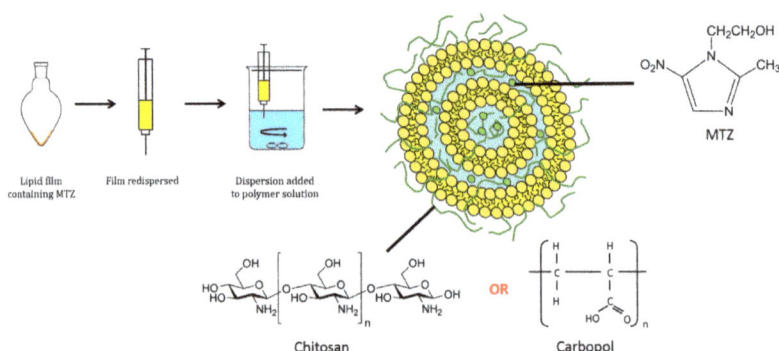

Figure A1. Schematic presentation of the preparation method.

References

1. Palmeira-de-Oliveira, R.; Palmeira-de-Oliveira, A.; Martinez-de-Oliveira, J. New strategies for local treatment of vaginal infections. *Adv. Drug Deliv. Rev.* **2015**, *92*, 105–122. [CrossRef] [PubMed]
2. Chiaffarino, F. Risk factors for bacterial vaginosis. *Eur. J. Obstet. Gynecol. Reproduct. Biol.* **2004**, *117*, 222–226. [CrossRef] [PubMed]
3. Nyirjesy, P. Vulvovaginal candidiasis and bacterial vaginosis. *Infect. Dis. Clin. N. Am.* **2008**, *22*, 637–652. [CrossRef] [PubMed]
4. Mashburn, J.J. Vaginal infections update. *J. Midwifery Women's Health* **2012**, *57*, 629–634. [CrossRef] [PubMed]
5. Van Der Pol, B. Diagnosing vaginal infections: It's time to join the 21st century. *Curr. Infect. Dis. Rep.* **2010**, *12*, 225–230. [CrossRef] [PubMed]
6. Brotman, R.R. Vaginal microbiome and sexually transmitted infections: An epidemiologic perspective. *J. Clin. Investig.* **2011**, *121*, 4610–4617. [CrossRef] [PubMed]
7. Swidsinski, A.; Mendling, W.; Loening-Baucke, V.; Swidsinski, S.; Dörffel, Y.; Scholze, J.; Lochs, H.; Verstraelen, H. An adherent *Gardnerella vaginalis* biofilm persists on the vaginal epithelium after standard therapy with oral metronidazole. *Am. J. Obstet. Gynecol.* **2008**, *198*, 97.e1–97.e6. [CrossRef] [PubMed]
8. Hussain, A.; Ahsan, F. The vagina as a route for systemic drug delivery. *J. Control. Release* **2005**, *103*, 301–313. [CrossRef] [PubMed]
9. Donders, G.; Bellen, G.; Ausma, J.; Verguts, L.; Vaneldere, J.; Hinoul, P.; Borgers, M.; Janssens, D. The effect of antifungal treatment on the vaginal flora of women with vulvo-vaginal yeast infection with or without bacterial vaginosis. *Eur. J. Clin. Microbiol. Infect. Dis.* **2011**, *30*, 59–63. [CrossRef] [PubMed]

10. Perioli, L.; Ambrogi, V.; Venezia, L.; Pagano, C.; Scuota, S.; Rossi, C. FG90 chitosan as a new polymer for metronidazole mucoadhesive tablets for vaginal administration. *Int. J. Pharm.* **2009**, *377*, 120–127. [CrossRef] [PubMed]

11. Baldrick, P. The safety of chitosan as a pharmaceutical excipient. *Regul. Toxicol. Pharmacol.* **2010**, *56*, 290–299. [CrossRef] [PubMed]

12. Kean, T.; Thanou, M. Biodegradation, biodistribution and toxicity of chitosan. *Adv. Drug Deliv. Rev.* **2010**, *62*, 3–11. [CrossRef] [PubMed]

13. Andersen, T. Novel Chitosan-Containing Liposomes as Mucoadhesive Delivery System for Vaginal Administration. Ph.D. Thesis, University of Tromsø The Arctic University of Norway, Tromsø, Norway, 2015.

14. Andersen, T.; Bleher, S.; Flaten, G.E.; Tho, I.; Mattsson, S.; Škalko-Basnet, N. Chitosan in mucoadhesive drug delivery: Focus on local vaginal therapy. *Mar. Drugs* **2015**, *13*, 222–236. [CrossRef] [PubMed]

15. Vanić, Ž.; Škalko-Basnet, N. Mucosal nanosystems for improved topical drug delivery: Vaginal route of administration. *J. Drug Deliv. Sci. Technol.* **2014**, *24*, 435–444. [CrossRef]

16. Li, N.; Zhuang, C.; Wang, M.; Sun, X.; Nie, S.; Pan, W. Liposome coated with low molecular weight chitosan and its potential use in ocular drug delivery. *Int. J. Pharm.* **2009**, *379*, 131–138. [CrossRef] [PubMed]

17. Zaru, M.; Manca, M.-L.; Fadda, A.M.; Antimisiaris, S.G. Chitosan-coated liposomes for delivery to lungs by nebulization. *Colloid Surf. B Biointerfaces* **2009**, *71*, 88–95. [CrossRef] [PubMed]

18. Bernkop-Schnürch, A.; Duennhaupt, S. Chitosan-based drug delivery systems. *Eur. J. Pharm. Biopharm.* **2012**, *81*, 463–469. [CrossRef] [PubMed]

19. Calderon, L.; Harris, R.; Cordoba-Diaz, M.; Elorza, M.; Elorza, B.; Lenoir, J.; Adriaens, E.; Remon, J.P.; Heras, A.; Cordoba-Diaz, D. Nano and microparticulate chitosan-based systems for antiviral topical delivery. *Eur. J. Pharm. Sci.* **2013**, *48*, 216–222. [CrossRef] [PubMed]

20. Berginc, K.; Suljaković, S.; Škalko-Basnet, N.; Kristl, A. Mucoadhesive liposomes as new formulation for vaginal delivery of curcumin. *Eur. J. Pharm. Biopharm.* **2014**, *87*, 40–46. [CrossRef] [PubMed]

21. Casettari, L.; Illum, L. Chitosan in nasal delivery systems for therapeutic drugs. *J. Control. Release* **2014**, *190*, 189–200. [CrossRef] [PubMed]

22. Jøraholmen, M.W.; Tho, I.; Vanic, Z.; Skalko-Basnet, N. Chitosan-coated liposomes for topical vaginal therapy: Assuring localized drug effect. *Int. J. Pharm.* **2014**, *472*, 94–101. [CrossRef] [PubMed]

23. Rabea, E.I.; Badawy, M.E.-T.; Stevens, C.V.; Smagghe, G.; Steurbaut, W. Chitosan and antimicrobial agent: Applications and mode of action. *Biomacromolecules* **2003**, *4*, 1457–1465. [CrossRef] [PubMed]

24. Seyfarth, F.; Schliemann, S.; Elsner, P.; Hipler, U.-C. Antifungal effect of high- and low-molecular-weight chitosan hydrochloride, carboxymethyl chitosan, chitosan oligosaccharide and N-acetyl-D-glucoamine against *Candida albicans*, *Candida krusei* and *Candida glabrata*. *Int. J. Pharm.* **2008**, *353*, 139–148. [PubMed]

25. Pu, Y.U.; Liu, A.; Zheng, Y.; Ye, B.I.N. In vitro damage of *Candida albicans* biofilms by chitosan. *Exp. Ther. Med.* **2014**, *8*, 929–934. [CrossRef] [PubMed]

26. Silva-Dias, A.; Palmeira-de-Oliveira, A.; Miranda, I.M.; Branco, J.; Colbrado, L.; Moneteiro-Soares, M.; Queiroz, J.A.; Pina-Vaz, C.; Rodrigues, A.G. Anti-biofilm activity of low-molecular weight chitosan hydrogel against *Candida* species. *Mol. Microbiol. Immunol.* **2014**, *203*, 25–33. [CrossRef] [PubMed]

27. Kandimalla, K.K.; Borden, E.; Omtri, R.S.; Boyapati, S.P.; Smith, M.; Lebby, K.; Mulpuru, M.; Gadde, M. Ability of chitosan gels to disrupt bacterial biofilms and their applications in the treatment of bacterial vaginosis. *J. Pharm. Sci.* **2013**, *102*, 2096–2101. [CrossRef] [PubMed]

28. Pavelić, Ž.; Škalko-Basnet, N.; Schubert, R. Liposomal gels for vaginal drug delivery. *Int. J. Pharm.* **2001**, *219*, 139–149. [CrossRef]

29. Bonacucina, G.; Martelli, S.; Palmieri, G.F. Rheological, mucoadhesive and release properties of carbopol gels in hydrophilic cosolvents. *Int. J. Pharm.* **2004**, *282*, 115–130. [CrossRef] [PubMed]

30. Ndesendo, V.M.K.; Pillay, V.; Choonara, Y.E.; Buchmann, E.; Bayever, D.N.; Meyer, L.C.R. A review of current intravaginal drug delivery approaches employed for the prophylaxis of HIV/AIDS and prevention of sexually transmitted infections. *AAPS PharmSciTech* **2008**, *9*, 505–520. [CrossRef] [PubMed]

31. Andersen, T.; Vanić, Ž.; Flaten, G.E.; Mattsson, S.; Tho, I.; Škalko-Basnet, N. Pectosomes and chitosomes as delivery systems for metronidazole: The one-pot preparation method. *Pharmaceutics* **2013**, *5*, 445–456. [CrossRef] [PubMed]

32. Das Neves, J.; Bahia, M.F.; Amiji, M.M.; Sarmento, B. Mucoadhesive nanomedicines: Characterization and modulation of mucoadhesion at the nanoscale. *Expert Opin. Drug Deliv.* **2011**, *8*, 1085–1104. [CrossRef] [PubMed]

33. Laffleur, F.; Bernkop-Schnurch, A. Strategies for improving mucosal drug delivery. *Nanomedicine (Lond.)* **2013**, *8*, 2061–2075. [CrossRef] [PubMed]

34. Lubrizol. Formulating controlled release tablets and capsules with carbopol polymers. In *Pharmaceutical Bulletin*; Lubrizol: Wickliffe, OH, USA, 2011; Volume 31.

35. Hainer, B.L.; Gibson, M.V. Vaginitis: Diagnosis and Treatment. *Am. Fam. Phys.* **2011**, *83*, 807–815.

36. Wain, A.M. Metronidazole vaginal gel 0.75% (MetroGel-Vagina®): A brief review. *Inf. Dis. Obstet. Gynecol.* **1998**, *6*, 3–7. [CrossRef] [PubMed]

37. Ensing, L.M.; Cone, R.; Hanes, J. Nanoparticle-based drug delivery to the vagina: A review. *J. Control. Release* **2014**, *190*, 500–514.

38. Zupančič, Š.; Potrč, T.; Baumgartner, S.; Kocbek, P.; Kristl, J. Formulation and evaluation of chitosan/polyethylene oxide nanofibers loaded with metronidazole for local infections. *Eur. J. Pharm. Sci.* **2016**, *95*, 152–160. [CrossRef] [PubMed]

39. Jøraholmen, M.W.; Škalko-Basnet, N.; Acharya, G.; Basnet, P. Resveratrol-loaded liposomes for topical treatment of the vaginal inflammation and infections. *Eur. J. Pharm. Sci.* **2015**, *79*, 112–121. [CrossRef] [PubMed]

40. Vanić, Ž.; Škalko-Basnet, N. Nanopharmaceuticals for improved topical vaginal therapy: Can they deliver? *Eur. J. Pharm. Sci.* **2013**, *50*, 29–41. [CrossRef] [PubMed]

41. Park, Y.; Kim, M.H.; Park, S.C.; Cheong, H.; Jang, M.K.; Nah, J.W.; Hahm, K.S. Investigation of the antifungal activity and mechanism of action of lmws-chitosan. *J. Microbiol. Biotechnol.* **2008**, *18*, 1729–1734. [PubMed]

42. Tayel, A.A.; Moussa, S.; El-Tras, W.F.; Knittel, D.; Opwis, K.; Schollmeyer, E. Anticandidal action of fungal chitosan against *Candida albicans*. *Int. J. Biol. Macromol.* **2010**, *47*, 454–457. [CrossRef] [PubMed]

43. Martinez, L.R.; Mihu, M.R.; Tar, M.; Cordero, R.J.B.; Han, G.; Friedman, A.J.; Friedman, J.M.; Nosanchuk, J.D. Demonstration of antibiofilm and antifungal efficacy of chitosan against candidal biofilms, using an in vivo central venous catheter model. *J. Infect. Dis.* **2010**, *201*, 1436–1440. [CrossRef] [PubMed]

44. Hawser, S.P.; Douglas, L.J. Resistance of *Candida albicans* biofilms to antifungal agents in vitro. *Antimicrob. Agents Chemother.* **1995**, *39*, 2128–2131. [CrossRef] [PubMed]

45. Mukherjee, P.K.; Chandra, J.; Kuhn, D.M.; Ghannoum, M.A. Mechanism of fluconazole resistance in *Candida albicans* biofilms: Phase-specific role of efflux pumps and membrane sterols. *Infect. Immun.* **2003**, *71*, 4333–4340. [CrossRef] [PubMed]

46. Pirotta, M.V.; Garland, S.M. Genital *Candida* species detected in samples from women in Melbourne, Australia, before and after treatment with antibiotics. *J. Clin. Microbiol.* **2006**, *44*, 3213–3217. [CrossRef] [PubMed]

47. Pradines, B.; Bories, C.; Vauthier, C.; Ponchel, G.; Loiseau, P.M.; Bouchemal, K. Drug-free chitosan coated poly(isobutylcyanoacrylate) nanoparticles are active against *Trichomonas vaginalis* and non-toxic towards pig vaginal mucosa. *Pharm. Res.* **2015**, *32*, 1229–1236. [CrossRef] [PubMed]

48. Melkumyan, A.R.; Priputnevich, T.V.; Ankirskaya, A.S.; Murav'eva, V.V.; Lubasovskaya, L.A. Effects of antibiotic treatment on the *Lactobacillus* composition of vaginal microbiota. *Bull. Exp. Biol. Med.* **2015**, *158*, 766–768. [CrossRef] [PubMed]

49. Bartlett, G.R. Phosphorus assay in column chromatography. *J. Biol. Chem.* **1959**, *234*, 466–468. [PubMed]

50. Sperstad, S.V.; Haug, T.; Paulsen, V.; Rode, T.M.; Strandskog, G.; Solem, S.T.; Styrvold, O.B.; Stensvåg, K. Characterization of crustins from the hemocytes of the spider crab, *Hyas araneus*, and the red king crab, *Paralithodes camtschaticus*. *Dev. Comp. Immunol.* **2009**, *33*, 583–591. [CrossRef] [PubMed]

marine drugs

MDPI

Article

Determination of Inorganic Cations and Anions in Chitooligosaccharides by Ion Chromatography with Conductivity Detection

Lidong Cao [1], Xiuhuan Li [1], Li Fan [2], Li Zheng [1], Miaomiao Wu [1], Shanxue Zhang [3] and Qiliang Huang [1,*]

[1] Institute of Plant Protection, Chinese Academy of Agricultural Sciences, No. 2 Yuanmingyuan West Road, Beijing 100193, China; caolidong@caas.cn (L.C.); lixiuhuan0822@163.com (X.L.); zhengli7seven@163.com (L.Z.); wumiaomiao2016@163.com (M.W.)
[2] Institute of Quality Standards & Testing Technology for Agro-Products, Chinese Academy of Agricultural Sciences, No. 12 Zhongguancun South Street, Beijing 100081, China; leefan66@126.com
[3] Hainan Zhengye Zhongnong High-Tech Co., LTD., No. 25 Nansha Road, Haikou 570206, Hainan, China; zhangshanxue@zyzn.net
* Correspondence: qlhuang@ippcaas.cn; Tel./Fax: +86-10-62816909

Academic Editors: Hitoshi Sashiwa and David Harding
Received: 8 January 2017; Accepted: 16 February 2017; Published: 22 February 2017

Abstract: Chitooligosaccharides (COSs) are a promising drug candidate and food ingredient because they are innately biocompatible, non-toxic, and non-allergenic to living tissues. Therefore, the impurities in COSs must be clearly elucidated and precisely determined. As for COSs, most analytical methods focus on the determination of the average degrees of polymerization (DPs) and deacetylation (DD), as well as separation and analysis of the single COSs with different DPs. However, little is known about the concentrations of inorganic cations and anions in COSs. In the present study, an efficient and sensitive ion chromatography coupled with conductivity detection (IC-CD) for the determination of inorganic cations Na^+, NH_4^+, K^+, Mg^{2+}, Ca^{2+}, and chloride, acetate and lactate anions was developed. Detection limits were 0.01–0.05 µM for cations and 0.5–0.6 µM for anions. The linear range was 0.001–0.8 mM. The optimized analysis was carried out on IonPac CS12A and IonPac AS12A analytical column for cations and anions, respectively, using isocratic elution with 20 mM methanesulfonic acid and 4 mM sodium hydroxide aqueous solution as the mobile phase at a 1.0 mL/min flow rate. Quality parameters, including precision and accuracy, were fully validated and found to be satisfactory. The fully validated IC-CD method was readily applied for the quantification of various cations and anions in commercial COS technical concentrate.

Keywords: chitooligosaccharides; ion chromatography; inorganic cations and anions; method validation; quantification

1. Introduction

Chitooligosaccharides (COSs) derive from the hydrolysis of chitosan, a cationic polysaccharide obtained by partial deacetylation of chitin, the second most abundant naturally occurring homopolysaccharide extracted from, among others, the exoskeleton of crustaceans and insects and fungal cell walls [1]. COSs are readily soluble in water due to their shorter chain lengths and free amino groups in D-glucosamine units. The greater solubility and low viscosity of COSs at neutral pH make COSs perform remarkable biological activities at the cellular or molecular level [2]. Moreover, COSs are promising as a drug candidate, and as a food ingredient, additive, and preservative that improve food quality and human health, because they are innately biocompatible, non-toxic, and non-allergenic to living tissues [3–5].

As a potential candidate for practical application in medicine and food, oral administration of COSs is inevitable. It has been claimed that COSs reach systemic circulation after oral administration [6]. Therefore, for the sake of human health, impurities in COSs, especially inorganic ions, must be clearly elucidated and precisely determined. COSs are produced by the enzymatic or alkaline heating deacetylation of chitosan or chitin [7]. The inherently present inorganic cations, such as sodium (Na^+), potassium (K^+), calcium (Ca^{2+}), magnesium (Mg^{2+}), and ammonium (NH_4^+) in raw material possibly persist in the final product of COSs. The free amino groups in COSs can undergo aerobic oxidation due to the lone electron pair of nitrogen atoms, leading to the decomposition of COSs during storage. This issue could be addressed by changing the neutral amino group to ammonium with the assistance of inorganic or organic acid during the manufacturing process. Hydrochloric, acetic, and lactic acid are the most commonly used acids. Thus, the counter anions of COSs should be qualitatively and quantitatively determined.

It is clear that mineral elements, such as Na^+, K^+, Ca^{2+}, and Mg^{2+}, are essential nutrients that function in the regulation of cardiac output and peripheral vascular resistance, which are the main determinants of blood pressure level [8]. There is a balanced concentration level for all these elements in order to maintain the essential functions of the human body [9,10]. Deficiency or excess of these elements over the required level can have implications for human health. As COSs are a promising candidate for the food and medicine fields, their inorganic cations and anions should be strictly monitored and controlled, which could improve the beneficial effects and avoid deleterious effects. Consequently, the development of an efficient method for the determination of inorganic cations and anions in COS samples is important from nutritional, toxicological, and technological points of view.

A variety of methods have been developed for analyzing such mineral inorganic cations in different matrices, including popularly used atomic spectroscopic methods such as atomic absorption spectroscopy (AAS) [11,12], inductively coupled plasma–mass spectrometry (ICP-MS) [13], atomic emission spectrometry (ICP-AES) [14], and optical emission spectrometry (ICP-OES) [15], as well as capillary electrophoresis (CE) [16]. However, some of these methods suffer from spectral and chemical interferences, asynchronous determination of mixed cations, laborious and prolonged procedures for sample preparation, as well as utilization of toxic concentrated acid and other reagents. Capillary electrophoresis has limits of poor reproducibility of migration times and peak areas, and moderate sensitivity [17]. Due to the sensitivity, stability of the separation system, good selectivity, and capacity of multi-element analysis in a single run, ion chromatography coupled with conductivity detection (IC-CD) has become the method of choice for the separation and determination of multiple cations and anions in various sample matrices [18–20]. This powerful and reliable technique has been widely used for analysis of Na^+, K^+, Ca^{2+}, Mg^{2+}, and anions in biological samples [21], food [22,23], biodiesel [24–26], oil [27], plant extract [28], and water [29–31], particularly at the level of trace concentrations.

As for COSs, most of the analytical methods focus on the determination of the average degrees of polymerization (DPs) and deacetylation (DD), as well as separation and analysis of the single COSs with different DPs [7]. Recently, we reported an efficient and sensitive analytical method based on high performance anion exchange chromatography with pulsed amperometric detection (HPAEC-PAD) for the simultaneous separation and determination of glucosamine and COSs, with DPs ranging from 2 to 6 without prior derivatization [32]. However, little is known about the concentrations of Na^+, NH_4^+, K^+, Mg^{2+}, Ca^{2+}, and anions in COSs. In the present study, an IC-CD method for determination of inorganic cations Na^+, NH_4^+, K^+, Mg^{2+}, Ca^{2+}, and chloride, acetate and lactate anions was developed. The detection is performed using conductimetry. Parameters such as linearity, sensitivity, precision, and accuracy were fully validated. Moreover, this validated method was used to quantify the inorganic cations and anions in COS technical concentrate. The procedure is simple and environmentally friendly because only water is used to dissolve the COS samples.

2. Results and Discussion

2.1. Optimization of Chromatographic Conditions

Ion chromatography is the most popular analytical method used for the determination of anions and cations in various sample matrices. Satisfactory separation depends mainly on the column, mobile phase, and flow rate. These three variables were screened during optimization of chromatographic conditions, which was carried out using mixed cations or anions standard solutions. The IonPac CS12A and IonPac AS12A analytical columns were used for cations and anions separation, respectively. The flow rate was set to 1.0 mL/min for both cations and anions optimization. The results demonstrate that the isocratic elution with 20 mM methanesulfonic acid solution enabled a satisfactory separation for Na^+, NH_4^+, K^+, Mg^{2+}, and Ca^{2+} within 15 min. While peak tailing and longer retention time occurred when 15 mM mobile phase was used. Isocratic elutions with 4, 7, 10, 15, and 20 mM sodium hydroxide solution were employed for anions separation. The results indicated that a 4 mM mobile phase can improve resolution for chloride, acetate, and lactate anions with 15 min. The resolutions for acetate and lactate were not satisfactory when isocratic elution with other concentrations were used. Representative IC-CD chromatograms of the mixed cations and anions standard are shown in Figures 1b and 2c, where the signals of Na^+, NH_4^+, K^+, Mg^{2+}, Ca^{2+}, and chloride, acetate, and lactate anions are clearly shown.

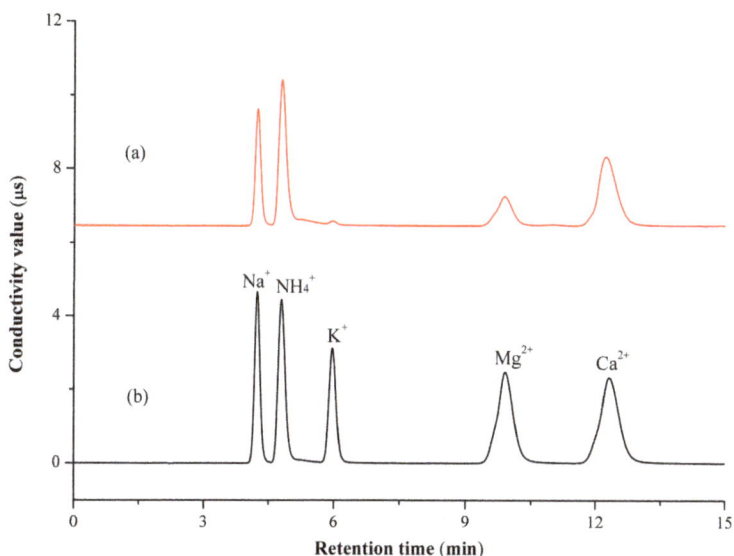

Figure 1. Representative ion chromatography with conductivity detection (IC-CD) chromatograms of cations in COS technical concentrate sample (**a**) and a mixed standard solution (**b**).

2.2. Calibration and Method Validation

Quality parameters such as sensitivity, linearity, precision, and accuracy were fully evaluated. Results showed that the linear ranges for cations and anions were 0.002–0.8 mM and 0.001–0.6 mM, respectively. All the calibration curves showed good linearity (R^2 = 0.9950–0.9999) in the tested range (Table 1). The limits of detection (LOD) and quantification (LOQ) were defined as the minimum amounts at which the analyte can be reliably detected and quantified. Typical signal-to-noise (S/N) ratios of the LOD and LOQ were 3 and 10, respectively. Diluted low concentrations of the cations and anions standard solutions were injected to determine the S/N ratio. Then, the LOD and LOQ were calculated. For cations, the LOD and LOQ ranged from 0.01 to 0.05 (corresponding to 0.25–1.25 pmol) and 0.03 to 0.15 µM (corresponding to 0.75–3.75 pmol), respectively. For anions, the LOD and LOQ were 0.5 µM (corresponding to 12.5 pmol) and 1.6 µM (corresponding to 40 pmol), respectively.

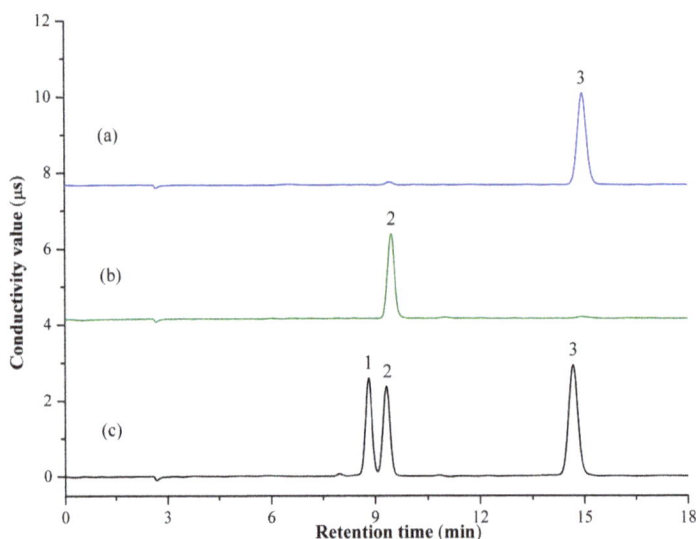

Figure 2. Representative IC-CD chromatograms of anions in COS technical concentrate sample (a), sample (b), and a mixed standard solution (c). Peaks: (1) lactate; (2) acetate; (3) chloride.

Table 1. Calibration parameters for cations and anions in standard solutions.

Analyte		Linear Range (mM)	Calibration Curve [a]	R^2	LOD (µM)	LOQ (µM)
Cation	Sodium	0.002–0.8	$y = 6.4217x + 0.0073$	0.9999	0.01	0.03
	Ammonium [b]	0.002–0.1	$y = 5.2160x + 0.0176$	0.9960	0.02	0.06
	Potassium	0.002–0.8	$y = 7.6307x - 0.0009$	0.9998	0.04	0.14
	Magnesium	0.002–0.8	$y = 13.3560x + 0.0130$	0.9999	0.05	0.15
	Calcium	0.002–0.8	$y = 14.4640x + 0.0256$	0.9999	0.02	0.05
Anion	Lactate	0.001–0.6	$y = 5.3829x - 0.0317$	0.9998	0.6	2.0
	Acetate	0.001–0.6	$y = 3.9231x + 0.0674$	0.9950	0.5	1.7
	Chloride	0.001–0.6	$y = 9.1727x - 0.1074$	0.9988	0.5	1.6

[a] y and x refer to the signal response (µS) and molar concentration (mM), respectively. [b] The calibration curve is $y = 2.9823x + 0.2796$ at the linear range 0.13–0.5 mM and the R^2 is 0.9961. LOD: limit of detection; LOQ: limit of quantification.

Precision was evaluated by measuring known amounts of the mixed cations and anions standards and the real sample determination. To establish repeatability (intraday) and intermediate (interday) precision, variations in terms of peak areas and retention times of the mixed standard solutions at three concentration levels were determined (Table 2). Repeatability was assessed using seven replicates in one day. Under repeatability conditions, retention times and integrated peak areas of all tested analytes were stable with 0.1–1.6 and 0.2–3.0 %RSD, respectively. Intermediate precision was assessed from nine determinations (three determinations daily over three days) using the same equipment, but performed on three consecutive days using three separately prepared batches of eluents. Under intermediate precision conditions, retention times and integrated peak areas of all tested analytes were stable with 0.3–3.7 and 0.2–6.2 %RSD, respectively. These values are slightly higher than what was found for repeatability. Method precision was also assessed by comparing the variations among seven replicates determinations of the same batch of COS technical concentrate with the Horwitz value (%RSDr). All the %RSD values of cations and anions determinations were less than the corresponding %RSDr (Table 3), indicating that the developed method is precise.

Table 2. Determination of method precision under repeatability (intraday) and intermediate precision (interday) conditions given as RSD(%) of peak area and retention time.

Analyte		Repeatability ($n = 7$)						Intermediate Precision ($n = 9$)					
		Peak Area			Retention Time			Peak Area			Retention Time		
		C1	C2	C3	C1	C2	C3	C1	C2	C3	C1	C2	C3
Cation	Sodium	0.39	0.21	0.60	0.46	0.07	0.61	0.94	6.44	4.57	0.49	0.39	0.40
	Ammonium	0.49	0.44	0.23	0.12	0.06	0.66	0.57	0.51	0.18	0.44	0.32	0.65
	Potassium	0.87	0.51	0.28	0.44	0.06	0.12	0.81	0.54	0.26	0.50	0.33	0.34
	Magnesium	1.13	0.70	0.19	0.67	0.13	0.21	1.39	0.82	0.18	1.14	1.30	0.64
	Calcium	2.79	0.75	0.37	0.63	0.08	0.17	3.08	1.20	0.47	1.44	0.79	0.63
Anion	Lactate	0.85	0.48	3.04	1.03	0.62	1.52	6.20	4.69	5.60	0.84	0.68	1.64
	Acetate	0.64	0.48	1.46	1.14	0.62	1.56	0.84	0.56	2.39	3.69	0.56	1.37
	Chloride	0.52	0.62	0.83	1.18	0.59	0.63	5.13	4.43	6.68	1.00	0.62	1.65

RSD: Relative standard deviation. For cations: C1 (mM): 0.04; C2 (mM): 0.08; C3 (mM): 0.12. For anions: C1 (mM): 0.13; C2 (mM): 0.19; C3 (mM): 0.25.

Table 3. Determination of each cation and anion in COS technical concentrates.

Analyte		COS Technical Concentrate A			COS Technical Concentrate B		
		Content (%) [a]	%RSD	%RSDr	Content (%) [a]	%RSD	%RSDr
Cation	Sodium	0.08	2.13	3.91	0.08	2.33	3.91
	Ammonium	0.39	1.18	3.08	0.36	2.71	3.14
	Potassium	0.01	4.91	5.24	0.01	4.95	5.24
	Magnesium	0.04	2.72	4.37	0.04	3.35	4.37
	Calcium	0.17	3.04	3.50	0.18	2.86	3.48
Anion	Acetate	17.64	1.30	1.74	–	–	–
	Chloride	–	–	–	11.57	0.34	1.85

[a] Mass percentage of each cation and anion in COS technical concentrate (means value of seven determinations).

The accuracy was evaluated through the standard addition method under optimized conditions, and it was found to be satisfactory with the recoveries ranging from 86.0% to 110.7% under three spiked concentration levels (Table 4). These validation results indicate that this IC-CD method is sensitive, precise, and accurate for the simultaneous quantitative determination of Na^+, NH_4^+, K^+, Mg^{2+}, Ca^{2+}, or chloride, acetate and lactate anions.

Table 4. Method accuracy for determining cations and anions in COS technical concentrates.

Analyte		Recovery (%)		
		Spiked C1	Spiked C2	Spiked C3
Cation	Sodium	102.13 ± 4.47	103.29 ± 3.45	98.79 ± 1.41
	Ammonium	89.68 ± 4.64	90.21 ± 2.13	87.80 ± 0.41
	Potassium	86.41 ± 0.37	90.51 ± 0.22	92.54 ± 1.82
	Magnesium	93.08 ± 2.19	97.02 ± 1.73	96.94 ± 3.02
	Calcium	91.90 ± 0.45	86.04 ± 0.24	93.42 ± 0.39
Anion	Lactate	110.65 ± 2.84	105.64 ± 4.79	107.14 ± 3.40
	Acetate	103.17 ± 3.04	108.97 ± 2.24	102.43 ± 6.03
	Chloride	97.97 ± 4.51	105.84 ± 3.87	108.15 ± 0.94

All values were given as mean recovery ($n = 3$) ± SD. SD: standard deviation. For cations: C1 (mM): 0.04; C2 (mM): 0.08; C3 (mM): 0.12. For anions: C1 (mM): 0.13; C2 (mM): 0.19; C3 (mM): 0.25.

2.3. Analysis of Cations and Anions in COS Technical Concentrates

During the manufacture of COS technical concentrate, the existing inorganic cations, such as Na^+, NH_4^+, K^+, Mg^{2+}, and Ca^{2+} in the raw-material or incomplete desalination process, may result in the inorganic cations impurities in the final product of COSs. For longer shelf life, commercial COSs are usually present in its ammonium salt form. Hydrochloric, acetic, and lactic acids are the most commonly used acids. Figures 1a and 2a,b show ion chromatograms obtained in the analysis of two COS technical concentrates where the signals of existing cations and anions are clearly indicated. Moreover, as presented in Table 3, all the inorganic cations of Na^+, NH_4^+, K^+, Mg^{2+}, and Ca^{2+} were detected in COS technical concentrate. Among the detected cations, NH_4^+ is the cation that has the highest concentration in both COS samples, followed by Ca^{2+}. The content of K^+ is the lowest. For anions, chloride and acetate were detected in COS Technical Concentrates A and B, with concentrations of 17.64% and 11.57%, respectively. As potential food and medicine field candidates, the ammonium lactate of COSs will be an ideal combination due to its biocompatibility, its low-toxicity, and the biodegradability of the lactic acid. Although the determination of lactate anion for a real COS sample was not performed, the recovery test clearly indicated that the proposed method is applicable for lactate anion analysis in COS samples.

3. Materials and Methods

3.1. Materials

Sodium hydroxide solution (50%, w/w) was purchased from Alfa Aesar Co., Ltd. (Tianjin, China). Methanesulfonic acid (99%) and sodium acetate (99%) were obtained from Sigma Aldrich Co. LLC. (Shanghai, China). Potassium chloride (99%), ammonium chloride (99%), magnesium sulfate (99%), calcium chloride (96%), and lactic acid (98%) were purchased from Sinopharm Chemical Reagent Beijing Co., Ltd. (Beijing, China). All stock standard solutions of anions and cations (20 mM) were prepared directly from the analytical reagent grade chemicals (as purchased) using deionized water, which was obtained using a MilliQ (Millipore, Bedford, MA, USA) water purification system. The working standard solutions were prepared as needed by appropriately diluting concentrated stock solutions with water. COS Technical Concentrate A with a number-average molecular weight (Mn) of 868 and a degree of deacetylation (DD) of 95% was provided by Hainan Zhengye Zhongnong High-Tech Co., Ltd. (Haikou, China). COS Technical Concentrate B with an Mn of 673 and a DD of 92% was obtained from Qingdao Zhongda Agricultural Science and Technology Co. Ltd. (Qingdao, China). The Mn was determined by matrix-assisted laser desorption/ionization time-of-flight mass spectrometry (MALDI-TOF-MS) analysis. The DD was determined by acid-base titration with bromocresol green as indicator. COS technical concentrate was accurately weighed and dissolved in water to prepare stock sample solution. The working sample solutions were prepared by dilution with water. All standards and samples solutions were stored in polyethylene bottles.

3.2. Chromatographic Analysis

All chromatographic analysis were performed using a Dionex ICS-3000 (Sunnyvale, CA, USA) system composed of an AS40 automated sampler, a GP40 gradient pump, and a CD20 conductivity detector. In order to reduce the background eluent conductivity, the detector was preceded by a Dionex self regenerating suppressor system. Suppression was achieved with a Dionex ASRS-300 (4 mm) for the anions and CSRS-300 (4 mm) for the cations. The ion separation was carried out with two different ion-exchange columns. Anions were separated on an IonPac AS12A column (250 mm × 4 mm i.d.) protected by an IonPac AG12A guard column (50 mm × 4 mm i.d.). Cations were determined using an IonPac CS12A column (250 mm × 4 mm i.d.) equipped with an IonPac CG12A guard column (50 mm × 4 mm i.d.). Data were collected using a Chromeleon 6.8 chromatogram workstation.

All eluents were degassed and pressurized under high-purity nitrogen to prevent dissolution of carbon dioxide and subsequent production of carbonate. An aqueous solution containing 20 mM methanesulfonic acid was used for elution of cations. An aqueous solution containing 4 mM sodium hydroxide served as eluent for anions. Elution was carried out at a flow rate of 1.0 mL/min and 25 μL was injected for both anion and cation determinations. The concentrations of each cations and anions in the samples were calculated using a calibration curve that produced the relationship between the amount of analyte and the peak area. All analyses were carried out in duplicate.

3.3. Calibration

To assess the linearity, calibration curves were plotted by a partial least squares method on the analytical data of peak area and concentration, using analyte standards covering the concentration range of 0.002–0.8 mM for Na^+, NH_4^+, K^+, Mg^{2+}, and Ca^{2+}, and 0.001–0.6 mM for chloride, acetate and lactate anions. The linear range of the curve was assessed by the value of linear correlation coefficient and the residuals. The dilute standard solution was further diluted to known low concentration with water for signal-to-noise (S/N) ratio determination. The limits of detection (LOD) and quantification (LOQ) were defined as the minimum concentrations resulting in signal-to-noise (S/N) ratios of 3 and 10, respectively.

3.4. Method Validation

The method precision was evaluated according to the repeatability (intraday) and intermediate (interday) precision and was expressed as a relative standard deviation (%RSD). For the mixed cations or anions standard solution, the precision in terms of retention time and peak area was determined. Repeatability was assessed using seven replicates in one HPLC run. Intermediate precision was evaluated from nine determinations (three determinations daily over three days) using the same equipment, but performed on three consecutive days using three separately prepared batches of mobile phase. Both repeatability and intermediate precision were determined at three concentrations levels for cations (0.04, 0.08, and 0.12 mM) and anions (0.13, 0.19, and 0.25 mM). In addition to the standard solutions, the precision was also evaluated by the real sample determination. For the COS technical concentrate sample, the contents of the containing cations and anions were measured under the prescribed conditions. The coefficients of variations of seven replicate determinations of the same batch of COS technical concentrate are compared with the Horwitz value (%RSDr) [33]. The Horwitz equations are described as follows:

$$\%RSD_R = 2^{(1-0.5\log_{10}C)} \tag{1}$$

$$\%RSDr = \%RSD_R \times 0.67 \tag{2}$$

where $\%RSD_R$ represents the inter-laboratory coefficient of variation (CV), $\%RSDr$ represents the repeatability CV, and C represents the concentration of the analyte in the sample as a decimal fraction.

The method accuracy was determined by spike-recovery test. A known amount of the cation or anion working standard solution was added to a predetermined amount of the COS technical concentrate, and the spiked sample was assayed. The total amount of each analyte was calculated from the corresponding calibration curve, and recovery was calculated using the following formula: recovery (%) = (observed amount − original amount)/spiked amount × 100%. The samples of COS technical concentrate were spiked with the analytes at three different concentrations. Three determinations were performed for each standard addition. Each determination was injected in duplicate.

3.5. Method Application

To determine the cations and anions, COS technical concentrate was accurately weighed and dissolved in water to prepare stock sample solution. The working sample solutions were prepared by dilution with water. For sample preparation, seven replicates were performed. For cations and anions determination, the concentration of the working sample solution was approximately 2000 and 50 mg/L, respectively.

4. Conclusions

The increasing interest of COSs in the food and medicine fields implies a need to control the quality of the product so that undesirable health effects are avoided. In the present work, an efficient, sensitive, and quick IC-CD method was established and demonstrated as suitable for separating, identifying, and quantifying inorganic cations of Na^+, NH_4^+, K^+, Mg^{2+}, Ca^{2+}, and chloride, acetate and lactate anions within 15 min. High sensitivity, satisfactory linearity, precision, and accuracy were achieved. The proposed method was readily applied for quantitative determination of the cations and anions stated above, providing a very useful method for the analysis of COSs for quality control and biological research purposes.

Acknowledgments: This project was supported by the National Natural Science Foundation of China (NSFC) (Nos. 31301701 and 31471805).

Author Contributions: Lidong Cao and Qiliang Huang conceived and designed the experiments; Xiuhuan Li, Li Fan and Li Zheng performed the experiments; Lidong Cao and Xiuhuan Li analyzed the data; Shanxue Zhang and Miaomiao Wu contributed reagents/materials/analysis tools; Lidong Cao wrote the paper. Xiuhuan Li took part in writing the paper. Lidong Cao and Xiuhuan Li contributed equally to this paper.

Conflicts of Interest: The authors declare no conflict of interest.

References

1. Jung, W.-J.; Park, R.D. Bioproduction of Chitooligosaccharides: Present and Perspectives. *Mar. Drugs* **2014**, *12*, 5328–5356. [CrossRef] [PubMed]
2. Zou, P.; Yang, X.; Wang, J.; Li, Y.; Zhang, Y.; Liu, G. Advances in characterisation and biological activities of chitosan and chitosan oligosaccharides. *Food Chem.* **2016**, *190*, 1174–1181. [CrossRef] [PubMed]
3. Aam, B.B.; Heggset, E.B.; Norberg, A.L.; Sørlie, M.; Vårum, K.M.; Eijsink, V.G.H. Production of chitooligosaccharides and their potential applications in medicine. *Mar. Drugs* **2010**, *8*, 1482–1517. [CrossRef] [PubMed]
4. Singh, P. Effect of chitosans and chitooligosaccharides on the processing and storage quality of foods of animal and aquatic origin. *Nutr. Food Sci.* **2016**, *46*, 51–81. [CrossRef]
5. Vela Gurovic, M.S.; Dello Staffolo, M.; Montero, M.; Debbaudt, A.; Albertengo, L.; Rodríguez, M.S. Chitooligosaccharides as novel ingredients of fermented foods. *Food Funct.* **2015**, *6*, 3437–3443. [CrossRef] [PubMed]
6. Chae, S.Y.; Jang, M.-K.; Nah, J.-W. Influence of molecular weight on oral absorption of water soluble chitosans. *J. Control. Release* **2005**, *102*, 383–394. [CrossRef] [PubMed]
7. Li, K.; Xing, R.; Liu, S.; Li, P. Advances in preparation, analysis and biological activities of single chitooligosaccharides. *Carbohydr. Polym.* **2016**, *139*, 178–190. [CrossRef] [PubMed]
8. Karppanen, H. Minerals and blood pressure. *Ann. Med.* **1991**, *2*, 299–305. [CrossRef]
9. Karppanen, H.; Karppanen, P.; Mervaala, E. Why and how to implement sodium, potassium, calcium, and magnesium changes in food items and diets? *J. Hum. Hypertens.* **2005**, *19*, S10–S19. [CrossRef] [PubMed]
10. Schmitt, S.; Garrigues, S.; de la Guardia, M. Determination of the mineral composition of foods by infrared spectroscopy: A review of a green alternative. *Crit. Rev. Anal. Chem.* **2014**, *44*, 186–197. [CrossRef] [PubMed]
11. De la Fuente, M.A.; Montes, F.; Guerrero, G.; Juárez, M. Total and soluble contents of calcium, magnesium, phosphorus and zinc in yoghurts. *Food Chem.* **2003**, *80*, 573–578. [CrossRef]
12. Ieggli, C.V.S.; Bohrer, D.; do Nascimento, P.C.; de Carvalho, L.M.; Garcia, S.C. Determination of sodium, potassium, calcium, magnesium, zinc, and iron in emulsified egg samples by flame atomic absorption spectrometry. *Talanta* **2010**, *80*, 1282–1286. [CrossRef] [PubMed]
13. Fantuz, F.; Ferraro, S.; Todini, L.; Piloni, P.; Mariani, P.; Salimei, E. Donkey milk concentration of calcium, phosphorus, potassium, sodium and magnesium. *Int. Dairy J.* **2012**, *24*, 143–145. [CrossRef]
14. Krejčová, A.; Černohorský, T.; Čurdová, E. Determination of sodium, potassium, magnesium and calcium in urine by inductively coupled plasma atomic emission spectrometry. The study of matrix effects. *J. Anal. At. Spectrom.* **2001**, *16*, 1002–1005. [CrossRef]
15. Edlund, M.; Visser, H.; Heitland, P. Analysis of biodiesel by argon-oxygen mixed-gas inductively coupled plasma optical emission spectrometry. *J. Anal. At. Spectrom.* **2002**, *17*, 232–235. [CrossRef]

16. Masár, M.; Sydes, D.; Luc, M.; Kaniansky, D.; Kuss, H.-M. Determination of ammonium, calcium, magnesium, potassium and sodium in drinking waters by capillary zone electrophoresis on a column-coupling chip. *J. Chromatogr. A* **2009**, *1216*, 6252–6255. [CrossRef] [PubMed]

17. Meng, H.-B.; Wang, T.-R.; Guo, B.-Y.; Hashi, Y.; Guo, C.-X.; Lin, J.-M. Simultaneous determination of inorganic anions and cations in explosive residues by ion chromatography. *Talanta* **2008**, *76*, 241–245. [CrossRef] [PubMed]

18. Liu, J.-M.; Liu, C.-C.; Fang, G.-Z.; Wang, S. Advanced analytical methods and sample preparation for ion chromatography techniques. *RSC Adv.* **2015**, *5*, 58713–58726. [CrossRef]

19. Zatirakha, A.V.; Smolenkov, A.D.; Shpigun, O.A. Preparation and chromatographic performance of polymer-based anion exchangers for ion chromatography: A review. *Anal. Chim. Acta* **2016**, *904*, 33–50. [CrossRef] [PubMed]

20. Michalski, R. Applications of ion chromatography for the determination of inorganic cations. *Crit. Rev. Anal. Chem.* **2009**, *39*, 230–250. [CrossRef]

21. Michalski, R.; Lyko, A. Research onto the contents of selected inorganic ions in the dialysis fluids and dialysates by using ion chromatography. *J. Liq. Chromatogr. Relat. Technol.* **2016**, *39*, 96–103. [CrossRef]

22. Kumar, S.D.; Narayan, G.; Hassarajani, S. Determination of anionic minerals in black and kombucha tea using ion chromatography. *Food Chem.* **2008**, *111*, 784–788. [CrossRef]

23. Rahimi-Yazdi, S.; Ferrer, M.A.; Corredig, M. Nonsuppressed ion chromatographic determination of total calcium in milk. *J. Dairy Sci.* **2010**, *93*, 1788–1793. [CrossRef] [PubMed]

24. De Caland, L.B.; Cardoso Silveira, E.L.; Tubino, M. Determination of sodium, potassium, calcium and magnesium cations in biodiesel by ion chromatography. *Anal. Chim. Acta* **2012**, *718*, 116–120. [CrossRef] [PubMed]

25. Cardoso Silveira, E.L.; de Caland, L.B.; Tubino, M. Simultaneous quantitative analysis of the acetate, formate, chloride, phosphate and sulfate anions in biodiesel by ion chromatography. *Fuel* **2014**, *124*, 97–101. [CrossRef]

26. Huang, Z.; Zhao, X.; Zhu, Z.; Pan, Z.; Wang, L.; Zhu, Y. Determination of anions and cations in biodiesel with on-line sample pretreatment column-switching ion chromatography. *J. Liq. Chromatogr. Relat. Technol.* **2015**, *38*, 1747–1752. [CrossRef]

27. Zhang, Y.; Thepsithar, P.; Jiang, X.; Tay, J.H. Simultaneous determination of seven anions of interest in raw *Jatropha curcas* oil by ion chromatography. *Energy Fuels* **2014**, *28*, 2581–2588. [CrossRef]

28. Cataldi, T.R.I.; Margiotta, G.; Del Fiore, A.; Bufo, S.A. Ionic content in plant extracts determined by ion chromatography with conductivity detection. *Phytochem. Anal.* **2003**, *14*, 176–183. [CrossRef] [PubMed]

29. Gros, N.; Gorenc, B. Performance of ion chromatography in the determination of anions and cations in various natural waters with elevated mineralization. *J. Chromatogr. A* **1997**, *770*, 119–124. [CrossRef]

30. Gros, N. Ion chromatographic analyses of sea waters, brines and related samples. *Water* **2013**, *5*, 659–676. [CrossRef]

31. Michalski, R. Ion chromatography as a reference method for determination of inorganic ions in water and wastewater. *Crit. Rev. Anal. Chem.* **2006**, *36*, 107–127. [CrossRef]

32. Cao, L.; Wu, J.; Li, X.; Zheng, L.; Wu, M.; Liu, P.; Huang, Q. Validated HPAEC-PAD method for the determination of fully deacetylated chitooligosaccharides. *Int. J. Mol. Sci.* **2016**, *17*, 1699. [CrossRef] [PubMed]

33. European Commission. Technical Material and Preparations: Guidance for Generating and Reporting Methods of Analysis in Support of Pre- and Post-Registration Data Requirements for Annex II (Part A, Section 4) and Annex III (Part A, Section 5) of Directive 91/414. Available online: https://ec.europa.eu/food/sites/food/files/plant/docs/pesticides_ppp_app-proc_guide_phys-chem-ana_tech-mat-preps.pdf (accessed on 20 February 2017).

marine drugs

MDPI

Article

Preparation of Chito-Oligomers by Hydrolysis of Chitosan in the Presence of Zeolite as Adsorbent

Khalid A. Ibrahim [1,2], Bassam I. El-Eswed [3], Khaleel A. Abu-Sbeih [4], Tawfeeq A. Arafat [5], Mahmoud M. H. Al Omari [6], Fouad H. Darras [6] and Adnan A. Badwan [7,*]

1   Department of Chemical Engineering, Faculty of Engineering, Al-Hussein Bin Talal University, P.O. Box 20, Ma'an 71111, Jordan; khalida@ahu.edu.jo
2   College of Engineering, King Saud University, P.O. Box 800, Riyadh 11421, Saudi Arabia
3   Department of Basic Sciences, Zarqa College, Al-Balqa Applied University, P.O. Box 313, Zarqa 13110, Jordan; bassameswed@bau.edu.jo
4   Department of Chemistry, Faculty of Science, Al-Hussein Bin Talal University, P.O. Box 20, Ma'an 71111, Jordan; abusbeih@ahu.edu.jo
5   Department of Pharmaceutical Medicinal Chemistry and Pharmacognosy, Faculty of Pharmacy and Medical Technology, Petra University, P.O. Box 961343, Amman 11196, Jordan; tarafat@uop.edu.jo
6   Research and Innovation Center (RIC), The Jordanian Pharmaceutical Manufacturing Co., P.O. Box 94, Naor 11710, Jordan; momari@jpm.com.jo (M.M.H.A.O.); fdarras@jpm.com.jo (F.H.D.)
7   The Jordanian Pharmaceutical Manufacturing Co., P.O. Box 94, Naor 11710, Jordan
*   Correspondence: jpm@go.com.jo; Tel.: +962-6-572-7207; Fax: +962-6-572-7641

Academic Editors: Hitoshi Sashiwa and David Harding
Received: 31 December 2015; Accepted: 15 February 2016; Published: 23 July 2016

Abstract: An increasing interest has recently been shown to use chitin/chitosan oligomers (chito-oligomers) in medicine and food fields because they are not only water-soluble, nontoxic, and biocompatible materials, but they also exhibit numerous biological properties, including antibacterial, antifungal, and antitumor activities, as well as immuno-enhancing effects on animals. Conventional depolymerization methods of chitosan to chito-oligomers are either chemical by acid-hydrolysis under harsh conditions or by enzymatic degradation. In this work, hydrolysis of chitosan to chito-oligomers has been achieved by applying adsorption-separation technique using diluted HCl in the presence of different types of zeolite as adsorbents. The chito-oligomers were retrieved from adsorbents and characterized by differential scanning calorimetry (DSC), liquid chromatography/mass spectroscopy (LC/MS), and ninhydrin test.

Keywords: chitosan; chito-oligomers; zeolite; depolymerization; hydrolysis

1. Introduction

Chitosan—among various renewable polymers—is one of the most commercially important biocompatible polymers from an environmental or biomedical point of view [1–3]. It is a linear copolymer of (1→4)-linked 2-acetamido-2-deoxy-β-D-glucan (GlcNAc) and 2-amino-2-deoxy-β-D-glucan (GlcN) units in varying proportions (Figure 1) [4,5]. Naturally, chitosan is produced by the hydration of a nitrogenous polysaccharide chitin (Figure 1), which is considered as the main building component of crustacean shells [6,7].

Recently, various N-containing products have been prepared from chitin [8–12]. For example, chitin has been converted to 3-acetamido-5-acetylfuran by using N-methyl-2-pyrrolidone [8] and ionic liquids [9] as solvents. Additionally, it has been converted to its corresponding amide/amino substituted sugar alcohols, smaller C_2–C_4 polyols and N-acetylmonoethanolamine over transition metal catalysts and hydrogen in water [10]. Two major products, namely hydroxyethyl-2-amino-2-

deoxyhexopyranoside and hydroxyethyl-2-acetamido-2-deoxyhexopyranoside, have been obtained by acid-catalyzed liquefaction of chitin in ethylene glycol [11]. Furthermore, N-containing carbon materials have been prepared by carbonization of chitin, which used as adsorbents to remove toxic heavy metals and in styrene epoxidation [12]. These findings offer opportunities to convert biomass such as chitin into value-added, renewable N-containing materials [13].

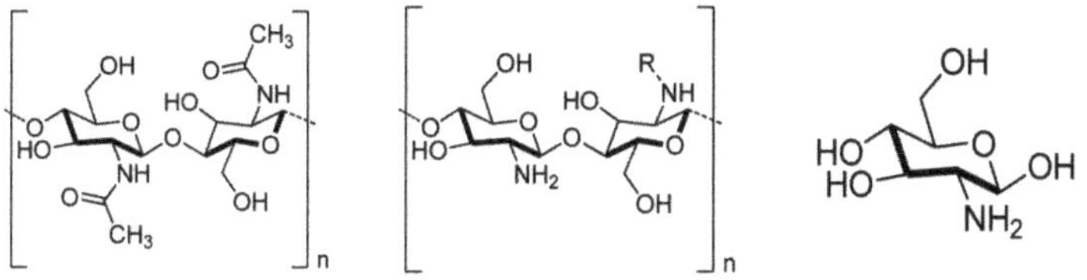

R = H-CO-CH₃

(A)   (B)   (C)

Figure 1. Chemical structure of (**A**) chitin; (**B**) chitosan; and (**C**) glucosamine.

The cationic nature of chitosan made it a unique polysaccharide, which is distinguished among other polysaccharides [14]. It is mainly obtained at the synthetic scale by deacetylation of chitin, and the process of deacetylation is carried out to different degrees depending upon the targeted applications, so numerous products with different degrees of deacetylation (DD) can be obtained. The physiological properties of chitosan, especially solubility, are determined by its molecular weight and degree of deacetylation. Chitosan is a water insoluble polysaccharide while, because of its cationic nature, is soluble in dilute acidic solutions [15]. The insoluble nature of chitosan in neutral pH restricted its use in solution for physiological applications in the medical and food industries [16].

Chito-oligomers are the hydrolysates of chitosan, mainly made up of -1,4 linked D-glucosamine and partially of -1,4 linked *N*-acetyl-D-glucosamine. Previously, many studies showed that chito-oligomers have significant potential in medicine and food fields due to their wide bioactivity, such as antibacterial, antifungal, antitumor activity, radical scavenging, antimicrobial activity, immunity modulatory effect, and wound healing [15,17–19]. Glucosamine is a monomer of chito-oligomers (Figure 1) which has a growing market due to its use for the treatment of osteoarthritis [20]. However, the in-depth knowledge of the mode of action of chito-oligomers is still limited because their biological activity has often been determined using heterogeneous and/or relatively poorly characterized oligomer mixtures [15].

Different methods have been described in the literature to prepare chito-oligomers from chitosan by enzymatic and chemical methods [15,16,19,21,22]. Enzymatic methods are selective and simple but their commercial applications are limited due to the cost, low yield, and limited availability of chitosan-specific enzymes [21,22]. The chemical methods include depolymerization of chitosan by a hydrolysis reaction, mainly using concentrated HCl [23], nitrous acid [24], fluorolysis in anhydrous hydrogen fluoride [25], and oxidative-reductive reaction by hydrogen peroxide [26]. Additionally, a few total chemical syntheses of chito-oligomers involving multiple protection and deprotection steps have also been reported [27,28]. Preparation of chito-oligomer mixtures from chitosan by physical methods (hydrothermal, microwave, ultrasonication, and gamma-ray) was reviewed by Yin *et al.* [29]. This subject was investigated in a limited number of studies and still uncovered [30].

Zeolites are well-known as valuable crystalline solids with framework structures containing discrete micropores of molecular dimensions that accommodate exchangeable extra-framework cation sites [31,32]. In terms of host-guest interactions, zeolites can be viewed as host frameworks with structurally intact and immutable three-dimensional (3D) structures [33,34]. Zeolites are widely used in commercial applications as petroleum refining, petrochemical industry, and fine chemical industry, as sorbents for small-molecule separation processes and as ion-exchangers in detergents [35–37].

Due to all of the properties mentioned above, zeolites are employed as adsorbent materials in this work to shift the equilibrium of the hydrolysis reaction toward the formation of chito-oligomers in an adsorption-separation technique.

Concentrated HCl hydrolysis of chitosan to chito-oligomers is the best known and applied chemical method [16,19,22,23,38]. Table 1 summarizes the reaction conditions, reagents, degree of acetylation (DA) of the chitosan used, degree of polymerization (DP), and characterization methods of the produced chito-oligomers (see Table 1). All reported methods started from almost deacetylated chitosan hydrolysis; the general protocol employed concentrated HCl for hydrolysis of chitosan, which has several disadvantages like harsh conditions, many purification steps to remove the strong acid and low yields of chito-oligomers obtained from such reaction conditions.

Table 1. Summary of reagent, reaction conditions, and final product characterization of a number of described hydrolysis reactions of chitosan using concentrated HCl (12 M).

DA of Starting Chitosan	Reaction Conditions	Final Product	Characterization	Reference
DA ~0%	72 °C, 1.5 h	DP = 2–12	Mass spectroscopy	[16]
DA < 10%	70 °C, 4 h	DP = 6–16	HPLC-light scattering detector	[19]
DA = 12%	72 °C, 0.5 h	DP = 3–16	Mass spectroscopy	[22]
DA ~0%	72 °C, 0.5–3 h	DP < 40	Size exclusion chromatography-refractive index detector	[23]

DA: Degree of acetylation; DP: Degree of polymerization.

Due to all of the mentioned disadvantages of using classical hydrolysis procedures of chitosan to its chito-oligomers, this work presents a novel application of adsorption reaction techniques to hydrolyze chitosan to chito-oligomers using diluted HCl in the presence of zeolite as adsorbent. The chito-oligomers obtained from the protocol used in the present work were subjected to a number of identification and characterization tests to understand the nature of the chito-oligomers obtained and to evaluate the hydrolysis technique applied in this work against the conventional hydrolysis methods used before.

2. Results and Discussion

2.1. Hydrolysis of Chitosan to Its Chito-Oligomers

As a result of the designed separation technique, starting from 2.0 g chitosan and 4.0 g of zeolite (different types: 10Å, 5Å, 3Å, Table 2), in 100 mL 1.8 M HCl (85 mL H_2O and 15 mL conc. HCl), the reaction mixture was stirred in a water bath under reflux for two hours, three solid samples were then collected and named accordingly S1, S2, and S3 as described briefly in Scheme 1.

Three different tests (Table 3) were performed as preliminary tests to determine the chito-oligomers content of the hydrolysis reactions products S1, S2, S3 obtained (Scheme 1). The first test was the water solubility test which is considered as the first indication of hydrolyzing the water insoluble chitosan to its water soluble chito-oligomers. Such data can give clear cut evidence of the hydrolysis of chitosan under the abovementioned reaction conditions. The second test performed is the light absorbance of the solution that results from the reaction of soluble obtained sample (S1, S2, and S3) with ninhydrin; the high absorbance is an indication to the high chito-oligomers content. Comparing the light absorbance data obtained from S1, S2, and S3 with the light absorbance of a chitosan (starting material) and glucosamine (pure monomer) samples gave an evidence of the presence or absence of chito-oligomers in the tested samples. The third test was the % loss upon ignition which is an indication of the % organic matter in the different samples collected within the work-up of the chitosan hydrolysis reaction (S1, S2, S3) in order to distinguish it from the zeolite and salt contents of the samples obtained from the hydrolysis reaction.

Table 2. Properties of zeolites used as adsorbents in the present work.

Property	HZSM-5 *	Molecular Sieves ** Beads 0.3 nm	Molecular Sieves ** Beads 1.0 nm
Pore diameter (Å)	5	3	10
Composition	Aluminosilicate Na < 700 ppm	Potassium sodium aluminum silicate	Sodium aluminum silicate
Main application	Petroleum industry-hydrocarbon isomerization	Water adsorption	Chloroform, carbon tetrachloride, benzene adsorption
SiO_2/Al_2O_3 molar ratio	400–570 50 [39] 76 [40]	2	3
Total acid density	0.2–0.26 [40] 0.11–0.64 [41]	-	-
BET total surface area (m^2/g)	300 364 [42] 389 [39] 392 [40] 826–1142 [41]	800	
Micropore surface area (m^2/g)	303 [39] 359 [40] 728–1036 [41]	-	-
Mesopore surface area (m^2/g)	85 [39] 33 [40] 98–106 [41]	-	-
Total pore volume (cm^3/g)	0.22 [40] 0.19–0.25 [41]	0.30	
Micropore volume (cm^3/g)	0.09 [42] 0.12 [39] 0.16 [40] 0.10–0.14 [41]	-	-
Mesopore volume (cm^3/g)	0.12 [42] 0.04 [39] 0.06 [40] 0.09–0.1 [41]	-	-

*: Values given without references were obtained from Acros Organics specifications; **: Values given without references were obtained from Merck specifications.

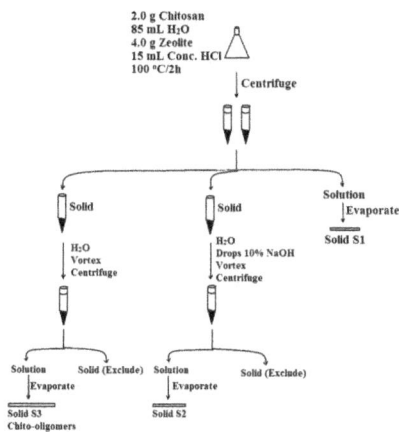

Scheme 1. Preparation protocol of chito-oligomers.

Table 3. Properties of glucosamine, chitosan, and the solids obtained from the hydrolysis of chitosan (S1, S2, and S3) using diluted HCl in the presence of zeolites.

Sample	Product Number	Solubility	Absorbance of Ninhydrin Test	% Loss Upon Ignition	DSC Peaks (°C)
Glucosamine		+	2.6	68	210–220
Chitosan		−	0.003	75	350
HCl(5Å-Z)-Acid retrieval	1 (S3)	+	2.9	68	200–230 230–250
HCl(5Å-Z)-Base retrieval	1 (S2)	+	0.004	13	No peaks were detected
HCl(5Å-Z)-First filtrate	1 (S1)	−	0.014	30	Not determined
HCl(10Å-Z)-Acid retrieval	2 (S3)	+	2.3	67	110–120 180–190 230–250
HCl(10Å-Z)-Base retrieval	2 (S2)	+	0.011	21	No peaks were detected
HCl(3Å-Z)-Acid retrieval	3 (S3)	+	2.4	81	110–140 150–160 180 220–250
HCl(3Å-Z)-Base retrieval	3 (S2)	+	0.006	10	No peaks were detected
HCl-Acid retrieval	4 (S3)	+			240–250

Surprisingly, very little content of the chito-oligomers was observed in the product sample S1 (Scheme 1) which is obtained from evaporation of the solution that results after centrifuging the reaction mixture. For example, 1 (S1) was found to be water insoluble, has a low absorbance in the ninhydrin test (0.014), and has a low % loss upon ignition (30%) (Table 3).

The solid product of the centrifugation of the reaction mixture was then neutralized to pH = 8 using 10% NaOH solution. After shaking, centrifugation, and evaporation of clear solution, solid product S2 (Scheme 1) was obtained. The product 1 (S2) was found to be water-soluble, has low absorbance in the ninhydrin test (0.004), and has a low % loss upon ignition (13%) (Table 3). These results indicated that there is very little content of chito-oligomers in this product and most of the solid is expected to be NaCl salt from the neutralization reaction.

As a third trial to find the chito-oligomers, the solid product of the centrifugation of the reaction mixture was strongly stirred (cortex) with distilled water (instead of 10% NaOH solution, as in the above paragraph), followed by centrifugation to remove the adsorbent material. Evaporating the obtained solution gave a light beige solid S3 (Scheme 1). S3 was found to be water- soluble, showed high absorbance in the ninhydrin test (absorbance = 2.4 using 3Å zeolite, 2.9 using 5Å zeolite, and 2.3 using 10Å zeolite) compared with chitosan with absorbance = 0.003, and it has also high % loss upon ignition (81% using 3Å zeolite, 68% using 5Å zeolite, 67% using 10Å zeolite) as an indication of organic compound content (Table 3).

Actually, the three types of zeolites employed in the present work were used in order to investigate the effect of pore dimension of zeolite on the hydrolysis reactions and products. The detailed properties of zeolites are given in Table 2. However, the type of zeolite did not affect chito-oligomers obtained significantly as indicated by the results in Table 3 and as indicated also by their mass spectra. This may be due to that chito-oligomers are adsorbed on the surface rather than in the pores of zeolites.

As shown in Table 2, HZSM-5 (5 Å) is distinguished by a high SiO_2/Al_2O_3 molar ratio, so it is a hydrophobic zeolite [43]. It was reported to be stable in acidic solution (1 M HCl) [43,44] and stable in aqueous hot water (150°–500°, 5–17 bar) [45] relative to other kinds of zeolites. In our experiments, it was observed that molecular sieve zeolites (3 and 10 Å) partially decomposed into solution as reflected by their orange color while HZSM-5 zeolite (5 Å) remained stable.

The preliminary results of study of S3 product using water solubility, the ninhydrin test, and the % loss upon ignition indicated that after centrifugation of the acidic hydrolysis reaction mixture;

chitosan was hydrolyzed to chito-oligomers which were completely adsorbed on zeolite. The adsorbed chito-oligomers could be retrieved by sonicating the solid product in deionized water. Thus, the hydrolyzed product S3, which contains chito-oligomers and decomposed zeolites, was chosen for further characterization and further study in the following sections.

2.2. Characterization of Chito-Oligomers

2.2.1. DSC Study

The DSC diagrams of chitosan and glucosamine are given in Figure 2. Chitosan has an exothermic peak at 350 °C while glucosamine has a sharp endothermic peak at 210 °C, followed by decomposition. The hydrolyzed product 1 (S5/5 Å-Zeolite) was found to have a broad endothermic peak at a temperature close to that of glucosamine (Figure 2), followed by decomposition. A similar behavior was observed in the case of hydrolyzed products 2 (S3/10 Å-Zeolite) and 3 (S3/3 Å-Zeolite), but an additional endothermic peak at 120 °C, corresponding to the onset of the evaporation of bound water, was observed as well (Figure 2) [46].

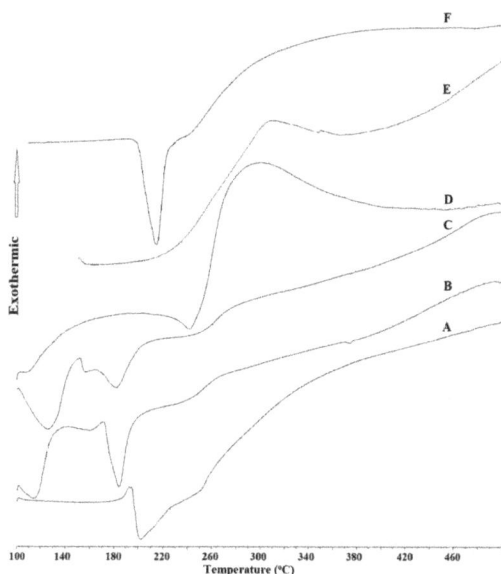

Figure 2. DSC of the hydrolyzed products (**A**) 1 (S3/5 Å-Zeolite); (**B**) 2 (S3/10 Å-Zeolite); (**C**) 3 (S3/3 Å-Zeolite) and (**D**) 4 (S3/No zeolite); (**E**) chitosan; and (**F**) glucosamine.

2.2.2. Mass Spectra

The theoretical molar mass values of chito-oligomers are: DP1 = 179.16, DP2 = 340.31, DP3 = 501.45, DP4 = 662.60, DP5 = 823.74, DP6 = 984.89, DP7 = 1146.03, DP8 = 1307.18, DP9 = 1468.32. The masses obtained experimentally are shown in Figure 3 for the products 4 (S3/No zeolite) and 1 (S3/5 Å-Zeolite). Similar values were obtained for all S3 products which gave positive tests in Table 2. Note that there are peaks resulting from the dehydration of the chito-oligomers (marked by * and # in Figure 3) and are due to the loss of one and two water molecules. The dehydration peaks were a result of the analysis method because these peaks were also observed in the case of the glucosamine standard (Figure 3). Interestingly, the intensity of the *m/z* peaks of the chito-oligomers decreases in the order of increasing DP. The highest DP chio-oligomer detected was that of DP = 9.

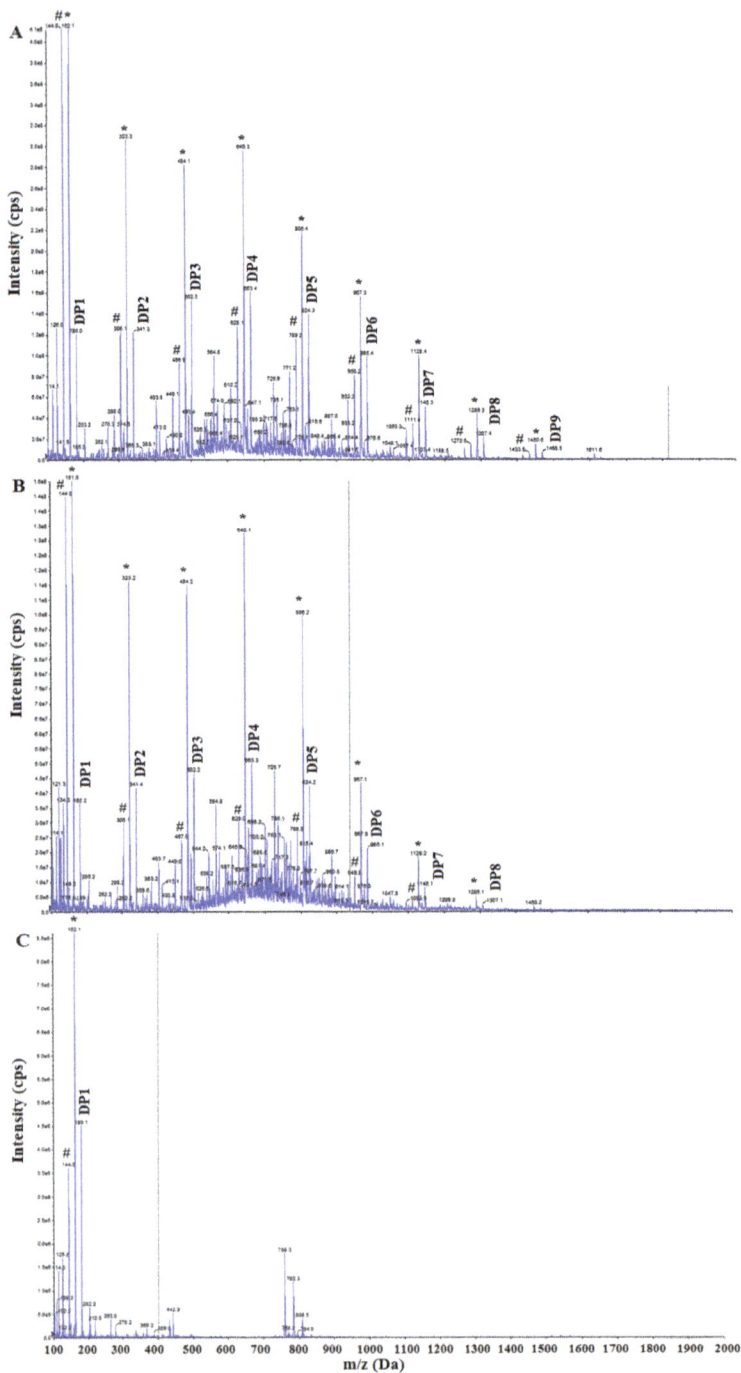

Figure 3. Mass spectra of hydrolyzed products (**A**) 1 (S3/5 Å-Zeolite); (**B**) 4 (S3/No zeolite); and (**C**) glucosamine.

2.2.3. Recrystallization of Chito-Oligomers

The aim of recrystallization was to remove the high molar mass chitosan from the chito-oligomers. The former are insoluble in neutral or basic medium while the latter are soluble. Thus, addition of base is necessary to precipitate the high molar mass chitosan. Ammonia was selected because the formed ammonium chloride replaces the chito-oligomers ammonium salts and desorbs them from chitosan and zeolite surfaces. The properties of recrystallized products are given in Table 4.

Table 4. Properties of recrystallized chito-oligomers.

Sample	Recrystallization Reagents	Solubility	% Yield	DSC Peaks (°C)
1 (S3/5 Å-Zeolite)	NH_3	+	47	
1 (S3/5 Å-Zeolite)	Ethanol-NH_3	+	38	220, 240, 280
2 (S3/10 Å-Zeolite)	NH_3	+	80	200–210 (NH_4Cl)
2 (S3/10 Å-Zeolite)	Ethanol-NH_3	+	96	220
3 (S3/3 Å-Zeolite)	NH_3	+	68	180–200 (NH_4Cl)
3 (S3/3 Å-Zeolite)	Ethanol-NH_3	+	56	220
4 (S3 No zeolite)	NH_3	+	12	NH_4Cl
4 (S3 No zeolite)	Ethanol-NH_3	+	6.0	NH_4Cl

The % yield of products (based on starting 2.0 g chitosan) after recrystallization in the absence of zeolites for the hydrolyzed product 4 (S3/No zeolite) were much lower than those in the presence of zeolites. Furthermore, The DSC diagrams of these samples showed only those peaks of ammonium chloride (200 °C, 340–360 °C broad). However, the mass spectra of these samples were very interesting (Figure 4) because chito-oligomers with DP from 1 to 3 were obtained as major products.

The low % yield of chito-oligomers in the absence of zeolite suggests that zeolites function as adsorbents for the produced chito-oligomers and causing shift of the hydrolysis equilibrium reaction forward and, thus, increasing the yield of the hydrolyzed chito-oligomers.

It is worth to mention that the % yields obtained cannot be compared with those in the literature because neither our study nor previous studies give cut evidences for the purity of chito-oligomers products.

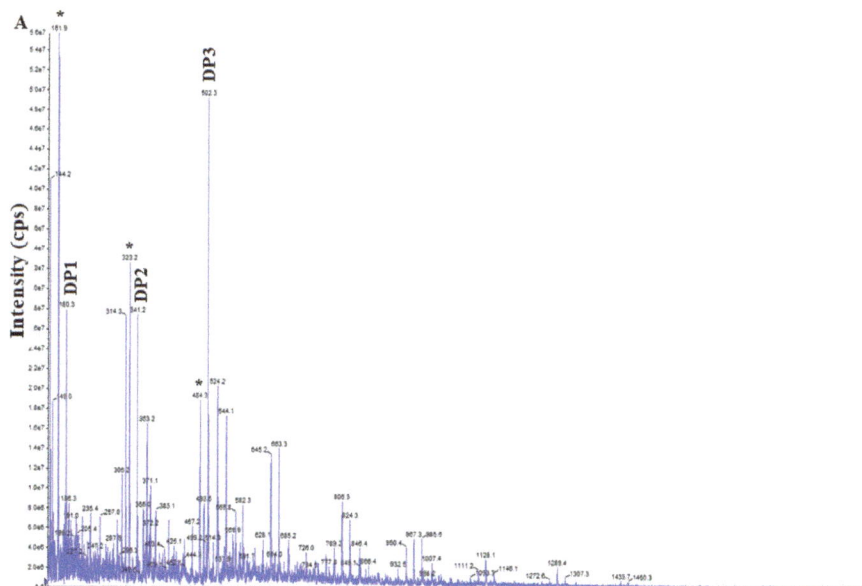

Figure 4. *Cont.*

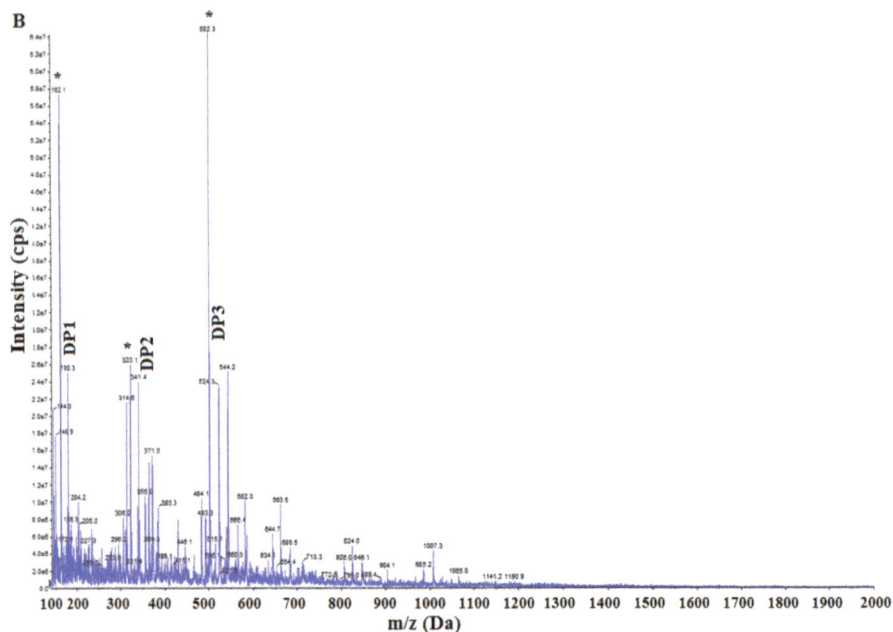

Figure 4. Mass spectra of recrystallized hydrolyzed product 4 (S3/No zeolite) using (**A**) ammonia-ethanol and (**B**) using ammonia alone.

3. Experimental Section

3.1. Materials

Chitosan (molecular weight ~250 kDa and DDA ~93%) was obtained from Homg Ju, HZSM-5 zeolite (5 Å) and molecular sieves (10 Å and 3 Å) were obtained from Across Organics and Merck, respectively. HCl (37%) was from Merck and sodium hydroxide was from GCC. The ammonia solution (25%) was obtained from Scharlue. Ethanol (Absolute) was purchased from GCC.

3.2. Methods

3.2.1. Hydrolysis of Chitosan

A 2.0 g sample of chitosan was mixed with 85 mL deionized water using a magnetic stirrer. A 4.0 g of zeolite (HZSM-5, molecular sieves 10 Å and 3 Å) was then added followed by the addition of 15 mL of concentrated HCl solution to give a final concentration of 1.8 M HCl solution. The mixture was then heated under reflux at 100 °C with stirring for 2 h. The mixture was then cooled to room temperature and divided into two portions and centrifuged (Hermle Z 320) at 4500 rpm for 15 min. The two solutions in the 50 mL centrifuge tubes were combined and evaporated for dryness (S1, Scheme 1) using a rotary evaporator. Distilled water was then added to the solid obtained in the two centrifuge tubes and the tubes were shaken using vortex (Labinco). The solution in one centrifuge tube was neutralized to pH = 8 using 10% NaOH solution. The resultant two mixtures were centrifuged at 4500 rpm for 15 min. The two resultant solutions were separated and evaporated in a rotary evaporator. The solids obtained from base and acid retrieval were designated S2 and S3, respectively (Scheme 1).

3.2.2. The Ninhydrin Test

A 0.10 g sample of the solids obtained from the hydrolysis procedure (2.2.1) was dissolved in 20 mL deionized water. A 2.0 mL portion was mixed with a buffer of KH_2PO_4 (0.2 M, pH 6) and 2.0 mL 0.8% ninhydrin. The solution was then heated to 80 °C in a water bath for 1 h. The absorbance of the resultant solution was measured using a UV-VIS Spectrophotometer (PG Instruments Ltd. T80).

3.2.3. The % Loss upon Ignition Measurement

A 0.1 g sample measured to the nearest of 0.0001 g was ignited in a crucible using a Bunsen burner and the total mass of the crucible and the residue was then measured. The empty crucible was previously ignited and its mass measured.

3.2.4. DSC Measurements

DSC measurements of the samples; in the range of 100–500 °C; were carried out in aluminum crucibles using Mettler Toledi DSC-1 Star System. The rate of heating was 10 °C/min.

3.2.5. Mass Spectra

The mass spectra of the samples were measured using AB Sciex LC-MSMS API-3200 LC-MS instrument. A 0.1 g sample was dissolved in 10.0 mL deionized water and introduced into the LC-MS instrument with a rate of 10 μL/min. The ionization source was turbo spray; the ion spray voltage was 5500. Typically, very little fragmentation or chemical reactions occur in this kind of soft ionization technique. The mass spectra were recorded in the range from 100 to 2000 *m/z* using MCA positive polarity mode where 60 scans were collected in 2.0 min. The step size was 0.1 Da. The peak area was obtained for all the peaks directly from the instrument.

3.2.6. Recrystallization Using Ammonia

A 1.0 g sample of the crude solid (S3, Scheme 1) obtained in procedure 2.2.1 was dissolved in 50 mL deionized water followed by the addition of 1.5 mL of 25% NH_3. The mixture was then centrifuged at 4500 rpm for 15 min, the solid precipitated was excluded and the solution was evaporated in the rotary evaporator.

3.2.7. Recrystallization Using Ethanol and Ammonia

A 1.0 g sample of the crude solid (S3, Scheme 1) obtained in procedure 2.2.1 was dissolved in 25 mL deionized water. Then 25 mL absolute ethanol was added followed by the addition of 0.5 mL of 25% NH_3. The mixture was then centrifuged at 4500 rpm for 15 min, the solid precipitated was excluded and the solution was evaporated in the rotary evaporator.

4. Conclusions

The method developed in the present study offers a simple process for preparation and separation of chito-oligomers from chitosan. The zeolites function as adsorbents for the produced chito-oligomers and cause a shift of the hydrolysis equilibrium reaction forward and, thus, increase the yield of the hydrolyzed products. More research is required on the hydrolysis of chitin/chitosan to its corresponding chito-oligomers using adsorbents other than zeolites. Studying the kinetics of the hydrolysis reaction in the presence of zeolites and the biological evaluation of the obtained chito-oligomers is also necessary to inspect the potential of these chito-oligomers.

Acknowledgments: The scientific and technical support for this project obtained from the Jordanian Pharmaceutical Manufacturing Company (JPM Co.) is gratefully acknowledged.

Author Contributions: Khalid A. Ibrahim, Bassam I. El-Eswed and Khaleel A. Abu-Sbeih conducted the experiments and analyzed the data. Mahmoud M.H. Al Omari and Fouad H. Darras participated in the data interpretation and preparation of the manuscript. Adnan A. Badwan and Tawfeeq A. Arafat contributed to the conception and design of the study and approved the final draft.

Conflicts of Interest: The authors declare no conflict of interest.

References

1. Gopalakrishnan, L.; Ramana, L.N.; Sethuraman, S.; Krishnan, U.M. Ellagic acid encapsulated chitosan nanoparticles as anti-hemorrhagic agent. *Carbohydr. Polym.* **2014**, *111*, 215–221. [CrossRef] [PubMed]
2. Ferris, C.; Casas, M.; Lucero, M.J.; de Paz, M.V.; Jimenez-Castellanos, M.R. Synthesis and characterization of a novel chitosan-N-acetyl-homocysteine thiolactone polymer using MES buffer. *Carbohydr. Polym.* **2014**, *111*, 125–132. [CrossRef] [PubMed]
3. Thakur, V.K.; Thakur, M.K.; Gupta, R.K. Rapid synthesis of graft copolymers from natural cellulose fibers. *Carbohydr. Polym.* **2013**, *98*, 820–828. [CrossRef] [PubMed]
4. Stefan, J.; Lorkowska-Zawicka, B.; Kaminski, K.; Szczubialka, K.; Nowakowska, M.; Korbut, R. The current view on biological potency of cationically modified chitosan. *J. Physiol. Pharmacol.* **2014**, *65*, 341–347. [PubMed]
5. Jiang, T.; Deng, M.; James, R.; Nair, L.S.; Laurencin, C.T. Micro- and nanofabrication of chitosan structures for regenerative engineering. *Acta Biomater.* **2014**, *10*, 1632–1645. [CrossRef] [PubMed]
6. Lai, G.J.; Shalumon, K.T.; Chen, S.H.; Chen, J.P. Composite chitosan/silk fibroin nanofibers for modulation of osteogenic differentiation and proliferation of human mesenchymal stem cells. *Carbohydr. Polym.* **2014**, *111*, 288–297. [CrossRef] [PubMed]
7. Lai, P.; Daear, W.; Lobenberg, R.; Prenner, E.J. Overview of the preparation of organic polymeric nanoparticles for drug delivery based on gelatine, chitosan, poly(d,l-lactide-co-glycolic acid) and polyalkylcyanoacrylate. *Colloids Surf. B Biointerfaces* **2014**, *118*, 154–163. [CrossRef] [PubMed]
8. Chen, X.; Chew, S.L.; Kerton, F.M.; Yan, N. Direct conversion of chitin into a N-containing furan derivative. *Green Chem.* **2014**, *16*, 2204–2212. [CrossRef]
9. Chen, X.; Liu, Y.; Kerton, F.M.; Yan, N. Conversion of chitin and N-acetyl-D-glucosamine into a N-containing furan derivative in ionic liquids. *RSC Adv.* **2015**, *5*, 20073–20080. [CrossRef]
10. Bobbink, F.D.; Zhang, J.; Pierson, Y.; Chen, X.; Yan, N. Conversion of chitin derived N-acetyl-D-glucosamine (NAG) into polyols over transition metal catalysts and hydrogen in water. *Green Chem.* **2015**, *17*, 1024–1031. [CrossRef]
11. Pierson, Y.; Chen, X.; Bobbink, F.D.; Zhang, J.; Yan, N. Acid-catalyzed chitin liquefaction in ethylene glycol. *ACS Sustain. Chem. Eng.* **2014**, *2*, 2081–2089. [CrossRef]
12. Gao, Y.; Chen, X.; Zhang, J.; Yan, N. Chitin-derived mesoporous, nitrogen-containing carbon for heavy-metal removal and styrene epoxidation. *ChemPlusChem* **2015**, *80*, 1556–1564. [CrossRef]
13. Yan, N.; Chen, X. Don't waste seafood waste. *Nature* **2015**, *524*, 155–157. [CrossRef] [PubMed]
14. Jayakumar, R.; Menon, D.; Manzoor, K.; Nair, S.V.; Tamura, H. Biomedical applications of chitin and chitosan based nanomaterials—A short review. *Carbohydr. Polym.* **2010**, *82*, 227–232. [CrossRef]
15. Aam, B.B.; Heggset, E.B.; Norberg, A.L.; Sorlie, M.; Varum, K.M.; Eijsink, V.G. Production of chitooligosaccharides and their potential applications in medicine. *Mar. Drugs* **2010**, *8*, 1482–1517. [CrossRef] [PubMed]
16. Trombotto, S.; Ladaviere, C.; Delolme, F.; Domard, A. Chemical preparation and structural characterization of a homogeneous series of chitin/chitosan oligomers. *Biomacromolecules* **2008**, *9*, 1731–1738. [CrossRef] [PubMed]
17. Muzzarelli, R.A.A. Chitins and chitosans for the repair of wounded skin, nerve, cartilage and bone. *Carbohydr. Polym.* **2009**, *76*, 167–182. [CrossRef]
18. Xia, W.; Liu, P.; Zhang, J.; Chen, J. Biological activities of chitosan and chitooligosaccharides. *Food Hydrocoll.* **2011**, *25*, 170–179. [CrossRef]
19. Li, K.; Xing, R.; Liu, S.; Qin, Y.; Li, B.; Wang, X.; Li, P. Separation and scavenging superoxide radical activity of chitooligomers with degree of polymerization 6–16. *Int. J. Biol. Macromol.* **2012**, *51*, 826–830. [CrossRef] [PubMed]

20. Einbu, A.; Varum, K.M. Depolymerization and de-*N*-acetylation of chitin oligomers in hydrochloric acid. *Biomacromolecules* **2007**, *8*, 309–314. [CrossRef] [PubMed]

21. Tian, M.; Chen, F.; Ren, D.; Yu, X.; Zhang, X.; Zhong, R.; Wan, C. Preparation of a series of chitooligomers and their effect on hepatocytes. *Carbohydr. Polym.* **2010**, *79*, 137–144. [CrossRef]

22. Cabrera, J.C.; Van Cutsem, P. Preparation of chitooligosaccharides with degree of polymerization higher than 6 by acid or enzymatic degradation of chitosan. *Biochem. Eng. J.* **2005**, *25*, 165–172. [CrossRef]

23. Domard, A.; Cartier, N. Glucosamine oligomers: 1. Preparation and characterization. *Int. J. Biol. Macromol.* **1989**, *11*, 297–302. [CrossRef]

24. Tommeraas, K.; Varum, K.M.; Christensen, B.E.; Smidsrod, O. Preparation and characterisation of oligosaccharides produced by nitrous acid depolymerisation of chitosans. *Carbohydr. Res.* **2001**, *333*, 137–144. [CrossRef]

25. Defaye, J.; Garcia Fernandez, J.M. Protonic reactivity of sucrose in anhydrous hydrogen fluoride. *Carbohydr. Res.* **1994**, *251*, 17–31. [CrossRef]

26. Nordtveit, R.J.; Vårum, K.M.; Smidsrød, O. Degradation of fully water-soluble, partially *N*-acetylated chitosans with lysozyme. *Carbohydr. Polym.* **1994**, *23*, 253–260. [CrossRef]

27. Aly, M.R.; Ibrahim, E.S.; El Ashry, E.S.; Schmidt, R.R. Synthesis of chitotetraose and chitohexaose based on dimethylmaleoyl protection. *Carbohydr. Res.* **2001**, *331*, 129–142. [CrossRef]

28. Kuyama, H.; Nakahara, Y.; Nukada, T.; Ito, Y.; Nakahara, Y.; Ogawa, T. Stereocontrolled synthesis of chitosan dodecamer. *Carbohydr. Res.* **1993**, *243*, C1–C7. [CrossRef]

29. Yin, H.; Du, Y.; Zhang, J. Low molecular weight and oligomeric chitosans and their bioactivities. *Curr. Top. Med. Chem.* **2009**, *9*, 1546–1559. [CrossRef] [PubMed]

30. Jung, W.; Park, R. Bioproduction of chitooligosaccharides: Present and perspectives. *Mar. Drugs* **2014**, *12*, 5328–5356. [CrossRef] [PubMed]

31. Roth, W.J. Chapter 7 Synthesis of delaminated and pillared zeolitic materials. *Stud. Surf. Sci. Catal.* **2007**, *168*, 221–239.

32. Sebastián, V.; Casado, C.; Coronas, J. Special Applications of Zeolites. In *Zeolites and Catalysis*; Wiley-VCH Verlag: Weinheim, Germany, 2010; pp. 389–410.

33. Corma, A. Inorganic solid acids and their use in acid-catalyzed hydrocarbon reactions. *Chem. Rev.* **1995**, *95*, 559–614. [CrossRef]

34. Bejblova, M.; Prochazkova, D.; Cejka, J. Acylation reactions over zeolites and mesoporous catalysts. *ChemSusChem* **2009**, *2*, 486–499. [CrossRef] [PubMed]

35. Čejka, J.; Wichterlová, B. Acid-catalyzed synthesis of mono- and dialkyl benzenes over zeolites: Active sites, zeolite topology, and reaction mechanisms. *Catal. Rev.* **2002**, *44*, 375–421. [CrossRef]

36. Perego, C.; Ingallina, P. Combining alkylation and transalkylation for alkylaromatic production. *Green Chem.* **2004**, *6*, 274–279. [CrossRef]

37. Zones, S.I. Translating new materials discoveries in zeolite research to commercial manufacture. *Micropor. Mesopor. Mater.* **2011**, *144*, 1–8. [CrossRef]

38. Domard, A.; Cartier, N. Glucosamine oligomers: 4. Solid state-crystallization and sustained dissolution. *Int. J. Biol. Macromol.* **1992**, *14*, 100–106. [CrossRef]

39. Lu, R.; Tangbo, H.; Wang, Q.; Xiang, S. Properties and characterization of modified HZSM-5 zeolites. *J. Nat. Gas Chem.* **2003**, *12*, 56–62.

40. Zhu, X.; Lobban, L.; Mallinson, R.; Resasco, D. Tailoring the mesopore structure of HZSM-5 to control product distribution in the conversion of propanal. *J. Catal.* **2010**, *271*, 88–98. [CrossRef]

41. Al-Dughaither, A.; de Lasa, H. HZSM-5 Zeolites with different SiO_2/Al_2O_3 ratios: Characterization and NH_3 desorption kinetics. *Ind. Eng. Chem. Res.* **2014**, *53*, 15303–15316. [CrossRef]

42. Gayubo, A.; Alonso, A.; Valle, B.; Aguayo, A.; Bilbao, J. Selective production of olefins from bioethanol on HZSM-5 zeolites catalysts treated with NaOH. *Appl. Catal. B Environ.* **2010**, *97*, 299–306. [CrossRef]

43. Zhou, C.; Zhu, J. Adsorption of nitrosamines in acidic solution by zeolites. *Chemosphere* **2005**, *58*, 109–114. [CrossRef] [PubMed]

44. Beyer, H. Dealumination techniques for zeolites. In *Molecular Sieves*; Springer-Verlag: Berlin, Germany, 2002; Volume 3, pp. 203–255.

45. Ravenelle, R.; Schübler, F.; D'Amico, A.; Danilina, N.; van Bokhoven, J.; Lercher, J.; Jones, C.; Sievers, C. Stability of zeolites in hot liquid water. *J. Phys. Chem.* **2010**, *114*, 19582–19595. [CrossRef]
46. Mourya, V.K.; Inamdar, N.N.; Tiwari, A. Carboxymethyl chitosan and its applications. *Adv. Mater. Lett.* **2010**, *1*, 11–33. [CrossRef]

marine drugs

MDPI

Article

Biological Potential of Chitinolytic Marine Bacteria

Sara Skøtt Paulsen [1], Birgitte Andersen [1], Lone Gram [1,*] and Henrique Machado [1,2]

[1] Department of Biotechnology and Biomedicine, Technical University of Denmark, DK-2800 Kgs. Lyngby, Denmark; saskp@dtu.dk (S.S.P.); ba@bio.dtu.dk (B.A.); henma@biosustain.dtu.dk (H.M.)

[2] Novo Nordisk Foundation Center for Biosustainability, Technical University of Denmark, DK-2800 Kgs. Lyngby, Denmark

* Correspondence: gram@bio.dtu.dk; Tel.: +45-2368-8295

Academic Editor: Hitoshi Sashiwa

Received: 18 November 2016; Accepted: 8 December 2016; Published: 16 December 2016

Abstract: Chitinolytic microorganisms secrete a range of chitin modifying enzymes, which can be exploited for production of chitin derived products or as fungal or pest control agents. Here, we explored the potential of 11 marine bacteria (*Pseudoalteromonadaceae*, *Vibrionaceae*) for chitin degradation using in silico and phenotypic assays. Of 10 chitinolytic strains, three strains, *Photobacterium galatheae* S2753, *Pseudoalteromonas piscicida* S2040 and S2724, produced large clearing zones on chitin plates. All strains were antifungal, but against different fungal targets. One strain, *Pseudoalteromonas piscicida* S2040, had a pronounced antifungal activity against all seven fungal strains. There was no correlation between the number of chitin modifying enzymes as found by genome mining and the chitin degrading activity as measured by size of clearing zones on chitin agar. Based on in silico and in vitro analyses, we cloned and expressed two ChiA-like chitinases from the two most potent candidates to exemplify the industrial potential.

Keywords: chitin; chitinases; antifungal; marine bacteria; *Pseudoalteromonadaceae*; *Vibrionaceae*

1. Introduction

Chitin, the β-1,4-linked homopolymer of *N*-acetylglucosamine (GlcNAc), is the most abundant polymer in the marine environment, and the second in nature after cellulose. Chitin is the structural basis for exoskeletons of crustaceans and insects, and a component of the fungal cell wall. The global production of chitin is estimated to be 10^{11} tons per year, however, chitin does not accumulate as it is hydrolyzed by marine microorganisms [1,2]. The hydrolysis is mediated by chitinolytic enzymes and allows the microorganisms to utilize chitin as a carbon and nitrogen source and chitin turnover is important for the biogeochemical C- and N-cycles. Chitin and chitinolytic enzymes are also of biotechnological interest with potential applications in the food, medical and agricultural sectors [3]. Also, chitin in the form of shellfish waste can be considered as a resource potentially used as a carbon-source in microbial fermentations. Shellfish waste constitutes an environmental problem of increasing magnitude [4,5], and the discovery of inexpensive processes, which can degrade chitin into chitooligosaccharides, chitosan and GlcNAc, may address this problem [6].

As mentioned, the cell wall of fungi contains chitin and some chitinolytic microorganisms can inhibit the growth of fungi by chitin degradation [7–9]. Fungal plant diseases are of great concern in agriculture and cause large losses at an estimated 5%–10% of the world's food production [10]. Fungal contamination and mycotoxin production is also a problem in the built environment [11]. Potentially, natural fungicides, such as chitinases, could replace the chemical fungicides in plant biocontrol [12] and toxic chemicals indoors, and since bacterial chitinases can inhibit fungal growth, they are of particular interest for this purpose [7,13].

Chitin is a recalcitrant insoluble polysaccharide and is degraded into soluble oligosaccharides or GlcNAc. Chitinases (EC 3.2.1.14) hydrolyze the β-1,4 glycosidic bonds between the GlcNAc residues to produce chitooligosaccharides. Chitinases are glycosyl hydrolases (GH) and are divided into GH families 18 and 19. Bacterial chitinases usually belong to family 18, although a few belonging to family 19 have been described [14]. The GH18 and GH19 chitinases differ in sequence similarity, three-dimensional structure and catalytic mechanism. Bacteria often secrete many chitinases, and it is believed that they do so in order to efficiently hydrolyze the different forms of chitin they encounter [7,15,16]. The GH19 chitinases are believed to be the primary enzymes involved in breakdown of fungal chitin, but this is also mediated by other types of chitinases [7,17].

Recently, a new extracellular enzyme involved in breakdown of chitin was discovered. Lytic polysaccharide monooxygenases (LPMOs) were first described in 2010 [18] and are metalloenzymes that oxidize the glycosidic bonds in the crystalline surface of chitin and facilitate access of chitinases. LPMOs were first classified as carbohydrate binding module family 33 and they were believed only to be involved in substrate recognition [19]. LPMOs are now reclassified in auxiliary activity group 10 (AA10) in the CAZY-database [20] and their facilitating activity in chitin degradation has become clear [21]. Another facilitating enzyme, which has so far only been described in some marine bacteria, is chitooligosaccharide deacetylase (COD) [22]. COD (EC 3.5.1.105) is secreted in low concentrations and produces a signal molecule, GlcNac-GlcN, which acts as an inducer for chitinase production [23]. Deacetylation by COD is of particular interest for the industrial production of chitosan oligomers. A combination of COD and another chitin deacetylase, NodB, is used for commercial production of defined chitosan oligomers [24].

The purpose of the present work was to determine the potential for chitin degradation in a collection of marine bacteria. The bacteria were isolated due to their production of antibacterial compounds [25] and we have recently shown that chitin influences the production of secondary metabolites, such as antibacterial compounds, in some of these bacteria [26,27]. We therefore rationalized that they likely would have a high potential for chitin degradation. We envision that these bacteria and their chitinolytic activities will be of interest to the biotech and building industries and in agricultural production. We used in silico genome-wide analysis combined with phenotypic testing to unravel the potential and to exemplify the industrial perspective we cloned and heterologously expressed two chitinases.

2. Results

2.1. Chitin Degrading Activity and in Silico Analysis

Chitin degradation by 11 bacterial strains was determined on agar plates containing crystalline chitin, colloidal chitin or chitosan of both shrimp and crab origin. All but one strain, *P. fuliginea* S3431, degraded chitin. At low temperatures (4 °C and 15 °C), the clearing zones in the chitin containing media were hardly visible. The most pronounced chitin hydrolysis was seen at 35 °C and 25 °C. There was no particular difference between degradation of crab and shrimp chitin. Degradation of crystalline chitin was slower than that of colloidal chitin. None of the strains had chitosanase activity.

The bacterial genomes have previously been mined for chitinase genes and ChiS [28], and we here extend this genome mining including also COD, LPMO and a subdivision of the chitinase genes (Table 1). Two to six putative chitinase genes were found in the 10 chitin degrading strains, and all but one strain, *V. galatheae* S2757, also encoded one to two putative LPMOs. The gene encoding the ChiS sensor was only present in strains belonging to the *Vibrionaceae* family, whereas the CdsS/CdsR pair was present in all *Pseudoalteromonas* strains, including strain S3431 that did not degrade chitin.

Table 1. Chitinolytic activity and chitinase genes in 11 marine bacteria. Strain S3431 is included as a non-chitin degrading control. Chitinase activity was graded according to clearing zone size (+ is zone size of 0–6.99 mm, ++ is zone size of >7 mm and − is no clearing zone). Activity was evaluated at 25 °C on colloidal shrimp chitin. Chitinases can be grouped into glycosyl hydrolases (GH) families 18 and 19. GH18 chitinases are further sub-grouped into ChiA, ChiD and unspecified (U).

Strain	Species	Chitinase Activity	# Of Chitinolytic Enzymes						CSS Type
			GH18 (ChiA)	GH18 (ChiD)	GH18 (U)	GH19	COD	LPMO	
S2753	*P. galatheae*	++	1	1 *	0	1	0	1	ChiS
S2052	*V. coralliilyticus*	+	4 **	1	0	1	1	2	ChiS
S2604	*V. nigripulchritudo*	+	2	1	0	3	1	1	ChiS
S2394	*V. neptunius*	+	2	1	0	1	1	2	ChiS
S2757	*V. galatheae*	+	1	0	0	1	1	0	ChiS
S1110	*V. fluvialis*	+	1	0	0	1	0	1	ChiS
S2040	*P. piscicida*	++	0	1 *	1	1	1	1	CdsS
S2724	*P. piscicida*	++	1	1	1	1	1	2	CdsS
S3137	*P. ruthenica*	+	1	1	0	1	0	1	CdsS
S2471	*P. rubra*	+	2	1	2	1	0	2	CdsS
S3431	*P. fuliginea*	−	0	0	0	0	0	0	CdsS

CSS: chitin sensing system, ChiS: chitin catabolic cascade sensor histidine kinase in *Vibrionaceae*, COD: chitooligosaccharide deacetylase, CdsS: chitin sensor kinase in Pseudoalteromonas, LPMO: lytic polysaccharide monooxygenases; * Have been classified as ChiD, due to phylogeny, ** One chitinase has been classified as ChiA, due to phylogeny.

We analysed the chitinases in Pfam, NCBI and SignalP which enabled the classification of chitinases into GH18 and GH19 groups (Table 1). The GH18 chitinases were sub-divided into ChiA, ChiD and an unspecified group. GH18 and, interestingly, also GH19 chitinases were found in all strains, except S3431.

COD genes were found in six strains and in both families. The COD genes were pairwise compared to other known COD genes and chitin deacetylases of bacterial and fungal origin. We included allontoinase genes since BLASTp analysis revealed that the *Vibrio* COD genes had 99% sequence similarity to allantoinases. Allantionases are enzymes that catalyze the hydrolytic cleavage of the hydantoin ring in allantoin, which is present in purine-derived compounds [29]. *Vibrio* COD genes had 96%–99% sequence similarity to chitin deacetylases (CDA) and the *Pseudoalteromonas* COD genes had 98%–99% sequence similarity to polysaccharide deacetylases (Figure 1). The *Vibrio* COD genes only had a low (10%–14%) similarity to other known COD genes from same genus. However, they share approximately 53% similarity with a chitin deacetylase, which was cloned from a metagenomic sample. The *Pseudoalteromonas* CODs had low similarity to the known CODs and to the metagenomic chitin deacetylase, 10% and 24%, respectively. The putative COD genes in our study were 10% similar to a *Shewanella* COD, which had more than 60% similarity to the known COD genes from *Vibrio* species. Low similarity was also found when comparing to fungal CDA. The Vibrio COD genes had low similarity to allantoinases from *Streptomyces coelicolor* and *Bacillus lichenformis*, however approximately 69% similarity to an allantoinase analog (PuuE) from *Pseudomonas fluorescens*.

	1	2	3	4	5	6	7	8	9	10	11	12	13	14	15	16	17
1		96.74	93.81	79.55	28.31	28.61	54.49	13.15	13.15	13.61	13.12	10.02	39.34	9.34	6.81	6.74	69.84
2	10		91.86	79.22	28.61	28.61	53.33	13.15	13.15	13.61	13.12	10.02	39.34	9.34	7.02	7.37	69.52
3	19	25		81.49	28.01	28.01	53.33	12.93	12.93	13.61	13.35	10.02	38.14	9.07	7.87	7.16	69.84
4	63	64	57		28.61	28.61	53.33	13.80	13.80	14.03	14.03	11.14	37.24	7.69	7.45	6.11	64.76
5	238	237	239	237		95.83	23.01	11.67	11.21	11.67	12.81	9.61	21.65	7.18	6.99	5.87	27.30
6	237	237	239	237	13		24.15	11.44	10.98	11.90	13.27	9.61	21.37	6.91	6.78	6.08	27.89
7	157	161	161	161	271	267		11.50	11.50	9.96	10.40	9.07	34.69	6.62	7.22	5.22	52.72
8	383	383	384	381	386	387	400		99.06	84.31	79.81	66.74	10.79	6.07	8.47	9.13	12.67
9	383	383	384	381	388	389	400	4		83.61	79.35	66.51	10.79	6.07	8.68	9.54	12.67
10	381	381	381	380	386	385	407	67	70		78.42	65.57	10.11	5.86	8.26	9.96	12.44
11	384	384	383	380	381	379	405	87	89	93		63.81	10.54	6.03	7.85	8.90	12.13
12	395	395	395	391	395	395	411	142	143	147	156		9.03	6.32	6.63	7.48	9.77
13	202	202	206	209	275	276	241	397	397	400	399	403		7.84	6.12	5.44	39.76
14	330	330	331	336	349	350	367	433	433	434	436	430	341		5.34	5.31	9.97
15	438	437	433	435	439	440	437	443	442	444	446	451	445	443		31.74	7.68
16	443	440	441	446	449	448	454	438	436	434	440	445	452	446	314		6.54
17	95	96	95	111	245	243	165	386	386	387	391	397	200	325	433	443	

Figure 1. Pairwise comparison of protein sequence identities of putative COD genes from this study (numbers 1: *Vibrio coralliilyticus* (KIY71281), 2: *Vibrio neptunius* S2394 (KIY93856), 3: *Vibrio galatheae* S2757 (KIY81897), 4: *Vibrio nigripulchritudo* S2604 (KIY75235), 5: *Pseudoalteromonas piscicida* S2040 (KIY92988) and 6: *Pseudoalteromonas piscicida* S2724 (KIY89714)), known COD genes (number 8: *Vibrio alginolyticus* H-8 (BAB21759), 9: *Vibrio parahaemolyticus* KN1699 (BAG70715), 10: *Vibrio* sp. SN84 (BAG82921), 11: *Vibrio cholerae* O1 (AAF94439) and 12: *Shewanella woodyi* ATCC 51908 (ACA84860), chitin deacetylases (number 7: *Metagenomic* CDA (AEJ31921), 13: *Schizosaccharomyces pombe* (CAB10114), 14: *Colletotrichum Lindemuthianum* (2IW0), 15: *Streptomyces coelicolor* A3(2), (NP_630347) and 16: *Bacillus Lichenformis* (AAU22686)) and an allantoinase analog PuuE (number 17: *Pseudomonas fluorescens*, *puuE* (ACA50280)). Values above the diagonal line refers to percent sequence identity and values below the diagonal refer to number of differences.

2.1.1. Phylogenetic Analysis of Chitinases

A phylogenetic tree comparing all the complete, translated chitinase genes identified in the 10 chitinolytic strains with other known chitinase genes from the GH18 and GH19 families was constructed. The genes clearly clustered within their respective families GH18 ChiA, GH18 ChiD and GH19, and, thus, were correctly annotated (Figure 2). The unclassified GH18 chitinases did not cluster with any known chitinases. The unclassified GH18 chitinase genes clustered together, with the exception of KJZ10112 from *P. rubra* S2471 that clustered alone, suggesting a new chitinase group.

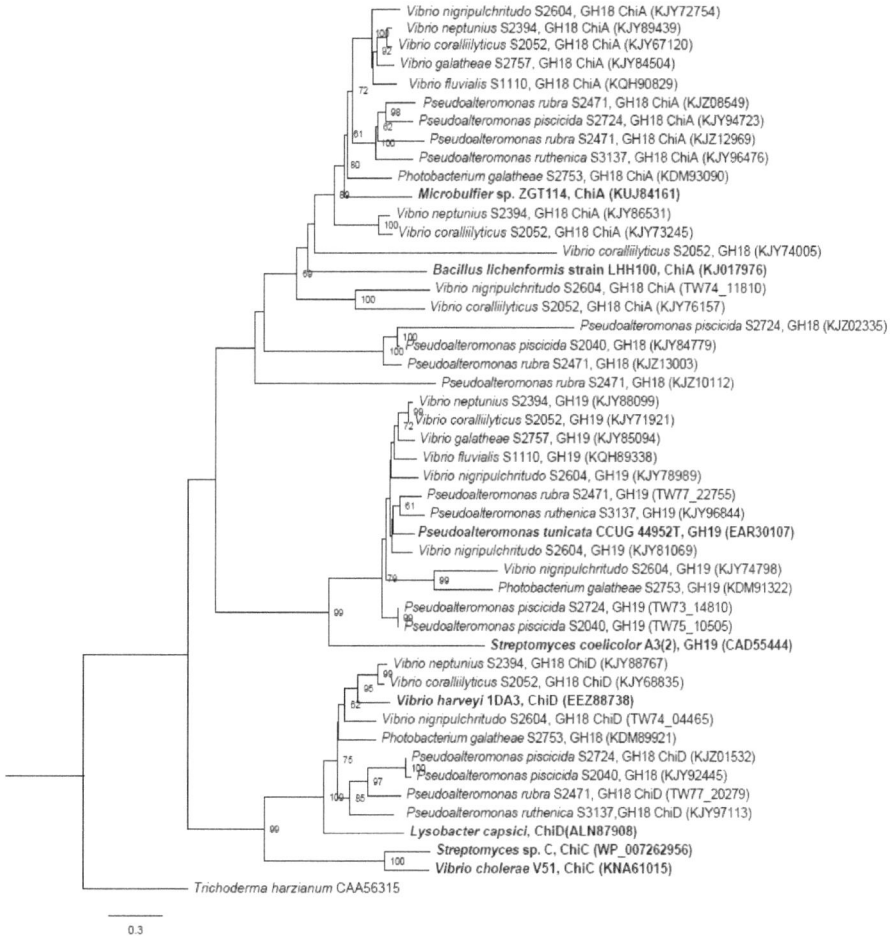

Figure 2. Phylogenetic relationship of the RAST-annotated chitinases from this study and known chitinases from the NCBI database (marked in bold). Branch support values (bootstrap proportions, with 1000 replicates in the analysis) are associated with nodes indicating that the support was <50%. The bar marker indicates the number of amino acid substitutions. Identifiers include species name and GH family, subfamily and accession numbers.

The unclassified GH18 chitinases contained signal peptides and catalytic GH18 domains (Figure 3). Three of four unclassified chitinases also contained a chitin binding motif (CBM). These domains provide evidence for their classification as chitinases.

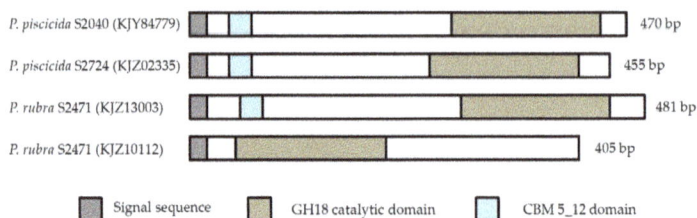

Figure 3. Domain structure of unclassified GH18 chitinases. Protein domains, as identified in Pfam, are shown. Blank areas in part of the proteins indicate that no match with characterized protein domains were found.

2.1.2. Cloning of Chitinases from S2753 and S2724

Chitinases from two strains with pronounced chitin degradation were cloned into a pBAD_His vector and expressed in an *E. coli* BL21 (DE3) host (Table 2). Chitinases are extracellular enzymes since a signalP detected signal peptides was found in all chitinases, as seen in the protein domain structure (Figure 4). The GH18 chitinase from S2724 (KJZ02335) was cloned without its signal peptide, as this region was left out from the RAST annotation.

Table 2. Name, length and expected protein sizes of the cloned chitinases.

Protein	Name	Length (Amino Acids)	Expected Size (kDa)
KJY84504	PG_ChiA	834	87.7
KDM89921	PG_ChiD	846	87.9
KJY85094	PG_GH19	539	59.4
KJY94723	PP_ChiA	822	87.5
KJZ01532	PP_ChiD	850	90.4
KJZ02335	PP_GH18	455	49.3
TW73_14810 *	PP_GH19	479	53.1

* Annotated as pseudo-gene in the NCBI database.

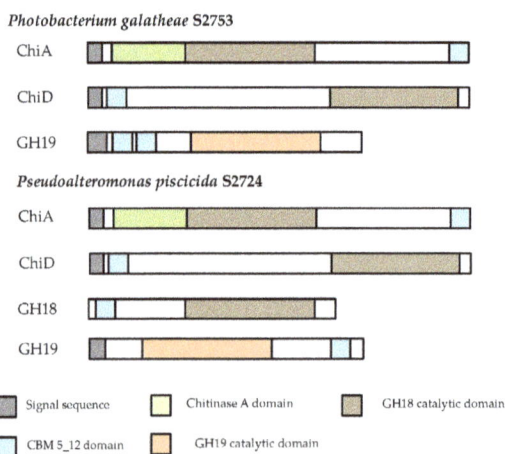

Figure 4. Domain structure of the cloned chitinases. Protein domains, as identified in Pfam, are shown. Blank areas in part of the proteins indicate that no match with characterized protein domains were found.

SDS-page analysis was used to determine if the chitinases were secreted or accumulated inside the cells. PG_ChiA and PP_ChiA were the only enzymes secreted (Figure 5).

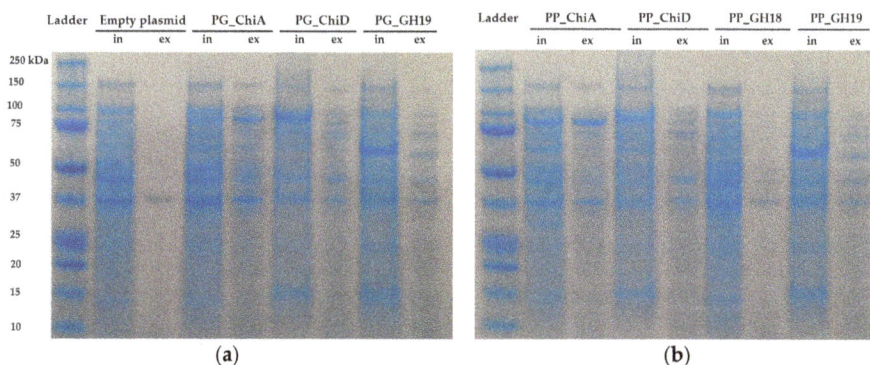

(a)   (b)

Figure 5. SDS-PAGE of induced *E. coli* BL21 harboring the cloned chitinases from (**a**) *P. galatheae* S2753 and (**b**) *P. piscicida* S2724 including both intracellular proteins (in) and extracellular protein extracts (ex). Expected protein sizes are PG_ChiA: 88 kDa, PG_ChiD: 88 kDa, PG_GH19: 59 kDa, PP_ChiA: 88 kDa, PP_ChiD: 90 kDa, PP_GH18: 49 kDa and PP_GH19: 53 kDa.

To test the substrate specificities of the chitinases, we tested the intracellular and extracellular extracts as well as the induced actively growing *E. coli* clones on crystalline and colloidal chitin and chitosan. Only ChiA clones and extracts degraded colloidal chitin but not crystalline chitin and not chitosan (Figure 6). Since we tested clone extracts, the protein concentrations of the chitinases were not determined, as other proteins are also present in the extracts.

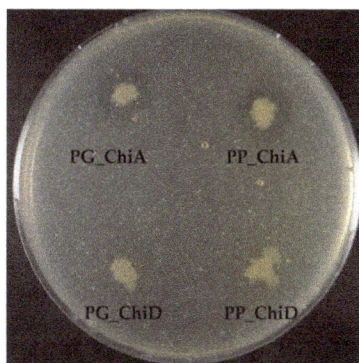

Figure 6. Colloidal chitin degradation by *E. coli* BL21 (DE3) expressing chitinase genes. The two ChiA-type chitinases are secreted, whereas the two ChiD-type chitinases accumulate inside the cells.

2.1.3. Antifungal Activity

Some chitinases have antifungal activity, and we tested the extracts from the chitinase clones against seven different fungi. None of the extracts were antifungal. We also tested the wild-type marine bacteria in two settings: one where the bacteria were spotted after fungal inoculation and one where the bacteria were spotted prior to fungal inoculation (Figure 7).

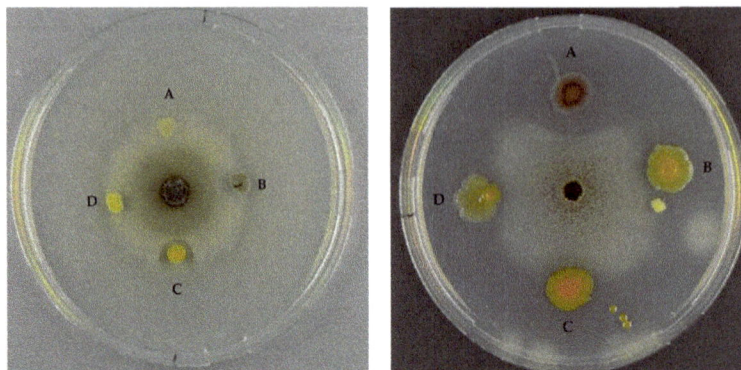

Figure 7. Left panel: Antifungal assay of *S. chartarum* by (**A**) S3137 (no inhibition); (**B**) S3431 (inhibition); (**C**) S2040 (inhibition); and (**D**) S2724 (inhibition). **Right panel**: *A. niger* and (**A**) S2471 (inhibition); (**B**) S2724 (inhibition); (**C**) S2040 (inhibition); and (**D**) S3137 (inhibition). Left panel bacteria were spotted after fungal inoculation, right panel bacteria were spotted prior to fungal inoculation.

In the first setting, one strain, S2040, had a pronounced antifungal effect against all seven fungi (Table 3, columns A). The remaining 10 strains were antifungal against *A. niger* and *B. cinerea*, however, they were not capable of retaining the antifungal effect over time. In the second setting (where the bacteria were spotted prior to the fungi), the same scenario as described above occurred. In addition, the two fungi with the slowest growth rates, *S. chartarum* and *B. cinerea* were inhibited by the bacterial presence (Table 3, columns P). In this setting, two additional bacteria, S3137 and S2471, retained their antifungal activity over time against *P. chrysogenum* and *A. niger*, respectively.

Table 3. Inhibition of fungi by 11 marine bacteria. Bacteria were spotted after (A) or prior (P) to inoculation of fungi. (+) describes an initial antifungal effect which was not retained after 14 days and (++) describes an antifungal effect which was retained after 14 days. (−) no antifungal effect, (sg) sparse growth of the fungi, (nt) not tested.

Strain	Activity against Fungal Strains													
	Penicillium chrysogenum		*Stachybotrys chartarum*		*Chaetomium globosum*		*Neosartorya hiratsukae*		*Aspergillus niger*		*Fusarium oxysporum*		*Botrytis cinerea*	
	A	P	A	P	A	P	A	P	A	P	A	P	A	P
S2753	+	nt	+	sg	−	nt	+	nt	+	nt	+	nt	+	sg
S2052	+	+	+	sg	−	−	−	+	+	+	−	−	+	sg
S2604	−	+	−	sg	−	+	+	+	+	+	−	−	+	sg
S2394	−	nt	−	sg	−	nt	−	nt	+	nt	−	nt	+	sg
S2757	−	−	−	sg	−	+	−	−	+	−	−	−	+	sg
S1110	+	+	−	sg	−	−	−	−	+	+	−	−	+	sg
S2040	++	++	++	sg	++	++	++	++	++	++	++	++	++	sg
S2724	+	+	+	sg	−	−	−	−	+	+	−	−	+	sg
S3137	+	++	−	sg	−	+	+	+	+	+	−	+	+	sg
S2471	+	−	−	sg	−	+	−	−	+	++	−	−	+	sg
S3431	+	nt	+	nt	−	nt	−	nt	+	nt	−	nt	+	sg

3. Discussion

Chitin degradation is an important process in both marine and terrestrial environments. Chitin is also an important resource in different industrial and medical processes, and chitin degrading enzymes or microorganisms are of interest, e.g., as antifungal agents or as bio-insecticides. Hence, there is a growing demand for new enzymes with chitin-modifying properties. We analyzed 11 marine bacteria with antibacterial activity and found a remarkable potential for chitin degradation. In silico, we identified genes involved in the chitin degradation process using an iterative strategy. We found a total of 50 putative chitinase genes of which 11 were considered to be wrongly annotated. The strains contained from two to six putative chitinase genes, and the number of chitinase genes per strain did not correlate with phenotypic chitin degradation ability. The three strains causing the largest clearing zones in chitin agar, S2753, S2724 and S2040, harbored three, three and four chitinase genes, respectively, whereas the poorer chitin degraders, S2050 and S2604, both had six chitinase genes. The chitinases were grouped in different categories of ChiA, ChiD and unclassified belonging to GH18 and GH19 chitinases.

Chitinolytic bacteria harbor different chitinases, most likely due to a specificity for different substrates [16,30–32]. Here, all but one bacterial strain contained at least one chitinase belonging to GH18 ChiA, whereas GH18 ChiD and unclassified GH18 chitinases were only present in some strains. This is not surprising, since ChiA-type chitinases are the most dominant chitinases in bacteria and play a key role in chitin degradation [16,33,34]. Also, this was the dominant chitinase group found in un-cultured bacteria [35]. *P. piscicida* S2040 does not encode for a ChiA-like chitinase, and yet was one of the most potent chitin degraders. The genome of S2040 contained one chitinase of the ChiD-like type, one belonging to GH19 and one unclassified GH18 chitinase, and chitinases of the ChiA-like type are not responsible for the pronounced chitin degradation by this strain. Four chitinases from strains S2724, S2040 and S2471 did not cluster into any existing chitinase subgroups, suggesting they belong to a new subgroup of chitinases. The domain structure provides further evidence that these are indeed chitinases, as they contain the main characteristics of a chitinase (signal peptide, CBM and catalytic domain). However, one chitinase (KJZ10112) does not contain a CBM. This does not disqualify its classification as a chitinase. CBMs are important for the overall performance of the enzyme, but chitinases do not lose function without CBMs, they merely display weaker binding [36–38]. These four chitinases were only identified in the *Pseudoalteromonadaceae* family, and it is interesting to note that two of the chitinases, KJZ02335 and KJY84779, originate from two of the three most potent chitin degrading strains. These unknown chitinases may play a significant role in chitin degradation by *Pseudoalteromonas* species.

All strains had at least one gene encoding chitinases belonging to the GH19 family, which for many years were considered to be unique to higher plants. The first bacterial GH19 chitinase was found in 1996 [39] and the majority of GH19 chitinases have been found in *Streptomyces* species [14]. However, GH19 chitinases from *Vibrio*, *Aeromonas*, *Pseudoalteromonas*, *Chitiniphilus*, *Nocardiopsis* and *Burkholderia* species have also been described [8,40–44]. The presence of the GH19 chitinase in all 10 chitinolytic strains could indicate that this gene is more widespread in the marine environment than hitherto believed.

Since the discovery of the function of LPMOs as facilitators of chitin degradation, the interest in these particular enzymes has increased. Our in silico analysis revealed that LPMOs are present in nine of the 10 genomes from chitinolytic bacteria. Recombinant LPMOs have not yet been characterized in marine bacteria and this potential should be further explored. These LPMOs could be of interest in enzyme cocktails as shown in a recent study where a LPMO from a *Streptomyces griseus* increased the chitin solubilization yields by up to 30-fold when combined with a *Serratia marcescens* GH18 chitinase [45].

COD genes have so far only been identified in *Vibrio* species and in *Shewanella woodyi* ATCC51908 [46,47]. Here, we found six putative COD genes, however, they only had little homology to known CODs. The *Vibrio* COD genes were similar to the allantoinase analog PuuE indicating a wrong annotation. Additional evidence for the wrong annotation is the observation that the neighboring genes to the putative *Vibrio* CODs are involved in purine degradation. PuuE type allantoinases have high similarity to polysaccharide deacetylases, which is also the observation in this study, where the *Vibrio* COD genes had approximately 50% identity to a CDA from a metagenomic sample [48]. The putative COD genes from the two *Pseudoalteromonas* strains were only 23%–27% identical to the metagenomic CDA and the PuuE allantoinase. PuuE proteins can be distinguished from polysaccharide deacetylases by two highly conserved segments, which are only present in PuuE proteins. These conserved segments are present in the *Vibrio* genes (data not shown), but not in the *Pseudoalteromonas* genes, which provide further evidence for the classification of *Vibrio* CODs as PuuE type allantoinases.

In contrast to the *Vibrio* COD genes, the *Pseudoalteromonas* COD genes have signal peptides. Thus, these CODs are likely secreted, and they may therefore potentially act as CDAs. CDAs of bacterial origin are of particular industrial interest, as they catalyze the conversion of chitin to chitosan, a highly coveted polymer [49]. Irrespective of the real function of the so-called *Pseudoalteromonas* COD, the enzymatic deacetylation of polymers has a high industrial potential.

Two of the seven cloned chitinases were secreted whereas the remaining five accumulated inside the cells. The signal peptide of ChiA type chitinases can be recognized by the secretion system in *E. coli*, and hence it is not surprising that ChiA is secreted by *E. coli*. [50,51]. However, cloned ChiA type chitinases are not always secreted [52], so the secretion does not seem to be specific to the chitinase type, but the secretion of the two ChiA-type chitinases from this study allows for an easy purification and exemplifies the potential industrial use of these enzymes. Extracellular and intracellular extracts were tested on different chitinous media, but only the extracts of the two ChiA enzymes degraded colloidal chitin, which may be due to substrate specificity and/or need for synergy between chitinases, as already mentioned.

We tested the extracts against seven fungi, covering both indoor contaminants and plant pathogens. GH19 chitinases are thought to be the main antifungal chitinases, but ChiA type chitinases have also displayed antifungal activity, like *Stm*ChiA from *Stenotrophomonas maltophilia* which was antifungal against *F. oxysporum* [52]. However, extracts from ChiA clones were not antifungal, nor were any of the other extracts. Since chitinase concentration in the extracts was unknown, an up-concentration could potentially result in a measurable effect.

All 11 bacterial strains, including the non-chitin degrader, were antifungal when tested as live cultures, however, with different fungal targets. Allowing pre-growth of the potential producer prior to fungal inoculation increased the antifungal activity of some of the bacteria, which is in agreement with the study by Giubergia and co-workers [27], in which bioactivity of a collection of *Vibrionaceae* increased 3-fold when the producer was allowed a 2-day pre-growth period. *P. piscicida* S2040 displayed pronounced antifungal effect towards all fungi, independently of the time of spotting, and hence would serve as a candidate for a broad-range antifungal bio-pesticide.

In summary, three of the ten strains were of interest due to their remarkable chitin degrading abilities and their antifungal activities. These strains could have potential in biodegradation of chitin-waste, and in biocontrol of unwanted fungal growth in agriculture and the building industry.

4. Materials and Methods

4.1. Strains and Plasmids

The bacterial strains used in this study (Table 4).

Table 4. Marine strains used in this study.

Strain	Species	Accession Number
S2753	*Photobacterium galatheae*	JMIB01
S2052	*Vibrio coralliilyticus*	JXXR01
S2604	*Vibrio nigripulchritudo*	JXXT01
S2394	*Vibrio neptunius*	JXXU01
S2757	*Vibrio galatheae*	JXXV01
S1110	*Vibrio fluvialis*	LKHR01
S2040	*Pseudoalteromonas piscicida*	JXXW01
S2724	*Pseudoalteromonas piscicida*	JXXX01
S3137	*Pseudoalteromonas ruthenica*	JXXZ01
S2471	*Pseudoalteromonas rubra*	JXYA01
S3431	*Pseudoalteromonas fuliginea*	JJNY01

The bacteria were isolated during the Galathea 3 expedition [25] and they have been whole-genome sequenced [28]. Genomes were assembled using CLC Genomics Workbench 7 (CLC bio, Aarhus, Denmark) and contig-based draft genomes were obtained. Gene annotation was performed using Rapid Annotation using Subsystem Technology (RAST), [53–55]. The genomes are available at the National Center for Biotechnology Information (NCBI) [56]. To compare annotations, the genomes were also downloaded from NCBI containing the NCBI-annotated genes. *Escherichia coli* Top10 was used for cloning and propagation of plasmids, and *E. coli* BL21 (DE3) was used for expression of chitinase genes from *P. galatheae* S2753 and *P. piscicida* S2724. The cloning and expression vector was pBAD_Myc_HisA. Plasmids were isolated using the QIAprep® Spin Miniprep kit (Qiagen, 27106, Hilden, Germany), and genomic DNA was extracted using the NucleoSpin® Tissue kit (Machery-Nagel, 740952, Düren, Germany). Strains and plasmids used for cloning can be seen in Table 5.

Table 5. Cloning and expression hosts and plasmids used in this study.

Strain/Plasmid	Details	Reference
Escherichia coli Top10	Cloning host	Invitrogen, C404010, Paisley, United Kingdom
Escherichia coli BL21 (DE3)	Expression host	Novagen, Madison, WI, USA
pBAD_Myc_HisA	Cloning and expression vector	Thermo Scientific, V44001, Waltham, MA, USA
PG_ChiA	pBAD_Myc_HisA vector containing the EA58_02560 gene	This study
PG_ChiD	pBAD_Myc_HisA vector containing the EA58_19900 gene	This study
PG_GH19	pBAD_Myc_HisA vector containing the EA58_12180 gene	This study
PP_ChiA	pBAD_Myc_HisA vector containing the TW73_17595 gene	This study
PP_ChiD	pBAD_Myc_HisA vector containing the TW73_14030 gene	This study
PP_GH18	pBAD_Myc_HisA vector containing the TW73_13265 gene	This study
PP_GH19	pBAD_Myc_HisA vector containing the TW73_14810 gene	This study

4.2. Preparation of Colloidal Chitin

Colloidal chitin was prepared from shrimp shell chitin (Sigma, C7170, Deisenhofen, Germany) or crab shell (Sigma, C9752, Deisenhofen, Germany) chitin as follows: 10 g chitin was hydrolyzed in 400 mL ice-cold 37% HCl for 6 h at 4 °C with stirring. The solution was transferred to 4 L cold dH$_2$O over night for settlement of chitin. The solution was neutralized using NaOH and adjusted to pH 7. Colloidal chitin was collected by centrifugation at $4000 \times g$ for 5 min and resuspended in dH$_2$O for a final concentration of 2%. The chitin solution was autoclaved at 121 °C for 15 min.

4.3. Chitinase and Chitosanase Activity Screening

The strains were tested for chitinase and chitosanase activity on plates containing different chitinous substrates. The basic media consisted of 2% Sea Salt (Sigma, S9883, Deisenhofen, Germany), 1,5% agar, 0.3% casamino acids and was supplemented with either 0.2% colloidal chitin from shrimp or crab, 0.2% crystalline shrimp chitin or 0.2% shrimp chitosan (Sigma, 50494, Deisenhofen, Germany) or crab chitosan (Sigma, 48165, Deisenhofen, Germany). Plates were spotted with one single colony from a streaked culture from freeze-stock on marine agar plates (Difco 2216). The plates were incubated at 4, 15,

25 and 35 °C for 11 days for colloidal chitin and chitosan and 35 days for crystalline chitin. The natural turbidity of the media allows for visual evaluation of chitin/chitosan degradation appearing as a clearing zone around the spotted bacteria. A qualitative grading of chitinase activity (clearing zone) was given, determined from the edge of the bacterial colony to the edge of the clearing zone, where a zone of 0–6.99 mm was graded one plus, +, and >7 mm was graded two plusses, ++.

4.4. In Silico Analysis

An annotation based search for chosen genes involved in chitin degradation was conducted using CLC Main Workbench 7 and included genes putatively encoding chitinases, chitin sensors (ChiS or CdsS), COD and LPMOs. 50 chitinase-encoding genes were found using RAST. The complete genes were translated to protein, and divided into GH18 and GH19 families by analysis using the Pfam protein family database [57,58] and further divided into subfamilies by a protein blast in the non-redundant database in NCBI [59]. Since chitinases are extracellular proteins they were checked for signal peptides using SignalP 4.1 [60]. 11 RAST-annotated chitinases were eliminated from the analysis, as they did not contain either GH18 or GH19 domains, and two others were deleted as they did not contain signal peptides or CBMs. All included and eliminated chitinases can be seen in Supplementary Materials Table S1. Since the COD proteins showed little homology to other known CODs, they were blasted against the non-redundant protein database to find proteins of high sequence similarity. The COD encoding genes were compared to already known CODs, chitin deacetylases and an allantoinase analog by pairwise comparison in CLC.

4.5. Phylogenetic Analysis of Chitinases

A phylogenetic tree was created consisting of the identified chitinase genes from the ten chitinolytic strains. To indicate correct groupings and potentially group the unidentified chitinase genes, known chitinase genes belonging to the GH18 subfamilies A, C and D and GH19 chitinases from the NCBI database were included. The phylogenetic tree was created from a multiple alignment of the translated chitinase proteins with a neighbor-joining construction method and a bootstrap analysis with 1000 replicates was included. The tree was visualized using FigTree [61]. A fungal chitinase from *Trichoderma harzianum* (accession number CAA56315) was used as root.

4.6. Construction of Plasmids for Chitinase Expression

Expression plasmids for the seven chitinase genes from *P. galatheae* S2753 and *P. piscicida* S2724 were constructed via USER cloning, as preciously described [62], using pBAD-Myc_HisA as the vector plasmid. Plasmid and chitinase genes were amplified by PCR using PfuX7 polymerase [63] and primers (Table 6) using following settings: initial denaturation at 98 °C for 2 min, 30 cycles of 98 °C for 20 s, 57 °C for 20 s, 72 °C for 1:30 min and a final extension at 72 °C of 2:30 min. For PG_GH19, PP_GH18 and PP_GH19 extension time was reduced to 50 s and final extension time reduced to 1:45 min. For plasmid DNA, extension time was prolonged to 2:30 min and final extension prolonged to 4:30 min. Each PCR reaction (50 μL) consisted of 5 μL Pfu buffer (200 mM Tris-HCl pH 8.8, 100 mM KCl, 60 mM $(NH_4)_2SO_4$, 20 mM $MgSO_4$, 1 mg/mL BSA (in nuclease-free water) and 1% Triton X-100), 5 μL dNTPs (2 mM), 1.2 μL $MgCl_2$ (50 mM), 5 μL forward primer (5 μM), 5 μL reverse primer (5 μM), 1 μL PfuX7 polymerase and 0.5–1 μL DNA.

In short, the chitinase containing plasmids were constructed in a reaction of 10 μL, consisting of 100 ng of each purified PCR product, 1 μL T4 DNA ligase buffer (New England BioLabs, #B0202S, Ipswich, MA, USA) and 1 μL USER™ enzyme (New England BioLabs, #M5505S, Ipswich, MA, USA). The reaction was incubated at 37 °C for 15 min, followed by 26 °C for 15 min and 10 °C for 10 min. 3 μL USER reaction was mixed with 40 μL chemically competent *E. coli* Top10. The mixture was incubated on ice for 30 min, followed by a 60 s heat shock at 42 °C, and 2 min incubation on ice. Cells were recovered in 1 mL LB for 1 h and subsequently harvested at 1600× *g* for 2 min. The pellet was plated on LB agar containing 100 μg/mL ampicillin and incubated at 37 °C overnight, and the following day colonies were grown in LB media with antibiotics and stored as glycerol stocks. Plasmids were purified and confirmed by sequencing (Macrogen, Amsterdam, The Netherlands).

Table 6. Primers used for chitinase gene amplification.

Primer	Sequence 5′–3′
pBAD_Myc_HisA_fw	AATTCGAAGCUTGGGCCCGAA
pBAD_Myc_HisA_rv	ATGGTTAATUCCTCCTGTTAGCC
PG_ChiA_fw	AATTAACCAUGTCTTTCAATAAGTTGAGTCCTATTGC
PG_ChiA_rv	AGCTTCGAATUCTGGCAGTTTGCTGCACCCA
PG_ChiD_fw	AATTAACCAUGCGTAAAACTCTGATTCAGACAGCTGT
PG_ChiD_rv	AGCTTCGAATUCTGAGCGTTCATAGCATCCAGCTTC
PG_GH19_fw	AATTAACCAUGAAACAAAAACTGTCCCCTCAATGGG
PG_GH19_rv	AGCTTCGAATUCTCAACGGTGACACCATAATATTTCTGG
PP_ChiA_fw	AATTAACCAUGAAACTTAATAAAATAACCAGCTATATAGGACTTG
PP_ChiA_rv	AGCTTCGAATUGTTAGTTACTGCCTTCCATACATCAGC
PP_ChiD_fw	AATTAACCAUGAAACCAACTTCTATATTACGATTGGCTTGG
PP_ChiD_rv	AGCTTCGAATUATTTCCTTGATTCATCTGCGTTAATTTATCGC
PP_GH18_fw	AATTAACCAUGGAAGTTGCACTGGCGGTTGACT
PP_GH18_rv	AGCTTCGAATUCTGACATTGATAGCTTGGTGTTACACCA
PP_GH19_fw	AATTAACCAUGAACAGTCTAAAATTAGCGACCGCAGTT
PP_GH19_rv	AGCTTCGAATUGTTAACCGCTAACCAAGGACCCG

4.7. Protein Expression and SDS-Page Analysis

Protein expression was initiated by transformation of the purified plasmids into electrocompetent *E. coli* BL21 (DE3) cells. Transformants were incubated at 37 °C overnight and one colony was used to incubate 10 mL LB media supplemented with 100 µg/mL ampicillin and incubated at 37 °C, 250 rpm, overnight. ON culture was used to incubate fresh LB media supplemented with ampicillin. At OD_{600} between 0.5 and 0.8 the cultures were induced with arabinose (total concentration 0.02%) and the cultures were further incubated at 37 °C for 4 h. 2 mL cell suspension was centrifuged for 1 min at $12,000 \times g$. The supernatant was separated from the pellet, and the pellet was resuspended in 1 mL lysis buffer (50 mM potassium phosphate, pH 7.4, 500 mM sodium acetate, 0.1 mM EDTA and 20% glycerol). The lysis mixtures were kept on ice and lysed by sonication (Soniprep, amplitude 10 µm) for 2×30 s, and hereafter centrifuged for 1 min at $12,000 \times g$. For SDS-page, protein concentration was estimated using the method of Bradford [64] with BSA as standard. Protein concentration of extracellular extracts ranged from 0.1 to 0.3 mg/mL and intracellular were diluted to a total concentration of 0.5 mg/mL. Fifty µL of protein extract were incubated with 10 µL loading dye (300 mM Tris HCl, pH 6.8, 0.01% bromophenol blue, 15% *v/v* glycerol and 6% SDS) at 95 °C for 5 min and 20 µL of each solution was loaded on a precast 4%–12% Bis-Tris gel (NuPAGE™ Novex™, Thermo Scientific, NP0321, Waltham, MA, USA). The gel was run for 1.5 h at 90 V and stained with Coomassie Brilliant Blue G-250. To identify the proteins on the gel, the sizes were estimated using the compute PI webtool [65].

4.8. Antifungal Activity

Seven different fungi were used for antifungal activity testing of both wildtype bacteria as well as the cloned chitinase enzyme extracts. The following fungi were chosen to cover indoor mold and plant-pathogenic fungi: *Penicillium chrysogenum* (IBT 33843), *Stachybotrys chartarum* (IBT 7709), *Chaetomium globosum* (IBT 7029), *Neosartorya hiratsukae* (IBT 28630), *Aspergillus niger* (IBT 32191), *Fusarium oxysporum* (IBT 41964) and *Botrytis cinerea* (IBT 41856) and were from the IBT Culture Collection at Department of Biotechnology and Biomedicine, Technical University of Denmark. The antifungal activity of wildtype bacterial strains was tested by adding a 20 µL spore suspension to a puncture well on the center of a MA plate. Due to different growth rates, the fungi were allowed to grow for 4–8 days, after which colony mass of each wildtype bacteria was spotted approximately 2 mm from the edge of the fungal colony. The plates were left for 4 days and antifungal activity was observed. We also tested the antifungal activity of the wild-type strains in a setting where the bacteria were spotted 2 days prior to the inoculating the fungi. The bacteria were spotted 2 cm from center of the plates. In both the above settings, plates were checked again after approximately 14 days to see if the bacteria retained their inhibitive effect. Antifungal effect was graded qualitatively, where one plus (+) describes antifungal effect which did not retain the effect after 14 days, and two pluses (++) describes the antifungal effect which was retained after 14 days.

For the chitinase extracts, the fungi were inoculated as described above and after 3–4 days of growth, holes were punched 2 mm from the edge of the fungal colony and 50 µL of each enzyme solution, prepared as in Section 4.7, was added to the wells. Plates were incubated for 2–4 days and checked for antifungal activity. All experiments were conducted at room temperature.

Supplementary Materials: The following are available online at http://www.mdpi.com/1660-3397/14/12/230/s1, Table S1: List of included and excluded chitinases.

Acknowledgments: H.M. was supported by a Ph.D. grant from the People Programme (Marie Curie Actions) of the European Union's Seventh Framework Programme FP7-People-2012-ITN, under grant agreement No. 317058, 'BACTORY'. B.A. was supported by a grant from the VILLUM Foundation. S.S.P. received funding from the Commission on Health, Food and Welfare under the Danish Council for Strategic Research and the MaCuMBA Project under the European Union's Seventh Framework Programme (FP7/2007-2013) under grant agreement No. 311975. The publication reflects the views only of the author, and the European Union cannot be held responsible for any use which may be made of the information contained therein. The present work was carried out as part of the Galathea 3 expedition under the auspices of the Danish Expedition Foundation. This is Galathea 3 contribution No. P122.

Author Contributions: H.M., S.S.P. and L.G. conceived and designed the experiments; S.S.P. performed the experiments with assistance from H.M. and B.A.; S.S.P. and H.M. analyzed the data; S.S.P. wrote the paper with support from H.M., L.G. and B.A.

Conflicts of Interest: The authors declare no conflict of interest. The founding sponsors had no role in the design of the study; in the collection, analyses, or interpretation of data; in the writing of the manuscript, and in the decision to publish the results.

References

1. Zobell, C.E.; Rittenberg, S.C. The Occurrence and Characteristics of Chitinoclastic Bacteria in the Sea. *J Bacteriol.* **1938**, *35*, 275–287. [PubMed]

2. Gooday, G.W. Physiology of microbial degradation of chitin and chitosan. *Biodegradation* **1990**, *1*, 177–190. [CrossRef]

3. Hayes, M.; Carney, B.; Slater, J.; Brück, W. Mining marine shellfish wastes for bioactive molecules: Chitin and chitosan—Part B: Applications. *Biotechnol. J.* **2008**, *3*, 878–889. [CrossRef] [PubMed]

4. Yan, N.; Xi, C. Don't waste seafood waste. *Nature* **2015**, *524*, 155–157. [CrossRef] [PubMed]

5. Blanco, M.; Sotelo, C.G.; Chapela, M.J.; Pérez-Martin, R.I. Towards sustainable and efficient use of fishery resources: Present and future trends. *Trends Food Sci. Technol.* **2007**, *18*, 29–36. [CrossRef]

6. Kim, S.; Mendis, E. Bioactive compounds from marine processing byproducts—A review. *J. Food Res.* **2006**, *39*, 383–393. [CrossRef]

7. Kawase, T.; Yokokawa, S.; Saito, A.; Fujii, T.; Miyashita, K.; Watanabe, T. Comparison of Enzymatic and Antifungal Properties between Family 18 and 19 Chitinases from *S. coelicolor* A3(2). *Biosci. Biotechnol. Biochem.* **2014**, *8451*, 988–998.

8. García-Fraga, B.; Da Silva, A.F.; López-Seijas, J.; Sieiro, C. A novel family 19 chitinase from the marine-derived *Pseudoalteromonas tunicata* CCUG 44952T: Heterologous expression, characterization and antifungal activity. *Biochem. Eng. J.* **2015**, *93*, 84–93. [CrossRef]

9. Hoster, F.; Schmitz, J.E.; Daniel, R. Enrichment of chitinolytic microorganisms: Isolation and characterization of a chitinase exhibiting antifungal activity against phytopathogenic fungi from a novel *Streptomyces* strain. *Appl. Microbiol. Biotechnol.* **2005**, *66*, 434–442. [CrossRef] [PubMed]

10. Pitt, J.I.; Hocking, A.D. Introduction. In *Fungi and Food Spoilage*; Springer Science: New York, NY, USA, 2009; p. 1.

11. Andersen, B.; Dosen, I.; Lewinska, A.M.; Nielsen, K.F. Pre-contamination of new gypsum wallboard with potentially harmful fungal species. *Indoor Air* **2016**. [CrossRef] [PubMed]

12. Brzezinska, M.S.; Jankiewicz, U.; Burkowska, A.; Walczak, M. Chitinolytic microorganisms and their possible application in environmental protection. *Curr. Microbiol.* **2014**, *68*, 71–81. [CrossRef] [PubMed]

13. Liu, D.; Cai, J.; Xie, C.; Liu, C.; Chen, Y. Purification and partial characterization of a 36-kDa chitinase from *Bacillus thuringiensis* subsp. *colmeri*, and its biocontrol potential. *Enzyme Microb. Technol.* **2010**, *46*, 252–256. [CrossRef]

14. Beier, S.; Bertilsson, S. Bacterial chitin degradation-mechanisms and ecophysiological strategies. *Front. Microbiol.* **2013**, *4*, 149. [CrossRef] [PubMed]

15. Svitil, A.; Chadhain, S.; Moore, J.; Kirchman, D. Chitin Degradation Proteins Produced by the Marine Bacterium *Vibrio harveyi* Growing on Different Forms of Chitin. *Appl. Environ. Microbiol.* **1997**, *63*, 408–413. [PubMed]

16. Orikoshi, H.; Nakayama, S.; Miyamoto, K.; Hanato, C.; Yasuda, M.; Inamori, Y.; Tsujibo, H. Roles of four chitinases (ChiA, ChiB, ChiC, and ChiD) in the chitin degradation system of marine bacterium *Alteromonas* sp. strain O-7. *Appl. Environ. Microbiol.* **2005**, *71*, 1811–1815. [CrossRef] [PubMed]

17. Watanabe, T.; Kanai, R.; Kawase, T.; Tanabe, T.; Mitsutomi, M.; Sakuda, S.; Miyashita, K. Family 19 chitinases of *Streptomyces* species: Characterization and distribution. *Microbiology* **1999**, *145*, 3353–3363. [CrossRef] [PubMed]

18. Vaaje-Kolstad, G.; Westereng, B.; Horn, S.; Liu, Z.; Zhai, H.; Sørlie, M.; Eisjink, V. An Oxidative Enzyme Boosting the enzymatic conversion of recalcitrant polysaccharides. *Science* **2010**, *330*, 219–222. [CrossRef] [PubMed]

19. Fushinobu, S. Metalloproteins: A new face for biomass breakdown. *Nat. Chem. Biol.* **2013**, *10*, 88–89. [CrossRef] [PubMed]

20. Levasseur, A.; Drula, E.; Lombard, V.; Coutinho, P.M.; Henrissat, B. Expansion of the enzymatic repertoire of the CAZy database to integrate auxiliary redox enzymes. *Biotechnol. Biofuels* **2013**, *6*, 41. [CrossRef] [PubMed]

21. Hamre, A.G.; Eide, K.B.; Wold, H.H.; Sørlie, M. Activation of enzymatic chitin degradation by a lytic polysaccharide monooxygenase. *Carbohydr. Res.* **2015**, *407*, 166–169. [CrossRef] [PubMed]

22. Kadokura, K.; Rokutani, A.; Yamamoto, M.; Ikegami, T.; Sugita, H.; Itoi, S.; Hakamata, W.; Oku, T.; Nishio, T. Purification and characterization of *Vibrio parahaemolyticus* extracellular chitinase and chitin oligosaccharide deacetylase involved in the production of heterodisaccharide from chitin. *Appl. Microbiol. Biotechnol.* **2007**, *75*, 357–365. [CrossRef] [PubMed]

23. Hirano, T.; Kadokura, K.; Ikegami, T.; Shigeta, Y.; Kumaki, Y.; Hakamata, W.; Oku, T.; Nishio, T. Heterodisaccharide 4-*O*-(*N*-acetyl-beta-D-glucosaminyl)-D-glucosamine is a specific inducer of chitinolytic enzyme production in *Vibrios* harboring chitin oligosaccharide deacetylase genes. *Glycobiology* **2009**, *19*, 1046–1053. [CrossRef] [PubMed]

24. Hamer, S.N.; Cord-Landwehr, S.; Planas, A.; Waegeman, H.; Moerschbacher, B.M.; Kolkenbrock, S. Enzymatic production of defined chitosan oligomers with a specific pattern of acetylation using a combination of chitin oligosaccharide deacetylases. *Sci. Rep.* **2015**, *5*, 8716. [CrossRef] [PubMed]

25. Gram, L.; Melchiorsen, J.; Bruhn, J.B. Antibacterial activity of marine culturable bacteria collected from a global sampling of ocean surface waters and surface swabs of marine organisms. *Mar. Biotechnol.* **2010**, *12*, 439–451. [CrossRef] [PubMed]

26. Wietz, M.; Månsson, M.; Gram, L. Chitin stimulates production of the antibiotic andrimid in a *Vibrio coralliilyticus* strain. *Environ. Microbiol. Rep.* **2011**, *3*, 559–564. [CrossRef] [PubMed]

27. Giubergia, S.; Phippen, C.; Gotfredsen, C.H.; Nielsen, K.F.; Gram, L. Influence of niche-specific nutrients on secondary metabolism in *Vibrionaceae*. *Appl. Environ. Microbiol.* **2016**, *82*, 4035–4044. [CrossRef] [PubMed]

28. Machado, H.; Sonnenschein, E.C.; Melchiorsen, J.; Gram, L. Genome mining reveals unlocked bioactive potential of marine Gram-negative bacteria. *BMC Genom.* **2015**, *16*, 158. [CrossRef] [PubMed]

29. Ramazzina, I.; Cendron, L.; Folli, C.; Berni, R.; Monteverdi, D.; Zanotti, G.; Percudani, R. Logical Identification of an Allantoinase Analog (*puuE*) Recruited from Polysaccharide Deacetylases. *J. Biol. Chem.* **2008**, *283*, 23295–23304. [CrossRef] [PubMed]

30. Boer, H.; Munck, N.; Natunen, J.; Wohlfahrt, G.; Söderlund, H.; Renkonen, O.; Koivula, A. Differential recognition of animal type β4-galactosylated and -fucosylated chito-oligosaccharides by two family 18 chitinases from *Trichoderma harzianum*. *Glycobiology* **2004**, *14*, 1303–1313. [CrossRef] [PubMed]

31. Vaaje-Kolstad, G.; Bunæs, A.C.; Mathiesen, G.; Eijsink, V.G.H. The chitinolytic system of *Lactococcus lactis* ssp. lactis comprises a nonprocessive chitinase and a chitin-binding protein that promotes the degradation of α- And β-chitin. *FEBS J.* **2009**, *276*, 2402–2415. [CrossRef] [PubMed]

32. Suzuki, K.; Sugawara, N.; Suzuki, M.; Uchiyama, T.; Katouno, F.; Nikaidou, N.; Watanabe, T. Chitinases A, B, and C1 of *Serratia marcescens* 2170 produced by recombinant *Escherichia coli*: Enzymatic properties and synergism on chitin degradation. *Biosci. Biotechnol. Biochem.* **2002**, *66*, 1075–1083. [CrossRef] [PubMed]

33. Uchiyama, T.; Katouno, F.; Nikaidou, N.; Nonaka, T.; Sugiyama, J.; Watanabe, T. Roles of the Exposed Aromatic Residues in Crystalline Chitin Hydrolysis by Chitinase A from *Serratia marcescens* 2170. *J. Biol. Chem.* **2001**, *276*, 41343–41349. [CrossRef] [PubMed]

34. Kawase, T.; Yokokawa, S.; Saito, A.; Fujii, T.; Nikaidou, N.; Miyashita, K.; Watanabe, T. Comparison of enzymatic and antifungal properties between family 18 and 19 chitinases from *S. coelicolor* A3(2). *Biosci. Biotechnol. Biochem.* **2006**, *70*, 988–998. [CrossRef] [PubMed]

35. Beier, S.; Jones, C.M.; Mohit, V.; Hallin, S.; Bertilsson, S. Global phylogeography of chitinase genes in aquatic metagenomes. *Appl. Environ. Microbiol.* **2011**, *77*, 1101–1106. [CrossRef] [PubMed]

36. Watanabe, T.; Ito, Y.; Yamada, T.; Hashimoto, M.; Sekine, S.; Tanaka, H. The Roles of the C-Terminal Domain and Type and Type III Domains of Chitinase A1 from *Bacillus circulans* WL-12 in chitin degradation. *J. Bacteriol.* **1994**, *176*, 4465–4472. [CrossRef] [PubMed]

37. Katouno, F.; Taguchi, M.; Sakurai, K.; Uchiyama, T.; Nikaidou, N.; Nonaka, T.; Sugiyama, J.; Watanabe, T. Importance of Exposed Aromatic Residues in Chitinase B from *Serratia marcescens* 2170 for Crystalline Chitin Hydrolysis. *J. Biochem.* **2004**, *168*, 163–168. [CrossRef] [PubMed]

38. Limón, M.C.; Lora, J.M.; García, I.; De La Cruz, J.; Llobell, A.; Benítez, T.; Pintor-Toro, J.A. Primary structure and expression pattern of the 33-kDa chitinase gene from the mycoparasitic fungus *Trichoderma harzianum*. *Curr. Genet.* **1995**, *28*, 478–483. [CrossRef] [PubMed]

39. Ohno, T.; Armand, S.; Hata, T.; Nikaidou, N.; Henrissat, B.; Mitsutomi, M.; Watanabe, T. A modular family 19 chitinase found in the prokaryotic organism *Streptomyces griseus* HUT 6037. *J. Bacteriol.* **1996**, *178*, 5065–5070. [CrossRef] [PubMed]

40. Honda, Y.; Taniguchi, H.; Kitaoka, M. A reducing-end-acting chitinase from *Vibrio proteolyticus* belonging to glycoside hydrolase family 19. *Appl. Microbiol. Biotechnol.* **2008**, *78*, 627–634. [CrossRef] [PubMed]

41. Ueda, M.; Kojima, M.; Yoshikawa, T.; Mitsuda, N.; Araki, K.; Kawaguchi, T.; Miyatake, K.; Arai, M.; Fukamizo, T. A novel type of family 19 chitinase from *Aeromonas* sp. No. 10S-24. *Eur. J. Biochem.* **2003**, *270*, 2513–2520. [CrossRef] [PubMed]

42. Huang, L.; Garbulewska, E.; Sato, K.; Kato, Y.; Nogawa, M.; Taguchi, G.; Shimosaka, M. Isolation of genes coding for chitin-degrading enzymes in the novel chitinolytic bacterium, *Chitiniphilus shinanonensis*, and characterization of a gene coding for a family 19 chitinase. *JBIOSC* **2012**, *113*, 293–299. [CrossRef] [PubMed]

43. Tsujibo, H.; Orikoshi, H.; Baba, N.; Miyamoto, K.; Yasuda, M.; Miyahara, M. Identification and Characterization of the Gene Cluster Involved in Chitin Degradation in a Marine Bacterium, *Alteromonas* sp. *Appl. Environ. Microbiol.* **2002**, *68*, 263–270. [CrossRef] [PubMed]

44. Kong, H.; Shimosaka, M.; Ando, Y.; Nishiyama, K. Species-specific distribution of a modular family 19 chitinase gene in *Burkholderia gladioli*. *FEMS Microbiol. Ecol.* **2001**, *37*, 135–141. [CrossRef]

45. Nakagawa, Y.S.; Kudo, M.; Loose, J.S.M.; Ishikawa, T.; Totani, K. A small lytic polysaccharide monooxygenase from *Streptomyces griseus* targeting alpha and beta chitin. *FEBS J.* **2015**, *282*, 1065–1079. [CrossRef] [PubMed]

46. Li, X.; Wang, L.X.; Wang, X.; Roseman, S. The chitin catabolic cascade in the marine bacterium *Vibrio cholerae*: Characterization of a unique chitin oligosaccharide deacetylase. *Glycobiology* **2007**, *17*, 1377–1387. [CrossRef] [PubMed]

47. Hirano, T.; Uehera, R.; Shiraishu, H.; Hakamata, W.; Nishio, T. Chitin Oligosaccharide Deacetylase from *Shewanella woodyi* ATCC51908. *J. Appl. Glycosci.* **2015**. [CrossRef]

48. Liu, J.; Jia, Z.; Li, S.; Li, Y.; You, Q.; Zhang, C.; Zheng, X.; Xiong, G.; Zhao, J.; Qi, C.; et al. Identification and characterization of a chitin deacetylase from a metagenomic library of deep-sea sediments of the Arctic Ocean. *Gene* **2016**, *590*, 79–84. [CrossRef] [PubMed]

49. Zhao, Y.; Park, R.D.; Muzzarelli, R.A.A. Chitin deacetylases: Properties and applications. *Mar. Drugs* **2010**, *8*, 24–46. [CrossRef] [PubMed]

50. De Fuente-Salcido, N.M.; Casados-Vázquez, L.E.; García-Pérez, A.P.; Barboza-Pérez, U.E.; Bideshi, D.K.; Salcedo-hernández, R.; Almendarez, B.E.G.; Barboza-Corona, J.E. The endochitinase ChiA Btt of *Bacillus thuringiensis* 2803 and its potential use to control the phytopathogen *Colletotrichum gloeosporioides*. *Microbiol. Open* **2016**, *5*, 819–829. [CrossRef] [PubMed]

51. Lobo, M.D.P.; Silva, F.A.D.; de Castro Landim, P.G.; da Cruz, P.R.; de Brito, T.L.; de Medeiros, S.C.; Oliveira, J.T.A.; Vasconcelos, I.M.; Pereira, H.D.; Grangeiro, T.B. Expression and efficient secretion of a functional chitinase from *Chromobacterium violaceum* in *Escherichia coli*. *BMC Biotechnol.* **2013**, *13*, 46. [CrossRef] [PubMed]

52. Suma, K.; Podile, A.R. Chitinase A from *Stenotrophomonas maltophilia* shows transglycosylation and antifungal activities. *Bioresour. Technol.* **2013**, *133*, 213–220. [CrossRef] [PubMed]

53. Overbeek, R.; Olson, R.; Pusch, G.D.; Olsen, G.J.; Davis, J.J.; Disz, T.; Edwards, R.A.; Gerdes, S.; Parrello, B.; Shukla, M.; et al. The SEED and the Rapid Annotation of microbial genomes using Subsystems Technology (RAST). *Nucleic Acids Res.* **2014**, *42*, 206–214. [CrossRef] [PubMed]

54. Aziz, R.K.; Bartels, D.; Best, A.A.; DeJongh, M.; Disz, T.; Edwards, R.A.; Formsma, K.; Gerdes, S.; Glass, E.M.; Kubal, M.; et al. The RAST Server: Rapid annotations using subsystems technology. *BMC Genom.* **2008**, *9*, 75. [CrossRef] [PubMed]

55. Rapid Annotation Using Subsystem Technology Server. Available online: http://rast.nmpdr.org/ (accessed on 14 December 2016).

56. The National Center for Biotechnology Information. Available online: https://www.ncbi.nlm.nih.gov/ (accessed on 14 December 2016).

57. Pfam. Available online: http://pfam.xfam.org/ (accessed on 14 December 2016).

58. Finn, R.D.; Coggill, P.; Eberhardt, R.Y.; Eddy, S.R.; Mistry, J.; Mitchell, A.L.; Potter, S.C.; Punta, M.; Qureshi, M.; Sangrador-vegas, A.; et al. The Pfam protein families database: Towards a more sustainable future. *Nucleic Acids Res.* **2016**, *44*, 279–285. [CrossRef] [PubMed]

59. BLAST. Available online: https://blast.ncbi.nlm.nih.gov/Blast.cgi (accessed on 13 December 2016).

60. SignalP. Available online: http://www.cbs.dtu.dk/services/SignalP/ (accessed on 13 December 2016).

61. FigTree. Available online: http://tree.bio.ed.ac.uk/software/figtree/ (accessed on 13 December 2016).

62. Nour-Eldin, H.H.; Hansen, B.G.; Nørholm, M.H.H.; Jensen, J.K.; Halkier, B.A. Advancing uracil-excision based cloning towards an ideal technique for cloning PCR fragments. *Nucleic Acids Res.* **2006**, *34*, e122. [CrossRef] [PubMed]

63. Nørholm, M.H.H. A mutant Pfu DNA polymerase designed for advanced uracil-excision DNA engineering. *BMC Biotechnol.* **2010**, *10*, 21.

64. Bradford, M.M. A Rapid and Sensitive Method for the Quantitation Microgram Quantities of Protein Utilizing the Principle of Protein-Dye Binding. *Anal. Chem.* **1976**, *254*, 248–254. [CrossRef]

65. Compute PI. Available online: http://web.expasy.org/compute_pi/ (accessed on 13 December 2016).

Review

An Overview of the Protective Effects of Chitosan and Acetylated Chitosan Oligosaccharides against Neuronal Disorders

Cui Hao [1],*, Wei Wang [2], Shuyao Wang [2], Lijuan Zhang [1] and Yunliang Guo [1],*

[1] Institute of Cerebrovascular Diseases, Affiliated Hospital of Qingdao University, Qingdao 266003, China; 18661801189@163.com

[2] Key Laboratory of Marine Drugs, Ministry of Education, Ocean University of China, Qingdao 266003, China; wwwakin@ouc.edu.cn (W.W.); shuyaowang224@126.com (S.W.)

* Correspondence: haocui2010@hotmail.com (C.H.); gekeli@qdu.edu.cn (Y.G.); Tel.: +86-532-8291-7322 (C.H. & Y.G.)

Academic Editor: Hitoshi Sashiwa
Received: 14 November 2016; Accepted: 15 March 2017; Published: 23 March 2017

Abstract: Chitin is the second most abundant biopolymer on Earth and is mainly comprised of a marine invertebrate, consisting of repeating β-1,4 linked *N*-acetylated glucosamine units, whereas its *N*-deacetylated product, chitosan, has broad medical applications. Interestingly, chitosan oligosaccharides have therapeutic effects on different types of neuronal disorders, including, but not limited to, Alzheimer's disease, Parkinson's disease, and nerve crush injury. A common link among neuronal disorders is observed at a sub-cellular level, such as atypical protein assemblies and induced neuronal death. Chronic activation of innate immune responses that lead to neuronal injury is also common in these diseases. Thus, the common mechanisms of neuronal disorders might explain the general therapeutic effects of chitosan oligosaccharides and their derivatives in these diseases. This review provides an update on the pathogenesis and therapy for neuronal disorders and will be mainly focused on the recent progress made towards the neuroprotective properties of chitosan and acetylated chitosan oligosaccharides. Their structural features and the underlying molecular mechanisms will also be discussed.

Keywords: chitosan; acetylated chitosan oligosaccharides; neuronal disorder; neuroprotection; molecular mechanism

1. Introduction

Neurodegeneration, the progressive loss of structure and function including the death of neurons in the central nervous system (CNS), is a major cause of cognitive and motor dysfunction [1]. While neuronal degeneration is well-known in Alzheimer's and Parkinson's diseases, it is also observed in neurotrophic infections, neoplastic disorders, prion diseases, multiple sclerosis, amyotrophic lateral sclerosis, stroke, and traumatic brain and spinal cord injuries, in addition to neuropsychiatric disorders and genetic disorders [1–3]. A common link among these diseases is observed at a sub-cellular level, such as atypical protein assemblies and induced neuronal death. Chronic activation of innate immune responses that lead to neuronal injury is also common in these diseases [1]. A large collection of evidence indicates that oxidative stress induced by reactive oxygen species (ROS) plays an important role in neurodegenerative diseases [4]. Moreover, high concentrations of glutamate can lead to neuronal injury and cell death through two different mechanisms: an accumulation of oxidative stress [5,6] and a massive influx of extracellular Ca^{2+} [2,7,8]. Thus, the common mechanisms of neuronal damage and neurodegeneration may offer the hope of discovering therapeutics that could treat many neurodegenerative diseases simultaneously. Indeed, chitosan oligosaccharides and their derivatives seem to have effects on different types of neurodegenerative diseases.

Chitosan, derived from chitin, is composed of randomly distributed β-(1→4)-linked D-glucosamine and N-acetyl-D-glucosamine (Figure 1) [9,10]. Chitin is often present in crustaceans, fungi, yeasts, diatoms sponges, corals, molluscs, and worms [11–16]. It is usually obtained by treating the chitin shells of shrimp and other crustaceans with sodium hydroxide [17,18]. Chitosan has received considerable attention as a functional, renewable, nontoxic, and biodegradable biopolymer for diverse applications, especially in pharmaceutics [19], food [20], and cosmetics [21]. In the medical field, chitosan has been developed not only as artificial skin and a wound healing accelerator, but also as a new physiological material due to its antitumor, immunoenhancing, and antimicrobial properties [22].

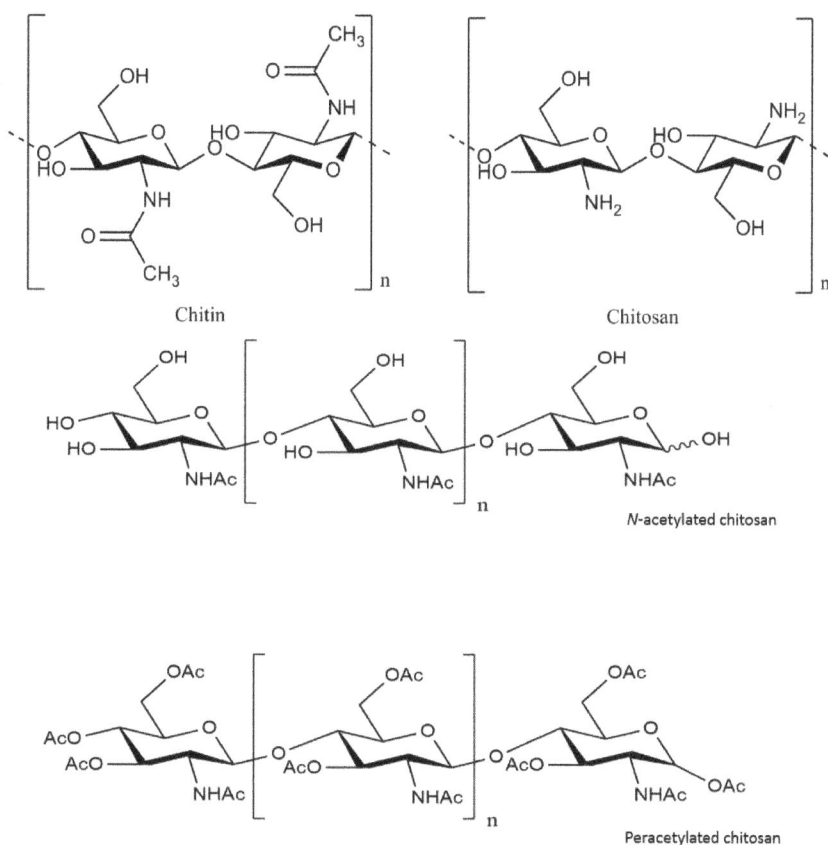

Figure 1. Chemical structure of chitosan and its derivatives.

On average, the molecular weight of commercially produced chitosan is between 3800 and 20,000 Daltons. Chitosan is soluble in acid and relatively insoluble in water. Chitooligosaccharides (COS), i.e., the oligosaccharides of chitosan, are readily soluble in water due to their shorter chain lengths [19]. A large number of studies have shown that the COSs have various biological activities, including antioxidant, antimicrobial, and antitumor activities [19,20]. Recently, it has been reported that the COSs possess good neuroprotective properties, such as β-amyloid and acetylcholinesterase inhibitory activities, anti-neuroinflammation, and anti-apoptosis effects [8,23–26], which suggest that the COSs and their derivatives might merit further investigation as potential neuroprotective agents against neurodegeneration.

This review provides an update on the pathogenesis and therapy for neuronal disorders and will mainly focus on the recent progress made towards the neuroprotective properties of chitosan and acetylated chitosan oligosaccharides. The structural features and their underlying molecular mechanisms will also be discussed.

2. Update on Pathogenesis and Therapy for Neuronal Disorders

2.1. The Pathogenesis of Neuronal Disorders

Neuronal disorders such as Alzheimer's disease, Parkinson's disease, amyotrophic lateral sclerosis, and frontotemporal lobar dementia, are among the most pressing problems for aging populations in the world [1,3]. While neuronal degeneration is well-known in Alzheimer's and Parkinson's diseases, it is also observed in neurotrophic infections, traumatic brain and spinal cord injuries, stroke, neoplastic disorders, prion diseases, multiple sclerosis, and amyotrophic lateral sclerosis, as well as neuropsychiatric disorders and genetic disorders [1–3]. A common link between these diseases is the chronic activation of innate immune responses including those mediated by microglia, the resident CNS macrophages. Such activation can trigger neurotoxic pathways leading to progressive degeneration [1,2]. Moreover, glutamate is one of the major endogenous excitatory neurotransmitters, which plays an important physiological role in the central nervous system [3,7]. However, in a variety of pathological conditions, accumulated high concentrations of glutamate can lead to neuronal injury and cell death through two different mechanisms [2,5–7]. One of the mechanisms occurs when glutamate-induced toxicity is mediated by the competitive inhibition of cystine uptake, which leads to oxidative stress [5,6]. Another mechanism presents when the excitotoxicity of glutamate is mediated by several types of excitatory amino acid receptors, resulting in a massive influx of extracellular Ca^{2+} [7,8]. In addition, the toxicity of misfolded protein aggregates is also reported to be responsible for the pathogenesis of Alzheimer's disease [27,28]. Aβ42 aggregates can cause oxidative stress, [29] and oxidative stress, in turn, increases the β-amyloid cleavage enzyme (BACE-1) activity and Aβ production [30].

Parkinson's disease (PD) is the second most common neurodegenerative disease, which is characterized by the loss of dopaminergic (DA) neurons in the substantia nigra pars compacta and the formation of Lewy bodies and Lewy neurites in surviving DA neurons, in most cases [31]. Numerous studies have shown that dysfunctional mitochondria may also play key roles in DA neuronal loss [31,32]. Both the genetic and environmental factors that are associated with PD contribute to mitochondrial dysfunction and PD pathogenesis [33]. Neuronal death could be due to metabolic disturbances related to alpha-synuclein accumulation, ubiquitin-proteasome system dysfunction, or oxidative stress [31,34,35]. On the other hand, oxidative stress induced by ROS may also play an important role in neurodegenerative disease, such as Parkinson's disease [36].

Huntington's disease (HD) is an autosomal dominant triplet repeat genetic disease, which results in progressive neuronal degeneration in the neostriatum and neocortex, and the associated functional impairments in motor, cognitive, and psychiatric domains [37]. Dopaminergic nigral neurons remain intact in HD and the dopamine level in the HD striatum is higher than normal [38]. Thus, HD is regarded as a relatively dopamine-predominant disease [38], but the mechanism by which this leads to neuronal cell death and the question of why striatal neurons are targeted, both remain unknown [37]. Besides that, prion diseases such as Creutzfeldt-Jakob disease are transmissible fatal neurodegenerative disorders in which infectivity is associated with the accumulation of PrP (Sc), a disease-related isoform of the normal cellular prion protein [39]. The link between PrP (Sc) and neurotoxicity is unclear, and alternative pathological processes need to be considered [39]. New insights into the mechanisms of neurotoxicity in prion diseases support the concept that PrP (Sc) itself is not directly neurotoxic, but neuronal prion propagation results in the production of a toxic intermediate or the depletion of a key constituent [39]. In summary, the pathogenesis process of neuronal disorders is a multi-factor and multi-step process, in which environmental and host factors both play important roles.

2.2. Current Treatments and Therapies for Neuronal Disorders

Neurodegenerative diseases are often characterized by the progressive degeneration of the structure and function of the nervous system. The mechanisms and strategies used to protect them against neuronal injury, apoptosis, dysfunction, and degeneration are known as neuroprotection [23,40]. The goal of neuroprotection is to limit neuronal dysfunction or death after CNS injury, in an attempt to maintain the highest possible integrity of cellular interactions in the brain, thus minimizing the disturbance to the neural function [41]. According to the mechanisms of neuroprotection, the neuronal disorders can be treated by using different neuroprotection agents [42–57], such as antioxidants [42,43] and anti-inflammatory factors [44,45].

Currently, there is no cure for Huntington's disease. The majority of therapeutics currently used in HD are designed to ameliorate the primary symptomatology of the HD condition itself (psychiatric agents for the control of behavioral symptoms, motor sedatives, cognitive enhancers, and neuroprotective agents), and thus improve the quality of life of the patient [58]. For prion diseases, such as Creutzfeldt-Jakob disease, PrP (Sc) is associated with both pathology and infectivity, and therapeutic approaches to date have largely aimed at preventing its accumulation, but this strategy has only produced modest results in animal models [39]. Passive immunization with anti-prion protein antibodies prevents peripheral prion replication and blocks the progression to clinical disease in peripherally infected mice [39]. Moreover, some disease-modifying therapies have been under development for Alzheimer's disease, such as BACE inhibitors, and anti-β-amyloid antibodies are in Phase 2 and 3 trials [59].

Furthermore, a large collection of evidence indicates that oxidative stress induced by ROS plays an important role in neurodegenerative disease. ROS are normal byproducts of aerobic respiration and their level is strictly controlled by various cellular antioxidant compounds and enzymes, while their overproduction leads to cell death [36]. Accordingly, tackling free radicals offers a promising therapeutic target in neurodegenerative disease. Many categories of natural and synthetic compounds have been reported to possess a neuroprotective activity. However, these synthetic neuroprotective agents are believed to have certain side effects, such as dry mouth, tiredness, drowsiness, sleepiness, anxiety or nervousness, difficulty in balancing, etc. [41]. Therefore, the development of novel anti-neuronal disorder agents with a low toxicity and high efficiency is of great importance.

3. The Potential Protective Effects of Chitosan and Its Derivatives against Neuronal Disorders

3.1. Potential Applications of Chitosan and Its Derivatives in Alzheimer's Disease Therapy

Chitosan oligosaccharides (COSs) are a degradation product of chitosan, which is derived from the deacetylation of chitin; the main component of the exoskeleton of crustaceans. Recently, it has been reported that the COSs possess good neuroprotective properties, such as β-amyloid and acetylcholinesterase inhibitory activities, anti-neuroinflammation, and anti-apoptosis effects [23–26]. Hao and co-workers discovered that the pretreatment of PC12 cells with the peracetylated chitosan oligosaccharides (PACOs) (Figure 1) [8,60] markedly inhibited glutamate-induced cell death in a concentration-dependent manner [8]. PACOs pretreatment significantly reduced lactate dehydrogenase release, reactive oxygen species production, and attenuated the loss of mitochondrial membrane potential. Further studies have indicated that the PACOs inhibited glutamate-induced cell death by preventing apoptosis through depressing the elevation of the Bax/Bcl-2 ratio and caspase-3 activation, which suggested that PACOs might be promising antagonists against glutamate-induced neural cell death [8].

Moreover, it was reported that orally administered COS at 200, 400, or 800 mg/kg doses were effective at reducing the learning and memory deficits in Aβ1-42-induced rats [61]. The neuroprotective effects of COS were closely associated with its ability to inhibit oxidative stress. COS was also shown to suppress the inflammatory response and decrease measures of inflammation via a decrease in the release of proinflammatory cytokines [62]. Thus, COSs have beneficial effects on the cognitive impairments seen in an Aβ1-42-induced model of Alzheimer's disease via inhibiting oxidative stress and neuroinflammatory responses. In addition, Dai et al. found that COS attenuated Aβ1-42-induced neurotoxicity in the cortical neurons of rats, and COSs may have anti-Aβ fibrillogenesis and fibril-destabilizing properties. Their findings highlight the potential role of COSs as novel therapeutic agents for the prevention and treatment of Alzheimer's Disease (AD) [61].

3.2. The Inhibitory Effects of Chitosan and Its Derivatives against Parkinson's Disease

Numerous studies have shown that dysfunctional mitochondria may play key roles in DA neuronal loss [31,32]. Both genetic and environmental factors that are associated with PD contribute to mitochondrial dysfunction and PD pathogenesis [33]. Thus, tackling mitochondrial dysfunction offers a promising therapeutic target in neurodegenerative disease. Wang et al. discovered that chitosan (CS) could significantly increase the cell viability and decrease the lactate dehydrogenase (LDH) release induced by Dibutyltin (DBT) in a dose-dependent manner [63]. CS could inhibit cell apoptosis, mitochondrial membrane potential (MMP) disruption, and ROS generation in PC12 cells [63]. Therefore, CS may inhibit DBT-induced apoptosis in PC12 cells through interfering with the mitochondria-dependent pathway [63].

Moreover, COSs were also reported to possess good protective effects against glutamate-induced neurotoxicity in cultured hippocampal neurons [26]. COS pretreatment could inhibit glutamate-induced neuron cell apoptosis in a concentration-dependent manner. COSs depressed glutamate-induced elevation in intracellular calcium concentration Ca^{2+}, and antagonized the glutamate-evoked activation of caspase-3 [26]. Thus, COSs may prevent cultured hippocampal neurons from glutamate-induced neuronal cell death by interfering with an increase in Ca^{2+}. In summary, chitosan and chitooligosaccharides can also be used for the therapy of Parkinson's disease.

3.3. The Inhibition Effects of Chitosan and Its Derivatives against Huntington's Disease

Huntington's disease (HD) is often regarded as a relatively dopamine-predominant disease [38], and there is no effective cure for HD. The majority of therapeutics currently used in HD are designed to ameliorate the primary symptomatology of the HD condition itself, and thus improve the quality of life of the patient [58]. However, the neurotoxicity of glutamate and ROS-induced neuronal damage may also play important physiological roles in the development of Huntington's disease. Thus, reagents that can inhibit the neurotoxicity of glutamate and ROS may be used for HD therapy.

Xu et al. discovered that chitooligosaccharides possessed protective effects against Cu(II)-induced neurotoxicity in the cortical neurons of rats [64]. Pretreatment with COSs could significantly attenuate the toxicity of Cu(II) to rat cortical neurons in a dose-dependent manner. COSs were found to depress Cu(II)-induced elevation in intracellular reactive oxygen species (ROS). Thus, COSs may protect against Cu(II)-induced neurotoxicity by interfering with the production of intracellular ROS [64]. Therefore, COSs may be used for HD therapy through attenuating the neurotoxicity of glutamate and the production of ROS in neurons.

3.4. The Inhibitory Effects of Chitosan and Its Derivatives against Other Neuronal Disorders

Chitosan has been demonstrated to seal compromised nerve cell membranes, thus serving as a potent neuroprotector following acute spinal cord trauma [65]. Cho et al. found that the topical application of chitosan after the complete transection or compression of the guinea pig spinal cord, facilitated the sealing of neuronal membranes in ex vivo tests, and restored the conduction of nerve impulses through the length of spinal cords in vivo, using somatosensory evoked potential recordings [65]. Moreover, chitosan preferentially targeted damaged tissues, serving as a suppressor of reactive oxygen species (ROS) generation, and the resultant lipid peroxidation of membranes, as shown in ex vivo spinal cord samples [65]. Therefore, chitosan treatment can be used as a novel medical approach to reduce the catastrophic loss of behavior after acute spinal cord and brain injuries.

Moreover, Gong and et al. explored the effects of chitooligosaccharides on nerve regeneration after peripheral nerve injuries, and discovered that COS treatment could significantly improve the number of regenerated myelinated nerve fibers, the muscle action potentials, the cross-sectional area of muscle fibers, and the thickness of regenerated myelin sheaths in the nerves [66]. Thus, COSs accelerated peripheral nerve regeneration after a crush injury to the common peroneal nerves of a rabbit. Therefore, the COSs merit further studies as potential neuroprotective agents to improve the peripheral nerve regeneration after an injury [66]. Furthermore, Jiang et al. reported that COS treatment could promote peripheral nerve regeneration with the desired functional recovery in the sciatic nerve crush injury model of a rat, which raises the possibility of developing COS as a potential neuroprotective agent for peripheral nerve repair applications [67].

4. The Mechanisms of Neuroprotective Effects of Chitosan and Its Derivatives

According to its mechanism, neuroprotection can be categorized into several mechanisms, such as: antioxidant (free radical trapper/scavenger) [42,43]; anti-inflammatory [44,45]; anti-excitotoxic [46]; apoptosis inhibitor [47]; gene expression modulator [48]; ion channel modulator [49,50]; metal ion chelator [51,52]; neurotrophic factor [53–55]; matrix metalloprotease inhibitor [56]; and combined mechanism (combining two mechanisms or more) [57].

4.1. Anti-Oxidative Stress Action

A large amount of evidence indicates that oxidative stress induced by reactive oxygen species plays an important role in neurodegenerative disease. ROS are normal byproducts of aerobic respiration and their level is strictly controlled by various cellular antioxidant compounds and enzymes, while their overproduction leads to cell death [4]. Accordingly, tackling free radicals offers a promising therapeutic target in neurodegenerative disease. Hao and co-workers indicated that PACOs pretreatment significantly reduced lactate dehydrogenase release and reactive oxygen species production in PC12 cells. Further studies indicated that the PACOs may inhibit glutamate-induced cell death by preventing apoptosis through depressing the elevation of the Bax/Bcl-2 ratio and caspase-3 activation [8]. Moreover, Khodagholi et al. found that chitosan could prevent oxidative stress-induced amyloid β formation in NT2 neuron cells [68], and the chitosan nanoparticles could also effectively, and statistically, reduce damage to the membrane integrity, secondary oxidative stress, and lipid peroxidation. Thus, chitosan may be able to attenuate neuronal damage through inhibiting the production of reactive oxygen species and ROS-induced cell death.

Xu et al. found that COSs showed protective effects against Cu(II)-induced neurotoxicity in the primary cultured cortical neurons of a rat [64]. The toxicity of Cu(II) to cortical neurons was obviously attenuated in a concentration-dependent manner by pretreated COSs. The data derived from lactate dehydrogenase (LDH) release and the Hoechst 33342 assay support the results from the MTT assay. After the 2′,7′-dichlorofluorescin (DCFH) assay, COSs were found to depress Cu(II)-induced elevation in intracellular reactive oxygen species, Therefore, COSs protect against Cu(II)-induced neurotoxicity in primary cortical neurons by interfering with an increase in intracellular reactive oxygen species (ROS) [64].

4.2. Suppressing Effect on Abeta Aggregation

β-Amyloid peptide (Aβ), the major component of senile plaques in patients with Alzheimer's disease (AD), is believed to facilitate the progressive neurodegeneration that occurs in this disease. The β-amyloid (Aβ) peptides can be cleaved from amyloid precursor proteins (APPs) by proteolysis enzymes such as β- and γ-secretase [69–71]. In APP proteolysis, it seems that the key enzyme is β-secretase, which is also known as the BACE-1, since it initiates the formation of Aβ [72].

Dai et al. reported that COS could inhibit the formation of Aβ1-42 fibrils and disaggregate preformed fibrils, suggesting that COS may have anti-Aβ fibrillogenesis and fibril-destabilizing properties. Pretreatment with COSs markedly inhibited cell death induced by Aβ exposure, and the ROS generation was also attenuated by COSs [61,73]. Moreover, Je et al. reported that chitosan derivatives could effectively inhibit the activity of BACE-1, and the aminoethyl derivative (AE-chitosan) demonstrated the strongest inhibitory activity compared to other derivatives [74]. Byun et al. indicated that the deacetylated chitosan could obviously inhibit the formation of β-amyloid through blocking the activity of BACE-1 [75]. Thus, the suppression of β-amyloid formation by chitosan and its derivatives may be able to enhance the medications for AD.

4.3. Anti-Neuroinflammatory

A growing number of studies are discovering intriguing links between chronic inflammation and a number of neurodegenerative disorders [76]. The neuroinflammation process plays a pivotal role in the initiation and progression of various neurodegenerative diseases [76]. A chronic inflammatory response associated with beta-amyloid (Abeta) and interleukin-1beta (IL-1beta) was reported to be responsible for the pathology of Alzheimer's disease [77]. Kim et al. discovered that a water-soluble chitosan (WSC) inhibited the production of pro-inflammatory cytokine in human astrocytoma cells activated by Aβ peptide 25–35 (Aβ25–35) and interleukin-1β (IL-1β) [77]. The secretion and expression of pro-inflammatory cytokines, TNF-alpha and IL-6, and the expression of inducible nitric oxide synthase (iNOS), were all significantly inhibited by pretreatment with WSC in human astrocytoma cells [77].

Fang et al. investigated the protective effect and mechanism of chitosan oligonucleotides on retinal ischemia and reperfusion (I/R) injury, and found that pretreatment with COSs, especially at a high dosage, effectively ameliorated the I/R-induced reduction of the b-wave ratio in ERGs and the retinal thickness, and the survival of RGCs at 24 h [78]. COSs decreased the expression of inflammatory mediators, p53 and Bax, increasing Bcl-2 expression and thereby reducing retinal oxidative damage and the number of apoptotic cells. More importantly, COSs attenuated IκB degradation and p65 presence in the retina, thus decreasing NF-κB/DNA binding activity after I/R. In conclusion, COSs prevented retinal I/R injury through their inhibition of oxidative stress and inflammation [78].

4.4. Anti-Apoptosis Action

The elimination of cells by apoptosis or programmed cell death is a fundamental event in development, while many human diseases such as acquired immunodeficiency syndrome and neurodegenerative disorders can be directly or indirectly attributed to cell apoptosis [79,80]. In neurodegenerative disorders, apoptosis might be pathogenic, and targeting it might mitigate neurodegenerative disorders [81]. Many researchers have reported that COS and its derivatives may be able to inhibit neuronal cell apoptosis in brain cells.

Wang et al. found that pretreatment with chitosan (CS) significantly increased the cell viability and decreased LDH release induced by DBT in a dose-dependent manner [63]. Meanwhile, DBT-induced cell apoptosis, the disruption of mitochondrial membrane potential (MMP), and the generation of intracellular ROS were attenuated by CS [63]. CS also inhibited the DBT-inducted activation of caspase-9 and -3 at mRNA and protein expression levels. Thus, CS could protect the PC12 cells from apoptosis induced by DBT through the inhibition of the mitochondria-dependent pathway [63]. Moreover, Koo et al. reported that high molecular weight water-soluble chitosan could protect against the cell apoptosis induced by serum starvation in human astrocytes [82]. Thus, the derivatives of chitosan may be able to inhibit neuronal disorders through blocking glutamate-induced neural cell death.

4.5. Anti-Excitotoxic Action

Glutamate is one of the major endogenous excitatory neurotransmitters and plays an important physiological role in the central nervous system [3]. However, in a variety of pathological conditions, accumulated high concentrations of glutamate can lead to neuronal injury and cell death, through two different mechanisms. One of the mechanisms occurs when glutamate-induced toxicity is mediated by the competitive inhibition of cysteine uptake, which leads to oxidative stress [5,6]. Another mechanism presents when the excitotoxicity of glutamate is mediated by several types of excitatory amino acid receptors, resulting in a massive influx of extracellular Ca^{2+} [7,8]. Based on both mechanism, it is predictable that proper antagonists would be able to prevent glutamate-induced neural injury and cell death.

Zhou et al. discovered that one chitooligosaccharide (M.W. 800) possessed good protective effects against glutamate-induced neurotoxicity in cultured hippocampal neurons [26]. They found that COS pretreatment could inhibit glutamate-induced cell apoptosis in cultured hippocampal neurons in a concentration-dependent manner. COSs were found to depress glutamate-induced elevation in intracellular calcium concentration Ca^{2+}, and could antagonize the glutamate-evoked activation of caspase-3 [26]. Thus, COSs may prevent cultured hippocampal neurons from glutamate-induced neuronal cell death by interfering with an increase in Ca^{2+} and inhibiting caspase-3 activity. Moreover, Dai and co-workers found that COSs may act as inhibitors of Aβ aggregation and this effect shows dose-dependency. The addition of COS could attenuate Aβ1-42-induced neurotoxicity in the cortical neurons of rats [73]. Thus, COS may be able to inhibit neuronal cell damage through interfering with glutamate-induced neurotoxicity, both in vitro and in vivo.

4.6. Other Mechanisms

The pathogenesis of AD has been linked to a deficiency in the brain neurotransmitter acetylcholine (ACh) [83,84]. The inhibition of the acetylcholinesterase (AChE) enzyme, which catalyzes the breakdown of ACh, may be one of the most realistic approaches to the symptomatic treatment of AD [83,85,86]. Yoon et al. synthesized COS derivatives with different substitution groups. Among three COS derivatives, diethylaminoethyl-COS (DEAE-COS) has the strongest AChEIs activity, with half maximal inhibitory concentration (IC_{50}) values of 9.2 ± 0.33 μg/mL. dimethyl aminoethyl-(DMAE-) and DEAE-COS were identified as competitive AChEIs, according to the Line weaver–Burk plot [87]. These findings suggest that chemical modification will enhance the utilization of COS as AChEIs, and their inhibitory activity depends on the hydrophobic nature of the group that is introduced to them [87].

Furthermore, Gong and co-workers investigated the effects of chitooligosaccharides on nerve regeneration after crush injuries to peripheral nerves, and found that the compound muscle action potentials, the number of regenerated myelinated nerve fibers, the thickness of regenerated myelin sheaths, and the cross-sectional area of tibialis posterior muscle fibers were significantly improved in the nerves that received COS treatment [66]. Thus, the COSs could become potential neuroprotective agents for the improvement of peripheral nerve regeneration after the injury and deserve further consideration [66].

In summary, the potential neuroprotective effects of chitosan and its derivatives against neuronal disorders discussed in this paper are summarized in Table 1 and their anti-neuronal disorder mechanisms are presented in Figure 2.

Table 1. Potential neuroprotective effects of chitosan and its derivatives against neuronal disorders.

Specific Polysaccharides	Anti-Neuronal Disorder Effects	Mechanisms	References
Chitosans (CSs)	Anti-Parkinson's disease; Anti-spinal cord injury	Anti-apoptosis action; Anti-oxidative stress	[63,65,68]
Chitooligosaccharides (COSs)	Anti-Alzheimer's disease	β-amyloid inhibitory activities; Anti-neuroinflammation; Anti-apoptosis action	[23–26,61,62,73–75]
	Anti-Parkinson's disease	Anti-exitotoxic action; Anti-apoptosis action	[26]
	Anti-Huntington's disease	Anti-exitotoxic action; Anti-oxidative stress	[64]
	Anti- nerve crush injury	Promoting nerve regeneration; Anti-neuroinflammation	[66,67,78]
Peracetylated chitosan oligosaccharides	Anti-Alzheimer's disease; Anti-Parkinson's disease	Anti-oxidative stress; Anti-apoptosis action	[8]
Water-soluble chitosans	Anti-Alzheimer's disease	β-amyloid inhibitory activities; Anti-apoptosis	[77,82]
COS derivatives	Anti-Alzheimer's disease	Anti-AChE and BACE-1 enzyme activities	[74,75,87]

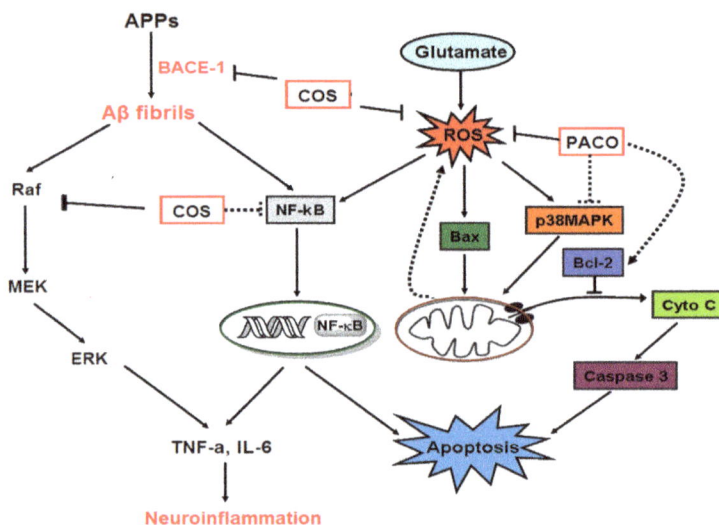

Figure 2. The schematic drawing of anti-neuronal disorder mechanisms of chitosan and its derivatives [8].

5. Progress of the Clinical Studies on Chitosan and Its Derivatives

To further analyze the potential of chitosan and its derivatives as novel antagonists against neurodegeneration, we also summarize the recent progress in the clinical studies on chitosan and its derivatives as medicinal materials (Figure 3).

As of 8 November 2016, a total of 53 clinical studies on chitosan or chitin were included on the clinicaltrials.gov website [88]. Most of them (16 out of 53) were designed to investigate the therapeutic effects of chitosan-based dressings for wound repair and for minimizing the bacterial re-colonization of wounds, especially in diabetic neuropathic foot ulcers (four cases). Moreover, positively charged chitosan can attract the negatively charged blood cells and platelets to promote clots, so 10 clinical trials were hemostasis studies, including those of a postpartum hemorrhage, dental surgery, and other surgical applications. These two types of clinical trials constitute almost half of the total clinical trials about chitosan and its derivatives (Figure 3). Other clinical trials mainly focused on the application of chitosan in treating chronic kidney disease, dry eye syndrome (DES), bone-related diseases, metabolic diseases, and immune-related diseases (Figure 3).

53 clinical studies on chitosan or chitin

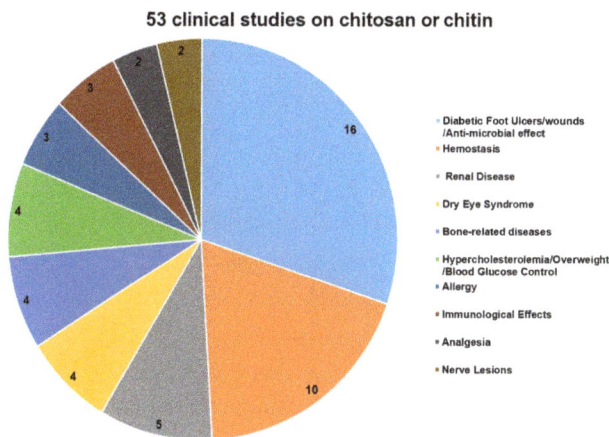

Figure 3. A chart of the clinical studies on chitosan and its derivatives.

Furthermore, only two clinical studies were performed to evaluate whether the additional use of a nerve tube in the primary microsurgical repair of traumatic sensory nerve lesions, influences convalescence and functional results. The results showed that chitosan was biocompatible and had positive effects on the survival and orientation of Schwann cells, as well as the survival and differentiation of neuronal cells and the prevention of painful neuromas. Although there has been much research on the neuroprotective effects of chitosan and its derivatives, studies on neuropathy-related clinical trials are rare (Figure 3). Therefore, chitosan and its derivatives merit further investigation in animal experiments or clinical trials as potential anti-neuronal disorder agents.

6. Conclusions

Recently, marine polysaccharides and their derivatives have been reported to possess various biological activities, such as antioxidant, antimicrobial, and antitumor activities [19,20]. As one of the bioactive compounds derived from the sea, chitosan and its derivatives have been reported to have good neuroprotective properties, such as β-amyloid and acetylcholinesterase inhibitory activities, anti-neuroinflammation, and anti-apoptosis effects [8,23–26]. Moreover, the accumulation of chitin fragments may contribute to the dementia of Alzheimer's disease and chitinase could possibly be used for the treatment of Alzheimer's disease [89]. Herein, our review presents an overview of the recent progress in research on the neuroprotective effects and mechanisms of chitosan and its derivatives against different neuronal disorders. According to the presented data, it seems that chitosan and its derivatives have the potential to be developed into novel neuroprotective agents in the future. However, further studies are needed in order to explore their activities in animal models and/or clinical trials. Nevertheless, chitosan and its derivatives merit further investigation as potential therapeutic candidates for neurodegenerative disorders.

Acknowledgments: This work was supported by National Natural Science Foundation of China (31500646, 81302811, 81672585), NSFC-Shandong Joint Fund (U1406402), the Promotive Research Fund for Excellent Young and Middle-aged Scientists of Shandong Province (BS2015YY040), Qingdao science and technology development project (15-9-1-67-JCH) and Development Plan of Traditional Chinese Medicine Science and Technology of Shandong Province (2015-183).

Conflicts of Interest: The authors declare no conflict of interest.

Abbreviations

Aβ	β-Amyloid peptide
AChE	acetylcholinesterase
AD	Alzheimer's disease
APPs	amyloid precursor proteins
BACE-1	β-amyloid cleavage enzyme
CNS	central nervous system
COS	chitooligosaccharide
CS	chitosan
DA	dopaminergic
DBT	Dibutyltin
DCFH	2′,7′-dichlorofluorescin
DEAE	diethylaminoethyl
DMAE	dimethyl aminoethyl
HD	Huntington's disease
IL-1	interleukin-1
iNOS	inducible nitric oxide synthase
LDH	lactate dehydrogenase
MMP	mitochondrial membrane potential
PACOs	peracetylated chitosan oligosaccharides
PD	Parkinson's disease
ROS	reactive oxygen species

References

1. Amor, S.; Peferoen, L.A.; Vogel, D.Y.; Breur, M.; van der Valk, P.; Baker, D.; van Noort, J.M. Inflammation in neurodegenerative diseases—An update. *Immunology* **2014**, *142*, 151–166. [CrossRef] [PubMed]
2. Bleich, S.; Romer, K.; Wiltfang, J.; Kornhuber, J. Glutamate and the glutamate receptor system: A target for drug action. *Int. J. Geriatr. Psychiatry* **2003**, *18* (Suppl. 1), S33–S40. [CrossRef] [PubMed]
3. Choi, D.W. Glutamate neurotoxicity and diseases of the nervous system. *Neuron* **1988**, *1*, 623–634. [CrossRef]
4. Droge, W. Free radicals in the physiological control of cell function. *Physiol. Rev.* **2002**, *82*, 47–95. [CrossRef] [PubMed]
5. Murphy, T.H.; Miyamoto, M.; Sastre, A.; Schnaar, R.L.; Coyle, J.T. Glutamate toxicity in a neuronal cell line involves inhibition of cystine transport leading to oxidative stress. *Neuron* **1989**, *2*, 1547–1558. [CrossRef]
6. Zablocka, A.; Janusz, M. The two faces of reactive oxygen species. *Postep. Hig. Med. Doswiadczalnej* **2008**, *62*, 118–124. [PubMed]
7. Monaghan, D.T.; Bridges, R.J.; Cotman, C.W. The excitatory amino acid receptors: Their classes, pharmacology, and distinct properties in the function of the central nervous system. *Annu. Rev. Pharmacol. Toxicol.* **1989**, *29*, 365–402. [CrossRef] [PubMed]
8. Hao, C.; Gao, L.; Zhang, Y.; Wang, W.; Yu, G.; Guan, H.; Zhang, L.; Li, C. Acetylated chitosan oligosaccharides act as antagonists against glutamate-induced PC12 cell death via Bcl-2/Bax signal pathway. *Mar. Drugs* **2015**, *13*, 1267–1289. [CrossRef] [PubMed]
9. Kumar, M.N.; Muzzarelli, R.A.; Muzzarelli, C.; Sashiwa, H.; Domb, A.J. Chitosan chemistry and pharmaceutical perspectives. *Chem. Rev.* **2004**, *104*, 6017–6084. [CrossRef] [PubMed]
10. Zargar, V.; Asghari, M.; Dashti, A. A Review on Chitin and Chitosan Polymers: Structure, Chemistry, Solubility, Derivatives, and Applications. *ChemBioEng Rev.* **2015**, *2*, 204–226. [CrossRef]
11. Brunner, E.; Richthammer, P.; Ehrlich, H.; Paasch, S.; Simon, P.; Ueberlein, S.; van Pee, K.H. Chitin-based organic networks: An integral part of cell wall biosilica in the diatom *Thalassiosira pseudonana*. *Angew. Chem. Int. Ed. Engl.* **2009**, *48*, 9724–9727. [CrossRef] [PubMed]
12. Ehrlich, H. Chitin and collagen as universal and alternative templates in biomineralization. *Int. Geol. Rev.* **2010**, *52*, 661–699. [CrossRef]
13. Ehrlich, H.; Ilan, M.; Maldonado, M.; Muricy, G.; Bavestrello, G.; Kljajic, Z.; Carballo, J.L.; Schiaparelli, S.; Ereskovsky, A.; Schupp, P.; et al. Three-dimensional chitin-based scaffolds from Verongida sponges (Demospongiae: Porifera). Part I. Isolation and identification of chitin. *Int. J. Biol. Macromol.* **2010**, *47*, 132–140. [CrossRef] [PubMed]

14. Bo, M.; Bavestrello, G.; Kurek, D.; Paasch, S.; Brunner, E.; Born, R.; Galli, R.; Stelling, A.L.; Sivkov, V.N.; Petrova, O.V.; et al. Isolation and identification of chitin in the black coral *Parantipathes larix* (Anthozoa: Cnidaria). *Int. J. Biol. Macromol.* **2012**, *51*, 129–137. [CrossRef] [PubMed]

15. Anitha, A.; Sowmya, S.; Kumar, P.T.S.; Deepthi, S.; Chennazhi, K.P.; Ehrlich, H.; Tsurkan, M.; Jayakumar, R. Chitin and chitosan in selected biomedical applications. *Prog. Polym. Sci.* **2014**, *39*, 1644–1667. [CrossRef]

16. Wysokowski, M.; Petrenko, I.; Stelling, A.; Stawski, D.; Jesionowski, T.; Ehrlich, H. Poriferan Chitin as a Versatile Template for Extreme Biomimetics. *Polymers* **2015**, *7*, 235–365. [CrossRef]

17. Hong, K.; Meyers, S.P. Preparation and Characterization of Chitin and Chitosan—A Review. *J. Aquat. Food Prod. Technol.* **1995**, *4*, 27–52. [CrossRef]

18. Younes, I.; Rinaudo, M. Chitin and chitosan preparation from marine sources. Structure, properties and applications. *Mar. Drugs* **2015**, *13*, 1133–1174. [CrossRef] [PubMed]

19. Zhang, J.; Xia, W.; Liu, P.; Cheng, Q.; Tahirou, T.; Gu, W.; Li, B. Chitosan modification and pharmaceutical/biomedical applications. *Mar. Drugs* **2010**, *8*, 1962–1987. [CrossRef] [PubMed]

20. Shahidi, F.; Abuzaytoun, R. Chitin, chitosan, and co-products: Chemistry, production, applications, and health effects. *Adv. Food Nutr. Res.* **2005**, *49*, 93–135. [CrossRef] [PubMed]

21. Nan, W.; Sun, A. Application of Chitosan and Oligochitosan in the Field of Cosmetics. *Chem. Ind. Eng. Prog.* **2003**, *32*, 3026–3031.

22. Bellich, B.; D'Agostino, I.; Semeraro, S.; Gamini, A.; Cesaro, A. "The Good, the Bad and the Ugly" of Chitosans. *Mar. Drugs* **2016**, *14*, 99. [CrossRef]

23. Pangestuti, R.; Kim, S.K. Neuroprotective properties of chitosan and its derivatives. *Mar. Drugs* **2010**, *8*, 2117–2128. [CrossRef] [PubMed]

24. Lee, S.H.; Park, J.S.; Kim, S.K.; Ahn, C.B.; Je, J.Y. Chitooligosaccharides suppress the level of protein expression and acetylcholinesterase activity induced by Abeta25–35 in PC12 cells. *Bioorg. Med. Chem. Lett.* **2009**, *19*, 860–862. [CrossRef] [PubMed]

25. Nidheesh, T.; Salim, C.; Rajini, P.S.; Suresh, P.V. Antioxidant and neuroprotective potential of chitooligomers in *Caenorhabditis elegans* exposed to Monocrotophos. *Carbohydr. Polym.* **2016**, *135*, 138–144. [CrossRef] [PubMed]

26. Zhou, S.; Yang, Y.; Gu, X.; Ding, F. Chitooligosaccharides protect cultured hippocampal neurons against glutamate-induced neurotoxicity. *Neurosci. Lett.* **2008**, *444*, 270–274. [CrossRef] [PubMed]

27. Soto, C. Unfolding the role of protein misfolding in neurodegenerative diseases. *Nat. Rev. Neurosci.* **2003**, *4*, 49–60. [CrossRef] [PubMed]

28. Khanam, H.; Ali, A.; Asif, M.; Shamsuzzaman. Neurodegenerative diseases linked to misfolded proteins and their therapeutic approaches: A review. *Eur. J. Med. Chem.* **2016**, *124*, 1121–1141. [CrossRef] [PubMed]

29. Butterfield, D.A.; Swomley, A.M.; Sultana, R. Amyloid beta-peptide (1–42)-induced oxidative stress in Alzheimer disease: Importance in disease pathogenesis and progression. *Antioxid. Redox Signal* **2013**, *19*, 823–835. [CrossRef] [PubMed]

30. Chami, L.; Checler, F. BACE1 is at the crossroad of a toxic vicious cycle involving cellular stress and beta-amyloid production in Alzheimer's disease. *Mol. Neurodegener.* **2012**, *7*, 52. [CrossRef] [PubMed]

31. Hu, Q.; Wang, G. Mitochondrial dysfunction in Parkinson's disease. *Transl. Neurodegener.* **2016**, *5*, 14. [CrossRef] [PubMed]

32. Jana, S.; Sinha, M.; Chanda, D.; Roy, T.; Banerjee, K.; Munshi, S.; Patro, B.S.; Chakrabarti, S. Mitochondrial dysfunction mediated by quinone oxidation products of dopamine: Implications in dopamine cytotoxicity and pathogenesis of Parkinson's disease. *Biochim. Biophys. Acta* **2011**, *1812*, 663–673. [CrossRef] [PubMed]

33. Migliore, L.; Coppede, F. Genetics, environmental factors and the emerging role of epigenetics in neurodegenerative diseases. *Mutat. Res.* **2009**, *667*, 82–97. [CrossRef] [PubMed]

34. Recchia, A.; Debetto, P.; Negro, A.; Guidolin, D.; Skaper, S.D.; Giusti, P. Alpha-synuclein and Parkinson's disease. *FASEB J.* **2004**, *18*, 617–626. [CrossRef] [PubMed]

35. Lim, K.L. Ubiquitin-proteasome system dysfunction in Parkinson's disease: Current evidence and controversies. *Expert Rev. Proteom.* **2007**, *4*, 769–781. [CrossRef] [PubMed]

36. Andersen, J.K. Oxidative stress in neurodegeneration: Cause or consequence? *Nat. Med.* **2004**, *10*, S18–S25. [CrossRef] [PubMed]

37. Kumar, A.; Kumar Singh, S.; Kumar, V.; Kumar, D.; Agarwal, S.; Rana, M.K. Huntington's disease: An update of therapeutic strategies. *Gene* **2015**, *556*, 91–97. [CrossRef] [PubMed]

38. Spokes, E.G. Neurochemical alterations in Huntington's chorea: A study of post-mortem brain tissue. *Brain* **1980**, *103*, 179–210. [CrossRef] [PubMed]

39. Mallucci, G.; Collinge, J. Update on Creutzfeldt-Jakob disease. *Curr. Opin. Neurol.* **2004**, *17*, 641–647. [CrossRef] [PubMed]

40. Kostrzewa, R.M.; Segura-Aguilar, J. Novel mechanisms and approaches in the study of neurodegeneration and neuroprotection. A review. *Neurotox. Res.* **2003**, *5*, 375–383. [CrossRef] [PubMed]

41. Tucci, P.; Bagetta, G. How to study neuroprotection? *Cell. Death Differ.* **2008**, *15*, 1084–1085. [CrossRef]

42. Pellicciari, R.; Costantino, G.; Marinozzi, M.; Natalini, B. Modulation of glutamate receptor pathways in the search for new neuroprotective agents. *Farmaco* **1998**, *53*, 255–261. [CrossRef]

43. Behl, C.; Moosmann, B. Antioxidant neuroprotection in Alzheimer's disease as preventive and therapeutic approach. *Free Radic. Biol. Med.* **2002**, *33*, 182–191. [CrossRef]

44. Agnello, D.; Bigini, P.; Villa, P.; Mennini, T.; Cerami, A.; Brines, M.L.; Ghezzi, P. Erythropoietin exerts an anti-inflammatory effect on the CNS in a model of experimental autoimmune encephalomyelitis. *Brain Res.* **2002**, *952*, 128–134. [CrossRef]

45. Gao, H.M.; Liu, B.; Zhang, W.; Hong, J.S. Novel anti-inflammatory therapy for Parkinson's disease. *Trends Pharmacol. Sci.* **2003**, *24*, 395–401. [CrossRef]

46. Volbracht, C.; van Beek, J.; Zhu, C.; Blomgren, K.; Leist, M. Neuroprotective properties of memantine in different in vitro and in vivo models of excitotoxicity. *Eur. J. Neurosci.* **2006**, *23*, 2611–2622. [CrossRef] [PubMed]

47. Yu, X.; An, L.; Wang, Y.; Zhao, H.; Gao, C. Neuroprotective effect of *Alpinia oxyphylla* Miq. fruits against glutamate-induced apoptosis in cortical neurons. *Toxicol. Lett.* **2003**, *144*, 205–212. [CrossRef]

48. Kietzmann, T.; Knabe, W.; Schmidt-Kastner, R. Hypoxia and hypoxia-inducible factor modulated gene expression in brain: Involvement in neuroprotection and cell death. *Eur. Arch. Psychiatry Clin. Neurosci.* **2001**, *251*, 170–178. [CrossRef] [PubMed]

49. Heurteaux, C.; Guy, N.; Laigle, C.; Blondeau, N.; Duprat, F.; Mazzuca, M.; Lang-Lazdunski, L.; Widmann, C.; Zanzouri, M.; Romey, G.; et al. TREK-1, a K+ channel involved in neuroprotection and general anesthesia. *EMBO J.* **2004**, *23*, 2684–2695. [PubMed]

50. Schwartz, G.; Fehlings, M.G. Evaluation of the neuroprotective effects of sodium channel blockers after spinal cord injury: Improved behavioral and neuroanatomical recovery with riluzole. *J. Neurosurg.* **2001**, *94* (Suppl. 2), 245–256. [CrossRef] [PubMed]

51. Youdim, M.B.; Fridkin, M.; Zheng, H. Novel bifunctional drugs targeting monoamine oxidase inhibition and iron chelation as an approach to neuroprotection in Parkinson's disease and other neurodegenerative diseases. *J. Neural Transm.* **2004**, *111*, 1455–1471. [CrossRef] [PubMed]

52. Gaeta, A.; Hider, R.C. The crucial role of metal ions in neurodegeneration: The basis for a promising therapeutic strategy. *Br. J. Pharmacol.* **2005**, *146*, 1041–1059. [PubMed]

53. Tremblay, R.; Hewitt, K.; Lesiuk, H.; Mealing, G.; Morley, P.; Durkin, J.P. Evidence that brain-derived neurotrophic factor neuroprotection is linked to its ability to reverse the NMDA-induced inactivation of protein kinase C in cortical neurons. *J. Neurochem.* **1999**, *72*, 102–111. [CrossRef] [PubMed]

54. Moalem, G.; Gdalyahu, A.; Shani, Y.; Otten, U.; Lazarovici, P.; Cohen, I.R.; Schwartz, M. Production of neurotrophins by activated T cells: Implications for neuroprotective autoimmunity. *J. Autoimmun.* **2000**, *15*, 331–345. [CrossRef] [PubMed]

55. Akerud, P.; Canals, J.M.; Snyder, E.Y.; Arenas, E. Neuroprotection through delivery of glial cell line-derived neurotrophic factor by neural stem cells in a mouse model of Parkinson's disease. *J. Neurosci.* **2001**, *21*, 8108–8118. [PubMed]

56. Woo, M.S.; Park, J.S.; Choi, I.Y.; Kim, W.K.; Kim, H.S. Inhibition of MMP-3 or -9 suppresses lipopolysaccharide-induced expression of proinflammatory cytokines and iNOS in microglia. *J. Neurochem.* **2008**, *106*, 770–780. [CrossRef] [PubMed]

57. Chandrasekaran, K.; Mehrabian, Z.; Spinnewyn, B.; Chinopoulos, C.; Drieu, K.; Fiskum, G. Neuroprotective effects of bilobalide, a component of Ginkgo biloba extract (EGb 761) in global brain ischemia and in excitotoxicity-induced neuronal death. *Pharmacopsychiatry* **2003**, *36* (Suppl. 1), S89–S94. [PubMed]

58. Handley, O.J.; Naji, J.J.; Dunnett, S.B.; Rosser, A.E. Pharmaceutical, cellular and genetic therapies for Huntington's disease. *Clin. Sci. (Lond.)* **2006**, *110*, 73–88. [CrossRef] [PubMed]

59. Selkoe, D.J.; Hardy, J. The amyloid hypothesis of Alzheimer's disease at 25 years. *EMBO Mol. Med.* **2016**, *8*, 595–608. [CrossRef] [PubMed]

60. Han, Z.; Zeng, Y.; Lu, H.; Zhang, L. Determination of the degree of acetylation and the distribution of acetyl groups in chitosan by HPLC analysis of nitrous acid degraded and PMP labeled products. *Carbohydr. Res.* **2015**, *413*, 75–84. [CrossRef] [PubMed]

61. Dai, X.; Chang, P.; Zhu, Q.; Liu, W.; Sun, Y.; Zhu, S.; Jiang, Z. Chitosan oligosaccharides protect rat primary hippocampal neurons from oligomeric beta-amyloid 1–42-induced neurotoxicity. *Neurosci. Lett.* **2013**, *554*, 64–69. [CrossRef] [PubMed]

62. Jia, S.; Lu, Z.; Gao, Z.; An, J.; Wu, X.; Li, X.; Dai, X.; Zheng, Q.; Sun, Y. Chitosan oligosaccharides alleviate cognitive deficits in an amyloid-beta1–42-induced rat model of Alzheimer's disease. *Int. J. Biol. Macromol.* **2016**, *83*, 416–425. [CrossRef] [PubMed]

63. Wang, X.; Miao, J.; Yan, C.; Ge, R.; Liang, T.; Liu, E.; Li, Q. Chitosan attenuates dibutyltin-induced apoptosis in PC12 cells through inhibition of the mitochondria-dependent pathway. *Carbohydr. Polym.* **2016**, *151*, 996–1005. [PubMed]

64. Xu, W.; Huang, H.C.; Lin, C.J.; Jiang, Z.F. Chitooligosaccharides protect rat cortical neurons against copper induced damage by attenuating intracellular level of reactive oxygen species. *Bioorg. Med. Chem. Lett.* **2010**, *20*, 3084–308. [CrossRef] [PubMed]

65. Cho, Y.; Shi, R.; Borgens, R.B. Chitosan produces potent neuroprotection and physiological recovery following traumatic spinal cord injury. *J. Exp. Biol.* **2010**, *213 Pt 9*, 1513–1520. [CrossRef] [PubMed]

66. Gong, Y.; Gong, L.; Gu, X.; Ding, F. Chitooligosaccharides promote peripheral nerve regeneration in a rabbit common peroneal nerve crush injury model. *Microsurgery* **2009**, *29*, 650–656. [CrossRef] [PubMed]

67. Jiang, M.; Zhuge, X.; Yang, Y.; Gu, X.; Ding, F. The promotion of peripheral nerve regeneration by chitooligosaccharides in the rat nerve crush injury model. *Neurosci. Lett.* **2009**, *454*, 239–243. [CrossRef] [PubMed]

68. Khodagholi, F.; Eftekharzadeh, B.; Maghsoudi, N.; Rezaei, P.F. Chitosan prevents oxidative stress-induced amyloid beta formation and cytotoxicity in NT2 neurons: Involvement of transcription factors Nrf2 and NF-kappaB. *Mol. Cell Biochem.* **2010**, *337*, 39–51. [CrossRef] [PubMed]

69. Evin, G. Future Therapeutics in Alzheimer's Disease: Development Status of BACE Inhibitors. *BioDrugs* **2016**, *30*, 173–194. [CrossRef] [PubMed]

70. Lukiw, W.J. Emerging amyloid beta (Ab) peptide modulators for the treatment of Alzheimer's disease (AD). *Expert Opin. Emerg. Drugs* **2008**, *13*, 255–271. [CrossRef] [PubMed]

71. Okamura, N.; Suemoto, T.; Shiomitsu, T.; Suzuki, M.; Shimadzu, H.; Akatsu, H.; Yamamoto, T.; Arai, H.; Sasaki, H.; Yanai, K.; et al. A novel imaging probe for in vivo detection of neuritic and diffuse amyloid plaques in the brain. *J. Mol. Neurosci.* **2004**, *24*, 247–255. [CrossRef]

72. Hampel, H.; Shen, Y. Beta-site amyloid precursor protein cleaving enzyme 1 (BACE1) as a biological candidate marker of Alzheimer's disease. *Scand. J. Clin. Lab. Investig.* **2009**, *69*, 8–12. [CrossRef] [PubMed]

73. Dai, X.; Hou, W.; Sun, Y.; Gao, Z.; Zhu, S.; Jiang, Z. Chitosan Oligosaccharides Inhibit/Disaggregate Fibrils and Attenuate Amyloid beta-Mediated Neurotoxicity. *Int. J. Mol. Sci.* **2015**, *16*, 10526–10536. [CrossRef] [PubMed]

74. Je, J.Y.; Kim, S.K. Water-soluble chitosan derivatives as a BACE1 inhibitor. *Bioorg. Med. Chem.* **2005**, *13*, 6551–6555. [CrossRef] [PubMed]

75. Byun, H.-G.; Kim, Y.-T.; Park, P.-J.; Lin, X.; Kim, S.-K. Chitooligosaccharides as a novel β-secretase inhibitor. *Carbohydr. Polym.* **2005**, *61*, 198–202. [CrossRef]

76. Kim, Y.S.; Joh, T.H. Microglia, major player in the brain inflammation: Their roles in the pathogenesis of Parkinson's disease. *Exp. Mol. Med.* **2006**, *38*, 333–347. [CrossRef] [PubMed]

77. Kim, M.S.; Sung, M.J.; Seo, S.B.; Yoo, S.J.; Lim, W.K.; Kim, H.M. Water-soluble chitosan inhibits the production of pro-inflammatory cytokine in human astrocytoma cells activated by amyloid beta peptide and interleukin-1beta. *Neurosci. Lett.* **2002**, *321*, 105–109. [CrossRef]

78. Fang, I.M.; Yang, C.M.; Yang, C.H. Chitosan oligosaccharides prevented retinal ischemia and reperfusion injury via reduced oxidative stress and inflammation in rats. *Exp. Eye Res.* **2015**, *130*, 38–50. [CrossRef] [PubMed]

79. Twomey, C.; McCarthy, J.V. Pathways of apoptosis and importance in development. *J. Cell. Mol. Med.* **2005**, *9*, 345–359. [CrossRef] [PubMed]

80. Fadeel, B.; Orrenius, S. Apoptosis: A basic biological phenomenon with wide-ranging implications in human disease. *J. Intern. Med.* **2005**, *258*, 479–517. [CrossRef] [PubMed]
81. Vila, M.; Przedborski, S. Targeting programmed cell death in neurodegenerative diseases. *Nat. Rev. Neurosci.* **2003**, *4*, 365–375. [CrossRef] [PubMed]
82. Koo, H.N.; Jeong, H.J.; Hong, S.H.; Choi, J.H.; An, N.H.; Kim, H.M. High molecular weight water-soluble chitosan protects against apoptosis induced by serum starvation in human astrocytes. *J. Nutr. Biochem.* **2002**, *13*, 245–249. [CrossRef]
83. Tabet, N. Acetylcholinesterase inhibitors for Alzheimer's disease: Anti-inflammatories in acetylcholine clothing! *Age Ageing* **2006**, *35*, 336–338. [CrossRef] [PubMed]
84. Terry, A.V., Jr.; Buccafusco, J.J. The cholinergic hypothesis of age and Alzheimer's disease-related cognitive deficits: Recent challenges and their implications for novel drug development. *J. Pharmacol. Exp. Ther.* **2003**, *306*, 821–827. [CrossRef] [PubMed]
85. Ibrahim, F.; Andre, C.; Thomassin, M.; Guillaume, Y.C. Association mechanism of four acetylcholinesterase inhibitors (AChEIs) with human serum albumin: A biochromatographic approach. *J. Pharm. Biomed. Anal.* **2008**, *48*, 1345–1350. [CrossRef] [PubMed]
86. Martinez, A.; Castro, A. Novel cholinesterase inhibitors as future effective drugs for the treatment of Alzheimer's disease. *Expert Opin. Investig. Drugs* **2006**, *15*, 1–12. [CrossRef] [PubMed]
87. Yoon, N.Y.; Ngo, D.-N.; Kim, S.-K. Acetylcholinesterase inhibitory activity of novel chitooligosaccharide derivatives. *Carbohydr. Polym.* **2009**, *78*, 869–872. [CrossRef]
88. Clinical Trials. Available online: https://clinicaltrials.gov (accessed on 8 November 2016).
89. Stern, R. Go Fly a Chitin: The Mystery of Chitin and Chitinases in Vertebrate Tissues. *Front. Biosci. (Landmark Ed.)* **2017**, *22*, 580–595. [CrossRef] [PubMed]

marine drugs

MDPI

Article

Controlling Properties and Cytotoxicity of Chitosan Nanocapsules by Chemical Grafting

Laura De Matteis [1,*], **Maria Alleva** [1], **Inés Serrano-Sevilla** [1,2], **Sonia García-Embid** [1], **Grazyna Stepien** [1], **María Moros** [3] **and Jesús M. de la Fuente** [2,*]

[1] Instituto de Nanociencia de Aragón (INA), Universidad de Zaragoza, Edificio I+D, calle Mariano Esquillor s/n, 50018 Zaragoza, Spain; mariaalleva1234@gmail.com (M.A.); inessersev@gmail.com (I.S.-S.); sonia.garcia.embid@gmail.com (S.G.-E.); gstepien@unizar.es (G.S.)

[2] Instituto de Ciencia de Materiales de Aragón (ICMA), CSIC-Universidad de Zaragoza, Edificio I+D, calle Mariano Esquillor s/n, 50018 Zaragoza, Spain

[3] Istituto di Scienze Applicate e Sistemi Intelligenti "E. Caianiello", Consiglio Nazionale delle Ricerche, Pozzuoli 80078, Italy; m.moros@isasi.cnr.it

* Correspondence: lauradem@unizar.es (L.D.M.); jmfuente@unizar.es (J.M.d.l.F.);
Tel.: +34-876-555-433 (L.D.M.); +34-606-949-073 (J.M.d.l.F.)

Academic Editors: David Harding and Hitoshi Sashiwa
Received: 28 July 2016; Accepted: 20 September 2016; Published: 30 September 2016

Abstract: The tunability of the properties of chitosan-based carriers opens new ways for the application of drugs with low water-stability or high adverse effects. In this work, the combination of a nanoemulsion with a chitosan hydrogel coating and the following poly (ethylene glycol) (PEG) grafting is proven to be a promising strategy to obtain a flexible and versatile nanocarrier with an improved stability. Thanks to chitosan amino groups, a new easy and reproducible method to obtain nanocapsule grafting with PEG has been developed in this work, allowing a very good control and tunability of the properties of nanocapsule surface. Two different PEG densities of coverage are studied and the nanocapsule systems obtained are characterized at all steps of the optimization in terms of diameter, Z potential and surface charge (amino group analysis). Results obtained are compatible with a conformation of PEG molecules laying adsorbed on nanoparticle surface after covalent linking through their amino terminal moiety. An improvement in nanocapsule stability in physiological medium is observed with the highest PEG coverage density obtained. Cytotoxicity tests also demonstrate that grafting with PEG is an effective strategy to modulate the cytotoxicity of developed nanocapsules. Such results indicate the suitability of chitosan as protective coating for future studies oriented toward drug delivery.

Keywords: chitosan; hydrogel; surface grafting; nanocapsules; stability

1. Introduction

In the last few decades, many kinds of nanocarriers have been developed for delivery and targeting of therapeutic or diagnostic agents, thanks to some important advantages that they offer depending on their physico-chemical properties [1,2].

Depending on the specific needs, the nanocarrier type and formulation process must be chosen on the basis of therapy goals and administration route [3]. The most common nanocarriers can be classified as follows: solid (inorganic or organic) nanoparticles [4,5], nanospheres (polymeric matrices or hydrogels) [6–8], or nanocapsules (usually liposomes, emulsion-based, or protein-based nanocapsules) [9–11].

According to Vrignaud and co-workers, nanocapsules are vesicular systems, composed of an oily or an aqueous core that can be considered as a reservoir in which the drug is confined to a cavity, surrounded by a polymeric shell [12]. Nanocapsules can be obtained combining nanoemulsion and a polymeric coating. Nanoemulsion particles are stable colloidal suspensions obtained by mixing an organic phase containing oil and a lipophilic surfactant with an aqueous one containing a hydrophilic surfactant, resulting in a particle size ranging from 20 to 600 nm. The characteristics of the obtained particles depend on the spontaneity of the emulsification process that is affected by the nature of the single components of the reaction mixture and also by the rate of the mixing process [13,14]. Depending on the desired application, a coating is necessary to further stabilize the nanoscaled particles resulting from this spontaneous process and improve surface properties. The most commonly used coatings are natural polymers, which are deposited on the nanoemulsion template surface to produce a rigid and dense shell [15]. The shell can be easily tailor-made to achieve desired characteristics and its surface chemistry can be tuned to obtain a proper functionalization for biological targeting [16–18].

Natural polymers are among the most used for these kind of coatings since they usually provide a high colloidal stability in water suspensions. Active research is now focused on the use of hydrophilic biopolymers as carrier coatings because of their biocompatibility and biodegradability [19,20]. Chitosan (CS) has been used for the development of sustained release carriers, mucoadhesive formulations, and peptide drug absorption systems [21–24]. It is currently employed to prepare nanomaterials with mucoadhesive properties since its positive charges allow the interaction of particles with the negative charge of mucin, resulting in a better interaction with mucosal tissues and with epithelial cells. Moreover, it is known that the positive charge of the polymer can promote the paracellular transport by tight-junction regulation [25,26].

In this work, core-shell nanocapsules made of a nanoemulsion core and a chitosan shell were synthesized and characterized with the aim of obtaining a multipocket nano-reservoir carrier to be used in future applications for sustained release of different drugs. The secondary effects—toxicity, poor solubility, and bioavailability—of new drugs lead to the need of their encapsulation to protect them from degradation and to enhance their stability and solubility [27,28].

Thanks to the presence of chitosan amino groups, the surface of the obtained nanocarrier can also be grafted with specific moieties in order to tune the net charge to introduce specific functional groups and/or improve the carrier stability in biological media and physiologic solutions for intravenous administration [29–33]. The development of a smart nanocarrier is strictly related with controlling its surface properties since they are responsible for specific recognition of targeted sites but also for non-specific adsorption of serum proteins. A decrease in protein adsorption leads to a reduced uptake by the mononuclear phagocytic system, leading to a prolonged circulation time in the blood stream and to a higher residence time of the encapsulated drug. Poly (ethylene glycol) (PEG) coatings are known to prevent aggregation and serum protein adsorption by steric and hydration repulsions leading to more stable colloidal suspensions of nanocapsules in physiological media [33–35]. In this work, the surface of the developed chitosan-coated nanocapsules was grafted covalently with PEG molecules through a novel, simple, and reproducible strategy based on the use of a homobifunctional crosslinker, bis (sulfosuccinimidyl) suberate (BS^3), that links aminated PEG molecules to amino groups on nanocapsule surface. Nanocapsule behavior in different media was evaluated in terms of aggregation degree before and after grafting and the effect of PEG on their cytotoxicity was also assessed.

2. Results

2.1. Chitosan-Coated Nanocapsules

A nanoemulsion method was developed to obtain small nanoparticles (smaller than 200 nm) to be used as a template for the following polymer coating and reinforcement. The aim was to obtain capsules with a lipophilic core and a hydrophilic cationic shell of chitosan hydrogel.

The synthesis of the nanocapsules was carried out in two steps. The formation of the nanoemulsion template particles was carried out simply by adding a water-miscible organic solution of Span® 85/oleic acid (Croda International PLC, Cowick Hall Snaith, Goole, East Yorkshire, UK) to a Tween® 20

(Croda International PLC, Cowick Hall Snaith, Goole, East Yorkshire, UK) aqueous solution under stirring. Optimal ratios between components have been found adapting a method reported by Bouchemal et al. [14]. The particles are immediately and spontaneously formed. Chitosan has been chosen for the coating of the nanoemulsion template since it is one of the richest in amino groups' natural polymers.

The presence of such functional groups is responsible for the ability of the polymer to gelify in presence of multi-anions leading to the formation of hydrogel. Moreover, the amino groups exposed on hydrogel shell surface allow the easy functionalization of the nanocapsule with the desired moiety.

Chitosan is directly added to the nanoemulsion that is subsequently mixed with a sodium sulfate solution to obtain the coating with a chitosan shell. This treatment has been used in several works to obtain chitosan particles [36–38]. In our case, the method allowed to obtain a hydrogel polymer shell as a result of the interaction of chitosan polyelectrolyte structure with sodium sulfate. Sodium sulfate acts as a bridge promoting interactions between polymeric chains.

The result of the synthesis is a water-stable suspension of chitosan-coated nanocapsules (CS-NCs).

A schematic representation of the hypothesized nanocapsule structure is reported in Figure 1A. Moreover, to investigate the morphology of the obtained material, an electron microscopy characterization was carried out on chitosan-coated nanocapsules using both Bright Field Transmission Electron Microscopy (BF-TEM) and Environmental Scanning Electron Microscopy (ESEM) (Figure 1B,C respectively). Due to the sensitive nature of the sample common in soft materials, previous fixation, dehydration, dyeing and resin embedding were necessary.

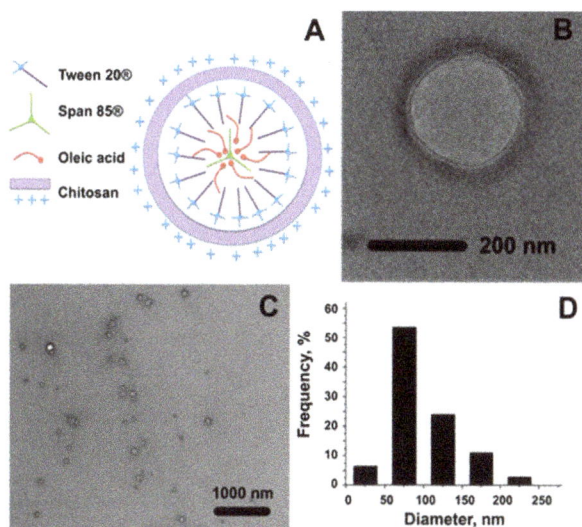

Figure 1. Morphological characterization of chitosan-coated nanocapsules (CS-NCs). (**A**) Schematic representation of a section of a nanoemulsion-based and chitosan-coated nanocapsule; (**B**) Bright Field Transmission Electron Microscopy (BF-TEM) image and (**C**) Environmental Scanning Electron Microscopy (ESEM) image of nanocapsules from a section of the epoxy resin block; (**D**) representation of frequency count analysis of size distribution.

The sample turned out to be composed of spherical capsules of a quite homogeneous size. The polymeric shell can be appreciated in BF-TEM and ESEM images (Figure 1B,C). ESEM estimation of the diameter distribution is reported in Figure 1D. The number of capsules of different sizes, as percentages over the total number of measured capsules, is reported in the graph as a function of the diameter. The calculated mean diameter of NCs is 104 nm.

As optimization of the process, the importance of the sonication treatment during hydrogel shell formation was evaluated by substituting it with a gentle stirring while adding nanocapsules to Na_2SO_4 solution. The elimination of the sonication step in the synthesis process could represent an advantage in terms of applicability of the process for future applications to industrial production with high levels of scale-up. Nanocapsules obtained with the modified process have been characterized in terms of hydrodynamic diameter and surface charge and they have been compared with the sonicated ones. In Figure 2, Dynamic Light Scattering (DLS) measures of the hydrodynamic diameter of the nanoemulsion template just before chitosan coating (A), nanocapsules obtained by sonication (sCS-NCs) (B), and non-sonicated nanocapsules obtained by stirring (nsCS-NC) (C) are reported and compared.

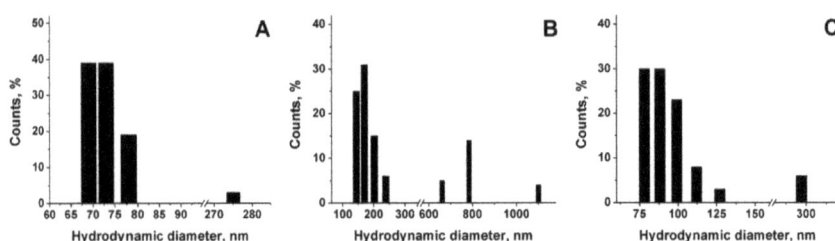

Figure 2. Dynamic Light Scattering (DLS) measures reported as percentages of frequency counts at each hydrodynamic diameter: nanoemulsion before chitosan coating (**A**); nanocapsules obtained by sonication (sCS-NCs) (**B**); and non-sonicated nanocapsules obtained by stirring (nsCS-NC) (**C**).

Hydrodynamic diameter of chitosan–coated nanocapsules always increases with respect to the nanoemulsion template (Figure 2A). Nevertheless, in the case of sonicated capsules the final hydrodynamic diameter is much higher than the non-sonicated sample (Figure 2B). The value is also higher than the diameter distribution obtained from SEM images referring to sonicated nanocapsules. Moreover, the presence of a significant percentage of big aggregates is observed in the sCS-NC. The increase in the diameter observed can be attributed to aggregation phenomena, indicating a lower stability of this sample in water suspension. In the case of non-sonicated samples, a certain amount of aggregates are detected too, even though such aggregates have a smaller diameter than the ones obtained by sonication. In any case, the presence of a low percentage (i.e., 5%–10%) of aggregates could be considered acceptable for future purposes. The hydrodynamic diameter of the nanocapsules stored in water suspension has been found to be reproducible over at least two months, indicating the suitability of chitosan to successfully stabilize the nanoemulsion.

It is supposed that the hydrogel formation treatment affects the properties of the polymer-coated surface of the nanocapsules depending on the degree of the interaction that could be established between $-NH_2$ and $-OH$ on polymer chains and Na_2SO_4 molecules. Moreover, it should be taken into account that the presence of salts can also promote hydrophobic interactions between polymer chains themselves. As a consequence of the establishment of such hydrophobic interactions, the exposure amino groups on the surface of the nanocapsule would be favored. To evaluate how the sonication during hydrogel formation can improve the interactions and so affect the nanocapsule surface properties, sonicated and non-sonicated nanocapsules have been compared in terms of surface potential. In particular, analysis of Z-potential and amino group spectrophotometric determination analysis are reported. Both capsules showed a positive potential when measured in a 10 mM KCl solution, being slightly more positive the surface of sCS-NC (+21.1 mV) than the surface of nsCS-NC (+13.8 mV). The observed variation in the surface potential depending on the sonication treatment could be explained in this case by the presence of a higher number of amino groups exposed on the outer surface of chitosan shell in the case of sCS-NC. This hypothesis was confirmed by the quantification of amino groups of the chitosan shell of nanocapsules by the spectrophotometric method of Orange II dye, previously reported and already optimized for inorganic nanoparticles [39]. Briefly, the method is based on a pH-dependent interaction between positively charged amino groups and –SO^{3-} group of Orange II dye. This spectrophotometric method is simple, inexpensive, and easy, and its most important advantage over other methods for amino quantification consists in the use of a molecule with low steric hindrance—an especially important aspect for porous materials materials. Moreover, the relationship between amino groups and reactant is in this case of 1:1, allowing a direct and reliable quantification [40]. In the present work, the method has been slightly modified to use syringe filters as support for the separation of nanocapsules from the solution during all the washing steps. Results from the spectrophotometric assay for the measurement of amino groups through the interaction with Orange II dye also proved the presence of a high number of positively charged amino groups in the case of sCS-NC (0.35 μmol·mg^{-1}) while a value of amino groups of only 0.2 μmol·mg^{-1} was obtained in the case of nsCS-NC.

It should be taken into account that results from Orange II interaction only represented the amount of amino groups available for the interaction with dye molecules and that the amount of these groups could be lower than the total amount of –NH$_2$ of the nanocapsules. Data from the Orange II assay were in good agreement with Z potential analysis. sCS-NC presented a number of moles mg^{-1} of –NH$_2$ groups almost double with respect to the moles mg^{-1} of –NH$_2$ of nsCS-NC.

Thermo Gravimetric Analysis (TGA) of chitosan-based nanocapsules sonicated or not during the synthesis process is reported in Figure S3 of the Supplementary Materials. The weight loss corresponding to chitosan shell can be appreciated at ~200 °C and it corresponded to 50% weight of the sCS-NC. Surprisingly in the case of nsCS-NC this percentage is even higher, where chitosan represents 65% of the total weight of the sample. This evidence demonstrated that the observed decrease of amino groups on the surface of non-sonicated nanocapsules was not due to a decrease in the chitosan total amount. Consequently, it is reasonable to suppose that the organization of chitosan layer and interactions between polymer chains themselves and with the surface of the nanoemulsion template are slightly different depending on the condition applied during the synthesis process.

2.2. Grafting of the Surface of Chitosan-Coated Nanocapsule

PEG coating has been chosen in this work for the further optimization of the carrier. As previously stated in the introduction and well documented in the literature, the prevention of unspecific adsorption of serum proteins on carrier surface represents a very important goal to allow a prolonged circulation time in the blood stream. From this point of view, pegylation is reported to be one of the most successful strategies [33–35].

Nanocapsules obtained without sonication during the synthesis have been selected for the process of grafting of the surface due to their better stability in water suspension. The surface was grafted with α-methoxy-ω-amino poly (ethylene glycol) (5000 Da) (aminated PEG) through a previous reaction with the homobifunctional linker bis (sulfosuccinimidyl) suberate (BS^3) (Scheme 1). As the colloidal stability of nanocapsules strongly depends on their surface properties, which in turn greatly affect nanocapsule cytotoxicity in animal cells, the possibility of obtaining nanocapsule surface with a low-density coverage of PEG molecules (5 nmol/mg initially added) (ldCS-NC) and a high-density coverage (100 nmol/mg initially added) (hdCS-NC) was explored. Consecutively, the final properties of the obtained materials were studied. A schematical representation of the strategy used for grafting is reported in Scheme 1.

The determination of the accessible amino groups was fundamental for the optimization of the grafting protocol. In fact, the method utilized here implied the use of a homobifunctional linker in which the main drawback lays in the possibility of crosslinking of amino groups from different nanocapsules and production of aggregates if the amount of reactants is not strictly controlled. For this reason, the value of accessible amino groups has been used as the starting point to adjust the amount of reactants in the following steps of the process. In particular, the amount of BS^3 has been always maintained below the total moles of functional groups of nanocapsules to reduce the possibility of crosslinking between capsules. To complete the process, the amount of aminated PEG exceeded twice the used amount of BS^3 in all the experiments to assure the reaction with all the reactive groups of BS^3 on the surface.

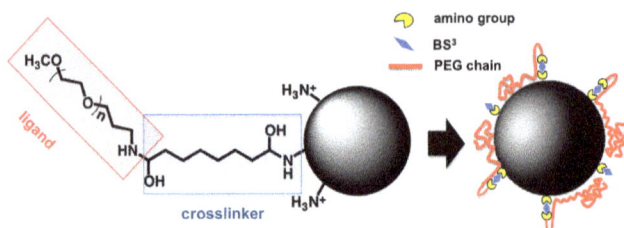

Scheme 1. Chitosan-coated nanocapsule grafted with aminated polyethylene glycol (PEG) through BS^3 linking.

The presence of PEG on nanocapsule surface has been evaluated by Fourier Transform Infrared Spectroscopy (FTIR) analysis (data reported in Supplementary Materials). The properties of chitosan-coated nanocapsules before and after the grafting have been compared and the FTIR spectra of all the single components of nanoemulsion-based nanocapsules are reported respectively in Figures S4 and S5 of the Supplementary Materials.

In the FTIR spectrum of nsCS-NC (Figure S4A), it is possible to recognize the presence of the four starting components, meaning that the final composition is compatible with the hypothesized structure of the nanocapsule (Figure 1A). Moreover, reported in Figure S6 of the Supplementary Materials a comparison of sCS-NCs and nsCS-NC demonstrates that the sonication process is not affecting the chemical interaction between nanoemulsion template and chitosan shell.

The grafting process consists of three steps. The first one is the incubation with the linker to provide the surface with the sulfosuccinimidyl ester group sensitive to the linking of PEG. During the second step the ligand is added and finally the grafted nanocapsules are incubated with Tris-HCl buffer to quench the free sulfosuccinimidyl ester groups eventually not linked to any PEG molecule. This last step is fundamental to avoid the crosslinking between nanocapsules due to the reaction between free sulfosuccinimidyl ester groups and amino group on a different capsule.

To determine if the grafting with PEG successfully occurred, the FTIR spectrum of the intermediate step of nanocapsules after the incubation with BS^3 is also reported (Figure S4B). It should be noted that at this intermediate step the sample is incubated with Tris-HCl buffer after the linking of BS^3 to avoid the crosslinking with amino groups on other nanocapsule surface through the reactive groups still available for ligand reaction. In the case of samples incubated with ligand in the second step, Tris-HCl was added at the end of the process (after incubation with the ligand) but in this case it can be supposed that BS^3, which should have reacted previously with PEG, is not available to react with Tris. This hypothesis was confirmed by comparing FTIR spectra of nanocapsules after the incubation with BS^3 (Figure S4B) and spectra in Figure S4C,D referring to nanocapsules grafted with a low density and high density of PEG, respectively. Peaks at 3180 and 3100 cm^{-1}, as well as peaks at 1630 and 1550 cm^{-1} referring to Tris are present in the intermediate step and (in ldCS-NC with lower intensities) but they completely disappeared in hdCS-NC spectrum.

Differences in the spectra could also be appreciated, comparing peaks and intensities ratios between 1150 and 1030 cm^{-1}. In this region the C-O-C peak (1105 cm^{-1}) presents an increased intensity, compared to other peaks in the same region, especially in the hdCS-NC spectrum. Moreover, in this sample peak at 1055 cm^{-1} is lost. This peak is supposed to refer to R-SO^{3-} group that is released from BS^3 in the linking reaction. In the BS^3 intermediate step the peak is still appreciable and it can be observed as a low intensity peak in ldCS-NCs too. It is possible that in these samples not all BS^3 reactive groups are quenched after Tris-HCl incubation and that some of them are still present on the surface. In any case this state did not represent a problem for the quality of the sample since any appreciable crosslinking and consequent aggregation are observed.

The successful grafting on nanocapsule surface was proved also by measuring the Z potential of the material surface before and after the modification with PEG. Using both a low amount and high amount of PEG, a decrease in the potential was observed, indicating a decrease in the free amino groups on the surface and an increase in the total electronegativity of the surface.

A confirmation of this evidence is the small difference in the decrease of amino groups (measured by Orange II spectrophotometric assay) registered between low-density and high-density PEG-covered nanocapsules. The measured amounts of amino groups are 0.1 and 0.08 µmol/mg respectively, corresponding to 50% and 40% of the amounts of amino groups on ungrafted nanocapsules. These values are considered as the apparent disappearing of amino groups probably due to a screening effect for the presence of PEG molecules on the surface. Orange molecule interaction with $-NH_3^+$ groups would be hampered by the presence of the polymer chains since, in the case of low-density PEG coverage, it can be supposed that polymer molecules stand in a conformation that lays around nanocapsule surface, probably thanks to a possible hydrogen bond interaction of the PEG chain with positively charged amino groups on the nanocapsule surface (see Scheme 1). The tendency to form such interactions could also be responsible for a lower than expected efficiency of linking in the case of high-density coverage sample.

The effect of the grafting on the behavior of nanocapsules in physiological medium was further evaluated by measuring the degree of aggregation of grafted and not-grafted nanocapsules in water and phosphate saline buffer (PBS) by means of hydrodynamic diameter measurement. Ungrafted nanocapsules are sensitive to the presence of salts in the medium and they tend to aggregate during incubation in physiological media like PBS. The grafting with PEG chains is a common strategy to improve the stability of nanoparticles in physiological and in vitro culture media [33]. Both phosphate ions and proteins can adsorb onto nanocapsule surface due to the presence of amino groups, producing the crosslinking between different capsules. Moreover, the presence of salts can produce aggregation due to a salting-out-like effect. The presence of PEG on nanocapsule surface would screen the amino groups on the surface from interaction with salts in the medium. All samples were measured in water and PBS and their diameters were compared to assess the effect of PEG on prevention of aggregation (Figure 3).

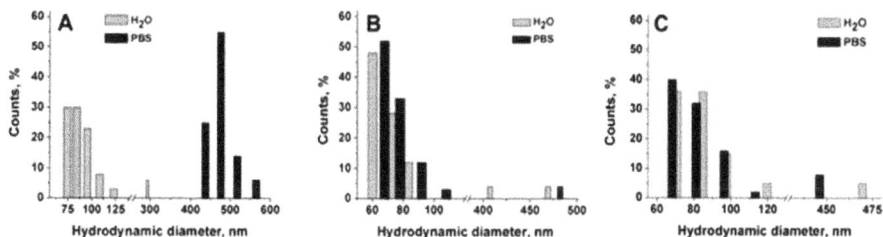

Figure 3. Hydrodynamic diameters of chitosan-coated nanocapsules before (**A**) and after grafting with different amounts of PEG ((**B**) and (**C**), respectively referring to low-density coverage (ldCS-NC) and high-density coverage (hdCS-NC)).

In Figure 3A, a strong aggregation effect in PBS is reported for ungrafted chitosan-based nanocapsules. The mean hydrodynamic diameter changed from 103 nm in water to 471 nm in PBS. On the contrary, the diameter and the PDI of grafted nanocapsules are maintained in PBS, even in the case of ldCS-NCs, confirming that the grafting successfully occurred in both cases and that it was effective for nanocapsule stabilization.

To further demonstrate the stabilizing effect of PEG grafting on nanocapsule surface, an aggregation test has been carried out by measuring the hydrodynamic diameter of grafted and ungrafted nanocapsules in the presence of increasing concentrations of Bovine Serum Albumin (BSA). As in the case of the above reported stability assay in presence of PBS, the high density of amino groups on nanocapsule surface can be considered responsible for the adsorption of proteins and aggregation observed.

The comparison of the behaviors of ungrafted nanocapsules, ldPEG-CS NC and hdPEG-CS NC, has been reported in Figure 4.

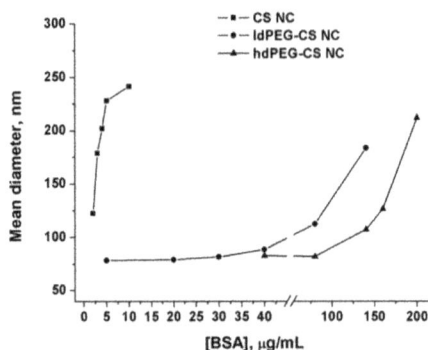

Figure 4. Hydrodynamic diameters of chitosan-coated nanocapsules before and after grafting with different amounts of PEG measured in presence of different concentrations of Bovine Serum Albumin (BSA).

Data reported in the graph demonstrated without any doubt the strong stabilizing effect of PEG on nanocapsules in the presence of proteins. The grafting with PEG, and so the strong decrease of the free amino groups on the surface lead to a very good stability of the nanocapsules in a protein-rich medium. Both ldPEG-CS NC and hdPEG-CS NC showed a significant increase in the hydrodynamic diameter only at very high concentrations of proteins (higher than 0.1 mg/mL). Comparatively, the concentration of BSA one order of magnitude lower is enough to produce the same increase in the diameter in the case of ungrafted NCs.

2.3. Cell Viabily and Internalization Assays

The cytotoxicity of chitosan-based nanocapsules before and after the grafting was tested on Vero cells through MTT spectrophotometric assay. Cells were incubated for 24 h with different concentrations of nsCS-NC, selected as ungrafted nanocapsules over the sonicated ones due to their better stability in water, and with the same concentrations of ldCS-NC and hdCS-NC. In Figure 5, the comparison between the grafted nanocapsules and the ungrafted ones is reported in terms of percentage of viability of cell culture after 24 h incubation.

It can be observed from Figure 5 that the grafting significantly improved the safety of the nanocapsules, especially when using high concentrations of nanocapsules (greater than 0.15 mg/mL). Both types of grafted nanocapsules were less toxic to the cells, especially the high density ones (at high concentrations).

Using an inverted microscope, it can be observed that non-sonicated nanocapsules lead to bigger vesicles inside cells than both types of grafted nanocapsules (Figure 6), which could be related with the higher toxicity).

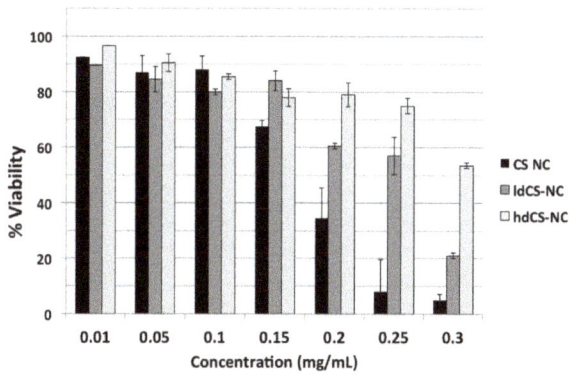

Figure 5. Viability test with chitosan-coated nanocapsules before and after grafting with low and high-density coverage of PEG.

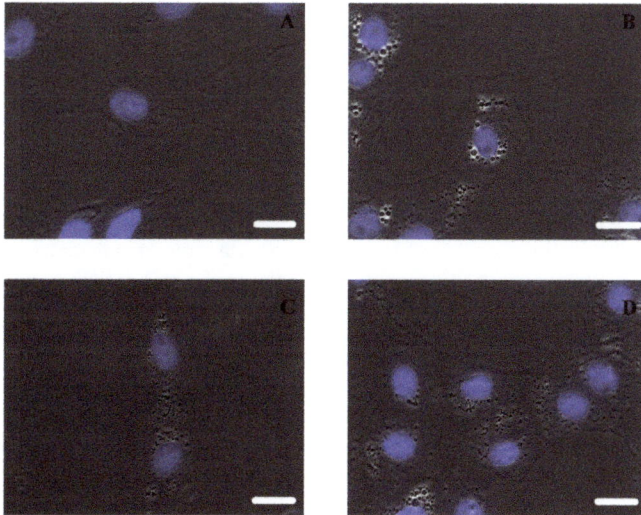

Figure 6. Inverted microscope images from (**A**) control cells; and cells incubated for 24 h with (**B**) non-sonicated nanocapsules; (**C**) ldCS-NC; and (**D**) hdCS-NC. Scale bars correspond to 20 μm.

The test proved not only that aminated PEG was definitively linked to nanocapsule surface, but also that the conformation of polymer chains obtained using the developed method of grafting for ldCS-NC hdCS-NC is effective for significantly improving the cytotoxicity of chitosan-based nanocapsules. Moreover, it should be noted that in the case of PEG-grafted nanocapsules the decrease in the cytotoxicity is still associated with a very high degree of internalization, indicating that the developed material can be considered a very effective carrier for drug delivery applications.

3. Discussion

Among all kinds of nanocarriers, nanocapsules are frequently the material of choice for biomedical applications since they offer the advantage of providing a good stability of encapsulated drugs and a favorable pharmacokinetic. The aim of the work was to develop nanocapsules to be used in the future as potential drug reservoir and whose composition could allow maximum flexibility of application together with great feasibility for the administration by different routes (intravenous injection, oral administration, and inhalation).

In this work, chitosan—a biocompatible polymer—has been used to obtain a potentially smart nanocarrier whose surface properties could be properly tuned depending on the desired application. A nanoemulsion-based nanocapsule has been used as template to be coated with a chitosan shell.

The process of chitosan hydrogel coating of a nanoemulsion template was studied in terms of exposure of amino groups on the surface depending on the interaction produced during the sonication process. It could be hypothesized that the sonication promoted interactions and produced stronger hydrophobic interaction between polymer chains, leading to a condensation of the structure that would lead to a higher exposure of amino groups and so to a higher availability for the interaction with the dye. On the other hand, it was also supposed that a different amount of polymer could be incorporated on the surface of the nanoemulsion template when sonication is used during the synthesis, leading to a different total amount of glucosamine units (and so amines) in the shell. The characterization of the material and the comparison between the sonicated and not sonicated one alloowed finally to confirm that the sonication process lead to a higher amount of amino groups exposed on nanocapsule surface.

Chitosan shell offers the possibility of an easy functionalization through amino groups on its surface so chitosan-based nanocapsules have been further coated with PEG since this is known to be a very effective strategy to stabilize surfaces in biologically relevant media. The possibility to tune the surface properties by varying the PEG density coverage has been explored and results obtained are compatible with a conformation of PEG molecules laying adsorbed on nanoparticle surface after covalent linking through their aminated terminal.

It is reported in literature that the attempt to produce high-density coverage of surfaces with PEG can result in a loss of efficiency of grafting. In fact, similarly to what reported for the grafting of other surfaces, it is reasonable to suppose that not all PEG molecules react immediately with BS^3 linker on nanocapsule surface. Once they are linked, their tendency is to assume a mushroom-like conformation until the density of PEG molecules on the surface does not reach a value high enough to produce a change into a brush-like conformation [41]. In the case of high-density PEG-covered nanocapsules, the decrease in amino groups on the surface was found to be not proportional to the reactants, differently to what happened in the case of low-density ones, indicating that eventually a conformation of PEG molecules laying adsorbed on nanoparticle surface was hampering the grafting of masked amino groups.

Despite the fact that the obtained grafting degree was slightly lower than the one expected, a highly improved stability in physiological medium (resistance to salting out effect) was observed with both low and high PEG-covered nanocapsules (although at different degrees) indicating that the presence of PEG on the surface will allow the use of the grafted capsules in physiological media for biomedical applications.

We also demonstrated that the presence of PEG on chitosan surface of nanocapsules was effective in avoiding the unspecific adsorption of proteins. This issue is of the utmost importance as, following intravenous injection, non-passivated nanoparticles tend to adsorb plasma proteins (protein corona), changing the charge and size of them. This binding of plasma components will be responsible of the final fate of the nanocapsules, and can greatly influence the biodistribution and therapeutic efficacy [42]. To date, PEG is the most widely used polymer to prevent this protein corona formation, although the final effect depends on different parameters such as the molecular weight or grafting density [43].

Following the encouraging results obtained in physiological media and protein-rich media, preliminary tests with cell cultures have been carried out showing very interesting results.

The obtained coating demonstrated effectiveness in tuning the cytotoxicity of chitosan-coated nanocapsules. Moreover, the nanocapsules showed a high degree of internalization in Vero cells, a useful property for the potential application as drug reservoirs directed to cytoplasm release. Finally, the tunability of all the properties of the synthesized chitosan-based nanocapsules was obtained by a very easy and reproducible chemical grafting, indicating that the developed nanocarrier was very promising for further studies oriented toward drug encapsulation for biomedical application.

4. Materials and Methods

Tween® 20 (Croda International PLC, Cowick Hall Snaith, Goole, East Yorkshire, UK), absolute ethanol, sodium sulfate anhydrous (99%–100.5%), and sodium chloride (99%), were purchased from Panreac. Span® 85 (sorbitanetrioleate) (Croda International PLC, Cowick Hall Snaith, Goole, East Yorkshire, UK), oleic acid (90%), chitosan (medium molecular weight), Orange II sodium salt, and Bovine Serum Albumine (BSA) were obtained from Sigma-Aldrich. Bis(sulfosuccinimidyl) suberate (BS3) was purchased from Pierce Biotechnology and α-methoxy-ω-amino poly(ethylene glycol) (PEG-MW 5000 Dalton) from IRIS Biotech GmbH. Water (double processed tissue culture) used in all nanocapsule synthesis was from Sigma. Millipore Biomax 300 kDa Ultrafiltration Discs were purchased from Merck Millipore.

Vero cells (monkey kidney epithelial cells) were purchased from the American Type Culture Collection (ATCC, Manassas, VA, USA, number CCL-81). Dulbecco's modified Eagle's medium (DMEM), Phosphate-Buffered Saline (PBS), Dulbecco's Phosphate-Buffered Saline (DPBS) were purchased from Lonza. MTT (3-(4,5-dimethylthiazolyl-2)-2,5-diphenyltetrazolium bromide), penicillin, streptomycin, glutamine solutions, and 4′,6-diamidino-2-fenilindolphenylindol (DAPI) were purchased from Invitrogen (Thermo Fisher Scientific, Waltham, MA, USA).

For the preparation of chitosan-based nanocapsules an organic solution containing 400 mg oleic acid and 86 mg Span® 85 in 40 mL of absolute ethanol was added to the aqueous one, containing 136 mg Tween® 20 solved in 80 mL water, under magnetic stirring during 15 min for the formation of the nanoemulsion. Then 25 mg from a 5 mg/mL chitosan solution in acetic acid 1% (v/v) were added and again the mixture was left under stirring 15 min. Finally, the chitosan-coated nanoemulsion was added to 200 mL of 50 mM Na_2SO_4 under sonication (or stirring in the case of optimized nanocapsules). Capsules were separated from Na_2SO_4 through ultracentrifugation (30 min, 69673 G, 10 °C), washed with 100 mL of water, centrifuged again, and resuspended in water. The concentration of the nanocapsules in water suspension was obtained by measuring the weight of 1 mL of sample after freeze-drying.

For the grafting of nanocapsule surface suspensions, 20 mg of nanocapsules at a concentration of 2 mg/mL in borate buffer 10 mM pH 8.3 were added with different amounts of the linker bis(sulfosuccinimidyl) suberate (BS3) (20–100 nmol/mgNC) and they were kept under stirring for 30 min. Then a double amount of α-methoxy-ω-amino poly (ethylene glycol) (MeO-PEG-NH$_2$) was added and the mixture was kept under stirring for 2 h at 37 °C. Finally, 20 mL of Tris-HCl buffer 10 mM pH 8.0 was added to quench the linker that eventually did not react with PEG. Grafted nanocapsules were filtered using an Amicon Ultrafiltration unit using Millipore Biomax 300 kDa Ultrafiltration Discs to separate them from unreacted PEG. After a washing with fresh water nanocapsules, they were concentrated to a final volume of 2 mL.

Several techniques have been used for the characterization of chitosan nanocapsules.

DLS analysis has been carried out using a Brookhaven 90Plus DLS instrument, by means of the Photo-Correlation Spectroscopy (PCS) technique. Nanoparticle hydrodynamic diameter and polydispersity index (PDI) have been measured in water at the concentration of 0.05 mg/mL.

Electrophoretic mobility (Z Potential) of nanoparticles at different pH values has been determined by measuring the potential of a 0.05 mg/mL nanoparticle suspension in 10 mM KCl with a Plus Particle Size Analyzer (Brookhaven Instruments Corporation).

Nanocapsule composition was analyzed by Fourier Transform Infrared Spectroscopy analysis in a JASCO FT/IR—4100 Fourier transform infrared spectrometer in a frequency range of 600–4000 cm^{-1} with a resolution of 2 cm^{-1} and a scanning number of 32.

Thermogravimetric analysis was performed in a TASTD 2960 thermogravimetric analyzer, by heating the sample at 10 °C/min under air atmosphere.

Environmental Scanning Electron Microscopy (ESEM) images were obtained using a QUANTA-FEG 250 microscope in Scanning Transmission Electron Microscopy (STEM) mode. Bright Field Transmission Electron Microscopy (BF-TEM) analysis was carried out in a FEI Tecnai T20 microscope operating at 200 kV. Due to the sensitive nature of the sample, previous fixation, dehydration, and epoxy resin embedding were necessary for both techniques. This process can be briefly achieved as follows. A fresh sample was synthesized and it was fixed with glutaraldehyde 0.25% in phosphate buffer 10 mM at pH 7.4 for 2 h. It was washed three times with buffer and incubated with 1% osmium tetroxide in PBS for further fixation and staining. Finally, the sample was accurately washed with water and resuspended in 5% gelatin. The sample was centrifuged to obtain a pellet and was incubated overnight at 4 °C. The obtained solid sample was cut in very small pieces before undergoing the subsequent steps. The dehydration of the samples was carried out using the following steps: incubation in ethanol 30%, ethanol 50%, and incubation overnight in ethanol 70%; incubation in ethanol 90% for 1 h; and finally incubation three times in absolute ethanol. After that, samples were incubated overnight in a 1:1 mixture absolute ethanol/epoxy resin (r.t.). The mixture was then removed, changed with absolute epoxy resin and samples were left for impregnation for 8 h at room temperature. After another change of the medium, the final incubation in epoxy resin was carried out overnight at 60 °C to obtain the polymerization.

From different ESEM images (Figures S1 and S2 of Supplementary Materials), an estimation of the diameter distribution has been obtained using Digital Micrograph® (Gatan Inc., Pleasanton, TX, USA) and OriginLab® (OriginLab, Northampton, MA, USA) softwares to measure the diameters of more than 100 nanocapsules and for the frequency count statistical analysis respectively.

Nanocapsule amino content was measured by the Orange II spectrophotometric assay [34,35]. 0.2 mg of nanocapsules were put in contact with 1 mL of 2 mM Orange II sodium salt acidic solution (pH 3) and kept under stirring for 30 min at 37 °C. Capsule suspension was passed through a syringe membrane filter (Millex syringe-driven filter unit, PVDF filter with 0.22 μm pores, purchased from Merck Millipore) to adsorb nanocapsules in the membrane and keep them retained in order to separate them from the Orange II solution. After that, an acidic solution (pH 3) was passed several times through the same filter until all the unbound dye was removed from the nanocapsules (verified by measuring spectrophotometrically the supernatant content). Then they were washed with an alkaline solution (pH 12) to desorb the bound dye from the amino groups on nanocapsules. The washing fractions were collected, the pH was adjusted at 3 and the amount of unadsorbed and desorbed dye was measured at a wave length of 480 nm with a Varian Cary 50 UV/V is spectrophotometer after carrying out a calibration curve.

Resistance of nanocapsules to aggregation has been determined by incubating nanocapsules (3 mL of a 0.15 mg/mL suspension) at different concentrations of albumin from bovine serum (BSA). The hydrodynamic diameter of the capsules under the incubation conditions was measured after 10 min using a Brookhaven 90Plus DLS instrument.

In vitro cell viability test was carried out to determine the cytotoxicity of nanocapsules using 3-(4,5-dimethylthiazol-2)-2,5-diphenyltetrazolium bromide (MTT) colorimetric assay. Vero cells were grown at 37 °C in a 5% CO_2 atmosphere in Dulbecco's modified Eagle's medium (DMEM) supplemented with 10% fetal bovine serum (FBS), penicillin (100 U/mL), streptomycin (100 μg/mL), and glutamine (2 mM). 7500 cells were seeded using a standard 96-well plate (five replicates per sample). After 24 h of incubation in a humidified atmosphere containing 5% CO_2, the medium was replaced with new medium containing five different concentrations of nanocapsules and a negative control containing no capsules (non-treated cells). After 24 h of incubation, the medium was replaced with fresh medium containing MTT dye solution (0.5 mg/mL in DMEM). After 2 h of incubation at 37 °C and 5% CO_2, the medium was removed and the formed crystals were dissolved in 200 μL of DMSO. The absorbance was read on a ThermoScientificMultiskan GO TM microplate reader at 570 nm. The relative cell viability (%) related to control cells without nanocapsules was calculated using the percentage ratio between absorbance of the sample and the absorbance of the control. Experiments were carried out in triplicate.

To perform optical microscopy analysis, 3×10^4 cells were seeded on glass coverslips in a 24-well plate at 37 °C. 24 h later, nanocapsules were added at 50 µg/mL in DMEM and incubated for 24 h at 37 °C. Non-internalized nanocapsules were removed, washing with DPBS twice. Cells were fixed with 4% paraformaldehyde for 20 min at 4 °C, washed twice with DPBS, and incubated for 10 min at room temperature with 4′,6-diamidino-2-fenilindolphenylindole (DAPI) for nucleus labeling. The coverslips were mounted on glass microscope slides using ProLong® (Thermo Fisher Scientific Inc., Waltham, MA, USA) Gold Antifade. Optical microscopy was performed using an inverted microscope (Nikon Eclipse Ti-E), and images were analyzed using NIS-Elements Advanced Research software.

Supplementary Materials: The following are available online at http://www.mdpi.com/1660-3397/14/10/175/s1, Figure S1: ESEM image of nanocapsules, Figure S2: ESEM image of nanocapsules, Figure S3: TGA analysis of nanocapsules, Figure S4: FTIR analysis of grafted nanocapsules, Figure S5: FTIR analysis of starting compounds, Figure S6: FTIR analysis of nanocapsules.

Acknowledgments: Authors would like to acknowledge the public funding from Fondo Social de la DGA (grupos DGA), Ministerio de la Economía y Competitividad del Gobierno de España for the public founding of ProyectosI+D+i—ProgramaEstatal de Investigación, Desarrollo e InnovaciónOrientada a los Retos de la Sociedad (project n. SAF2014-54763-C2-2-R), the European Seventh Framework Program (NAREB Project 604237), LLP/Erasmus fellowship 2013/2014, INA fellowship "Iniciación a la Investigación" 2014 and 2015, and the European Union's Horizon 2020 research and innovation program for MCSA Fellowship (Grant Agreement No. 660228). The authors also acknowledge José Antonio Ainsa and Ainhoa Lucía for the fruitful discussions as well as Rodrigo Fernandez-Pacheco and Alfonso Ibarra from Advanced Microscopy Laboratories of the Universidad de Zaragoza and Iñigo Echaniz for their technical support. The costs to publish in open access have been covered by funds from NAREB Project (see before).

Author Contributions: L.D.M. and J.M.d.l.F. conceived and designed the experiments; L.D.M., M.A., I.S.-S. and S.G.-E. performed the experiments; L.D.M., M.M., G.S. and J.M.d.l.F. analyzed the data; L.D.M. wrote the paper; all the authors revised the paper besides contributing to the work performance.

Conflicts of Interest: The authors declare no conflict of interest. The founding sponsors had no role in the design of the study; in the collection, analyses, or interpretation of data; in the writing of the manuscript, and in the decision to publish the results.

References

1. Wilczewska, A.Z.; Niemirowicz, K.; Markiewicz, K.H.; Car, H. Nanoparticles as drug delivery systems. *Pharmacol. Rep.* **2012**, *64*, 1020–1037. [CrossRef]
2. Martinez, J.O.; Brown, B.S.; Quattrocchi, N.; Evangelopoulos, M.; Ferrari, M.; Tasciotti, E. Multifunctional to multistage delivery systems: The evolution of nanoparticles for biomedical applications. *Chin. Sci. Bull.* **2012**, *57*, 3961–3971. [CrossRef] [PubMed]
3. Mishra, B.; Patel, B.B.; Tiwari, S. Colloidal nanocarriers: A review on formulation technology, types and applications toward targeted drug delivery. *Nanomedicine* **2010**, *6*, 9–24. [CrossRef] [PubMed]
4. Zhang, L.; Li, Y.; Jimmy, C.Y. Chemical modification of inorganic nanostructures for targeted and controlled drug delivery in cancer treatment. *J. Mater. Chem. B* **2014**, *2*, 452–470. [CrossRef]
5. Hans, M.L.; Lowman, A.M. Biodegradable nanoparticles for drug delivery and targeting. *Curr. Opin. Solid State Mater. Sci.* **2002**, *6*, 319–327. [CrossRef]
6. Maya, S.; Sarmento, B.; Nair, A.; Rejinold, N.S.; Nair, S.V.; Jayakumar, R. Smart stimuli sensitive nanogels in cancer drug delivery and imaging: A review. *Curr. Pharm. Des.* **2013**, *19*, 7203–7218. [CrossRef] [PubMed]
7. Kawaguchi, H. Thermoresponsive microhydrogels: Preparation, properties and applications. *Polym. Int.* **2014**, *63*, 925–932. [CrossRef]
8. Karimi, M.; Sahandi Zangabad, P.; Ghasemi, A.; Amiri, M.; Bahrami, M.; Malekzad, H.; Ghahramanzadeh Asl, H.; Mahdieh, Z.; Bozorgomid, M.; Ghasemi, A.; et al. Hambling, Temperature-responsive smart nanocarriers for delivery of therapeutic agents: Applications and recent advances. *ACS Appl. Mater. Interfaces* **2016**, *8*, 21107–21133. [CrossRef] [PubMed]
9. Allen, T.M.; Cullis, P.R. Liposomal drug delivery systems: From concept to clinical applications. *Adv. Drug Deliv. Rev.* **2013**, *65*, 36–48. [CrossRef] [PubMed]

10. Natarajan, J.V.; Nugraha, C.; Ng, X.W.; Venkatraman, S. Sustained-release from nanocarriers: A review. *J. Control. Release* **2014**, *193*, 122–138. [CrossRef] [PubMed]
11. Shimanovich, U.; Bernardes, G.J.L.; Knowles, T.P.J.; Cavaco-Paulo, A. Protein micro-and nano-capsules for biomedical applications. *Chem. Soc. Rev.* **2014**, *43*, 1361–1371. [CrossRef] [PubMed]
12. Vrignaud, S.; Benoit, J.P.; Saulnier, P. Strategies for the nanoencapsulation of hydrophilic molecules in polymer-based nanoparticles. *Biomaterials* **2011**, *32*, 8593–8604. [CrossRef] [PubMed]
13. Solans, C.; Izquierdo, P.; Nolla, J.; Azemar, N.; Garcia-Celma, M.J. Nano-emulsions. *Curr. Opin. Colloid Interface Sci.* **2005**, *10*, 102–110. [CrossRef]
14. Bouchemal, K.; Briançon, S.; Perrier, E.; Fessi, H. Nano-emulsion formulation using spontaneous emulsification: Solvent, oil and surfactant optimisation. *Int. J. Pharm.* **2004**, *280*, 241–251. [CrossRef] [PubMed]
15. Anton, N.; Benoit, J.P.; Saulnier, P. Design and production of nanoparticles formulated from nano-emulsion templates—A review. *J. Control. Release* **2008**, *128*, 185–199. [CrossRef] [PubMed]
16. Soppimath, K.S.; Aminabhavi, T.M.; Kulkarni, A.R.; Rudzinski, W.E. Biodegradable polymeric nanoparticles as drug delivery devices. *J. Control. Release* **2001**, *70*, 1–20. [CrossRef]
17. Kumari, A.; Yadav, S.K.; Yadav, S.C. Biodegradable polymeric nanoparticles based drug delivery systems. *Colloids Surf. B Biointerfaces* **2010**, *75*, 1–18. [CrossRef] [PubMed]
18. Zimmer, A.; Kreuter, J. Microspheres and nanoparticles used in ocular delivery systems. *Adv. Drug Deliv. Rev.* **1995**, *16*, 61–73. [CrossRef]
19. Sundar, S.; Kundu, J.; Kundu, S.C. Biopolymeric nanoparticles. *Sci. Technol. Adv. Mater.* **2010**, *11*, 1–13. [CrossRef]
20. Huang, S.; Fu, X. Naturally derived materials-based cell and drug delivery systems in skin regeneration. *J. Control. Release* **2010**, *142*, 149–159. [CrossRef] [PubMed]
21. Younes, I.; Rinaudo, M. Chitin and chitosan preparation from marine sources. Structure, properties and applications. *Mar. Drugs* **2015**, *13*, 1133–1174. [CrossRef] [PubMed]
22. Prabaharan, M. Chitosan-based nanoparticles for tumor-targeted drug delivery. *Int. J. Biol. Macromol.* **2015**, *72*, 1313–1322. [CrossRef] [PubMed]
23. Nagpal, K.; Singh, S.K.; Mishra, D.N. Chitosan nanoparticles: A promising system in novel drug delivery. *Chem. Pharm. Bull.* **2010**, *58*, 1423–1430. [CrossRef] [PubMed]
24. Chen, C.K.; Wang, Q.; Jones, C.H.; Yu, Y.; Zhang, H.; Law, W.C.; Lai, C.K.; Zeng, Q.; Prasad, P.N.; Pfeifer, B.A.; et al. Synthesis of pH-responsive chitosan nanocapsules for the controlled delivery of doxorubicin. *Langmuir* **2014**, *30*, 4111–4119. [CrossRef] [PubMed]
25. Felt, O.; Buri, P.; Gurny, R. Chitosan: A unique polysaccharide for drug delivery. *Drug Dev. Ind. Pharm.* **1998**, *24*, 979–993. [CrossRef] [PubMed]
26. Onoue, S.; Yamada, S.; Chan, H.K. Nanodrugs: Pharmacokinetics and safety. *Int. J. Nanomed.* **2014**, *9*, 1025–1037. [CrossRef] [PubMed]
27. Lehner, R.; Wang, X.; Marsch, S.; Hunziker, P. Intelligent nanomaterials for medicine: Carrier platforms and targeting strategies in the context of clinical application. *Nanomedicine* **2013**, *9*, 742–757. [CrossRef] [PubMed]
28. Nicolas, J.; Mura, S.; Brambilla, D.; Mackiewicz, N.; Couvreur, P. Design, functionalization strategies and biomedical applications of targeted biodegradable/biocompatible polymer-based nanocarriers for drug delivery. *Chem. Soc. Rev.* **2013**, *42*, 1147–1235. [CrossRef] [PubMed]
29. Kyzas, G.Z.; Bikiaris, D.N. Recent modifications of chitosan for adsorption applications: A critical and systematic review. *Mar. Drugs* **2015**, *13*, 312–337. [CrossRef] [PubMed]
30. Moghimi, S.M.; Hunter, A.C.; Murray, J.C. Long-circulating and target-specific nanoparticles: Theory to practice. *Pharmacol. Rev.* **2001**, *53*, 283–318. [PubMed]
31. Zhang, Y.; Chan, H.F.; Leong, K.W. Advanced materials and processing for drug delivery: The past and the future. *Adv. Drug Deliv. Rev.* **2013**, *65*, 104–120. [CrossRef] [PubMed]
32. Sapsford, K.E.; Algar, W.R.; Berti, L.; Gemmill, K.B.; Casey, B.J.; Oh, E.; Stewart, M.H.; Medintz, I.L. Functionalizing nanoparticles with biological molecules: Developing chemistries that facilitate nanotechnology. *Chem. Rev.* **2013**, *113*, 1904–2074. [CrossRef] [PubMed]
33. Rabanel, J.M.; Hildgen, P.; Banquy, X. Assessment of PEG on polymeric particles surface, a key step in drug carrier translation. *J. Control. Release* **2014**, *185*, 71–87. [CrossRef] [PubMed]

34. Klibanov, A.L.; Maruyama, K.; Torchilin, V.P.; Huang, L. Amphipathic polyethyleneglycols effectively prolong the circulation time of liposomes. *FEBS Lett.* **1990**, *268*, 235–237. [CrossRef]

35. Maruyama, K.; Yuda, T.; Okamoto, A.; Ishikura, C.; Kojima, S.; Iwatsuru, M. Effect of molecular weight in amphipathic polyethyleneglycol on prolonging the circulation time of large unilamellar liposomes. *Chem. Pharm. Bull.* **1991**, *39*, 1620–1622. [CrossRef] [PubMed]

36. Tavares, I.S.; Caroni, A.L.P.F.; Dantas-Neto, A.A.; Pereira, M.R.; Fonseca, J.L.C. Surface charging and dimensions of chitosan coacervated nanoparticles. *Colloids Surf. B Biointerfaces* **2012**, *90*, 254–258. [CrossRef] [PubMed]

37. Berthold, A.; Cremer, K.; Kreuter, J. Preparation and characterization of chitosan microspheres as drug carrier for prednisolone sodium phosphate as model for antiinflammatory drugs. *J. Control. Release* **1996**, *39*, 17–25. [CrossRef]

38. Mao, H.-Q.; Roy, K.; Troung-Le, V.L.; Janes, K.A.; Lin, K.Y.; Wang, Y.; August, J.T.; Leong, K.W. Chitosan-DNA nanoparticles as gene carriers: Synthesis, characterization and transfection efficiency. *J. Control. Release* **2001**, *70*, 399–421. [CrossRef]

39. Arenal, R.; De Matteis, L.; Custardoy, L.; Mayoral, A.; Tence, M.; Grazú, V.; De La Fuente, J.M.; Marguina, C.; Ibarra, M.R. Spatially-resolved EELS analysis of antibody distribution on bio-functionalized magnetic nanoparticles. *ACS Nano* **2013**, *7*, 4006–4013. [CrossRef] [PubMed]

40. Noel, S.; Liberelle, B.; Robitaille, L.; De Crescenzo, G. Quantification of primary amine groups available for subsequent biofunctionalization of polymer surfaces. *Bioconjug. Chem.* **2011**, *22*, 1690–1699. [CrossRef] [PubMed]

41. Thierry, B.; Griesser, H.J. Dense PEG layers for efficient immunotargeting of nanoparticles to cancer cells. *J. Mater. Chem.* **2012**, *22*, 8810–8819. [CrossRef]

42. Moros, M.; Mitchell, S.G.; Grazú, V.; Tence, M.; De La Fuente, J.M. The fate of nanocarriers as nanomedicines in vivo: Important considerations and biological barriers to overcome. *Curr. Med. Chem.* **2013**, *20*, 2759–2778. [CrossRef] [PubMed]

43. Duncan, R. Polymer therapeutics as nanomedicines: New perspectives. *Curr. Opin. Biotechnol.* **2011**, *22*, 492–501. [CrossRef] [PubMed]

marine drugs

MDPI

Article

Novel Spray Dried Glycerol 2-Phosphate Cross-Linked Chitosan Microparticulate Vaginal Delivery System—Development, Characterization and Cytotoxicity Studies

Emilia Szymańska [1,*], Marta Szekalska [1], Robert Czarnomysy [2], Zoran Lavrič [3], Stane Srčič [3], Wojciech Miltyk [4] and Katarzyna Winnicka [1,*]

[1] Department of Pharmaceutical Technology, Faculty of Pharmacy, Medical University of Białystok, Mickiewicza 2c, Białystok 15-222, Poland; marta.szekalska@umb.edu.pl
[2] Department of Synthesis and Technology of Drug, Faculty of Pharmacy, Medical University of Białystok, Kilińskiego 1, Białystok 15-089, Poland; robert.czarnomysy@umb.edu.pl
[3] Department of Pharmaceutical Technology, Faculty of Pharmacy, University of Ljubljana, Aškerčeva cesta 7, Ljubljana SI-1000, Slovenia; zoran.lavric@ffa.uni-lj.si (Z.L.); stanko.srcic@ffa.uni-lj.si (S.S.)
[4] Department of Pharmaceutical Analysis, Faculty of Pharmacy, Medical University of Białystok, Mickiewicza 2d, Białystok 15-222, Poland; wmiltyk@umb.edu.pl
* Correspondence: esz@umb.edu.pl (E.S.); kwin@umb.edu.pl (K.W.); Tel.: +48-85-748-5893 (E.S.); +48-85-748-5615 (K.W.); Fax: +48-85-748-5616 (E.S. & K.W.)

Academic Editors: Hitoshi Sashiwa and David Harding
Received: 9 August 2016; Accepted: 14 September 2016; Published: 28 September 2016

Abstract: Chitosan microparticulate delivery systems containing clotrimazole were prepared by a spray drying technique using glycerol 2-phosphate as an ion cross-linker. The impact of a cross-linking ratio on microparticle characteristics was evaluated. Drug-free and drug-loaded unmodified or ion cross-linked chitosan microparticles were examined for the in vitro cytotoxicity in VK2/E6E7 human vaginal epithelial cells. The presence of glycerol 2-phosphate influenced drug loading and encapsulation efficacy in chitosan microparticles. By increasing the cross-linking ratio, the microparticles with lower diameter, moisture content and smoother surface were observed. Mucoadhesive studies displayed that all formulations possessed mucoadhesive properties. The in vitro release profile of clotrimazole was found to alter considerably by changing the glycerol 2-phosphate/chitosan ratio. Results from cytotoxicity studies showed occurrence of apoptotic cells in the presence of chitosan and ion cross-linked chitosan microparticles, followed by a loss of membrane potential suggesting that cell death might go through the mitochondrial apoptotic pathway.

Keywords: chitosan; microparticles; glycerol 2-phosphate; spray drying; mucoadhesiveness; cytotoxicity

1. Introduction

Microparticulate drug delivery systems—multiunit carriers composed of spherical particles with average diameter 1–500 μm—have gained particular interest in the pharmaceutical field, due to their ability to assure prolonged drug release profile, providing stability of labile substances or reducing toxicity of active agents [1,2]. In addition, a large surface area of the microparticles enables assuring more uniform drug absorption, which might be favorable, especially with respect to topical delivery vehicles [3]. In designing microparticulate dosage forms intended for local delivery (buccal, nasal, vaginal, pulmonal), a particularly important aspect is to provide an intimate contact between drug carrier and the mucosal surface. To prolong residence time of the multiunit drug carriers at the site of administration, mucoadhesive polymers are employed [4–6]. Among natural polymers, chitosan is one of the most extensively used in technology of microparticle preparation [7–9].

Chitosan—a natural multifunctional polysaccharide, comprised of glucosamine and N-acetylglucosamine units—is obtained by deacetylation of chitin possessed from exoskeleton of insects, crustaceans or fungi [10,11]. Due to its biocompatibility, biodegradability, intrinsic antimicrobial and penetration enhancement properties, chitosan is considered a useful compound in pharmaceutical technology [12–14]. Chitosan's hydration and gel formation abilities give it the opportunity to prolong release of the drug at the administration site [15]. Additionally, owing to mucoadhesive properties arising from the cationic behavior and the presence of free amine and hydroxyl groups, chitosan is capable of interacting with mucin by electrostatic and hydrogen bonds [16]. Recently, chitosan microparticles have been shown to be useful in oral [17], vaginal [18], nasal [19] and ocular drug delivery [9].

A number of novel techniques have been investigated to develop micro- and nanoparticulate drug carrier preparation, including pressurized gyration [20], microfluidic [21], spray drying [22] and electrohydrodynamic (electrospraying or electrospinning) processing [23–25]. Among various encapsulation techniques, spray drying is an advanced, easy to scale-up method of chitosan microparticle preparation, in which dry particles are obtained from a fluid state by evaporating the solvent. Spray drying offers a very flexible control over microparticle properties such as size, flow characteristics and drug encapsulation efficacy [26,27]. This method is uncomplicated, but, in order to receive microparticles with desirable properties, it requires an understanding of the process and careful adjustment of the spray-drying conditions [28].

Despite these advantages, a serious inconvenience of chitosan microparticles is their high solubility in an acidic environment, and, as a consequence, limited ability to control the drug release rate. In order to avoid too rapid dissolution of the polymer formulation and escape of the drug from the microparticle matrix, chemical or physical cross-linking of the chitosan backbone is being practiced [29–31]. Glycerol 2-phosphate (β-glycerophosphate disodium, βGP) is a non-toxic divalent ion that interacts with chitosan through electrostatic forces, creating ionic cross-linked networks [32]. It should be noted that in the presence of βGP (used in a proper concentration), chitosan becomes thermally sensitive and may undergo sol/gel transition around body temperature. Recently, chitosan/βGP material has attracted considerable attention as a promising tool for a variety of applications, such as local drug carriers or injectable delivery systems for tissue engineering [33,34]. In addition, data previously published by our group showed that ionic interaction between βGP and chitosan improved mechanical properties of hydrogels and enabled prolonged drug release profile [35,36].

The aim of this work was to design and prepare vaginal mucoadhesive microparticles using chitosan or chitosan/βGP by the spray drying method. Obtained microparticles would be utilized to prepare multiple-unit dosage form intended for topical administration. Particular effort was made to investigate whether the presence of βGP cross-linked chitosan could improve the physicochemical properties of designed microparticulate delivery systems or influence the drug release profile. Clotrimazole—CLO—an imidazole derivative commonly used as the drug of choice for the fungal infections of the urogenital tract—was employed as a model agent. The prepared microparticles were analysed for drug loading, encapsulation efficacy, production yield, surface morphology, hydration capability and the in vitro release. Drug-polymer-cross-linking agent interactions in the solid state were investigated by differential scanning calorimetric analysis. In addition, to examine the mucoadhesive properties of the obtained microparticles, the ex vivo residence time, maximum detachment force and work of adhesion in the presence of porcine vaginal mucosa were evaluated.

Furthermore, this study concentrated on the cytotoxicity examination of microparticles with chitosan or chitosan cross-linked with different ratios of βGP by 3-(4,5-dimethylthiazol-2-yl)-2,5-diphenyltetrazolium bromide MTT assay followed by three independent methods of apoptosis evaluation: fluorescent microscopy, flow cytometry assessment of annexin V conjugated to green-fluorescent FITC dye (annexin V-FITC binding and analysis of mitochondrial membrane potential. The in vitro cytotoxicity profile of both drug-free and drug-loaded microparticles using human vaginal mucosa epithelium VK2/E6E7 was reported for the first time.

2. Results and Discussion

2.1. Characterization of Microparticles

Chitosan and ion cross-linked chitosan microparticles for vaginal delivery of CLO were prepared for the first time by spray drying method using βGP as a cross-linking agent. Preliminary experiments were performed in order to select suitable amount of excipients and optimal operating conditions of the process. The characteristics of formulated microparticles F1–F6 were summarized in Table 1.

Table 1. Characteristics of chitosan (CS) and βGP/CS microparticles with clotrimazole (CLO).

Formu-lation	βGP:CS Ratio (w/w)	CLO:CS Ratio (w/w)	Production Yield (%) [a] *	Encapsulation Efficacy (%) [b] *	CLO Loading (%) [c] *	Diameter Range (Mean Diameter *) (μm)	Moisture Content (%) *	Residence Time (min) *
			1.0% (w/v) chitosan solution					
F1	-	1:4	75.4 ± 1.3	59.5 ± 3.1	11.2 ± 2.8	n.t.	14.3 ± 3.5	n.t.
F2	1:2	1:4	78.0 ± 1.6	58.2 ± 3.9	8.3 ± 2.7	n.t.	8.9 ± 2.6	n.t.
F3	-	1:3	73.5 ± 0.9	67.2 ± 9.1	16.8 ± 4.1	1.52–5.28 (3.39 ± 1.87)	10.4 ± 2.9	60 ± 2
F4	1:3	1:3	78.8 ± 1.2	59.6 ± 1.0	11.9 ± 2.7	1.49–6.33 (3.96 ± 2.45)	7.6 ± 2.2	116 ± 4
F5	1:2	1:3	79.5 ± 1.1	73.2 ± 9.0	13.3 ± 3.2	1.23–5.25 (3.12 ± 2.13)	6.5 ± 2.4	132 ± 5
F6	1.5:2	1:3	80.6 ± 1.9	68.4 ± 11.3	10.9 ± 2.1	0.93–3.93 (2.54 ± 1.59)	7.9 ± 2.7	154 ± 5

[a] the weight of obtained microparticles/overall weight of components in feed solution ×100; [b] actual CLO content in microparticles/theoretical CLO content ×100; [c] actual CLO content in microparticles/examined amount of microparticles ×100; * mean ± S.D; $n = 3$; n.t.—not tested.

The production yield was in the range from 73.5% ± 0.9% (F3) to 80.6% ± 1.9% (F6) and was found to be higher for βGP/chitosan microparticles. The increase in drug/polymer ratio from 1:4 to 1:3 resulted in a substantial improvement in CLO loading and encapsulation efficacy. The presence of βGP had a negative impact on the drug-loading of microparticles regardless of the amount of chitosan or drug used. Nevertheless, by increasing the ratio of βGP:chitosan (from 1:3 to 1:2), the rise in the encapsulation efficacy was noticed (from 59.6% ± 1.0% to 73.2% ± 9.0%). The moisture's content was found to be in the range from 6.5% ± 2.4% (formulation F5) to 14.3% ± 3.5% (F1). Microparticles with βGP/chitosan exhibited lower water content compared to non-cross-linked formulations (Table 1).

2.2. SEM and DSC Studies

To assess the surface morphology and size of microparticles F3–F6 (with CLO:chitosan ratio 1:3), SEM studies were carried out. The representative SEM photographs of drug-loaded microparticles with non-cross-linked chitosan or βGP/chitosan were presented in Figure 1. In all investigated formulations, the predominance presence of spherical forms was observed. In the case of microparticles F5 and F6 with βGP/chitosan, some elliptical shape particles that resemble biconcave discs in structure were also noticed (Figure 1C). The microparticles F3 with unmodified chitosan were found to be uniform with ridged surface. Notably, with increasing the amount of βGP in chitosan microparticles, the wrinkles on the particles' surface disappeared and the smoothest and most intact surface in formulations F5 and F6 was observed under identical magnification (×20,000). The diameter range of drug-loaded microparticles was diversified and varied between 0.93 (F6) and 6.33 μm (F4) (Table 1). Interestingly, the mean diameter of βGP/chitosan microparticles decreased from 3.96 ± 2.45 μm (F4) to 2.54 ± 1.59 μm (F6) with increasing βGP cross-linking ratio.

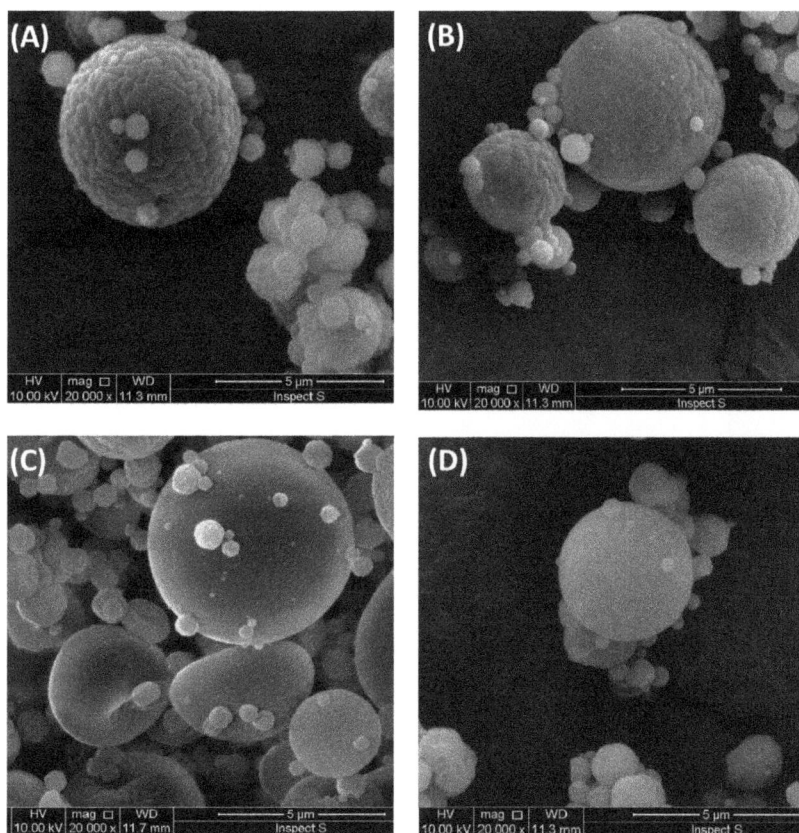

Figure 1. Representative SEM images of clotrimazole (CLO)-loaded microparticles prepared with: (**A**) unmodified chitosan (F3); (**B**) βGP/chitosan 1:3 (*w/w*) (F4); (**C**) βGP/chitosan 1:2 (*w/w*) (F5); (**D**) βGP/chitosan 1.5:2 (*w/w*) (F6); and original magnification ×20,000.

The differential scanning calorimetry DSC thermogram of a CLO sample revealed onset of melting occurring at 142 °C and thermal decomposition starting at about 286 °C (Figure 2). Pure chitosan exhibited a broad endotherm between 10 °C and 140 °C (ΔH 188 J/g) corresponding to its water loss. Chitosan decomposition process occurred around 296 °C [37]. The DSC curve of pure βGP exhibited numerous peaks in the range 80 °C to 150 °C—characteristic for water loss and process of βGP melting [38]—and a decomposition peak at about 264 °C.

The results from DSC studies of physical mixtures (Figure 2C,H,J) displayed no interactions between the drug and chitosan or βGP, whereas a decrease in CLO peak intensity in the thermograms of F1 and F2 microparticles (Figure 2I,K) was noticed indicating a partial loss of drug crystallinity within the polymer matrix. An observed shift to a lower CLO melting temperature (T_{melt}) in F1 (130 °C) and F2 microparticles (120 °C) thermal profiles, followed by a decrease in temperature of thermal decomposition (266 and 255 °C for F1 and F2, respectively), might be caused by increasing interactions of CLO with the chitosan matrix and more rigid structure formation. This effect was enhanced with incorporation of βGP cross-linking agent to microparticles. On the contrary, thermal stressing of F2 microparticle samples led to an increase in melting temperature of CLO, which might reflect either alterations in polymorphic forms of the drug or thermal relaxation of chitosan matrix during heating/cooling program cycle. No melting peak of βGP was present in the thermogram of F2 microparticles, which may indicate that βGP was molecularly dispersed within the polymer matrix. The obtained results might suggest that CLO encapsulation in chitosan or chitosan/βGP matrix by using a spray-drying method exerted a stabilizing effect on the drug's molecular dispersion. The above observations are in agreement with previously reported data in which the preparation method was found to influence the state of the drug in the solid dispersion [39].

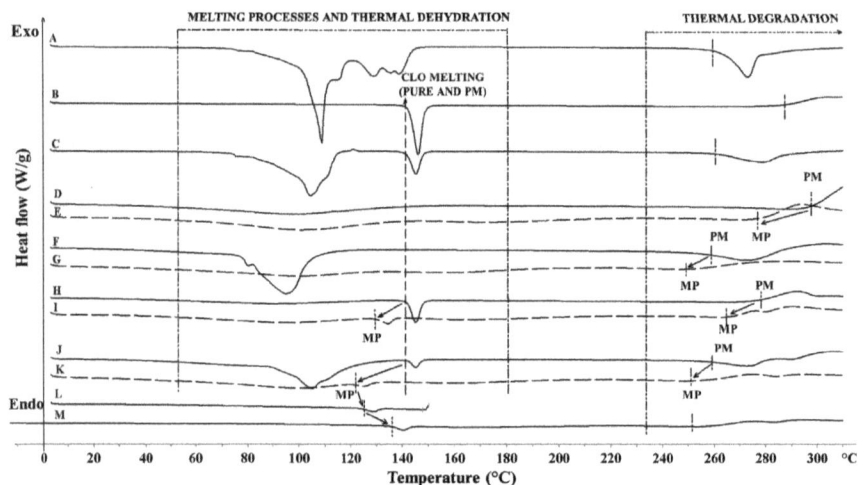

Figure 2. DSC thermograms of: (**A**) βGP powder; (**B**) CLO powder; (**C**) CLO/βGP 1:1 (*w/w*) physical mixture (PM); (**D**) chitosan powder; (**E**) placebo microparticles (MP) P1; (**F**) chitosan/βGP 1:1 (*w/w*) physical mixture (PM); (**G**) placebo microparticles (MP) P2; (**H**) CLO/chitosan 1:1 (*w/w*) physical mixture (PM); (**I**) microparticles F1 (MP); (**J**) CLO/chitosan/βGP (F2) physical mixture (PM); (**K**) microparticles (MP) F2; (**L**) MP F2 heated to 110 °C; (**M**) MP F2 heated for 1 h at 110 °C.

2.3. Mucoadhesive and in Vitro Drug Release Studies

In order to evaluate microparticles' characteristics in contact with mucosal surfaces, mucoadhesive studies were accomplished using texture analyser. Porcine vaginal mucosa was applied for the imitation of vaginal mucoadhesion due to its resemblance to the human vagina in terms of anatomical structure, pH, permeability and vaginal secretion [40]. The effect of βGP cross-linking at various ratios, the amount of the drug on the force of detachment (F_{max}), and work required to overcome the microparticles-porcine vaginal mucosa interaction (W_{ad}) was presented in Figure 3.

All examined formulations exhibited mucoadhesive properties, expressed as F_{max} with the median range from 0.17 N (F2) to 0.23 N (F1 and F4) and W_{ad} between 271.8 µJ (F2) and 673.5 µJ (F1). Several studies revealed that the presence of cross-linker reduced chitosan capability of interacting with mucosa as a result of a drop in positively charged free amino groups [41,42]. The preserved ability of βGP/chitosan microparticles to interact with mucosal material could be explained by incomplete chitosan cross-linking with βGP partially retaining protonation of polymers' functional groups that were still capable of physicochemical interactions with mucosa [43]. Basically, the presence of higher content of CLO crystalline particles in formulations (F1 vs. F3 and F2 vs. F5) did not significantly alter the mucoadhesiveness of the microparticles. Based on the plotted trend line (Figure 3A), it was also noticed that unmodified chitosan and βGP/chitosan microparticles possessed comparable values ($p > 0.05$) of F_{max}. The obtained data is in agreement with results previously reported by our group, indicating that βGP modification of chitosan in semi-solid drug carriers maintained polymers' ability to interact with mucosal tissue [35]. Nonetheless, the obtained W_{ad} values were found to demonstrate a downward trend (Figure 3B) with raising the amount of ion cross-linker in chitosan microparticles.

This drop in work required to separate microparticles mucoadhesive material might be attributed to the presence of a lower amount of water in βGP/chitosan formulations (compared with unmodified chitosan microparticles) responsible for formation of less cohesive structure with mucus. It should also be pointed out that the rough surface of unmodified chitosan microparticles (as shown in Figure 1A) might favor an intimate contact between mucin and polymer compared to more intact βGP/chitosan formulations.

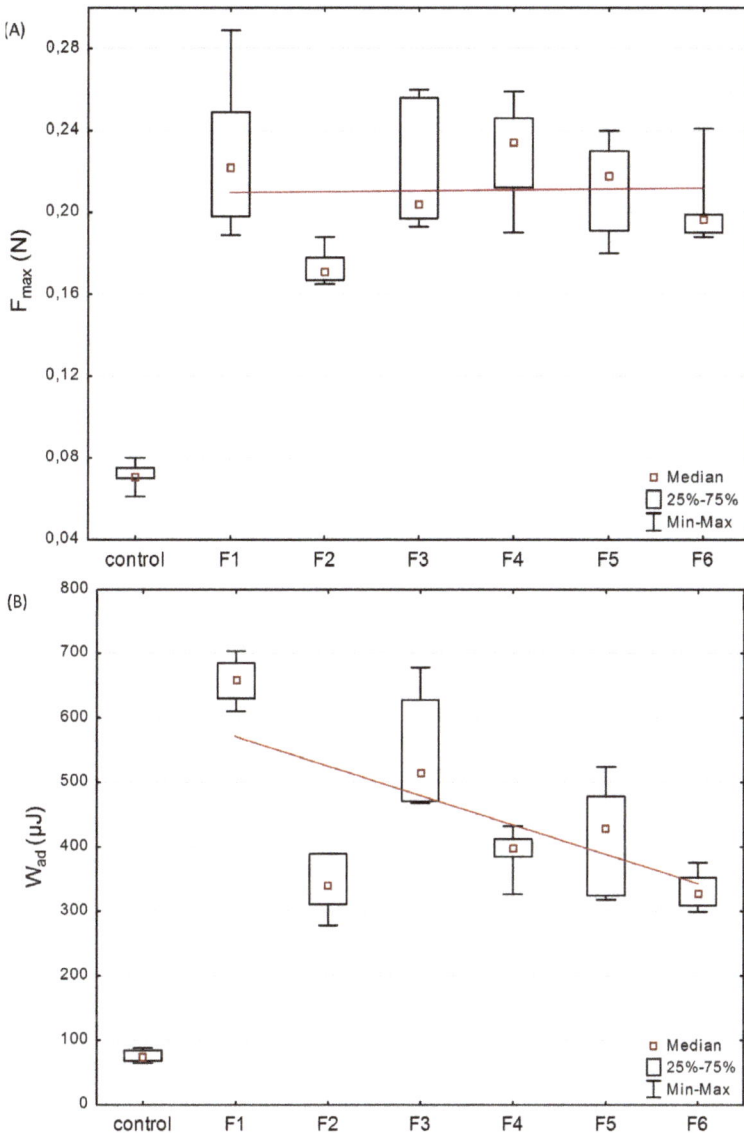

Figure 3. Box-plot graphs presenting mucoadhesive properties: (**A**) maximum force of detachment (F_{max}); and (**B**) work of adhesion (W_{ad}) of formulations F1–F6 and control—cellulose paper (median; $n = 6$; trend line plotted through median values).

Additionally, in order to investigate microparticles' behavior in contact with mucosal surface, the residence time to the porcine vaginal mucosa using a self-constructed apparatus was determined [44]. A capability of increased drug residence time after administration in the vaginal cavity appears to be essential in reducing the dosage frequency and improving the patient compliance. It was observed that examined microparticles F3–F6 adhered immediately to the mucosal surface. Notably, the contact time of formulations with βGP/chitosan with vaginal tissue was considerably higher than those obtained for non-cross-linked chitosan microparticles (Table 1). The residence time was found to correlate with βGP concentration and formulation F6, with the highest βGP:chitosan ratio exhibiting the longest period of contact time with mucosal surface. Nevertheless, it should be noted that, throughout the test, a progressive dissolution and surface erosion of chitosan and βGP/chitosan matrix in acidic pH was observed, and, thus, prolonged residence time of βGP/chitosan microparticles could be attributed to a slower matrix dissolution rate compared to unmodified chitosan formulation.

To examine the influence of βGP cross-linking on CLO release profiles from chitosan microparticles (with CLO:chitosan ratio 1:3), formulation F3 (with unmodified chitosan) and formulations F4–F6 (with different βGP:chitosan ratios) were selected for the in vitro release study. The dissolution test was carried out with using the enhancer cell equipped with Cuprophan membrane. This model reflects the conditions of the vaginal cavity more suitably as it enables maintaining microparticles in a reservoir where drug carriers slowly hydrate and form gel matrix from which drug is released [45]. As shown in Figure 4, all designed chitosan microparticles exhibited prolonged CLO release rates as compared to control—simple dispersion of CLO in the release medium. However, it was found that the drug was certainly faster released from formulation F3 with non-cross-linked chitosan (the plateau was reached within 24 h) as a result of a high degree of protonation of amino groups responsible for electrostatic repulsion between charged chains and expansion of the polymer network. As expected, the amount of ion cross-linker had a remarkable effect on the CLO release from βGP/chitosan microparticles. It was noticed that release behavior from formulation F4 (with 1:3 βGP:chitosan ratio) was comparable ($p > 0.05$) with the data obtained for unmodified chitosan formulation F3, whereas microparticles F5 and F6 (with βGP:chitosan in a ratio 1:2 and 1.5:2 (w/w), respectively) retarded the release of CLO markedly. The observed phenomenon might be attributed to a more intact structure of microparticles F5 and F6, as indicated by the SEM picture (Figure 1C,D).

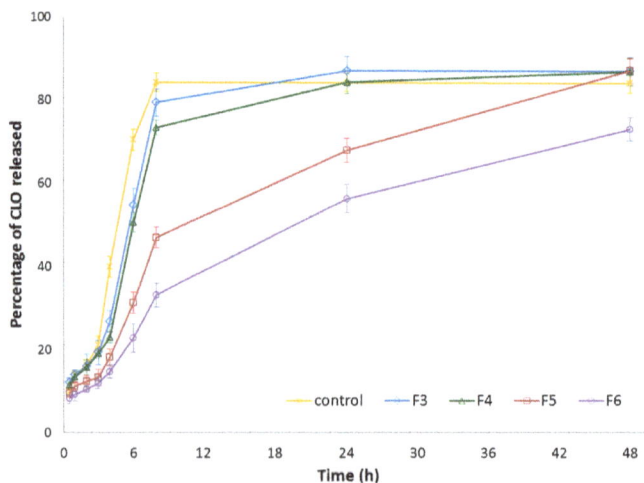

Figure 4. Percentage of CLO released from microparticles with unmodified chitosan (F3) or βGP/chitosan (F4–F6) compared to 1.0% (w/v) CLO dispersion in the release medium (control) (mean ± SD; $n = 3$).

Due to the fact that the drug release behavior from chitosan microparticles is strongly associated with their water uptake properties, swelling index studies of formulations F3–F6 were additionally carried out and reported in Figure 5. As expected, all examined formulations displayed hydration capability at acidic pH resulting from protonation of chitosan amino groups and repulsion of the polymer chain. Microparticles with unmodified chitosan (F3) were found to swell rapidly and dissolve almost immediately after direct contact with acidic buffer solution; thus, it was impossible to measure their accurate degree of swelling. It was noticed that the swelling capacity of formulations with βGP/chitosan decreased with increasing a degree of cross-linking. Furthermore, βGP:chitosan ratio 1:2 (F5) and 1.5:2 (*w*/*w*) (F6) appeared to be essential for preserving microparticles integrity in acidic environment up to 2 and 3 h, respectively. Higher amounts of βGP might have interfered with the process of chitosan chain relaxation, thus leading to a suppression of the relaxation mechanism. As a result, slower disintegration of βGP/chitosan formulations and drug dissolution rate in acidic pH was observed.

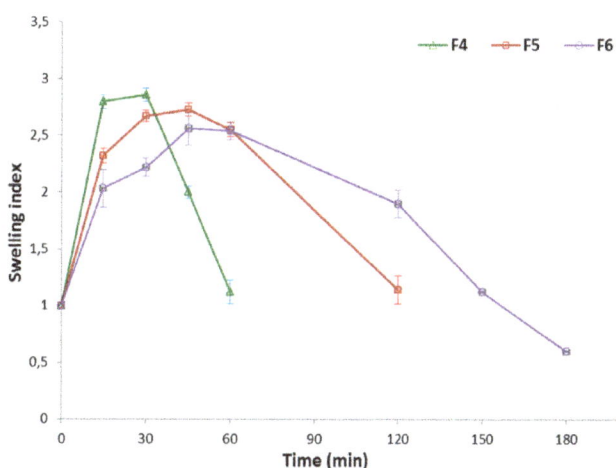

Figure 5. Swelling index study of microparticles F4–F6 (mean ± SD; *n* = 3).

2.4. In Vitro Cytotoxicity Studies

Chitosan has been widely used for the preparation of microparticulate drug delivery systems. The biocompatibility and non-toxicity of chitosan material was proved in vitro on several types of cell lines and in vivo after intravenous or long-term oral administration [46–48]. Despite the fact that chitosan is extensively studied as an excipient in microparticulate dosage forms [5,19,42], its safety profile has still not been fully determined. The polymer biocompatibility may be influenced by the applied processing method or the presence of an active agent. In addition, because of unique physicochemical properties emerging throughout chitosan microparticle preparation (e.g., large surface area), its toxicity profile could remarkably differ in comparison to native polymer material [49]. Therefore, the safety profile of drug-free and CLO-loaded chitosan microparticles prepared by spray drying in VK2/E6E7 cells was evaluated in the present work. As any modification of chitosan structure should be followed by an exhaustive evaluation of its cytotoxicity, a particular effort was made to investigate whether the presence of βGP cross-linker in chitosan microparticles impacted the VK2/E6E7 cell viability. Formulations with unmodified chitosan (placebo P3, drug-loaded F3) and βGP cross-linked chitosan at two different ratios 1:2 (P5, F5) and 1.5:2 (P6, F6) were selected for the cytotoxicity investigations. To exclude the influence of ion cross-linker itself and the pH of the medium on the cell viability, several controls were employed concomitantly with experimental samples and summarized in Table 2.

Table 2. Abbreviations used in the in vitro cytotoxicity evaluation.

Abbreviation		Description
K0		non treated cells
K1	Cells treated with:	acetic buffer pH 4.5
K2		aqueous solution of βGP *
K3		aqueous solution of βGP **
P3/F3		unmodified chitosan placebo/drug-loaded microparticles
P5/F5		βGP:chitosan (1:2) placebo/drug-loaded microparticles
P6/F6		βGP:chitosan (1.5:2) placebo/drug-loaded microparticles

The amount of βGP (*w*/*w*) added corresponded with those added to chitosan microparticles (*) P5, F5; (**) P6, F6.

The MTT assay, a quantitative and quick colorimetric test was chosen to preliminary screen the cytotoxicity range of drug-free and CLO-loaded chitosan and βGP/chitosan microparticles. All investigated formulations showed dose-dependent and time-dependent cytotoxicity in the concentration range of 0.01 to 0.5 mg/mL as presented in Figure 6. At the lowest applied concentration of chitosan 0.01 mg/mL, microparticles were found to exert low cytotoxicity after 4 h incubation (the range of median of cell viability 66.5%–81.2%), whereas a substantial loss of viable cells (up to median range 31.6%–41.8%) ($p < 0.05$) after 48 h incubation was noticed. Cells exposed to chitosan and βGP/chitosan formulations at concentration 0.5 mg/mL exhibited high cytotoxic effects (median cell viability below 25%) regardless of the incubation time. In addition, a slightly lower inhibitory effect of drug-free microparticles on VK2/E6E7 cell viability compared to CLO-loaded formulations after 24 h incubation was demonstrated. As shown in Figure 6A, βGP/chitosan microparticles caused greater reduction in VK2/E6E7 cells viability (median 66.5%, 68.3% for P5, P6 and 69.4%, 72.5% for F5 and F6, respectively) compared to microparticles with unmodified chitosan (median 80.8% (P3) and 81.2% (F3)) within 4 h incubation, suggesting that ion cross-linked chitosan microparticles exerted stronger impact on the VK2/E6E7 metabolic activity at early time points. Nevertheless, the differences in MTT cytotoxicity between non-cross-linked and ion cross-linked microparticles after 48 h incubation were found to be statistically irrelevant ($p > 0.05$) (Figure 6C).

The MTT assay reflects the metabolic activity of the cells that may be not directly related to cell death as an outcome; therefore, to elucidate the nature of cell death induced by chitosan and βGP/chitosan microparticles in VK2/E6E7 cells, flow cytometric analysis after annexin V-FITC and propidium iodide staining was performed. Based on the preliminary results from the MTT assessment, two microparticle concentrations (which corresponded to chitosan concentration)—0.01 and 0.1 mg/mL—were applied in these studies. The incubation of VK2/E6E7 cells with chitosan and βGP/chitosan microparticles induced visible phosphatidylserine exposure after 24 h of incubation (Figure 7A).

At a concentration of 0.1 mg/mL, the number of early and late apoptotic cells was significantly higher ($p < 0.05$) than at a concentration of 0.01 mg/mL. The differences in the nature of cell death evoked by drug-free and drug-loaded microparticles were found to be statistically insignificant ($p > 0.05$), which suggest that the inhibition of the cell viability is not influenced by the presence of the drug. In addition, the obtained results demonstrated comparable viability of VK2/E6E7 cells within exposures to unmodified and βGP cross-linked chitosan microparticles. Interestingly, the above findings did not correspond with the cytotoxicity studies previously reported by our group in which lesser degrees of cytotoxicity had been noticed in VK2/E6E7 incubated with solutions of chitosan cross-linked with βGP (prepared by simple dissolution of the polymer in acetate buffer without further processing) [35], which might indicate that the applied spray drying process could have influenced the cytotoxicity profile of βGP/chitosan.

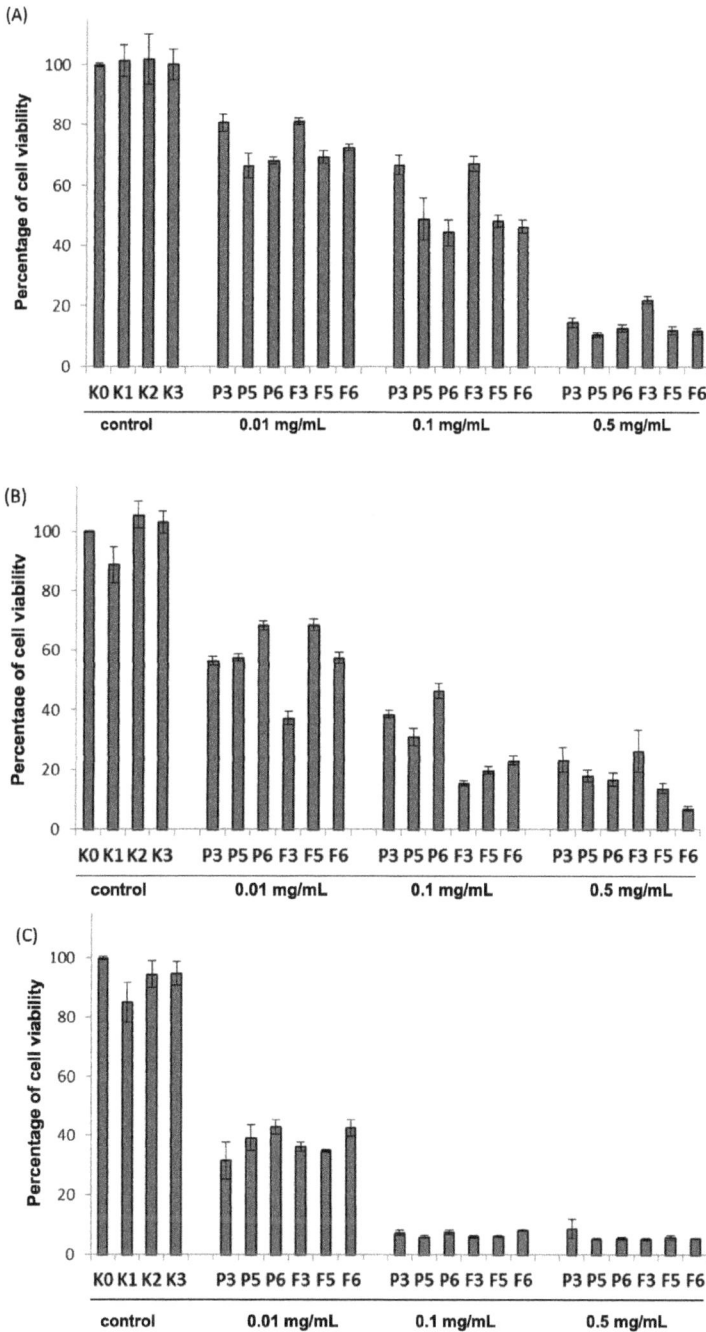

Figure 6. Column charts presenting the viability of VK2/E6E7 cells with controls (as described in Table 2) and different concentrations (expressed in mg/mL) of drug-free (P3, P5, P6) or CLO-loaded (F3, F5, F6) microparticles with unmodified (P3, F3) or chitosan cross-linked with βGP at different ratios (P5–P6, F5–F6) incubated for: (**A**) 4 h; (**B**) 24 h; and (**C**) 48 h measured by using MTT assay (*n* = 6).

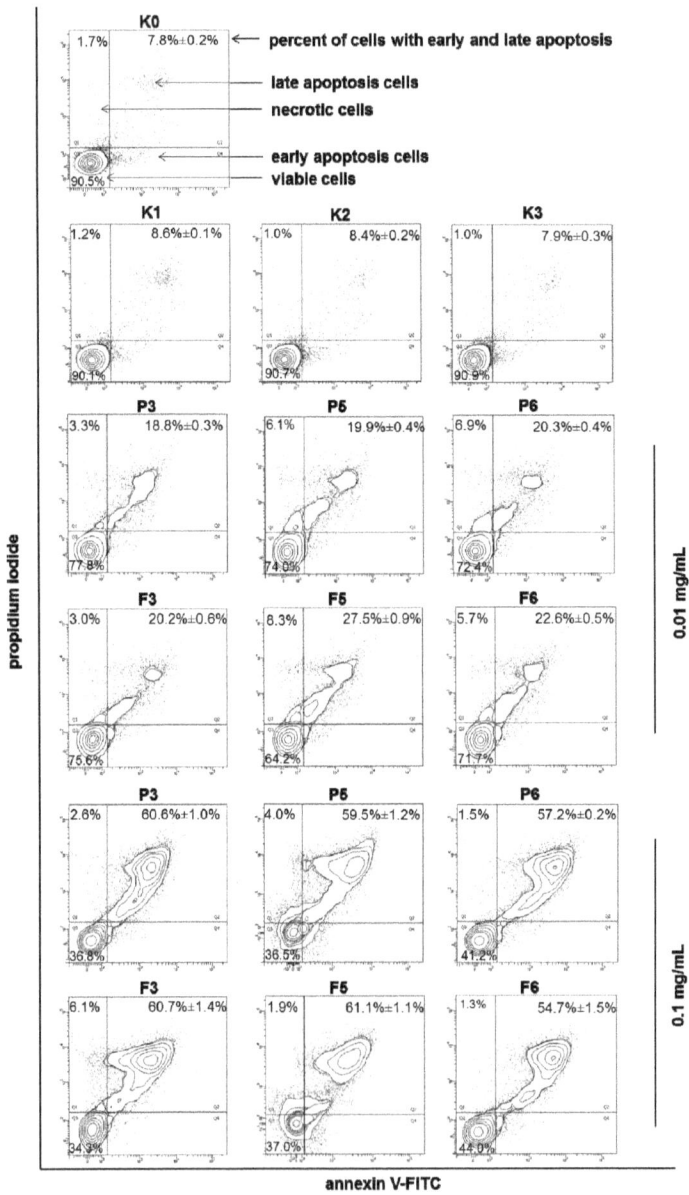

(A)

Figure 7. *Cont.*

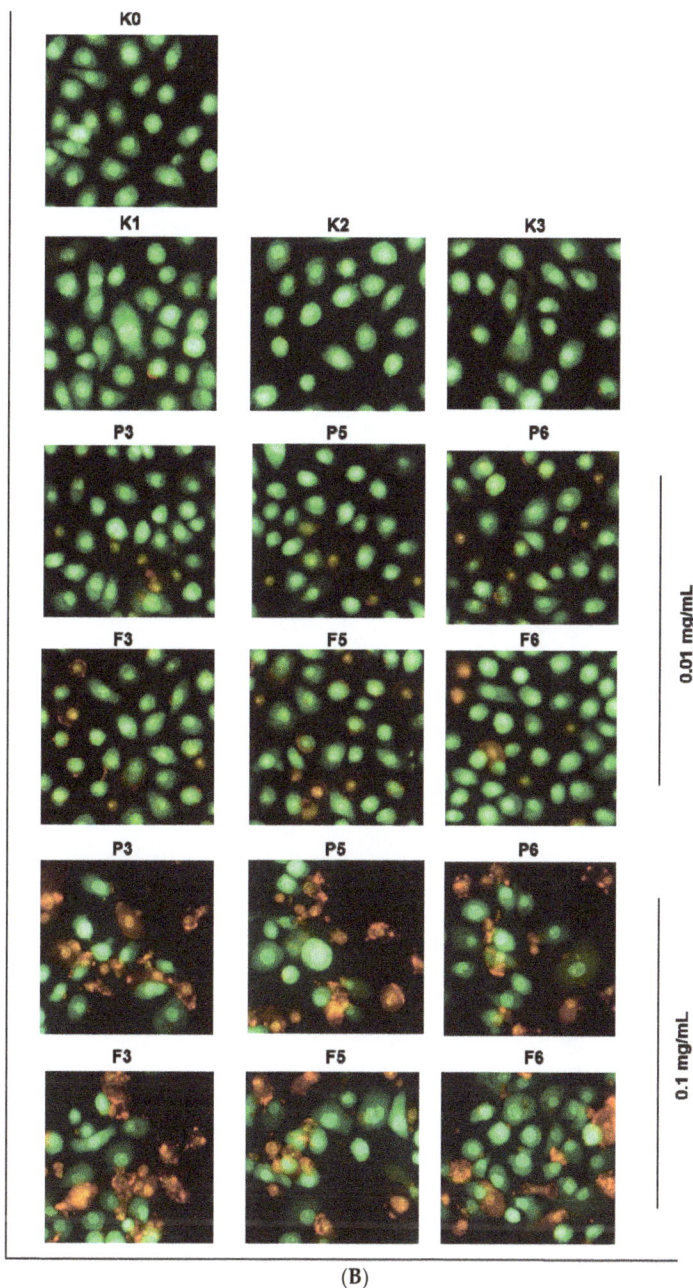

(B)

Figure 7. Representative (**A**) flow cytometry dot plots for Annexin V-FITC assay (mean ± SD; *n* = 3); (**B**) fluorescence microscopy images (magnification ×200) of VK2/E6E7 cells incubated with 0.01 or 0.1 mg/mL of drug-free (P) or CLO-loaded (F) chitosan (P3, F3) and βGP/chitosan microparticles (P5–P6, F5–F6) and controls (as described in Table 2) for 24 h.

VK2/E6E7 cells were also subjected to chitosan and βGP/chitosan microparticle incubation followed by acridine orange-ethidium bromide double staining. Acridine orange—a nucleic acids-selective fluorescent dye—binds to DNA, staining it green, and interacts with RNA, making it appear orange-red. Ethidium bromide is solely absorbed by nonviable cells and emits orange fluorescence by intercalation to chromatin of necrotic cells [50]. The mixture of both dyes is commonly used to detect specific features of apoptosis: apoptotic bodies formation and nuclear envelope disruption. Representative fluorescence microscopy images presented in Figure 7B clearly displayed morphological changes after 24 h incubation with 0.01 and 0.1 mg/mL chitosan and βGP/chitosan microparticles. The results of flow cytometry and fluorescence microscopy assessments revealed that chitosan and βGP/chitosan microparticles induced the VK2/E6E7 cell death via an apoptotic mode.

The cytotoxicity induced by chitosan microparticulate delivery systems is not entirely unexpected as Prego et al. [51] reported a cytotoxic effect of chitosan nanocapsules prepared by the solvent displacement method on human epithelial colorectal adenocarcinoma cells. Investigations performed on tripolyphosphate cross-linked chitosan microparticles obtained by ionotropic gelation also revealed a cytotoxic effect of the polymer on retinal cells in vitro in a dose-dependent manner [9]. In contrast, Pai et al. [52] found that rifampicin and rifabutin-loaded chitosan microparticles (prepared by combination of ionotropic gelation and spray drying technique) exerted no toxic effect in vivo on Sprague-Dawley rats. In addition, the in vitro cytotoxicity of tenofovir-loaded, unmodified and thiolated chitosan nanoparticles confirmed no cytotoxicity in vaginal epithelial cells [53].

Cytotoxicity was additionally evaluated by the fluorescent mitochondrial probe JC-1—a marker of mitochondrial membrane integrity. Permealization of mitochondrial membrane followed by a decrease in its potential (MMP) is a very useful marker to evaluate induction of apoptosis [54]. In a normal cell, JC-1 is present as a monomer in cytosol (where emits green fluorescence) and as aggregates in mitochondria (emitting red fluorescence), whereas in apoptotic cell with disrupted mitochondrial membrane, the dye retains its monomeric form in mitochondria and produces green fluorescence only. The effects of chitosan and βGP/chitosan microparticles on the MMP of VK2/E6E7 cells after 24 h incubation were displayed in Figure 8.

The percentage of cells with the loss of MMP rose significantly with an increase in chitosan microparticles concentration from 0.01 to 0.1 mg/mL. The JC-1 assay confirmed that mitochondrial functions did not vary markedly ($p > 0.05$) between cells exposed to unmodified chitosan and βGP/chitosan microparticles. The observed mitochondrial membrane disruption in the presence of chitosan microparticulate delivery systems could be explained by the electrostatic interactions of a polymer's positively charged functional groups in the polymer backbone (partially maintained also in case of βGP/chitosan microparticles as a consequence of incomplete chitosan cross-linking with βGP) with membrane phospholipids. These results are in accordance with those obtained in the annexinV-FITC/PI assessment, pointing out that the apoptosis evoked by chitosan and βGP/chitosan microparticles might go via the mitochondrial pathway.

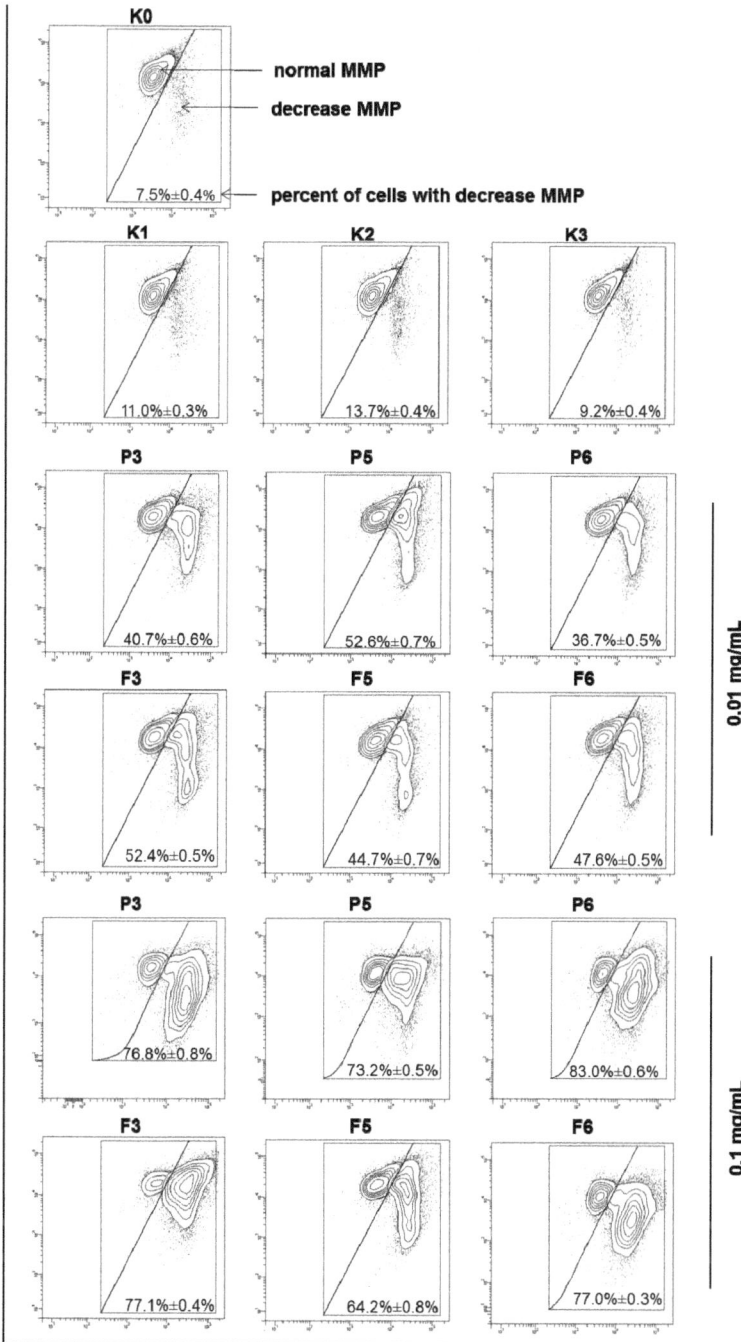

Figure 8. Representative dot-plots presenting the loss of mitochondrial membrane potential (MMP) of VK2/E6E7 cells after 24 h incubation with 0.01 or 0.1 mg/mL of drug-free (P) or CLO-loaded (F) chitosan (P3, F3) and βGP/chitosan microparticles (P5–P6, F5–F6) and controls (as described in Table 2) measured by JC-1 fluorescence (mean ± SD; *n* = 3).

3. Materials and Methods

3.1. Materials

High quality chitosan (Chitoscience® product line) with an individual certificate of analysis was obtained from Heppe Medical Chitosan GmbH (Haale, Germany). The number of average molecular weight (80 kDa) and weight average molecular weight (232 kDa) were assessed by Agilent 1260 Infinity GPC/SEC at 35 °C with a refractive index detector (Agilent Technologies, Santa Clara, CA, USA) and PSS Novema columns (PSS Standards Polymer Service GmBh, Mainz, Germany). A sample of 2 mg of chitosan was diluted in 0.3 M formic acid (1 mL) overnight prior analysis. The average viscosity 32 ± 3 mPa·s of 1.0% chitosan in 1.0% acetic acid (w/v) was measured at 25 °C by Haake Viscotester (Thermo Scientific, Darmstadt, Germany), whereas the moisture content 6.5% $\pm$ 0.2% was evaluated by using a moisture analyser Radwag WPS 50SX (Radom, Poland). The deacetylation degree 79.9% was determined by titration method according to Czechowska-Biskup et al. [55].

Clotrimazole was a gift from Ziaja Ltd. (Gdańsk, Poland). Glycerol 2-phosphate disodium salt hydrate, dimethyl sulfoxide (DMSO, anhydrous), Cremophor EL, penicillin, streptomycin, 3-(4,5-dimethylthiazol-2-yl)-2,5-diphenyltetrazolium bromide (MTT), acridine orange, and ethidium bromide were purchased from Sigma Aldrich (Steinheim, Germany). FITC Annexin V Apoptosis Detection Kit II and JC-1 MitoScreen Kit were products of BD Biosciences (San Jose, CA, USA). Porcine vaginal mucosa from white pigs weighing approximately 200 kg was obtained from the veterinary service of an indigenous slaughterhouse (Turośń Kościelna, Poland) and prepared according to Squier et al. [56]. Normal human vaginal epithelial cells (VK2/E6E7; CRL 2616) were obtained from the American Type Culture Collection (Manassas, VA, USA). Keratinocyte serum-free medium (ker-SFM) with a supplement kit was from Life Technologies Sp. z o.o. (Warsaw, Poland). Methanol (HPLC grade) was obtained from Merck (Darmstadt, Germany). Water for HPLC was distilled and passed through a reverse osmosis system Milli-Q Reagent Water System (Billerica, MA, USA). Acetic buffer (pH 4.5) for mucoadhesive, swelling and in vitro release studies was prepared according to Ph. Eur. 8.0 [57] with the following composition (per liter of distilled water): anhydrous sodium acetate 63 g and acetic acid 90 mL; ionic strength (0.14 M acetic acid/0.08 M sodium acetate). All other solvents and chemicals were purchased from Chempur (Piekary Śląskie, Poland).

3.2. Preparation of Microparticles

Drug-free and drug-loaded microparticles with chitosan or βGP/chitosan were prepared by using a Bűchi Mini Spray Dryer (Flawil, Switzerland) with a standard 0.5 mm nozzle. In order to select the optimal process parameters and obtain microparticles with desired properties, preliminary studies were conducted and the experimental parameters of the spray drying were set as follows: inlet temperature 155 °C–158 °C, outlet temperature 90 °C–92 °C, aspirator flow 37 m^3/h, feed flow 1.5 mL/min, and spray flow 600 L/h. Microparticles were prepared using two drug/polymer mass ratios (1:4, 1:3) and different cross-linking agent/polymer ratios (1:3, 1:2, 1.5:2). Based on the preformulation experiments, concentration of chitosan solution 1.0% in 1.0% acetic acid (w/v) was selected for microparticles preparation.

For drug-free formulation, chitosan was solubilized in 1.0% (w/v) acetic acid solution and chilled in ice bath for 60 min. The appropriate amount of βGP (corresponding to polymer amount (w/w)) was next added to the cold chitosan base with continuous stirring. To ensure the high microbiological purity of microparticles, the feed solutions were prepared under aseptic conditions in a laminar flow cabinet Lamil Plus-13 with HEPA filters (Karstulan Metalli Oy, Karstula, Finland). The solutions were next sterilized using 0.22 µm membrane syringe filters (Millipore, Billerica, MA, USA) and subjected to spray drying. As CLO is sparingly soluble in water [58], to prepare drug-loaded microparticles, the drug was gradually suspended in a small amount of previously obtained chitosan or βGP/chitosan base, and then diluted with the remaining amount of the polymer base by transferring suspension to the baker with mechanical stirring (300 rpm, 1 h).

3.3. Morphology and Microparticle Size Analysis

Morphology and measurement of the microparticles size were performed using scanning electron microscopy SEM Hitachi SH 200 (Tokyo, Japan) after sputter-coating with gold (6.0 nm) in an argon atmosphere (Leica EM AC 2000, Wetzlar, Germany). The longest dimension from edge to edge of a single microparticle was measured for particles present in at least three areas of observation ($\times$5000 magnification).

3.4. Drug Loading, Encapsulation Efficacy and Production Yield

CLO was extracted from accurately weighted amount of microparticles (10 mg) with a mixture of 1.0% acetic acid and methanol (1:4, v/v; 5 mL) under agitation at 150 rpm for 4 h in a water bath (25 °C) [59]. After filtration through 0.45 µm nylon Millipore filter (Millipore, Billerica, MA, USA), CLO concentration was determined by the HPLC method according to Hájková et al. [60] in the subsequent conditions: Zorbax Eclipse XDB–C18, 150 mm $\times$ 4.6 mm, 5 µm column (Agilent Technologies, Cary, NC, USA); mobile phase: methanol:phosphate buffer pH 7.4 (4:1, v/v), flow rate: 1.0 mL/min; UV detection 210 nm; retention time 5.4 min; the standard calibration curve was linear in the range from 1 to 100 µg/mL ($R^2 = 0.996$).

3.5. Moisture Content Determination

In order to determine the percentage of moisture content, accurately weighted microparticles (50 mg) were placed in the aluminium pan of moisture analyser Radwag WPS 50SX (Radom, Poland), heated from 30 to 120 °C and the average of three measurements for each formulation was computed.

3.6. Differential Scanning Calorimetry

DSC measurements of CLO, chitosan, βGP, their physical binary mixtures in the ratio 1:1 (w/w) and a ternary mixture of CLO, βGP and chitosan with a ratio of 1:2:4 (w/w) (equaling the ratio of F2 formulation), placebo (P1 and P2) and drug-loaded microparticles (F1 and F2) were done on DSC-1 calorimeter (Mettler Toledo, Urdorf, Switzerland) equipped with an HSS-8 sensor. About 5 mg samples were placed in aluminum pans and hermetically sealed with lids. Analysis was performed under nitrogen atmosphere with a flow rate of 50 mL/min and heating rate of 10 °C/min. A simple heating program (from 0 °C to 310 °C) was used for all samples. Additionally, in order to assess changes in F2 microparticles under the influence of high temperature, a heating/cooling program cycle consisting of heating to 110 °C (kept at that temperature for 0 h or 1 h), cooling to 0 °C and heating again to 310 °C was used. Recorded thermograms were analyzed in the StarE program (version, Mettler Toledo, Urdorf, Switzerland).

3.7. Determination of the Mucoadhesive Properties

TA.XT.Plus Texture Analyser (Stable Microsystems, Godalming, UK) equipped with a 5 kg load cell, cylinder probe and the mucoadhesion measuring system A/MUC was applied for examination of mucoadhesive properties. The mucoadhesion of chitosan and βGP/chitosan microparticles were evaluated on porcine vaginal mucosa, which was attached to the upper probe with α-cyanoacrylate glue and lowered on the surface of microparticles with a constant speed of 0.5 mm/s [61]. Each formulation of microparticles (100 mg) was located on the platform below the texture analyser probe and moisturized with 100 µL of acetate buffer pH 4.5. The tests were conducted at 37 °C $\pm$ 0.5 °C. An acquisition rate of 200 points/s and a trigger force of 0.003 N were chosen for all measurements. After keeping contact for 100 s under an initial contact force 0.5 N, the factors were determined during preliminary studies, and the two surfaces were detached at a constant speed (0.1 mm/s). The maximum detachment force F_{max} (N) as a function of displacement was recorded directly from Texture Exponent 32 software (version 5.0, Stable Microsystems, Godalming, UK), whereas the work of mucoadhesion W_{ad} (expressed in µJ), was calculated from the area under the force vs. distance curve. Cellulose paper was used as a negative control.

3.8. Determination of the Residence Time

The in vitro residence time of microparticles was determined using a self-constructed apparatus according to Nakamura et al. [44] (by modifying USP Disintegration tester Electrolab ED-2L, Mumbai, India). Segments of porcine vaginal mucosa (2 cm long) were attached to the internal side of a beaker above the level of 500 mL acetic buffer pH 4.5 at 37 °C ± 0.5 °C. Microparticles (100 mg) were moisturized with acetic buffer (pH 4.5) and put in contact with the mucosal membrane. A plexiglass cylinder (weight 280 g, 6 cm diameter), vertically fixed to the apparatus, was allowed to move up and down to enable the complete immersion of microparticles in the buffer solution [61]. The time required for entire detachment of microparticles from the vaginal tissue was examined visually.

3.9. Swelling Index Test

Each formulation of microparticles was accurately weighted (50 mg) and placed in a Petri plate containing 2 mL acetic buffer (pH 4.5). At the predetermined time intervals (0.25, 0.5, 0.75, 1, 2, and 3 h), microparticles were removed, wiped off with filter paper, and reweighted. The swelling index was calculated by using the subsequent formula:

$$SI = \frac{W_2 - W_1}{W_1} \tag{1}$$

where SI was the swelling index, W_1 was the initial weight of microparticles, and W_2 was the weight of microparticles after the specific swelling time interval. Each test was performed in triplicate.

3.10. In Vitro Drug Release Studies

In vitro release of CLO from microparticles was assessed through natural cellulose membrane Cuprophan (Molecular Weight Cut-Off 10,000 Da, Medicell, London, UK) using an Enhancer cell (Agilent Technologies, Cary, NC, USA). The precisely weighted amount of each formulation of microparticles (referred to 10 mg of CLO) was added to 2 mL of release medium and the mixture was immediately located in the enhancer cell assembly making certain that no entrapped air was present at the membrane-drug carrier interface. Simple CLO dispersion in the release medium (1.0% (w/v)) was used as a control. A USP II dissolution tester (Agilent 708-DS, Agilent Technologies, Cary, NC, USA) equipped with mini paddles and mini vessels (250 mL) was applied to measure the CLO release from the drug reservoir [35]. The dissolution medium—acetic buffer pH 4.5 [57] with addition of 1.0% Cremophor EL—was maintained at 37 °C ± 0.5 °C and the rotation speed was 75 rpm. CLO solubility in the dissolution medium was 0.781 ± 0.091 mg/mL at 37 °C; in order to provide sink conditions, the volume of the medium was 100 mL. Samples (1 mL) were withdrawn at the predetermined time intervals (0.25, 0.5, 1, 2, 3, 4, 6, 8, 24 and 48 h), filtered through 0.45 μm cellulose acetate filters, diluted with mobile phase, and analyzed with the HPLC technique (as described in Section 3.4). Withdrawn samples were substituted with equal volumes of the fresh medium.

3.11. Cell Culture

VK2/E6E7 human vaginal mucosa cell line was maintained according to the attached procedure protocol in ker-SFM with 0.1 ng/mL human recombinant epidermal growth factor, 0.05 mg/mL bovine pituitary extract, 44.1 mg/mL calcium chloride (final concentration 0.4 mM), 50 U/mL penicillin and 50 μg/mL streptomycin. Cells were cultured in Costar flasks (Sigma Aldrich, Steinheim, Germany) and grown in a 5% CO_2 atmosphere at 37 °C to attain subconfluence of 70%–80%. After discarding culture medium, confluent cells were rinsed with 0.25% (w/v) trypsin-0.03% (w/v) EDTA solution, counted in hemocytometer and seeded in 6-well or 24-well plates (Nunc, Thermo Scientific, Darmstadt, Germany) in 2 mL or 1 mL ker-SFM, respectively. Cultures were re-fed with 1–2 mL media per well every two to three days.

3.12. MTT Assay

Prior cytotoxicity studies, different amounts of freshly prepared sterile drug-loaded microparticles with unmodified and βGP cross-linked chitosan (F3, F5, F6) and placebo formulations (P3, P5, P6), which corresponded to a βGP:chitosan ratio of microparticles with CLO, were suspended in sterile acetate buffer pH 4.5 for 4 h under aseptic environment in a laminar flow cabinet Lamil Plus 13 (Karstulan Metalli Oy, Karstula, Finland). Next, the suitable amount of prepared samples were added to 24-well plates containing 1×10^5 per well VK2/E6E7 confluent cells (to give final chitosan concentrations of 0.01 to 0.5 mg/mL) [62] and incubated for 4, 24 or 48 h at 37 °C in a 5% CO_2 humidified atmosphere. The pH of the medium with samples varied in the range between 6.9 (for P3 and F3) to 7.32 (for P6 and F6 formulations). The cytotoxicity assay was performed according to the method of Carmichael et al. [63] with modifications using 3-(4,5-dimethylthiazol-2-yl)-2,5-diphenyltetrazolium bromide (MTT). Absorbance of converted dye in living cells was evaluated at a wavelength of 570 nm. Cell viability of VK2/E6E7 cells cultured in the presence of studied compounds was calculated as a percent of control cells. The control comprising cells not exposed to the chitosan formulations. After treatment of VK2/E6E7 cells with the samples, the ratio of survived to dead cells in tested and control cells was calculated for each concentration of unmodified or chitosan/βGP preparations.

3.13. Flow Cytometry Assessment of Annexin V-FITC/Propidium Iodide Binding

Apoptosis was determined by BD FACSCanto II flow cytometer (BD Biosciences Systems, San Jose, CA, USA) using an Apoptosis Detection Kit II according to the manufacturer's instructions. Annexin V-FITC binds exclusively with phosphatidylserine, which is exposed at the cell surface during apoptosis [64], whereas propidium iodide (PI) selectively stains cells with a disrupted cell membrane and could be used to recognize late apoptotic and dead cells. The combination of annexin V-FITC with PI enables discrimination of viable cells that are unlabeled, apoptotic cells which are stained with annexin V-FITC, and necrotic cells—labeled with both annexin V-FITC and PI. Briefly, after 24 h incubation with the analyzed compounds, the cells were collected, washed twice with PBS, and resuspended in 100 μL binding buffer [65]. For each analysis, 10,000 VK2/E6E7 cells were counted. Cells cultured in a microparticle-free medium were used as controls. Forward scatter and side scatter signals were identified on a logarithmic scale histogram. FITC was detected in the FL1 channel (FL1 539, Threshold value 52). Results were analyzed with FACSDiva software (version 7.0, BD Biosciences Systems, San Jose, CA, USA).

3.14. Fluorescent Microscopy Assay

The morphology of VK2/E6E7 cells incubated for 24 h in the presence of unmodified chitosan or βGP cross-linked chitosan microparticles was observed by inverted fluorescence microscope Nikon Eclipse Ti-U (Nikon, Tokyo, Japan). Prior to the analysis, the cell suspension (250 μL) was stained with 10 μL of the dye mixture (10 μM acridine orange and 10 μM ethidium bromide), which was prepared in PBS. The control was cells incubated without analyzed compounds. The representative images were captured using a Nikon Digital Sight DS-Fi1c camera (Nikon, Tokyo, Japan) at a total of ×200 magnification.

3.15. Analysis of Mitochondrial Membrane Potential

Disruption of the mitochondrial membrane potential (MMP) was assessed using the lipophilic cationic probe 5,5',6,6'-tetrachloro-1,1',3,3'-tetraethylbenzimidazolcarbocyanine iodide (JC-1 MitoScreen kit; BD Biosciences, San Jose, CA, USA) as described previously [66]. Briefly, after 24 h incubation with the analyzed compounds, unfixed cells were washed and resuspended in PBS supplemented with JC-1. Cells were then incubated for 15 min at room temperature in the dark, washed, and resuspended in PBS for immediate BD FACSCanto II flow cytometry analysis. The percentage of cells with disrupted MMP was calculated in the FACSDiva software (version 7.0, BD Biosciences Systems, San Jose, CA, USA).

3.16. Statistical Analysis

Quantitative variables were expressed as the mean ± standard deviation and (or) the median. A statistical analysis was accomplished using nonparametric methods: the Kruskal-Wallis and Mann-Whitney-U test with using the Statistica 12.5 software (StatSoft, Kraków, Poland). Differences between groups were reflected to be significant at $p < 0.05$.

4. Conclusions

To our best knowledge, this work demonstrated for the first time the preparation of chitosan microparticulate delivery carriers by spray drying while using βGP as cross-linking agent. A particular effort was made to evaluate the impact of βGP, applied in different ratios, on chitosan microparticles' characteristics. Unmodified chitosan microparticles had a rough surface, whereas smoother and more intact microparticles F5 and F6 (with βGP:chitosan ratio 1:2 and 1.5:2 (w/w) respectively) were observed. The microparticles' size and drug loading were found to decrease with raising the amount of βGP. Microparticles with βGP/chitosan ratio 1:2 (w/w) exhibited the highest loading efficacy. In addition, the amount of ion cross-linker had a remarkable prolonged effect on the drug release behavior from βGP/chitosan microparticulate dosage forms. All investigated formulations exhibited beneficial mucoadhesive properties and displayed prolonged residence time to porcine vaginal mucosa. A drop in work of adhesion observed for βGP/chitosan microparticulate might be a result of more intact particle surface and lower water content, leading to formation of a less cohesive structure with mucosal tissue.

The data from the cytotoxicity studies indicated that chitosan and βGP/chitosan microparticles affected cell viability and accelerated the rate of apoptosis in human vaginal epithelial cells in a dose-dependent manner. The cytotoxic effect of unmodified and βGP cross-linked chitosan was hardly affected by the presence of drugs. Notably, mitochondrial integrity measured by JC-1 assay appeared to play a substantial role in apoptosis induced by chitosan microparticles in VK2/E6E7 cells. The findings from the cytotoxicity studies added novel insight into the potential safety profile of spray-dried chitosan microparticles and could help to better understand the toxicity of chitosan microparticulate delivery systems.

Acknowledgments: Preparation and characteristics of microparticles were accomplished with the use of equipment purchased by the Medical University of Białystok, Poland as part of the OP DEP 2007–2013, Priority Axis I.3, contract No. POPW.01.03.00-20-008/09 and supported by Medical University of Białystok grant (number N/ST/ZB/16/006/2215). The authors wish to thank Anna Basa from University of Białystok (Białystok, Poland) and Katja Richter from Heppe Medical Chitosan GmbH (Haale, Germany) for their contribution to this work. The equipment of Center BioNanoTechno: Leica EM AC 2000 (Wetzlar, Germany) and SEM Hitachi SH 200 (Tokyo, Japan) was partly supported by EU funds via project number POPW.O1.03.00-20-004/11.

Author Contributions: Emilia Szymańska conceived and designed the experiments, evaluated mucoadhesive, swelling, in vitro release studies and statistical analysis, analyzed the data, and wrote the paper. Marta Szekalska accomplished preparation of microparticles. Zoran Lavrič and Stane Srčič performed DSC measurements and Wojciech Miltyk conducted the VK2/E6E7 2616 cell culture. Robert Czarnomysy and Emilia Szymańska were responsible for fluorescence microscopy and flow cytometry assessments. Katarzyna Winnicka contributed to the conception and design of the experiments, participated in the data analysis and approved the final draft. All authors revised the manuscript and approved the final version.

Conflicts of Interest: The authors declare no conflict of interest. The founding sponsors had no role in the design of the study; in the collection, analyses, or interpretation of data; in the writing of the manuscript, and in the decision to publish the results.

References

1. Martín-Villena, M.J.; Fernández-Campos, F.; Calpena-Campmany, A.C.; Bozal-de Febrer, N.; Ruiz-Martínez, M.A.; Clares-Naveros, B. Novel microparticulate systems for the vaginal delivery of nystatin: Development and characterization. *Carbohydr. Polym.* **2013**, *94*, 1–11. [CrossRef] [PubMed]
2. Singh, M.N.; Hemant, K.S.Y.; Ram, M.; Shivakumar, H.G. Microencapsulation: A promising technique for controlled drug delivery. *Res. Pharm. Sci.* **2010**, *5*, 65–77. [PubMed]
3. Sezer, A.D.; Cevher, E. Topical drug delivery using chitosan nano- and microparticles. *Expert Opin. Drug Deliv.* **2012**, *9*, 1129–1146. [CrossRef] [PubMed]

4. Mishra, M.; Mishra, B. Mucoadhesive microparticles as potential carriers in inhalation delivery of doxycycline hyclate: A comparative study. *Acta Pharm. Sin. B* **2012**, *2*, 518–526. [CrossRef]

5. Patel, J.K.; Patil, P.S.; Sutariya, V.B. Formulation and characterization of mucoadhesive microparticles of cinnarizine hydrochloride using supercritical fluid technique. *Curr. Drug Deliv.* **2016**, *10*, 317–325. [CrossRef]

6. Szekalska, M.; Amelian, A.; Winnicka, K. Alginate microspeheres obtained by the spray drying technique as mucoadhesive carriers of ranitidine. *Acta Pharm.* **2015**, *65*, 15–27. [CrossRef] [PubMed]

7. Cerchiara, T.; Abruzzo, A.; Parolin, C.; Vitali, B.; Bigucci, F.; Gallucci, M.C.; Nicoletta, F.P.; Luppi, B. Microparticles based on chitosan/carboxymethylcellulose polyelectrolyte complexes for colon delivery of vancomycin. *Carbohydr. Polym.* **2016**, *143*, 124–130. [CrossRef] [PubMed]

8. Sinha, V.R.; Singla, A.K.; Wadhawan, S.; Kaushik, R.; Kumria, R.; Bansal, K.; Dhawan, S. Chitosan microspheres as a potential carrier for drugs. *Int. J. Pharm.* **2004**, *274*, 1–33. [CrossRef] [PubMed]

9. Wassmer, S.; Rafat, M.; Fong, W.G.; Baker, A.N.; Tsilfidis, C. Chitosan microparticles for delivery of proteins to the retina. *Acta Biomater.* **2013**, *9*, 7855–7864. [CrossRef] [PubMed]

10. Dash, M.; Chiellini, F.; Ottenbrite, R.M.; Chiellini, E. Chitosan—A versatile semi-synthetic polymer in biomedical applications. *Prog. Polym. Sci.* **2011**, *36*, 981–1014. [CrossRef]

11. Ifuku, S. Chitin and chitosan nanofibers: Preparation and chemical modifications. *Molecules* **2014**, *19*, 18367–18380. [CrossRef] [PubMed]

12. Kong, M.; Chen, X.G.; Xing, K.; Park, H.J. Antimicrobial properties of chitosan and mode of action: A state of the art review. *Int. J. Food Microbiol.* **2010**, *144*, 51–63. [CrossRef] [PubMed]

13. Shaji, J.; Jain, V.; Lodha, S. Chitosan: A novel pharmaceutical excipient. *Int. J. Pharm. Appl. Sci.* **2010**, *1*, 11–28.

14. Yeh, T.H.; Hsu, L.W.; Tseng, M.T.; Lee, P.L.; Sonjae, K.; Ho, Y.C.; Sung, H.W. Mechanism and consequence of chitosan-mediated reversible epithelial tight junction opening. *Biomaterials* **2011**, *32*, 6164–6173. [CrossRef] [PubMed]

15. Nilsen-Nygaard, J.; Strand, S.P.; Vårum, K.M.; Draget, K.I.; Nordgård, C.T. Chitosan: Gels and interfacial properties. *Polymers* **2015**, *7*, 552–579. [CrossRef]

16. Andersen, T.; Bleher, S.; Eide Flaten, G.; Tho, I.; Mattsson, S.; Škalko-Basnet, N. Chitosan in mucoadhesive drug delivery: Focus on local vaginal therapy. *Mar. Drugs* **2015**, *13*, 222–236. [CrossRef] [PubMed]

17. Kavianinia, I.; Plieger, P.G.; Cave, N.J.; Gopakumar, G.; Dunowska, M.; Kandile, N.G.; Harding, D.R. Design and evaluation of a novel chitosan-based system for colon-specific drug delivery. *Int. J. Biol. Macromol.* **2016**, *85*, 539–546. [CrossRef] [PubMed]

18. Sánchez-Sánchez, M.P.; Martín-Illana, A.; Ruiz-Caro, R.; Bermejo, P.; Abad, M.J.; Carro, R.; Bedoya, L.M.; Tamayo, A.; Rubio, J.; Fernández-Ferreiro, A.; et al. Chitosan and kappa-carrageenan vaginal acyclovir formulations for prevention of genital herpes. In vitro and ex vivo evaluation. *Mar. Drugs* **2015**, *13*, 5976–5992. [CrossRef] [PubMed]

19. Caetano, L.A.; Almeida, A.J.; Gonçalves, L.M. Effect of experimental parameters on alginate/chitosan microparticles for BCG encapsulation. *Mar. Drugs* **2016**, *14*, 90–120. [CrossRef] [PubMed]

20. Brako, F.; Raimi-Abraham, B.; Mahalingam, S.; Craig, D.Q.M.; Edirisinghe, M. Making nanofibres of mucoadhesive polymer blends for vaginal therapies. *Eur. Polym. J.* **2015**, *70*, 186–196. [CrossRef]

21. Parhizkar, M.; Sofokleous, P.; Stride, E.; Edirisinghe, M. Novel preparation of controlled porosity particle/fibre loaded scaffolds using a hybrid micro-fluidic and electrohydrodynamic technique. *Biofabrication* **2014**, *6*, 045010–045024. [CrossRef] [PubMed]

22. Zhang, T.; Zhang, C.; Agrahari, V.; Murowchick, J.B.; Oyler, N.A.; Youan, B.B. Spray drying tenofovir loaded mucoadhesive and pH-sensitive microspheres intended for HIV prevention. *Antivir. Res.* **2013**, *97*, 334–346. [CrossRef] [PubMed]

23. Bohr, A.; Wan, F.; Kristensen, J.; Dyas, M.; Stride, E.; Baldursdottír, S.; Edirisinghe, M.; Yang, M. Pharmaceutical microparticle engineering with electrospraying: The role of mixed solvent systems in particle formation and characteristics. *J. Mater. Sci. Mater. Med.* **2015**, *26*, 61–74. [CrossRef] [PubMed]

24. Eltayeb, M.; Stride, E.; Edirisinghe, M.; Harker, A. Electrosprayed nanoparticle delivery system for controlled release. *Mater. Sci. Eng. C Mater. Biol. Appl.* **2016**, *66*, 138–146. [CrossRef] [PubMed]

25. Parhizkar, M.; Reardon, P.J.; Knowles, J.C.; Browning, R.J.; Stride, E.; Pedley, B.R.; Harker, A.H.; Edirisinghe, M. Electrohydrodynamic encapsulation of cisplatin in poly (lactic-co-glycolic acid) nanoparticles for controlled drug delivery. *Nanomedicine* **2016**, *12*, 1919–1929. [CrossRef] [PubMed]

26. Gupta, N.V.; Natasha, S.; Getyala, A.; Bhat, R.S. Bioadhesive vaginal tablets containing spray dried microspheres loaded with clotrimazole for treatment of vaginal candidiasis. *Acta Pharm.* **2013**, *63*, 359–372. [CrossRef] [PubMed]

27. Hani, U.; Shivakumar, H.G.; Gowrav, M.P. Formulation design and evaluation of a novel vaginal delivery system of clotrimazole. *Int. J. Pharm. Sci. Res.* **2014**, *5*, 220–227.

28. Cal, K.; Sollohub, K. Spray drying technique. I: Hardware and process parameters. *J. Pharm. Sci.* **2010**, *99*, 575–586. [CrossRef] [PubMed]

29. Anal, A.K.; Stevens, W.F.; Remuñán-López, C. Ionotropic cross-linked chitosan microspheres for controlled release of ampicillin. *Int. J. Pharm.* **2006**, *7*, 166–173. [CrossRef] [PubMed]

30. Jain, A.; Thakur, K.; Sharma, G.; Kush, P.; Jain, U.K. Fabrication, characterization and cytotoxicity studies of ionically cross-linked docetaxel loaded chitosan nanoparticles. *Carbohydr. Polym.* **2016**, *137*, 65–74. [CrossRef] [PubMed]

31. Mourya, V.K.; Inamdar, N.N. Chitosan—Modifications and applications: Opportunities galore. *React. Funct. Polym.* **2008**, *68*, 1013–1051. [CrossRef]

32. Ding, K.; Yang, Z.; Zhang, Y.L.; Xu, J.Z. Injectable thermosensitive chitosan/β-glycerophosphate/collagen hydrogel maintains the plasticity of skeletal muscle satellite cells and supports their in vivo viability. *Cell Biol. Int.* **2013**, *37*, 977–987. [CrossRef] [PubMed]

33. Rossi, S.; Ferrari, F.; Bonferoni, M.C.; Sandri, G.; Faccendini, A.; Puccio, A.; Caramella, C. Comparison of poloxamer- and chitosan-based thermally sensitive gels for the treatment of vaginal mucositis. *Drug Dev. Ind. Pharm.* **2014**, *40*, 352–360. [CrossRef] [PubMed]

34. Supper, S.; Anton, N.; Seidel, N.; Riemenschnitter, M.; Curdy, C.; Vandamme, T. Thermosensitive chitosan/glycerophosphate-based hydrogel and its derivatives in pharmaceutical and biomedical applications. *Expert Opin. Drug Deliv.* **2014**, *11*, 249–267. [CrossRef] [PubMed]

35. Szymańska, E.; Sosnowska, K.; Miltyk, W.; Rusak, M.; Basa, A.; Winnicka, K. The effect of β-glycerophosphate cross-linking on chitosan cytotoxicity and properties of hydrogels for vaginal application. *Polymers* **2015**, *7*, 2223–2244. [CrossRef]

36. Szymańska, E.; Winnicka, K.; Wieczorek, P.; Sacha, P.T.; Tryniszewska, E.A. Influence of unmodified and beta-glycerophosphate cross-linked chitosan on anti-Candida activity of clotrimazole in semi-solid delivery systems. *Int. J. Mol. Sci.* **2014**, *15*, 17765–17777. [CrossRef] [PubMed]

37. Mucha, M.; Pawlak, A. Thermal analysis of chitosan and its blends. *Thermochim. Acta* **2005**, *427*, 69–76. [CrossRef]

38. Sigma Aldrich. Available online: http://www.sigmaaldrich.com/catalog/product/sigma/g9422?lang=en&region=SI (assessed on 3 August 2015).

39. Bromana, E.; Khoob, C.; Taylor, L.S. A comparison of alternative polymer excipients and processing methods for making solid dispersions of a poorly water soluble drug. *Int. J. Pharm.* **2001**, *222*, 139–151. [CrossRef]

40. D'Cruz, O.J.; Erbeck, D.; Uckun, F.M. A study of the potential of the pig as a model for the vaginal irritancy of benzalkonium chloride in comparison to the nonirritant microbicide PHI-443 and the spermicide vanadocene dithiocarbamate. *Toxicol. Pathol.* **2005**, *33*, 465–476. [CrossRef] [PubMed]

41. Abruzzo, A.; Bigucci, F.; Cerchiara, T.; Saladini, B.; Gallucci, M.C.; Cruciani, F.; Vitali, B.; Luppi, B. Chitosan/alginate complexes for vaginal delivery of chlorhexidine digluconate. *Carbohydr. Polym.* **2013**, *91*, 651–658. [CrossRef] [PubMed]

42. Fernandes, M.; Gonçalves, I.C.; Nardecchia, S.; Amaral, I.F.; Barbosa, M.A.; Martins, M.C. Modulation of stability and mucoadhesive properties of chitosan microparticles for therapeutic gastric application. *Int. J. Pharm.* **2013**, *454*, 116–124. [CrossRef] [PubMed]

43. Chenite, A.; Buschmann, M.; Wang, D.; Chaput, C.; Kandani, N. Rheological characterisation of thermogelling chitosan/glycerol-phosphate solutions. *Carbohydr. Polym.* **2001**, *46*, 39–47. [CrossRef]

44. Nakamura, F.; Ohta, R.; Machida, Y.; Nagai, T. In vitro and in vivo nasal mucoadhesion of some water-soluble polymers. *Int. J. Pharm.* **1996**, *134*, 173–181. [CrossRef]

45. Gupta, J.; Tao, J.Q.; Garg, S.; Al-Kassas, R. Design and development of an in vitro assay for evaluation of solid vaginal dosage forms. *Pharmacol. Pharm.* **2011**, *2*, 289–298. [CrossRef]

46. Baldrick, P. The safety of chitosan as a pharmaceutical excipient. *Regul. Toxicol. Pharmacol.* **2010**, *56*, 290–299. [CrossRef] [PubMed]

47. Gades, M.D.; Stern, J.S. Chitosan supplementation and fecal fat excretion in men. *Obes. Res.* **2003**, *11*, 683–688. [CrossRef] [PubMed]

48. Hirano, S.; Iwata, M.; Yamanaka, K.; Tanaka, H.; Toda, T.; Inui, H. Enhancement of serum lysozyme activity by injecting a mixture of chitosan oligosaccharides intravenously in rabbits. *Agric. Biol. Chem.* **1991**, *55*, 2623–2629.

49. Rodrigues, S.; Dionísio, M.; López, C.R.; Grenha, A. Biocompatibility of chitosan carriers with application in drug delivery. *J. Func. Biomater.* **2012**, *3*, 615–641. [CrossRef] [PubMed]

50. Martin, D.; Lenardo, M. Morphological, biochemical, and flow cytometric assays of apoptosis. *Curr. Protoc. Mol. Biol.* **2001**, *49*. [CrossRef]

51. Prego, C.; Torres, D.; Alonso, M.J. Chitosan nanocapsules as carriers for oral peptide delivery: Effect of chitosan molecular weight and type of salt on the in vitro behaviour and in vivo effectiveness. *J. Nanosci. Nanotechol.* **2006**, *6*, 2921–2928. [CrossRef]

52. Pai, R.V.; Jain, R.R.; Bannalikor, A.S.; Menon, M.D. Development and evaluation of chitosan microparticles based dry powder inhalation formulations of rifampicin and rifabutin. *J. Aerosol Med. Pulm. Drug Deliv.* **2015**, *29*, 179–195. [CrossRef] [PubMed]

53. Meng, J.; Zhang, T.; Agrahari, V.; Ezoulin, M.J.; Youan, B.B. Comparative biophysical properties of tenofovir-loaded, thiolated and nonthiolated chitosan nanoparticles intended for HIV prevention. *Nanomedicine* **2014**, *9*, 1595–1612. [CrossRef] [PubMed]

54. Lemasters, J.J.; Qian, T.; Bradham, C.A.; Brenner, D.A.; Cascio, W.E.; Trost, L.C.; Nishimura, Y.; Nieminen, A.L.; Herman, B. Mitochondrial dysfunction in the pathogenesis of necrotic and apoptotic cell death. *J. Bioenerg. Biomembr.* **1999**, *31*, 305–319. [CrossRef] [PubMed]

55. Czechowska-Biskup, R.; Jarosińska, D.; Rokita, B.; Ulański, P.; Rosiak, J.M. Determination degree of deacetylation of chitosan: Comparison of methods. *Prog. Chem. Appl. Chitin Its Deriv.* **2012**, *17*, 5–20.

56. Squier, C.A.; Mantz, M.J.; Schlievert, P.M.; Davis, C.C. Porcine vagina ex vivo as a model for studying permeability and pathogenesis in mucosa. *J. Pharm. Sci.* **2008**, *97*, 9–21. [CrossRef] [PubMed]

57. The European Directorate for the Quality of Medicines. *The European Pharmacopeia*, 8th ed.; Council of Europe: Strasburg, France, 2014; Volume 1.

58. Pedersen, M. Effect of hydrotropic substances on the complexation of clotrimazole with β-cyclodextrin. *Drug Dev. Ind. Pharm.* **1993**, *19*, 439–448. [CrossRef]

59. Martinac, A.; Filipović-Grčić, J.; Perissutti, B.; Voinovich, D.; Pavelić, Z. Spray-dried chitosan/ethylcellulose microspheres for nasal drug delivery: Swelling study and evaluation of in vitro drug release properties. *J. Microencapsul.* **2005**, *22*, 549–561. [CrossRef] [PubMed]

60. Hájková, R.; Sklenářová, H.; Matysová, L.; Švecová, P.; Solich, P. Development and validation of HPLC method for determination of clotrimazole and its two degradation products in spray formulations. *Talanta* **2007**, *73*, 483–489. [CrossRef] [PubMed]

61. Szymańska, E.; Winnicka, K.; Amelian, A.; Cwalina, U. Vaginal chitosan tablets with clotrimazole-design and evaluation of mucoadhesive properties using porcine vaginal mucosa, mucin and gelatine. *Chem. Pharm. Bull.* **2014**, *62*, 160–167. [CrossRef] [PubMed]

62. Huang, M.; Khor, E.; Lim, L.Y. Uptake and cytotoxicity of chitosan molecules and nanoparticles: Effects of molecular weight and degree of deacetylation. *Pharm. Res.* **2004**, *21*, 344–353. [CrossRef] [PubMed]

63. Carmichael, J.; Degraff, W.; Gazdar, A.; Minna, J.; Mitchell, J. Evaluation of a tetrazolium-based semiautomated colorimetric assay: Assessment of chemosensitivity testing. *Cancer Res.* **1987**, *47*, 936–942. [PubMed]

64. Westermann, B. Molecular machinery of mitochondrial fusion and fission. *J. Biol. Chem.* **2008**, *283*, 13501–13505. [CrossRef] [PubMed]

65. Winnicka, K.; Bielawski, K.; Bielawska, A. Synthesis and cytotoxic activity of G3 PAMAM-NH$_2$ dendrimer-modified digoxin and proscillaridin A conjugates in breast cancer cells. *Pharmacol. Rep.* **2010**, *62*, 414–423. [CrossRef]

66. Czarnomysy, R.; Bielawska, A.; Muszyńska, A.; Bielawski, K. Effects of novel alkyl pyridine platinum complexes on apoptosis in Ishikawa endometrial cancer cells. *Med. Chem.* **2015**, *11*, 540–550. [CrossRef] [PubMed]

marine drugs

Article

Development and Characterization of VEGF165-Chitosan Nanoparticles for the Treatment of Radiation-Induced Skin Injury in Rats

Daojiang Yu [1,2,†], Shan Li [3,†], Shuai Wang [1,†], Xiujie Li [1], Minsheng Zhu [2], Shai Huang [2], Li Sun [2], Yongsheng Zhang [4], Yanli Liu [3,*] and Shouli Wang [2,5,6,*]

1   Department of Plastic Surgery, the Second Affiliated Hospital, Soochow University, Suzhou 215004, China; ydj51087@163.com (D.Y.); wsyy0514@163.com (S.W.); 18362720093@163.com (X.L.)
2   Department of Pathology, School of Biology & Basic Medical Sciences, Soochow University, Suzhou 215123, China; zhuminsheng714@126.com (M.Z.); hshine1992@126.com (S.H.); 20144221021@stu.suda.edu.cn (L.S.)
3   College of Pharmaceutical Science, Soochow University, Suzhou 215123, China; m18862163651@163.com
4   Department of Pathology, the Second Affiliated Hospital of Soochow University, Suzhou 215004, China; shengyongzh@163.com
5   Institute of Radiology & Oncology, Soochow University, Suzhou 215006, China
6   Suzhou Key Laboratory of Tumor Microenvironment Pathology, Suzhou 215123, China
*   Correspondence: liuyanli@suda.edu.cn (Y.L.); wangsoly112@hotmail.com (S.W.);
    Tel.: +86-512-69561421 (Y.L.); +86-512-65880129 (S.W.); Fax: +86-512-65882089 (Y.L.); +86-512-65880103 (S.W.)
†   These authors contributed equally to this work.

Academic Editors: Hitoshi Sashiwa and David Harding
Received: 12 August 2016; Accepted: 31 August 2016; Published: 11 October 2016

Abstract: Radiation-induced skin injury, which remains a serious concern in radiation therapy, is currently believed to be the result of vascular endothelial cell injury and apoptosis. Here, we established a model of acute radiation-induced skin injury and compared the effect of different vascular growth factors on skin healing by observing the changes of microcirculation and cell apoptosis. Vascular endothelial growth factor (VEGF) was more effective at inhibiting apoptosis and preventing injury progression than other factors. A new strategy for improving the bioavailability of vascular growth factors was developed by loading VEGF with chitosan nanoparticles. The VEGF-chitosan nanoparticles showed a protective effect on vascular endothelial cells, improved the local microcirculation, and delayed the development of radioactive skin damage.

Keywords: radiation-induced skin injury; VEGF; chitosan; nanoparticles; apoptosis

1. Introduction

Radiotherapy is one of the standard treatment options for patients with cancer, as ionizing radiation can kill tumor cells by generating free radicals (FR) or reactive oxygen species (ROS). However, the beneficial effects of radiation are modest in the vast majority of patients. The failure of targeted radiotherapy lies in the non-discriminative killing of both cancer and normal cells. Radiation-induced skin injury is the most common complication, and late effects, which are the most severe, are characterized by subcutaneous fibrosis and morbidity [1,2]. Recent advances in cancer radiation biology resulted in the identification of new and promising targets for tumor radiosensitization in addition to normal tissue radioprotection in radiotherapy [3]. Growth factors have been used for a long time to rescue progenitor cells and hematopoietic cells following irradiation. Current research has focused on microcirculation, an important breakthrough in the treatment of radioactive skin damage with typical features of radiation-induced vasculopathies, including necrosis and inflammation within the arterial wall [4]. Therefore, reducing the damage to microcirculation and promoting the reconstruction of impaired microvessels are key issues in the treatment of skin radiation injury.

Vascular endothelial growth factor (VEGF), a master regulator of angiogenesis, has the ability to start a complex cascade of events leading to endothelial cell activation, assembly of new vascular structures, mural cell recruitment, and vessel stabilization [5–7]. However, incorporation of growth factors is difficult due to their short half-life of only several minutes in circulation [8]. Many studies have reported the development of VEGF-loaded nanoparticles for wound-healing angiogenesis, bone regeneration, or inhibition of the graft shrinkage [9–12]. Chitosan (CS) nanoparticles—as non-viral vectors that can deliver cytokines with low toxicity, favorable biodegradability, and lack of immunologic effects—have been studied extensively by our and other research groups [13–16]. Here, we successfully developed CS nanoparticles loaded with VEGF165, a powerful vascular growth factor [17], in a model of radiation-induced skin injury in rats. The results showed that the VEGF-CS nanoparticles protected vascular endothelial cells, improved the local microcirculation, and alleviated radiation-induced skin injury in rats.

2. Results and Discussion

2.1. The Characteristics of CS Nanoparticles Formulated with VEGF

VEGF has been formulated into various sustained release delivery systems because of its short half-life. The long-term VEGF delivery is usually based on its encapsulation in biodegradable polymers, which are designed to release the loaded VEGF in a sustained manner following the degradation of the polymer. To improve the sustained release, VEGF has been conjugated with CS, which secures the loaded growth factor and releases it in a biologically relevant manner. Here, we used CS nanoparticles prepared by the ionic interaction between a positively charged amino group of CS and a negatively charged counter-ion of tripolyphosphate sodium (TPP). CS has been used to improve or control drug release based on its ability to form a hydrogel spontaneously upon contact with multivalent polyanions.

VEGF-CS nanoparticles were prepared by adding a sodium tripolyphosphate solution to a chitosan solution under stirring as previous reported [15]. Figure 1a shows a representative scanning electron microscope (SEM) image of particles with spherical structures and almost uniform size distribution. The average particle size was 386.9 nm, the Zata potential was 25.3 V, and the encapsulation rate reached up to 85% (data not shown). VEGF was released from nanoparticles for 30 days. In the first 2 days, the amount of VEGF released was 10%–20%, with a gradual increase for more than 20 days (Figure 1b). This result demonstrated that CS nanoparticles achieve a controlled release of VEGF, suggesting that VEGF-CS nanoparticles are a promising delivery system for the treatment of radiation-induced skin injury.

(a)   (b)

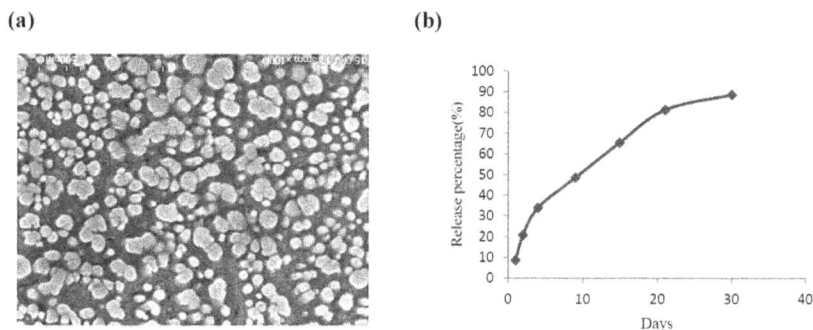

Figure 1. Characteristics of vascular endothelial growth factor (VEGF)-chitosan (CS) nanoparticles. (a) Representative shapes of VEGF-CS nanoparticles via scanning electron microscope (SEM) technology; (b) Release percentage of VEGF-chitosan nanoparticles in vitro.

2.2. Establishment of a Radiation-Induced Skin Injury Model

All rats survived after receiving 45 Gy X-ray irradiation, with varying degrees of depression, reduced activity, and weight loss. Acute skin reactions were detected after 1 week of irradiation. Hair loss was detected at 2 weeks after irradiation, while local skin erythema, blisters, and eczema appeared at approximately 3 weeks after irradiation. Skin erosion, necrosis, and ulceration were observed at 4 weeks after irradiation, and chronic ulcers developed gradually starting at 5 weeks (Figure 2a). In the histological study, as shown in Figure 2b, injury can be found in the irradiated skin group. The number of vessels decreased after 1 week of irradiation, and inflammatory cells began to infiltrate. The vascular basement membrane was partly disintegrated and incomplete, and necrosis of endothelial cells and microthrombosis were observed. After 2 weeks of irradiation, the blood vessels disappeared, and a large number of inflammatory cells infiltrated the irradiation area.

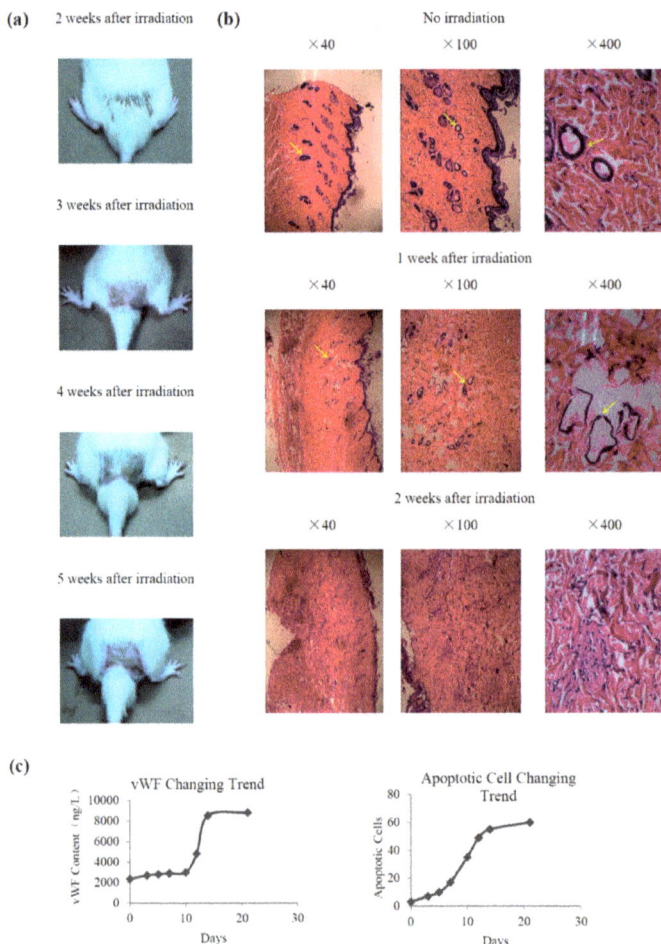

Figure 2. Establishment of a radiation-induced skin injury model. (**a**) Macroscopic images of radiation-induced skin injury in rats at 2–5 weeks; (**b**) Histological changes associated with radiation-induced skin injury in rats at 1 and 2 weeks after irradiation and in the control group; (**c**) Changes of von Willebrand factor (vWF) content and number of apoptotic vascular endothelial cells. Arrows pointing vessels.

Apoptosis of vascular endothelial cells and changes of von Willebrand factor (vWF), which closely resembled those of vascular disease, were a sensitive index of vascular endothelial cell injury [18–20]. Figure 2c shows that the contents of vWF and the number of apoptotic cells increased after 1 week of irradiation and nearly reached a peak after 2 weeks, then remained stable, with significantly higher values than those of the non-irradiated group ($p < 0.05$).

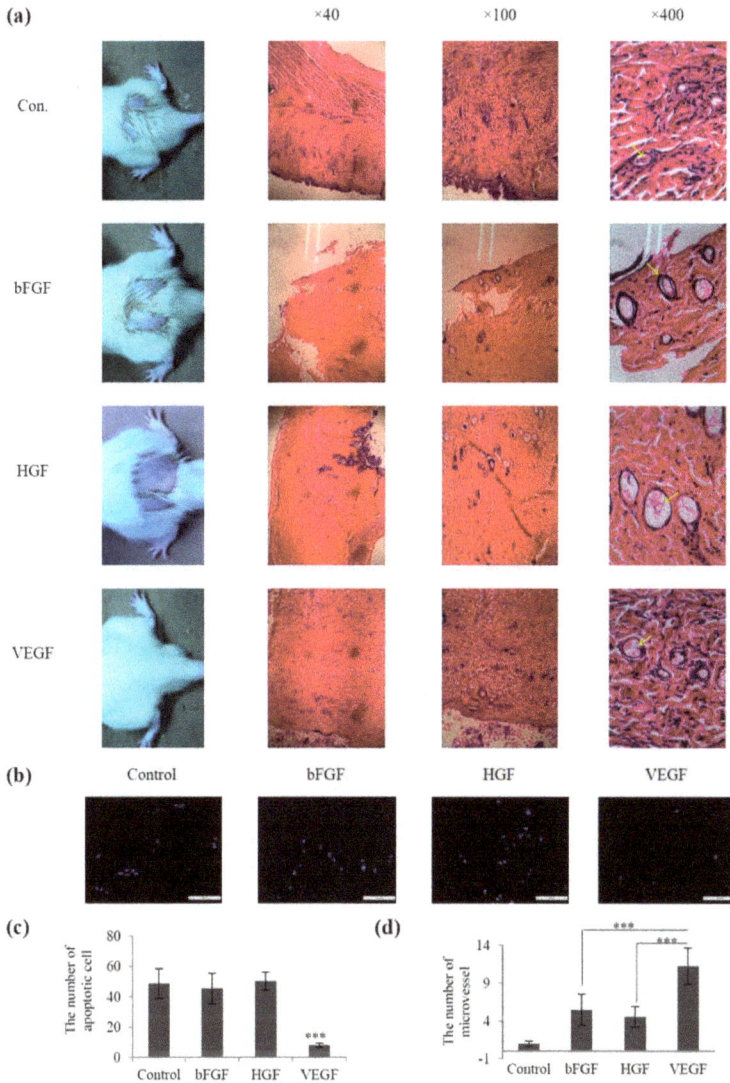

Figure 3. Screening of growth factor treatment for radiation-induced skin injury. (**a**) Macroscopic images and histological changes in irradiation-induced skin injury groups treated with different growth factors; (**b**) Representative images of apoptosis staining-positive cells upon basic fibroblast growth factor (bFGF), hepatocyte growth factor (HGF), and VEGF treatment ($\times 200$); (**c**) Quantification of apoptotic cells upon bFGF, HGF, and VEGF treatment (**d**) Number of microvessels in skin sections of all groups observed under the microscope. Data represent the mean $\pm$ SD from three independent experiments. *** $p < 0.005$ (**c–d**). Arrows pointing vessels.

2.3. Screening of Vascular Growth Factor Treatment in Radiation-Induced Skin Injury

The same amounts of growth factors such as basic fibroblast growth factor (bFGF), hepatocyte growth factor (HGF), and VEGF were used for radiation-induced skin injury treatment. Figure 3a shows that after 2 weeks of treatment, all rats displayed hair loss in the control group, which was treated with equal volumes of normal saline. The bFGF- and HGF-treated groups showed slight hair loss, while the VEGF-treated group had no significant hair loss. Furthermore, the development of hair loss and ulcers was delayed in the VEGF group compared with that in the other groups. Morphological studies showed that the VEGF-treated group showed a nearly normal strata structure, complete re-epithelialization, and few hair follicles, whereas the other groups showed thickening of the epidermal layer and poorly vascularized granulation tissue. In addition, the number of apoptotic cells was lower in the VEGF-treated group than in the other groups (Figure 3b,c). The number of microvessels was higher in the VEGF group than in the other groups, and the difference was statistically significant ($p < 0.05$) (Figure 3d). Taken together, these data suggested that the rats injected with VEGF had better results than the other groups. This observation coincides with the results of other research groups, who used EVGF165 for the treatment of wounds in diabetic mice [21].

2.4. Effect of VEGF-CS Nanoparticles on Healing of Irradiation-Induced Skin Injury

VEGF165 promotes tissue repair in a rat model of radiation-induced injury [22] or relieves endothelial injury after deep vein thrombectomy [23]. In the present study, we explored the effects of different dose of VEGF165 loaded in CS nanoparticles and different delivery systems on irradiation-induced skin injury. As shown in Figure 4a, there were no differences in hair loss and ulcer development time between the group receiving a single treatment of VEGF165 (700 ng/mL) and the control group (injected with saline) ($p > 0.05$). However, when VEGF165 was injected daily (100 ng/mL) for 1 week, it delayed hair loss and ulcer development, and the difference was statistically significant compared with the control group ($p < 0.01$). A single injection of VEGF165 bound to nanoparticles (700 ng VEGF165) had a significantly better effect regarding the delay in hair loss and ulcer formation than that of the group receiving VEGF165 injection ($p < 0.01$). These results were confirmed by hematoxylin and eosin (HE) staining, which showed few microvessels in groups A and B, while rich microvessels were observed in groups C and D (Figure 4b). Figure 4c shows a significantly greater number of microvessels in the group treated with VEGF165 daily (100 ng/mL) for 1 week and that treated with a single dose of VEGF165 nanoparticles (loaded 700 ng VEGF165) than in the control group and the group receiving a single treatment of VEGF165 700 ng/mL ($p < 0.01$). These results were confirmed by measuring the content of vWF (Figure 4d) and suggested the potential value of VEGF-CS nanoparticles for the treatment of irradiation-induced skin injury.

2.5. Mechanism Underlying the Effect of VEGF165-CS on Alleviating Radiation-Induced Skin Injury

To explore the possible mechanism by which VEGF165-CS nanoparticles alleviated radiation-induced skin injury, we further examined the expression of VEGF165 and caspase3, which plays an important role in apoptosis in endothelial cells. As shown in Figure 6a, caspase3 expression was lower in groups C and D than in groups A and B. This result suggested that the effects of VEGF165 on inhibiting apoptosis and protecting endothelial cells in the form of nanoparticles were superior to those of a single injection and the control group. In addition, the expression of VEGF165 in groups C and D was significantly higher than that in groups A and B (Figure 6b). Furthermore, extensive vascular tissue and a complete lumen were observed in group D.

(a)

(b)

(c)

(d)

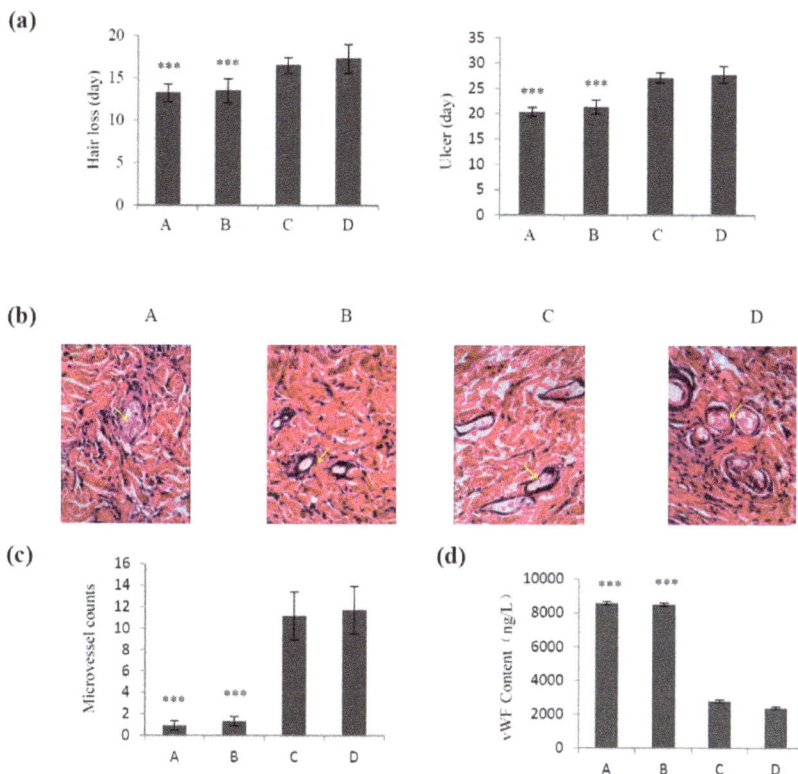

Figure 4. Effect of CS-VEGF nanoparticles on healing of irradiation-induced skin injury. The 64 rats bearing irradiation-induced skin injury were divided into four groups (A–D): A, single treatment with 1 mL normal saline as control; B, single treatment with 1 mL VEGF165 (700 ng/mL); C, treatment with 1 mL VEGF165 (100 ng/mL) daily for 1 week; and D, single treatment with 1 mL heavy suspension of VEGF165-CS nanoparticles. (**a**) Hair loss and ulcer development in each group after irradiation; (**b**) Hematoxylin and eosin staining ($\times 100$) of skin tissues; (**c**) Number of microvessels; (**d**) vWF content in each group. *** $p < 0.005$ (**c**–**d**). Arrows pointing to vessels.

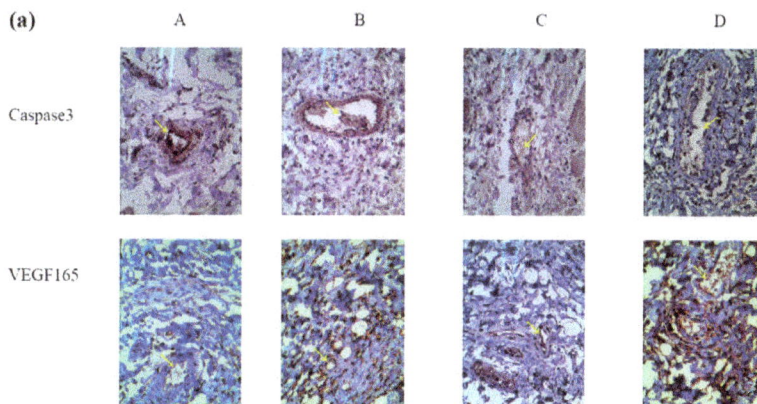

(a)

Figure 5. *Cont.*

(b)

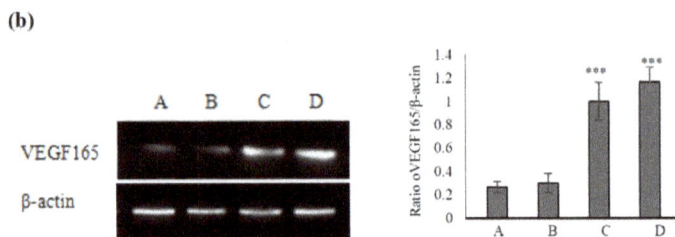

Figure 6. Mechanism of VEGF165-CS nanoparticle-mediated alleviation of radiation-induced skin injury. The 64 rats bearing irradiation-induced skin injury were divided into four groups (A–D): A, single treatment with 1 mL normal saline as control; B, single treatment with 1 mL VEGF165 (700 ng/mL); C, treatment with 1 mL VEGF165 (100 ng/mL) daily for 1 week; and D, single treatment with 1 mL heavy suspension of VEGF165-CS nanoparticles. (**a**) Immunohistochemical staining of caspase3 and VEGF165 in each group; (**b**) Western blot assessment of VEGF165 expression in each group (normalized to β-actin). Data represent the mean ± SD from three independent experiments. *** $p < 0.005$. Arrows pointing vessels or VEGF165 and caspase3 expression.

3. Experimental Procedures

3.1. Materials

Male Sprague Dawley (SD) rats (4 weeks old) were purchased from Shanghai SLAC Laboratory Animal Co., Ltd. (Shanghai, China). Chitosan with a degree of deacetylation of 95% and average molecular weight of 200 kDa, sodium tripolyphosphate, NaOH, and bFGF protein were obtained from Gibco (Shanghai, China). VEGF165 was purchased from Sigma-Aldrich (St. Louis, MO, USA). HGF was purchased from Invitrogen (Life Technologies, Carlsbad, CA, USA).

3.2. Methods

3.2.1. The Establishment of a Model

Seventy-two healthy male rats weighting 220 ± 20 g SD were used. Rats were anesthetized by intraperitoneal injection of 10% chloral hydrate (0.35 mL/100 g). Rats were fixed with adhesive tape on a plastic plate. A 3 cm thick piece of lead was used to shield the rats and localize the radiation field (45 × 40 mm). A single dose of 45 Gy was administered to the buttock skin of each rat at a dose rate of 600 cGy/min using a 4 MeV electron beam accelerator (Phillips, Amsterdam, the Netherlands). To observe the hair and skin changes in the irradiated area, the animals were sacrificed and skin tissues and blood samples were collected. All animal procedures were approved by the Ethics Review Committee for Animal Experimentation of Soochow University.

3.2.2. Enzyme-Linked Immunosorbent Assay (ELISA)

To detect the changes of vWF content, blood samples were analyzed by ELISA. Rats were sacrificed and blood samples were collected into citrated tubes, centrifuged at 3000 g (relative centrifugal force, RCF) for 10 min, and stored at 4 °C for subsequent analysis by ELISA according to the manufacturer's data sheets (Shenggong, Shanghai, China).

3.2.3. Cell Apoptosis Analysis

Cell apoptosis analysis was performed using the One Step TUNEL Apoptosis Detection Kit (Merck Millipore, Shanghai, China). Specimens were frozen and sections were fixed with 4% paraformaldehyde, and incubated in an ice bath. Then, the cells were stained with Hoechst 33258 (10 µg/mL) for 30 min. Nuclear condensation and fragmentation were observed under a fluorescence microscope (Eclipse TE2000-U, Nikon, Japan.)

3.2.4. Screening of Vascular Growth Factors

The same approach was used to establish an irradiation-induced skin injury model as that described above. The injury animals were assigned to four groups ($n = 16$) and injected subcutaneously with 1 mL (100 ng/mL) HGF, VEGF, bFGF, or an equal volume of normal saline as control once a day for 1 week. The observation parameters were the same as those described above.

3.2.5. The Preparation and Characterizations of CS Nanoparticles

CS nanoparticles were created by modified ionic gelation with negatively charged TPP ions [24]. Briefly, CS with a molecular weight of 200 kDa and a degree of deacetylation of 95% was dissolved to a concentration of 2 mg/mL in 1% acetic acid solution, and NaOH solution was added with slow stirring, followed by an equal volume of protein solution (20 ng/mL). TPP at a concentration of 1.5 mg/mL was prepared with deionized water and added dropwise under constant stirring to the mixture [25]. The physical size and zeta potential of the nanoparticles were measured using a 3000HSA Zetasizer (Malvern Instruments, Malvern, UK). The morphology of nanoparticles was observed by scanning electron microscopy (SEM).

3.2.6. Release of VEGF from VEGF-CS Nanoparticles

The constant temperature oscillation method was used to measure VEGF association efficiency. CS nanoparticles (5 mg) and 3 mL PBS buffer were mixed in a 5 mL test tube. After constant shaking (100 r/min) at 37 °C, 3 mL of the supernatant were collected at different time periods and added with fresh medium at the same time. Then the release of protein content and the cumulative release rate of each time point were determined.

3.2.7. Application of VEGF-CS Nanoparticles in Vivo

Sixty-four SD rats bearing irradiation-induced skin injury were randomly assigned to four groups according to the different VEGF165 delivery systems as follows: Group A, single injection of 1 mL normal saline only; Group B, single injection of 1 mL VEGF165 (700 ng/mL); Group C, injection of 1 mL VEGF165 (100 ng/mL) daily for 1 week; Group D, single injection of 1 mL suspension of VEGF165-CS nanoparticles (loaded 700 ng VEGF165). The observation parameters were as mentioned above.

3.2.8. Hematoxylin and Eosin Staining and Immunohistochemistry

The skin specimens were fixed in 10% formalin for 2 days, embedded in paraffin, sectioned at 5 μm, and mounted. After the sections were stained with HE, the morphology and vascular density of skin tissues were observed under a light microscope (Olympus CX31, ×40 magnification). Immunohistochemistry staining was performed by the Streptavidin–Peroxidase kit method according to the manufacturer's guide. In brief, paraffin sections were deparaffinized and incubated at 4 °C overnight with anti-VEGF165 and Caspase3 (Abcam, Inc., Cambridge, MA, USA) in PBS containing 1% BSA. Then, the slides were incubated with peroxidase-conjugated IgG (Shanghai Genomics, Shanghai, China) and counterstained with hematoxylin.

3.2.9. Western Blot Determination of VEGF165 Expression

Skin tissues were harvested and lysed in TNES buffer. Equivalent 25 μg aliquots of proteins were electrophoresed and electrotransferred onto polyvinylidine difluoride membranes. After the blots were blocked, the membranes were incubated with 1:1000 anti-VEGF165 (Santa Cruz Biotechnology, Santa Cruz, CA, USA) for 3 h at room temperature, and visualized with peroxidase-conjugated secondary antibodies using an enhanced chemiluminescence detection system (Santa Cruz, CA, USA).

4. Conclusions

In the present study, we established a model of radioactive skin damage, and demonstrated that VEGF can inhibit apoptosis in vascular endothelial cells and accelerate wound healing in irradiated areas. However, the biological function of VEGF depends on its method of delivery. We developed a CS nanoparticle delivery system modified by TPP and successfully loaded VEGF165. Assessment of microvessels, vWF content, and apoptosis demonstrated that VEGF165-CS nanoparticles delayed the development of radiation-induced skin injury and promoted healing. The design of VEGF165-CS nanoparticles with controlled local release of VEGF165 for long periods provides an excellent delivery system for the treatment of radiation-induced skin injury.

Acknowledgments: This work was supported in part by Chinese Nature Science Foundation (81272738, 81372867, 81472297) and Project Founded by the Priority Academic Program Development of Jiangsu Higher Education Institutions.

Author Contributions: Shouli Wang and Yanli Liu designed the experiments and wrote the manuscript. Daojiang Yu, Shan Li, Shuai Wang, Xiujie Li, Minsheng Zhu, Shai Huang, Li Sun and Yongsheng Zhang performed the experiments and analyzed the data. All authors endorse the full content of this work.

Conflicts of Interest: The authors declare no conflict of interest.

References

1. Wang, X.J.; Lin, S.; Kang, H.F.; Dai, Z.J.; Bai, M.H.; Ma, X.L.; Ma, X.B.; Liu, M.J.; Liu, X.X.; Wang, B.F. The effect of *RHIZOMA COPTIDIS* and *COPTIS CHINENSIS* aqueous extract on radiation-induced skin injury in a rat model. *BMC Complement. Altern. Med.* **2013**, *13*, 105. [CrossRef] [PubMed]
2. Kim, J.H.; Kolozsvary, A.J.; Jenrow, K.A.; Brown, S.L. Mechanisms of radiation-induced skin injury and implications for future clinical trials. *Int. J. Radiat. Biol.* **2013**, *89*, 311–318. [CrossRef] [PubMed]
3. Kumar, S.; Singh, R.K.; Meena, R. Emerging targets for radioprotection and radiosensitization in radiotherapy. *Tumour Biol.* **2016**. [CrossRef] [PubMed]
4. Lucas, J.; Mack, W.J. Effects of ionizing radiation on cerebral vasculature. *World Neurosurg.* **2014**, *81*, 490–491. [CrossRef] [PubMed]
5. Carmeliet, P. Angiogenesis in health and disease. *Nat. Med.* **2003**, *9*, 653–660. [CrossRef] [PubMed]
6. Caron, C.; DeGeer, J.; Fournier, P.; Duquette, P.M.; Luangrath, V.; Ishii, H.; Karimzadeh, F.; Lamarche-Vane, N.; Royal, I. CdGAP/ARHGAP31, a Cdc42/Rac1 GTPase regulator, is critical for vascular development and VEGF-mediated angiogenesis. *Sci. Rep.* **2016**, *6*. [CrossRef] [PubMed]
7. Gianni-Barrera, R.; Trani, M.; Reginato, S.; Banfi, A. To sprout or to split? VEGF, notch and vascular morphogenesis. *Biochem. Soc. Trans.* **2011**, *39*, 1644–1648. [CrossRef] [PubMed]
8. George, M.L.; Eccles, S.A.; Tutton, M.G.; Abulafi, A.M.; Swift, R.I. Correlation of plasma and serum vascular endothelial growth factor levels with platelet count in colorectal cancer: Clinical evidence of platelet scavenging? *Clin. Cancer Res.* **2000**, *6*, 3147–3152. [PubMed]
9. Geng, H.; Song, H.; Qi, J.; Cui, D. Sustained release of VEGF from PLGA nanoparticles embedded thermo-sensitive hydrogel in full-thickness porcine bladder acellular matrix. *Nanoscale Res. Lett.* **2011**, *6*, 312. [CrossRef] [PubMed]
10. Subbiah, R.; Hwang, M.P.; Van, S.Y.; Do, S.H.; Park, H.; Lee, K.; Kim, S.H.; Yun, K.; Park, K. Osteogenic/angiogenic dual growth factor delivery microcapsules for regeneration of vascularized bone tissue. *Adv. Healthc. Mater.* **2015**, *4*, 1982–1992. [CrossRef] [PubMed]
11. Mohandas, A.; Anisha, B.S.; Chennazhi, K.P.; Jayakumar, R. Chitosan-hyaluronic acid/VEGF loaded fibrin nanoparticles composite sponges for enhancing angiogenesis in wounds. *Colloids Surf. B Biointerfaces* **2015**, *127*, 105–113. [CrossRef] [PubMed]
12. Jiang, X.; Xiong, Q.; Xu, G.; Lin, H.; Fang, X.; Cui, D.; Xu, M.; Chen, F.; Geng, H. VEGF-Loaded Nanoparticle-Modified BAMAs Enhance Angiogenesis and Inhibit Graft Shrinkage in Tissue-Engineered Bladder. *Ann. Biomed. Eng.* **2015**, *43*, 2577–2586. [CrossRef] [PubMed]
13. Al Rubeaan, K.; Rafiullah, M.; Jayavanth, S. Oral insulin delivery systems using chitosan-based formulation: A review. *Expert Opin. Drug Deliv.* **2016**, *13*, 223–237. [CrossRef] [PubMed]

14. Lauzon, M.A.; Daviau, A.; Marcos, B.; Faucheux, N. Nanoparticle-mediated growth factor delivery systems: A new way to treat Alzheimer's disease. *J. Control. Release* **2015**, *206*, 187–205. [CrossRef] [PubMed]

15. Xu, Q.; Guo, L.; Gu, X.; Zhang, B.; Hu, X.; Zhang, J.; Chen, J.; Wang, Y.; Chen, C.; Gao, B.; et al. Prevention of colorectal cancer liver metastasis by exploiting liver immunity via chitosan-TPP/nanoparticles formulated with IL-12. *Biomaterials* **2012**, *33*, 3909–3918. [CrossRef] [PubMed]

16. Wang, S.L.; Yao, H.H.; Qin, Z.H. Strategies for short hairpin RNA delivery in cancer gene therapy. *Expert Opin. Biol. Ther.* **2009**, *9*, 1357–1368. [CrossRef] [PubMed]

17. Losordo, D.W.; Vale, P.R.; Symes, J.F.; Dunnington, C.H.; Esakof, D.D.; Maysky, M.; Ashare, A.B.; Lathi, K.; Isner, J.M. Gene therapy for myocardial angiogenesis: Initial clinical results with direct myocardial injection of phVEGF165 as sole therapy for myocardial ischemia. *Circulation* **1998**, *98*, 2800–2804. [CrossRef] [PubMed]

18. Hassler, S.N.; Johnson, K.M.; Hulsebosch, C.E. Reactive oxygen species and lipid peroxidation inhibitors reduce mechanical sensitivity in a chronic neuropathic pain model of spinal cord injury in rats. *J. Neurochem.* **2014**, *131*, 413–417. [CrossRef] [PubMed]

19. Ruggeri, Z.M. Structure of von Willebrand factor and its function in platelet adhesion and thrombus formation. *Best Pract. Res. Clin. Haematol.* **2001**, *14*, 257–279. [CrossRef] [PubMed]

20. Lip, G.Y.; Blann, A. von Willebrand factor: A marker of endothelial dysfunction in vascular disorders? *Cardiovasc. Res.* **1997**, *34*, 255–265. [CrossRef]

21. Yoon, C.S.; Jung, H.S.; Kwon, M.J.; Lee, S.H.; Kim, C.W.; Kim, M.K.; Lee, M.; Park, J.H. Sonoporation of the minicircle-VEGF(165) for wound healing of diabetic mice. *Pharm. Res.* **2009**, *26*, 794–801. [CrossRef] [PubMed]

22. Wang, T.; Liao, T.; Wang, H.; Deng, W.; Yu, D. Transplantation of bone marrow stromal cells overexpressing human vascular endothelial growth factor 165 enhances tissue repair in a rat model of radiation-induced injury. *Chin. Med. J.* **2014**, *127*, 1093–1099. [PubMed]

23. Tang, J.J.; Meng, Q.Y.; Cai, Z.X.; Li, X.Q. Transplantation of VEGFl65-overexpressing vascular endothelial progenitor cells relieves endothelial injury after deep vein thrombectomy. *Thromb. Res.* **2016**, *137*, 41–45. [CrossRef] [PubMed]

24. Mohammadpourdounighi, N.; Behfar, A.; Ezabadi, A.; Zolfagharian, H.; Heydari, M. Preparation of chitosan nanoparticles containing *Naja naja* oxiana snake venom. *Nanomedicine* **2010**, *6*, 137–143. [CrossRef] [PubMed]

25. Vimal, S.; Taju, G.; Nambi, K.S.N.; Majeed, S.A.; Babu, V.S.; Ravi, M.; Hameed, A.S.S. Synthesis and characterization of CS/TPP nanoparticles for oral delivery of gene in fish. *Aquaculture* **2012**, *358*, 14–22. [CrossRef]

marine drugs

MDPI

Article

Chitosan Oligosaccharide Reduces Propofol Requirements and Propofol-Related Side Effects

Zhiwen Li [1], Xige Yang [1], Xuesong Song [1], Haichun Ma [1,*] and Ping Zhang [2,*]

[1] Department of Anesthesiology, the First Hospital of Jilin University, Changchun 130021, China;
    li_zhiwen1159@sina.com (Z.L.); yangxige_xg@sina.com (X.Y.); songxs1981@sina.com (X.S.)
[2] Department of Hepatobiliary and Pancreatic Surgery, the First Hospital of Jilin University,
    Changchun 130021, China
*   Correspondence: mahc_cc@126.com (H.M.); zhangp_1978@126.com (P.Z.);
    Tel.: +86-431-88782955 (H.M.); +86-431-85667038 (P.Z.)

Academic Editors: Hitoshi Sashiwa and David Harding
Received: 2 November 2016; Accepted: 29 November 2016; Published: 21 December 2016

Abstract: Propofol is one of the main sedatives but its negative side effects limit its clinical application. Chitosan oligosaccharide (COS), a kind of natural product with anti-pain and anti-inflammatory activities, may be a potential adjuvant to propofol use. A total of 94 patients receiving surgeries were evenly and randomly assigned to two groups: 10 mg/kg COS oral administration and/or placebo oral administration before being injected with propofol. The target-controlled infusion of propofol was adjusted to maintain the values of the bispectral index at 50. All patients' pain was evaluated on a four-point scale and side effects were investigated. To explore the molecular mechanism for the functions of COS in propofol use, a mouse pain model was established. The activities of Nav1.7 were analyzed in dorsal root ganglia (DRG) cells. The results showed that the patients receiving COS pretreatment were likely to require less propofol than the patients pretreated with placebo for maintaining an anesthetic situation ($p < 0.05$). The degrees of injection pain were lower in a COS-pretreated group than in a propofol-pretreated group. The side effects were also more reduced in a COS-treated group than in a placebo-pretreated group. COS reduced the activity of Nav1.7 and its inhibitory function was lost when Nav1.7 was silenced ($p > 0.05$). COS improved propofol performance by affecting Nav1.7 activity. Thus, COS is a potential adjuvant to propofol use in surgical anesthesia.

Keywords: chitosan oligosaccharide; propofol; surgery patients; mouse; dorsal root ganglia; voltage-gated sodium channel gene 1.7

1. Introduction

Propofol (2,6-diisopropylphenol), as a sedative agent, has been used widely in the induction of surgical anesthesia [1]. However, propofol-induced side effects become apparent [2], including hypotension and respiratory depression [3]. Propofol-induced injection pain is a major issue for propofol as an anesthetic in surgery [4,5]. Various alternative and folk remedies have also been used effectively for many years [6–8]. Remifentanil preventing propofol-induced injection pain has been proved effective. However, the combination therapy will be affected by the time interval between remifentanil and propofol injection, as well as the dosage of remifentanil [4]. Lidocaine is often used before being injected with propofol. Lidocaine pretreatment or mixed with propofol has also been used successfully for preventing propofol-induced pain [9]. Although the effectiveness is obvious, the side effects of the medicine are also palpable [10,11].

Thus, it is critical to explore a new agent for preventing or treating pain disorders. Chitosan oligosaccharide (COS) is a polysaccharide mainly obtained from crustacean shells and consists of

2-amino-2-deoxy-D-glucan combined with glycoside linkages. COS is made from chitin, which is a homopolymer of 1-4 linked 2-acetamido-2-deoxy-β-D-glucopyranose. COS will be formed when chitin is deacetylated >50%. COS can be applied in many primary industries, including microbial control in agriculture, maintenance of overall fruit and vegetable quality [12] and nutritional dietary additive [13]. Chitosan has many medical and pharmaceutical uses with anti-inflammation and antioxidant activities and fewer side effects [14,15]. The analgesic effect of COS on pain has been proved due to its absorption of proton ions [16]. Thus, COS may be a potential adjuvant to propofol use. To understand the functions of COS, it is necessary to explore the molecular mechanism for the role of COS in propofol therapy.

Voltage-gated sodium channels (Navs) are important indicators of the development of mammalian hyperalgesia [17]. Navs are localized in a mammalian central nervous system [18,19] and DRG (dorsal root ganglia) [20,21]. Navs participate in the pain caused by inflammatory responses [22]. Carrageenan and complete Freund's adjuvant (CFA) have been used widely to produce mechanical and thermal hyperalgesia in an inflammatory animal model [23–26]. Thus, these models provide convenient tools in exploring the molecular mechanism of a pain cause. There are many members of Navs with different functions. Three main voltage-gated sodium channels, Nav1.7, Nav1.8, and Nav1.9, are preferentially expressed in dorsal root ganglia (DRG) cells [27]. These channels are involved in different pain. Nav1.9 and Nav1.8 play important roles in the development of cold pain [28]. Previous work showed that little change could be found for inflammation-induced hypersensitivity in the mice lacking Nav1.8 or Nav1.9 [28]. Comparatively, a great reduction in hypersensitivity could be found in Nav1.7 knockout mice [29]. Furthermore, Nav1.7 is essential for burn-induced heat hypersensitivity [30]. An alpha-subunit gene, SCN9A, encodes the Nav1.7 sodium channel [31,32]. An earlier study indicated that SCN9A is essential for human nociception [33]. Sodium channel Nav1.7 is associated with the reduction of neuropathic pain, which is caused by chronic constriction injury of the sciatic nerve in animal models. Behavior tests indicated that the thresholds for thermal and mechanical hyperalgesia were greatly reduced in neuropathic pain models. Meanwhile, the levels of Nav1.7 were significantly increased in DRG cells [34]. In contrast, loss-of-function mutations of Nav1.7 caused congenital insensitivity to pain [35]. Intrathecal injection of Nav1.7 shRNA reduced the levels of Nav1.7 and inactivated astrocytes and microglia of DRG. Nav1.7 can improve the pain tolerance in an animal model [36]. Given the key role of Nav1.7 in human pain, the effects of dual therapy on Nav1.7 were investigated.

To uncover the more specific functions of the combined therapy of COS and propofol, the present study was performed to examine the effects of a combined therapy on the level of Nav1.7 in a mouse pain model.

2. Results

2.1. Chitosan Oligosaccharide (COS) Pretreatment Reduces Propofol Dose during Anesthesia

According to an earlier study, COS is a kind of hemostatic agent, which can reduce pain by blocking nerve endings [37]. COS pretreatment may prevent propofol-induced injection pain and reduce propofol dose. Thus, the effects of COS on propofol doses were measured. As Figure 1 shows, there was no statistical significance of differences for propofol dose at intubating conditions ($p > 0.05$). In contrast, the effect-site concentration of propofol was greatly lower in CG (COS-pretreated group) than that in PG (placebo-pretreated group) ($p < 0.05$). The results suggest that COS pretreatment reduces propofol dose during anesthesia.

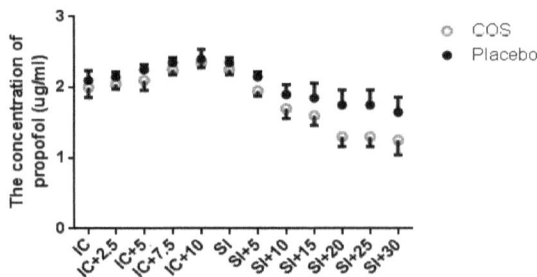

Figure 1. The effects of chitosan oligosaccharide (COS) on propofol requirements. All the selected subjects were evenly assigned to two groups before being injected with propofol: 10 mg/kg COS oral administration and 10 mg/kg placebo oral administration. After five min, propofol was started with step increases of 0.5 µg/mL/2.5 min until the patient lost consciousness. Propofol target-controlled infusion (TCI) was adjusted to maintain the values of bispectral index (BIS) at 50.

2.2. The Incidence of Propofol-Induced Injection Pain in the Subjects Undergoing Surgery

Propofol induces high-incidence pain during intravenous injection. However, few non-pharmacological methods have been applied to control propofol-induced injection pain. COS may be a potential natural product to control the pain. The effects of COS on propofol-induced injection pain were measured. As Table 1 shows, the incidence of propofol-induced pain at a four-point scale in the subjects undergoing surgery was higher in PG than in CG ($p < 0.05$). Furthermore, there was no toxic symptom of COS in all subjects. The results suggest that COS may inhibit the propofol-induced injection pain and can be a potential adjuvant to propofol use.

Table 1. Intravenous COS pretreatment reduces propofol-induced pain.

Grade, *n* (%)	CG (*n* = 47)	PG (*n* = 47)	Chi-Square Statistic	*p* Values
No pain	36 (76.6)	6 (12.77)		
Mild pain	4 (8.51)	12 (25.53)	39.25	0.000
Moderate pain	5 (10.64)	16 (34.04)		
Severe pain	2 (4.26)	13 (27.66)		
Total pain	11 (23.4)	41 (87.23)	38.736	0.000

Note: Chi-square test was performed. 4 × 4 contingency test was used for the comparison of four-grade pain and 2 × 2 contingency test was used for the comparison of total pain. BMI, body mass index. There is statistical significance of differences if $p < 0.05$.

2.3. COS Pretreatment Reduces the Side Effects of Propofol

Besides propofol-induced injection pain, propofol can cause some other side effects. For instance, propofol use induces sedation and may have a significant effect on the pattern of upper airway obstruction [38]. Hypotension has been reported to be a common adverse effect caused by propofol, but there is no reliable method to determine which patients have the risk for propofol-induced hypotension [39]. Therefore, it is necessary to find a new method to control these side effects caused by propofol. Based on this idea, the effects of COS on these side effects were measured. Table 2 shows the most common side effects, which were found in both groups. The patients had lower inadequate ventilation in CG than in PG ($p < 0.05$). Similarly, the patients had a lower incidence of tachycardia and hypotension in CG than in PG ($p < 0.05$). Other side effects showed the similar incidences between two groups. However, there is no statistical significance of differences for bradypnea ($p > 0.05$), and no nausea or vomiting was found in both groups after seven-day surgery, although the symptoms were widely reported in propofol use [40,41].

Table 2. The effects of COS on the side effects caused by propofol.

Side Effects	CG ($n = 47$)	PG ($n = 47$)	Chi-Square Statistic	p Values
Apnea	2 (4.26)	8 (17.02)	2.798	0.094
Bradypnea (breaths < 6/min)	1 (2.13)	6 (12.77)	2.470	0.116
Obstructive respiration	0 (0)	5 (10.64)	4.451	0.035
Tachycardia (HR > 30% above BL)	1 (2.13)	8 (17.02)	4.424	0.035
Hypertension(MAP > 30%above BL)	0 (0)	6 (12.77)	2.470	0.116
Bradycardia (HR > 30% under BL)	0 (2.13)	5 (10.64)	4.451	0.035
Hypotension (MAP > 30% under BL)	0 (0)	7 (14.89)	5.557	0.018
Burning and stinging	2 (4.26)	10 (21.28)	6.114	0.013

Note: Chi-square test was performed. HR: heart rate, MAP: mean arterial pressure, BL = baseline (measurement before induction). There is statistical significance of differences if $p < 0.05$.

2.4. Analysis of Mechanic Hyperalgesia

Intraplantar injection of 0.9% NaCl solution did not induce mechanical hyperalgesia and is regarded as a control group (Figure 2) Intraplantar injection of CFA increased mechanical hyperalgesia of a mouse model by reducing its thresholds for pain (Figure 2). Propofol and COS treatment decreased CFA-induced hyperalgesia (Figure 2). The combination treatment of COS and propofol attenuated the hyperalgesia more than propofol used alone ($p < 0.05$). However, Nav1.7-silenced groups attenuated hyperalgesia significantly though COS and/or propofol no longer attenuated hyperalgesia (Figure 2). There is no statistical significance of differences among the Nav1.7-silenced groups treated or untreated by COS and/or propofol ($p > 0.05$).

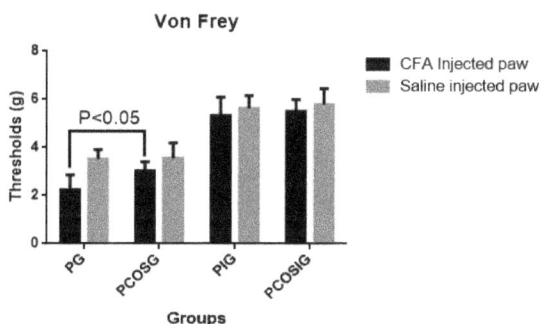

Figure 2. The threshold in an inflammatory pain model. There were 32 mouse pain models evenly assigned into four groups: PG group (received 10 mg/kg propofol treatment), PCOSG group (received both 10 mg/kg COS and propofol treatment), PIG group (Nav1.7-silenced model mouse received 10 mg/kg propofol treatment) and PCOSIG group (Nav1.7-silenced model mouse received both 10 mg/kg COS and propofol treatment). All data were presented as mean ± S.D. and $n = 8$ in each group. There is statistical significance of differences if $p < 0.05$.

2.5. Analysis of Thermal Hyperalgesia

Thermal hyperalgesia was found in CFA-induced mouse pain models but not in the mice only treated with 0.9% NaCl solution (Figure 3A). COS reduced thermal hyperalgesia by increasing its latency (Figure 3A). COS attenuated the mechanic hyperalgesia caused by propofol ($p < 0.05$). COS pretreatment resulted in insensitivity to the pain in a mouse model ($p < 0.05$). Comparatively, Nav1.7 silence also attenuated mechanic hyperalgesia significantly but COS no longer attenuated mechanic hyperalgesia (Figure 3A). There is no statistical significance of differences among Nav1.7-silenced groups treated or untreated by COS ($p > 0.05$).

For thermal pain, there is statistical significance of differences for the jumping times between COS-treated and non-treated groups ($p < 0.05$, Figure 3B), suggesting that COS has better effects on thermal hyperalgesia than propofol used alone ($p < 0.05$). Notably, Nav1.7 silence reduced jumping times significantly but propofol and/or COS was not able to reduce jumping times (Figure 3B). There was no statistical significance of differences between the groups treated by propofol and the combination therapy of propofol and COS when Nav1.7 was silenced ($p > 0.05$).

For a cold-plate test, Nav1.7 silence could not reduce the rearing times, and COS and propofol could not maintain reduction of rearing times on the cold plate (Figure 3C). There is no statistical significance of differences among Nav1.7-silenced groups treated or untreated by COS and/or propofol ($p > 0.05$), suggesting that Nav1.7 is also not associated with cold pain.

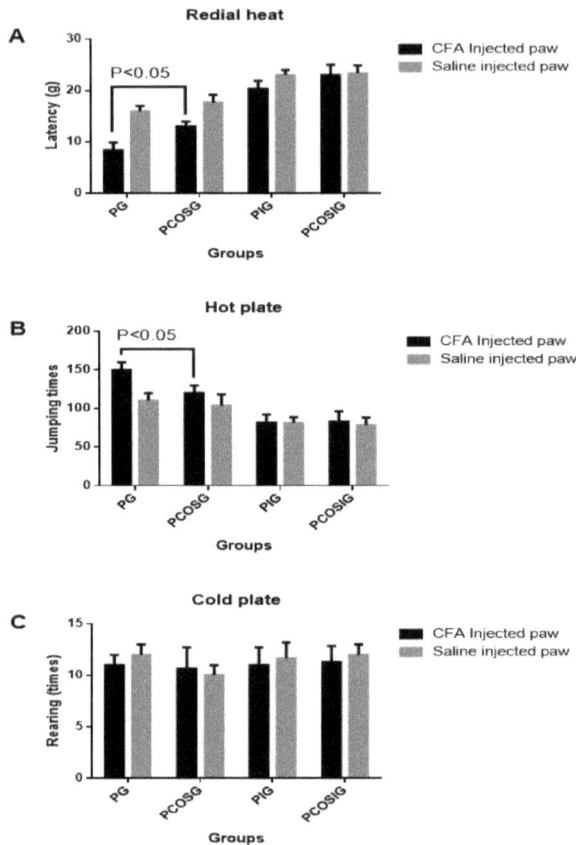

Figure 3. Analysis of hyperalgesia in different groups. (**A**), radial heat analysis of complete Freund's adjuvant (CFA)-induced thermal hyperalgesia; (**B**), the latency times for jumping responses after exposure to the 50-centigrade plate within 10 min; (**C**), rearing times after exposure to the 50-centigrade plate within 10 min. The mice received 10 mg/kg COS in dietary before 2 h propofol injection. All data were presented as mean ± S.D. and *n* = 8 in each group. There is statistical significance of differences if $p < 0.05$.

2.6. The Protein Level of Voltage-Gated Sodium Channels (Nav)1.7 in Dorsal Root Ganglia (DRG) Neurons

The protein level of Nav1.7 was analyzed by Western blot. The results showed that Nav1.7 was at a low level when the mice were injected with 0.9% NaCl solution (Figure 4). CFA increased the protein level of Nav1.7 ($p < 0.01$) and there was statistical significance of differences between control and model groups (Figure 4). There was no change in protein level when the mice were treated with propofol and COS (Figure 4) ($p > 0.05$), suggesting that propofol or COS cannot affect the protein level of Nav1.7.

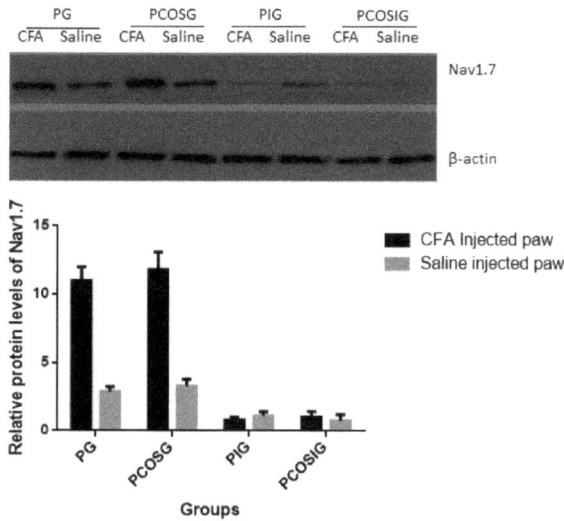

Figure 4. The protein levels of Nav1.7 in the dorsal root ganglia (DRG) neurons of the mice from different groups. All data were presented as mean $\pm$ S.D. and $n = 8$ in each group. There is statistical significance of differences if $p < 0.05$.

2.7. COS Reduces the Activity of Nav1.7

To investigate the effects of COS on propofol performance for blocking Nav1.7 activities, the electrophysiological properties of Nav1.7 were compared by using whole-cell patch-clamp recordings. As shown in Figure 5A, propofol blocked Nav1.7 activities in a concentration-dependent manner and COS improved propofol blocking the channels (Figure 5B). Resting channels were measured at a holding potential of -120 mV by test pulses to 0 mV applied at 0.1 Hz. The IC50 values for propofol were 231 ± 12 μM (Hill coefficient 1.8 ± 0.4, $n = 10$) and the values of the combination of COS and propofol were 165 ± 18 μM (Hill coefficient 1.1 ± 0.2, $n = 10$). Figure 5C showed that there was statistical significance of differences for the blocking potencies of resting Na$^+$ channels between the propofol and combined groups ($p = 0.02$, unpaired t-test). Figure 5D showed that COS enhanced the tonic block of inactivated Na$^+$ channels when compared to the group only treated with propofol (propofol, IC50 value 188 ± 10 μM; Hill coefficient 1.6 ± 0.2, $n = 10$; propofol and COS, IC50 value 121 ± 8 μM; Hill coefficient 1.3 ± 0.1, $n = 10$; $p = 0.02$, unpaired t-test).

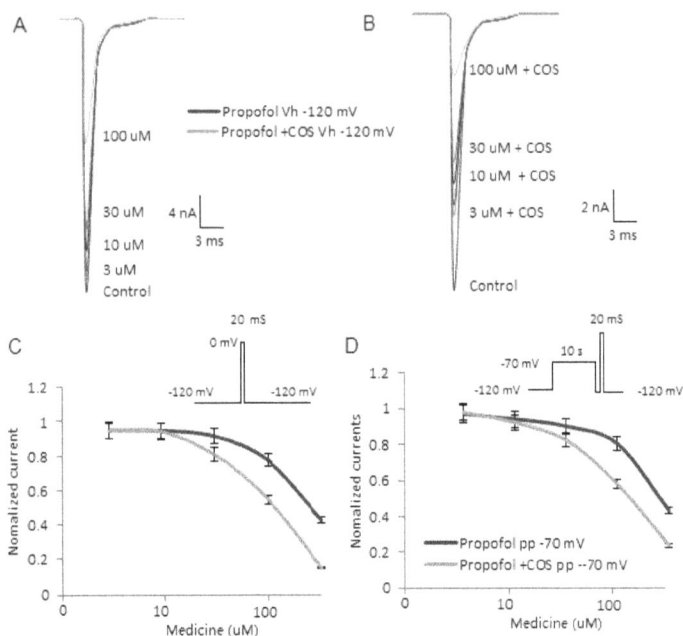

Figure 5. Nav1.7 channels were blocked by propofol and COS. (**A**), Representative traces of Nav1.7 currents in the DRGs treated by different concentrations of propofol. The cells were held at a holding potential of −120 mV and test pulses were stepped to 0 mV and applied at 0.1 Hz; (**B**), Representative traces of Nav1.7 currents in DRGs treated by COS and different concentrations of propofol. The cells were held at a holding potential of −120 mV and test pulses were stepped to 0 mV and applied at 0.1 Hz; (**C**), a tonic block of resting Nav1.7 channels by propofol and/or the combination of COS and propofol. Resting channels were measured at a holding potential of −120 mV; (**D**), a tonic block of inactivated Na+ channels by propofol and/or the combination of COS and propofol. Inactivated channels were induced by a 10 s pre-pulse to −70 mV followed by a 100 ms pulse at −120 mV and a test pulse to 0 mV. Peak amplitudes of Nav1.7 currents were normalized with respect to the peak amplitude in control solution and plotted against the concentration of propofol or a combination of propofol and COS.

2.8. COS Also Promotes Propofol-Produced Stabilization of Fast and Slow Inactivation

Fast inactivation was caused by 50 ms pre-pulses ranging from −120 to 0 mV in a five-mV step (Figure 6A), and the remaining fraction of channels was measured with a 20 ms pre-pulse to 0 mV. Figure 6B showed that 100 µM propofol caused a ten-mV hyperpolarization shift of steady-state fast inactivation from $V_{1/2}$ of −75 ± 2 mV (n = 10) in control to $V_{1/2}$ of −85 ± 5 mV (n = 10) (p < 0.01). COS stabilized the fast inactivation and caused a ten-mV hyperpolarization shift of the steady-state fast inactivation of propofol (propofol: $V_{1/2}$ of −85 ± 5 mV; propofol and COS: $V_{1/2}$ of −95 ± 6 mV; n = 10) (p < 0.05). There is statistical significance of differences when compared with the combination treatment of COS and propofol (p < 0.05).

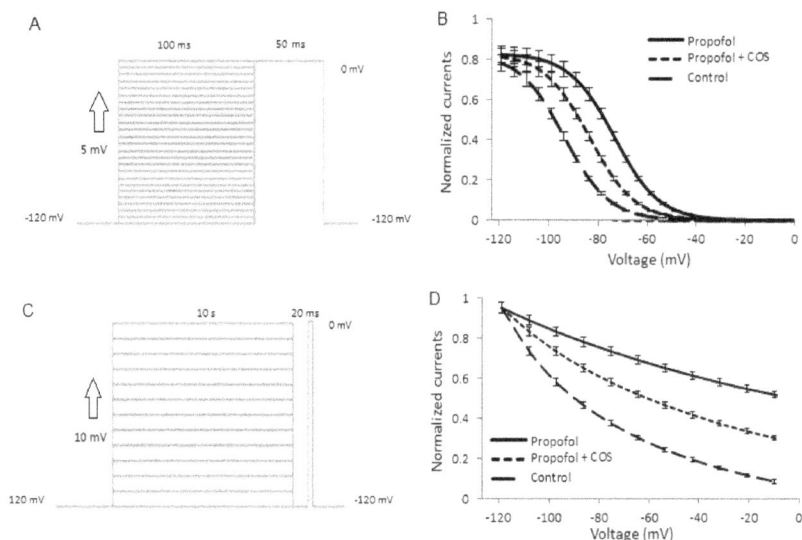

Figure 6. Voltage dependency of fast and slow inactivation of Nav 1.7 in the presence of propofol and COS. (**A**), fast inactivation was caused by 50 ms pre-pulses ranging from −120 to 0 mV in five-mV step, and the remaining fraction of channels was measured with a 20 ms pre-pulse to 0 mV; (**B**), Voltage-dependency of fast inactivation of Nav1.7 in the presence of control solution, or propofol, and/or combination treatment of propofol and COS; (**C**), The voltage protocol of slow inactivation. Slow inactivation was caused by 10 s pre-pulses ranging from −120 to −30 mV in steps of 10 mV followed by a 100-ms interpulse at −120 mV, which allows the recovery from fast inactivation; (**D**), Voltage-dependency of slow inactivation of Nav1.7 in control solution, or propofol, and/or combination treatment of propofol and COS. The lines were fitted by using a Boltzmann equation.

Slow inactivation was caused by 10 s pre-pulses ranging from −120 to −10 mV in ten-mV step, followed by a 100 ms pulse at −120 mV, which allows recovery from fast inactivation, and followed by a test pulse to −10 mV. Propofol at 100 µM induced a small shift of the voltage dependency of slow inactivation of Nav1.7 (control: $V_{1/2}$ of −20 ± 1 mV; propofol: $V_{1/2}$ of −65 ± 2 mV $n = 10$; Figure 6C). In contrast, combination treatment caused the shift of slow inactivation when compared with only propofol used (propofol: $V_{1/2}$ of −65 ± 2 mV, combined: $V_{1/2}$ of −90 ± 4 mV, $n = 10$; Figure 6C). Neither propofol nor a combination of propofol and COS caused an apparent shift of the voltage-dependency of activation (data not shown).

2.9. COS Promotes Propofol Blocking Veratridine-Induced Persistent Sodium Current of Nav1.7

To understand the activity of the propofol and the combination of propofol and COS on the persistent Nav1.7 currents, tonic activation was created by adding 50 µM veratridine. Figure 7A shows that veratridine caused a prominent persistent current, which was stimulated by 50 ms pulses in cells at a holding potential of −120 mV. Figure 7B shows that COS promoted propofol blocking the persistent current. The calculated IC50 values of propofol were at 202 ± 27 µM, $n = 8$, and a combination of COS and propofol at 126 ± 47 µM ($p = 0.03$, unpaired t-test).

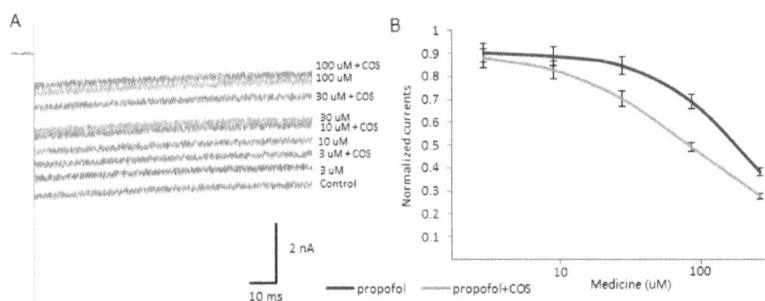

Figure 7. Propofol and COS inhibit persistent Na$^+$ currents in DRGs. (**A**), Representative traces of Nav1.7 currents in the presence of 50 μM veratridine. Cells were held at −120 mV and currents were activated at 0.1 Hz; (**B**), the block of the persistent Na$^+$ currents in presence of propofol or COS. Peak current amplitudes of the persistent Nav1.7 currents were normalized and fitted with the Hill equation.

3. Discussion

Present findings indicated that COS greatly inhibited the incidence and severity of propofol-induced injection pain if the patients received 10 mg/kg COS via oral administration before being injected with propofol (Table 1). No toxic symptom or fewer side effects were observed in all the patients treated with COS (Table 2). The results suggest that COS may be a potential natural adjuvant to improve propofol performance.

From pain analyses, an animal pain model was successfully established after CFA injection. The mouse model had mechanical and thermal hyperalgesia because of inflammatory pain, which was tested by a von Frey filament assay and hot/cold plate assay. Propofol is one kind of medicine mainly used for decreasing human pain. Present findings indicated that Nav1.7 was increased in CFA-induced hyperalgesia, which suggested that Nav1.7 plays a critical role in inflammatory pain. Subsequent work showed that COS and propofol reduced pain thresholds.

Injection pain is a normal unwanted adverse effect for propofol use. The side effects can be reduced when combined with COS because they can produce more analgesic efficacy [42]. Another study also used COS as an anesthesia supplement of propofol injection, which was successfully used in topical local anesthesia for surgery on a child [43]. All the results suggest that propofol and COS may have synergistic functions. However, the complementary functions remain unclear. Since many Navs play important roles in pain [44,45] and neural disorders [46,47], we want to explore the effects of combined medicine on the level of Navs. The mutant SCN9A gene-encoding Nav1.7 caused insensitivity to pain in mammals [35]. Furthermore, many pyrrolo-benzo-1,4-diazine derivatives were synthesized to inhibit the activity of Nav1.7, and showed anti-nociceptive oral efficacy in an inflammatory pain model [48].

CFA increasing the expression of Nav1.7 was also reported in an earlier study [49]. CFA increased the colocalization of protein kinase B/Akt with Nav1.7 in L4/5 DRG neurons while Akt pathway induced the upregulation of Nav1.7 [50]. Thus, the level of Nav1.7 was higher than in an animal model than in a healthy control. However, no evidence has shown that propofol and COS can reduce the level of Nav1.7 yet (Figure 4). According to a previous report, opioid receptor activation will reduce the level of Nav1.7 [51] while propofol can increase the expression of an opioid receptor [52]. Present work revealed a functional role of COS for controlling pain, which was not associated with the changes of Nav1.7 level (Figure 4). The present findings showed that the combined treatment was better than only one kind of medicine used for decreasing the mechanic and thermal pain ($p < 0.05$)(Figures 2 and 3).

The main aim of our work was to evaluate whether COS and propofol functionally interact with the sodium channel Nav1.7. Our data suggested that COS was a potential adjuvant to improve propofol performance, concentration- and state-dependent inhibitors of Nav1.7. Our results also suggested that propofol and COS interacted and modulated Nav1.7. Therefore, the findings showed that COS reinforced the inhibitory properties of propofol on Nav1.7 activity.

Previous work showed that steady-state plasma concentration of propofol during sedation was in the order of 22–44 μM [53]. It can intensively (97%–98%) bind plasma proteins [54]. In most cases, only the unbound fraction is able to interact with Na$^+$ channels. Therefore, a higher concentration was used in pain therapy [55]. Propofol is mainly eliminated by hepatic conjugation to inactive metabolites, which are secreted from the kidney [56]. On the other hand, the persons have a reduced clearance for propofol and may have increased levels of plasma propofol [57]. Additionally, the terminal half-life of propofol ranges from one to three days [58].

COS showed as a preventive agent by improving propofol performance in a pain model. COS improves propofol performance by suppressing pain symptoms and inhibiting Nav1.7 activity (Figures 6–8). Furthermore, COS caused an obvious hyperpolarization shift of the steady-state fast inactivation of Nav1.7 (Figure 6). There is statistical significance of differences when compared to the combination of COS and propofol ($p < 0.05$, unpaired t-test). COS has no systemic adverse effects on the mouse model. Clinically relevant plasma levels of propofol will cause related effects on Nav1.7. Therapeutic levels of COS are low in the present experiment (10 mg/Kg).

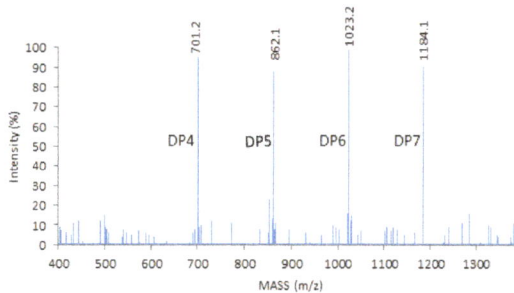

Figure 8. MALDI–TOF MS analysis of COS. The main products for the degree of polymerization (DP) were DP4, 5, 6 and 7 when potassium adducts ions were calculated in MALDI-TOF.

One important thing should be mentioned here: −120 mV hyperpolarized potentials were artificial and did not present the membrane properties of DRGs in vivo. With a physiological resting membrane potential around −50 mV, and with an ongoing DRG activity, the data from inactivated channels can be used to evaluate the function of Na$^+$ channel blockers. A tonic block of Nav1.7 channels by propofol and COS may be a better means for pain therapy. Present findings showed that COS were potential adjuvants to induce a higher tonic block as compared to use of only propofol.

There are some limitations for the present study: (1) Most studies, if not all, examined the effect of COS in addition to propofol, and the possible effects of COS alone have not been studied. This seems to make the mechanisms of COS effects vague and mysterious. Propofol has been proved to be an important sedative. However, we are not sure whether only COS can be a kind of sedative although it has been reported to have anti-pain functions. To avoid unknown risks, the test was not performed in the patients receiving surgeries. We are influenced by the design for human experiment and the test was not performed in the animal models with only COS treatment; (2) Low-molecular-weight COS cannot be injected in most cases although it has been used widely as healthy products in China; (3) Detail molecular mechanism for the inhibitory function of COS and propofol for Nav1.7 remains unknown; (4) Nav1.7 is only one critical effector for evaluating the functions of COS, and many other Nav members should be analyzed in the future.

4. Materials and Methods

4.1. COS Preparation and MALDI–TOF (Matrix-Assisted Laser-Desorption Ionization–Time-of-Flight) MS Analysis

Low-molecular-weight, water-soluble COS was purchased from GlycoBio Company (Dalian, China). The COS was marine natural products and prepared from marine resources according to a previous report [59]. A 1 µL sample solution was mixed with 2 µL 2,5-dihydroxybenzoic acid (15 mg/mL) in 30% ethanol. Mass spectra were made on an Agilent 6530 Accurate-Mass (Santa Clara, CA, USA) in a positive ion mode. In the measurement, a nitrogen laser (Spectra-Physics, Mountain View, CA, USA) (at 337 nm, 3 ns pulse width, 3 Hz) was performed. All spectra were examined in a reflector mode by using external calibration. MALDI–TOF MS analysis of COS showed that the degree of polymerization (DP) of the main products were DP4, 5, 6 and 7 when potassium adduct ions were summed together in MALDI-TOF (Figure 8).

4.2. Participants

Before the present study, all protocols were approved by the Ethical Committee of the First Hospital of Jilin University (Changchun, China). The subjects with the physical status of American Association of Anesthesiology (ASA) I or II received surgery at our hospital from 3 May to 12 October. Including criteria was used according to previously reported [60]. Excluding criteria includes following items: (1) the patients could not express themselves clearly; (2) they took other anti-pain medicine within one day of surgery; (3) the patients refused to sign an informed consent for present experiments. Finally, a total of 188 patients were selected.

4.3. Patient Grouping

All the selected subjects were evenly assigned to two groups before being injected with propofol: 10 mg/kg COS (CG) treatment and 10 mg/kg placebo (PG) treatment daily. COS and placebo were administered orally. To avoid the intervention of baseline characters for final results, demographic data were investigated including age, gender, BMI (body mass index), lifestyle and ASA. After 2 h pretreatment, the patients received 2 mg/kg/h saline treatment. After five min, propofol TCI was started with step increases of 0.5 µg/mL/2.5 min until the patient lost consciousness. *Cis*-atracurium was injected at 0.2 mg/kg to promote tracheal intubation. Meanwhile, propofol TCI was adjusted to maintain BIS values at 50. The pain was evaluated by clinical experts according to a four-point scale (no pain, mild pain, moderate pain and severe pain) from propofol injection to the time when the patients lost consciousness. Side effects were recorded from day 4 to 7 after the surgery. Table 3 showed that the baseline characters were similar between CG and PG groups, including age, gender, BMI, lifestyle and ASA ($p < 0.05$). The results suggest that the baseline clinical characters will not affect the final results of COS and propofol treatment.

Table 3. Baseline characters of patients receiving surgery between CG and PG groups.

Baseline Characters	CG (n = 47)	PG (n = 47)	T Value/Chi-Square Statistic	p Values
Age	38.9 ± 15.6	40.1 ± 16.8	0.141	0.235
Gender, male (%)	32 (68.09)	34 (72.34)	0.203	0.652
BMI	22.7 ± 4.8	23.9 ± 6.5	0.037	0.326
Smoking, n (%)	28 (59.57)	26 (55.32)	0.174	0.677
Drinking, n (%)	29 (61.7)	25 (53.19)	0.696	0.404
Spouse, n (%)	44 (93.62)	46 (97.87)	0.261	0.409
ASA				
I	35 (74.47)	37 (78.72)	0.237	0.626
II	12 (25.53)	10 (21.28)		

Note: *t*-test and Chi-squared test were performed. CG, the patients received COS oral administration before being injected with propofol. PG, the patients received placebo oral administration before being injected with propofol. ASA, American Society of Anesthesiologists. There is statistical significance of differences if $p < 0.05$.

4.4. Animals

To explore the molecular mechanism, an animal pain model was established. All the protocols were established according to the guidance for the use of laboratory animals (National Academy Press) and approved by the Ethical Committee of the First Hospital of Jilin University (Changchun, China). Four-week-old C57BL/6 male mice were purchased from Shanghai SLAC Laboratory Animal Co., Ltd. (Shanghai, China). A total of 32 mice (20–25 g) were anesthetized with 2% isoflurane (Cat. No. CDS019936, Sigma, St. Louis, MO, USA) and injected with complete Freund's adjuvant (CFA, Cat. No. F5881, 10 μL 0.5 mg/mL heat-killed *M. tuberculosis*) (Sigma, St. Louis, MO, USA) in the plantar of one hind paw to cause inflammatory pain symptoms. Meanwhile, another hind paw was injected with 10 μL 0.9% NaCl as a control. Animal behaviors were observed after one-day pain induction.

4.5. Nav1.7 Gene Silencing

pTZU6+1 vector was from Chongqing Medical University (Chongqing, China). shRNA for Nav1.7 gene silencing was constructed by using the primers: sense, 5′-ACCTCGACCTCAGA GCTTCGTTCACTTTGGAGTGAACGAAGCTCTGAGGTCTT-3′; antisense, 5′-CAAAAAGACCTCAG AGCTTCGTTCACTCCAAAGTGAACGAAGCTCTGAGGTCG-3′. Restriction sites, *Sal*I and *Xba*I, were added on either end of the oligos and linked with pTZU6+1, and pTZU6+1-Nav1.7 were reconstructed. The reconstructed plasmids were injected into mice via tail veins. Eight hours after injection, propofol injection was performed and animal behaviors of mechanical and thermal hyperalgesia were analyzed.

4.6. Animal Grouping

The mice received 10 mg/kg COS treatment before 2 h propofol injection and the dosage was used according to a previous report [61]. There were 32 pain-model mice evenly assigned into four groups: PG group (received 10 mg/kg propofol treatment), PCOSG group (received both 10 mg/kg COS and propofol treatment), PIG group (Nav1.7-silenced model mouse received 10 mg/kg propofol treatment) and PCOSIG group (Nav1.7-silenced model mouse received both 10 mg/kg COS and propofol treatment).

4.7. Animal Behavior of Mechanical and Thermal Hyperalgesia

Mechanic pain sensitivity was measured immediately by testing the responding forces to the stimulation by Electronic von Frey monofilaments (Nanjing Jisheng Medical Technology Company, Nanjing, China) after propofol injections. The thermal pain was examined by an algesiometer (Shanghai AoBopharmtech, Shanghai, China). Hot- and cold-induced pains were tested by a Hot/Cold Plate Analgesia Meter (YLS-6B, Huaibei Zhenghua Biologic Apparatus Facilities Ltd. Co., Huaibei, China).

4.8. Western Blot

According to a previous report, CFA infection increases the expression of Nav1.7 [49]. Nav1.7 can be upregulated in L4/5 DRG neurons in a certain evoking situation [50]. Therefore, L4-5 DRG samples from different groups were obtained. Protein was isolated using a plasma membrane protein isolation kit (Cat. No. ab65400, Abcam Trading (Shanghai) Company Ltd., Shanghai, China). Rabbit anti-mouse monoclonal Nav1.7 antibody (Cat. No. 62758, dilution 1:5000, Abcam Trading (Shanghai) Company Ltd., Shanghai, China) was used as the first antibody. Polyclonal Goat Anti-Rabbit IgG H&L (Cat. No. ab6721, dilution 1:3000, Abcam Trading (Shanghai) Company Ltd., Shanghai, China) was used as a secondary antibody. A rabbit anti-mouse β-actin polyclonal antibody (1:2000 dilution; Cat. No. 4967, Cell Signaling Technology, Danvers, MA, USA) was used as a loading control. All protein bands were visualized by using an enhanced chemiluminescence substrate (Sangon Biotech, Co., Ltd., Shanghai, China). The image intensity of protein bands was quantified by using NIH ImageJ software (Bethesda, MD, USA).

4.9. Electrophysiology Analysis of Nav1.7

Primary DRG cells were cultured in DMEM media and treated with different concentrations of propofol and/or 10 µg/mL COS for 24 h. To investigate the activities of Nav1.7, the electrophysiological properties of Nav1.7 were compared in primary DRG cells by using whole-cell patch-clamp recordings. The following test solution was prepared (mM): 100 NaCl, 50 choline chloride, 5 KCl, 1 MgCl$_2$, 1 CaCl$_2$, 10 HEPES, and 15 glucose. The pH value was adjusted to 7.0 with tetraethylammonium hydroxide (Sigma, St. Louis, MO, USA). The pipette solution consists of the following components (mM): 140 CsF, 10 NaCl, 1 ethyleneglycol-bis (2-aminoethylether)-*N,N,N',N'*-tetraacetic acid, 10 HEPES and pH value was adjusted to 7.0 with CsOH.

The membrane currents were recorded by using a patch clamp and an EPC10 amplifier (HEKA Instruments Inc., Bellmore, NY, USA). Data were obtained and stored with Patchmaster v20 × 60 software (HEKA Instruments Inc., Bellmore, NY, USA). Patch pipettes were pulled from glass capillaries (Science Products, Hofheim, Germany) by using a DMZ-Universal Puller (Zeitz, Germany) and then heat polished to give a resistance of 2.0 to 2.5 MΩ when it was filled with pipette solution. Currents were filtered at 5 kHz. The series resistance was compensated by 60%–80% to minimize voltage errors, and the capacitance artifacts were canceled using the amplifier circuitry. Linear leak subtraction based on resistance estimates from hyperpolarized pulses was applied before the pulse test.

4.10. Statistical Analysis

M Data were represented as mean ± S.D. Chi-square test was used for the comparison between two groups. The comparisons of independent groups of data were performed with the ANOVA test by using IBM SPSS Statistics 20.0 (Brea, CA, USA). Data analysis, curve fitting, and statistical analyses were also performed using the same software. IC50 values were calculated by normalizing peak current amplitudes at different concentrations to the value obtained in control solution. Data were fitted with Hill equation $y = y_{max} \times (IV50n/IC50n \times Cn)$, where y_{max} is the maximal amplitude, IC50 is the concentration at which $y/y_{max} = 0.5$, and n is the Hill coefficient. To obtain inactivation curves, peak currents evoked by a test pulse were measured, normalized, and plotted against the conditioning repulse potential. Data were fitted by the Boltzmann equation [62], $y = 1/(1 + \exp(Epp - h0.5)/kh)$, where Epp is the membrane potential of test pulse, h0.5 is the voltage at which y equals 0.5, and kh is a slope factor.

5. Conclusions

Taken together, propofol and chitosan oligosaccharide (COS) can synergistically reduce inflammation pain symptoms. While propofol causes some adverse effects, COS improves the propofol performance with fewer side effects by reducing inflammation and inhibiting the activity of voltage-gated sodium channel (Nav)1.7. Our data demonstrate that both substances block the Na$^+$ channel Nav1.7 and potentially contribute to pain relief. Thus, this study identified a potential adjuvant for the pain therapy with low-dose propofol.

Acknowledgments: We are very grateful to the anonymous reviewers for their important and strategic comments, which have significantly improved the quality of the present paper.

Author Contributions: Z.L. and X.Y. conceived and designed the experiments; X.Y. performed the experiments; X.S. and H.M. analyzed the data; P.Z. contributed reagents/materials/analysis tools; P.Z. wrote the paper.

Conflicts of Interest: The authors declare no conflict of interest.

References

1. Sapate, M.; Andurkar, U.; Markandeya, M.; Gore, R.; Thatte, W. To study the effect of injection dexmedetomidine for prevention of pain due to propofol injection and to compare it with injection lignocaine. *Braz. J. Anesthesiol.* **2015**, *65*, 466–469. [CrossRef] [PubMed]

2. Marik, P.E. Propofol: Therapeutic indications and side-effects. *Curr. Pharm. Des.* **2004**, *10*, 3639–3649. [CrossRef] [PubMed]

3. Mays, N. Reducing unwarranted variations in healthcare in the English NHS. *BMJ* **2011**, *342*, d1849. [CrossRef] [PubMed]

4. Lee, S.H.; Lee, S.E.; Chung, S.; Lee, H.J.; Jeong, S. Impact of time interval between remifentanil and propofol on propofol injection pain. *J. Clin. Anesth.* **2016**, *34*, 510–515. [CrossRef] [PubMed]

5. Madan, H.K.; Singh, R.; Sodhi, G.S. Comparsion of Intravenous Lignocaine, Tramadol and Keterolac for Attenuation of Propofol Injection Pain. *J. Clin. Diagn. Res.* **2016**, *10*, UC05–UC08. [CrossRef] [PubMed]

6. Berberian, P.; Obimba, C.; Glickman-Simon, R.; Sethi, T. Herbs for Low-Back Pain, Acupuncture for Psychological Distress, Osteopathic Manipulative Therapy for Chronic Migraine, Honey Dressings for Burns, Vegetarian Diet and Risk of Colorectal Cancer. *Explore* **2015**, *11*, 410–414. [CrossRef] [PubMed]

7. Schroder, S.; Beckmann, K.; Franconi, G.; Meyer-Hamme, G.; Friedemann, T.; Greten, H.J.; Rostock, M.; Efferth, T. Can medical herbs stimulate regeneration or neuroprotection and treat neuropathic pain in chemotherapy-induced peripheral neuropathy? *Evid. Based Complement. Altern. Med.* **2013**, *2013*, 423713. [CrossRef] [PubMed]

8. Tatsumi, S.; Mabuchi, T.; Abe, T.; Xu, L.; Minami, T.; Ito, S. Analgesic effect of extracts of Chinese medicinal herbs Moutan cortex and Coicis semen on neuropathic pain in mice. *Neurosci. Lett.* **2004**, *370*, 130–134. [CrossRef] [PubMed]

9. Euasobhon, P.; Dej-Arkom, S.; Siriussawakul, A.; Muangman, S.; Sriraj, W.; Pattanittum, P.; Lumbiganon, P. Lidocaine for reducing propofol-induced pain on induction of anaesthesia in adults. *Cochrane Database Syst. Rev.* **2016**. [CrossRef]

10. Joo, J.D.; In, J.H.; Kim, D.W.; Jung, H.S.; Kang, J.H.; Yeom, J.H.; Choi, J.W. The comparison of sedation quality, side effect and recovery profiles on different dosage of remifentanil patient-controlled sedation during breast biopsy surgery. *Korean J. Anesthesiol.* **2012**, *63*, 431–435. [CrossRef] [PubMed]

11. McCleskey, P.E.; Patel, S.M.; Mansalis, K.A.; Elam, A.L.; Kinsley, T.R. Serum lidocaine levels and cutaneous side effects after application of 23% lidocaine 7% tetracaine ointment to the face. *Dermatol. Surg.* **2013**, *39*, 82–91. [CrossRef] [PubMed]

12. Romanazzi, G.; Feliziani, E.; Banos, S.B.; Sivakumar, D. Shelf life extension of fresh fruit and vegetables by chitosan treatment. *Crit. Rev. Food Sci. Nutr.* **2017**, *57*, 579–601. [CrossRef] [PubMed]

13. Swiatkiewicz, S.; Swiatkiewicz, M.; Arczewska-Wlosek, A.; Jozefiak, D. Chitosan and its oligosaccharide derivatives (chito-oligosaccharides) as feed supplements in poultry and swine nutrition. *J. Anim. Physiol. Anim. Nutr. (Berl.)* **2015**, *99*, 1–12. [CrossRef] [PubMed]

14. Chung, M.J.; Park, J.K.; Park, Y.I. Anti-inflammatory effects of low-molecular weight chitosan oligosaccharides in IgE–antigen complex-stimulated RBL-2H3 cells and asthma model mice. *Int. Immunopharmacol.* **2012**, *12*, 453–459. [CrossRef] [PubMed]

15. Guo, M.; Ma, Y.; Wang, C.; Liu, H.; Li, Q.; Fei, M. Synthesis, anti-oxidant activity, and biodegradability of a novel recombinant polysaccharide derived from chitosan and lactose. *Carbohydr. Polym.* **2015**, *118*, 218–223. [CrossRef] [PubMed]

16. Aranaz, I.; Mengíbar, M.; Harris, R.; Paños, I.; Miralles, B.; Acosta, N.; Galed, G.; Heras, Á. Functional characterization of chitin and chitosan. *Curr. Chem. Biol.* **2009**, *3*, 203–230. [CrossRef]

17. Lai, J.; Porreca, F.; Hunter, J.C.; Gold, M.S. Voltage-gated sodium channels and hyperalgesia. *Annu. Rev. Pharmacol. Toxicol.* **2004**, *44*, 371–397. [CrossRef] [PubMed]

18. Shah, B.S.; Stevens, E.B.; Pinnock, R.D.; Dixon, A.K.; Lee, K. Developmental expression of the novel voltage-gated sodium channel auxiliary subunit $\beta 3$, in rat CNS. *J. Physiol.* **2001**, *534*, 763–776. [CrossRef] [PubMed]

19. Whitaker, W.; Faull, R.; Waldvogel, H.; Plumpton, C.; Burbidge, S.; Emson, P.; Clare, J. Localization of the type VI voltage-gated sodium channel protein in human CNS. *Neuroreport* **1999**, *10*, 3703–3709. [CrossRef] [PubMed]

20. Yin, R.; Liu, D.; Chhoa, M.; Li, C.M.; Luo, Y.; Zhang, M.; Lehto, S.G.; Immke, D.C.; Moyer, B.D. Voltage-gated sodium channel function and expression in injured and uninjured rat dorsal root ganglia neurons. *Int. J. Neurosci.* **2015**, *126*, 182–192. [CrossRef] [PubMed]

21.   Rabert, D.K.; Koch, B.D.; Ilnicka, M.; Obernolte, R.A.; Naylor, S.L.; Herman, R.C.; Eglen, R.M.; Hunter, J.C.; Sangameswaran, L. A tetrodotoxin-resistant voltage-gated sodium channel from human dorsal root ganglia, hPN3/SCN10A. *Pain* **1998**, *78*, 107–114. [CrossRef]

22.   Cohen, C.J. Targeting voltage-gated sodium channels for treating neuropathic and inflammatory pain. *Curr. Pharm. Biotechnol.* **2011**, *12*, 1715–1719. [CrossRef] [PubMed]

23.   Suh, H.R.; Chung, H.J.; Park, E.H.; Moon, S.W.; Park, S.J.; Park, C.W.; Kim, Y.I.; Han, H.C. The effects of Chamaecyparis obtusa essential oil on pain-related behavior and expression of pro-inflammatory cytokines in carrageenan-induced arthritis in rats. *Biosci. Biotechnol. Biochem.* **2015**, *80*, 203–209. [PubMed]

24.   Yang, Y.; Li, Y.X.; Wang, H.L.; Jin, S.J.; Zhou, R.; Qiao, H.Q.; Du, J.; Wu, J.; Zhao, C.J.; Niu, Y.; et al. Oxysophocarpine Ameliorates Carrageenan-induced Inflammatory Pain via Inhibiting Expressions of Prostaglandin E2 and Cytokines in Mice. *Planta Med.* **2015**, *81*, 791–797. [CrossRef] [PubMed]

25.   Jiang, Y.L.; He, X.F.; Shen, Y.F.; Yin, X.H.; Du, J.Y.; Liang, Y.I.; Fang, J.Q. Analgesic roles of peripheral intrinsic met-enkephalin and dynorphin A in long-lasting inflammatory pain induced by complete Freund's adjuvant in rats. *Exp. Ther. Med.* **2015**, *9*, 2344–2348. [CrossRef] [PubMed]

26.   Qian, B.; Li, F.; Zhao, L.X.; Dong, Y.L.; Gao, Y.J.; Zhang, Z.J. Ligustilide Ameliorates Inflammatory Pain and Inhibits TLR4 Upregulation in Spinal Astrocytes Following Complete Freund's Adjuvant Peripheral Injection. *Cell. Mol. Neurobiol.* **2015**, *36*, 143–149. [CrossRef] [PubMed]

27.   Strickland, I.T.; Martindale, J.C.; Woodhams, P.L.; Reeve, A.J.; Chessell, I.P.; McQueen, D.S. Changes in the expression of NaV1.7, NaV1.8 and NaV1.9 in a distinct population of dorsal root ganglia innervating the rat knee joint in a model of chronic inflammatory joint pain. *Eur. J. Pain* **2008**, *12*, 564–572. [CrossRef] [PubMed]

28.   Leo, S.; D'Hooge, R.; Meert, T. Exploring the role of nociceptor-specific sodium channels in pain transmission using Nav1.8 and Nav1.9 knockout mice. *Behav. Brain Res.* **2010**, *208*, 149–157. [CrossRef] [PubMed]

29.   Nassar, M.A.; Stirling, L.C.; Forlani, G.; Baker, M.D.; Matthews, E.A.; Dickenson, A.H.; Wood, J.N. Nociceptor-specific gene deletion reveals a major role for Nav1.7 (PN1) in acute and inflammatory pain. *Proc. Natl. Acad. Sci. USA* **2004**, *101*, 12706–12711. [CrossRef] [PubMed]

30.   Shields, S.D.; Cheng, X.; Üçeyler, N.; Sommer, C.; Dib-Hajj, S.D.; Waxman, S.G. Sodium channel Nav1.7 is essential for lowering heat pain threshold after burn injury. *J. Neurosci.* **2012**, *32*, 10819–10832. [CrossRef] [PubMed]

31.   De Rooij, A.M.; Gosso, M.F.; Alsina-Sanchis, E.; Marinus, J.; Hilten, J.J.V.; Maagdenberg, A.M.V.D. No mutations in the voltage-gated NaV1.7 sodium channel alpha1 subunit gene SCN9A in familial complex regional pain syndrome. *Eur. J. Neurol.* **2010**, *17*, 808–814. [CrossRef] [PubMed]

32.   Diss, J.K.; Calissano, M.; Gascoyne, D.; Djamgoz, M.B.; Latchman, D.S. Identification and characterization of the promoter region of the Nav1.7 voltage-gated sodium channel gene (SCN9A). *Mol. Cell. Neurosci.* **2008**, *37*, 537–547. [CrossRef] [PubMed]

33.   Cox, J.J.; Reimann, F.; Nicholas, A.K.; Thornton, G.; Roberts, E.; Springell, K.; Karbani, G.; Jafri, H.; Mannan, J.; Raashid, Y.; et al. An SCN9A channelopathy causes congenital inability to experience pain. *Nature* **2006**, *444*, 894–898. [CrossRef] [PubMed]

34.   Liu, C.; Cao, J.; Ren, X.; Zang, W. Nav1.7 protein and mRNA expression in the dorsal root ganglia of rats with chronic neuropathic pain. *Neural Regen. Res.* **2012**, *7*, 1540–1544. [PubMed]

35.   Minett, M.S.; Pereira, V.; Sikandar, S.; Matsuyama, A.; Lolignier, S.; Kanellopoulos, A.H.; Mancini, F.; Iannetti, G.D.; Bogdanov, Y.D.; Santana-Varela, S.; et al. Endogenous opioids contribute to insensitivity to pain in humans and mice lacking sodium channel Nav1.7. *Nat. Commun.* **2015**, *6*, 8967. [CrossRef] [PubMed]

36.   Gandini, R.; Merolla, S.; Chegai, F.; Del Giudice, C.; Stefanini, M.; Pampana, E. Foot Embolization During Limb Salvage Procedures in Critical Limb Ischemia Patients Successfully Managed With Mechanical Thromboaspiration: A Technical Note. *J. Endovasc. Ther.* **2015**, *22*, 558–563. [CrossRef] [PubMed]

37.   SudheesháKumar, P. Flexible, micro-porous chitosan–gelatin hydrogel/nanofibrin composite bandages for treating burn wounds. *RSC Adv.* **2014**, *4*, 65081–65087.

38.   Capasso, R.; Rosa, T.; Tsou, D.Y.; Nekhendzy, V.; Drover, D.; Collins, J.; Zaghi, S.; Camacho, M. Variable Findings for Drug-Induced Sleep Endoscopy in Obstructive Sleep Apnea with Propofol versus Dexmedetomidine. *Otolaryngol. Head Neck Surg.* **2016**, *154*, 765–770. [CrossRef] [PubMed]

39.   Au, A.K.; Steinberg, D.; Thom, C.; Shirazi, M.; Papanagnou, D.; Ku, B.S.; Fields, J.M. Ultrasound measurement of inferior vena cava collapse predicts propofol-induced hypotension. *Am. J. Emerg. Med.* **2016**, *34*, 1125–1128. [CrossRef] [PubMed]

40.   Bang, Y.S.; Kim, Y.U.; Oh, D.; Shin, E.Y.; Park, S.K. A randomized, double-blind trial evaluating the efficacy of palonosetron with total intravenous anesthesia using propofol and remifentanil for the prevention of postoperative nausea and vomiting after gynecologic surgery. *J. Anesth.* **2016**, *30*, 935–940. [CrossRef] [PubMed]

41.   Bataille, A.; Letourneulx, J.F.; Charmeau, A.; Lemedioni, P.; Leger, P.; Chazot, T.; Guen, M.L.; Diemunsch, P.; Fischler, M.; Liu, N. Impact of prophylactic combination of dexamethasone-ondansetron on postoperative nausea and vomiting in obese adult patients undergoing laparoscopic sleeve gastrectomy during closed-loop propofol-remifentanil anaesthesia: A randomised double-blind placebo study. *Eur. J. Anaesthesiol.* **2016**, *33*, 898–905. [PubMed]

42.   Cho, S.Y.; Jeong, C.W.; Jeong, C.Y.; Lee, H.G. Efficacy of the combination of cold propofol and pretreatment with remifentail on propofol injection pain. *Korean J. Anesthesiol.* **2010**, *59*, 305–309. [CrossRef] [PubMed]

43.   Zhang, J.; Wang, Y.; Li, B.; Zhang, W. Remifentail infusion for paediatric bronchoscopic foreign body removal: Comparison of sevoflurane with propofol for anaesthesia supplementation for bronchoscope insertion. *Anaesth. Intensive Care* **2010**, *38*, 905–910. [PubMed]

44.   Rivara, M.; Zuliani, V. Novel sodium channel antagonists in the treatment of neuropathic pain. *Expert Opin. Investig. Drugs* **2015**, *25*, 215–226. [CrossRef] [PubMed]

45.   Hockley, J.R.; Winchester, W.J.; Bulmer, D.C. The voltage-gated sodium channel Na 1.9 in visceral pain. *Neurogastroenterol. Motil.* **2016**, *28*, 316–326. [CrossRef] [PubMed]

46.   Mackenzie, F.E.; Parker, A.; Parkinson, N.J.; Oliver, P.L.; Brooker, D.; Underhill, P.; Lukashkina, V.A.; Lukashkin, A.N.; Holmes, C.; Brown, S.D. Analysis of the mouse mutant Cloth-ears shows a role for the voltage-gated sodium channel Scn8a in peripheral neural hearing loss. *Genes Brain Behav.* **2009**, *8*, 699–713. [CrossRef] [PubMed]

47.   Teixeira, C.E.; Baracat, J.S.; Arantes, E.C.; de Nucci, G.; Antunes, E. Effects of β-adrenoceptor antagonists in the neural nitric oxide release induced by electrical field stimulation and sodium channel activators in the rabbit corpus cavernosum. *Eur. J. Pharmacol.* **2005**, *519*, 146–153. [CrossRef] [PubMed]

48.   Yang, S.W.; Ho, G.D.; Tulshian, D.; Bercovici, A.; Tan, Z.; Hanisak, J.; Brumfield, S.; Matasi, J.; Sun, X.; Sakwa, S.A.; et al. Bioavailable pyrrolo-benzo-1,4-diazines as Nav 1.7 sodium channel blockers for the treatment of pain. *Bioorg. Med. Chem. Lett.* **2014**, *24*, 4958–4962. [CrossRef] [PubMed]

49.   Huang, C.-P.; Chen, H.-N.; Su, H.-L.; Hsieh, C.-L.; Chen, W.-H.; Lai, Z.-R.; Lin, Y.-W. Electroacupuncture reduces carrageenan-and CFA-induced inflammatory pain accompanied by changing the expression of Nav1.7 and Nav1.8, rather than Nav1.9, in mice dorsal root ganglia. *Evid. Based Complement. Altern. Med.* **2013**. [CrossRef] [PubMed]

50.   Liang, L.; Fan, L.; Tao, B.; Yaster, M.; Tao, Y.-X. Protein kinase B/Akt is required for complete Freund's adjuvant-induced upregulation of Nav1.7 and Nav1.8 in primary sensory neurons. *J. Pain* **2013**, *14*, 638–647. [CrossRef] [PubMed]

51.   Chattopadhyay, M.; Mata, M.; Fink, D.J. Continuous delta-opioid receptor activation reduces neuronal voltage-gated sodium channel (NaV1.7) levels through activation of protein kinase C in painful diabetic neuropathy. *J. Neurosci.* **2008**, *28*, 6652–6658. [CrossRef] [PubMed]

52.   Li, Z.; Pei, Q.; Cao, L.; Xu, L.; Zhang, B.; Liu, S. Propofol increases micro-opioid receptor expression in SH-SY5Y human neuroblastoma cells. *Mol. Med. Rep.* **2012**, *6*, 1333–1336. [PubMed]

53.   Richards, M.J.; Skues, M.A.; Jarvis, A.P.; Prys-Roberts, C. Total i.v. anaesthesia with propofol and alfentanil: Dose requirements for propofol and the effect of premedication with clonidine. *Br. J. Anaesth.* **1990**, *65*, 157–163. [CrossRef] [PubMed]

54.   Altmayer, P.; Buch, U.; Buch, H.P. Propofol binding to human blood proteins. *Arzneimittelforschung* **1995**, *45*, 1053–1056. [PubMed]

55.   Ludbrook, G.L.; Upton, R.N.; Grant, C.; Gray, E.C. Brain and blood concentrations of propofol after rapid intravenous injection in sheep, and their relationships to cerebral effects. *Anaesth. Intensive Care* **1996**, *24*, 445–452. [PubMed]

56.   Raoof, A.A.; Obbergh, L.J.V.; de Goyet, J.V.; Verbeeck, R.K. Extrahepatic glucuronidation of propofol in man: Possible contribution of gut wall and kidney. *Eur. J. Clin. Pharmacol.* **1996**, *50*, 91–96. [CrossRef] [PubMed]

57.   Shafer, A.; Doze, V.A.; Shafer, S.L.; White, P.F. Pharmacokinetics and pharmacodynamics of propofol infusions during general anesthesia. *Anesthesiology* **1988**, *69*, 348–356. [CrossRef] [PubMed]

58. Kotani, Y.; Shimazawa, M.; Yoshimura, S.; Iwama, T.; Hara, H. The experimental and clinical pharmacology of propofol, an anesthetic agent with neuroprotective properties. *CNS Neurosci. Ther.* **2008**, *14*, 95–106. [CrossRef] [PubMed]

59. Younes, I.; Rinaudo, M. Chitin and chitosan preparation from marine sources. Structure, properties and applications. *Mar. Drugs* **2015**, *13*, 1133–1174. [CrossRef] [PubMed]

60. Becher, R.D.; Peitzman, A.B.; Sperry, J.L.; Gallaher, J.R.; Neff, L.P.; Sun, Y.; Miller, P.R.; Chang, M.C. Damage control operations in non-trauma patients: Defining criteria for the staged rapid source control laparotomy in emergency general surgery. *World J. Emerg. Surg.* **2016**, *11*, 10. [CrossRef] [PubMed]

61. Yousef, M.; Pichyangkura, R.; Soodvilai, S.; Chatsudthipong, V.; Muanprasat, C. Chitosan oligosaccharide as potential therapy of inflammatory bowel disease: Therapeutic efficacy and possible mechanisms of action. *Pharmacol. Res.* **2012**, *66*, 66–79. [CrossRef] [PubMed]

62. Wang, S.Y.; Wang, G.K. A mutation in segment I-S6 alters slow inactivation of sodium channels. *Biophys. J.* **1997**, *72*, 1633–1640. [CrossRef]

![marine drugs logo] *marine drugs*

MDPI

Article

Chitin Oligosaccharide (COS) Reduces Antibiotics Dose and Prevents Antibiotics-Caused Side Effects in Adolescent Idiopathic Scoliosis (AIS) Patients with Spinal Fusion Surgery

Yang Qu, Jinyu Xu, Haohan Zhou, Rongpeng Dong, Mingyang Kang and Jianwu Zhao *

Department of Orthopedics, The Second Hospital of JiLin University, Changchun 130041, China;
quyang_22d@163.com (Y.Q.); xujy0913@sina.com (J.X.); zhouhh1609@sina.com (H.Z.);
drpeng1507@sina.com (R.D.); kmy_fge@126.com (M.K.)
* Correspondence: jianwu@jlu.edu.cn; Tel.: +86-431-8879-6940

Academic Editors: Hitoshi Sashiwa and David Harding
Received: 12 January 2017; Accepted: 8 March 2017; Published: 14 March 2017

Abstract: Antibiotics are always considered for surgical site infection (SSI) in adolescent idiopathic scoliosis (AIS) surgery. However, the use of antibiotics often causes the antibiotic resistance of pathogens and side effects. Thus, it is necessary to explore natural products as drug candidates. Chitin Oligosaccharide (COS) has anti-inflammation and anti-bacteria functions. The effects of COS on surgical infection in AIS surgery were investigated. A total of 312 AIS patients were evenly and randomly assigned into control group (CG, each patient took one-gram alternative Azithromycin/Erythromycin/Cloxacillin/Aztreonam/Ceftazidime or combined daily), experiment group (EG, each patient took 20 mg COS and half-dose antibiotics daily), and placebo group (PG, each patient took 20 mg placebo and half-dose antibiotics daily). The average follow-up was one month, and infection severity and side effects were analyzed. The effects of COS on isolated pathogens were analyzed. SSI rates were 2%, 3% and 8% for spine wounds and 1%, 2% and 7% for iliac wound in CG, EG and PG ($p < 0.05$), respectively. COS reduces the side effects caused by antibiotics ($p < 0.05$). COS improved biochemical indexes and reduced the levels of interleukin (IL)-6 and tumor necrosis factor (TNF) alpha. COS reduced the antibiotics dose and antibiotics-caused side effects in AIS patients with spinal fusion surgery by improving antioxidant and anti-inflammatory activities. COS should be developed as potential adjuvant for antibiotics therapies.

Keywords: chitin oligosaccharide; antimicrobial prophylaxis; adolescent idiopathic scoliosis; spinal fusion

1. Introduction

In most surgeries, surgical infection caused by pathogens is a common problem, which increases patients' morbidity and prolongs the duration of hospital stay [1]. The risk of post-operative surgical infection is still increasing, especially in the surgery for adolescent idiopathic scoliosis (AIS). Azithromycin is the normally used antibiotics for preventing surgical site infections (SSIs). Erythromycin combined with the other methods has been proved an effective and safe method in seroma therapy in general surgery and traumatology [2]. However, long-term utilization of low-dose erythromycin after surgery was not recommended for surgical patients [3]. There are many ways for inhibiting the risk and the progression of SSI. Earlier results suggested that cloxacillin is effective for AIS therapy in methicillin-sensitive staphylococcal infections [3]. Aztreonam is a monobactam antibiotic mainly used to fight against an infection caused by Gram-negative aerobic bacteria. Aztreonam has been proven effective for prophylaxis and therapy of urinary tract infections after prostate surgery [4]. Antibiotic prophylaxis with ceftazidime has been proven effective for preventing surgical infections [5].

Although these antibiotics have been proved to be effective for preventing surgical infection after surgery, the trouble it that drug resistance is the main threat for public health with the widespread use of antibiotics and it is a big challenge to control drug-resistant pathogens [3,6]. The resistance of enterobacteriaceae to antibiotics has been steadily increasing [7]. The resistance of pathogens to cloxacillin [8,9], aztreonam [10,11] and ceftazidime [12,13] has been widely reported. The problem is becoming more pronounced because drug resistance accumulates faster than new antibiotics have been developed.

Furthermore, some adverse effects of these antibiotics limit their application. Side effects have been reported in azithromycin-treated patients (gastric upset, nausea, and headache) and there are some manifestations like vomiting or diarrhea and abdominal pain symptoms [14]. The application of cloxacillin will result in the side effects including hypotension, tachycardia, flushing, palpitation, headache and nausea [15]. cloxacillin induces seizure in a hemodialysis patient [16] and often causes fever, chills, malaise, abdominal pain and anxiety [17]. The most frequent side effects of azithromycin are associated with a gastrointestinal system, such as abdominal pain, diarrhea, and nausea [14]. The most frequent side effects for aztreonam are acute renal failure, skin rash, and eosinophilia [18]. Long-term use of ceftazidime will result in side effects such as fever, rash, and pancytopenia. Gastrointestinal side effects limit erythromycin treatment and compliance [19]. Therefore, it is critical to find a novel way to control drug-resistant bacteria and related infections [20], and it is necessary to explore natural products for antibiotic therapy.

Chitin, a long-chain polymer of an N-acetylglucosamine, is the most abundant natural resource on the earth. Chitin is the major component of crustacean (crab [21], lobster [22] and shrimp [23]) shells, the cell walls of fungi [24] and insects [25], and the radulae of Mollusca [26]. The safety of chitin has been proven as a carrier for drug delivery [27]. Chitin can be digested and broken into Chitin Oligosaccharide (COS) by chitinase [28]. COS shows diverse pharmacological potential. COS has functions against cancer and inflammation [29] and is a potential anti-inflammatory drug by activating tumor necrosis factor (TNF) alpha [30]. From the information, COS may be effective for controlling surgical infection. The main aim of the present work is to investigate the effects of COS on surgical infection and related molecular mechanisms in AIS patients. Meanwhile, antibiotics-caused side effects and the levels of related oxidant-related molecules and inflammatory cytokines were measured.

2. Results

2.1. The Bacteria Isolated from AIS Patients

The following bacteria were isolated from infected wounds of AIS patients and confirmed by 16s rRNA: *Pseudomonas aeruginosa*, *Burkholderia cepacia*, *Burkholderia cenocepacia*, *Acinetobacter baumannii*, *Acinetobacter lwoffii*, *Klebsiella pneumoniae*, *Escherichia coli*, *Staphylococcus aureus* and *Providencia stuartii*. These isolates represent the most frequently encountered resistance types. The resistance to aminoglycosides and fluoroquinolones is widely found in isolated *P. aeruginosa* [31]. *P. aeruginosa* can cause acute and chronic infections. New treatment strategies against *P. aeruginosa* infection are in especially high demand for its increasing resistance to the antibiotic Azithromycin. The overexpression of PA3297 (a DEAH-box helicase; DEAH, Asp-Glu-Ala-His) was found to be caused by the interaction between Azithromycin and ribosomes. The mutant PA3297 will increase unprocessed 23TS-5S rRNA in the presence of Azithromycin, which increases the sensitivity to Azithromycin-reduced effects, suggesting the bacterial reactions to counteract the effects of Azithromycin [32]. Lipopolysaccharides/lipooligosaccharides are abundant in the cell surface and increase antibiotics resistance in most Gram-negative bacteria. *A. baumannii* can develop resistance to Azithromycin via the loss of lipopolysaccharides (LPS) with mutant genes, which are associated with lipid synthesis [33]. Long-term therapy of Ceftazidime for surgical infection will increase beta-lactam resistance in Burkholderia and results in treatment failure [34]. *A. lwoffii* shows multi-drug resistance (MDR), which may be caused by a bla (NDM-1)-bearing plasmid, pNDM-BJ01, and mutant pNDM-BJ02 [35]. MDR is common in *E. coli* and its resistance to antibiotics has been reported to be associated with virulence [36]. The MDR of *S. aureus* has been found to be associated with class 1 and 2 integrons and gene cassettes [37].

2.2. Effects of COS on Isolated Bacteria

COS was proved to prevent MDR bacteria growth in a dose-dependent way after one-day culture (Figure 1). This effect was dose dependent for the concentration range investigated (0, 10, 20 and 30 mg/L). Generally, the growth rates of these bacteria were decreasing with increasing concentration of COS. The inhibitory functions were prominent for the pathogens *P. aeruginosa*, *B. cenocepacla*, *A. baumannii*, *K. pneumoniae*, *E. coli*, *S. aureus* and *P. stuartli*.

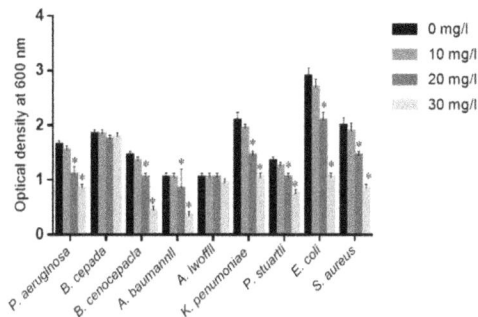

Figure 1. The effects of chitin oligosaccharide (COS) on the growth of clinically isolated pathogens. All pathogens were cultured for 24 h with different concentrations of COS at 37 °C. All data were presented as average value ± standard derivation (S.D.) * $p < 0.05$ vs. 0 mg/L COS. There is a statistical significance of differences if $p < 0.05$.

2.3. COS Reduces the Antibiotics Resistance of MDR Bacteria

The effects of COS on the dose of antibiotics for fighting against MDR bacteria were measured. The antibiotics (Azithromycin, Erythromycin, Cloxacillin, Aztreonam and Ceftazidime) were selected for detecting antibiotic-resistant pathogens (Table 1).

As Table 1 showed, COS addition decreased the minimum inhibitory concentration (MIC) of most antibiotics, including Azithromycin, Erythromycin, Cloxacillin, Aztreonam and Ceftazidime. Among these antibiotics, the MIC of *P. aeruginosa* was decreased from 19 to 0.5 μg/mL for Ceflazidime, suggesting the bacteria could be inhibited by reducing antibiotics resistance via COS. The MIC of *P. aeruginosa* was decreased from 250 to 6 μg/mL for erythromycin with the increase of COS dose from 0 to 30 mg/L. COS could reduce the antibiotics resistance of *B. cepacia* for Azithromycin, Erythromycin, Cloxacillin, Aztreonam and Ceftazidime from 30, 60, 120, 250 and 60 to 0.1, 9, 2, 2 and 0.5 μg/mL, respectively. The inhibitory functions of COS for antibiotic resistance were prominent for other pathogens except for *P. stuartli* by reducing the MICs of most antibiotics. COS had similar inhibitory functions for antibiotics resistance if combined antibiotics were used (data were not shown). A lesser effect was observed with Gram-positive isolates.

Table 1. MICs of antibiotics alone and with increasing concentrations (0, 10, 20 and 30 mg/L) of COS for a range of antibiotics-resistant clinical isolates (Antibiotic MIC, mg/L).

Antibiotics	COS	PA	BC	BC1	AB	AL	KP	EC	SA	PS
Azithromycin	0	7	30	3	17	0.2	500	8	0.1	8
	10	3	16	2	4	0.1	250	4	0.1	2
	20	2	5	1	0.5	0.1	60	1	0.1	1
	30	0.5	0.1	0.5	0.02	0.1	32	0.5	0.1	0.1
Erythromycin	0	250	60	68	9	0.2	250	30	0.1	20
	10	126	48	38	4	0.1	120	18	0.1	16
	20	30	17	16	1	0.1	65	8	0.1	0.1
	30	6	9	8	0.3	0.1	20	4	0.1	0.1
Cloxacillin	0	1000	120	39	2	0.1	2	4	0.1	0.1
	10	1000	60	16	2	0.1	2	4	0.1	0.1
	20	1000	17	8	1	0.1	1	4	0.1	0.1
	30	1000	2	2	1	0.1	1	2	0.1	0.1
Aztreonam	0	34	250	129	678	30	4000	1000	512	0.1
	10	14	100	32	312	15	2000	510	234	0.1
	20	9	8	3	250	4	500	250	249	0.1
	30	3	2	1	126	1	250	64	244	0.1
Ceftazidime	0	19	60	8	778	2	4230	4123	12	8
	10	8	8	4	546	2	2078	3124	8	4
	20	1	2	1	312	0.5	1536	2341	2	2
	30	0.5	0.25	0.25	156	0.1	1324	1097	1	1

Note: MIC, minimal inhibitory concentration; COS, chitin oligosaccharide; PA, *P. aeruginosa*; BC, *B. cepacia*; BC1, *B. cenocepacla*; AB, *A. baumannii*; AL, *A. lwoffii*; KP, *K. pneumoniae*; EC, *E. coli*; SA, *S. aureus*; PS, *P. stuartli*. A maximum dose for an adult is 10 mg/kg for Azithromycin, 50 mg/kg for erythromycin, 100 mg/kg for Cloxacillin, 200 mg/kg for Aztreonam, and 150 mg/kg for Ceftazidime daily.

2.4. The Measurement of Resistance to COS

As Table 2 showed, MIC analysis indicated that *P. aeruginosa* had no resistance to COS although the concentration of COS was increased and the bacteria were cultured for three months in the media with COS. There was no statistical significance of differences among the groups with different concentrations of COS, suggesting no resistance to COS in the pathogens. Similarly, the increasing concentration of COS could not increase the drug resistance of all other pathogens ($p > 0.05$).

Table 2. MICs of antibiotics for clinical isolates with increasing concentrations of COS (mg/L).

COS	PA	BC	BC1	AB	AL	KP	EC	SA	PS
0	10 ± 1.2	5 ± 0.6	8 ± 1	20 ± 2.6	4 ± 0.6	106 ± 11	9 ± 1.1	2 ± 0.1	11 ± 1.3
10	12 ± 1.5	5 ± 0.9	9 ± 1.1	18 ± 2.2	4 ± 0.8	100 ± 10	11 ± 0.8	2 ± 0.2	9 ± 1.6
20	11 ± 1.4	6 ± 0.7	10 ± 1.2	17 ± 2.4	3 ± 0.4	95 ± 9	12 ± 0.9	3 ± 0.2	12 ± 1.5
30	12 ± 1.3	5.5 ± 0.6	10 ± 1.3	21 ± 1.6	5 ± 0.6	112 ± 12	14 ± 1.5	4 ± 0.4	13 ± 1.1
p values	0.56	0.73	0.4	0.23	0.18	0.09	0.15	0.10	0.27

Note: PA, *P. aeruginosa*; BC, *B. cepacia*; BC1, *B. cenocepacla*; AB, *A. baumannii*; AL, *A. lwoffii*; KP, *K. pneumoniae*; EC, *E. coli*; SA, *S. aureus*; PS, *P. stuartli*. There is statistical significance of differences if $p < 0.05$.

2.5. Baseline Characters

A total of 312 AIS subjects were selected in the present study and the average follow-up period was one month (which ranged from two to ten weeks). Seventy-one (31.4%) patients (group B) received antimicrobials until drain removal (range 3–5 days). There was no statistical significance of differences among three groups, including average age, gender, body mass index (BMI), lifestyles and the symptoms of AIS (scoliosis curve type, mean number of levels fused per patient, intra-operative transfusion, post-operative transfusion and duration of drain left in situ). There was no statistical significance of differences in all parameters found among three groups ($p > 0.05$, Table 3). To avoid instrument interference for surgery, different instruments were also compared among three groups. No statistical significance of differences was found either (Table 4, $p > 0.05$). Different surgical levels would cause different results, and thus the fused levels, surgical duration, blood loss and spine drains were also compared here. The results showed that there was no statistical significance of differences in these factors among three groups (Table 5, $p > 0.05$).

Table 3. Baseline characters ($n = 104$ for each group).

Parameters	CG, Mean (SD; Range)	EG, Mean (SD; Range)	PG, Mean (SD; Range)	*p*
Age, years	12.03 (SD 2.64; 10–16)	11.80 (SD 2.28; 10–16)	12.37 (SD 2.70; 11–26)	0.86
Gender, male/female	74/30	76/28	70/34	0.65
BMI	18.7 (SD 2.8; 13.42–28.27)		18.93 (SD 3.62; 12.29–30.49)	0.28
Smoking	23	20	18	0.65
Scoliosis curve type (MT/DM/DT/TM/TL)	58/22/19/0/5	55/25/17/2/5	59/21/16/4/4	0.26
Mean number of levels fused per patient	9.26 (SD 2.06; 7–13)	9.38 (SD 2.21; 7–13)	9.57 (SD 2.23; 7–13)	0.78
Surgical duration (minutes)	253.88 (SD 62.15; 198–487)	287.95 (SD 61.85; 178–483)	279.94 (SD 76.7; 196–437)	0.53
Mean number of anchor points per patient	10.1 (SD 2.49; 67–20)	11.1 (SD 2.69; 7–21)	9.2(SD 2.06; 7–18)	0.39
Post-operative transfusion, mL	Blood: 198.26 (SD 149.80; 0–745) FFP: 21.32 (SD 77.23; 0–545)	Blood: 218.26 (SD 169.80; 0–721) FFP: 20.42 (SD 76.63; 0–521)	Blood: 207.68 (SD 282.43; 0–986) FFP: 22.48 (SD 75.37; 0–530)	0.21
Post-operative spine drain, mL	Total drain: 218.01 (SD 221.76; 5–1500)	Total drain: 473.02 (SD 355.40; 5–1320)	Total drain: 463.01 (SD 345.90; 5–1640)	0.41
Post-operative duration of drain in situ, days	3.2 (SD 0.68; 2–7)	3.0 (SD 0.61; 2–7)	3.3 (SD 0.75; 2–7)	0.36

Note: All patients were randomly and evenly assigned into three groups according to different therapies after operation: control group (CG, the patients received one-gram alternative drugs Azithromycin/Erythromycin/Cloxacillin/Aztreonam/Ceftazidime or combined in one gram daily), experiment group (EG, the patients received 20 mg COS and half-dose antibiotics daily) and placebo group (PG, the patients received 20 mg placebo and half-dose antibiotics daily). SD, standard deviation. ANOVA test and chi-square test were performed for comparing the statistical significance of difference among three groups. There is statistical significance of differences if $p < 0.05$.

Table 4. Comparison of instruments among three groups.

Instrument	CG	EG	PG	*p*-Value
All pedicle screw constructs	35	30	34	0.73
Pedicle screws and hook constructs	42	48	46	0.69
All hook constructs	10	10	10	1.00
Pedicle screws, hooks and sublaminar wire construct	11	10	8	0.77
Pedicle screw and sublaminar wire construct	6	6	6	1.00

Note: chi-square test was performed for comparing the statistical significance of difference among three groups. There is statistical significance of differences if $p < 0.05$.

Table 5. Surgical operation.

Levels Fusion		T6-L2	T2-L1	T2-L2	T5-L4	T7-L2	T6-12	T4-11
	CG	35	24	12	9	10	8	6
Numbers	EG	33	26	13	8	11	7	6
	PG	34	25	14	7	9	8	7
p-value					1			
Surgical	CG	250 ± 45	263 ± 51	298 ± 46	305 ± 62	527 ± 71	296 ± 43	248 ± 32
duration	EG	242 ± 49	251 ± 59	277 ± 53	311 ± 53	512 ± 63	280 ± 37	239 ± 28
(minutes)	PG	256 ± 40	266 ± 52	302 ± 43	299 ± 47	531 ± 58	301 ± 26	254 ± 21
p-value		0.91	0.85	0.69	0.27	0.53	0.44	0.75
Blood loss (mL)	CG	537 ± 214	587 ± 196	716 ± 243	436 ± 178	NA	NA	NA
	EG	498 ± 253	502 ± 245	642 ± 215	451 ± 210	NA	NA	NA
	PG	399 ± 128	546 ± 237	597 ± 204	407 ± 185	NA	NA	NA
p-value		0.12	0.26	0.37	0.55			
Spine	CG	270 ± 123	346 ± 188	164 ± 49	105 ± 31	541 ± 103	90 ± 21	100 ± 17
drain	EG	244 ± 105	320 ± 171	154 ± 33	96 ± 18	501 ± 83	84 ± 13	93 ± 11
(mL)	PG	252 ± 98	305 ± 124	137 ± 41	99 ± 27	527 ± 66	89 ± 17	102 ± 22
p-value		0.18	0.29	0.08	0.57	0.62	0.71	0.41

Note: chi-square test was performed for comparing the statistical significance of difference among three groups. There is statistical significance of differences if $p < 0.05$.

2.6. COS Were Potential Adjuvants of Antibiotics for Preventing Surgical Infection

All patients were randomly and evenly assigned into three groups according to different therapies after operation: control group (CG, the patients received one-gram alternative drugs Azithromycin/Erythromycin/Cloxacillin/Aztreonam/Ceftazidime or combined in one gram daily), experiment group (EG, the patients received 20 mg COS and half-dose antibiotics daily) and placebo group (PG, the patients received 20 mg placebo and half-dose antibiotics daily). The overall rates of SSI were 1%, 2% and 10% for spine wounds and 0%, 1% and 3% for the iliac crest wounds in CG, EG and PG, respectively. The patients had lower SSI rates for spine wounds in CG and EG when compared with PG (Table 6, $p < 0.05$). Although the patients also had lower SSI rates for iliac wounds among three groups, there was no statistical significance of differences (Table 6, $p > 0.05$). The results may be caused by a small population in the present work and result in a limited number of the patients with iliac wound. The present findings indicated that COS were potential adjuvants in antibiotics therapy for preventing surgical infection since only half-dose antibiotics were used in EG.

Table 6. The rates of surgical site infections.

Wound Category	CG	EG	PG	P1	P2	P3
Spine wound	1	2	10	1	0.004	0.005
Iliac wound	0	1	3	1	0.245	0.607

Note: chi-square test was performed for comparing the statistical significance of difference for the rates of surgical site infections among three groups. The infection was not considered if the infection duration was less than three days. The infection would be counted if the duration was more than three days. P1, CG vs. EG; P2, CG vs. PG; P3, EG vs. PG. There is statistical significance of differences if $p < 0.05$.

2.7. COS Prevented the Side Effects Caused by Antibiotics

Although most antibiotics can control most surgical infection well, the adverse effects are obvious, which limit their clinical use. Just as we proposed, the side effects were higher in CG than in EG and or PG (Table 7, $p < 0.05$). Fewer side effects were found in EG and PG groups. Most side effects included gastric upset, nausea, headache, vomiting, diarrhea, abdominal pain, seizure, chills, malaise, anxiety and fever in CG.

Table 7. Side effects of AIS patients in different groups.

Side Effects	CG	EG	PG	P1	P2	P3
gastric upset	15	3	1	0.003	0.000	0.614
nausea	13	1	0	0.001	0.000	1.000
headache	15	3	1	0.003	0.000	0.614
vomiting	20	4	2	0.000	0.000	0.679
diarrhea	7	0	0	0.210	0.210	1.000
abdominal pain	10	0	1	0.004	0.005	1.000
seizure	3	0	1	0.245	0.614	1.000
chills	6	1	1	0.124	0.124	1.000
malaise	5	1	1	0.214	0.214	1.000
anxiety	6	0	1	0.038	0.124	1.000
fever	8	0	1	0.012	0.041	1.000

Note: AIS, adolescent idiopathic scoliosis; chi-square test was performed for comparing the statistical significance of difference among three groups. P1, CG via EG; P2, CG via PG; P3, EG via PG. There is statistical significance of differences if $p < 0.05$.

2.8. COS Improves the Biochemical Parameters of AIS Patients

Before the experiment, serum biochemical index analysis showed that there was no statistical significance of differences for serum levels of SOD (superoxide dismutase), GSH (reduced glutathione), ALT (alanine aminotransferase) and AST (aspartate amino-transaminase) among three groups (Table 8) ($p > 0.05$). Comparatively, serum biochemical index analysis showed that serum ALT and AST, reached the highest level in PG when compared with the other two groups after an average of one-month follow-up (Table 8) ($p < 0.05$). In contrast, the serum SOD and GSH reached the lowest level in PG as compared to two other groups (Table 8) ($p < 0.05$). COS increased the levels of SOD and GSH, and reduced the serum levels of ALT and AST.

Table 8. Biochemical parameters of enzyme activities in AIS patients.

Stages	Group (n = 10)	SOD (U/mL)	GSH (pg/mL)	ALT (pg/mL)	AST (pg/mL)
Before experiment	CG	26.24 ± 3.16	22.15 ± 2.04	44.12 ± 10.13	100.32 ± 25.67
	EG	24.25 ± 4.16	23.23 ± 1.93	47.79 ± 8.40	108.26 ± 19.64
	PG	25.34 ± 2.32	21.28 ± 1.74	43.79 ± 12.36	103.42 ± 24.48
	p-value	0.56	0.74	0.12	0.68
After experiment	CG	22.73 ± 3.28 Δ,●	21.09 ± 3.06 Δ	48.31 ± 8.25 Δ,●	118.36 ± 26.82 Δ,●
	EG	28.29 ± 5.13 *,●	25.44 ± 2.01 *,●	38.19± 6.25 *,●	76.34 ± 14.38 *,●
	PIG	15.32 ± 5.71 *,Δ	14.26 ± 4.39 *,Δ	53.69 ± 18.31 *,Δ	89.46 ± 27.52 *,Δ
	p-value	0.01	0.01	0.01	0.01

Note: AIS, adolescent idiopathic scoliosis; SOD, superoxidedismutase; GSH, glutathione; AST, aspartate transaminase; ALT, alanine transaminase. * $p < 0.05$ vs. the control group (CG); Δ $p < 0.05$ vs. the experimental group (EG); ● $p < 0.01$ vs. the placebo group (PG). There is statistical significance of differences if $p < 0.05$.

2.9. COS Reduced Relative mRNA Levels of Inflammatory Cytokines (IL-6 and TNF Alpha)

qRT-PCR analysis indicated that the mRNA levels of IL-6 and TNF alpha were higher in PG than in CG, and EG (Figure 2) ($p < 0.05$). Furthermore, the mRNA levels of IL-6 and TNF alpha were higher in CG than in EG (Figure 1) ($p < 0.05$). The results suggest that COS reduces the mRNA levels of IL-6 and TNF alpha.

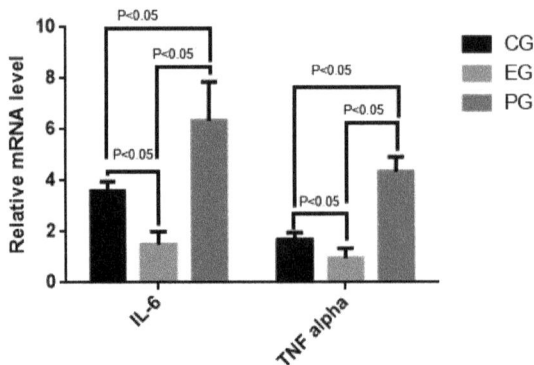

Figure 2. Real-time qRT-PCR analysis of the effects of COS on relative mRNA levels of interleukin (IL-6) and tumor necrosis factor (TNF) alpha. All data were presented as average value $\pm$ S.D. There is statistical significance of differences if $p < 0.05$.

2.10. COS Reduced Relative Protein Levels of Inflammatory Cytokines (IL-6 and TNF Alpha)

Just as qRT-PCR analysis, Western blot results indicated that the protein levels of IL-6 and TNF alpha were higher in PG than in CG, and EG (Figure 3) ($p < 0.05$). Furthermore, the protein levels of IL-6 and TNF alpha were higher in CG than in EG (Figure 3) ($p < 0.05$). The results suggest that COS reduces the protein levels of IL-6 and TNF alpha.

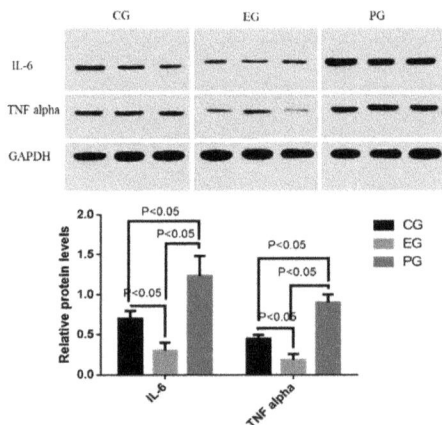

Figure 3. Western blot analysis of the effects of COS on relative protein levels of IL-6 and TNF alpha. All data were presented as average value $\pm$ S.D. There is statistical significance of differences if $p < 0.05$.

3. Discussion

Present results showed that COS reduced the dose and side effects of antibiotics, and SSI rates, which may be associated with antioxidant and anti-inflammation activities of COS. Infection will result in the increase of reactive oxygen species (ROS) [38] by affecting ROS-related molecules SOD [39], GSH [40], ALT [41] and AST [42]. COS improved antioxidant activities by increasing the activities of the anti-oxidant enzymes SOD and GSH, and decreasing the levels of oxidative-stress-related biomarkers ALT and AST (Table 8, $p < 0.05$). IL-6 and TNF alpha are involved with the infection inflammation and their levels will be increased. For instance, IL-6 trans-signaling plays an important role in the angiogenesis of the peritoneal membrane [43], in which IL-6 is an important inflammatory cytokine. Vascular inflammation is an important reason for causing atherosclerosis. High-level TNF alpha will induce vascular inflammation. TNF alpha neutralizing antibodies have been administered to treat many inflammatory disorders [44]. COS may improve anti-inflammatory activity by reducing levels of IL-6 and TNF alpha (Figures 2 and 3, $p < 0.05$). All the results may contribute to the reduction of SSI rates after COS treatment (Table 6, $p < 0.05$).

Other reports showed that COS performed its anti-inflammatory functions via activating nuclear factor-kappa B [45], cyclooxygenase-2 [46], and inducible nitric oxide synthase [45]. Chitin is essential structural polysaccharide of many fungal pathogens and plays an important role in human immune responses. COS inhibited LPS-induced inflammation and contributed to human immune response when the pathogens were killed. Furthermore, NOD2 and TLR9 were regarded as chitin receptors and regulated inflammatory conditions by preventing the expression of chitinases, whose activity was critical to inflammatory conditions. COS has an important role in infectious and allergic disorders [47]. However, chitin has no such effects because it is hard to be dissolved in solution and must be digested into COS by chitinase. In this way, COS can be absorbed well by humans.

Antibiotics have been widely used in the therapy of surgical infection. However, the development of MDR bacteria is a big challenge for an anti-infection therapy. COS showed potential activities against MDR microorganisms. Present work proved that long-term use of COS decreased the MIC values of most MDR bacteria, which were isolated clinically. The reduction of MIC of most antibiotics will be beneficial to effectively control surgical infection. On the other hand, COS had been demonstrated to be effective against various antibiotics-resistant bacteria. COS provided a new way to make use of present antibiotics in low dosage. COS promoted therapeutic progression in treating surgical infection, which was caused by *Burkholderia* and *Acinetobacter* species. These pathogens are aggressively treated and often develop antibiotics resistance after long-term use of antibiotics.

Our subsequent work proved that COS was safe and tolerable in AIS patients. COS can be widely used in food and medical fields for its lesser side effects and safety for human health. Present work also proved that long-term use of high-concentrations of COS will not increase the resistance of most pathogens for COS. Although chitin cannot be used well, it cannot be solved in solution. COS can be produced with high purification via chitinase. The method is superior to the normal chemical method, which can cause environmental contamination.

There were some limitations for present work: (1) the population was still in a small sample size when different antibiotics and pathogens were considered; (2) we did not explore the molecular mechanisms for the development of antibiotics resistance of the clinically isolated pathogens; (3) AIS is a very complex disorder and many other affected factors may be not considered in the present work; (4) the anti-inflammatory and antioxidant mechanisms of COS remained widely unknown and could not be used alone as an experimental group; (5) present findings showed that fewer side effects were found in EG and PG groups compared to CG. Most of the side effects included gastric upset, nausea, headache, vomiting, diarrhea, abdominal pain, seizure, chills, malaise, anxiety and fever in CG. Half-dose antibiotics were used in EG and PG groups, and thus side effects would be reduced since they were caused by these antibiotics; and (6) the bacterial culture experiments showed that COS decreased MIC values for most of the bacteria. All of these pathogens were clinically isolated from human subjects. However, the effects of COS on these bacteria inside human subjects were not performed. Further work is still needed in the future.

4. Materials and Methods

4.1. Materials

COS was purchased from Qingdao BZ-Oligo Co., Ltd. (Qingdao, China). The COS was made from crab shells and obtained by chitinase digestion, chemical derivatization and column chromatography. Briefly, chitins were treated with alkali to increase solubility of the substrates. Chitinase was expressed in *Pichia pastoris* and purified [48]. In addition, 1 mL (1 mg/mL) of purified chitinase was added to one liter of chitin hydrolysis (1% chitin, w/v). The digesting reaction was performed at 39 °C and pH 5 for half an hour. The digesting solution was ultrafiltrated with a membrane NMWL of 3 kDa (Millipore Corporation, Billerica, MA, USA). Filtrated solution was further purified via gel filtration chromatography (Amersham Pharmacia Biotech Inc., Piscataway, NJ, USA). The COS with the degrees of polymerization (DP) ranging from 2 to 10 was obtained, and followed by spray drying.

4.2. Participants

Before the experiments, all protocols were approved by the ethical committee of our hospital. From 3 March 2014 to 16 May 2015, a total of 312 AIS patients who had undergone posterior spinal fusion were collected.

4.3. Inclusion Criteria

All the patients meet the diagnostic criteria and classification criteria of AIS [49]; and the ages ranged from 10 to 16 years old. For the patients in the coronal range of 25° to 45°, they require surgical treatment.

4.4. Exclusion Criteria

Those who do not meet the diagnostic criteria; had experienced surgery more than twice; patients did not meet the inclusion age; coronal plane angle: angle jump angle less than 25°; patients wo had received other relevant treatment, which may affect the effect of present experimental results; patients had other diseases, including cardiovascular system, liver, kidney, hematopoietic system, endocrine system diseases and cancer; patients had allergic disorders; patients had brain diseases; and patients who did not sign the informed consent.

4.5. Patients Grouping

All patients were randomly and evenly assigned into three groups according to different therapies as above mentioned: CG, EG and PG groups. The average follow-up was one month (from two weeks to ten weeks), and infection severity and side effects were analyzed. Analysis was performed to evaluate differences in post-operative variables among three groups.

4.6. Baseline Measurement

The baseline characters (age, gender, lifestyle, infection rate and daily calorie uptake) were compared among three groups. The following parameters (related with surgical risks) were also considered: blood loss in the surgery, surgical length, vertebral levels fused and anchor points, post-operative drain collection and its duration were recorded. For patients who received antimicrobials until drain removal, the number of days that the drug was administered was recorded. Serious adverse events were recorded including gastric upset, nausea, headache, vomiting, diarrhea, abdominal pain, seizure, chills, malaise, anxiety and fever.

4.7. Measurement of Surgical Infection

Presently, there is no uniform standard for the diagnosis of orthopedic infection. Postoperative infection of orthopedic surgery can be determined if the patients had purulent secretion exudation, clinical or surgical or pathological or imaging diagnosis of deep incision. There are abscesses, sinus secretions, joint punctures, and intraoperative lesions in the fluid culture that can be diagnosed as pathogens and other infections. In addition, white blood cell count (WBC), neutrophils, C-reactive protein (CRP), erythrocyte sedimentation rate (ESR), body temperature and other clinical abnormalities can also help to diagnose postoperative infection. Increased WBC is often considered as an indicator of the diagnosis of orthopedic infection. Postoperative infection was measured with clinical infection, such as skin, intravenous tissue, or muscles over the fascial layer. For each patient with SSI, the organism was isolated, and its antimicrobial sensitivity and subsequent management were recorded. The species of infected bacteria were identified by 16S rRNA.

4.8. Measurement of Anti-Bacteria Activities of COS

To explore the inhibitory functions of COS on isolated pathogens, the effects of COS on these pathogens were analyzed according to an earlier report [50]. The effects of COS on bacterial growth were measured by using different concentrations of COS (from 0 to 30 mg/L). All bacteria were cultured in tryptone soya broth for 20 h and transferred into a new 50 mL tube with 5 mL tryptone soya broth with different concentrations (from 0 to 30 mg/L) and cultured at 37 °C for 24 h. Cellular concentrations were measured at OD_{600nm} (1 OD = 10^7 cells/mL). Anti-bacterial functions of COS were further determined by using MIC [51]. A single colony was cultured for 20 h in tryptone soya broth and diluted in Phosphate salt buffer until the OD_{600nm} was 0.01 (10^7 CFU/mL). Serial-diluted antibiotics were added to the above broth or the broth with different concentrations of COS in a 96-well plate and cultured at 37 °C for 20 h. MICs were calculated as the lowest concentration when no growth was observed.

4.9. Measurement for the Resistance to COS after Long-Term Culture

All isolated pathogens were sub-cultured (one time within 24 h) in tryptone soya broth with different concentrations of COS (from 0 to 30 mg/L) for three months. MIC was analyzed on every tenth day to find if there was resistance of pathogens to COS with different concentrations of COS.

4.10. Biochemical Analysis

Oxidative stress was analyzed because oxidative stress is an important risk factor for the development of AIS. The serum activity of SOD was measured by the formazan-WST (water-soluble tetrazolium salt) method [52]. The serum concentration of GSH was determined by using Dithiobis-2-nitrobenzoic acid (DTNB) [53]. The serum concentrations of AST and ALT were evaluated by using the Hitachi 7170A/7180 Biochemical Analyzer (Hitachi, Japan). Serum levels of IL-6 and TNF alpha were measured by using Human IL-6 Quantikine ELISA Kit D6050 and Human TNF-alpha Quantikine ELISA Kit DTA00C ELISA kits (R&D Systems Inc., Minneapolis, MN, USA), respectively.

4.11. qRT-PCR

Total hepatic RNA was exacted and purified by using a RNA purification kit. cDNA was produced based on the instructions of the RT-PCR kit. qRT-PCR was performed to assay the mRNA levels of IL-6 (Forward primer, 5'-cacaacagaccagtatatac-3'; Reverse primer, 5'-gtatttctggaagtttcag-3') and TNF alpha (Forward primer, 5'-gtggcgggggccaccacgctc-3'; Reverse primer, 5'-cgagttttgagaagatgatc-3') genes. GAPDH (Glyceraldehyde 3-phosphate dehydrogenase) was used as an internal control to standardize the copy number (Ct value) of each sample. qRT-PCR was performed on the CFX96 Touch Real-Time PCR Detection System (Bio-Rad, Hercules, CA, USA). The mean Ct value represented the mRNA levels of individual genes.

4.12. Western Blot Analysis

Protein was extracted and the concentration was determined by Bradford protein assay kit (Beyotime Biotechnology, Beijing, China). In addition, 30-µg proteins were taken from each group, separated by 12% SDS-PAGE, and then transferred to a polyvinylidene difluoride (PVDF) membranes (Millipore Corporation, Bedford, MA, USA), which was blocked by 5% non-fat milk. The membrane was treated with primary antibodies at 4 °C for 10 h. Secondary antibodies were added and incubated for one hour. Protein bands were shown after one-hour exposure with GE's Amersham ECL+

Mar. Drugs **2017**, *15*, 70

Chemiluminescent CCD camera (City, US State abbrev. if applicable, Country). The protein level was indicated as the value according to the relative ratio to control GAPDH.

4.13. Statistical Analysis

All data were presented as mean values ± S.D. Analysis of variance (ANOVA) test was used to compare the changes in post-operative infection rate among three groups. Chi-square tests and student's *t*-tests were used for analyzing the difference between two groups. Statistical analysis was carried out by using SPSS 20.0 (SPSS Inc., Chicago, IL, USA). There was statistical significance of differences if p was <0.05.

5. Conclusions

COS reduced clinical pathogens resistantance to normal antibiotics and improved the management of antibiotics. COS can reduce the dose of antibiotics and control the risk of the development of antibiotics resistance in some clinical pathogens. The exact COS concentration and clinical pathogens still need to be studied in the future. COS should be developed as a kind of potential adjuvant of antibiotics therapy for clinical infection.

Acknowledgments: We are very grateful to the anonymous reviewers for their critical and strategic comments, which have significantly improved the quality of our manuscript.

Author Contributions: Y.Q. and J.X. conceived and designed the experiments; H.Z. and R.D. performed the experiments; M.K. and J.Z. analyzed the data; Y.Q. contributed reagents/materials/analysis tools; and Y.Q. and J.Z. wrote the paper.

Conflicts of Interest: The authors declare no conflict of interest.

References

1. Kasatpibal, N.; Whitney, J.D.; Dellinger, E.P.; Nair, B.G.; Pike, K.C. Failure to Redose Antibiotic Prophylaxis in Long Surgery Increases Risk of Surgical Site Infection. *Surg. Infect. (Larchmt)* **2016**, *3*, 1449.
2. Salgado, M.; Fernandez, F.; Aviles, C.; Cordova, C. Erythromycin Seromadesis in Orthopedic Surgery. *J. Orthop. Case Rep.* **2016**, *6*, 92–94. [PubMed]
3. Healey, K.R.; Zhao, Y.; Perez, W.B.; Lockhart, S.R.; Sobel, J.D.; Farmakiotis, D.; Kontoyiannis, D.P.; Sanglard, D.; Taj-Aldeen, S.J.; Alexander, B.D.; et al. Prevalent mutator genotype identified in fungal pathogen Candida glabrata promotes multi-drug resistance. *Nat. Commun.* **2016**, *7*, 11128. [CrossRef] [PubMed]
4. Moyano Calvo, J.L.; Arellano Ganan, R.; Sempere Gutierrez, A.; Sanz Sacristan, J.; Teba del Pino, F.; Melon Rey, F.J.; Herrero Torres, L.; Pereira Sanz, I. Prophylaxis in prostatic surgery with Aztreonam. Our experience. *Arch. Esp. Urol.* **1992**, *45*, 519–521. [PubMed]
5. Boffi, L.; Panebianco, R. A comparative study of 2 schedules of antibiotic prophylaxis using Ceftazidime in the prevention of infections in elective surgery of the biliary surgery. Preliminary results. *Clin. Ter.* **1992**, *140*, 265–271. [PubMed]
6. Wang, Y.; Li, H.; Chen, B. Pathogen distribution and drug resistance of nephrology patients with urinary tract infections. *Saudi Pharm. J.* **2016**, *24*, 337–340. [CrossRef] [PubMed]
7. Gomes, C.; Martinez-Puchol, S.; Palma, N.; Horna, G.; Ruiz-Roldan, L.; Pons, M.J.; Ruiz, J. Macrolide resistance mechanisms in Enterobacteriaceae: Focus on Azithromycin. *Crit. Rev. Microbiol.* **2017**, *43*, 1–30. [CrossRef] [PubMed]
8. Yin, O.Q.; Tomlinson, B.; Chow, M.S. Effect of multidrug resistance gene-1 (ABCB1) polymorphisms on the single-dose pharmacokinetics of cloxacillin in healthy adult Chinese men. *Clin. Ther.* **2009**, *31*, 999–1006. [CrossRef] [PubMed]
9. Nwobu, R.A.; Dosunmu-Ogunbi, O.; Rotimi, V.O. Phage-types and resistance pattern of Staphylococcus aureus, isolated from clinical specimens, to penicillin and cloxacillin in a Lagos hospital. *Cent. Afr. J. Med.* **1986**, *32*, 155–158. [PubMed]

10. Marshall, S.; Hujer, A.M.; Rojas, L.J.; Papp-Wallace, K.M.; Humphries, R.M.; Spellberg, B.; Hujer, K.M.; Marshall, E.K.; Rudin, S.D.; Perez, F.; et al. Can Ceftazidime/avibactam and Aztreonam overcome beta-lactam resistance conferred by metallo-beta-lactamases in Enterobacteriaceae? *Antimicrob. Agents Chemother.* **2017**. [CrossRef] [PubMed]

11. Braz, V.S.; Furlan, J.P.; Fernandes, A.F.; Stehling, E.G. Mutations in NalC induce MexAB-OprM overexpression resulting in high level of Aztreonam resistance in environmental isolates of Pseudomonas aeruginosa. *FEMS Microbiol. Lett.* **2016**, *363*. [CrossRef] [PubMed]

12. Cummings, J.E.; Slayden, R.A. Transient In Vivo Resistance Mechanisms of Burkholderia pseudomallei to Ceftazidime and Molecular Markers for Monitoring Treatment Response. *PLoS Negl. Trop. Dis.* **2017**, *11*, e0005209. [CrossRef] [PubMed]

13. Shields, R.K.; Chen, L.; Cheng, S.; Chavda, K.D.; Press, E.G.; Snyder, A.; Pandey, R.; Doi, Y.; Kreiswirth, B.N.; Nguyen, M.H.; et al. Emergence of Ceftazidime-avibactam resistance due to plasmid-borne blaKPC-3 mutations during treatment of carbapenem-resistant Klebsiella pneumoniae infections. *Antimicrob. Agents Chemother.* **2017**, *61*, e02097-16. [CrossRef] [PubMed]

14. Hopkins, S. Clinical toleration and safety of Azithromycin. *Am. J. Med.* **1991**, *91*, 40S–45S. [CrossRef]

15. Richa; Tandon, V.R.; Sharma, S.; Khajuria, V.; Mahajan, V.; Gillani, Z. Adverse drug reactions profile of antimicrobials: A 3-year experience, from a tertiary care teaching hospital of India. *Indian J. Med. Microbiol.* **2015**, *33*, 393–400.

16. El Nekidy, W.; Dziamarski, N.; Soong, D.; Donaldson, C.; Ibrahim, M.; Kadri, A. Cloxacillin-induced seizure in a hemodialysis patient. *Hemodial. Int.* **2015**, *19*, E33–E36. [CrossRef] [PubMed]

17. Wanjiru, M.M. *Isolation and Characterization of Bacteria Pathogens in Blood and Stool Samples among Patients Presenting with Typhoid Fever Symptoms in Alupe, Busia County*; Kenyatta University: Nairobi County, Kenya, 2013.

18. Pazmino, P. Acute renal failure, skin rash, and eosinophilia associated with Aztreonam. *Am. J. Nephrol.* **1988**, *8*, 68–70. [PubMed]

19. Anastasio, G.D.; Robinson, M.D.; Little, J.M., Jr.; Leitch, B.B.; Pettice, Y.L.; Norton, H.J. A comparison of the gastrointestinal side effects of two forms of erythromycin. *J. Fam. Pract.* **1992**, *35*, 517–523. [PubMed]

20. Lee, S.; Jang, J.; Jeon, H.; Lee, J.; Yoo, S.M.; Park, J.; Lee, M.S. Latent Kaposi's sarcoma-associated herpesvirus infection in bladder cancer cells promotes drug resistance by reducing reactive oxygen species. *J. Microbiol.* **2016**, *54*, 782–788. [CrossRef] [PubMed]

21. Aklog, Y.F.; Egusa, M.; Kaminaka, H.; Izawa, H.; Morimoto, M.; Saimoto, H.; Ifuku, S. Protein/CaCO(3)/Chitin Nanofiber Complex Prepared from Crab Shells by Simple Mechanical Treatment and Its Effect on Plant Growth. *Int. J. Mol. Sci.* **2016**, *17*, 1600. [CrossRef] [PubMed]

22. Sayari, N.; Sila, A.; Abdelmalek, B.E.; Abdallah, R.B.; Ellouz-Chaabouni, S.; Bougatef, A.; Balti, R. Chitin and chitosan from the Norway lobster by-products: Antimicrobial and anti-proliferative activities. *Int. J. Biol. Macromol.* **2016**, *87*, 163–171. [CrossRef] [PubMed]

23. Mhamdi, S.; Ktari, N.; Hajji, S.; Nasri, M.; Sellami Kamoun, A. Alkaline proteases from a newly isolated Micromonospora chaiyaphumensis S103: Characterization and application as a detergent additive and for chitin extraction from shrimp shell waste. *Int. J. Biol. Macromol.* **2017**, *94*, 415–422. [CrossRef] [PubMed]

24. Aranda-Martinez, A.; Lopez-Moya, F.; Lopez-Llorca, L.V. Cell wall composition plays a key role on sensitivity of filamentous fungi to chitosan. *J. Basic Microbiol.* **2016**, *56*, 1059–1070. [CrossRef] [PubMed]

25. Li, T.; Chen, J.; Fan, X.; Chen, W.; Zhang, W. MicroRNA and dsRNA targeting chitin synthase A reveal a great potential for pest management of a hemipteran insect Nilaparvata lugens. *Pest. Manag. Sci.* **2016**. [CrossRef] [PubMed]

26. Peters, W. Occurrence of chitin in Mollusca. *Comp. Biochem. Physiol. B Comp. Biochem.* **1972**, *41*, 541–544. [CrossRef]

27. Geetha, P.; Sivaram, A.J.; Jayakumar, R.; Gopi Mohan, C. Integration of in silico modeling, prediction by binding energy and experimental approach to study the amorphous chitin nanocarriers for cancer drug delivery. *Carbohydr. Polym.* **2016**, *142*, 240–249. [CrossRef] [PubMed]

28. Ohnuma, T.; Numata, T.; Osawa, T.; Inanaga, H.; Okazaki, Y.; Shinya, S.; Kondo, K.; Fukuda, T.; Fukamizo, T. Crystal structure and chitin oligosaccharide-binding mode of a 'loopful' family GH19 chitinase from rye, Secale cereale, seeds. *FEBS J.* **2012**, *279*, 3639–3651. [CrossRef] [PubMed]

29. Azuma, K.; Osaki, T.; Minami, S.; Okamoto, Y. Anticancer and anti-inflammatory properties of chitin and chitosan oligosaccharides. *J. Funct. Biomater.* **2015**, *6*, 33–49. [CrossRef] [PubMed]
30. Yoon, H.J.; Moon, M.E.; Park, H.S.; Im, S.Y.; Kim, Y.H. Chitosan oligosaccharide (COS) inhibits LPS-induced inflammatory effects in RAW 264.7 macrophage cells. *Biochem. Biophys. Res. Commun.* **2007**, *358*, 954–959. [CrossRef] [PubMed]
31. Michalska, A.D.; Sacha, P.T.; Ojdana, D.; Wieczorek, A.; Tryniszewska, E. Prevalence of resistance to aminoglycosides and fluoroquinolones among Pseudomonas aeruginosa strains in a University Hospital in Northeastern Poland. *Braz. J. Microbiol.* **2014**, *45*, 1455–1458. [CrossRef] [PubMed]
32. Tan, H.; Zhang, L.; Weng, Y.; Chen, R.; Zhu, F.; Jin, Y.; Cheng, Z.; Jin, S.; Wu, W. PA3297 Counteracts Antimicrobial Effects of Azithromycin in Pseudomonas aeruginosa. *Front. Microbiol.* **2016**, *7*, 317. [CrossRef] [PubMed]
33. Garcia-Quintanilla, M.; Carretero-Ledesma, M.; Moreno-Martinez, P.; Martin-Pena, R.; Pachon, J.; McConnell, M.J. Lipopolysaccharide loss produces partial colistin dependence and collateral sensitivity to Azithromycin, rifampicin and vancomycin in Acinetobacter baumannii. *Int. J. Antimicrob. Agents* **2015**, *46*, 696–702. [CrossRef] [PubMed]
34. Chantratita, N.; Rholl, D.A.; Sim, B.; Wuthiekanun, V.; Limmathurotsakul, D.; Amornchai, P.; Thanwisai, A.; Chua, H.H.; Ooi, W.F.; Holden, M.T.; et al. Antimicrobial resistance to Ceftazidime involving loss of penicillin-binding protein 3 in Burkholderia pseudomallei. *Proc. Natl. Acad. Sci. USA* **2011**, *108*, 17165–17170. [CrossRef] [PubMed]
35. Hu, H.; Hu, Y.; Pan, Y.; Liang, H.; Wang, H.; Wang, X.; Hao, Q.; Yang, X.; Yang, X.; Xiao, X. Novel plasmid and its variant harboring both a blaNDM-1 gene and type IV secretion system in clinical isolates of Acinetobacter lwoffii. *Antimicrob. Agents Chemother.* **2012**, *56*, 1698–1702. [CrossRef] [PubMed]
36. Liu, S.W.; Xu, X.Y.; Xu, J.; Yuan, J.Y.; Wu, W.K.; Zhang, N.; Chen, Z.L. Multi-drug resistant uropathogenic Escherichia coli and its treatment by Chinese medicine. *Chin. J. Integr. Med.* **2016**. [CrossRef] [PubMed]
37. Mostafa, M.; Siadat, S.D.; Shahcheraghi, F.; Vaziri, F.; Japoni-Nejad, A.; Vand Yousefi, J.; Rajaei, B.; Harifi Mood, E.; Ebrahim zadeh, N.; Moshiri, A.; et al. Variability in gene cassette patterns of class 1 and 2 integrons associated with multi drug resistance patterns in Staphylococcus aureus clinical isolates in Tehran-Iran. *BMC Microbiol.* **2015**, *15*, 152. [CrossRef] [PubMed]
38. Gobert, A.P.; Wilson, K.T. Polyamine- and NADPH-dependent generation of ROS during Helicobacter pylori infection: A blessing in disguise. *Free Radic. Biol. Med.* **2016**. [CrossRef] [PubMed]
39. Broxton, C.N.; Culotta, V.C. SOD Enzymes and Microbial Pathogens: Surviving the Oxidative Storm of Infection. *PLoS Pathog.* **2016**, *12*, e1005295. [CrossRef] [PubMed]
40. Yuan, C.; Fu, X.; Huang, L.; Ma, Y.; Ding, X.; Zhu, L.; Zhu, G. The synergistic antiviral effects of GSH in combination with acyclovir against BoHV-1 infection in vitro. *Acta Virol.* **2016**, *60*, 328–332. [CrossRef] [PubMed]
41. Ikuabe, P.O.; Ebuenyi, I.D.; Harry, T.C. Limited Elevations in Antituberculosis Drug-Induced Serum Alanine Aminotransferase (Alt) Levels in a Cohort of Nigerians on Treatment for Pulmonary Tuberculosis and Hiv Infection in Yenagoa. *Niger. J. Med.* **2015**, *24*, 103–107. [PubMed]
42. Costa, M.M.; Franca, R.T.; Da Silva, A.S.; Paim, C.B.; Paim, F.; do Amaral, C.H.; Dornelles, G.L.; da Cunha, J.P.; Soares, J.F.; Labruna, M.B.; et al. Rangelia vitalii: Changes in the enzymes ALT, CK and AST during the acute phase of experimental infection in dogs. *Rev. Bras. Parasitol. Vet.* **2012**, *21*, 243–248. [CrossRef] [PubMed]
43. Catar, R.; Witowski, J.; Zhu, N.; Lucht, C.; Derrac Soria, A.; Uceda Fernandez, J.; Chen, L.; Jones, S.A.; Fielding, C.A.; Rudolf, A.; et al. IL-6 Trans-Signaling Links Inflammation with Angiogenesis in the Peritoneal Membrane. *J. Am. Soc. Nephrol.* **2016**. [CrossRef] [PubMed]
44. Khalili, H.; Lee, R.W.; Khaw, P.T.; Brocchini, S.; Dick, A.D.; Copland, D.A. An anti-TNF-alpha antibody mimetic to treat ocular inflammation. *Sci. Rep.* **2016**, *6*, 36905. [CrossRef] [PubMed]
45. Izumi, R.; Azuma, K.; Izawa, H.; Morimoto, M.; Nagashima, M.; Osaki, T.; Tsuka, T.; Imagawa, T.; Ito, N.; Okamoto, Y.; et al. Chitin nanofibrils suppress skin inflammation in atopic dermatitis-like skin lesions in NC/Nga mice. *Carbohydr. Polym.* **2016**, *146*, 320–327. [CrossRef] [PubMed]
46. Jeon, I.H.; Mok, J.Y.; Park, K.H.; Hwang, H.M.; Song, M.S.; Lee, D.; Lee, M.H.; Lee, W.Y.; Chai, K.Y.; Jang, S.I. Inhibitory effect of dibutyryl chitin ester on nitric oxide and prostaglandin E(2) production in LPS-stimulated RAW 264.7 cells. *Arch. Pharm. Res.* **2012**, *35*, 1287–1292. [CrossRef] [PubMed]

47. Wagener, J.; Malireddi, R.S.; Lenardon, M.D.; Köberle, M.; Vautier, S.; MacCallum, D.M.; Biedermann, T.; Schaller, M.; Netea, M.G.; Kanneganti, T.-D. Fungal chitin dampens inflammation through IL-10 induction mediated by NOD2 and TLR9 activation. *PLoS Pathog.* **2014**, *10*, e1004050. [CrossRef] [PubMed]

48. Youxi, Z.; Huihui, J.; Zhiming, R.; Yizhi, J.; Yanling, C.; Yanhe, M. High level expression of Saccharomyces cerevisiae chitinase (ScCTS1) in Pichia pastoris for degrading chitin. *Int. J. Agric. Biol. Eng.* **2015**, *8*, 142–150.

49. Pasquini, G.; Cecchi, F.; Bini, C.; Molino-Lova, R.; Vannetti, F.; Castagnoli, C.; Paperini, A.; Boni, R.; Macchi, C.; Crusco, B.; et al. The outcome of a modified version of the Cheneau brace in adolescent idiopathic scoliosis (AIS) based on SRS and SOSORT criteria: A retrospective study. *Eur. J. Phys. Rehabil. Med.* **2016**, *52*, 618–629. [PubMed]

50. Khan, S.; Tøndervik, A.; Sletta, H.; Klinkenberg, G.; Emanuel, C.; Onsøyen, E.; Myrvold, R.; Howe, R.A.; Walsh, T.R.; Hill, K.E. Overcoming drug resistance with alginate oligosaccharides able to potentiate the action of selected antibiotics. *Antimicrob. Agents Chemother.* **2012**, *56*, 5134–5141. [CrossRef] [PubMed]

51. Armstrong, E.S.; Mikulca, J.A.; Cloutier, D.J.; Bliss, C.A.; Steenbergen, J.N. Outcomes of high-dose levofloxacin therapy remain bound to the levofloxacin minimum inhibitory concentration in complicated urinary tract infections. *BMC Infect. Dis.* **2016**, *16*, 710. [CrossRef] [PubMed]

52. Tan, A.S.; Berridge, M.V. Superoxide produced by activated neutrophils efficiently reduces the tetrazolium salt, WST-1 to produce a soluble formazan: A simple colorimetric assay for measuring respiratory burst activation and for screening anti-inflammatory agents. *J. Immunol. Methods* **2000**, *238*, 59–68. [CrossRef]

53. Smith, I.K.; Vierheller, T.L.; Thorne, C.A. Assay of glutathione reductase in crude tissue homogenates using 5,5'-dithiobis(2-nitrobenzoic acid). *Anal. Biochem.* **1988**, *175*, 408–413. [CrossRef]

marine drugs

MDPI

Article

Fluorescent Property of Chitosan Oligomer and Its Application as a Metal Ion Sensor

Hun Min Lee [1], Min Hee Kim [1], Young Il Yoon [2,*] and Won Ho Park [1,*]

[1] Department of Advanced Organic Materials and Textile System Engineering, Chungnam National University, Daejeon 34134, Korea; hun1062@naver.com (H.M.L.); vvvkmhvvv@nate.com (M.H.K.)

[2] Laboratory of Molecular Imaging and Nanomedicine (LOMIN), National Institute of Biomedical Imaging and Bioengineering (NIBIB), National Institutes of Health (NIH), Bethesda, MD 20892, USA

* Correspondence: youngil.yoon@nih.gov (Y.I.Y.); parkwh@cnu.ac.kr (W.H.P.); Tel.: +1-301-827-0575 (Y.I.Y.); +82-42-821-6613 (W.H.P.)

Academic Editors: David Harding and Hitoshi Sashiwa
Received: 26 December 2016; Accepted: 29 March 2017; Published: 4 April 2017

Abstract: An aqueous solution was successfully prepared using a low-molecular-weight chitosan oligomer and FITC, and its structural and fluorescent properties were observed by using ^{1}H NMR, ^{13}C NMR, FT-IR, XRD, UV-Vis, and PL spectrometry. Its application as a metal ion sensor was also evaluated. The fluorescence in the water-soluble chitosan oligomer was a result of the carbamato anion (NHCOO-), and a synthesized FITC-labeled chitosan oligomer exhibited an effective detection effect for copper ion as well as energy transfer by the ion near FITC that caused a fluorescence decrease (quenching). The chitosan oligomer was confirmed to be applicable as a selective and sensitive colorimetric sensor to detect Cu^{2+}.

Keywords: chitosan oligomer; fluorescent property; metal ion sensor

1. Introduction

Among all natural polymers, chitosan is a promising biopolymer that is commercially available. It is a well-known polysaccharide that is mainly produced from chitin [1], and in the past few decades, naturally occurring chitosan has attracted a significant amount of interest due to its large quantities in nature, biodegradability, and extensive applicability [2]. A water-soluble chitosan oligomer is composed of β-(1,4)-2-amido-2-deoxy-D-glucan and β-(1,4)-2-acetoamido-2-deoxy-D-glucan (acetylglucosamine), and a substance with a low molecular weight can be obtained through acidic or enzymatic hydrolysis of chitosan. To date, many researchers have examined chitosan oligomers as promising materials for biomedical applications due to their good biocompatibility, biodegradability, antimicrobial activity, and wound healing effects [3–5].

Dye-labeled chitosan can be also employed in bio-imaging systems because it has little toxicity. Some dyes, such as Alexa Fluor, Cibacron Blue, and fluorescein isothiocyanate (FITC), have been used to create a dye-labeled chitosan particulate system [6]. However, a significant part of such approaches uses chemical reagents in the synthesis of dye-labeled chitosan due to its low solubility in water [7]. Numerous studies have addressed the utilization of chitosan as a fluorescence probe [8,9]. This approach has limited applicability in medical and pharmaceutical applications because most of these may be environmentally toxic or biologically hazardous [10].

Copper ions (Cu^{2+}) have been classified as a potentially carcinogenic substance because they induce DNA damages [11]. Malondialdehyde and 4-hydroxynonenal are produced through a reaction of the ions with lipid hydroxyperoxide, and this can result in impairment to tissues [12]. Therefore, effective detection of these ions is required for various fields [13].

This study focused on the fluorescent property of a low-molecular-weight chitosan oligomer and an environmentally friendly approach using a water-soluble chitosan derivative to sense the metal ion (Cu^{2+}). The simple synthesis and fluorescent properties of the FITC-labeled chitosan oligomer were carried out in distilled water.

2. Materials and Methods

2.1. Materials

A chitosan (CHI) oligomer was provided by Kittolife Co., Pyeongtaek, Korea. Its degree of deacetylation (DD), molecular weight, and Cl ion content were 97%, ~1000 Da, and 3.2% respectively. Fluorescein isothiocyanate (FITC) was purchased from Sigma-Aldrich Co., Saint Louis, MO, USA. Ethanol (EtOH) was obtained from Samchun Chemical Co., Yeosu, Korea. Metal cations such as Na^+, Cr^+, Ni^+, Sn^+, Li^+, Mg^{2+}, Al^{2+}, Co^+, Ni^{2+}, Cu^{2+}, Zn^{2+}, Cd^{2+}, Hg^{2+}, Pb^{2+}, and Fe^{3+} were supplied from Alfa Aesar Co., Haverhill, MA, USA. The chemicals were used without further purification or additional processes.

2.2. Preparation of CHI Oligomer-FITC Complexes

To conduct the one-step synthesis of CHI oligomer-FITC complexes, 100 mL solutions of 0.01%–0.05% (*w/v*) FITC in EtOH were added to a 10 mL solution of 2% (*w/v*) CHI oligomer in distilled water (DW). To obtain the desired products, the solutions were stirred at room temperature for 24 h in a darkroom. After the reaction, the unreacted FITC was removed through two centrifugal separations at 4000 rpm for 10 min using EtOH. The end products were obtained as a powder by using a vacuum dryer (VO-10x, Jeio Tech Co., Seoul, Korea) at room temperature for 24 h.

2.3. Structural Analyses of the CHI Oligomer and CHI Oligomer-FITC Complexes

To verify the presence of carbamato anion (NHCOO-) in the CHI oligomer, ^{13}C nuclear magnetic resonance (^{13}C NMR) spectra were recorded on a NMR spectrometer (300 MHz, FT-NMR, Bruker, Billerica, MA, USA). The Fourier transform vacuum infrared (FT-IR) spectra of the CHI oligomer-FITC complexes were collected using a FT-IR spectrometer (VERTEX 80v, Bruker) with a frequency range of 675–4000 cm^{-1}, and X-ray diffraction (XRD) patterns were obtained at room temperature with 2θ = 5°–80° using an X-ray diffractometer (D8 DISCOVER, Bruker AXS, Billerica, MA, USA) to confirm structural changes in the complexes. Also, the degree of substitution (DS) of the complexes is ascertained by recording ^{1}H nuclear magnetic resonance (1H NMR) spectra on a NMR spectrometer (300 MHz, FT-NMR, Bruker). The DS (%) values were expressed as $(I_{AR}/9)/(I_{H2-H6}/6) \times 100$ [7]. I_{AR} and I_{H2-H6} indicated peak areas of aromatic protons in FITC and C2-C6 protons in the chitosan backbone, respectively.

2.4. Fluorescent Analyses of the CHI Oligomer and CHI Oligomer-FITC Complexes

To analyze the UV-Vis absorbance and photoluminescence (PL) properties of the CHI oligomer, aqueous solutions (0.1%–7% (*w/v*)) of the CHI oligomer in DW were prepared after stirring at room temperature for 1 h. The absorption spectra of the solutions were conducted on a UV-Vis spectrophotometer (UV-2450PC, Shimadzu Co., Kyoto, Japan) with a measurable range of 190–1100 nm and a resolution of 0.1 nm. The photoluminescence spectra of the solution were collected using a luminescence spectrophotometer (Varian Cary Eclipse, Varian, Palo Alto, CA, USA) equipped with a xenon flash lamp excitation source. The absorption spectra were obtained at 25 °C and the emission spectra of CHI oligomer and CHI oligomer-FITC were obtained at 475 nm and 520 nm, respectively, using an excitation wavelength at 395 nm with resolution of 1 nm and scan rate of 600 nm/min. Also, the changes in the fluorescent spectra of the CHI oligomer-FITC complexes with or without metal ions were measured using the same UV-Vis and luminescence spectrophotometers.

2.5. Adsorption Behaviors of Metal Ions onto the CHI Oligomer-FITC Complex

To examine applicability of the CHI oligomer-FITC complex as a colorimetric sensor, the reactions between the complex and the metal ions such as Na^+, Cr^+, Ni^+, Sn^+, Li^{2+}, Mg^{2+}, Al^{2+}, Co^+, Ni^{2+}, Cu^{2+}, Zn^{2+}, Cd^{2+}, Hg^{2+}, Pb^{2+}, and Fe^{3+} were monitored. The solutions containing metal ions (10^{-3} M, 0.1 mL) were added into the CHI oligomer-FITC complex solutions (1 mL). After that, the changes in color of the mixed solutions were checked after exposure to UV radiation (Ex. = 395 nm).

3. Results and Discussion

3.1. Structural and Fluorescent Analyses of the CHI Oligomer

The low-molecular-weight CHI oligomer with a high water solubility showed unique structural and fluorescent characteristics. In general, fluorescence is caused by molecules with delocalized electrons at conjugated double bonds [14]. The CHI oligomer featured a molecular structure that had no delocalized electron, but the fluorescent property of the oligomer was observed by the fluorescence photometer and the naked eye. Recently, several polymers without conjugated double bonds were reported to exhibit fluorescence [15], and it turned out that the carbamato anion (NHCOO-) formed by the reaction between carbon dioxide and amine induced this phenomenon [16]. The ^{13}C NMR spectrum in Figure 1A shows a typical carbamato anion (NHCOO-) peak in CHI oligomer at 174 ppm. It was induced by the reaction between the amino group in the CHI oligomer and carbon dioxide in air. The fluorescent intensity was monitored depending on the concentration of the CHI oligomer (Figure 1B,C). The excitation and emission wavelengths of the CHI oligomer were 400 and 470 nm, respectively.

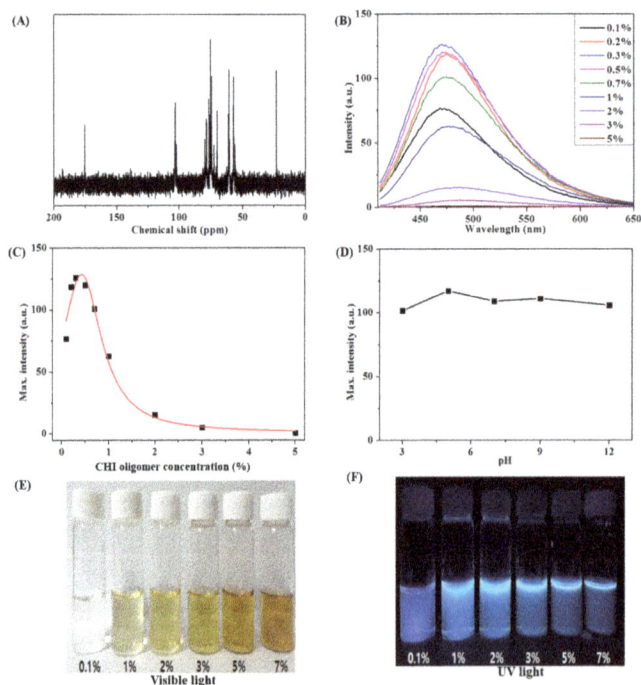

Figure 1. Analysis results of a CHI oligomer. ^{13}C NMR spectra (**A**); PL spectra (**B**), change in fluorescent intensities at 470 nm by the concentration of the CHI oligomer (**C**); change in fluorescent intensities at 470 nm by pH conditions (**D**); images under visible (**E**) or ultraviolet light (Ex. = 395 nm) (**F**).

The photoluminescence (PL) spectra revealed that the maximal CHI oligomer concentration showing the highest PL intensity was 0.3% (w/v). At 0.5% (w/v) and over, the PL intensities showed a tendency to decrease due to self-quenching among the CHI oligomer molecules [17]. The fluorescent pattern of the CHI oligomer was confirmed to be similar to that given in the results above under UV radiation (Ex. = 395 nm) (Figure 1E,F). In addition, observation at various pH values from 3 to 12 revealed that the PL intensities of the CHI oligomer were unaffected by the pH 3–12 conditions (Figure 1D).

3.2. Structural and Fluorescent Analyses of the CHI Oligomer-FITC Complexes

A reaction of a primary amine with an isothiocyanate (NCS), one of various amine modification methods, has been widely known to lead to a thiourea as its product [18]. Herein, to develop a sensitive and selective imaging probe that not only has high water solubility but is also immune to the pH conditions, the CHI oligomer–FITC complexes were prepared through a reaction of the CHI oligomer with FITC. The so-called "FITC labeling" reaction took place with the primary amine of the CHI oligomer and the NCS of FITC. The reaction mechanism is illustrated in Figure 2.

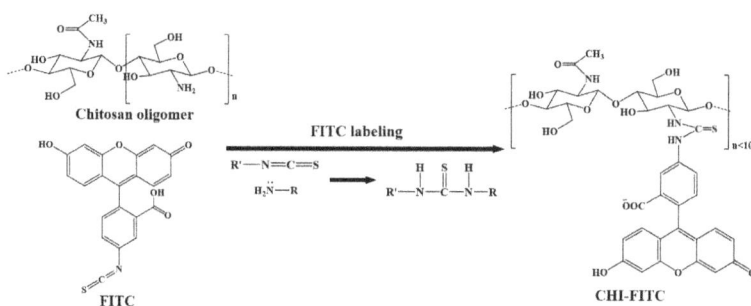

Figure 2. Schematic representation of a reaction mechanism between the CHI oligomer and FITC.

The optimal condition of this reaction was confirmed by obtaining the ^{1}H NMR, XRD, and FI-IR spectra [19–21]. The concentration of the CHI oligomer was fixed, and the concentration of FITC varied from 0 to 7×10^{-2}% (w/v). The peaks at 6.5–8 ppm, indicating aromatic rings, were observed with an increase in the FITC concentration (Figure 3A). Also, the DS (%) was calculated by the proportion of areas of 6.5 to eight peaks corresponding to the nine protons of FITC to areas of three to four peaks corresponding to the six protons of the CHI oligomer (Figure 3B). The DS (%) values increased to four depending on the FITC concentration. Therefore, the sample with DS (%) = 4 was named as CHI-FITC-4. In the same way, the samples with DS (%) = 2 or DS (%) = 3 were named as CHI-FITC-2 or CHI-FITC-3, respectively. Figure 3C shows the XRD patterns for the CHI-FITC complexes. A spectrum of the CHI oligomer showed $2\theta = 20°$ as a typical crystalline peak. However, according to the increase in the DS (%) value, the crystalline peak gradually decreased due to the introduction of a bulky FITC group. Due to the substitution of the amino group to thiourea by FITC, the steric hindrance between the CHI oligomers increased considerably, and hydrogen bonds were broken between the CHI oligomers. Also, through an FT-IR analysis, the peaks at 1458, 1535, and 1594 cm^{-1} related to the stretching vibration of the aromatic ring were clearly observed in the CHI-FITC-4 (Figure 3D). After the reaction of the CHI oligomer with FITC, a peak at 2015 cm^{-1} corresponding to the NCS vibration disappeared completely. Figure 3 verifies that the CHI oligomer–FITC complex was successfully synthesized. In addition, the fluorescent intensities of CHI-FITC-4 were evaluated from the PL spectra (Figure 4A,B), and the intensities were dependent on the concentration of CHI-FITC-4.

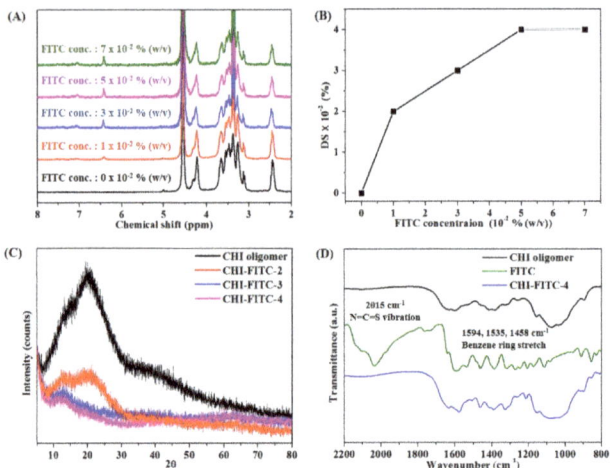

Figure 3. Results of the analysis of CHI-FITC complexes. ^{1}H NMR spectra (**A**); DS (**B**); XRD spectra (**C**); FT-IR spectra (**D**).

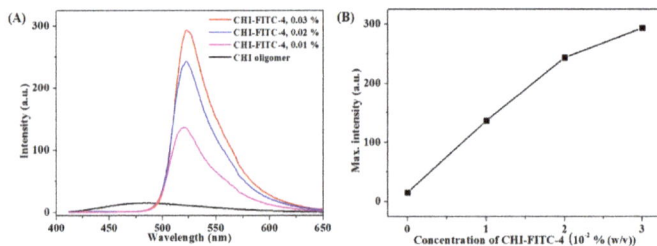

Figure 4. Results of the analysis of CHI-FITC-4 depending on the concentration. PL spectra (**A**); PL intensity at 520 nm (**B**).

3.3. Colorimetric Sensing of CHI-FITC-4 against Metal Ions

Selectivity and sensitivity tests were carried out to assess the applicability of CHI-FITC-4 as a colorimetric sensor for metal ions. To begin, the selectivity of CHI-FITC-4 was investigated using metal ions such as Na^+, Cr^+, Ni^+, Sn^+, Li^{2+}, Mg^{2+}, Al^{2+}, Co^+, Ni^{2+}, Cu^{2+}, Zn^{2+}, Cd^{2+}, Hg^{2+}, Pb^{2+}, and Fe^{3+}. Figure 5A,B show that the CHI-FITC-4 sample had selectivity for Cu^{2+}. In particular, a decrease in the fluorescent intensity could be clearly distinguished from those of other metal ions under UV radiation (Ex. = 395 nm). As quantitative criteria, the UV-vis and PL spectra of CHI-FITC-4 with metal ions were monitored (Figure 5C–F). The absorbance (λ_{max} = 492 nm) and fluorescent intensity (λ_{max} = 520 nm) of CHI-FITC-4 with Cu^{2+} decreased to approximately 60% and 80%, respectively. According to the concentration range of 0.1–7.0 mM for Cu^{2+}, a sensitive colorimetric assay of CHI-FITC-4 was implemented, and the color changes were observed under UV radiation (Ex = 395 nm) (Figure 6A,B). The UV-vis spectra indicated that the limit of detection (LOD) of CHI-FITC-4 for Cu^{2+} was close to 60 µM, and its correlation coefficient (R^2) was 0.99 (Figure 6C,D). In addition, the fluorescent intensities at 520 nm of CHI-FITC-4 with Cu^{2+} displayed a tendency for decay according to the concentration (0.1–7.0 mM) of Cu^{2+} (Figure 6E,F). This phenomenon could be explained by the fluorescent quenching mechanism. The fluorescence intensity of CHI-FITC-4 decreased due to energy transfer resulting from the formation of a selective complex with Cu^{2+}, as shown in Figure 7. The energy transfer occurred due to Cu^{2+} being located near the FITC group, resulting in a decrease in fluorescence (quenching) [22].

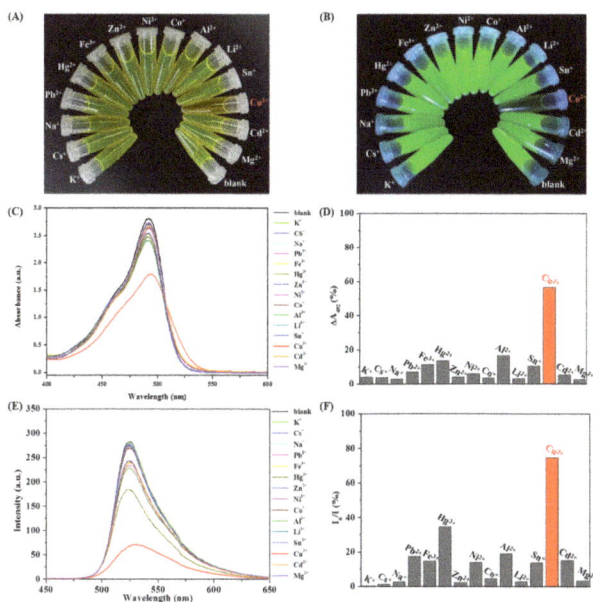

Figure 5. Colorimetric assay to sense metal ions using CHI-FITC-4. Images under visible (**A**) or ultraviolet radiation (**B**); UV-vis spectra (**C**); Change comparison of UV-vis intensity at 492 nm (**D**); PL spectra (**E**); Change comparison of PL intensity at 520 nm (**F**).

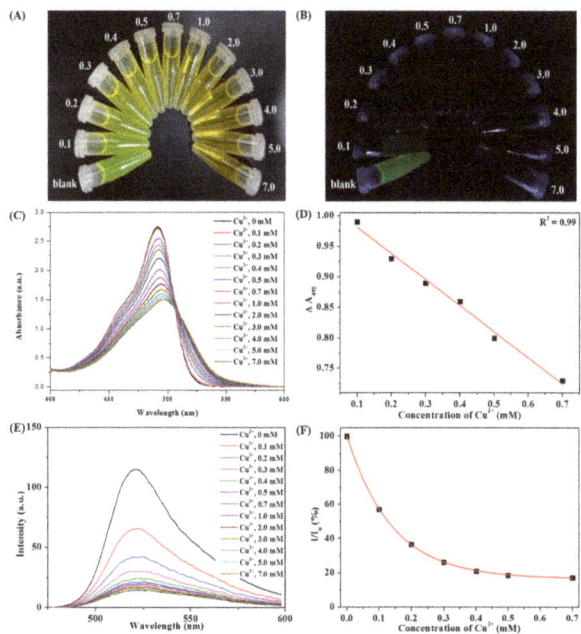

Figure 6. Colorimetric assay to sense Cu^{2+} using CHI-FITC-4. Images under visible (**A**) or ultraviolet radiation (**B**); UV-vis spectra (**C**); Change comparison of UV-vis intensity at 492 nm (**D**); PL spectra (**E**); Change comparison of PL intensity at 520 nm (**F**).

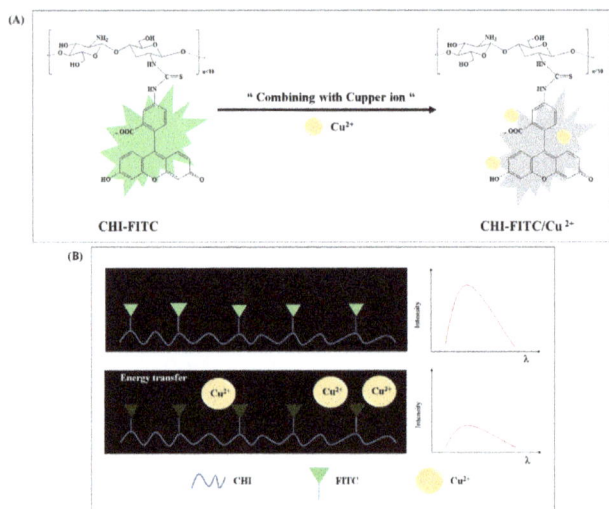

Figure 7. Schematic representation of the reaction mechanism between CHI-FITC and Cu^{2+} (**A**) and the energy transfer phenomenon (**B**).

4. Conclusions

In this study, aqueous solutions of low-molecular-weight CHI oligomers were successfully prepared, and their structural and fluorescent properties were then meticulously observed using ^{1}H NMR, ^{13}C NMR, FT-IR, XRD, UV-Vis, and PL spectrometry. The presence of the carbamato anion (NHCOO-) as a fluorophore was verified as the origin of the fluorescent properties of the CHI oligomer. With respect to the concentration and pH of the aqueous solutions of the CHI oligomer, the optimal concentration of the oligomer was 0.3% (w/v), and the solutions were not affected in the range of pH of 3–12. Depending on the addition of the metal ions, the color changes of the synthesized CHI oligomer–FITC complex showed a remarkable difference. It had an excellent selectivity to detect copper ions (Cu^{2+}), and its limit of detection (LOD) was 60 µM. When Cu^{2+} was combined with FITC, the energy transfer between FITC and Cu^{2+} led to fluorescent quenching of the CHI oligomer–FITC complex. In conclusion, the CHI oligomer–FITC complex has great potential as a promising colorimetric sensor to detect Cu^{2+}.

Acknowledgments: This research was supported by the Basic Science Research Program through the National Research Foundation of Korea (NRF) funded by the Ministry of Science, ICT & Future Planning (2015R1A2A2A01007954).

Author Contributions: Won Ho Park and Young Il Yoon conceived and designed the experiments; Hun Min Lee performed the experiments; Min Hee Kim and Hun Min Lee analyzed the data; Young Il Yoon and Won Ho Park wrote the paper.

Conflicts of Interest: The authors declare no conflict of interest.

References

1. Rinaudo, M. Chitin and chitosan: Properties and applications. *Prog. Polym. Sci.* **2006**, *31*, 603–632. [CrossRef]
2. Ilium, L. Chitosan and its use as a pharmaceutical excipient. *Pharm. Res.* **1998**, *15*, 1326–1331. [CrossRef]
3. Agrawal, P.; Strijkers, G.J.; Nicolay, K. Chitosan-based systems for molecular imaging. *Adv. Drug Deliv. Rev.* **2010**, *62*, 42–58. [CrossRef] [PubMed]
4. Kumar, M.R. A review of chitin and chitosan applications. *React. Funct. Polym.* **2000**, *46*, 1–27. [CrossRef]

5. Pan, X.; Ren, W.; Gu, L.; Wang, G.; Liu, Y. Photoluminescence from chitosan for bio-imaging. *Aust. J. Chem.* **2014**, *67*, 1422–1426. [CrossRef]

6. Cormode, D.P.; Skajaa, T.; Fayad, Z.A.; Mulder, J.M. Nanotechnology in medical imaging probe design and applications. *Arterioscler. Thromb. Vasc. Biol.* **2009**, *29*, 992–1000. [CrossRef] [PubMed]

7. Gonil, P.; Sajomsang, W.; Ruktanonchai, U.R.; Ubol, P.N.; Treetong, A.; Opanasopit, P.; Puttipipatkhachorn, S. Synthesis and fluorescence properties of *N*-substituted 1-cyanobenz[f]isoindole chitosan polymers and nanoparticles for live cell Imaging. *Biomacromolecules* **2014**, *15*, 2879–2888. [CrossRef] [PubMed]

8. Huang, H.; Liu, F.; Chen, S.; Zhao, Q.; Liao, B.; Long, Y.; Zeng, Y.; Xia, X. Enhanced fluorescence of chitosan based on size change of micelles and application to directly selective detecting Fe^{3+} in humanserum. *Biosens. Bioelectron.* **2013**, *42*, 539–544. [CrossRef] [PubMed]

9. Qaqish, R.; Amiji, M. Synthesis of a fluorescent chitosan derivative and its application for the study of chitosan-mucin interactions. *Carbohydr. Polym.* **1999**, *38*, 99–107. [CrossRef]

10. Keana, T.; Thanou, M. Biodegradation, biodistribution and toxicity of chitosan. *Adv. Drug Deliv. Rev.* **2010**, *62*, 3–11. [CrossRef] [PubMed]

11. Aruoma, O.I.; Halliwell, B.; Gajewski, E.; Dizdaroglu, M. Copper-ion-dependent damage to the bases in DNA in the presence of hydrogen peroxide. *Biochem. J.* **1991**, *273*, 601–604. [CrossRef] [PubMed]

12. Kobal, A.B.; Horvat, M.; Prezelj, M.; Briski, A.S.; Krsnik, M.; Dizdarevic, T.; Mazej, D.; Falnoga, I.; Stibilj, V.; Arneric, N.; et al. The impact of long-term past exposure to elemental mercury on antioxidative capacity and lipid peroxidation in mercury miners. *J. Trace Elem. Med. Biol.* **2004**, *17*, 261–274. [CrossRef]

13. Gumpu, M.B.; Sethuraman, S.; Krishnan, U.M.; Rayappan, J.B.B. A review on detection of heavy metal ions in water-an electrochemical approach. *Sens. Actuators B Chem.* **2015**, *213*, 515–533. [CrossRef]

14. Jameson, D.M.; Croney, J.C.; Moens, P.D. Fluorescence: Basic concepts, practical aspects, and some anecdotes. *Methods Enzymol.* **2003**, *360*, 1–43. [PubMed]

15. Pan, X.; Wang, G.; Lay, C.L.; Tan, B.H.; He, C.; Liu, Y. Photoluminescence from amino-containing polymer in the presence of CO_2: Carbamato anion formed as a fluorophore. *Sci. Rep.* **2013**, *3*, 1–6. [CrossRef] [PubMed]

16. Zhang, X.; Lee, S.; Liu, Y.; Lee, M.; Yin, J.; Sessler, J.L.; Yoon, J. Anion-activated, thermoreversible gelation system for the capture, release, and visual monitoring of CO_2. *Sci. Rep.* **2014**, *4*, 1–8. [CrossRef] [PubMed]

17. Dulkeith, E.; Morteani, A.C.; Niedereichholz, T.; Klar, T.A.; Feldmann, J. Fluorescence quenching of dye molecules near gold nanoparticles: Radiative and nonradiative effects. *Phys. Rev. Lett.* **2002**, *89*, 1–4. [CrossRef] [PubMed]

18. Serdyuk, O.V.; Heckel, C.M.; Tsogoeva, S.B. Bifunctional primary amine-thioureas in asymmetric organocatalysis. *Org. Biomol. Chem.* **2013**, *11*, 7051–7071. [CrossRef] [PubMed]

19. Lavertu, M.; Xia, Z.; Serreqi, A.N.; Berrada, M.; Rodrigues, A.; Wang, D.; Buschmann, M.D.; Gupta, A. A validated 1H NMR method for the determination of the degree of deacetylation of chitosan. *J. Pharm. Biomed. Anal.* **2003**, *32*, 1149–1158. [CrossRef]

20. Clark, G.L.; Smith, A.F. X-ray diffraction studies of chitin, chitosan, and derivatives. *J. Phys. Chem.* **1936**, *40*, 863–879. [CrossRef]

21. Duarte, M.L.; Ferreira, M.C.; Marvão, M.R.; Rocha, J. An optimised method to determine the degree of acetylation of chitin and chitosan by FTIR spectroscopy. *Int. J. Biol. Macromol.* **2002**, *31*, 1–8. [CrossRef]

22. Qiao, Y.; Zheng, X. Highly sensitive detection of copper ions by densely grafting fluorescein inside polyethyleneimine core-silica shell nanoparticles. *Analyst* **2015**, *140*, 8186–8193. [CrossRef] [PubMed]

MDPI AG

St. Alban-Anlage 66

4052 Basel, Switzerland

Tel. +41 61 683 77 34

Fax +41 61 302 89 18

http://www.mdpi.com

Marine Drugs Editorial Office

E-mail: marinedrugs@mdpi.com

http://www.mdpi.com/journal/marinedrugs